A study book for the NEBOSH International General Certificate

Technical Member (TechIOSH)

This category recognises the qualifications and competence needed by professionals working in a range of operational health and safety roles. To become a Technical member, you must have a suitable level of experience in a health and safety role, as well as an IOSH accredited qualification.

Students on a course of study towards Tech IOSH will find this **Essential Health and Safety Guide** particularly useful in developing the knowledge requirements for an IOSH accredited qualification at this level.

RMS Publishing

Suite 3, Victoria House, Lower High Street, Stourbridge DY8 1TA

© RMS Publishing ™

First Published September 2007.

Second Edition March 2012.

Third Edition May 2014.

Fourth Edition August 2015.

Fifth Edition January 2016.

Sixth Edition December 2019.

Cover design by Smudge Creative Design, Birmingham.

Printed and bound in Great Britain by Stephens and George Print Group, Merthyr Tydfil.

ISBN-13: 978-1-906674-88-5

CONTENTS

PREFACE

Publication users

This Essential Health and Safety Guide to the NEBOSH International General Certificate qualification provides an excellent reference for all those with management responsibilities for health and safety, wherever in the world they may be working. It focuses on international standards and management systems to provide a broad understanding of health and safety principles and practices that are relevant globally.

The practical and informative approach of the publication will provide managers, supervisors and others who have associated responsibilities for health and safety with a perspective of the significance of health and safety risk in the workplace and an understanding of the control measures that can be applied to manage risk. In addition, this publication would be particularly useful for corporate risk managers, human resources managers, designers, engineers, facilities managers, process or service managers, suppliers, contractors and site managers.

The general nature of this guide makes it highly applicable to a range of work activities and industries, providing an introduction to health and safety in the workplace.

This guide is an excellent study book and source of information for those undergoing the following learning programmes:

- The NEBOSH International General Certificate in Occupational Health and Safety Qualification.

- To provide an introduction to the management of general occupational health and safety risks for those studying the NEBOSH International Technical Certificate in Oil and Gas Operational Health and Safety Qualifications.

Successful completion of the NEBOSH qualifications will enable applicants to apply for first level membership of many national professional bodies related to health and safety, for example, Associate Membership (AIOSH) of the Institution of Occupational Safety and Health (IOSH). The qualification also meets the academic requirements for Technical Membership (Tech IOSH) of IOSH.

Scope and contents

Scope

Topics covered by the guide include:

- International standards for health and safety.

- Implementation of health and safety management systems.

- Identification of workplace hazards.

- Methods of risk control.

This guide contains an emphasis on practical solutions to workplace health and safety issues. Full colour photographs, tables and sample documents are provided to aid in the understanding of how these risks can be managed.

All statistics shown throughout this publication are the latest available at time of going to press.

Supporting study features

The authors have taken particular care to reinforce learning and understanding, following many years' experience in producing such publications. Where relevant within each Element of study there are sections included entitled: Review; Consider and Case Study.

 Review
After a review of a particular Element take an opportunity to answer a question related to the learning outcome to check your understanding.

 Consider
Tasks to try or think about, at times relating to your own workplace or job role, in support of your learning process.

 Case study
Applicable case studies to reflect on in context with your learning.

The additional sections at the back of the publication are designed to help the student through study tips, revision techniques and exam practice together with practical assessment guidance.

International standards

This guide refers to significant international conventions, recommendations, codes, guidance and standards in the form of a summary and in context with the topics covered. Examples of how the topics relate to health and safety globally support the information provided.

Assessment

The International General Certificate qualification is assessed by two unit assessments:

- Unit IG1: Management of health and safety – question paper.

- Unit IG2: Risk assessment – practical assessment.

In order that users may check their understanding of the topic expressed in this guide, a number of study questions have been included at the end of each element. Although elements 5 to 11 are only assessed by the risk assessment (IG2), RMS has chosen to include questions for all elements to help reinforce learning across the full syllabus.

Any resemblance to NEBOSH examination questions is coincidental; the questions have been created to illustrate the type of question which may be asked at examination and to check understanding of the relevant element learning outcomes.

Revision and examination guidance

There are ways of making study time more effective to give a learner the greatest possible chance of success in their exam. This section is included to aid learners with useful study tips, time management and revision techniques.

Syllabus

This 6th Edition has been thoroughly revised and updated to embrace the current specification of the syllabus for the NEBOSH International General Certificate in Occupational Health and Safety (IG), although the publication will suit all those interested in the topic in general and studying for other qualifications.

It has been designed to reflect the order and content of the IG syllabus which is structured in a very useful way, focusing on core management of health and safety principles (Elements 1 to 4) then hazards and their control (Elements 5 to 11); and in this way, the learner studying for this qualification can be confident that this guide reflects the themes of the syllabus and forms an excellent study book for that purpose.

LEARNING OUTCOMES AND ASSESSMENT CRITERIA

	Sub-elements	Learning outcome The learner will be able to:	Assessment criteria
Element 1	1.1-2.2	Justify health and safety improvements using moral, financial and legal arguments.	1.1 Discuss the moral, financial and legal reasons for managing health and safety in the workplace 1.2 Explain how health and safety is regulated and the consequences of non-compliance.
	1.3	Advise on the main duties for health and safety in the workplace and help their organisation manage contractors.	1.3 Summarise the main health and safety duties of different groups of people at work and explain how contractors should be selected, monitored and managed.
Element 2	2.1-2.2	Work within a health and safety management system, recognising what effective policy, organisational responsibilities and arrangements should look like.	2.1 Give an overview of the elements of a health and safety management system and the benefits of having a formal/certified system 2.2 Discuss the main ingredients of health and safety management systems that make it effective – policy, responsibilities, arrangements.

	Sub-elements	Learning outcome The learner will be able to:	Assessment criteria
Element 3	3.1 - 3.3	Positively influence health and safety culture and behaviour to improve performance in their organisation.	
	3.4	Do a general risk assessment in their own workplace profiling and prioritising risks, inspecting the workplace, recognising a range of common hazards, evaluating risks (taking account of current controls), recommending further control measures, planning actions.	3.4 Explain the principles of the risk assessment process
	3.5	Recognise workplace changes that have significant health and safety impacts and effective ways to minimise those impacts.	3.5 Discuss typical workplace changes that have significant health and safety impacts and ways to minimise those impacts
	3.6 – 3.8	Develop basic safe systems of work (including taking account of typical emergencies) and knowing when to use permit-to-work systems for special risks.	3.6 Describe what to consider when developing and implementing a safe system of work for general activities 3.7 Explain the role, function and operation of a permit to-work system 3.8 Discuss typical emergency procedures (including training and testing) and how to decide what level of first aid is needed in the workplace
Element 4	4.2	Take part in incident investigations	4.2 Explain why and how incidents should be investigated, recorded and reported
	4.1, 4.3, 4.4	Help their employer to check their management system effectiveness – through monitoring, audits and reviews.	4.1 Discuss common methods and indicators used to monitor the effectiveness of management systems. 4.3 Explain what an audit is and why and how it is used to evaluate a management system. 4.4 Explain why and how regular reviews of health and safety performance are needed.
Elements 5 - 11		Do a general risk assessment in their own workplace - profiling and prioritising risks, inspecting the workplace, recognising a range of common hazards, evaluating risks (taking account of current controls), recommending further control measures, planning actions.	5-11 Produce a risk assessment of a workplace which considers a wide range of identified hazards (drawn from elements 5 – 11) and meets best practice standards ('suitable and sufficient')

Production of the publication

Managing Editor

Ian Coombes, Managing Director ACT, CMIOSH; former member of NEBOSH Council, former member of NEBOSH Board of Trustees, former member of NEBOSH Qualifications and Technical Council and former NEBOSH examiner. Former member of IOSH Professional Affairs Committee and chairman of the Initial Professional Development sub-committee. Member of the Safety Groups UK (SGUK) management committee.

Acknowledgements

RMS Publishing Ltd wishes to acknowledge the following contributors and thank them for their assistance in the preparation of the publication:

Geoff Littley CMIOSH, NEBOSH Diploma, CSPA; experienced health and safety professional. Areas of work include manufacturing, NHS Trusts, Local Authorities, construction and power generation in the UK and Africa. A lead tutor for NEBOSH Diploma and General and Construction Certificate courses. Also an experienced examiner for NEBOSH Diploma Units A and C and both General and Construction Certificate courses.

Kevin Coley, CMIOSH; a NEBOSH examiner with many years' health and safety experience in the private and public sector. Kevin has been a production manager in a large foundry, so has a heavy engineering background as well as a senior safety manager for large government bodies.

Barrie Newell, Former Director ACT, FCMI, Lead Auditor OHSAS 18001; former member of the NEBOSH Diploma and Certificate Panels, former senior manager in the chemical industry with over 20 years' experience in the management of high risk facilities processing highly flammable and toxic chemicals, including HAZOP implementation. Implemented waste management systems including, waste reduction, recycling, reuse, incineration, including energy recovery and disposal to land fill.

Jonathan Backhouse, CertEd BA(Hons) MA MRes DipNEBOSH EnvDipNEBOSH MIFireE CFIOSH; highly qualified and experienced Chartered Safety and Health Practitioner with a vast amount of experience as an occupational safety and health professional, fire risk assessor and qualified teacher (working in UK, Africa, Europe, Middle East and the USA). An author of three books and over a dozen articles, actively involved within IOSH, current roles include Tees Branch Vice-Chair, Mentor and Peer Review Interviewer.

Victoria L Kieran, Chartered Health and Safety Practitioner and Qualified Teacher, CMIOSH DipNEBOSH SpDipEM; Lecturer and Mentor on a range of recognised health and safety courses including the NEBOSH Diploma in Occupational Safety.

Brian W. Stones, CEng, FIMM, CFIOSH; providing training, education and consultancy for over 30 years.

Stephen Whitehouse; an experienced health and safety professional with more than 20 years' experience working with the energy sector.

Julie Skett, Support and Development Manager. Rhiannon Davies, Andy Taylor, Robert Sannwald-Dunn and Rosie Hierons, layout and formatting.

FOREWORD

Despite the increased global recognition of the importance of health and safety at work, accidents and work-related ill health continue to have a significant effect on workers. The ILO has identified that countries that have growing industrialisation and development are experiencing an increase in harm to workers.

The ILO estimates that over 6,000 people are killed per day as a direct result of either workplace accidents or exposures to hazardous substances daily as a result of occupational accidents or disease. More than 2.3 million work-related deaths occur each year. Worldwide, there are around 340 million occupational accidents and 160 million victims of work-related illnesses annually.

This scale of harm is reported to have an effect equivalent to approximately 4% of the gross national product (GDP) of the countries affected.

Improving health and safety at work is part of an organisation's corporate social responsibility and where this has been accepted, significant improvements in health and safety at work have been accomplished. The practical and informative approach taken by this publication will help the reader gain a perspective of the significance of health and safety risk in the workplace and an understanding of how the risk can be managed.

The first four elements in this guide indicates how risks are being managed in countries in different parts of the world. The supporting references to international standards in the form of conventions, recommendations and legislative approach taken in different countries provides the reader with a structure to what corporate social responsibility for health and safety means. The elements which cover the control of international workplace risks provides a practical insight into how risks can be managed.

Application of the information provided by this publication will enable readers to make a significant contribution to health and safety in the workplace.

PUBLICATIONS ALSO AVAILABLE FROM RMS:

Publication	Edition	ISBN
A Study Book for the NEBOSH National General in Occupational Health and Safety	Eleventh	978-1-906674-93-9
A Study Book for the NEBOSH National Certificate in Fire Safety and Risk Management	Sixth	978-1-906674-66-3
The Management of Construction Health and Safety Risk	Fourth	978-1-906674-37-3
The Management of Environmental Risks in the Workplace	Fourth	978-1-906674-24-3
The Management of Health and Well-being in the Workplace	First	978-1-906674-14-4
A Guide to International Oil and Gas Operational Safety	Second	978-1-906674-65-6
Study Books for the NEBOSH National Diploma in Occupational Safety and Health:		
(Unit A) Managing Health and Safety	Sixth	978-1-906674-55-7
(Unit B) Hazardous Agents in the Workplace	Sixth	978-1-906674-56-4
(Unit C) Workplace and Work Equipment Safety	Sixth	978-1-906674-57-1
Study Books for the NEBOSH International Diploma in Occupational Safety and Health:		
(Unit IA) International Management of Health and Safety	Fourth	978-1-906674-52-6
(Unit IB) International Control of Hazardous Agents in the Workplace	Fourth	978-1-906674-53-3
(Unit IC) International Workplace and Work Equipment Safety	Fourth	978-1-906674-54-0
Essential Revision Guides for the NEBOSH Diploma in Occupational Health and Safety:		
(Unit A) Managing Health and Safety	First	978-1-906674-76-2
(Unit IA) Managing Health and Safety	First	978-1-906674-78-6
(Unit B & IB) Hazardous substances/agents	First	978-1-906674-80-9
(Unit C & IC) Workplace and work equipment safety	First	978-1-906674-82-3

LIST OF ABBREVIATIONS

ABS	Anti-lock Braking System
AC	Alternating Current
ACGIH	American Conference of Governmental Industrial Hygienists
ACM	Asbestos-Containing Material
ACOP	Approved Code of Practice
ACSNI	Advisory Committee on the safety of Nuclear Installations
AETR	European Agreement concerning the work of Crews of Vehicles Engaged in International Road Transport
AIB	Asbestos Insulating Boards
AIDS	Acquired Immune Deficiency Syndrome
APF	Assigned Protection Factor
BBVs	Blood Borne Viruses
BLVs	Biological Limit Values
BMGVs	Biological Monitoring Guidance Values
BOLVs	Binding Occupational Limit Values
BSI	British Standards Institute
CAR	Control of Asbestos Regulations 2012
CAT	Cable Avoidance Tool
CBM	Condition-based Maintenance
CBT	Cognitive Behavioural Therapy
CCTV	Closed Circuit Television
CE	Conformité Européene
CEN	European Standards
CFPA-E	Confederation of Fire Protection Associations Europe
CLP	Classification, Labelling and Packaging
CO_2	Carbon Dioxide
COPD	Chronic Obstructive Pulmonary Disease
COSHH	Control of Substances Hazardous to Health
CPR	Cardio Pulmonary Resuscitation
dB	Decibel
DC	Direct Current
DNA	Deoxyribonucleic acid
DOSH	Department of Occupational Safety and Health
DSE	Display Screen Equipment
DVD	Digital Versatile Disc
EC	European Commission
EHS	Environment, Health and Safety
EHSMS	Environment, Health and Safety Management System
EPL	Equipment Protection Level
ESA	European Safety Agency
ETA	Event Tree Analysis
EU	European Union
FLTs	Fork-lift Trucks
FMEA	Failure Mode and Effect Analysis
FTA	Fault Tree Analysis
GAD	Generalised Anxiety Disorder
GHS	Globally Harmonised System
GNP	Gross National Product
GCS	Ground Control System
HASAWA	Health and Safety at Work etc., Act 1974
HAVS	Hand Arm Vibration Syndrome
HAZOP	Hazard and Operability Studies
HIV	Human Immunodeficiency Virus

HSE	Health and Safety Executive
HSG	Health and Safety Guidance
Hz	Hertz
IAEA	International Atomic Energy Agency
ICAO	International Civil Aviation Organisation
ICRP	International Commission on Radiological Protection
IEC	International Electrotechnical Commission
ILO	International Labour Organisation
ILO-OSH	International Labour Organisation – Occupational Safety and Health
ILV	Indicative Limit Values
IOELV	Indicative Occupational Exposure Limit Value
IQ	Intelligence Quotient
IR	Infared
ISO	International Organisation for Standardisation
ISRS	International Safety Rating System
IT	Information Technology
KPI	Key Performance Indicator
LEV	Local Exhaust Ventilation
LGV	Large Goods Vehicle
LITE	Load, Individual, Task and Environment
LOTO	Lock Out Tag Out
LPG	Liquefied Petroleum Gas
LTEL	Long-term Exposure Limit
MAC	Maximum Allowable (Acceptable) Concentrations
mS	milliSeconds
MEWP	Mobile Elevated Work Platform
MHSWR	Management of Health and Safety at Work Regulations 1999
MSDs	Musculoskeletal Disorders
MW	Megawatt
NEBOSH	National Examination Board in Occupational Safety and Health
NGO	Non-Governmental Organisation
NHS	National Health Service
NICE	National Institute for Health and Care Excellence
NIHL	Noise-induced Hearing Loss
NRR	Noise Reduction Rating
OCD	Obsessive Compulsive Disorder
OEL	Occupational Exposure Limit
OES	Occupational Exposure Standards
OH&S	Occupational Health and Safety
OHSAS	Occupational Health and Safety Assessment Series
OSHA	Occupational Safety and Health Administration
OSHAD-SF	Occupational Safety and Health System Framework
OHSMS	Occupational Health and Safety Management System
OSH	Occupational Safety and Health
Pa	Pascal
PAT	Portable Appliance Testing
PCV	Passenger Carrying Vehicle
PDCA	Plan-Do-Check-Act
PEL	Permissible Exposure Limits
PNA	Predicted Noise Attenuation
PPE	Personal Protective Equipment
PPM	Planned Preventive Maintenance
PTS	Permanent Threshold Shift
PTSD	Post-traumatic Stress Disorder
PTW	Permit-to-work

PUWER	Provision and Use of Work Equipment Regulations 1998
RCD	Residual Current Device
REACH	Registration, Evaluation, Authorisation and Restriction of Chemicals
RIDDOR	Reporting of Injuries, Diseases and Dangerous Occurrences Regulations
RPA	Radiation Protection Adviser
RPE	Respiratory Protective Equipment
Rz	Roughness
SDS	Safety Data Sheets
SLC	Sound Level Conversion
SMART	Specific, Measurable, Achievable, Reasonable and Time-bound
SNR	Single Number Rating
SRV	Slip Resistance Rating
STEL	Short-term Exposure Limit
Sv	Sievert
SWL	Safe Working Load
TILE	Task, Individual, Load, and Environment
TLVs	Threshold Limit Values
TTS	Temporary Threshold Shift
TWA	Time Weighted Average
UAV	Unmanned Aerial Vehicle
UK	United Kingdom
UN	United Nations
UNRTDG	UN Recommendations on the Transport of Dangerous Goods
USA	United States of America
USSR	Union of Soviet Socialist Republics
UV	Ultraviolet
VLOS	Visual Line of Site
VWF	Vibration White Finger
WAH	Work at Height Regulations 2005
WAOSHR	Western Australian Occupational Safety and Health Regulations 1996
WBGT	Wet Bulb Globe Temperature
WBVS	Whole Body Vibration Syndrome
WEL	Workplace Exposure Limit
WHO	World Health Organisation
WHSWR	Workplace (Health, Safety and Welfare) Regulations 1992
WRULD	Work-related Upper Limb Disorder
2HC	Two-Hand Controls

This page is intentionally left blank

Element 1

Why we should manage workplace health and safety

Contents

1.1   Morals and money

GENERAL ARGUMENT

The main reasons why organisations should manage workplace health and safety are moral, financial and regulatory.

Moral and financial reasons provide a strong motivation to promote good health and safety standards and are discussed in this section. The regulatory reasons are discussed in *Section 1.2.*

CASE STUDY

Effect of a serious accident/incident at work

A 32 year-old engineer fell 5 metres from a ladder while doing maintenance work on a building. He suffered major injuries to his head, arm and back when he hit the ground. He was taken to hospital where he remained unconscious for 24 hours and had surgery to repair the fractures he had suffered. He spent many days in hospital receiving treatment for his injuries.

The engineer's manager had to tell his family that he had been injured and was in hospital as a result of the fall at work. His manager said "It was really difficult to tell his family knowing that he was one of my workers. I knew his family well and felt that I had let them down by letting him get injured. It was hard to tell his wife and although his three young children did not understand I could see they were very upset to know he was in hospital."

His wife spent all her available time at the hospital with him. She shared the pain and emotional stress of his injuries. His condition affected everyone in the family. His wife said "One of his sons did not recognise him because of his injuries. I think it was very hard for them to see him like that."

The engineer used to bicycle to work every day. Although he had received treatment for his injuries and a long programme of exercises to improve his mobility the damage to his spine meant that he had difficulty in walking. He could no longer use a bicycle and would not be able to walk without a walking stick. This meant he could not return to his previous job.

The engineer said "I was once fit and healthy, now look at me, all because I didn't have the right equipment to work at height safely."

CONSIDER

Consider the case study and identify three human effects or consequences that came from the incident.

MORAL EXPECTATIONS OF GOOD STANDARDS OF HEALTH AND SAFETY

Summary

Workplace injuries and ill-health can result in a great deal of pain and suffering for those affected. A worker should not have to expect that, by coming to work, their life is at risk. They should also not expect to be affected by hazardous substances that could shorten their life or cause long-term harm. Nor should others be adversely affected by work activities. Ensuring good health and safety standards at work may therefore be seen as the right way for organisations to conduct themselves and harming people through work activities as the wrong way.

Taking care not to harm people through work activities is a widely accepted custom of conduct and the right thing to do. This is reflected in many of the world's religions and cultures. This moral reason to prevent harm is usually further reinforced by societal expectations of behaviour, which requires the consideration of others that may be affected by interaction with them. In particular this includes work activities and how they may harm those involved or affected by the activity. This societal expectation is often expressed in both civil law and criminal law as, without the potential for litigation or regulatory action, many employers would not act upon their moral obligation to provide protection. In many countries, it is a specific legal requirement to safeguard the health and safety of workers and others that might be affected by an employer's work activities.

Discussion

The moral reason for achieving good standards of health and safety in work activities is founded on the desire to prevent harm to those that may be affected by the work activities. This is a very important reason as, globally, so much harm is regularly caused to workers and others. The size of the health and safety problem internationally is difficult to fully quantify in terms of deaths, injuries and incidence of work related ill-health as data is reported in different ways in different countries. The ILO has estimated that globally there are 2.3 million work-related deaths or diseases each year. Men suffer two thirds of those deaths. In addition, it is reported that there are 340 million accidents/incidents at work each year. The ILO estimates that 160 million people are suffering from work-related illnesses. The biggest groups of work-related diseases are cancers, circulatory diseases and communicable diseases. The ILO has estimated that the cost of accidents/incidents and diseases is equivalent to 4% of the gross national product (GNP) of a country. Some examples of the international experience of various regulatory agencies with regard to injuries and ill-health are given below.

In the European Union (EU), the European Agency for Safety and Health at Work identified that in 2015 there were 3,876 work-related deaths in the EU, equating to 1.83 deaths per 100,000 workers. In addition, in the same year there were 3.2 million serious accidents that resulted in more than four days absence from work, equating to 1,601 serious accidents per 100,000 workers. Among the EU Member States, the highest incidence of fatal accidents at work in 2015 was recorded in Romania (5.56 deaths from accidents at work per 100,000 persons employed). No other European EU Member State reported an incidence rate above the level of four fatal accidents per 100,000 persons employed. The Association of Workers' Compensation Boards of Canada states that 852 workplace deaths (303 due to injury other than industrial disease related) were recorded in Canada in 2015, which equates to 3.6 work-related deaths per work day based on 5 work days per week and 25 days (5 weeks) holiday per year. In addition, 'accepted time loss injuries' were reported to be 231,725 in 2015. Accepted time loss injuries are injuries where an employee is compensated by a Canadian Worker's Compensation Board for a loss of wages following a work related accident/incident or receives compensation for a permanent disability, for example, hearing loss from excessive noise in the workplace. This does not include minor first-aid injuries where there is no loss of wages or injuries to the self-employed.

In the USA the Bureau of Labour Statistics records, 5190 people suffered a work-related death in 2016, which equates to over 14 each working day. The reported statistic for 2012 was 4,628 work-related deaths, equating to over 12.68 deaths per work day or a rate of 3.3 per 100,000 full time workers. In 2015 there were 2.9 million workplace injuries and illnesses reported, 2.9 per 100 full time workers. In 2013 there were 1,162,210 people who received work-related injuries that involved days absent from work. The reported statistic for 2012 was approximately 1,153,980 injuries, the statistic for 2013 equates to a rate of 109.4 per 10,000 full time workers.

A report on the occupational health and safety situation in the Arab states of the Persian Gulf region in 2007 produced by the ILO confirmed that in Kuwait during 2006, 31 deaths and 2,818 work-related accidents/incidents occurred. Similarly, in Bahrain during the same period, 19 deaths and 2,247 occupational accidents/incidents occurred. The Bahrain Ministry of Labour reported that the number of occupational deaths in all industry types almost doubled from 19 in 2006 to 36 in 2008, despite the introduction in 2007 of a ban on construction work taking place between 12.00 and 16.00 hours in July and August, the peak time for fatal accidents/incidents induced by heat stress.

The School of Public Health of the Shanghai Medical University reported that 235 work-related deaths occurred in the construction industry in one new economic development area in eastern China between 1991 and 1997. These deaths represented 55% of all occupational deaths in that development area. The average annual death rate was 51.5 per 100,000 construction workers.

Data derived by the Japan Construction Safety and Health Association confirmed that work-related deaths for 2017 were 978. The Association also confirmed that there were 120,460 accidents/incidents in 2017, an increase of 2.2% on 2016. The Department of Occupational Safety and Health in Malaysia recorded that in 2014 there were 207 work-related deaths, 145 permanent disability injuries and 2,456 non-permanent injuries reported to them.

	Work-related deaths	Accidents/incidents
ILO - global estimate	2.3 million (annually)	340 million (annually)
EU	3,876 (2015)	3.2 million (2015)
USA	5,190 (2015)	2.9 million (2015)
UK	144 (2017/18)	70,114 (2017/18) RIDDOR
Canada	852 (2015)	231,725 (2015)
Japan	978 (2017)	120,460 (2017)
Malaysia	117 (to June 2018)	1,692 (to June 2018)
Kuwait	31 (2006)	2,818 (2006)
Bahrain	19 (2006)	2,247 (2006)

Figure 1-1: Number of work-related deaths and accidents/incidents. Source: Various.

In the United Kingdom (UK) the Health and Safety Executive (HSE) statistics revealed that 144 workers were killed at work during the year 2017/18 representing 0.44 deaths per 100,000 workers. The construction industry (42%) and agriculture (27%) accounted for most of these. Falls from height, being struck by a vehicle and being struck by a falling object accounted for approximately 50% of these deaths. In the same period there were 71,062 non-fatal injuries reported. Slips, trips and falls on the level accounted for 31% of all non-fatal injuries. The second most common injuries were sustained whilst lifting, handling or carrying, representing 21% and falls from height were 8% of the non-fatal injuries.

In addition to work-related injuries, it was reported in 2017/18 that in the UK 1.4 million people were suffering from an illness (long-standing as well as new cases) they believed was caused or made worse by their current or past work. Of these illnesses 0.5 million of them were new conditions that started during the year. In 2016 2,595 people died from mesothelioma, a type of cancer caused by exposure to asbestos fibres that usually starts in the membrane located between the lungs and chest wall. Thousands more people died from other occupational cancers and diseases such as chronic obstructive pulmonary disease (COPD), which is a work related lung disease caused by breathing in certain dusts, fumes, chemicals or gases.

In the UK a total of 31.2 million days were lost due to work-related harm, 25.7 million days lost were due to work-related ill-health and 5.5 million due to workplace injury. The total cost of work-related ill-health and injury in the UK was estimated at £14.9 billion in 2016/17, illness represented £9.7 billion (65%) of these costs and injuries cost £5.2 billion. In 2016/17 it was calculated that there was an average cost of £1.6 million per fatality and £8,400 per non-fatal injury.

Irrespective of the country involved, the types of injuries arising from accidents/incidents tend to follow broad trends. Falling from height, being struck by a vehicle and being struck by a falling object are primary causes of work-related deaths, whereas manual handling and slips, trips and falls are significant causes of other injuries leading to lost time. It can be seen by the statistics outlined that harm to people from work-related activities is very large and provides a strong moral reason to ensure good health and safety in the workplace. Societal expectations of good standards of health and safety vary across the world. In established market economies the societal expectation of good standards of health and safety is at its strongest, causing organisations to have good standards that prevent injury and ill-health. In developing market economies the societal expectations of good standards of health and safety may be more of

an aspiration and have less influence, leading to a smaller number of organisations achieving good standards.

Societal expectations tend to influence health and safety standards in two ways:

1) Strategically, the general mass of the public influence societal expectations concerning tolerance or intolerance of specific workplace risks or situations, for example, intolerance of risks of work-related major disasters.

2) Locally, these influences tend to surround tolerance or intolerance of the practices of a specific organisation. This influence is often strongest following an accident/incident in an organisation and has led to the closure of some smaller organisations.

Globally there is broad agreement that society should expect and demand a safe and healthy work environment that does not harm workers, contractors, the self-employed and the general public. This is reflected in the principles set out in ILO conventions that have been developed and adopted by societies throughout the world, in particular the Occupational Safety and Health Convention C155 (Convention C155), 1981, which creates the following principles:

> "Part II - Principles of National Policy - Article 4
>
> 1) Each Member shall, in the light of national conditions and practice, and in consultation with the most representative organisations of employers and workers, formulate, implement and periodically review a coherent national policy on occupational safety, occupational health and the working environment.
>
> 2) The aim of the policy shall be to prevent accidents and injury to health arising out of, linked with or occurring in the course of work, by minimising, so far as is reasonably practicable, the causes of hazards inherent in the working environment."

Figure 1-2: Principles of national policy on safety and health. Source: Occupational Safety and Health Convention C155, ILO.

The principles established in Article 4 of ILO Convention C 155 use the term 'reasonably practicable' in relation to National Policy, this is a term that is often used in National health and safety legislation and is explained later in this element in 1.3 "The role of national governments and international bodies in formulating a framework for regulation of health and safety". The ILO Occupational Safety and Health Convention C155, 1981, expects each country to meet these principles in a way that reflects the country's state of development and societal expectations. Many countries of the world have established requirements for good health and safety standards. These requirements tend to confirm that employers need to pay particular attention to five important factors in order to prevent work-related injury and ill-health. Employers need to provide five health and safety requirements:

1) A safe and healthy place of work.

2) Safe and healthy plant and equipment.

3) Safe and healthy systems of work.

4) Training and supervision.

5) Competent workers.

These health and safety requirements are often reflected in national civil and criminal laws related to workplace health and safety. This further endorses the societal expectation to prevent harm from work-related activities. The ILO Occupational Safety and Health Convention C155, 1981, sets out broad requirements for member countries to follow to ensure health and safety requirements are set into national laws. An example of this is Article 16, which requires:

"1. Employers shall be required to ensure that, so far as is reasonably practicable, the workplaces, machinery, equipment and processes under their control are safe and without risk to health.

2. Employers shall be required to ensure that, so far as is reasonably practicable, the chemical, physical and biological substances and agents under their control are without risk to health when the appropriate measures of protection are taken."

Many countries have in place requirements reflecting the ILO Occupational Safety and Health Convention C155, 1981, that ensure health and safety is included in National laws, in particular criminal laws. The UK established similar requirements in the Health and Safety at Work etc Act 1974, before their formalisation in Convention C155 in 1981. The Health and Safety at Work etc Act 1974 set out specific legal requirements to provide a safe place of work, safe plant and equipment, safe systems of work, training and supervision. A number of countries have enacted similar legislation in more recent years, for example, Section 15 of the Occupational Safety and Health Act 1994 of Malaysia. National criminal laws enable society, in the form of the State, to punish those that do not meet their responsibilities to prevent harm from work activities.

The five important health and safety requirements have also been adopted within the civil laws of many countries, they create a separate 'duty of care' towards those that might be affected by work activities.

This duty of care is a legal requirement imposed on employers and requires that they exercise a reasonable standard of care while performing any work-related activities that could foreseeably harm others.

In the UK the employers' duty of care in civil law requires employers to take ***'reasonable care of those that might foreseeably be affected by their acts or omissions'***. The duty of care provides protection of workers and others (for example, visitors, members of the public, etc.) who might foreseeably be affected. It follows that workers have the right to work in a workplace and in a way that provides reasonable protection from harm. If an employer fails to meet this duty the employer may be required to compensate the person for the harm done. The civil and criminal responsibility to meet the five important health and safety requirements could be considered as society's formalisation of the moral responsibilities held by an individual or organisation towards another individual or organisation.

REVIEW

List two good reasons, with a description of each, for preventing accidents in the workplace.

THE FINANCIAL COST OF INCIDENTS

Summary

Accidents/incidents and ill-health at work cost organisations a great deal of money. Especially when the costs arising from damage accidents/incidents are added and particularly as they may interrupt production, downgrade the quality of products or impair the environment. Many governments realise that poor occupational health and safety performance gives rise to additional costs to the State, for example, payments to the incapacitated worker and the costs of medical treatment. Employers also sustain costs in the event of accidents/incidents and ill-health at work. These can include legal fees, fines, compensation payments, investigation time and lost goodwill from the workforce, customers and wider community.

Insured and uninsured costs

Direct costs

Direct costs are those costs that directly relate to the accident/incident or incidence of ill-health. Direct costs may be insured or uninsured depending on the cause of the loss. In addition, many smaller costs may not be covered by insurance policies because the insurance company expects the employer to meet some of the costs of losses. Sometimes this is referred to as a 'threshold' or 'excess' amount that the employer is required to pay before the insurance company pays their part of the compensation.

Insured direct costs often include:

- Claims for compensation made against the employer by employees and other people. This may be covered by employer's liability insurance or public liability insurance or similar insurance policies.
- Damage to buildings, tools, equipment, vehicles and materials, specifically covered by insurance policies, this may depend on the type of accident/incident, for example damage by a fire may be extensively covered by insurance. However, vehicles may only have insurance cover for accidental damage on public roads.
- Medical costs related to harm caused to a worker may be insured. This can include treatment, hospital and physician charges, occupational therapy, prescription medicines and medical equipment.
- Legal costs of defending a claim for compensation or prosecution are often insured.
- Long term loss of business may be insured, usually related to a major disaster like a fire.

Uninsured direct costs often include:

- Lost time of an injured worker or worker suffering ill-health.
- Continued payments to the worker or the worker's family as part of an employer's worker absence scheme.
- Additional payments to other workers or contractors to do the work the worker would have done.
- Wage costs due to decreased productivity once the injured worker returns to work. This is may be due to restricted movement or nervousness/cautiousness on the part of the injured worker and time spent discussing the accident/incident with other employees etc.
- Damage to equipment, tools, property, plant or materials not covered by specific insurance.

- Most first-aid costs.
- Time and materials to clean up after the accident/incident or incidence of ill-health.
- Increases in insurance premiums resulting from accidents/incidents and ill-health.
- Charges related to citations, notices and other forms of enforcement agency intervention.
- Fines imposed by a criminal court following prosecution.

Indirect costs

Indirect costs are those costs that indirectly relate to the accident/incident or incidence of ill-health. Some of these costs may be caused by a single accident/incident or incidence of ill-health or may be the result of an accumulation of occurrences over time. Indirect costs can also be insured or uninsured, though not many can be insured.

Insured indirect costs include:

- Claims for consequential loss related to effects on the supply chain, from customers or suppliers.

Uninsured indirect costs include:

- Lost time by other workers who stop work or reduce performance.
  - Out of curiosity
  - Out of sympathy
- Workers and managers may feel that health and safety is a low priority, which could weaken morale and lead to reduced productivity.
- Loss of important experience that the worker had.
- Lost time by supervisors or other managers.
  - Assisting the injured worker.
  - Investigating the causes of the accident/incident or ill-health.
  - Arranging for the injured workers' production to be continued by other workers.

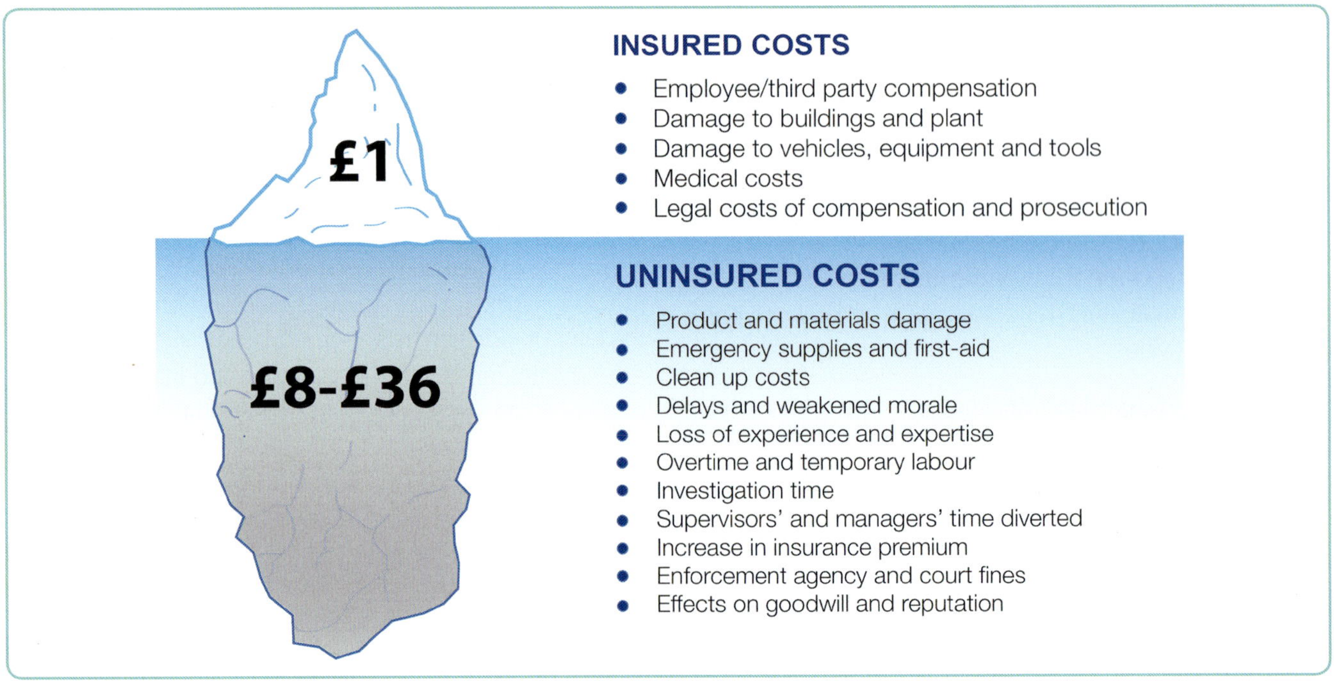

INSURED COSTS
- Employee/third party compensation
- Damage to buildings and plant
- Damage to vehicles, equipment and tools
- Medical costs
- Legal costs of compensation and prosecution

£1

£8-£36

UNINSURED COSTS
- Product and materials damage
- Emergency supplies and first-aid
- Clean up costs
- Delays and weakened morale
- Loss of experience and expertise
- Overtime and temporary labour
- Investigation time
- Supervisors' and managers' time diverted
- Increase in insurance premium
- Enforcement agency and court fines
- Effects on goodwill and reputation

Figure 1-3: Main insured and uninsured costs, typical ratio of costs. Source: RMS/UK, HSE HSG96 (no longer available).

- Organising activities to resume the operation of the organisation.
- Selecting or training a new worker to replace an injured worker.
- Preparing reports, attending enforcing agency meetings.
- Attending hearings or inquest courts.
- Interference with production leading to failure to fill orders on time, loss of bonuses, penalty payments and similar losses that do not benefit from insurance.
- Effects on goodwill and the reputation of the organisation.

A number of case studies over recent years have illustrated the difference between insured costs and uninsured costs. It has been shown in the case studies that uninsured costs were often between 8 and 36 times greater than the insured costs, *see Figure 1-3* .

The Occupational Safety and Health Administration (OSHA) of the USA found that the ratio of indirect costs to direct costs varies widely, but for smaller claims of up to $3,000 they are often 4.5:1. OSHA also identified that workplace injuries alone cost USA businesses over $110 billion in 1993.

Similarly, in 1993, in the UK the HSE published the results of a series of case studies they conducted to illustrate just how much accidents/incidents at work could cost a company. The industries chosen were from a wide range of activities and they were studied over a period of time.

The table in *Figure 1-4* illustrates the losses identified in the case studies. As the studies lasted less than a year the annualised costs of the losses from accidents/incidents were calculated from the data obtained. The annualised losses were the total amounts of money the organisations would lose in one year if nothing was done to mitigate the risks.

CONSIDER

What sort of financial costs would a company incur as a result of having poor health and safety standards?

1.2 Regulating health and safety

The International Labour Organisation is a UN body that is made up of representatives from member countries of the UN. It sets out international treaty agreements in the form of conventions and member countries confirm their acceptance of the conventions by ratifying them. They agree to be bound by conventions that they have ratified. Member countries use the ILO conventions to structure and guide their approach to health and safety. This is particularly useful to emerging countries that have not gained experience in the practices necessary to manage health and safety in a complex industrialised country. It saves them developing their own approach through their own experience, which would be both painful and costly in terms of human life.

The ILO has established a number of conventions and recommendations, for example those that relate to the management of health and safety, control of chemicals and the work environment, and construction activities. ILO conventions set out what should be done by member countries at a national level and what should be done at employer level. Workers' rights and responsibilities are included within the conventions.

The ILO Occupational Safety and Health Convention C155, 1981, establishes broad requirements for member countries to follow to ensure health and safety requirements are set into national laws.

Article 8 of Convention C155 requires:

"Each Member shall, by laws or regulations or any other method consistent with national conditions and practice and in consultation with the representative organisations of employers and workers concerned, take such steps as may be necessary to give effect to Article 4 of this Convention."

Article 4 of Convention C155 requires:

"A coherent national policy on occupational safety, occupational health and the working environment."

		Total loss	Annualised loss	Representing
1	Construction site	£245,075	£700,000	8.5% tender price
2	Creamery	£243,834	£975,336	1.4% operating costs
3	Transport company	£48,928	£195,712	1.8% of operating costs/37% of profits
4	Oil platform	£940,921	£3,763,684	14.2% of potential output

Figure 1-4: Sample costs of accidents/incidents. Source: The costs of accidents at work, HSG 96, HSE Books (no longer available).

National Governments of member countries establish laws in a form that suits their culture, perspectives and level of economic development. In the past, national legislation has concentrated on specific problems that have occurred in specific industries. More recently, national legislation is being introduced in some countries that better reflects the broad requirements set out in ILO Occupational Safety and Health Convention C155, 1981, and encourages employers to manage the risks they create. A good example of this would be the Malaysian Occupational Safety and Health Act 1994. Legislation written in this way is designed to make employers put in arrangements to manage the risks without further specific national legislation detailing actions for each risk.

WHAT ENFORCEMENT AGENCIES DO AND THE CONSEQUENCES OF NON-COMPLIANCE

The ILO Occupational Safety and Health Convention C155 creates broad requirements for member countries to follow to ensure health and safety requirements are set into national laws and enforced. Member countries that confirm acceptance of the convention are expected to put in place mechanisms to enforce the law and establish penalties for non-compliance. In order that the enforcement agencies can identify compliance or non-compliance the ILO Convention C155 expects member countries to establish a system of inspection.

In addition, ILO Convention C155 expects employers to be provided with information and advice on compliance. The convention recognises that compliance with law will not happen without advice and information and that compliance has to be verified by inspection. Countries may decide that the advisory part of the role of enforcement agencies be set apart from the inspection and enforcement part of the role to maintain independence of compliance verification and action.

The Convention C155 sets out the following requirements:

"Article 9:

The enforcement of laws and regulations concerning occupational safety and health and the working environment shall be secured by an adequate and appropriate system of inspection.

The enforcement system shall provide for adequate penalties for violations of the laws and regulations.

Article 10

Measures shall be taken to provide guidance to employers and workers so as to help them to comply with legal obligations."

Some countries have had health and safety enforcement agencies established for many years. In the USA,

OSHA, the Occupational Safety and Health Administration, has been regulating occupational health and safety since 1971. In the UK, the Health and Safety Executive (HSE) enforces health and safety legislation under powers provided by the Health and Safety at Work etc. Act 1974. In Malaysia, part of the Ministry of Human Resource called the Department of Occupational Safety and Health (DOSH) is responsible for ensuring the safety, health and welfare of workers in both the public and private sector. DOSH enforces the Occupational Safety and Health Act 1994.

Enforcement agencies may take a number of approaches to their role. It is common for enforcement agencies to assist their government in the formulation of a framework for the regulation of health and safety. They will usually be involved in the drafting of legislation and guidance on how to comply with the legislation. The enforcement agencies will usually develop a regulatory plan and approach to regulation of compliance for their government to consider and confirm. A common approach to ensuring compliance would be, where possible, that the enforcement agency adopt a monitoring and advisory role in the workplace but, where necessary, have a range of powers that enable them to enforce compliance. When regulating compliance in workplaces the enforcement agency will often set out requirements orally or in writing as they see fit. If a formal approach is required they will usually issue some form of formal notice of non-compliance, an enforcement notice, or take steps to prosecute.

In order for enforcement agencies to carry out their role effectively they are usually provided with a range of powers, commonly this would include powers to:

- Enter premises at any reasonable time.
- Take a police officer or some other authorised person if the enforcement officer is likely to be obstructed in the execution of their duty.
- Examine and investigate.
- Direct that premises or part of premises be left undisturbed.
- Take photographs, measurements and recordings.
- Sample articles and substances in the premises and the atmosphere in the vicinity.
- Order the dismantling or testing of articles or substances that have or are likely to cause danger.
- Take possession of articles and substances for examination and evidence.
- Require answers to questions with a signed statement, if necessary.
- Inspect and copy statutory books and documents or any other relevant documents.
- Require facilities and assistance when carrying out their duty.
- Order a medical examination.

- Issue improvement notices/citations and prohibition notices.
- Take steps to prosecute.

The consequences of non-compliance will depend on the powers of the enforcement agencies. The consequences can include the enforcement agencies doing the following:

- Issuing an enforcement notice that requires the employer to improve health and safety. The fact that an enforcement notice has been issued will usually be placed on the employer's record held by the enforcement agency and may guide decisions on the consequences of future non-compliance.
- Issuing a legal agreement that requires specified actions to be taken by the employer to comply with legislation, for example 'Enforceable Undertakings' are used in parts of Australia. These legal agreements will usually involve the employer in training, education and supervision for a period. They will also usually include a commitment to good future health and safety standards and actions to prevent the specific non-compliance happening again.
- Issuing a notice that prohibits part or whole of an employer's work activities.
- Withdrawal of a licence, where a licence is required to carry out a specified activity.
- Issuing a formal caution, that specifies the non-compliance and warns that repeat offences may result in taking action to prosecute.
- Issuing a fixed penalty fine, given by the enforcement agency without the need to prosecute in a criminal court.
- Taking action to prosecute the employer in a criminal court. This can result in fines, prison sentences, court orders that require action to be taken, continued prohibition of unsafe actions till risks are improved and prevention of people involved holding a controlling role in companies. In some countries, for example the USA, the court may order the company and directors to be placed under probation for offences committed. Any further offence within a set time limit would result in the original offence being punished also.

Details of notices, fines, prosecutions and court actions may be made available for public scrutiny or publicised in some other way. In some countries, for example some states in Australia, the court may order that the employer publicise their offence.

THE ROLE OF INTERNATIONAL STANDARDS

The international organisation for standardisation

The International Organisation for Standardisation (ISO) is an international standard-setting body composed of representatives from various national standards bodies from throughout the world. Founded in 1947, the organisation produces world-wide industrial and commercial standards. While the ISO defines itself as a non-governmental organisation (NGO), its standards often become law, either through treaties or national standards. In practice, the ISO acts as a consortium with strong links to Governments. There are approximately 158 members of the ISO, each of which represents one country. ISO's main products are the international standards, for example, ISO 9001:2015 'Quality management systems – Requirements' and ISO 14001:2015 'Environmental management systems – Requirements with guidance for use'. More recently the ISO produced its international standard on health and safety management systems, ISO 45001:2018 'Occupational health and safety management systems – requirements with guidance for use'. The ISO also creates technical reports, technical specifications, publicly available specifications, technical corrigenda (minor amendments to standards), and guides.

ISO international standards are not binding on either Governments or industry even though they are internationally agreed standards. This is to allow for situations where certain types of standards may conflict with social, cultural or legislative expectations and requirements. This also reflects the fact that national and international experts responsible for creating these standards do not always agree and not all proposals become standards by unanimous vote. The individual member countries and their standards bodies can decide whether to adopt international standards written by the ISO and make them apply to their country.

International standards like ISO 45001 provide a common approach to achieving something. Because they are developed from wide technical and practical experience they provide a well-developed basis that should be suitable and effective in most situations. This helps to save those using them a great deal of time in developing their own standards, which is very important for smaller organisations. It also encourages international consistency, allowing global organisations to establish approaches, such as health and safety management systems, that can apply throughout the countries they operate in. This helps to enable organisations to operate with good standards and reduces unnecessary technical barriers to trade. The harmonisation of the views of the countries involved helps to establish effective global policy and regulatory compatibility.

INTERNATIONAL LABOUR ORGANISATION (ILO) CONVENTION C155

The International Labour Organisation

The International Labour Organisation (ILO) states that it is "devoted to advancing opportunities for women and men to obtain decent and productive work in conditions of freedom, equity, security and human dignity. Its main aims are to promote rights at work, encourage decent employment opportunities, enhance social protection and strengthen dialogue in handling work-related issues. In promoting social justice and internationally recognised human and labour rights, the organisation continues to pursue its founding mission that labour peace is essential to prosperity. Today, the ILO helps advance the creation of decent jobs and the kinds of economic and working conditions that give working people and business people a stake in lasting peace, prosperity and progress."

The ILO was founded in 1919. It is a 'tripartite' United Nations agency that brings together representatives of governments, employers and workers to jointly shape policies and programmes. The ILO derives international treaties, called Conventions, which countries that are members of the ILO sign to accept the Convention. This is called ratification of conventions. When the member countries accept the convention they are expected to meet the requirements in it. An example of a convention on health and safety is the Occupational Safety and Health Convention C155 (Convention C155), 1981. This convention requires member countries to adopt a coherent national occupational health and safety policy that includes action by governments and employers to promote occupational health and safety and to improve working conditions. Convention C155 requires member countries to establish national legislation to meet the requirements of the convention, including the creation of legal responsibilities for employers to ensure health and safety.

Occupational safety and health convention C155

The purpose of Convention C155 is to encourage countries to formulate, implement and periodically review a coherent national policy on occupational safety, health and working environment that reflects the globally agreed requirements established in the convention. The aim of the national policy is to "prevent accidents and injury to health arising out of, linked with or occurring in the course of work, by minimising, so far as is reasonably practicable, the causes of hazards inherent in the working environment."

The requirements of Convention C155 include action at the national level to establish strategies that support the development of good health and safety standards in the workplace, for example introduction of laws, guidance to employers, reporting of accidents and diseases, learning through holding inquiries into reports of harm and an enforcement system that provides adequate penalties for violation of laws. Other requirements include "action at the level of the undertaking" to establish employers' responsibilities for health and safety in the workplace, workers' rights/responsibilities and how workers/worker representatives co-operate with the employer.

The ILO also produces a number of other documents in support of Conventions:

- Recommendations, for example 'Occupational Safety and Health Recommendation R164' (Recommendation R164), 1981, which provides guidance on how Convention C155 should be met. Recommendations are not binding on member countries, but provide supporting advice to member countries that is useful when meeting conventions.
- Guidelines, for example 'Guidelines on Health and Safety Management Systems, ILO-OSH 2001'.
- Codes of practice, for example 'Ambient Factors in the Workplace'.
- Reports on issues that affect health and safety.

In addition, the ILO produces a number of informative books, including the very useful health and safety encyclopaedia.

Employer's responsibilities

According to Convention C155 employers have multiple responsibilities. These include ensuring that, so far as is reasonably practicable, the workplaces, machinery, equipment and processes under their control are safe and without risk to health.

Figure 1-5: Storage of gas cylinders.
Source: RMS.

They must also ensure that, so far as is reasonably practicable, the chemical, physical and biological substances and agents under their control are without risk to health when the appropriate measures of protection are taken.

This includes provision, where necessary, of adequate protective equipment and protective clothing to prevent risk of accidents/incidents or adverse effects on health. Employers must also provide measures to deal with emergencies and accidents/incidents, including adequate first-aid arrangements. Whenever two or more employers are engaged in activities at the same time and at one workplace, they must collaborate in applying the above requirements.

Recommendation R164 clarifies the employers responsibilities defined in Article 16 of Convention C155.

The term 'worker' used in ILO Convention C155, Recommendation R164 and this element is the same as the term 'employee'.

Clause 10 of Recommendation R164 clarifies that typical responsibilities established in national legislation by ILO member countries might include responsibilities to:

- Provide and maintain workplaces *(safe place of work)*, machinery and equipment *(safe plant and equipment)*.

- Use work methods that are safe and without risk to health *(safe system of work)*.

- Give necessary instructions and *training,* taking account of the functions and capacities of different categories of workers *(competent workers)*.

- Provide adequate *supervision* of work, of work practices and of application and use of occupational safety and health measures.

- Establish organisational arrangements regarding occupational health, safety and the working environment that are adapted to the size of the undertaking and the nature of its activities.

- Provide, without any cost to the worker, adequate personal protective clothing and equipment which are reasonably necessary when hazards cannot be otherwise prevented or controlled.

- Ensure that work organisation, particularly with respect to hours of work and rest breaks, does not adversely affect occupational safety and health.

- Take measures to eliminate excessive physical and mental fatigue.

- Undertake studies and research or otherwise keep up to date with the scientific and technical health and safety knowledge.

- Collaborate with other employers where their work activities take place at the same time and in the same workplace.

Figure 1-6: Office workplace hazards.
Source: Unsplash.

- Take measures to facilitate co-operation between the managers and workers, for example by the appointment of worker health and safety representatives and health and safety committees to enable consultation with workers or their representatives.

- Verify implementation of applicable health and safety standards, for example by regular workplace environmental monitoring and conducting health and safety audits.

- Keep health, safety and working environment records and provide reports as required by legislation or other competent authorities. For example, records of occupational accidents/incidents and ill-health.

- The employer, having regard to the nature of activities and size of the organisation, should also:

- Set out in writing their health and safety policy, responsibilities and arrangements to meet the policy

- Make provision for the availability of an occupational health service and a safety service, within the organisation, jointly with other employers or under arrangements with an outside body.

- Make provision for obtaining specialists advice on particular occupational health or safety problems or the supervision of the application of measures to meet them.

In addition, an employer is generally expected to carry out risk assessments in order to meet the specific responsibilities described earlier. In some countries the employers' responsibilities are defined in civil and criminal law.

An example of a country that has established general employers' responsibilities in criminal law is the United States of America (USA), in the Occupational Safety and Health Act 1970:

"Section 5 - Duties:

Each employer - shall furnish to each of his employees employment and a place of employment which are free from recognized hazards that are causing or are likely to cause death or serious physical harm to his employees. Shall comply with occupational safety and health standards promulgated under this Act."

Similarly, in Malaysia general responsibilities of employers and the self-employed were established by the Occupational Safety and Health Act 1994 include:

"Section 15 - General duties of employers and self-employed persons to their employees.

(1) It shall be the duty of every employer and every self-employed person to ensure, so far as is practicable, the safety, health and welfare at work of all his employees."

In common with a number of other countries and reflecting some of clause 10 of Recommendation R164, Malaysia clarifies this general duty in section 15 of their Occupational Safety and Health Act 1994 with specific duties of employers that included:

- Provision and maintenance of plant and systems of work.
- Safety and absence of risks to health in connection with the use or operation, handling, storage and transport of plant and substances.
- Information, instruction, training and supervision as is necessary.
- Place of work maintained in a condition that is safe and without risks to health and the provision and maintenance of the means of access to and egress.
- Provision and maintenance of a working environment and facilities for welfare at work.

Similar legislation has been in place in South Africa since 1993 through the Occupational Health and Safety Act 1993. Requirements of this Act include an interesting responsibility of employers to deal with hazards in a form of hierarchical control, section 8, the general duties of employers to employees, subsection 2b requires:

"Taking such steps as may be reasonably practicable to eliminate or mitigate any hazard or potential hazard to the safety or health of employees, before resorting to personal protective equipment."

From the examples provided it can be seen that the employers' responsibilities established in ILO Recommendation R164 are typical of the duties countries create in National legislation. Although there are many aspects to an employer's responsibilities the basic responsibilities will include:

- Safe place of work.
- Safe plant and equipment.
- Safe systems of work.
- Training and supervision.
- Competent workers.

Workers' responsibilities and rights

Similar to employers, in the Occupational Safety and Health Convention C155, 1981, the ILO sets out workers' responsibilities to protect themselves and those around them from harm. The term 'worker' used in ILO Convention C155 and in this section is the same as the term 'employee'. Workers are expected to co-operate with the employer with regard to obligations placed upon the employer, including reporting any situation that presents imminent risk or serious danger. In order to achieve this, workers should receive adequate information and training on measures taken by the employer to secure occupational health and safety. They or their representatives should be consulted by the employer on all aspects of occupational health and safety associated with their work. Workers should not be made to work in situations of continuing imminent or serious danger, until the employer takes remedial action.

Worker responsibilities

ILO Convention C155 emphasises that employers need to share responsibility for health and safety at work, therefore employers and employees are both given responsibilities, they reflect their different roles in achieving health and safety. Article 19 of Convention C155 provides the following responsibilities of workers.

- Workers, in the course of performing their work, must co-operate with their employer in meeting the employer's responsibilities.
- Workers must promptly report to their supervisor any situation which they have reasonable justification to believe presents an imminent and serious danger to their life or health. Until the employer has taken remedial action, if necessary, the employer cannot require workers to return to a work situation where there is continuing imminent and serious danger to life or health.

Recommendation R164 clarifies the worker responsibilities set out in Convention 155 and expects that workers:

"(a)Take reasonable care for their own safety and that of other persons who may be affected by their acts or omissions at work.

(b) Comply with instructions given for their own safety and health and those of others and with safety and health procedures.

(c) Use safety devices and protective equipment correctly and do not render them inoperative.

(d) Report forthwith to their immediate supervisor any situation which they have reason to believe could present a hazard and which they cannot themselves correct.

(e) Report any accident or injury to health which arises in the course of or in connection with work."

In the USA the Occupational Safety and Health Act 1970 defines workers' responsibilities in a general way in section 5:

"(b) Each employee shall comply with occupational safety and health standards and all rules, regulations, and orders issued pursuant to this Act which are applicable to his own actions and conduct."

In Malaysia the Occupational Safety and Health Act 1994 establishes a number of duties to define and clarify the responsibilities of workers:

"Section 24 - General duties of employees at work.

(1) It shall be the duty of every employee while at work:

(a) To take reasonable care for the safety and health of himself and of other persons who may be affected by his acts or omissions at work.

(b) To co-operate with his employer or any other person in the discharge of any duty or requirement imposed on the employer or that other person by this Act or any regulation made there under.

(c) To wear or use at all times any protective equipment or clothing provided by the employer for the purpose of preventing risks to his safety and health; and

(d) to comply with any instruction or measure on occupational safety and health instituted by his employer or any other person by or under this Act or any regulation made hereunder.

(2) A person who contravenes the provisions of this section shall be guilty of an offence and shall, on conviction, be liable to a fine not exceeding one thousand ringgit or to imprisonment for a term not exceeding three months or to both."

"Section 25 - Duty not to interfere with or misuse things provided pursuant to certain provisions.

A person who intentionally, recklessly or negligently interferes with or misuses anything provided or done in the interests of safety, health and welfare in pursuance of this Act shall be guilty of an offence and shall, on conviction, be liable to a fine not exceeding twenty thousand ringgit or to imprisonment for a term not exceeding two years or to both."

Legislation similar to that in Malaysia has been in place in South Africa since 1993 through the Occupational Health and Safety Act 1993. Requirements of this Act include worker responsibilities regarding reporting health and safety matters, Section 14, General duties of employees at work includes:

"(d) If any situation which is unsafe or unhealthy comes to his attention, as soon as practicable report such situation to his employer, or to the health and safety representative for his workplace or section thereof, as the case may be, who should report it to the employer.

(e) If he is involved in any incident which may affect his health or which has caused an injury to himself, report such incident to his employer or to anyone authorized thereto by the employer, or to his health and safety representative, as soon as practicable but not later than the end of the particular shift during which the incident occurred, unless the circumstances were such that the reporting of the incident was not possible, in which case he shall report the incident as soon as practicable thereafter."

A number of approaches have been adopted by different countries to establish workers' responsibilities, which has resulted in a range of legal duties being created for workers. The ILO Convention C155 and Recommendation R164 provide a useful summary of the international perspective on workers' responsibilities.

Employers will often include similar requirements to those provided in Article 16 of Recommendation R164 in the responsibilities they define for their workers. However, the responsibilities they define may be written in a simpler way and may include additional responsibilities to comply with specific rules the employer has made. The main health and safety responsibilities of workers, as they are defined by employers, are generally taken to include workers' responsibilities to:

- Take reasonable care for their own health and safety and that of other people that might be affected by their acts or omissions, for example, fellow workers or members of the public.

- Co-operate with their employer, including taking part in consultation processes.

- Follow rules, instructions and procedures.

- Use health and safety devices and protective equipment properly and not misuse them.

- Report accidents/incidents, work-related ill-health and any situation in the workplace that they believe could present a hazard and which they cannot themselves correct.

Employers may also define additional responsibilities that clarify what is required of workers when they are taking care of their own or other people's health and safety, for example, responsibilities to keep their own work area safe and healthy and not to drink alcohol or take drugs while at work.

Rights

ILO Convention C155 emphasises that workers must have certain rights to enable them to be involved in health and safety at work, these rights enable them to co-operate with the employer and provide them with protection from discrimination by their employer for things they do while carrying out their rights and responsibilities. Articles 17 and 19 to 21 of Convention C155 provides the following rights to workers:

- Workers and their representatives must be given appropriate training in health and safety.
- Workers or their representatives, in accordance with national law and practice, can enquire into all aspects of health and safety associated with their work and be consulted by the employer. For this purpose technical advisers may, by mutual agreement, be brought in from outside the organisation.
- Employers must make arrangements so that workers and/or their representatives can co-operate with the employer on health and safety. Co-operation between management and workers and/or their representatives must be an essential element of organisational and other measures taken.
- Representatives of workers must be given adequate information on measures taken by the employer to secure health and safety and may consult their representative organisations about this information provided they do not disclose commercial secrets.
- Occupational health and safety measures must not involve any expenditure for the workers.

Article 17 of Recommendation R164 establishes an expectation that, where a worker reports health and safety issues, they should not be discriminated against by their employer, *see Figure 1-7.*

"No measures prejudicial to a worker should be taken by reference to the fact that, in good faith, he complained of what he considered to be a breach of statutory requirements or a serious inadequacy in the measures taken by the employer in respect of occupational safety and health and the working environment."

Figure 1-7: Prejudicial action against workers that complain. Source: ILO, Occupational Safety and Health Recommendation R164, Article 17.

In Malaysia the Occupational Safety and Health Act 1994 establishes a number of rights of workers:

"Section 26 - Duty not to charge employees for things done or provided.
No employer shall levy or permit to be levied on any employee of his any charge in respect of anything done or provided in pursuance of this Act or any regulation made hereunder.
Section 27 - Discrimination against employee, etc.

1) *No employer shall dismiss an employee, injure him in his employment, or alter his position to his detriment by reason only that the employee:*

   a) *Makes a complaint about a matter which he considers is not safe or is a risk to health.*

   b) *Is a member of a safety and health committee established pursuant to this Act.*

   c) *Exercises any of this functions as a member of the safety and health committee*

2) *No trade union shall take any action on any of its members who, being an employee at a place of work:*

   a) *Makes a complaint about a matter which he considers is not safe or is a risk to health.*

   b) *Is a member of health and safety committee established pursuant to this Act.*

   c) *Exercises any of his functions as a member of the safety and health committee.*

3) *An employer who, or a trade union which, contravenes the provisions of this section shall be guilty of an offence and shall, on conviction, be liable to a fine not exceeding ten thousand ringgit or to a term of imprisonment not exceeding one year or to both.*

4) *Notwithstanding any written law to the contrary, where a person is convicted of an offence under this section the Court may, in addition to imposing a penalty on the offender, make one or both of the following orders:*

   a) *An order that the offender pays within a specific period to the person against whom the offender has discriminated such damages as it thinks fit to compensate that person.*

   b) *An order that the employee be reinstated or re-employed in his former position or, where that position is not available, in a similar position."*

Section 15 of the Trinidad and Tobago Occupational Safety and Health Act 2004 (as amended) follows a similar approach and includes circumstances where an employee may refuse to work on grounds of health and safety, their obligation to report to the employer or worker representative and the need for the health and safety committee to investigate the circumstances. Circumstances include the following, but do not relate to those employed in emergency services where it would be "inherent in the employee's work":

Section 15 says - An employee may refuse to work or do particular work where he has sufficient reason to believe that:

a) *"There is serious and imminent danger to himself or unusual circumstances have arisen which are hazardous or injurious to his health or life."*

Furthermore, under the same legislation Section 20 affords employees acting in accordance with their rights protection in law:

"No employer or person acting on behalf of an employer shall:

b) *Dismiss or threaten to dismiss a worker;*

c) *Discipline or suspend or threaten to discipline or suspend a worker;*

d) *Impose any penalty upon a worker, or intimidate or coerce a worker, because the worker has acted in compliance with this Act."*

It can be seen from the worker responsibilities and rights depicted in the ILO convention and recommendations, as well as the examples from the national legislation of member countries, that workers play a significant role in health and safety. Their co-operation with the employer is essential to make health and safety effective. This includes reporting unsafe and unhealthy situations. They are also expected to work in a way that takes care of themselves and others.

REVIEW

What are employer's health and safety responsibilities?

SOURCES OF INFORMATION ON NATIONAL STANDARDS

Sources of information on National Standards for health and safety will typically include National legislation, official government guidance and codes of practice to support the legislation, information provided during enforcement action by enforcement agencies, court actions and government/commercial standard setting organisations.

Legislation - acts, decrees, regulations and orders

An important source of information on National Standards for health and safety is legislation created by each country to regulate health and safety nationally. This can be in a variety of forms depending on the country and the structure by which legislation is created, for example it may be in the form of acts, decrees, regulations and orders. These are used to set out a framework of national standards of occupational health and safety by means of a legislative structure.

The national legislation will provide information on National Standards by identifying work-related risks and defining actions to reduce risks. The legislation may establish health and safety standards for specific risks, for example control limit levels for exposure to noise, or more general standards, for example requirements to have a health and safety management system.

The legal requirements for health and safety standards that are expected of employers are made as duties created by the National legislation. These duties are usually expressed in the form of absolute and qualified duties. Typical qualified duties are 'practicable', 'reasonably practicable' and 'reasonable'. National legislation will often use these duties to create four levels of duty, 'absolute' duties being the highest and 'reasonable' being the lowest. Whether a duty set by legislation is absolute or qualified may depend on the risk being controlled and how important the control measure is to reducing risk. The four levels of duty are:

- Absolute.
- Practicable.
- Reasonably practicable.
- Reasonable

The duties used by national organisations when setting health and safety standards will often reflect the absolute and qualified requirements in ILO conventions.

Absolute duties

Absolute duties do not allow choice and take no allowance of how much effort is required to meet them. An employer is expected to comply fully with an absolute duty whatever the cost or difficulty. Absolute duties usually relate to fundamental preventive measures that all employers must apply and to measures to deal with high risk work activities. Absolute requirements can be identified in national legislation by the use of words such as 'shall' and 'must'. The Irish Safety, Health and Welfare at Work (Construction) Regulations 2013 creates an absolute duty when making a requirement to appoint a project supervisor for health and safety matters on major construction projects.

"PART 2 - DESIGN AND MANAGEMENT

Duties of clients - appointments of project supervisors.

6. (1) Except as provided for in paragraphs (5) a client shall appoint, in writing, for every project -

(a) a competent project supervisor for the design process, and

(b) a competent project supervisor for the construction stage,

and the client shall obtain written confirmation of acceptance of each of the appointments."

Similarly, the Malaysian Occupational Safety and Health Act 1994 creates an absolute duty to formulate a health and safety policy, an essential item for successful health and safety.

"Section 16. Duty to formulate safety and health policy

Except in such cases as may be prescribed, it shall be the duty of every employer and every self-employed person to prepare and as often as may be appropriate revise a written statement of his general policy with respect to the safety and health at work of his employees and the organization and arrangements for the time being in force for carrying out that policy, and to bring the statement and any revision of it to the notice of all of his employees."

The Irish Safety, Health and Welfare at Work (Construction) Regulations 2013, Regulation 51, creates an absolute duty to take precautions related to excavations and similar construction work. This is a high risk activity that can result in death of workers if control measures are not used, therefore it is established as an absolute duty to take the specified measures.

"51. (1) A contractor responsible for a construction site shall ensure for that site that adequate precautions are taken in any excavation, shaft, earthwork, underground works or tunnel to -

(a) guard against danger to persons at work from a fall or dislodgement of earth, rock or other material by suitable shoring or otherwise,

(b) guard against dangers arising from the fall of materials or objects or the inrush of water, mud, sand or other material that might flow into the excavation, shaft, earthworks, underground works or tunnel,

(c) secure adequate ventilation at all workstations so as to maintain an atmosphere fit for respiration and to limit any fumes, gases, vapours, dust or other impurities to levels which are not dangerous or injurious to health,

(d) guard against the occurrence of fire or flooding,

(e) enable persons at work to reach safety in the event of fire or an inrush of water or materials,

(f) avoid risk to persons at work arising from possible underground dangers such as underground cables or other distribution or transmission systems, the circulation of fluids or the presence of pockets of gas, by undertaking appropriate investigations to locate them before excavation begins, and

(g) provide a safe means of access to and egress from each place of work."

These examples show the use of 'absolute' duties in National legislation to control the provision of fundamental preventive measures that all employers must apply and to measures to deal with high risk work activities.

Practicable duties

Practicable duties have to be met to an extent only limited by the current state of knowledge and invention, irrespective of cost or difficulty. This duty is often used for control of the provision of important or high risk precautions that involve applying technical solutions. As technical solutions may be influenced by technical characteristics of the work the duty allows the employer to adjust the solution to meet these characteristics and achieve health and safety. In addition technical solutions may change over time as "knowledge and invention" produces alternative and better solutions. Employers are expected to use these solutions to meet the duty, therefore the duty expects employers to keep up to date with alternative solutions and apply them promptly when they become generally available.

The Irish Safety, Health and Welfare at Work (Construction) Regulations 2013 creates in Regulation 55 a duty that must be met as far as is practicable when fencing excavations:

55. (1) A contractor responsible for a construction site shall ensure for that site that every accessible part of an excavation, shaft, pit or opening in the ground near to which persons are working and into or down which a person is liable to fall a distance liable to cause personal injury -

(a) has a suitable barrier placed as close as is practicable to the edge, or

(b) is securely covered.

In the UK the Provision and Use of Work Equipment Regulations (PUWER) 1998, Regulation 11, requires:

"The provision of fixed guards enclosing every dangerous part or rotating stock-bar where and to the extent that it is practicable to do so."

These examples show the use of 'practicable' duties in National legislation to enable technical solution to reflect the technical characteristics of the work activities and the solutions available.

Reasonably practicable duties

A duty that has to be carried out 'as far as is reasonably practicable' is one where employers are entitled to balance the costs of actions to comply with requirements against the risk. The costs of the action to comply can include time, trouble and money. The risk will depend on the likelihood and severity of harm from non-compliance. Where the risk is low the costs of taking action can be balanced according to this and may also be low. Where the risk is high the costs of action taken to reduce the risk and to comply with the duty may also be expected to be high. Only where the costs of taking a specific action greatly outweigh the risk should alternative actions with lower costs be used to comply.

Figure 1-8: Reasonably practicable.
Source: RMS/Shutterstock.

This type of duty ensures that the costs of actions taken to control risk are correctly balanced against the risk. It particularly suits situations where national standards have to fit a wide range of risk. This range could include the type and level of risk. The duty can be applied to a wide variety of situations because in meeting this duty solutions can be matched to the level of risk. For example, in Section 8 of the South African Occupational Health and Safety Act 1993 it creates a general duty that uses the term 'as far as is reasonably practicable' to define the extent of the duty:

"8. General duties of employers to their employees

1. Every employer shall provide and maintain, as far as is reasonably practicable, a working environment that is safe and without risk to the health of his employees."

Similarly, Section 8 of the Irish Safety, Health and Welfare Act 2005 requires:

"8) (1) Every employer shall ensure, so far as is reasonably practicable, the safety, health and welfare at work of his or her employees."

There are great advantages in using 'reasonably practicable' duties in National legislation to enable solutions to be balanced against risk. This enables a cost effective approach to the management of health and safety.

Reasonable

Reasonable requirements are those requirements that might normally be expected of an average person taking care of what they are doing. For example, on a national level, Section 24 of the Malaysian Occupational Safety and Health Act 1994 requires:

"It shall be the duty of every employee while at work:
(a) To take reasonable care for the safety and health of himself and of other persons who may be affected by his acts or omissions at work."

A similar approach has been taken in the UK when expressing employees' duties under the Health and Safety at Work etc. Act 1974.

These examples show the use of 'reasonable' duties in National legislation to ensure employees are involved in health and safety, but with a lower level of duty than employers. Employers usually have duties set at a higher level, having absolute, practicable and reasonably practicable duties.

REVIEW

What is meant by the terms: absolute, practicable, reasonably practicable and reasonable?

Codes of practice and official guidance

In many countries, the legislation that establishes requirements for health and safety also establishes agencies for standard setting and regulation of compliance with legislation. Because the legislation may not provide sufficient detail to enable employers to meet requirements the agencies provide national standards, approved codes of practice and official guidance to assist with compliance.

These standards and approved codes of practice are set out to establish minimum levels of compliance, and non-compliance is usually punishable by prosecution or other sanctions. For example, in the United States of America (USA) the Occupational Safety and Health Act 1970 established a requirement to meet national standards and established the Occupational Safety and Health Administration (OSHA) to set and enforce protective workplace health and safety standards. OSHA standards are rules that describe the methods employers are legally required to follow to protect their workers from hazards. Before OSHA can issue a standard, it must go through a very extensive and lengthy process that includes substantial public engagement, notice and comment. The agency must show that a significant risk to workers exists and that there are feasible measures employers can take to protect their workers.

Examples of OSHA standards include requirements to provide fall protection, prevent trenching collapse, prevent exposure to some infectious diseases, ensure the safety of workers who enter confined spaces, prevent exposure to such harmful substances as asbestos and lead, put guards on machines, provide respirators or other safety equipment, and provide training for certain dangerous jobs.

A similar national standard setting arrangement exists in the UK, where the Health and Safety Executive can provide a recognised interpretation of how an employer may comply with health and safety legislation in the form of an approved code of practice (ACOP). The ACOP is not a statutory instrument and it does not have direct legal binding force, therefore, it is not automatically a criminal offence to break it. However, it can be cited in evidence in a court considering a case of prosecution for breach of health and safety legislation and a person who breaks it is much more likely to be held to be in breach of law. Employers must either meet the standards contained in an ACOP or show that they have complied with an equal or better standard. Examples of UK ACOPs include 'Control of Asbestos Regulations 2012. Approved Code of Practice and Guidance', 'Safe Use of Work Equipment' which supplements the Provision and Use of Work Equipment Regulations (PUWER) 1998 and 'Workplace Health, Safety and Welfare' supplementing the Workplace (Health, Safety and Welfare) Regulations (WHSWR) 1992.

Different countries give different status to official guidance, in some it is legally binding and in others it provides more general advice. Enforcement agencies and Government Departments may therefore issue guidance notes that are purely advisory and have no standing in law. The advice in guidance note form is generally more practical than that contained in a standard or an ACOP, and may be referred to in legal cases as persuasive argument of what may have been done to prevent a breach of legislation or injury. This guidance may help explain standards/ACOPs, hazards/risks reduction strategies and provide assistance in developing effective health and safety programmes.

Enforcement agencies

The enforcement agencies may provide information on health and safety national standards directly through the various ways they use to influence the performance of organisations. These may include:

- The provision of information such as codes of practice, guidance notes, leaflets, statistics, website information.

- Campaigns for reducing risk and campaigns aimed at particular business sectors.

- Carrying out health and safety inspections, including inspections following receipt of complaints of poor workplace conditions. Follow up visits may be made to ensure compliance with the advice given.

- Carrying out investigation of accident/incident and incidence of ill-health.

- Examination of arrangements and systems in place to manage risk.

- Giving advice and assistance to employers on how to comply with their statutory duties.

- Enforcement action, such as the issue of improvement or prohibition notices and taking criminal proceedings.

Role of the courts

When a case is heard in court it can provide insight into the application of legislation to a specific situation. Over time, the cases heard can provide a better practical understanding of what the legislation expects and what the national standards for health and safety are.

CONSIDER

Consider sources of information on National standards available in your workplace, are they sufficient for you to carry out your job to an appropriate professional standard?

1.3 Who does what in organisations

EMPLOYERS

The employer has overall responsibility for the protection of workers' health and safety. The employer must clearly **allocate responsibility**, accountability and authority for the development, implementation and performance of the health and safety management system, including the policy and the achievement of objectives. This will require the employer to establish health and safety as a known and accepted line-management responsibility at all levels, and identify important roles with responsibility, authority and accountability for the identification, evaluation and control of health and safety hazards and risks. A clear delegation of responsibilities will assist in sharing out the health and safety workload, ensuring contributions from different levels and functions of the organisation. It will also help to establish clear lines of reporting and communication. The employer should ensure that health and safety roles and responsibilities of individuals are defined in the organisation's health and safety management system.

This will include details related to the most senior accountable person, line managers, supervisors, workers and their representatives as well as those persons who have specific roles and responsibilities for health and safety, including health and safety practitioners, occupational health advisers, occupational hygienists, engineers, fire officers and first-aiders.

Ensuring that all individuals in an organisation are aware of their roles and responsibilities will increase their motivation, indicate the commitment and leadership of senior management and help to improve the health and safety culture within the organisation as a whole. Defining clear roles and responsibilities will assist in the identification of individual competencies and training needs, particularly for specific roles, for example roles relating to first aid and fire arrangements. It is usual to express how health and safety is organised by means of an organisational chart, which shows lines of communication and feedback loops for both line managers and other people, for example health and safety practitioners.

ROLES AND RESPONSIBILITIES OF DIRECTORS/MANAGERS/SUPERVISORS

Directors

In many situations directors and senior managers (sometimes called top managers) are appointed by employers to ensure employers' health and safety responsibilities are met. The role of directors and senior managers will include ensuring that the aims, objectives and overall approach to health and safety are suitable and sufficient. Employers give directors and senior managers health and safety responsibilities (an obligation to achieve something) relating to their roles.

The directors and senior managers are accountable (personally answerable for what is done or not done) to their employer for meeting these responsibilities. The importance of the role and responsibilities of directors and senior managers was emphasised in the guidance document on occupational safety and health (OSH) roles and responsibilities that accompanied the Abu Dhabi Occupational Safety and Health System Framework (OSHAD-SF), see Figure 1-9.

> Actively involved and accountable senior managers, who drive health and safety activities, are more likely to deliver successful organizational OSH outcomes.

Figure 1-9: Senior management role.
Source: Abu Dhabi Occupational Safety and Health Systm Framework (OSHAD-SF) Guidance Document - OSH Roles and responsibilities.

In practice, the responsibilities of directors and senior managers will include ensuring a suitable management system is established and maintained. This will require the directors and senior managers to establish an appropriate health and safety policy, measurable health and safety objectives and programmes to meet the objectives. Their responsibilities will also include ensuring risk assessments have been carried out, management arrangements are made and control measures are in place and used. In addition, they will usually have a responsibility to lead by example and participate in health and safety leadership activities like visits to the workplace. It is usual for them to have specific responsibilities to ensure that health and safety needs are resourced, that they receive and consider reports on health and safety and that action is taken to ensure deficiencies are dealt with and objectives are met. They are also expected to be responsible for arranging competent health and safety assistance and conducting a regular (usually annual) review of health and safety performance. Their responsibilities will also include establishing the roles and responsibilities of middle managers, people with primary health and safety functions, and workers. Directors and senior managers are also responsible for establishing arrangements to hold these people to account for the responsibilities they are given.

In addition to the responsibilities directors and senior managers have in relation to the organisation they will be given individual responsibilities relating to their specific role in the organisation, for example, the management of a region, division or site.

The role of senior management in meeting legal requirements was emphasised in the guidance document on environment, health and safety (EHS) roles and responsibilities that accompanied the Abu Dhabi environment, health and safety management system (EHSMS) Regulatory Framework, see Figure 1-10.

> Senior management is in the best position to demonstrate its commitment to, and provide leadership in, the development and implementation of OSHMS programs to meet legal requirements.

Figure 1-10: Senior management role in meeting legal requirements.
Source: Abu Dhabi Occupational Safety and Health System Framework (OSHAD-SF) Guidance Document - OSH Roles and responsibilities.

In some countries directors and senior managers can be prosecuted for failing to carry out responsibilities and causing their employer to be in breach of National health and safety laws. For example, in the UK, Section 37 of the Health and Safety at Work etc. Act (HASAWA) 1974 and in Malaysia Section 52 of the Occupational Safety and Health Act 1994 provide the means to prosecute directors and senior managers in these circumstances.

Middle Managers

Middle managers play a particularly important role in ensuring the potential causes of accidents/incidents and ill-health are identified and controlled. In a similar way to senior managers of an organisation, middle managers have a responsibility to ensure health and safety is effectively established in their area of control. If they fail to do this they could be held accountable for their failings. The responsibilities of middle managers generally reflect the main responsibilities of employers established in the ILO Occupational Safety and Health Recommendation R164, 1981, discussed earlier in this element. This is because the middle managers, on behalf of the employer, have the responsibility to ensure that the employer's responsibilities are met. Middle managers will typically have responsibilities that include ensuring that:

- There is provision and maintenance of workplaces and equipment that are safe and without risk to health, as far as possible.

- Risk assessments are conducted, maintained and reviewed.

- Appropriate rules and procedures for health and safety are established.

- Work methods that are used are safe and without risk to health.

- Necessary instructions and training are given.

- There is adequate supervision of work.

- There is adequate personal protective clothing and equipment for situations where hazards cannot be otherwise prevented or controlled.

- They organise work so that it does not adversely affect occupational health and safety, particularly with respect to hours of work, rest breaks and physical/mental fatigue.

- There is co-operation between other managers, workers and contractors.

- Participate in consultation with workers or their representatives.

- Take part in accident/incident investigations and work-related ill-health.

- There is evaluation of compliance with arrangements, procedures and rules, including by the use of workplace environmental monitoring and health and safety audits.

- Appropriate corrective and improvement actions are taken when required.

- Appropriate health and safety records are kept and reports provided, for example, of accidents/incidents and ill-health.

The active involvement of managers in achieving health and safety is important for its success, this includes the role of the line manager in ensuring conformity with the policy. It is necessary for managers to arrange for the implementation of the policy and ensure work is being carried out in a safe and healthy way. It is essential therefore that they monitor the effectiveness of the policy and the processes used to achieve it. Where the policy and processes are found not to be effective the managers should be able to communicate this, provide feedback on their observations and suggest improvements. It must be recognised that middle managers can only achieve success in controlling health and safety within the limits of their current knowledge, skill and experience. It is, therefore, important to define their responsibilities, keep the responsibilities up to date and ensure they have the competencies to meet their responsibilities.

Supervisors

The term 'supervisor' is often used for a first line manager, i.e. an employee who manages one or more workers in a work group or team. Unlike middle management, supervisors tend to be practically involved with the workers they supervise, often contributing directly to the aims of the task. For example a supervisor in a hotel kitchen would carry out similar tasks to those under supervision. Supervisors will know the capabilities of the work team and will be readily able to assess and communicate the feasibility of the tasks placed on them to their managers.

Supervisors' responsibilities are usually similar to those of middle managers, but relate to their role of providing immediate supervision of work activities. Supervisors therefore have the responsibility for control of work in their area of responsibility. They should take part in risk assessments and in the development of safe systems of work, ensuring that members of their team are trained in the systems once they have been introduced. Supervisors will have specific responsibilities relating to communication with their team and consultation with workers or their representatives. They also have a responsibility to set a good example and control the behaviour of their team to ensure that rules and procedures are complied with.

Supervisors would usually have a responsibility to carry out the initial investigation into accidents/incidents and ill-health, and to provide reports to their manager and health and safety practitioner. They should carry out inspections of their working areas and take action to control any work conditions or actions that are observed which are unsafe or unhealthy. Supervisors also have a responsibility to report to their manager situations where they do not have the authority to take the necessary action which would ensure health and safety.

HOW TOP MANAGEMENT CAN DEMONSTRATE COMMITMENT

Top management, for example, directors/senior management, should be active in the management of health and safety. They should demonstrate their commitment by taking action to meet their responsibilities, for example, providing resources for a health and safety management system. In addition, they should consider appointing a person from the senior management team to ensure health and safety is given equal priority.

ISO 45001:2018 Clause 5.1 requires:

"Top management shall demonstrate leadership and commitment with respect to the OH&S management system by:

a) taking overall responsibility and accountability for the prevention of work-related injury and ill health, as well as the provision of safe and healthy workplaces and activities;

b) ensuring that the OH&S policy and related OH&S objectives are established and are compatible with the strategic direction of the organization;

c) ensuring the integration of the OH&S management system requirements into the organization's business processes;

d) ensuring that the resources needed to establish, implement, maintain and improve the OH&S management system are available;

e) communicating the importance of effective OH&S management and of conforming to the OH&S management system requirements;

f) ensuring that the OH&S management system achieves its intended outcome(s);

g) directing and supporting persons to contribute to the effectiveness of the OH&S management system;

h) ensuring and promoting continual improvement;

i) supporting other relevant management roles to demonstrate their leadership as it applies to their areas of responsibility;

j) developing, leading and promoting a culture in the organization that supports the intended outcomes of the OH&S management system;

k) protecting workers from reprisals when reporting incidents, hazards, risks and opportunities;

l) ensuring the organization establishes and implements a process(es) for consultation and participation of workers (see 5.4);

m) supporting the establishment and functioning of health and safety committees, [see 5.4 e) 1)]."

Making resources available for a health and safety management system

When determining the resources needed to establish, implement and maintain a health and safety management system top management should consider the:

- Amount of support needed from top management.

- Financial resources required.

- People required to establish, implement and maintain the system, in particular managers, health and safety practitioners and worker representatives. It may be necessary to get assistance from people outside the organisation where there is a need for people with specific knowledge, skill and experience.

- Time required from those involved in establishing, implementing and maintaining the system.

- Technical information and equipment that might be required. For example, access to information technology (IT) to enable the written content of the management system to be made available to managers.

- Organisational resources to coordinate the involvement of managers and others in the development of the system. Organisational resources are also likely to be required to ensure those using the health and safety management system have a clear understanding of the system.

The amount of resources needed for the introduction of an effective health and safety management system should not be underestimated and may require substantial investment. This was emphasised in the ILO publication 'Fundamental Principles of Occupational Health and Safety', *see Figure 1-11*.

If a simple analogy in manufacturing terms is considered, introduction of a health and safety management system can require a commitment to the equivalent level of resources that an organisation might make following the identification of the need to install a new piece of equipment, for example, a production line to fill bottles in a milk packaging company.

Figure 1-11: Providing resources for health and safety.
Source: ILO, Fundamental Principles of Occupational Health and Safety.

A detailed project to install the new equipment would have to be established. This would involve the allocation of money to purchase and install the equipment, the installation planned, performance checks organised and managers and workers trained in its use. Perhaps more importantly, the equipment would need to be maintained for the whole of its life cycle, with periodic reviews of quality and efficiency of the process. The reviews would need to include changes in health and safety legislation that occurred over the period of its use, for example, a reduction in exposure levels for noise.

In the same way, the introduction of a health and safety management system would need to have money allocated, its design agreed, implementation planned, measures put in place to check it is working properly and managers and workers would need to be trained in how to use it. The management system would need to be reviewed and maintained to ensure it was kept up to date and working effectively. Similarly, the reviews would need to take account of changes in health and safety legislation. It can be seen from this analogy that introducing a health and safety management system requires a commitment to provide significant resources. It is important that senior management demonstrate their commitment to health and safety by ensuring adequate resources are made available.

Defining roles and responsibilities

It is important that everyone in an organisation knows their health and safety role, how they can contribute to achieving health and safety objectives and what the employer expects of them. This will help the organisation to achieve and maintain a good health and safety performance. The top management of an organisation should identify who needs to do what with regards to the management of health and safety. In particular, it is essential that top management clearly and concisely define in writing the roles and responsibilities of managers and others that have a specific role in health and safety. This will include defining roles and responsibilities for the most senior manager (top manager), members of the senior management team and anyone in the senior management team taking a lead role in health and safety.

The roles and responsibilities of all managers should be documented and defined in sufficient detail for them to understand what they are responsible for. When defining responsibilities, simple, summary statements like "the manager will be responsible for health and safety in their area" should be avoided unless more detail is provided to support what is meant by the summary statement.

Responsibilities of managers should at least include requirements for:

- Planning for health and safety - for example, responsibilities for organising meetings and a programme for risk assessments.

- Implementation of health and safety - for example, responsibilities for communication between managers, provision of information to workers, specific actions required to keep the workplace safe and healthy.

- Evaluation of health and safety - for example, responsibilities for carrying out workplace inspections and investigations into accidents/incidents.

It is also important that managers are provided with appropriate authority (power to make decisions and take action independently) to carry out their role and meet their responsibilities. This is emphasised in ILO-OSH 2001, *see Figure 1-12*. In addition, it is essential that managers are made aware of their responsibilities, the extent of their authority and what they are accountable for.

Figure 1-12: Responsibilities, accountability and authority.
Source: ILO-OSH 2001, Clause 3.3.2.

Where different managers of the organisation work together or share responsibility for something it is important to ensure roles and responsibilities are clear and do not conflict with each other.

For example, operations and maintenance managers may share responsibility for the health and safety of work equipment that needs to be taken out of use for maintenance.

The roles and responsibilities of the operations and maintenance managers should clarify who is in control of the equipment at the different stages involved, including when it is being taken out of use, maintained and returned to use.

Other roles that relate directly to the effective management of health and safety that should be defined include:

- Health and safety practitioners.

- Occupational health specialists.

- Those that conduct governance audits, if they include health and safety.

- Human resources managers.

- Training managers.

- Purchasing and contracts managers.

- Workers and their representatives.

Appointing senior managers with specific responsibility for health and safety

Though the top manager (most senior manager) has overall responsibility for health and safety, the senior management team shares this responsibility. The members of the senior management team have a collective and individual responsibility to ensure health and safety. Therefore, an important part of management commitment is that the senior management team willingly accept their responsibility for health and safety. Senior management teams sometimes demonstrate their commitment to health and safety by selecting someone from the team to provide extra emphasis on health and safety and to ensure that health and safety is an integral part of the senior management team activities and decisions.

The senior manager appointed can be anyone from the senior management team. Some employers have the view that the most senior manager should take this role, other employers may feel that the attributes of another person in the team better suit the role. Another approach is to appoint a different member of the senior management team each year. In time this would ensure a higher involvement and a better understanding of health and safety by each member of the team. Whichever option is chosen for selecting and appointing senior managers with specific responsibility for health and safety, the demonstration of commitment to health and safety shown by making the appointment can have a strong effect on the health and safety culture of an organisation.

Appointing senior managers with specific responsibility for health and safety ensures that sufficient time and effort is committed to health and safety at senior management level and assists with the development of health and safety objectives and plans. The senior manager appointee usually has responsibility for the regular coordination and presentation of data on health and safety performance to the senior management team. They should also oversee the implementation of

the arrangements for health and safety and provide input to activities that require senior management commitment. For example, as part of their role they might lead the promotion of health and safety on behalf of the senior management team. They would also be involved in organising an annual senior management team to review health and safety performance. The role of the senior management appointee is illustrated in ILO-OSH 2001, *see Figure 1-13*.

"A person or persons at the senior management level should be appointed, where appropriate, with responsibility, accountability and authority for:

The development, implementation, periodic review and evaluation of the OSH management system.

(Periodic reporting to the senior management on the performance of the OSH management system.

Promoting the participation of all members of the organisation."

Figure 1-13: Senior management appointee.
Source: ILO-OSH 2001, Clause 3.3.3.

Appointing competent people to help the organisation meet its obligations

Top management should appoint sufficient competent people to provide primary health and safety functions and assist in ensuring the effective management of health and safety. This will usually include the appointment of at least one health and safety practitioner who is competent to provide the assistance required. The appointment of a health and safety practitioner does not remove line management responsibilities for health and safety, but provides support to line managers in fulfilling these responsibilities.

The health and safety practitioner assists with the development of the health and safety policy, aims, objectives and plans. They will also support the implementation of the arrangements for health and safety. In doing this they will provide advice and guidance to directors/senior management regarding compliance with health and safety legislation/standards, evaluating risks, devising and applying control measures and assisting with the management of health and safety. As part of their role they would support top management in the promotion of health and safety and consultation with workers or their representatives. The health and safety practitioner will often contribute to the evaluation of health and safety performance by participating in accident/incident investigation and carrying out audits. The health and safety practitioner usually has responsibility to maintain health and safety records and for the regular coordination and presentation of general data on health and safety performance to management, for example, data on the number and type of accidents/incidents. They may also be involved in senior management team reviews of health and safety performance and provide impartial, authoritative, advice for any liaison the organisation has with enforcing authorities.

To enable health and safety practitioners to be effective they should:

- Have sufficient information and support for their role.

- Have adequate time and other means available to fulfil their role. In deciding if the time and other means available are adequate consideration should be given to the size of the organisation, the nature and distribution of its risks and minimum National legal requirements.

- Co-operate with each other if more than one person is appointed.

Competent people are appointed by senior management, on behalf of the employer, to provide advice and assistance to the employer and management team in meeting their responsibilities. The appointment of a competent person does not remove senior management or line-management responsibilities for health and safety, but it provides support to managers in fulfilling these responsibilities. Article 3.3.2 of ILO-OSH 2001 emphasises the importance of ensuring health and safety is a clear line management responsibility, which is known and accepted at all levels.

The appointment of a sufficient number of competent people in an organisation helps to demonstrate the commitment of the senior management team to health and safety. Competent people, such as a health and safety practitioner, may be appointed to coordinate health and safety in general or to focus on specific aspects of health and safety, such as fire or health risks. Competent people periodic information on performance and will often assist the senior management team with the annual review of health and safety performance. In addition, competent persons may be appointed to conduct specific tasks such as accident/incident investigation or audits.

Sometimes there may be a need to seek professional help from a specialist. Specialists tend to focus on a number of limited specific topics or topic areas and have the expert knowledge and experience often gained through personal research leading to higher qualifications. An example of a specialist might be someone who could advise on the suitability of fire precautions for an underground rail system, or carry out subsurface geological surveys ahead of a major project such as determining the suitability of the site for the proposed area for the position of the footings for a transport bridge across a river. These specialists may be employed by the organisation or external, providing specialist services under contract when needed.

Directors/senior managers should ensure that those they appoint to assist them with meeting their health and safety obligations are competent to carry out their role and the tasks they are assigned. Competent people appointed should have:

- Knowledge and understanding of the work activities involved.

- Knowledge and understanding of the principles of risk assessment and prevention.

- Knowledge and understanding of relevant legal requirements, international standards and current best practice.

- The ability to apply knowledge and understanding to the work activities involved and the tasks required to fulfil their role.

- The ability to identify problems and assess the need for action.

- The ability to design, develop and implement strategies and plans.

- The ability to evaluate the effectiveness of actions taken to manage health and safety.

- The ability to promote and communicate health and safety risks, requirements for action and success achieved.

- Experience of applying knowledge and abilities to relevant work activities and health and safety situations.

- An awareness of their own limitations.

- The willingness and ability to supplement their existing knowledge and experience.

- A competence based qualification.

- Membership of a professional body or similar.

The appointment of competent people by top management should be clearly communicated to managers and workers as part of senior management demonstrating their commitment to health and safety. It is important that managers and workers understand that competent people are appointed to assist in matters of health and safety and not to remove responsibility from them. Reporting lines from the competent people to management should be clear and enable managers and workers to see that health and safety and the competent people are given appropriate status and priority. Competent people appointed to help the organisation to meet its health and safety obligations should report directly to top management in the organisation, which would emphasise an appropriate commitment to health and safety by top management on behalf of the employer.

Reviewing health and safety performance

Top management should be active in the management of health and safety reviews. Review of the organisation's performance in managing health and safety should be carried out periodically, for example, annually. This should include the review of the organisation's whole approach to health and safety, including its health and safety management system. This is best conducted with the involvement of top management in the review process as it supports the motivation of all levels of management and workers. It is also important because top management will decide the resources that are necessary for the continuing success of the organisation's health and safety performance.

RESPONSIBILITIES OF ORGANISATIONS WHO SHARE A WORKPLACE

Article 17 of Convention C155 and Recommendation 11 of Recommendation R164 establish similar responsibilities of organisations who share a workplace to work together on health and safety issues, Recommendation 11 states:

"Whenever two or more undertakings engage in activities simultaneously at one workplace, they should collaborate in applying the provisions regarding occupational safety and health and the working environment, without prejudice to the responsibility of each undertaking for the health and safety of its employees. In appropriate cases, the competent authority or authorities should prescribe general procedures for this collaboration."

Therefore organisations who work together in the same workplace have a responsibility to collaborate with each other to ensure health and safety for their own workers and that of the other employers. They should inform other employers with whom they share a workplace, about risks to the other employers' workers that may arise out of working together in a shared workplace.

Organisations should co-operate with each other by considering each other's activities when conducting risk assessments. They should also coordinate any control measures used to deal with any shared risks identified in the risk assessment.

Organisations have responsibilities regarding the health and safety of the workplace they occupy. When contractors are working in a workplace occupied by an employer it can be said that they share the workplace and have shared responsibilities. In a similar way, where a building is occupied by a number of organisations they would be responsible for the parts of the building that they occupy separately, but would share responsibility for areas they share with others.

It is necessary for all organisations and self-employed people who are involved in situations where they share the same workplace to satisfy themselves that the emergency arrangements are adequate. Both co-operation and coordination of effort will be required for the effective provision of emergency arrangements for shared risks, including response and evacuation procedures.

For example, when establishing fire evacuation procedures on a site occupied by multiple employers all the employers on the site should be considered. A procedure should be agreed that ensures each employer co-operates with the other employers to provide a coordinated response. Each employer should ensure that all relevant managers and workers are aware of the emergency procedures and fully take part in any shared practice emergency activities.

In the case of rented offices the organisation in control of the premises, usually the landlord, will usually have the overall responsibility for the integrity of the building and for the routes in the building that enable those renting the offices to get to the part they are renting. This will include the entrance lobby, lifts, stairs, lobby areas on each floor and corridors. These are often called the 'common areas' of the building as they are generally available for everyone to use. Responsibility for the integrity of the building will usually include its strength, stability and protection from fire. The organisation in control of the premises may also have responsibility for the central services provided throughout the building, for example, the air conditioning.

HOW CLIENTS AND CONTRACTORS SHOULD WORK TOGETHER

Clients and contractors

'Client' is the term used for someone who establishes a contract with someone else and the person with whom the client contracts is called a 'contractor'. In the context of health and safety at work the client is generally an employer and they might contract for a wide range of work activities to be carried out, for example, cleaning, repair, maintenance, refurbishment, catering or security activities. In addition to these activities the client may also establish contracts for large-scale construction works, such as an extension to a building, construction of a new building or installation of new equipment. Where the client is a government department they may contract to carry out major civil engineering projects, for example, construction of a road or bridge. A contractor will be either an employer or self-employed person.

The relationship of client and contractor enables the client to use the expertise and resources of a con-tractor for work activities and projects. This is very helpful to the client in situations where a work activity is not required frequently and would not justify employing extra workers to do the work, for example, periodic maintenance of equipment or major construction work. Some clients establish long-term contracts for contractors to provide services that the client would rather not do, often because they prefer to focus on the main activities of the organisation. This would often include cleaning, catering and security work, but might extend to a great many other activities.

The duties the client and the contractor have to each other

How the client organises a contract and the way the contractor meets a contract can significantly affect health and safety. For example, the client could put unacceptable demands in the contract or the way that the contractor works could create unacceptable risks. In some cases, the work activities may be organised such that the various work activities of the contractor conflict with each other or conflict with the work activities of the client, putting people at risk.

In most countries, because clients and contractors are employers they are obliged by National health and safety laws to protect their own workers and other people from health risks and personal injury. The principal responsibilities clients and contractors have to each other and the other's workers are described in the earlier sections of this element and show that a client, as an employer, has responsibilities towards contractors and contractors' workers as 'other people that may be affected by the employer's work activities'. They also show that the contractor, as an employer, has the same responsibilities towards the client and the client's workers. This means that when a client commissions a contractor to do work, the client retains responsibility to see that it is conducted in a safe and healthy manner, including that it does not affect the client's workers, contractor's workers and the public. It also means a contractor that agrees to a contract must provide appropriate health and safety standards when conducting the contracted work. These good standards will benefit all those that might be affected, including the contractor's and client's workers.

The relationship of client and contractor means that they both have an interest in the success of the contracted work activities, including the health and safety aspects of the activity. Because clients retain overall responsibility for the health and safety of contracted work activities they must establish contracts that can be achieved safely and healthily and must carefully select suitable contractors that can meet the contracts they are given. The client

and contractor share responsibility to ensure collaboration, building health and safety into the contract and the work methods used by the contractor.

The ILO Code of Practice – Safety and Health in Construction establishes specific duties for clients:

"2.7. General duties of clients

2.7.1 Clients should:

(a) co-ordinate or nominate a competent person to co-ordinate all activities relating to safety and health on their construction projects;

(b) inform all contractors on the project of special risks to health and safety of which the clients are or should be aware;

(c) require those submitting tenders to make provision for the cost of safety and health measures during the construction process.

2.7.2. In estimating the periods for completion of work stages and overall completion of the project, clients should take account of safety and health requirements during the construction process.

Effective planning and co-ordination of contracted work

It is essential that all contracted work is effectively planned and coordinated. To meet their separate responsibilities, it is important that the client and contractor provide each other with health and safety information that might affect the contracted work activity, for example, the client might provide information on known hazards in the workplace, like asbestos. The effective planning and coordination of contracts will include conducting appropriate risk assessments and foreseeing how the work activities and those involved in the work interact with each other. The information provided and risk assessments should be used to develop plans and statements of how the work is to be conducted and co-ordinated.

The ILO Code of Practice – Safety and Health in Construction establishes the following specific duties:

"2.4. Co-operation and co-ordination

2.4.1. Whenever two or more employers undertake activities at one construction site, they should co-operate with one another as well as with the client or client's representative and with other persons participating in the construction work being undertaken in the application of the prescribed safety and health measures.

2.4.2. Whenever two or more employers undertake activities simultaneously or successively at one construction site, the principal contractor, or other person or body with actual control over or primary responsibility for overall construction site activities, should be responsible for planning and co-ordinating safety and health measures and, in so far as is compatible with national laws and regulations, for ensuring compliance with such measures.

2.4.3. In so far as is compatible with national laws and regulations, where the principal contractor, or other person or body with actual control over or primary responsibility for overall construction site activities, is not present at the site, they should nominate a competent person or body at the site with the authority and means necessary to ensure on their behalf co-ordination and compliance with safety and health measures.

2.4.4. Employers should remain responsible for the application of the safety and health measures in respect of the workers placed under their authority."

These shared responsibilities mean that the client and contractor need to collaborate throughout the contract period to ensure health and safety is achieved, including prior to the contract being signed, before commencement of work and during the work. The client and contractor also share responsibility to evaluate the health and safety performance related to the contracted work activity, this will help to enable the successful management of the work. As the client retains overall responsibility for the contracted work activity, if the level of health and safety performance is not acceptable the client should take action to ensure the contractor meets the health and safety requirements of the contract.

Pre-selection and management of contractors

The client's selection of contractors for contracted work usually involves the balancing of a number of issues, one of which is financial. Selection of a contractor based primarily on the price the contractor wants to charge the client for the work may not meet the client's health and safety responsibilities. The price a contractor charges is not usually a good indicator of their likely health and safety performance during the contract. Therefore an approach where health and safety is considered before the price is considered is a preferred way to select contractors.

There are five main elements to a health and safety focused selection procedure for contractors. The extent to which each element is relevant will depend upon the degree of risk and nature of work to be contracted.

The elements are:

1) Identification of the likely hazards and risks related to the work defined in the contract specification.

2) Identification of suitable bidders from those with an interest in bidding (sometimes called development of a preferred list of bidders). This is done by assessing those contractors with an interest in bidding against specific health and safety factors. The assessment might consider factors such as:

- The status and adequacy of the contractor's health and safety policy.
- The status of the competent person appointed to assist the contractor with health and safety.
- Previous accident/incident and ill-health records of the contractor.
- Details of any prosecutions or significant civil claims against the contractor.
- Details of any enforcing agency action against the contractor, for example, enforcement notices.
- The competence of the contractor's workers.
- Whether the contractor has a management system certified by a recognised organisation.
- The level and type of insurance the contractor holds.
- References from previous clients of the contractor.

3) Providing the contract specification to suitable bidders.

4) Assessing the health and safety aspects of bids, including how they intend to control hazards and manage health and safety risks.

5) Deciding on the contractors to proceed to final selection based on other factors than health and safety, in particular, quality, environmental and financial factors.

After final selection, the contractor should be managed and monitored during work activities. After completion of the contract the contractor's health and safety performance should be reviewed and their status on the list of preferred contractors updated.

REVIEW

What factors should the employer consider when selecting an individual to fulfil the role of health and safety practitioner?

Sources of reference

Reference information provided, in particular web links, was correct at time of publication, but may have changed.

Global Database on Occupational Safety and Health Legislation (LEGOSH) http://www.ilo.org/dyn/legosh/en/f?p=14100:1:0::NO:::

Guidelines on Occupational Safety and Health Management systems, Malaysia, ISBN 978-983-2014-75-1, can be downloaded free from the Malaysian Department of Occupational Safety and Health website - http://www.dosh.gov.my/index.php?option=com_docman&task=cat_view&gid=12&Itemid=179&lang=en

Health and Safety at Work etc Act 1974, UK, www.legislation.gov.uk (please enter the name of the Act)

ILOLEX (ILO database of International Law) http://www.ilo.org/ilolex/index.htm (follow quick links on web page)

International Labour Organisation (ILO) www.ilo.org

International Organisation for Standardisation (ISO) www.iso.org

ISO 45001:2018 Occupational health and safety management systems – requirements with guidance for use, International Organization for Standardization

NORMLEX (ILO database of international labour standards, including ILO Conventions, and National laws) http://www.ilo.org/dyn/normlex/en/f?p=NORMLEXPUB:1:0::NO:::

Occupational Safety and Health System Framework (OSHAD SF), Abu Dhabi, https://www.oshad.ae/Lists/Publications/Attachments/13/12.%20%D8%A7%D9%84%D9%85%D9%84%D8%AE%D8%B5%20%D8%A7%D9%84%D8%AA%D9%86%D9%81%D9%8A%D8%B0%D9%8A%20-%20English.pdf

Occupational Health and Safety Act 1993, South Africa, http://www.acts.co.za/occupational-health-and-safety-act-1993/

Occupational Safety and Health Act 1970, United States of America, https://www.osha.gov/law-regs.html

Occupational Safety and Health Act 1994, Malaysia, www.dosh.gov.my (please follow the links via legislation to Acts)

Occupational Safety and Health Act 2004, Trinidad and Tobago, http://www.ttparliament.org/publications.php?mid=29

Occupational Safety and Health Convention C155, 1981, ILO, http://www.ilo.org/dyn/normlex/en/f?p=1000:12000:0::NO::: (please select C155 from the Technical list).

Occupational Safety and Health Recommendation R164, 1981, ILO, http://www.ilo.org/dyn/normlex/en/f?p=1000:12010:0::NO::: (please select R164 from the Technical list).

Provision and Use of Work Equipment Regulations 1998, UK, www.legislation.gov.uk (please enter the name of the Regulation)

Safety and Health in Construction Recommendation, 1988 (R175), http://www.ilo.org/dyn/normlex/en/f?p=NORMLEXPUB:12100:0::NO::P12100_INSTRUMENT_ID:312513

Safety and health in construction, ILO Code of Practice, ILO Geneva, ISBN: 92-2-107104-9, http://www.ilo.org/wcmsp5/groups/public/---ed_protect/---protrav/---safework/documents/normativeinstrument/wcms_107826.pdf

Using contractors, a brief guide, INDG368, HSE Books, http://www.hse.gov.uk/pubns/indg368.pdf

Workplace (Health, Safety and Welfare) Regulations, 1992, Approved Code of Practice and guidance, http://www.hse.gov.uk/pubns/books/l24.htm

Web links to these references are provided on the RMS Publishing website for ease of use – www.rmspublishing.co.uk

STUDY QUESTIONS

1. What is meant by the terms direct and indirect cost of a health and safety incident, with an example for each? (8)

2. What are the social reasons for preventing accidents/incidents and ill-health in the workplace? (8)

3. What are responsibilities and rights of workers in the Occupational Safety and Health Convention C155? (8)

4. What is meant by practicable duties placed on employers? (8)

5. Why are ISO international standards not in any way binding on either governments or industry merely by virtue of being international standards? (8)

For guidance on how to answer NEBOSH questions please refer to the 'study question answer guidance' section located at the back of this guide.

Element 2

How health and safety management systems work and what they look like

2

Contents

2.1 What are they and the benefits they bring

A health and safety management system is a set of interrelated elements established to effectively manage health and safety in an organisation. Health and safety management systems are usually structured from a small number of elements that provide the scope of the system, outline the aims and objectives of the organisation and identify the means by which these are achieved. The number of elements and the names given to them differ depending on the design of the management system.

Employers are usually free to establish their own structure for their health and safety management system, though some countries may prescribe the structure by National Government guidance or legal requirements, for example, Malaysia, Abu Dhabi and Egypt. The structure of many employers' management systems has therefore been developed over time from their experience, others may be based on available guidance that reflects the experience of a number of organisations.

ISO 45001: 2018 OCCUPATIONAL HEALTH AND SAFETY MANAGEMENT SYSTEM STANDARD

ISO 45001 is the world's first internationally agreed occupational health and safety management system (OHSMS) standard. The international standard has drawn on experience from BS OHSAS 18001:2007 and other approaches from around the world. This has led to an enhanced and more comprehensive standard that will correspond with the values of some of the more enlightened organisations and their strategies for managing health and safety risks.

In line with good practice expectations, top management leadership of health and safety is a significant aspect of ISO 45001. Reflecting the goals of many organisations, there is an increased focus on continually improving occupational health and safety performance in the standard.

A principal requirement of ISO 45001 is for managers to take an active and personal involvement in the management of OH&S in their organisation. Depending on the present level of management engagement and the prevailing culture of the organisation, occupational health and safety managers may need to get more 'buy-in' from managers, including top management if implementation of ISO 45001 is to be successful. Without such ownership and commitment from top management the necessary culture for ongoing improvement is unlikely to be achieved.

When conforming to ISO 45001 requirements management needs to take a bigger role in managing occupational health and safety issues. The focus of ISO 45001, is all about prevention, not cure – taking a proactive approach. Managers need to recognise the potential occupational health and safety issues in the work that they manage, long before they cause significant harm and the organisation is forced to deal with them retrospectively.

Summary

ISO 45001 utilises the Plan-Do-Check-Act (PDCA) cycle, see **Figure 2-1**, which shapes the OH&S management system and can be applied to all processes within the system to achieve continual improvement. The PDCA cycle can be applied to the OH&S management system and to each of its individual elements, as follows:

- Plan - determine and assess OH&S risks, OH&S opportunities and other risks and other opportunities taking into account issues identified in clause 4, establish OH&S objectives and processes necessary to deliver results in accordance with the organisation's OH&S policy;

- Do - implement the processes as planned;

- Check - monitor and measure activities and processes with regard to the OH&S policy and objectives, and report the results;

- Act - take actions to continually improve OH&S performance to achieve the intended outcomes of the OH&S management system.

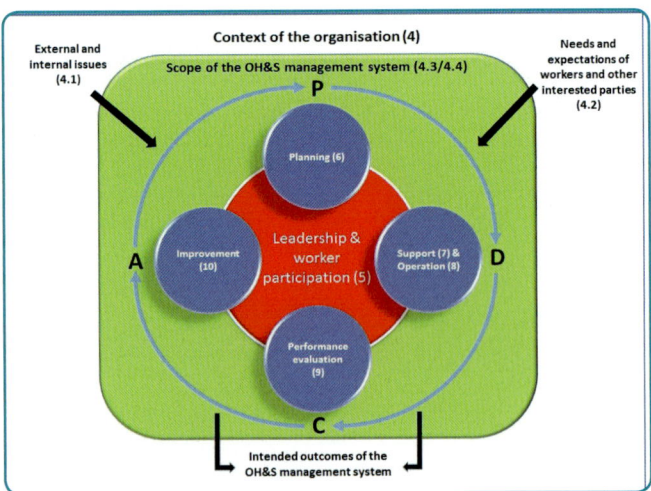

Figure 2-1: Relationship between PDCA and ISO 45001. Source: RMS.

Clause 4: Context of the organisation

This clause establishes the context of the OH&S management system and underpins the rest of standard. It gives an organisation the opportunity to identify and understand the external and internal factors and interested parties that affect the intended outcome(s) of the OH&S management system. It also, in part, addresses the concept of preventive action.

Firstly, the organisation will need to identify external and internal issues that are relevant to its purpose i.e. what the relevant issues are, both inside and are out, that have an impact on or affect its ability to achieve the intended outcome(s) of the OH&S management system. It should be noted that the term 'issue' covers not only problems or potential problems, but also important topics for the system to address, such as changing circumstances, legal requirements and other obligations.

Secondly, an organisation will need to identify and take into account the needs and expectations of the 'interested parties' relevant to their OH&S management system. The requirements of ISO 45001 explicitly refer to workers as interested parties to be taken into account, other interested parties include customers, owners, clients and visitors.

Next, the scope of the OH&S management system has to be determined taking the above into account. The scope is intended to clarify the boundaries to which the system will apply, especially if the organisation is part of a larger organisation. Finally, the last requirement of Clause 4 is to establish, implement, maintain and continually improve the OH&S management system in accordance with the requirements of the standard.

Clause 5: Leadership and worker participation

It is worth noting that leadership and worker participation have been placed together in the same clause, reflecting the fact that creating an effective OH&S management system requires the combined efforts of a range of people in the organisation, not just a health and safety specialist. This clause places requirements on 'top management' which is the person or group of people who directs and controls the organisation at the highest level. Note that if the organisation that is the subject of the OH&S management system is part of a larger organisation, then the term 'top management' refers to the smaller organisation.

Top management must take overall responsibility and accountability for the protection of workers' work-related health and safety and need to develop, lead and promote a culture that supports the OH&S management system. They must ensure that the requirements are integrated into the organisation's processes and that the policy and objectives are compatible with the strategic direction of the organisation. They also need to establish the OH&S policy and the standard defines the characteristics and properties that the policy is to include. In ISO 45001 there is a significant focus on top management being

able to demonstrate leadership and commitment to the management system and ensure active participation of workers in the development, planning, implementation and continual improvement of the OH&S management system. This includes using consultation and the identification and removal of obstacles or barriers to worker participation.

Top management need to ensure that the importance of effective OH&S management is communicated and understood by all parties and that the OH&S management system achieves its intended outcomes. Also contained within this clause is the requirement to establish, implement and maintain an OH&S policy in consultation with workers at all levels. This must include commitments to provide safe and healthy working conditions, fulfilling legal requirements, setting OH&S objectives and continual improvement.

Finally, top management need to assign and communicate responsibilities, accountabilities and authorities for relevant roles within the system, including workers at each level within the organisation assuming responsibility for those aspects of the OH&S management system over which they have control.

Clause 6: Planning

ISO 45001 requires that the planning part of the OH&S management system takes full account of the issues determined when meeting Clause 4 'context of the organisation' and includes consideration of opportunities as well as risks.

It is also important that planning is seen as an ongoing, proactive process that anticipates changing circumstances, continually determining risks and opportunities for improvement. The initial part of this process will be familiar as it involves the identification of hazards, including at the conceptual design stage of workplaces, facilities, products or the way activities are organised. This will include consideration of routine and non-routine activities as well as the range of people that may be affected, for example, workers, contractors, visitors and others not under the direct control of the organisation. A particularly discernible difference is that the ISO 45001 standard sets out comprehensive aspects that the process must fulfil, including consideration of human factors and social factors like workload, work hours, victimisation and bullying. This marks a significant shift from consideration of traditional safety hazards and will lead organisations to consider the causes of wider harm, such as stress.

This clause recognises the importance of legal requirements that might affect obligations to assess and control risk or to manage health and safety in a particular way. Organisations are required to establish a process to determine and update legal and other requirements (for example, customer, client or contracted requirements) which are applicable to its hazards, OH&S risks and OH&S management system.

A separate and specific requirement is to plan actions to address risks/opportunities and legal/other requirements as well as plans to prepare for and respond to emergency situations. There is also an expectation to plan how to integrate these actions into OH&S processes and, importantly, evaluate their effectiveness. Another important requirement of this clause is the need to establish and plan to achieve OH&S objectives which are measurable or capable of evaluation. The purpose of effective OH&S objectives is to demonstrate maintenance and continual improvement to the OH&S management system. The objectives will need to be set at relevant levels of the organisation and can be strategic, tactical or operational in nature.

Clause 7: Support

It is important that objectives do not just remain good intentions. The requirements of clause 7 – Support, details what actions the organisation will need to take to support the delivery of the OH&S system. As those who have experience of managing health and safety will know, objectives meet many barriers if they are not resourced. Clause 7 requires the organisation to determine and provide resources to support the establishment, implementation, maintenance, and continual improvement of the OH&S management system. This critical requirement covers all OH&S resource needed, including human resources, infrastructure and financial resources.

Other important support requirements relate to competence, awareness and communication. Organisations will need to formally determine the necessary competence of workers (at all levels) that affect or can affect OH&S performance and ensure they receive the appropriate education and training to meet these needs. Within this clause there is a requirement to retain documented information as evidence of competence.

Organisations also need to ensure that all workers (at all levels) are aware of the OH&S policy, OH&S objectives, the OH&S hazards and risks that are relevant to them and their contribution to the effectiveness of the system, as well as the implications of not conforming to system requirements. As would be expected when establishing effective management of health and safety, organisations also need to have a communication process to determine internal and external communications relevant to the OH&S management system. Finally, this clause refers to requirements for 'documented information', for example 'documents' and 'records', which includes the creation, updating and control of documented information.

Clause 8: Operation

This clause focuses on the 'Do' part of the PDCA cycle, the implementation of the actions identified in previous clauses. Operational planning and control of processes need to be established as necessary to enhance health and safety, by the elimination of hazards and reduction of OH&S risks. Operational control of processes can include a variety of methods, for example, procedures, method statements, systems of work, preventative maintenance regimes, inspection programmes and engineering or administrative controls, such as a permit to work.

Clause 8 specifies a hierarchy of controls for a systematic approach to the elimination of hazards and reduction of risks. This will be familiar to those who have received basic health and safety education – eliminate, substitute, engineering controls, administrative controls and finally, personal protective equipment.

This clause also sets out a required approach to the management of change. Change needs to be planned for in a systematic manner, to prevent the introduction of new hazards or risks. Importantly, at times of proposed changes to processes organisations will need to identify opportunities to reduce OH&S risks or create improvements to the effectiveness of the OH&S management system.

Procurement and outsourcing feature as specific issues in ISO 45001, requiring the organisation to establish processes to ensure that procured goods/services and outsourced functions/processes are controlled and they conform to the requirements of the OH&S management system. As part of procurement controls, contractors will need to be specifically considered as their activities can involve different types of hazard and levels of OH&S risks. The organisation must ensure that the requirements of its OH&S management system are met by its contractors and their workers. Procurement and contractors are considered in ISO 45001 and outsourcing has been brought within scope of the OH&S management system.

Processes relating to the control of emergency preparedness and response also feature in significant detail in this clause.

Clause 9: Performance evaluation

This clause requires the organisation to be active in various aspects of performance evaluation:

- Monitoring, measurement and analysis.
- Evaluation of compliance.
- Internal audit.
- Management review.

This establishes a comprehensive range of performance evaluation. Organisations will need to determine what information they need in order to evaluate OH&S performance and effectiveness. This will help them to identify what specifically needs to be measured and monitored, who is going to do it, when and how it will be done.

Requirements for internal audit and management review are specified. One notable inclusion in the inputs for the management review is the adequacy of resources for maintaining the effectiveness of the OH&S management system. The frequency of monitoring and measuring will need to be appropriate to the risk and its performance, as well as the size and nature of the organisation. It is worth noting that documented information that provides evidence of this must be retained.

Clause 10: Improvement

The first requirement in this clause is a general one to determine opportunities for improvement and implement actions. This is a proactive requirement to make improvements. Notwithstanding this, if incidents or nonconformities occur there is a requirement in Clause 10 to react to them and take action in a timely manner.

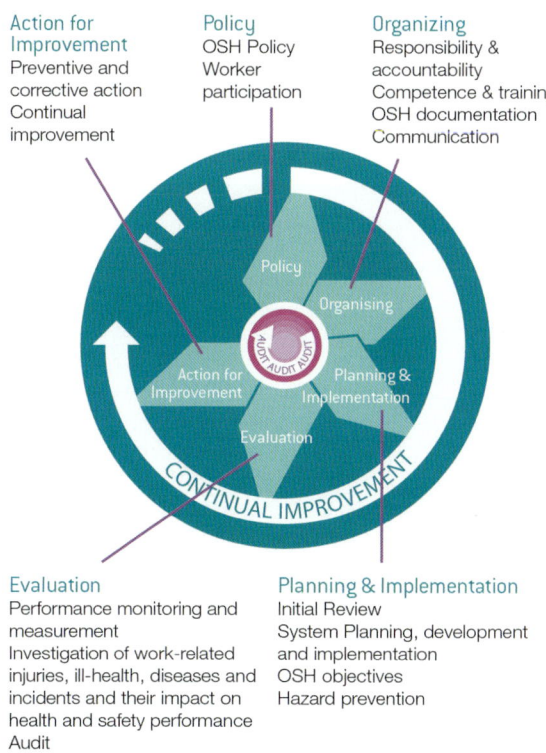

Action for Improvement
Preventive and corrective action
Continual improvement

Policy
OSH Policy
Worker participation

Organizing
Responsibility & accountability
Competence & training
OSH documentation
Communication

Evaluation
Performance monitoring and measurement
Investigation of work-related injuries, ill-health, diseases and incidents and their impact on health and safety performance
Audit
Management review

Planning & Implementation
Initial Review
System Planning, development and implementation
OSH objectives
Hazard prevention

Figure 2-2 Key elements of successful health and safety management. Source: ILO-OSH 2001.

Health and safety professionals will be pleased to note that consideration of corrective action to eliminate root causes is referred to specifically. This will need to be done with the participation of workers and involvement of other relevant parties, which might challenge the approach taken by some organisations where investigations have been led by health and safety professionals to the exclusion of workers and managers.

An additional requirement is to determine whether similar incidents or nonconformities exist, or could potentially occur, and the implications to existing risk assessments, potentially leading to appropriate corrective actions across the whole organisation if necessary.

Corrective actions to the management system will be preventive in nature, as would the current requirement to determine similar incidents or nonconformities that could potentially occur. The requirement for continual improvement has been extended to continually improve the suitability, adequacy and effectiveness the OH&S management system through:

- Enhancing performance
- Promoting a culture that is supportive of the OH&S management system
- Promoting worker participation
- Communicating the results of continual improvement

ILO GUIDELINES ON OCCUPATIONAL SAFETY AND HEALTH MANAGEMENT SYSTEMS

The ILO developed their 'Guidelines on Occupational Safety and Health Management Systems – ILO-OSH 2001' (ILO-OSH 2001) as voluntary guidance on establishing an occupational health and safety management system. The guidance provides practical recommendations, which are not legally binding and are not intended to replace National legislation or accepted standards. Their application and use does not require certification of an employer's management system by a third party organisation.

The ILO 'Guidelines on Occupational Safety and Health Management Systems - ILO-OSH 2001', **see Figure 2-2**, follow a structure that uses the following five key elements:

1) Policy.

2) Organising.

3) Planning and implementation.

4) Evaluation.

5) Actions for improvement.

Policy (Plan)

Organisations that are successful in achieving high standards of health and safety have health and safety policies that contribute to their business performance, while meeting their responsibilities in a way which fulfils both the spirit and the letter of the law. In this way they satisfy the expectations of shareholders, workers, customers and society at large. Their policies are aimed at achieving the preservation and development of physical and human resources and minimisation of financial losses and liabilities. Their health and safety policies influence all their activities and decisions, including those to do with the selection of resources and information, the design and operation of working systems, the design and delivery of products and services, and the control and disposal of waste.

Organising (Plan)

Organisations that achieve high health and safety standards are structured and operated so as to put their health and safety policies into practice in a way that is effective. They clearly define the roles and responsibilities of everyone in the organisation. This is helped by the creation of a positive culture that secures involvement and participation at all levels. This culture is sustained by effective communications and the promotion of competence that enables all managers and workers to make a responsible and informed contribution to the health and safety effort. The visible and active leadership of senior managers develops and maintains a culture supportive of health and safety management. The aim of these senior managers is not simply to avoid accidents/incidents, but to motivate and empower people to work safely and healthily. The vision, values and beliefs of leaders become the shared 'common knowledge' of all and result in behaviours that provide high health and safety standards.

Planning and implementing (Do)

Successful organisations adopt a planned and systematic approach to policy implementation. Their aim is to minimise the risks created by work activities, products and services. They use risk assessment methods to identify hazards, decide priorities and set objectives for hazard elimination and risk reduction. Performance standards are then established. Specific actions needed to promote a positive health and safety culture and to eliminate and control risks are identified. Wherever possible, risks are eliminated by the careful selection and design of facilities, equipment, substances and processes or minimised by the use of physical control measures. Where this is not possible, provision of a safe system of work and personal protective equipment are used to control risks. Risk control measures that are implemented as part of the management system should be supported by emergency prevention, preparedness and response arrangements. Arrangements for planning and implementation will generally also include specific risk control measures for procurement, contracting and the management of change.

Evaluation (Check)

Organisations that successfully manage health and safety establish procedures to evaluate health and safety performance on a regular basis. Responsibilities,

accountability and authority for evaluation are clearly allocated at different levels in the management structure. Performance evaluation is consistently used to determine the extent to which the policy and objectives have been met and risks controlled.

Monitoring, measuring and investigation

Performance is often monitored through active and reactive monitoring of health and safety. Active (sometimes called proactive) monitoring involves the monitoring of the things done to prevent accidents/incidents and ill-health. This includes an examination of both hardware (premises, equipment and substances) and software (people, procedures and systems), including individual behaviour. Failures to control risks are evaluated through reactive monitoring, which involves analysis of the quantity and type of these failures along with the thorough investigation of accidents/incidents, ill-health or other events with the potential to cause harm or loss.

Health and safety performance is evaluated against pre-determined standards. These reveal when and where action is needed to improve performance. The role of active and reactive monitoring is to determine the immediate causes of sub-standard performance and, more importantly, to identify the underlying and root causes of sub-standard performance and the implications for the design and operation of the health and safety management system.

Figure 2-3: Audit of a health and safety policy.
Source: Shutterstock.

Auditing

Successful organisations use formal auditing processes to determine the extent to which the health and safety management system, or elements of it, conform to standards. Auditing assesses what is in place, whether it is adequate and if it is effective. Auditing can provide an important formal assessment of whether health and safety objectives and targets are being met. Internal auditing is used to confirm conformity with internal, legal and other standards. From time to time the whole health and safety management system may be audited by an independent organisation to determine the extent to which it conforms with recognised health and safety management systems and appropriate legislation.

Successful organisations develop an audit policy and programme that includes auditor competency, audit scope, frequency of audits, audit methodology and reporting.

Review

Successful organisations understand that learning from all relevant experience and applying the lessons learned are important in the effective management of health and safety. They do this systematically through regular reviews of performance based on data from evaluation activities, including audits of the whole or part of the health and safety management system. In addition to using data from internal evaluation activities, some organisations find it useful to compare their health and safety performance against similar organisations with known high standards, this process is sometimes called benchmarking. Review of the health and safety management system and its elements is an essential activity and provides a good basis for improvement actions.

Action for improvement (Act)

Evaluation activities of successful organisations identify system non-conformities, for example the presence of previously unidentified hazards or incorrectly identified risk levels, procedures not being followed or planned action not taken. They take prompt corrective and preventive action related to system non-conformities, particularly those that can result in major accidents/incidents and serious ill-health. Corrective action is that action to correct the non-conformity and prevent further harm being caused by it, this tends to deal with the immediate causes of the non-conformity. Preventive action deals with the underlying and root causes of the non-conformity. This type of action reduces the likelihood of the non-conformity re-occurring in the same place or in other places.

These organisations establish arrangements to aid the continual improvement of health and safety management system elements and the system as a whole, taking into account such areas as the health and safety objectives of the organisation, changes in National legislation, voluntary programmes, collective agreements and any other new relevant information. Commitment to continual improvement therefore involves organisations in the regular review and further development of their health and safety policy and plans. It also includes the constant development of their approaches to implementation and their techniques of risk prevention and control.

Organisations that achieve high standards of health and safety assess their health and safety performance by reference to internal key performance indicators (KPIs) and by external comparison with the performance of other similar organisations, benchmarking. They often also record and account for their performance in their annual reports.

BENEFITS FOR A FORMAL/CERTIFIED HEALTH AND SAFETY MANAGEMENT SYSTEM

Organisations will often have well developed systems for managing design, production, quality and the financial aspects of their business. Health and safety should be managed with the same emphasis and degree of control as these aspects. A formal management system provides the organisation with a clear aim and specific objectives to work towards, preventing it from progressing in an unstructured way towards a perceived, but unclear goal. A systematic approach makes the management of health and safety both easier and more effective.

A formal management system allows the role of all individuals, from senior managers to workers, as well as those with specific responsibilities to be defined encouraging all members of the organisation to participate in the objectives of the management system.

The processes for identifying, evaluating and controlling risks that form part of such a management system will provide a consistent approach. As the system will also identify monitoring and review processes, the ongoing effectiveness of the system will be ensured which will assist with ongoing legal compliance, reduction of risks and the incidence of injury and ill-health.

The commitment demonstrated by such a formal system increases the morale of the workforce, increasing their commitment to the organisation. This can have a positive impact on health and safety and loyalty. For many businesses, public perception is important as is the view of third-party suppliers and customers who will sometimes only do business with those organisations who can demonstrate a tight control over their health and safety. In some cases, this will mean organisations who have obtained certification by a third party that confirms the scope and conformity of their management system

A formal health and safety management system that conforms to recognised standards like ISO 45001 or ILO-OSH 2001 will help organisations provide a safe and healthy workplace for workers and other people, as well as continually improve their OH&S performance.

Organisations who conform to recognised standards will also improve their resilience by ensuring they can anticipate, assess and prepare responses to risks, allowing them to adapt to changes so they continue to succeed and prosper. Formal management systems allow organisations to meet challenges by instilling good practices and validating, through certification, that they are properly established in the organisation.

By having a formal management system an organisation could have the following benefits:

- Demonstrates corporate responsibility and meets supply chain requirements.
- Ensures OH&S management is aligned with the strategic direction of the organisation.
- Improves integration with other management systems.
- Gives an organisation more order and structure.
- Organisations become more professional in managing health and safety risk.
- Greater health and safety efficiency and effectiveness.
- Improves occupational health and safety performance the through proactive management of risks.
- Ensures that risks are defined and managed.
- Eliminates hazards or minimise OH&S risks.
- Identifies problems and opportunities in good time.
- Reduces the errors that can lead to harm.
- Reduces work related injuries, ill health and death.
- Enables learning processes to avoid the same mistakes in the future.
- Increases involvement of the leadership team.
- Motivates and engages workers through consultation and participation.
- Establishes (or improves) a positive health and safety culture.
- Provides workers, customers and other stakeholders with added confidence.
- Reduces costs arising from harm due to ill health, injury and regulatory action.
- Lower insurance premiums.
- Gives a competitive edge.

Certification is the provision of written independent assurance (a certificate) that a management system meets specific requirements set out in recognised standards.

Organisations can choose to work in accordance with recognised standards, and they can also choose to be certified by an independent party (such as a certification body) or not. Whether an organisation seeks certification may depend on what the costs and benefits are of obtaining it. As the organisation would receive a conformity assessment as part of the certification process, this will usually incur a charge of fees from the certification organisation. The conformity assessment process will also usually require significant quantities of management time in order to provide evidence of conformity to the satisfaction of the certification organisation. Cost of the certification processes can depend on the industry sector (type and level of risk), turnover and location/number of sites of the organisation. In some situations, certification is a legal requirement or contractual requirement in order to supply or tender, particularly when dealing with international organisations or those in the public sector.

Therefore, the commercial advantage of having certification may outweigh the cost of obtaining it.

Undergoing the conformity assessment that forms part of the certification process has a number of benefits:

- Demonstrates that the organisation meets the requirements of a recognised standard.
- Provides objective proof that the organisation attaches great importance to health and safety.
- Ensures clear responsibilities, (communication) structures and processes throughout the entire organisation.
- Confirms a commitment to be open to independent scrutiny.
- Adds credibility to the organisation.
- Improves regard and reputation.
- Gives a positive image of the organisation.
- Communicates a positive message to workers, customers and other stakeholders.
- Confirms to regulators that recognised health and safety standards have been met.
- Meets customers' expectations.
- Communicates a readiness to ensure consistent and improving health and safety performance.
- Increases the reliability of the work done and raises awareness.
- Establishes a level of performance to be maintained over the certification period.
- Increases confidence in the organisation and its ability to manage health and safety risk.
- Helps to distinguish the organisation from others who do not have certification.

 REVIEW

What are the five key elements in the ILO-OSH 2001 Guidelines on Occupational Safety and Health Management Systems?

2.2 What good health and safety management systems look like

THE OCCUPATIONAL HEALTH AND SAFETY POLICY

Role of the health and safety policy

The overall role of a health and safety policy is to provide direction for an organisation, establishing a management commitment that will both guide the organisation to satisfy its aims and maintain the standards that it sets to meet the aims. This will influence the decisions made by an organisation because the decisions will need to meet the aims of the policy and the commitments made in it. Without active and guided management involvement in health and safety any attempt at organised accident/incident and ill-health prevention will tend to be restricted and predominantly reactive. Organisations that successfully manage health and safety establish a formal, usually written health and safety policy that defines what the organisation aims to achieve, outlines the responsibilities and arrangements to achieve this and demonstrates top management's commitment to health and safety.

Many employers are required to have a health and safety policy by National or local legislation, and it is a key element of any occupational safety and health management system, whether the system is certified or voluntary.

The ILO recognises the importance of having a written health and safety policy in their Occupational Safety and Health Recommendation R164, 1981.

"14. Employers should, where the nature of the operations in their undertakings warrants it, be required to set out in writing their policy and arrangements in the field of occupational safety and health, and the various responsibilities exercised under these arrangements, and to bring this information to the notice of every worker, in a language or medium the worker readily understands."

Figure 2-4: Health and safety policy.
Source: ILO, Occupational Safety and Health Recommendation R164.

As can be seen by the example from the Malaysian Occupational Safety and Health Act 1994 in **Figure 2-5** the need to have a health and safety policies are often included as requirements in National legislation.

"Section 16 - Duty to formulate safety and health policy.

Except in such cases as may be prescribed, it shall be the duty of every employer and every self-employed person to prepare and as often as may be appropriate revise a written statement of his general policy with respect to the safety and health at work of his employees and the organisation and arrangements for the time being in force for carrying out that policy, and to bring the statement and any revision of it to the notice of all of his employees."

Figure 2-5: Requirements for a health and safety policy.
Source: Malaysia, Occupational Safety and Health Act 1994.

To be effective the health and safety policy should be:

- Specific to the organisation and appropriate to its size and nature of its activities.

- Developed in consultation with workers and managers.

- Concise and clearly written.

- In formats that are suitable for workers and managers.

- In suitable languages.

- Endorsed or signed by the employer or most senior (top) manager of the organisation.

- Effectively communicated.

- Monitored through audits.

- Reviewed and revised as appropriate.

Figure 2-6: Senior manager signing health and safety policy.
Source: Unsplash.

Content of a health and safety policy

The features and content of the health and safety policy are often developed through experience and general good practice.

The health and safety policy should state the overall aims of the organisation in terms of health and safety performance by establishing a general statement of intent. The ILO document 'Guidelines on occupational safety and health management systems - ILO-OSH 2001' suggests the health and safety policy statement should include, as a minimum, the following aims (what the ILO calls "key principles and objectives") to which an organisation should be committed:

- Protecting the health and safety of all members of the organisation by preventing work-related injuries, ill-health, diseases and incidents.
- Complying with relevant National/International occupational health and safety legislation, voluntary programmes, collective agreements on health and safety and other requirements to which the organisation subscribes.
- Ensuring that workers and their representatives are consulted and encouraged to participate actively in all elements of the health and safety management system.
- Continually improving the performance of the health and safety management system.

The aims that the ILO suggests are a useful basis for setting an organisation's health and safety general statement of intent and may be adapted to suit a particular organisation's needs. The health and safety aims of an organisation establish broad intentions and commitments that may remain unchanged for a number of years.

Organisations may also wish to align the content of their policy with the requirements of ISO 45001:2018, which establishes that an OH&S policy is a set of principles stated as commitments in which top management outlines the long-term direction of the organisation to support and continually improve its OH&S performance. The OH&S policy provides an overall sense of direction, as well as a framework for the organisation to set its objectives and take actions to achieve the intended outcomes of the OH&S management system.

Clause 5.2 of ISO 45001:2018 specifies the content of a health and safety policy:

"Top management shall establish, implement and maintain an OH&S policy that:

includes a commitment to provide safe and healthy working conditions for the prevention of work-related injury and ill health and is appropriate to the purpose, size and context of the organization and to the specific nature of its OH&S risks and OH&S opportunities;

provides a framework for setting the OH&S objectives;

includes a commitment to fulfil legal requirements and other requirements;

includes a commitment to eliminate hazards and reduce OH&S risks;

includes a commitment to continual improvement of the OH&S management system;

includes a commitment to consultation and participation of workers, and, where they exist, workers' representatives."

The needs of different organisations

Organisations differ greatly in their aims, risks, structure and to what they feel capable of committing themselves. In the same way, a policy is therefore a 'personal' thing, setting out a particular organisation's position at that point in time. The policy enables the organisation to communicate its personal commitment, expectations it has of people and its approach to health and safety.

Guidance on effective health and safety policies, including the ILO document 'Guidelines on occupational safety and health management systems - ILO-OSH 2001', suggests that the health and safety policy must be appropriate to the size and nature of an organisa-tion's activities and the nature and scale of its health and safety risks. This means that a small organisation with low risks may have a simple policy, whereas a large complex organisation with higher risks would need a policy that had wider scope and greater depth. In some cases it may be necessary for large complex organisations to have different health and safety policies for different locations in order to better reflect the risks they are managing at the locations. In all but the very small organisations, for example those employing less than five employees, employers are usually expected to establish a written health and safety policy.

The policy should be proportionate to the needs of the organisation and clearly written with the reader in mind. The format, complexity and language used should be considered in relation to those that need to read and use it. Organisations should avoid writing their policy using complicated technical health and safety language or in the form of a complex legal contract using legal terms. Unnecessary information and detail should be avoided, it may be necessary to cross-reference the health and safety policy to other documents, for example, other parts of the health and safety management system.

When the health and safety policy has been developed the employer or the most senior accountable person (top manager) in the organisation should endorse the policy and require it to be implemented. In order to empha-sise the necessary commitment to the organisation's aims the statement should be signed by the most senior accountable person in the organisation and dated.

This shows who is the most senior person account-able for health and safety and that the senior person has accepted their responsibilities. It also gives authority to the policy and demonstrates senior management commitment to health and safety. By signing the policy, the personal commitment demonstrated should mean that managers are more likely to implement the policy and workers are more likely to believe in it and follow it. Strong management commitment, shown by the application of a signature, can emphasise that health and safety is equal to other business objectives. The addition of a date to the statement will indicate the last time the statement was reviewed, which will assist with ensuring the statement remains relevant and up to date. The health and safety policy statement is often posted in prominent positions in the organisation to reinforce the commitment to health and safety.

The effective communication of the policy is important to ensure it is implemented correctly. All affected by it must understand it and their obligations to it. In order to achieve this, a lot has to be done. Merely posting the policy on a notice board or distributing a copy to workers is not enough. Training and briefings will be necessary, as a minimum, to ensure effective communication. For new workers this is often done as part of the induction process. The health and safety policy should be communicated in a language and medium that managers and workers readily understand. The health and safety policy should be readily accessible to all persons in their workplace and made available to external interested parties, for example visitors, neighbours, customers and suppliers. The health and safety policy is often posted in prominent positions in the organisation as part of the communication process and can serve as a lasting reminder to managers and workers after completion of training.

The health and safety policy should be monitored to identify whether it has been implemented properly and if managers and workers are conforming to it.

 **REVIEW**

Explain the overall role of a policy

Health and safety objectives

The aims of the organisation that are set out in the health and safety policy may not change greatly over time. These should provide a framework for setting and reviewing health and safety objectives, which may change each year.

The ILO in their guidance document ILO-OSH 2001 recommends that objectives should be established, which are:

a) Specific to the organisation, and appropriate to and according to its size and nature of activity. Objectives for a small retail store are likely to be different to a large construction company. In particular they should reflect the risks of the different work activities of the organisations.

b) Consistent with the relevant and applicable National laws and regulations, and the technical and business obligations of the organisation with regard to health and safety.

c) Focused towards continually improving workers' health and safety protection to achieve the best health and safety performance.

d) Realistic and achievable.

e) Documented, and communicated to all relevant functions and levels of the organisation.

f) Periodically evaluated and if necessary updated.

There should be a direct link between the work that workers perform and the organisation's objectives. It is

important that the objectives **reflect the hazards and risks** of the organisation. Objectives that relate to specific risks are usually required to ensure focus and action to manage them. For example, objectives may be set to introduce better routes for vehicles operating on site to reduce the risk of accidents/incidents due to the collision of vehicles.

Objectives should also include consideration of **technological options** and take account of any technical advances that may have become available since the last review of objectives. For example, objectives may involve the introduction of mechanical handling equipment to assist with moving heavy loads or the introduction of quieter machinery to improve risks to hearing due to excessive noise. Good objectives will also reflect the organisation's **financial, operational and business** requirements integrating objectives so that there is no conflict of goals.

In line with other business objectives, **quantifiable performance targets** should be set for health and safety objectives. This will enable the monitoring and measurement of progress on completion of targets and objectives.

Objectives such as 'prevent accidents/incidents' and 'consult workers on health and safety' are not defined sufficiently to be confident the right action will follow. It is important that objectives are supported by and translated into quantifiable targets. In order to achieve this many organisations adopt the **'SMART'** approach. This ensures that targets are specific, measurable, achievable, reasonable and time bound.

An example of a health and safety objective could be "Managers will receive appropriate health and safety training". Set as a quantifiable performance target (SMART target) this could be:

"The training manager will ensure that 80% of managers will take the NEBOSH International General Certificate health and safety training course by the end of the year."

This contains specific requirements that are measurable and time bound, presuming that 80% of managers is both reasonable and achievable for the organisation setting the objective. This approach to setting targets helps managers responsible for fulfilling them to identify what needs to be done, by whom, with what outcome and by when. This is explained further in **Figure 2-7**.

The targets the organisation sets to achieve its objectives are used to direct the health and safety performance of the organisation over the period of time they relate to. The organisation should periodically measure its health and safety performance and compare performance against targets.

Specific	A specific objective has a much greater chance of being accomplished than a general goal or objective because it enables a greater understanding of what the objective is. To set a specific objective it is necessary to determine: • **Who** is involved • **What** is to be accomplished • **Where**: identify the location • **When**: a time frame should be established • **Which**: identify requirements and any constraints • **Why**: specific reasons, purpose or benefits of accomplishing the objective.
Measurable	If an objective is not measurable it is not possible to know whether progress is being made towards completion of the objective. Establish a reliable system to measure progress towards the achievement of the objective. Measurable objectives should have quantifiable indicators, for example a specific number of things to achieve. Avoid using words like "improve understanding" when setting an objective because understanding can be very difficult to measure. An example of a measurable objective could be "80% of managers will take the NEBOSH International General Certificate health and safety training course by the end of the year."
Achievable	Many objectives, if they have clearly been assigned to someone, are achievable eventually. However, the amount to be achieved in the time available may be unrealistic. For example, it is realistic for a person to want to lose two kilograms in weight. However, it is unrealistic to want to lose two kilograms in one week. Any barriers to the success of achieving the objective should be identified. Each barrier identified can then be analysed to determine how significant it is and its likely effect on achieving the objective. This provides an opportunity to check and ensure the objectives set are achievable.
Reasonable	Objectives should be reasonable in order for those who have to achieve them to be motivated to achieve them. Reasonable includes the objectives supporting the overall aims of the organisation. Objectives set for departments and similar should link well to higher level objectives and the overall aims of the organisation. Objectives should relate to the nature and size of the organisation. They also have to be reasonable in that they relate to the management of health and safety.
Time bound	It is important to ensure objectives are completed in good time. If the objective contains a requirement for it to be completed by a time it is more likely to get the attention necessary to achieve it. Those required to achieve the objective are more likely to plan to achieve it within the time and monitor progress towards completion. For short-term or project related objectives to be effective a defined time frame for completion should be set. Where objectives require something to be repeated it is usual to establish the number of times it is to be repeated over a defined time period.

Figure 2-7: SMART objectives.   Source: RMS

It can be useful to establish interim targets that lead towards completion of the final target.

For example, if the target requires 80% of managers to have taken the NEBOSH International General Certificate health and safety training course by the end of the end of a year period it may be useful to set performance targets that lead towards achieving this. A minimum target number of managers to receive the training may be set for each quarter of the year. This helps to ensure smooth progress towards the end of year target and avoids a rush of action to try to meet the target just before the end of the year period.

This also ensures the organisation gets the benefit of more managers operating with a better understanding of health and safety for a longer period in the year.

The organisation should set a range of quantifiable performance targets that reflect their specific intentions over a period of time.

The range of health and safety targets that may be set could relate to:

• Reduction in the number of accidents/incidents.

• Reduction in the number of work-related ill-health cases.

- Maintenance of health exposure levels below defined limits.

- Maintenance of conformity scores of audits.

- Improvement in the number of workplace inspections.

- Improvement in the number of managers receiving health and safety training.

- Improvement in the number of completed risk assessments.

- Improvement in the number of contractors receiving a health and safety induction.

- Improvement in the number of workers wearing the required personal protective equipment.

- Improvement in the number of management meetings that have active health and safety items on the agenda.

- Improvement in the number of consultation meetings with workers.

It is important that objectives and targets are documented. Documenting the objectives and targets will help to ensure they are seen as an obligation that must be fulfilled and that effort must be made to meet them. Objectives and targets should be communicated to those whose support will be needed for their fulfilment and to other relevant parties, such as workers, contractors and other stakeholders.

Managers responsible for meeting the objectives and targets should keep them in focus and monitor progress towards their completion, which should be reported periodically. This will provide an opportunity to identify if the planning to meet the objectives and the targets set were effective or whether the plan needs to be amended to better support completion of the targets and objectives.

Objectives and their targets must be reviewed periodically in order that they remain relevant and support continual improvement. This is usually done annually and enables an annual planning cycle to follow.

CONSIDER

How would you ensure that your policy was adequately communicated to those affected?

CONSIDER

Do you have a clear policy for health and safety; is it written down?

What did you achieve in health and safety last year and what do you want to achieve this year? Be realistic.

Does your policy prevent injuries, reduce risks and really affect the way you work? Be honest.

RESPONSIBILITIES OF WORKERS AT ALL LEVELS TO TAKE RESPONSIBILITY

Everyone in an organisation has a role and responsibilities to achieve the aims in the health and safety policy statement. It is important that roles and responsibilities are clearly defined and individuals are made aware of them. This will enable everyone to take an active and effective part in health and safety. The health and safety policy should indicate that the fulfilment of the policy and objectives requires workers, at all levels, to accept the responsibility for those aspects of health and safety they control. While responsibility and authority can be assigned, ultimately top management is still accountable for the functioning of the health and safety management system.

It is important that workers at all levels in an organisation understand their responsibilities and that health and safety is an important part of their job. This will establish ownership of health and safety by top management, line managers and individual workers as relevant to their role. ISO 45001:2018 Clause 5.3 requires "Workers at each level of the organisation shall assume responsibility for those aspects of the OH&S management system over which they have control."

The employer should clearly allocate responsibility, accountability and authority for the development, implementation and performance of the health and safety policy. This will require the employer to establish health and safety as a known and accepted line-management responsibility at all levels, and identify important roles with responsibility, authority and accountability for the identification, evaluation and control of health and safety hazards and risks. This will include details related to the most senior accountable person, line managers, supervisors, workers and their representatives, as well as others who have specific roles and responsibilities for health and safety, including health and safety practitioners, occupational health advisers, occupational hygienists, engineers, fire officers, first-aiders.

A clear delegation of responsibilities will assist in sharing out the health and safety workload, ensuring contributions from different levels and functions of the organisation. It will also help to establish clear lines of reporting and communication. Making individuals aware of their own roles and responsibilities will indicate to them that health and safety is a core part of their job.

Ensuring that all individuals in an organisation are aware of their roles and responsibilities will increase their motivation, indicate the commitment and leadership of senior management and help to improve the health and safety culture within the organisation as a whole. Defining clear roles and responsibilities will assist in the identification of individual competencies and training needs, particularly for specific roles, for example roles relating to first aid and fire arrangements.

Some health and safety responsibilities are common to workers at all levels, for example, this could include responsibilities to:

a) Take reasonable care for their own and others' health and safety who may be affected by their acts or omissions.

b) Comply with procedures and instructions given for their own and others' health and safety.

c) Use health and safety devices and protective equipment correctly.

d) Report to their immediate line manager any situation which they have reason to believe could present a hazard/unacceptable risk and which they cannot themselves correct.

e) Report any accident/incident or harm to health which arises in the course of or in connection with work they control.

The roles and responsibilities of individuals within the organisation, including typical responsibilities are covered in more detail in *Element 1.2 – Regulating health and safety – workers' responsibilities.*

Health and safety arrangements

The health and safety policy should state how it is supported by health and safety arrangements for achieving general and specific aims and objectives of the policy. This should explain where these arrangements are set out/defined, for example the arrangements could be an integral part of the organisation's health and safety management system.

The health and safety arrangements need to reflect the context, nature and size of the organisation. Though there will be some arrangements common to organisations, the range of arrangements to implement a health and safety policy for a construction company should be different from one for a retail outlet as the risks they manage are different.

The policy should summaries the arrangements for planning, organising, controlling hazards, consultation, communication, monitoring compliance with the arrangements and assessing the effectiveness of the arrangements to implement the policy.

The extent to which these arrangements are established in written form will also depend on the needs of the organisation, for example a small, low risk organisation may not need comprehensive written arrangements, whereas a complex high risk organisation may do. ISO 45001 encourages organisations to take an effective approach, Clause 7.5 Documented information states:

"The organization's OH&S management system shall include:

a) documented information required by this document;

b) documented information determined by the organization as being necessary for the effectiveness of the OH&S management system.

NOTE The extent of documented information for an OH&S management system can differ from one organization to another due to:

– the size of organization and its type of activities, processes, products and services;

– the need to demonstrate fulfilment of legal requirements and other requirements;

– the complexity of processes and their interactions;

– the competence of workers."

Typical general and specific arrangements might include the following.

General arrangements:

- Planning, including procedures for complying with legal requirements, setting objectives and targets and establishing programmes to meet the objectives.

- Allocation of resources for health and safety, including procedures for budgeting for health and safety.

- Organising, including procedures for worker communication and consultation.

- Training and competence, including procedures for induction and specific job training (such as managers, fork lift truck drivers, fire wardens and first-aiders).

- General controls for hazards, for example, risk assessments, safe systems of work, permits to work and provision of personal protective equipment.

- Hazard management, including procedures for reporting hazards and taking corrective and preventive actions.

- Cleaning, including arrangements for cleaning the workplace and welfare facilities.

- Inspections, including arrangements for inspecting the workplace and work equipment.

- Welfare facilities, including arrangements for washing, eating and provision of drinking water.

- Emergencies, including arrangements for someone to take charge of an emergency, to get assistance, to bring the emergency under control and reduce the effects of the emergency.

- Medical, including arrangements for health surveillance and first-aid.

- Accident/incident and ill-health reporting and investigation.

- Evaluation of performance, including arrangements for monitoring accidents/incidents and ill health, audits and periodic reviews.

Specific arrangements for hazards:

- Chemical and biological substances, including arrangements to identify and label substances, to prevent exposure in use, for storage and transport.

- Confined spaces, including arrangements for access, controlling work activities and rescue.

- Contractors, including procedures for contractor selection and management.

- Drugs, alcohol and smoking, including rules and measures to control these risks while at work.

- Electricity, including procedures for the maintenance and testing of equipment.

- Fire, including arrangements to prevent and control fires and procedures for raising a fire alarm and responding to a fire alarm.

- Maintenance, including safe systems of work and permit to work procedures.

- Manual handling, including arrangements to limit handling loads by the use of mechanical devices.

- Noise, vibration and radiation, including arrangements to set exposure limits, to purchase equipment with low exposure levels where possible, to prevent exposure and provide medical surveillance to identify possible effects of exposure.

- Stress and violence, including arrangements to identify and control causes and the provision of support and counselling.

- Transport, including arrangements to separate pedestrians from transport where possible and the safe control of reversing vehicles.

- Visitors, including procedures for their identification and supervision.

- Work at height, including arrangement to avoid the need to work at height and for safe working at height.

- Working alone, including arrangements for monitoring those working alone and for emergencies.

Worker/worker representative participation is an essential element of the management of health and safety in an organisation. The employer should therefore ensure that arrangements to enable workers and their representatives to participate in health and safety are defined and clarified to show that through their participation they support the line managers and do not take responsibility for health and safety away from them. They should be consulted, informed and trained on all aspects of health and safety that affect them or who they represent. This should include hazards, risks and their controls as well as emergency arrangements associated with their work. The employer should make arrangements for workers and their health and safety representatives to have the time and resources to actively participate in health and safety. The arrangements should establish a clear perspective of the role of health and safety committees if they exist, including clarification of the lines of communication with them and from them, including *feedback loops* to line management and health and safety practitioners.

KEEPING THE HEALTH AND SAFETY MANAGEMENT SYSTEM CURRENT

It is important that the health and safety management system is maintained so that it is current and remains effective. As part of this process the health and safety management system should be reviewed as necessary to ensure this. A number of circumstances may lead to a need to review the policy, for example, the passage of time, technological, organisational or legal changes and the results of monitoring or following a major incident.

As time passes the aims, objectives, processes and procedures defined in the health and safety management system should be reviewed. In general aims and objectives will be reviewed at least annually in order to ensure they remain relevant and this would usually involve a review of the adequacy and effectiveness of the health and safety management system to determine its ability to meet the aims and objectives.

ISO 45001 Clause 9.3 states that the management review must include consideration of:

"a) The status of actions from previous management reviews.

b) Changes in external and internal issues that are relevant to the health and safety management system, including legal requirements.

c) The extent to which the OH&S policy and the OH&S objectives have been met.

d) Information on the OH&S performance, including trends in incidents, audit results, consultation with workers.

e) Adequacy of resources for maintaining an effective OH&S management system.

f) Relevant communication(s) with interested parties.

g) Opportunities for continual improvement."

Figure 2-8: Review of health and safety policy.
Source: Unsplash.

Technological change is happening in workplaces all the time, and this can mean that operational controls may have been defined against circumstances that no longer exist, for example different equipment or substances may have been used. It might also be that the organisation has been able to take advantage of technological advances in something, for example, materials handling, yet the health and safety management system refers to the earlier way of working.

Changes in **organisation** have a specific bearing on the health and safety management system. For example, if the top manager changes this would require the new top manager to take ownership of the policy by reviewing it, making any adjustments they considered were necessary and emphasising their commitment by signing the policy.

Legislation changes periodically and it usually reflects a strengthening of society's expectations with regard to health and safety. This may mean that the aims of the policy may no longer conform to legal requirements or legislation may require additional emphasis on the control measures for specific risks in order to reflect the higher standards it requires.

Monitoring of the health and safety management system might be as a result of planned activities, enforcement action, professional advice or following an accident/incident/ill-health investigation. If monitoring methods are in place and are working they could identify a gap or reduced effectiveness, perhaps due to a specific risk that monitoring activities identified was not effectively controlled. This may mean parts of the management system, for example, assessment of risks, objectives and operational controls, may need to be clarified.

This process of reviewing the health and safety management system helps to maintain it so that it reflects the organisation's current needs and expectations and remains suitable and adequate. It also provides an opportunity to learn from what is working or not and ensures it remains effective.

 **REVIEW**

What might initiate a review of policy?

Sources of reference

Reference information provided, in particular web links, was correct at time of publication, but may have changed.

Guidelines for auditing management systems, ISO 19011:2018

Guidelines on Occupational Safety and Health Management systems, Malaysia, ISBN 978-983-2014-75-1, can be downloaded free from the Malaysian Department of Occupational Safety and Health website - http://www.dosh.gov.my/index.php/legislation/guidelines/general/597-04-guidelines-on-occupational-safety-and-health-management-systems-oshms/file

Health and Safety at Work etc Act 1974, UK, www.legislation.gov.uk (please enter the name of the Act)

The health and safety toolbox, How to control risks at work, HSG268, HSE Books, ISBN: 978-0-7176-6587-7 http://www.hse.gov.uk/pUbns/priced/hsg268.pdf

ILO Guidelines on Occupational Safety and Health Management Systems (ILO-OSH 2001), ISBN: 92-2-111634-4 (2nd edition), can be downloaded free from ILO web site. http://www.ilo.org/global/publications/ilo-bookstore/order-online/books/WCMS_PUBL_9221116344_EN/lang--en/index.htm

ISO 45001:2018 Occupational health and safety management systems – requirements with guidance for use, International Organization for Standardization

NORMLEX (ILO data base of international labour standards, including ILO Conventions, and National laws) http://www.ilo.org/dyn/normlex/en/f?p=NORMLEXPUB:1:0::NO:::

Occupational health and safety management systems – requirements with guidance for use: ISO 45001: 2018, ISBN: 978-0-580-86393-6

Occupational Safety and Health Act 1994, Malaysia, www.dosh.gov.my (please follow the links via legislation to Acts)

Occupational Safety and Health Recommendation R164, 1981, ILO, http://www.ilo.org/dyn/normlex/en/f?p=1000:12010:0::NO::: (please select R164 from the Technical list).

Web links to these references are provided on the RMS Publishing website for ease of use – www.rmspublishing.co.uk

STUDY QUESTIONS

1) (a) What are the five key elements in ILO 'Guidelines on Occupational Safety and Health Management Systems?' (5)

   (b) Explain the requirement for two of the key elements identified. (3)

2) What are the main benefits of introducing a recognised health and safety management system? (8)

3) What are the key aims an organisation should commit to in their health and safety policy statement? (8)

4) What is the purpose of the three key features of an effective health and safety policy? (8)

5) What methods could you use to communicate the health and safety policy to workers and others? (8)

6) (a) Who should sign the health and safety policy statement? (2)

   (b) Why should the policy be signed and dated? (6)

For guidance on how to answer NEBOSH questions please refer to the 'study question answer guidance' section located at the back of this guide.

This page is intentionally left blank

Element 3

Managing risk - understanding people and processes

Contents

3.1 Health and safety culture

MEANING OF THE TERM 'HEALTH AND SAFETY CULTURE'

The term 'safety culture' was first used in the International Atomic Energy Agency's (IAEA) initial report following the Chernobyl disaster, in 1986. One definition of the term 'safety culture' was given in 1993 by the UK, Health and Safety Executive (HSE) Advisory Committee on the Safety of Nuclear Installations (ACSNI), not long after the IAEA report on Chernobyl. The definition was included in the ACSNI Human Factors Study Group's third report - 'Organising for safety', *see Figure 3-1*.

> *"The safety culture of an organisation is the product of individual and group values, attitudes, perceptions, competencies, and patterns of behaviour that determine the commitment to, and the style and proficiency of, an organisation's health and safety management".*

Figure 3-1: Safety culture of an organisation.
Source: UK, ACSNI.

In more recent years the term 'safety culture' has been widened in its general use to 'health and safety' culture to reflect the increased interest in work-related ill-health. To better understand the term health and safety culture a useful approach is to distinguish between three inter-related aspects of health and safety culture, specifically:

- 'How people feel' about health and safety.
- 'What people do' about health and safety in the organisation.
- 'What the organisation has' in place for health and safety.

'How people feel' about health and safety encompasses the values, beliefs, attitudes and perceptions of individuals and groups at all levels of the organisation, which are often referred to as the health and safety climate of the organisation.

'What people do' within the organisation includes the health and safety related activities, actions and behaviours of individuals and groups at all levels.
For example, individuals making time for health and safety and giving it due priority when making decisions.

'What the organisation has' is reflected in the organisation's policies, operating procedures, management systems, control systems, communication and workflow systems. For example, health and safety is integrated in planning work activities and design of the workplace. The table in *Figure 3-2* provides examples of how these three aspects relate to a positive health and safety culture.

The UK, HSE Advisory Committee on the Safety of Nuclear Installations provided a useful perspective on what a positive health and safety culture is, *see Figure 3-3*.

> *"Organisations with a positive safety culture are characterised by communications founded on mutual trust, by shared perceptions of the importance of safety and by confidence in the efficacy of preventive measures."*

Figure 3-3: Positive safety culture.
Source: UK, ACSNI.

Aspect of culture	A positive health and safety culture
'How people feel'	For example, values and beliefs: • People in the organisation value health and safety, regard it as mandatory and operate in a safe and healthy way. • People in the organisation believe that health and safety makes commercial and professional sense, that individuals are not the sole cause of accidents/incidents and ill-health and that the next accident/incident or ill-health event is waiting to happen.
'What people do'	For example, problem solving methods: • People in the organisation use risk assessment and accident/incident and ill-health investigation/analysis, they also use cost-benefit analysis. • People in the organisation actively search for problems in advance of accidents/incidents or ill-health events.
'What the organisation has'	For example, working practices: • Health and safety is integral to design and operations practices. • Health and safety is the first item on meeting agendas up to Board level.

Figure 3-2: Aspects of a positive health and safety culture. Source: RMS.

RELATIONSHIP BETWEEN CULTURE AND HEALTH AND SAFETY PERFORMANCE

Because the culture of an organisation has such a strong influence on the way it does things, there is a direct relationship between the culture of an organisation and its health and safety performance. Each decision taken by an organisation is influenced by its culture. The way the decisions affect health and safety may be positive or negative, depending on the culture of the organisation. Organisations may have a culture that focuses only on the short-term or they may have a culture that values a longer term, more controlled, preventive approach to health and safety.

Organisations with a culture that supports a preventive approach have a greater chance of a good health and safety performance. For example, an organisation with a positive health and safety culture naturally includes health and safety in their decisions and gives it due priority. This means that the organisation is more likely to have good standards of health and safety and take early action to minimise risks. As organisations with a positive health and safety culture also tend to have managers and workers that are committed to health and safety they are more likely to behave in a way that leads to good health and safety performance.

The relationship between health and safety culture and health and safety performance can be illustrated by research that has been conducted. Research has shown that influences on the health and safety culture of an organisation made by the introduction of a health and safety programme significantly improves health and safety performance. For example, after the introduction of a health and safety programme in a range of forestry and logging organisations in Columbia, it was found by Painter and Smith (1986) that there were dramatic improvements in performance. The accident/incident frequency rate was reduced by 75% and the workers' compensation costs were reduced by 62%. In further research, Lauriski and Guymon (1989) found that after a health and safety management programme had been introduced at the Utah Power and Light Company, lost time injury rates were reduced by 60% over a period of five years. From 1980 to 1988, the accident/incident frequency rate was reduced from 40 to 8 per annum, while production more than doubled.

A reduction in accident/incident rates experienced by an organisation with a positive health and safety culture would be seen as a positive step forward and could lead to a further positive influence on the health and safety culture and probable future improvement in performance.

INDICATORS OF AN ORGANISATION'S HEALTH AND SAFETY CULTURE

Developing and promoting a positive health and safety culture is an important part of health and safety management. A number of indicators can be used to assess an organisation's health and safety culture. Some of these indicators relate to health and safety outcomes and include:

- The number of accidents/incidents occurring.
- The number of work-related illnesses occurring.
- Sickness and absenteeism rates.
- The number of complaints about working conditions.
- The number of workers leaving employment.

Where these outcomes are high, for example, if there are a lot of accidents/incidents, it usually indicates the presence of a negative health and safety culture. Although, where data on outcomes show they are low this may be as a result of poor reporting. Other indicators used to assess health and safety culture relate to what an organisation does to manage health and safety. Where there are a good range of indicators present and they are effective it usually indicates the presence of a positive health and safety culture.

These indicators include:

- Values and beliefs expressed in a health and safety policy.
- Visible leadership and commitment.
- Health and safety as the first item on the agenda of meetings.
- Health and safety performance reported in the annual report.
- Giving health and safety an equal priority to other matters.
- Directors/senior management visits to workplaces.
- The amount of risk assessments conducted.
- The amount of health and safety training conducted.
- Communication of health and safety matters.
- Manager and worker involvement.
- Compliance with health and safety rules and procedures.
- The amount of workplace and work equipment inspections.
- The use of investigation and analysis processes.
- Timely completion of corrective and improvement actions.

The incident investigation process used by an organisation can provide a particularly useful indication of how positive the health and safety culture is. If investigation is limited to only identifying the immediate causes of accidents/incidents it would indicate that the culture is not strongly positive. If underlying and root causes are identified routinely by the accident/incident investigation process it would indicate a more positive health and safety culture. In addition, the management of the various causes identified in the investigation process may indicate how well developed the health and safety culture is. If the procedure for accident/incident investigation requires the structured management of recommendations and these were completed appropriately it would indicate a positive health and safety culture. However, if the same types of causes are repeatedly resulting in accidents/incidents it may indicate a lack of commitment to the prevention and improvement actions identified in the investigation recommendations. This would indicate the presence of a negative health and safety culture.

When assessing an organisation's health and safety culture, it is usual to consider both types of indicator what an organisation does to manage health and safety, and the outcomes.

INFLUENCE OF PEERS ON HEALTH AND SAFETY CULTURE

The term 'peer' is usually used to refer to those people that work with an individual. Peers are usually at the same level in an organisation as the individual, they may also be members of the same group, for example, members of a health and safety committee or management team.

Many individuals are influenced by what people think of them and may act in a particular way to ensure that the people around them feel positive towards them. In group situations this can include the individual acting in the same way as the group to maintain a feeling of belonging in the group. This desire for belonging and to be liked can lead to an individual consciously or subconsciously behaving in a similar way to those they work with, their peers. An individual's peers will also try to influence the individual to behave in a similar way to the way they behave. This influence of peers may lead to peer pressure to behave in a particular way and can be a very strong influence when peers act as a group. This can have a significant effect on an organisation's health and safety culture as the behaviour of people in an organisation is an important part of what influences the culture.

The influence of peers may have the effect of promoting good health and safety, for example, within a work team. The influence of peers means that if the members of a team believe that working safely is the right way to do the job, each team member will watch over the activities of the others in the team to ensure they conform to the way the team works. The group as a whole will ensure any new member follows their example to ensure they work safely. However, peer pressure may have a negative influence. The influence of peers may lead to managers and workers receiving direct or indirect pressure not to put effort into achieving good health and safety standards. This may particularly be the case where the peers do not value health and safety or consider it prevents them from achieving things that they feel are important. For example, a manager or worker may feel that working safely will limit their ability to achieve productivity rewards and they may put pressure on their peers to ignore health and safety requirements.

It is therefore important to anticipate the influence of peers and plan to use this influence to promote health and safety, helping to establish a positive health and safety culture. This will usually involve obtaining the commitment of the senior manager and using it to influence other managers, who in turn influence those they work with. In this way the influence of peers can have a significant positive effect on an organisation's health and safety culture.

REVIEW

Explain the following aspects of health and safety culture: 'What people think', 'What people do', 'What the organisation has'.

3.2 Improving health and safety culture

Health and safety culture at work can be improved by the use of a range of measures and by the coordinated effort of everyone in the organisation. Measures that can improve health and safety culture and contribute to establishing a positive health and safety culture include securing management commitment, leadership, promoting health and safety standards, holding people to account for their behaviour, ensuring the competence of individuals, effective communication, encouraging co-operation, arranging consultation and the provision of training.

GAINING COMMITMENT OF MANAGEMENT

Securing management commitment is one of the most important steps in establishing a positive health and safety culture, which is recognised as an effective way of ensuring appropriate health and safety behaviour. The absence of management commitment would indicate that health and safety was a low priority in the organisation and that it was not important. This could lead to managers and workers behaving in a way that is not supportive to health and safety. For example, when decisions are made, health and safety may not be given the resources required.

Securing a commitment from management would help to ensure health and safety was properly integrated in the processes of the organisation. This includes preparing and implementing a health and safety policy that reflects positive values and beliefs. The policy should be supported by effective management systems and procedures that include health and safety. Management should be committed to giving an equal priority to health and safety issues as other objectives, such as production and quality. This will involve managers actively ensuring that their work and management practices give appropriate priority to health and safety. For example, providing a pleasant working environment with good welfare facilities will indicate a commitment to health and safety and the good standards expected of managers and workers.

By securing the commitment of management, health and safety would be identified as one of the core values of the organisation and individuals would be more motivated to behave safely and healthily.

PROMOTING HEALTH AND SAFETY STANDARDS BY LEADERSHIP, EXAMPLE AND DISCIPLINARY PROCEDURES

Management actions at all levels should send clear signals to other managers and workers of the importance of the health and safety standards set by the organisation. Managers should promote the health and safety standards by showing their commitment to them when they are carrying out their normal management activities and specific health and safety activities assigned to them. This will often include managers showing leadership by setting a good example. For example, managers should use the correct personal protective equipment, follow health and safety rules and show commitment by attending health and safety training. This will reinforce the need to conform to health and safety standards.

Managers should also show leadership by making health and safety important in meetings and decision making, for example, ensuring health and safety is dealt with as a priority item on the agenda of meetings. This will promote the importance of managers putting effort and priority into achieving good health and safety standards. Managers would then be more likely to give proper consideration to health and safety during meetings and in their other work. Managers would also have an opportunity to promote health and safety standards in these meetings by referring to them when operational issues are discussed and ensuring the standards are not compromised when making decisions. In situations where there is a pressure to compromise health and safety standards because of time constraints managers should emphasise the importance of maintaining standards and, where

it is appropriate, promoting the acceptance that a delay will be necessary to complete the task safely. This strong leadership will set a good example to other managers and the workers affected by the decision, which will encourage positive health and safety behaviour in general.

There are many other opportunities for managers to show leadership and set an example, including during visits made to the workplace and when health and safety audits or reviews are conducted. Senior managers show good leadership when they promote health and safety standards during their opening presentation at health and safety training sessions that other managers and workers attend. This will also emphasise the importance of the training being provided and show a commitment to good health and safety standards. The significance of strong leadership to health and safety was emphasised in the guidance document on occupational safety and health (OSH) roles and responsibilities that accompanied the Abu Dhabi Occupational Safety and Health System Framework (OSHAD-SF) Guidance Document - OSH Roles and responsibilities, *see Figure 3-4.*

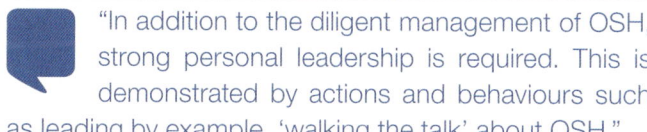

"In addition to the diligent management of OSH, strong personal leadership is required. This is demonstrated by actions and behaviours such as leading by example, 'walking the talk' about OSH."

Figure 3-4: Management leadership.
Source: Abu Dhabi Occupational Safety and Health System Framework (OSHAD-SF) Guidance Document - OSH Roles and responsibilities.

It is important that managers show leadership and support for health and safety standards during problem solving processes. This will demonstrate that health and safety is valued and that there is a management desire for good health and safety standards. It is therefore useful for managers to be involved in the investigation of accidents/incidents and ill-health and is particularly useful for them to be involved in preventive measures like risk assessment processes. Both of these processes provide an opportunity to promote health and safety standards by identifying if the standards were in place, encouraging standards to be met and enabling measures to be put in place for individuals to work safely and healthily. The manager's involvement in risk assessment processes would promote the importance of preventive measures and could encourage supervisors and workers to behave in a way that is more likely to reduce risks.

To have the maximum effect on health and safety behaviour the organisation's health and safety standards should also be promoted through the good example set by supervisors and influential individuals, such as worker representatives and committee members.

If regular checks are made by managers to ensure health and safety standards are being met it would also show positive leadership to supervisors and workers and reinforce the agreed standards. Where health and safety standards are not being met, the reasons for this should be established by identifying the underlying and root causes. A blame approach should be avoided where there are acceptable reasons for not meeting standards. However, disciplinary procedures should be used where they are appropriate, to reinforce the need to maintain health and safety standards.

COMPETENT WORKERS

If managers and workers are to be effective in contributing to the correct health and safety behaviour at work it is essential that they are competent. Ensuring competence improves health and safety behaviour because managers and workers become more aware of the health and safety aspects of work activities, their perception of risk and danger increases to an appropriate level and there should be an improvement in their attitude and motivation. In addition, competent managers and workers will have a better understanding of the rules and procedures and are more likely to follow them. The decisions made by competent managers are more likely to give proper regard to health and safety and lead to actions that support workers following rules and behaving appropriately.

Competence means more than just providing training and involves ensuring that the relevant knowledge, skill and experience are established. Studies have shown that human errors can occur through an absence of knowledge and skill. It is important that competencies held by managers and workers include the correct health and safety behaviour for their job. Where possible, correct health and safety behaviour should be learnt as part of the main job competencies, for example, when a manager is learning to manage contractors the correct health and safety behaviour expected of them and the contractors should be included. This will ensure that having the correct health and safety behaviour is seen as part of doing a job well and is more likely to become a normal part of doing the job.

Because individuals may not be capable of developing the competencies and behaving in the correct way needed for certain jobs, their suitability should be assessed. This will help to prevent situations where human errors and violations occur because a person is not capable of doing the job. For example, an individual may not have the aptitude, dexterity and physical ability needed to do certain jobs competently. Similarly, not everyone will be able to work at height or in confined spaces.

When new appointments of managers and workers are being made it should not be assumed that they have the competence that will provide the right health and safety behaviour, even if they are good at other aspects of their work, for example, quality. Managers and workers should be assessed for their range of competencies prior to appointment. Where competencies need to be developed it is important that managers and workers gain competence in a controlled way to ensure they learn the correct things and develop the right skills and behaviour. If this is not done, they may not develop the correct health and safety behaviour, instead they may just become very skilled at their bad habits. In order to minimise human errors, it may be necessary to arrange for individuals to gain experience under controlled conditions, for example, by close supervision. ILO-OSH 2001 emphasises the need to ensure competence in Clause 3.4.1, *see Figure 3-5.*

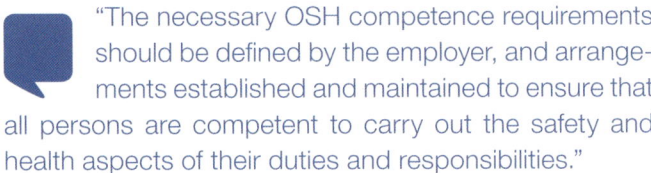

"The necessary OSH competence requirements should be defined by the employer, and arrangements established and maintained to ensure that all persons are competent to carry out the safety and health aspects of their duties and responsibilities."

Figure 3-5: Requirements for competence.
Source: ILO-OSH 2001, Clause 3.4.1.

Competence may be achieved and maintained by actions that include:

- Defining the competencies needed to carry out jobs and tasks safely and healthily.
- Assessing the competencies that managers and workers have and planning competence development.
- Providing the means to ensure that all managers and workers, including temporary workers, develop competencies they require in a controlled way, particularly for those involved in high risk work.
- Arranging for help, advice and supervision while they develop the correct competencies and health and safety behaviour.
- Ensuring individuals know the limits of their competency.
- Reviewing and refreshing competencies as needs change or where health and safety behaviour shows it is required.

GOOD COMMUNICATION WITHIN THE ORGANISATION

Studies of major accidents/incidents have shown that without effective communication, managers and workers are less likely to make the right decisions and errors or violations may occur. It is therefore important to provide effective communication within an organisation in order to help to ensure the correct health and safety behaviour takes place. ILO-OSH 2001, Clause 3.6, shows the type of communication arrangements and procedures required in an organisation, *see Figure 3-6*.

"Arrangements and procedures should be established and maintained for:

(a) Receiving, documenting and responding appropriately to internal and external communications related to OSH;

(b) Ensuring the internal communication of OSH information between relevant levels and functions of the organisation; and

(c) Ensuring that the concerns, ideas and inputs of workers and their representatives on OSH matters are received, considered and responded to."

Figure 3-6: Communication arrangements and procedures.
Source: ILO-OSH 2001, Clause 3.6.

Effective communication involves providing what is required in the right way and at the right time. This will mean using the best communication method for the information or instruction being provided and the circumstances. It is therefore necessary to consider the individual receiving the communication to ensure that the information is understood and the correct health and safety behaviour results. For example, when managers communicate with individuals it is important that managers are aware of the individual's needs, desires, capabilities and expectations in order to communicate with them effectively. In some cases, this will mean ensuring that the communication provides information on why action is to be taken as well as what is to be done. This is particularly the case with regard to health and safety. For example, a manager may communicate to maintenance workers that they must wear eye protection when doing a task, but without emphasising why they should wear eye protection they may not behave correctly and wear it.

It is also important that managers at all levels communicate their commendation of positive health and safety behaviour when they see it, as this will make it more likely that the positive behaviour will continue.

Communication within an organisation should include workers communicating to manager's information on concerns they have regarding health and safety. It is important that there are means for workers to do this in a positive way, so that they are encouraged to report hazards and deficiencies that could be a danger to themselves and others. To encourage workers to continue to report concerns managers should provide workers with feedback on concerns they communicate.

Information

Purpose	To improve awareness about health and safety generally and in relation to specific hazards, their controls and management performance to bring about these controls. In itself passive, it relies on the recipient to interpret.		
Subject	Legislation. Company policy statements.	Incident and ill health statistics. General hazards and controls	Names of appointed first-aiders
Means of communicating	Bulletins and news sheets. Notice-boards, propaganda, films.	Team briefing. Written material for visitors.	Site signs and labels.

Instruction

Purpose	To control the behaviour of workers, contractors and visitors with regard to general and specific health and safety arrangements. Typically, one-way communication, often no real check of understanding.		
Subject	Health and safety rules. Policy, arrangements and plans.	Use of PPE. Specific hazards, for example, smoking	Emergency procedures. Reporting incidents and ill health.
Means of communicating	Carried out formally using verbal, written and visual material - notice-boards, induction and job training, direct issue of document, 'tool-box talks'.		

The communication process

Communication may be defined as 'a process by which information is exchanged between individuals through a common system of symbols, signs or behaviour'.

Exchanging information may seem a simple task, but communicating in an effective way is often much more difficult than it appears. Poor communication, while common, can result in serious misunderstandings, leading to incorrect health and safety behaviour, accidents/incidents and work-related harm to health.

Barriers to effective communication

It is essential that communications are understood in order for them to be acted on and for individuals to have the correct health and safety behaviour. However, communication is not always effective as there are a number of barriers that may prevent it being understood, including:

- Noise and other similar distractions.
- Sensory impairment (poor hearing or eyesight).
- Complexity of the information.
- Language/dialect of the speaker.
- Illogically presented information
- Ambiguity of the information.
- Use of technical and local terms or abbreviations.
- The timeliness of the communication.
- Lengthy communication methods.
- Inattention of the person receiving the information.
- Lack of trust or respect.
- Capabilities of the person receiving the information to understand it, due to reduced health, fatigue, stress or limitations on their mental processing ability.

General principles of communication

Communication is a skill that people often take for granted. Like any other skill, some individuals are better at it than others. A one-way communication process involves sending information to an individual and not obtaining confirmation of effective communication. This type of process has the limitation that confirmation of receipt and understanding of the information is not achieved.

Therefore, communication is generally a two-way process and is most effective in this form. In a two-way communication process the needs of the person receiving the information are equally as important as the needs of the person sending the information. The person sending the information should not assume that the person receiving it has understood what was sent. The two-way process requires the person sending the information to confirm that the other person has received and understood it. To ensure the success of two-way communication the person sending the information should:

- Communicate to the person receiving the information at a time that is suitable for their circumstances.

- Communicate in a form capable of being understood by the person receiving the information.

- Keep the content of the communication concise and relevant to the information being communicated.

- Provide the information in an order that is logical to the person receiving it.

- Use clear and unambiguous terms.

- Ensure that what is communicated matches the

way it is communicated. If something is important, the communication of it should not be rushed. Non-verbal communication may communicate something different to what is said. Consider the need for respect and whether anger or humour may distract from what is communicated.

- Plan and encourage feedback to confirm understanding, including time, method and the form of feedback. The person receiving the communication may have to respond in a particular way, for example, repeating the information back to the person who gave the information.

- Use open ended questions to investigate understanding.

- Use closed questions, which tend to provide yes/no answers, to confirm understanding.

- Repeat the communication or use a different method of communication if understanding is unclear.

Communication methods

In ensuring effective communication, the correct method of communication should be selected and used for the particular information and circumstances in which it is being provided. This will mean considering the principles of good communication and the benefits and limitations of different methods. The different methods used to communicate health and safety typically include verbal, written and graphic methods.

Verbal communication

Verbal communication is widely used in relation to health and safety and involves direct speech between individuals or through another medium, for example, a telephone, radio, video link or electronic speaker system. Two-way forms of verbal communication provide the advantage of being able to communicate and obtain confirmation of understanding efficiently and promptly. This allows communication to be done in small amounts where necessary, ensuring understanding after each part of the information is communicated. This is particularly useful where information needs to be communicated accurately and understanding confirmed.

The barriers to communication outlined above are all relevant to verbal communication. In particular this method of communication can be influenced by the language spoken, dialects, accents and the unintended wrong use of words. These barriers can lead to misunderstandings or the wrong meaning being communicated. Where clarity of small amounts of important information is to be communicated the convention of spelling the letters of words and using a phonetic spelling alphabet is sometimes adopted. For example, the phonetic spelling alphabet of the International Civil Aviation Organisation (ICAO) assigns code words to letters of the English language, for example, A - Alpha, B - Bravo, C - Charlie, etc. This is particularly useful when communicating by radio or telephone and helps to avoid errors in verbal communication.

Where the information to be communicated is complex it may be preferable to use written communication to ensure clarity or a combination of written and verbal communication.

Written communication

Written communications are widely used for health and safety purposes, including policy documents, work instructions, procedures, site rules, permits-to-work, contractor contracts, meeting agendas and minutes, and reports. The written method of communication enables the information to be retained by the person receiving it and referred to if they forget what the information was. It may also enable an individual to receive information prior to needing it and means they can read it at a time that suits them. Obtaining confirmation of understanding of written communications can be as important as for verbal communication. However, written communications are often used as a form of one-way communication, for example, notices or instructions posted on an information board.

One advantage of the written form of communication is that it can provide information to a large number and range of people at the same time. However, this can make it difficult to take account of the range of people reading it and the need to ensure effective communication to each of them. In particular, it may be hard to obtain or manage feedback from all of the people sufficiently to ensure they understood what was communicated. For example, this difficulty would affect a situation where all workers in an organisation were issued with a new health and safety handbook. It could be difficult to confirm whether they had all read and understood the handbook.

When communicating in the written form it is important to consider the barriers to and principles of communication, this will help to make the communication effective. Particular attention should be paid to the use of appropriate structure, language, writing style and terms used to ensure the person receiving the communication understands it. Clarity and understanding of written communications can be tested by getting a sample of those who will receive the written communication to read it and checking they understood it. For example, procedures may be read by a sample of workers who have to follow them.

Graphical communication

Graphical communications may be used to replace or support other forms of communication. Verbal presentations and written reports will often include graphical images to illustrate or clarify complex relationships, for example, changes in accident/incident frequency rates over a period of time may be illustrated by a graph.

Graphical communication is used to good effect with health and safety signs, where the colour and pictorial representation of hazards and control measures are used. This avoids problems where workers may have limited language abilities and would struggle to read words. Graphical communication is also used in posters and moving image communications, for example, electronic induction packages and message boards.

Graphical communication in the form of drawings, photographs, still or moving digital images may be used to clarify information on hazards or risk control measures. They can provide an opportunity to make the communication of information more effective by containing examples of hazards and control measures from the actual workplace of those persons with whom they are communicating. The visual effect of graphical communication of this type may enable managers and workers to better understand practical aspects of health and safety. For example, a rule that requires a worker to wear a hard hat correctly could easily be demonstrated and understood by showing an image of a worker wearing a hard hat correctly.

Use and effectiveness of various communication measures

Notice boards

A traditional communication measure is to post health and safety information on notice boards. The advantage of this communication measure is that the communication is easily available to everyone in a particular work area where the notice board is located and to those that pass near it. They can be a cheap and effective technique of communication when used to make general statements or to keep workers aware of current information and proposed developments. Notice board information relies on an individual's ability to read, understand and apply the information correctly. Therefore, particular care should be taken to ensure appropriate language(s) are used for information placed on a notice board and they should not be used as a substitute for more effective two-way communication that can provide confirmation of receipt and understanding of information. Also, the information posted on the notice board must be kept up to date and maintained in a legible condition if it is to be effective. This usually requires someone being given responsibility for the control of items placed on and removed from notice boards.

Health and safety media

Moving image media

Moving images of relevant health and safety topics are often used to provide and refresh awareness and understanding of health and safety. They can provide information in a way that improves motivation and leads to better health and safety behaviour. The visual impact in them is also a strong stimulus that maintains attention while learning and enables training to provide practical examples of hazards and measures to minimise risk. A common use for moving image media is at site induction of new workers and contractors, where it can provide information about the site, its rules and correct health and safety behaviour.

Shocking images of injuries and work-related ill-health are sometimes used to illustrate what might happen if procedures are not followed. It has been found that their effect may not change attitude in the long term, but can gain the attention of those seeing the images and allow them to accept the importance of health and safety measures when this is explained after seeing the images. However, for some people the shocking images cause them to put up a barrier to receiving further information and can prevent the desired health and safety message being received.

Poster campaigns

Posters displaying health and safety information are sometimes seen as an inexpensive and visible way of showing commitment to health and safety. This approach can be ineffective if management place too much reliance on them and use them to inform workers to take care to avoid hazards that should have been managed by better control measures. To be effective, messages communicated by posters should be positive, aimed at the correct audience and be believable.

a) *Positive* - posters warning of the consequences if a particular action is not taken can be ineffective. To be effective, messages should emphasise the positive health and safety benefits of working safely and healthily to those reading the posters.

b) *Aimed at the correct audience* - posters quickly lose their impact and blend into the background, therefore posters should be changed regularly to maintain their impact as a means of communication. There can also be a reduction in impact when the message being communicated is perceived to be irrelevant by those reading the poster. Therefore, campaigns should be carefully targeted and posters positioned so they are seen by those the campaign is aimed at. For example, a poster reminding workers to wear the correct clothing when working with machinery should be sited close to the relevant machines.

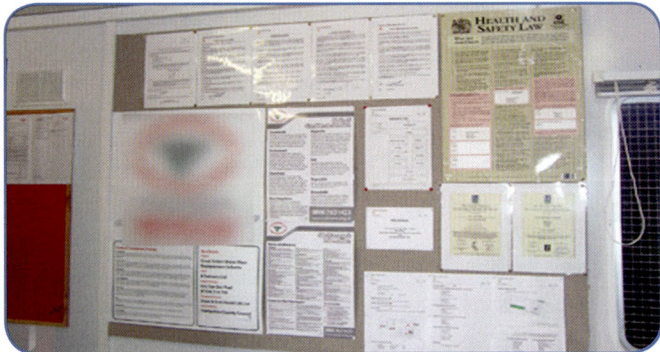

Figure 3-7: Health and Safety notice board.
Source: RMS .

Figure 3-8: Health and safety suggestion scheme.
Source: RMS.

c) **Believable** - messages should be convincing and realistic. Care must be taken to avoid offending or distracting the person reading the poster from the message by the use of inappropriate images of people. Similarly, pictures of horrific injuries can lead to a rejection of the intended message as the person reading it may focus on the image rather than the message. Some images may be so distressing that individuals refuse to look at the poster at all.

The advantages of using posters for communicating health and safety include:

- They have a relatively low cost.

- They provide visual impact to health and safety messages, allowing brief messages to be easily understood.

- They are flexible, enabling them to be displayed in the most appropriate positions and moved easily.

- They can be used to reinforce verbal instructions and provide a constant reminder of important health and safety issues.

- They can enable workers to be involved in health and safety locally, by encouraging them to suggest ideas for posters, selecting or designing posters and locating the posters where they will have most effect.

The disadvantages of using posters for communicating health and safety include:

- There is a need to change posters on a regular basis to maintain attention on them.

- They may become soiled, defaced and out of date.

- There is a possibility that they might be seen to trivialise serious matters.

- They might offend people if inappropriate stereotypes are used.

- It is not easy to assess whether the message has been understood.

- They may be used by some employers as an easy, if not particularly effective, way of discharging their responsibility for health and safety by moving the responsibility onto workers.

Toolbox talks

'Toolbox talks' are short communication sessions provided by supervisors and team leaders to explain or remind workers of important health and safety risks and control measures. The term toolbox talk comes from a scheme where short talks on health and safety were provided by supervisors in the workplace, sometimes near where the workers kept their tools. They were designed to take place in the workplace so that workers were not distracted from their work for too long. The term is now often used to describe any form of communication where workers are gathered together for a short talk in or near to the worker's workplace.

Because toolbox talks are planned and organised in advance, the technique provides fast and reasonably consistent communication of specific subjects. The toolbox talks often cover only one subject to ensure the attention of workers is focused and two-way communication is encouraged. Subjects covered in talks may provide a reminder of a particular risk related to work activities or discuss recent good or poor health and safety experience. The talks may be used to tell workers about proposed changes, for example, changes in personal protective equipment, work practices or procedures. Toolbox talks enable health and safety issues to be discussed by supervisors and workers, providing increased awareness/understanding of risks and control measures and reinforcement of rules. They can be regularly used to encourage workers to have appropriate health and safety behaviour.

Electronic messages and e-mails

Electronic messages and e-mails are a useful way to provide health and safety communication on a timely basis. For example, they can give information on proposed changes to procedures before the amend-ment and distribution of the whole procedure. Electronic messages are often used as a means of one-way communication. One of the difficulties in using electronic messages is that the person sending the message might assume that everyone to whom they sent the message received it, however this might not be correct as loss of power of battery operated equipment and poor signal can influence receipt of the message. Where the message is important to health and safety there should be a process that confirms receipt. Similar problems can exist with the use of e-mails, although software is available to check whether a person has received and opened an e-mail. However, confirmation of receipt of a memo or e-mail does not confirm that it has been read, understood or actioned.

Many people complain about the number of emails they receive. Where individuals receive a high volume of e-mails there is a risk that important health and safety e-mails may be missed. This situation is reported to be particularly difficult in situations where individuals are copied in to e-mails in bulk, causing people to receive e-mails that are of very low or no relevance to them. This could reduce the perceived value of e-mails received from these people and lead to important health and safety messages they send being missed.

Worker handbooks

Workers need to know the current rules and procedures so they can follow them and have the correct health and safety behaviour. Issuing a worker handbook will help to provide this information, but if it is not kept up to date it can lead to workers making errors.

Handbooks are often issued to new workers at induction. They are useful in communicating site rules and information such as accident/incident and ill-health reporting procedures. Similarly, they will often contain information relating to on site emergency arrangements such as fire and first-aid. To be effective, the organisation should establish a mechanism to recall and reissue the handbooks when changes occur. Some organisations operate a loose-leaf folder design to enable amendments to be made. If paper-based handbooks are provided, consideration needs to be given to where each person keeps the handbook so that it is available for use and updating. Some organisations have computer systems available and therefore choose to provide handbooks in an electronic format. This can enable changes to the electronic copy of the handbook to be made quickly. However, unless workers can be equally quickly informed of the changes, they are not likely to follow the new rules and procedures because they will not be aware of them. Toolbox talks or other means of communication may have to be used to inform workers affected by the changes. If toolbox talks are used, they would provide an opportunity to discuss the changes and ensure people knew why the changes were taking place and that they understood what the changes were. This could provide additional confidence that everyone would follow the new rules and procedures.

When choosing an appropriate communication process that uses a particular communication method and communication measure, it is important to consider the risk that might result from poor communication. For example, it may not be appropriate to rely on a one-way communication process, like notice boards, for important communications related to high risk work. In most communication situations, confirmation that the information communicated was received is important and experience shows that confirmation of understanding of the information is essential if the correct health and safety behaviour is to result from it. Therefore, the communication methods and measures that provide a two-way communication process are usually preferred.

CONSIDER

How effective is health and safety communication in your organisation?

What could be done to improve it?

Co-operation and consultation with the workforce and contractors

Those organisations that recognise the importance of establishing a positive health and safety culture have arrangements in place for consultation in order to develop good management and worker co-operation. This may be achieved by consulting workers directly or indirectly, consulting with worker health and safety representatives or by using health and safety committees. It is important to establish that there is a difference between informing people and consulting them. Informing is a one-way process involving the provision of relevant information, for example, by management to workers, whereas consulting is a two-way process where account is taken of the views of those being consulted before a decision is taken.

The significance of co-operation and consultation was emphasised in the guidance document on occupational safety and health (OSH) roles and responsibilities that accompanied the Abu Dhabi Occupational Safety and Health System Framework (OSHAD-SF) Guidance Document - OSH Roles and responsibilities, see *Figure 3-9.*

"One of the primary objectives of the OSHAD-SF is to foster a co-operative consultative relationship between employers and employees on the health, safety and welfare of such employees at work."

Figure 3-9: Co-operation and consultation.
Source: Abu Dhabi Occupational Safety and Health System Framework (OSHAD-SF) Guidance Document - OSH Roles and responsibilities

In some organisations co-operation and consultation would extend to the involvement of contractors, this is particularly important where contractors provide services in a workplace on a long-term contract, for example maintenance and security contractors.

Worker participation
Roles and benefits of worker participation

The main role of worker participation in health and safety is to provide the employer with a wider view of how risks affect workers, obtain feedback on the effectiveness of current control measures and gain their view on proposed control measures. Worker participation in health and safety through consultation has the benefit that it will draw on the worker's experience and knowledge. This can be used to improve the effectiveness of health and safety measures. This improved effectiveness and participation of workers in the development of new methods can provide better worker acceptance of health and safety measures and lead to long term improvements in worker health and safety behaviour.

"The active involvement of each individual in the workplace is essential for the success of any OSHMS, and helps to develop an 'OSH culture' in the workplace."

Figure 3-10: The need for involvement.
Source: Abu Dhabi Occupational Safety and Health System Framework (OSHAD-SF) Guidance Document - OSH Roles and responsibilities.

In addition, to providing an opportunity for workers to contribute to health and safety, the role of worker participation is to show management commitment and motivate workers to work safely and healthily. The benefit of worker participation of this type is that it has been identified as one of the most significant factors to influence health and safety behaviour and promote a positive health and safety culture in organisations. Workers feel valued and involved in decision making. This can lead to shared health and safety values and the motivation of those involved to work together to improve health and safety. Worker participation provides an effective way of ensuring worker feedback on health and safety, for example proposed changes to the layout of a workplace, options for content and delivery of training, hazard identification and procedures to control risks. Workers often have a good understanding of the risks in the workplace and by involving them it shows them that health and safety is taken seriously and their views on problems and solutions are valued.

The ILO in Occupational Safety and Health Convention C155 and Occupational Safety and Health Recommendation R164 establishes a requirement that employers consult workers or their representatives, see figure 3-12. The term 'worker' used in ILO Convention C155 is the same as the term 'employee'.

The employer should therefore consult workers on matters of health and safety that affect them, in particular before any changes in arrangements are made. National legislation may specify what workers need to be consulted on, the Abu Dhabi Occupational Safety and Health Framework Guidance Document – Communication and Consultation suggest that in general this could be:

i.   Any changes which may substantially affect worker health and safety, for example, new or different procedures, types of work, equipment, premises, ways of working (shift patterns, hours of work).

ii. Arrangements for getting competent health and safety people to help the organisation meet its legal obligations, for example a health and safety advisor.

iii. Information that must be given to workers on likely risks in their workplace and precautions they should take.

iv. Planning of health and safety training.

v.   Health and safety consequences of introducing new technology.

vi. Incident investigation reports, risk assessments and emergency plans.

vii. Occupational health risks like stress and musculoskeletal disorders, including how they affect workers and control measures.

This may be done by consulting individuals directly or indirectly to obtain the participation of workers in the consultation process. There are a range of ways to consult with workers directly, including: one-to-one discussions, regular tours/walkabouts carried out by managers, inclusion of health and safety on meeting agendas, toolbox talks, special worker meetings, working groups. Ways to consult workers indirectly could include

Figure 3-11: Involving your workforce in health and safety. Source: UK, HSE HSG263.

company internet sites/apps that provide information and ask for views, worker surveys, notice boards and suggestion schemes.

Alternatively, consultation may be by the participation of worker representatives, where groups of workers are represented by a specific individual. The representative would participate in consultation with managers on behalf of the group of workers. This could be as single worker representatives or in groups of representatives at a health and safety committee.

> "Article 19 (e) workers or their representatives and, as the case may be, their representative organisations in an undertaking, in accordance with national law and practice, are enabled to enquire into, and are consulted by the employer on, all aspects of occupational safety and health associated with their work; for this purpose technical advisers may, by mutual agreement, be brought in from outside the undertaking.
>
> Article 20 Co-operation between management and workers and/or their representatives within the undertaking shall be an essential element of organisational and other measures taken in pursuance of Articles 16 to 19 of this Convention."

Figure 3-12: Co-operation and consultation. Source: ILO, Occupational Safety and Health Convention C155.

Worker representatives

A significant number of countries have established National legal requirements for worker health and safety representatives. For example, Section 17 of the South African Occupational Health and Safety Act 1993 requires health and safety representatives be appointed where more than 20 people are employed and specifies that there must be at least one representative for every 100 workers in offices and shops and at least one representative for every 50 workers for other workplaces. Other legislation, for example, in the UK, is less prescriptive and provides rights to trade unions to appoint health and safety representatives.

One of the benefits of using worker representatives is that it is easier to provide them with time to be consulted, rather than consulting each individual worker or a number of small groups of workers. Consultation with worker representatives can also enable them to quickly get to know what the organisation's aims are and how it is dealing with health and safety issues.

The employer may not have a duty to adopt any suggestions made by workers or their representatives, but there is a value in obtaining their views. Because the consultation process can lead to different views and disagreement on health and safety matters, a grievance procedure should be in place to deal with any disputes or misunderstandings that may occur in the consultation process. In some countries, grievances that cannot be resolved internally may be referred to a government enforcement agency for resolution. The rights and functions of worker health and safety representatives may be defined in national legislation, typically their rights would reflect those set out in ILO Occupational Safety and Health Recommendation R164, see *Figure 3-13*. The term 'worker' used in ILO Recommendation R164 is the same as the term 'employee'.

"Article 12 (2) Workers' safety delegates, workers' safety and health committees, and joint safety and health committees or, as appropriate, other workers' representatives should:

(a) Be given adequate information on safety and health matters, enabled to examine factors affecting safety and health, and encouraged to propose measures on the subject.

(b) Be consulted when major new safety and health measures are envisaged and before they are carried out, and seek to obtain the support of the workers for such measures.

(c) Be consulted in planning alterations of work processes, work content or organisation of work, which may have safety or health implications for the workers.

(d Be given protection from dismissal and other measures prejudicial to them while exercising their functions in the field of occupational safety and health as workers' representatives or as members of safety and health committees.

(e) Be able to contribute to the decision-making process at the level of the undertaking regarding matters of safety and health.

(f) Have access to all parts of the workplace and be able to communicate with the workers on safety and health matters during working hours at the workplace.

(g) Be free to contact labour inspectors.

(h) Be able to contribute to negotiations in the undertaking on occupational safety and health matters.

(i) Have reasonable time during paid working hours to exercise their safety and health functions and to receive training related to these functions.

(j) Have recourse to specialists to advise on particular safety and health problems."

Figure 3-13: Rights of worker health and safety representatives and committees.
Source: ILO, Occupational Safety and Health Recommendation R164.

The South African Occupational Health and Safety Act 1993 provides for the following functions of health and safety representatives:

a) Review the effectiveness of health and safety measures.

b) Identify potential hazards and major incidents.

c) In collaboration with the employer, examine the causes of incidents in the workplace.

d) Investigate health and safety complaints.

e) Make representations relating to (a) to (d) to the employer or health and safety committee or where representations are unsuccessful to an inspector.

f) Make general representations to the employer on health and safety affecting employees they represent.

g) Inspect the workplace of those they represent at such intervals as may be agreed with the employer, provided reasonable notice of inspection is provided and the employer may be present.

h) Participate in consultations with inspectors.

i) Receive information from inspectors.

j) Attend meetings of the health and safety committee in connection with the functions.

Health and safety committees
Role and benefits of health and safety committees
The role of health and safety committees is to provide a forum for discussion, ideas and the development of recommendations that can be offered to the employer in order to maintain or improve health and safety. They provide an efficient method to consult managers that have to implement the organisation's health and safety policy and those affected by it, enabling the views of workers and line managers to be gained. In this way, a health and safety committee can promote co-operation on health and safety matters and support the normal worker/employer systems for the reporting and control of workplace health and safety problems.

One of the important benefits of a health and safety committee is that by encouraging everyone in the health and safety committee to participate it enables people with different views and experience to contribute to solving problems. Other worker and manager committee members not directly involved in the problem may be able to provide ideas and support to enable problems to be solved effectively. When a health and safety problem is discussed by the committee, the fact that a number of people have considered it and agreed recommended actions can mean that the actions are more likely to be taken. This can also provide a persuasive argument to

Element 3

help convince people in the workplace to accept recommended solutions to problems.

In addition to helping to solve problems, experience has shown that regular health and safety committee meetings can stimulate a more active (proactive) approach to health and safety by providing an opportunity to consider and discuss ideas for health and safety improvements. The health and safety committee enables workers to provide feedback on difficulties they may be having working with current control measures and provides a particularly useful method of gaining worker feedback on proposed changes and ideas for improvement. Worker members of the committee can discuss the proposals and ideas with the workers they represent and provide feedback to the committee. This feedback can assist in ensuring that current controls, changes and improvements are effective. Where worker feedback is found to be useful this should be communicated to the workers that provided the feedback to encourage their participation in the future.

ILO Occupational Safety and Health Recommendation R164 established a requirement to appoint health and safety committees, see **Figure 3-14**. The term 'worker' used in ILO Recommendation R164 is the same as the term 'employee'.

> "Article 12 (1) The measures taken to facilitate the co-operation referred to in Article 20 of the Convention should include, where appropriate and necessary, the appointment, in accordance with national practice, of workers' safety delegates, of workers' safety and health committees, and/or of joint safety and health committees; in joint safety and health committees workers should have at least equal representation with employers' representatives."

Figure 3-14: Appointment of health and safety committees.
Source: ILO, Occupational Safety and Health Recommendation R164.

In some countries there is a legal requirement for health and safety committees to be established where a certain number of people are employed. For example, section 30 of the Malaysian Occupational Safety and Health Act 1994 requires that a health and safety committee be established where 40 or more people are employed and in Ontario, Canada, a committee is required where 20 or more people are employed. In other countries, where two or more health and safety representatives are appointed to meet a legal responsibility to consult workers and they request a health and safety committee, one must be formed, for example, in South Africa and the UK. Where the requirement to have health and safety committees has been defined by law, legislation is often used to clarify the functions of the committee and set out parameters that will make the committee effective.

Effective health and safety committees

For any committee to operate effectively, it is necessary to determine clear objectives, functions and composition of the committee. An effective health and safety committee requires both worker representatives and management to show honest commitment to the same health and safety aims as the organisation. The objectives and functions of the health and safety committee should be clearly defined and communicated. The composition of the health and safety committee should be decided and the members of the committee should agree who will be the chairperson for meetings, how often they meet and the format of the agenda.

Functions of a health and safety committee

Though the functions of health and safety committees may be defined in national legislation, typical functions may include:

- To review the measures taken to ensure health and safety, for example, purchasing and maintenance programmes.

- To review accident/incident and occupational health trends.

- To examine health and safety audit reports.

- To consider enforcing agency reports and information releases.

- To consider reports which worker health and safety representatives may wish to submit.

- To assist in the development of health and safety rules, systems of work and procedures.

- To consider the effectiveness of the health and safety content of manager and worker training.

- To consider the adequacy of communication and publicity in the workplace.

- To consider new developments and proposed changes, for example, legislation and new technology.

- To provide a communication link with National enforcing agencies.

Composition

The membership and structure of the health and safety committee should be settled in consultation be-tween management and workers or any worker health and safety representatives concerned or in accord with any National legal requirements. The composition of the health and safety committee should be aimed at keeping the total size as small as possible and preferably balancing the number of management and worker representatives evenly.

A typical health and safety committee might consist of:

- Chairperson.
- Someone to provide administrative support to the committee and note taking.
- Management representatives, including a senior manager.
- Worker representatives.
- Health and safety practitioner.
- Other management, for example, project engineers, planning engineers, electrical engineers and operational supervisors.

Frequency of meetings

How often health and safety committee meetings are held would depend on the nature of the organisation's activities, the risks involved, how active the health and safety programme is, items on the agenda and other local considerations (such as whether there is a hierarchy of committees representing departments/locations/sites). Usually, meetings are held at a frequency varying between monthly and quarterly. Some countries, for example, parts of Canada, have set out legal requirements that specify the frequency of meeting must be no less than once every three months. A similar requirement has been established in Abu Dhabi, **see Figure 3-15**.

 "6.5 Meeting Frequency, Assistance and Training

(a) Section 3.2(f) of – OSHAD-SF - Element 04 - Consultation and Communication requires entities with more than 50 staff to meet at least 4 times a year.

(b) The meeting frequency for your OSH committee should be defined by the risks that are inherent within your undertaking and higher risk entities should meet on a more regular basis."

Figure 3-15: Frequency of committee meetings.
Source: Abu Dhabi OSHAD-SF Technical Guidline - Consultation and Communication.

Agenda and minutes

A typical format for an agenda of a health and safety committee is:

- Apologies for absence.
- Minutes of the previous meeting.
- Matters arising.
- Reports of the health and safety practitioner.
- Other reports, for example, fire officer, nurse and occupational hygienist.
- New items (and emergency items).
- Date of next meeting.

The specific items considered by the committee will be influenced by its functions and new items for consideration and discussion would be added as needed.

The Abu Dhabi OSHAD-SF Technical Guideline – Consultation and Communication confirms this and advises on typical minimum items to consider on an agenda, **see Figure 3-16.**

"There is no set rule as to what items should be included on the OSH committee agenda as these will differ between entities, however the following items should be considered as a minimum:

(i) statistics on accident records, ill health and sickness absence;

(ii) accident investigations and subsequent action;

(iii) OSH management systems updates/changes;

(iv) inspections and audits of the workplace;

(v) risk management programme;

(vi) OSH training;

(vii) emergency procedures;

(viii) changes in the workplace affecting the OSH and welfare of staff; and

(ix) adequacy of OSH communications and publicity in the workplace."

Figure 3-16: Items to consider in the agenda of committee meetings. Source: Abu Dhabi OSHAD-SF Technical Guideline - Consultation and Communication.

The minutes should be circulated as soon as possible after the meeting.

Consideration should be given to making copies of the minutes readily available to all workers. This could be done by posting sufficient copies on notice boards or in rest areas used by workers and by making an electronic copy available on the organisation's intranet.

Although some of the factors are discussed in more detail above, a summary of the factors that make health and safety committees effective is provided below. Effective health and safety committees have:

- A clear management commitment.
- Clear objectives and functions.
- An even balance between management and worker representatives.
- Agendas that are agreed,distributed in advance and adhered to in meetings.
- Minutes or notes of the meetings being produced promptly and distributed in good time for actions to be taken before the next meeting.

- Minutes that are provided personally to each member of the committee and the most senior manager of the organisation.

- Publicity given to discussions and recommendations, including the posting/displaying of copies of the minutes of meetings.

- Effective chairing of meetings, which enables points to be raised related to the agenda and the control of points raised that are not part of the agenda.

- Full participation by members.

- Access to the organisation's decision-making processes through the chairperson so that the committee's views are taken into account and decisions on recommendations by management are made without delay.

- Regular meetings at a frequency that reflects the matters to be discussed and the risks of the organisation.

- Dates of meetings arranged well in advance and published to members of the committee and those they represent.

- Meetings that are not cancelled or postponed, except in very exceptional circumstances.

- Appropriate topics for consideration and discussion.

- Access to health and safety expertise, for example, a health and safety practitioner.

- Sub-committees established where there is a need to focus in detail on specific items.

Health and safety committees can provide a very effective way of resolving health and safety problems and deciding on improvements. Involvement of managers and workers in this process helps to develop a positive health and safety culture and can greatly influence health and safety behaviour.

Figure 3-17: Using health and safety committees at work.
Source: ILO.

Improving an organisation's consultation arrangements

Many aspects of consultation have been discussed in this element. Some of the ways that arrangements for consultation with workers might be made more effective are summarised as follows:

- Define formal requirements and arrangements to consult.

- Provide consultation training to both management and workers.
- Plan direct consultation as part of the standard operation of the organisation, for example, at departmental meetings, team meetings, toolbox talks.
- Consult with worker health and safety representatives when logistical barriers make it difficult to consult directly with workers.
- Establish health and safety committees to enable planned consultation of workers and managers in a way that encourages discussion and recommendations to be made.
- Use questionnaires, surveys and suggestion schemes.
- Consult as part of the development process for safe systems of work and procedures.
- Consult as part of accident/incident or ill-health investigations or as part of the risk assessment process.
- Enable informal consultation to take place between supervisors and their team.

CONSIDER

How effective is the health and safety consultation process in your organisation? What could be done to improve it?

REVIEW

What actions could an organisation take that would help to establish a positive health and safety culture?

WHEN TRAINING IS NEEDED

Training should be carried out from the start of an individual's employment with an organisation, during induction, and throughout their career. Training may also be required when transferring between jobs or work departments or promotion to a management role. The training provided should not only consider routine aspects of work, but also work that takes place infrequently, such as maintenance and emergencies.

"The primary role of training in occupational safety and health is to promote action. It must therefore stimulate awareness, impart knowledge and help recipients to adapt to their own roles. Training for the acquisition of technical skills should therefore always include an OSH component."

Figure 3-18: The role of health and safety training.
Source: ILO, Fundamental Principles of Occupational Health and Safety.

The ILO 'Code of Practice for Ambient Factors in the Workplace', 2001 recommends that health and safety training should be provided in accordance with an individual's needs and that this may include a range of items. The items that are suggested are illustrated in *Figure 3-19*, however the items appropriate for an individual will depend on their and the work they do. The term 'hazardous ambient factors in the workplace' means hazardous needs airborne chemicals, ionising and non-ionising radiation, ultraviolet, infra-red and (in some circumstances) visible radiation, electric and magnetic fields, noise, vibration, high and low temperatures and humidity. The term 'worker' used in this code of practice is the same as the term 'employee'.

Although the ILO list, shown in *Figure 3-19*, relates to ambient factors in the workplace it illustrates the range of items typically included in health and safety training. A suitable training programme should be defined and implemented for all individuals, from worker to director level. Some of the training requirements identified by the ILO would typically be met by standard induction training or job specific health and safety training provided to the individual. The ILO recommends that training should take place during working hours and it is good practice that a record should be maintained of the analysis of training needs, training plans and training conducted.

 (a) Relevant health and safety legislation, such as the responsibilities and rights of employers and workers.

(b) The nature and degree of hazards or risks to health and safety.

(c) The correct and effective use of prevention, control and protection measures, especially engineering control measures, and the workers' own responsibility for using measures properly.

(d) Correct methods for the handling of substances, the operation of processes and equipment, and for storage, transport and waste disposal.

(e) Assessments, reviews and exposure measurement and the rights and duties of workers in this regard.

(f) The role of health surveillance, the rights and duties of workers in this regard, and access to information.

(g) Instructions on personal protective equipment, its importance, correct use, limitations, factors which may show inadequacy or malfunction of the equipment.

(h) Warning signs and symbols.

(i) Emergency and first-aid measures.

(j) Appropriate hygiene practices to prevent, for example, the transmission of hazardous substances to the home or family environment.

(k) Cleaning, maintenance, storage and waste disposal to the extent that these may cause exposure for the workers concerned.

(l) Procedures to be followed in an emergency.

Figure 3-19: Scope of training required in relation to ambient factors. Source: ILO, Code of Practice for Ambient Factors in the Workplace.

To ensure individuals retain knowledge, skill and the right experience refresher training should be given at regular intervals, where appropriate. When considering the need for refresher training, particular attention should be paid to infrequent tasks, complex tasks or tasks that are critical for ensuring health and safety. When risks are assessed and accidents/incidents and ill-health are investigated, they may identify a need for training or refresher training. Training will also be required when situations change that can affect health and safety, for example, the introduction of new technology, new systems of work or new work equipment.

Induction training for new managers and workers

Induction training is generally defined as the training given when an individual starts working for an organisation for the first time or works at a new location of the same organisation. The purpose of a health and safety induction is to orientate individuals to the organisation's approach to health and safety and provide them with essential information that will enable them to be safe and healthy in their work and contribute to the health and safety of the organisation. It provides an opportunity to explain the organisation's values and commitment to health and safety. The induction process should motivate individuals and make them aware of the health and safety behaviour the organisation expects of them. It provides confidence to the workforce that health and safety is an important part of doing work in the right way. It also provides clear communication on the line of reporting for supervision and explains where workers report to if they identify a conflict of interest between what is being expected and getting the work done.

Induction training is an important integral part of the prevention process. It is essential that systematic arrangements are in place to ensure a consistent standard of induction for health and safety. Induction training for new managers and workers can be done in two stages, by providing a general induction to the organisation/location in which they work and a specific induction to their job. When a job induction is carried out as well as a general induction it enables the information provided to be more specific to the individual's workplace. In smaller organisations the content of a general induction may be essentially the same as the content of a job induction, making a job induction not necessary. However, in larger organisations both types of induction may be needed.

For example, there may be general things to know about site rules and emergency arrangements, which can be provided in a general induction, but there may also be very specific additional rules and emergency arrangements related to the individual's job. These important additional items may not be relevant to all the people attending the general induction and may therefore best be covered in a job induction.

General induction training for new managers and workers should include:

- Review and discussion of the health and safety policy.

- Hazards of the workplace and control measures.

- Health and safety signs.

- Specific training requirements.

- Welfare facilities.

- Personal protective equipment (PPE) provisions - limitations, use and maintenance etc.

- Fire and emergency procedures.

- First-aid procedures and facilities.

- Accident/incident and ill-health reporting and investigation.

It is necessary for workers to know the relevant hazards and controls for each location in which they are working. For example, this will mean knowing areas they can and cannot go to, the use of walkways, where rest facilities are, who the first-aiders are and to whom they report hazards/accidents.

For job induction training to be meaningful, it has to be relevant to the trainee and focus on the controls that should be used as well as the hazards that may be encountered.

A job induction for a driver who makes both internal and external deliveries could include:

- Information on internal traffic routes.
- Speed limits.
- Parking/loading areas.
- Any hazardous substances on site.
- The need to carry out regular inspections of the vehicle and how to report defects.
- Company rules regarding the carrying of passengers, mobile phones, etc.
- How to report accidents and incidents.

Specific health and safety training

The specific health and safety training required by an individual will greatly depend on the individual's job, for example, whether they are a worker or manager. The need for specific training for a particular job should be defined through an analysis of training needs. Some examples of specific health and safety training that may be required include training related to:

- Safe systems of work. The actual training provided would depend on the individual's role. The training a worker needs may be more practically based than that of a manager, who needs to understand the health and safety requirements and how to ensure the work takes place safely and healthily.

- Equipment training, for example, specific training to develop competence in driving a fork-lift truck.

- Personal protective equipment training. For example, involving the practical aspects of selection, use and maintenance of respiratory protection equipment (RPE).

- Fire training, for example, how to act as a fire marshal for a specific part of a building.

- First-aid training, for example, how to give first-aid, this may be very specific training relating to the types of injury that might be experienced in a particular workplace.

- Workplace inspections and risk assessments. This type of training may involve learning the theory of the process and then the practical application of the technique in the workplace under controlled conditions.

When job specific health and safety training is provided it is common for competence to be formally assessed and recorded. For example, an individual may work under a higher than usual level of supervision for an initial period following the training to confirm that the important aspects of training have been transferred to the individual's work. This systematic approach helps to ensure the individual can apply what they learnt and that they have the correct health and safety behaviour.

Supervisor and manager training

General health and safety training should be provided to supervisors and managers to ensure responsibilities are known and to enable them to carry out the organisation's health and safety policy. This will enable them to fully understand and participate in the management of health and safety in their organisation. It will also help them to consistently and reliably respond to health and safety risks, which will ensure the organisation's standards for health and safety are maintained.

General health and safety training for supervisors and managers should include:

- The health and safety policy of the organisation.
- Causes and consequences of accidents/incidents and ill-health.
- The range of hazards and risks related to the organisation's activities.
- The legal framework the organisation works in and the legal responsibilities of the organisation, managers, workers and others.
- The structure and operation of the health and safety management system.
- How health and safety is organised and the responsibilities everyone has been given. Including methods available to guide health and safety behaviour, disciplinary procedures and commendation procedures.
- How health and safety is planned and implemented, including risk assessment, risk prevention and risk reduction techniques. This should be supported by examples of measures to control the hazards and risks related to the organisation's activities and an explanation of specific arrangements/processes/procedures for the management of risks, for example, permits-to-work.
- How health and safety performance is evaluated, including reactive and active monitoring, for example, accident/incident investigation, inspection procedures, auditing, reporting to more senior managers and review processes.
- How corrective and improvement actions are managed.

Refresher training

As time passes, a manager's or worker's approach to the health and safety aspects of their job can change from the way they were originally trained. This may be because they have forgotten the way they were trained, sometimes because of infrequent use, or because they have developed a different way of working that they prefer. This could mean that they work in unexpected ways when they work with other people and the practices they have developed could put themselves and others at unnecessary risk. It is therefore important that regular refresher training is used to reinforce the correct way of working and maintain an awareness of risks.

Many countries require annual refresher training for first-aiders and a similar approach may be taken for work activities where the consequences of the individual not working in the way they were initially trained could result in high risks. However, two or three years may be an acceptable interval between training and refresher training periods for other work activities. Refresher training may also be required where the causes of accidents/incidents or ill-health are linked to competence. Similarly, when an individual returns to work after a period of sustained absence, perhaps for ill-health reasons, it is often useful to provide them with refresher training to remind them of important health and safety aspects of their job. Some organisations have well established refresher training programmes for workers who need to retain skills, it is also important to ensure refresher training is provided to managers as well.

Training			
Purpose	To develop the knowledge and skill of people, their attitudes, perception and motivation with regard to health and safety to ensure acceptable actions. Training should use two-way communication - information/instruction given and understanding/skill checked. This may be by observation of a person's practical skill, for example, by driving a fork-lift truck or using a simulator, and/or by written or verbal assessment of understanding.		
Subject	Accident investigation Conducting risk assessments Conducting inspections/audits How to comply with instructions	How to set up your display screen workstation How to use work equipment, for example, rough terrain fork-lift truck	Use of personal protective equipment Manual handling techniques. Emergency procedures
Means of communicating	On/off the job Internal/external trainers	Explanation, demonstration, discussion and practice	One-to-one or group Written, oral and visual material

Changes that create a need for training

As an individual's job or the work process changes training is required to update techniques and ensure awareness of the correct methods. Without effective training the individual will not be fully prepared for their new job or may try to apply their old knowledge to new processes. The training should be designed to prevent the individual transferring old methods to the new job/process. This may require a significant amount of practice under controlled conditions to ensure the new methods have been learnt. Whether the individual is a manager or worker the absence of training could lead to incorrect behaviour and risk of harm.

Where new legislation/standards are introduced it can have the effect of introducing new requirements that the organisation has to meet. This will make it necessary to train managers and workers in these requirements and the action taken by the organisation to meet them. For example, a change in legislation could involve a reduction in occupational exposure limits and may require managers and workers to use more stringent exposure controls. This could mean managers and workers would need training to under-stand the importance of the new exposure limit and what specific actions they need to take to ensure the limit is not exceeded. As the introduction of new health and safety measures can take some time to complete, training may need to be provided in stages as the work to introduce the measures is completed and the effect on work activities is known.

The introduction of new technology will often require the adoption of new work practices, for example, the introduction of mechanical aids to improve manual handling risks will require new methods of working. The use of new technology will often include a need to develop skills to interpret equipment control layouts and data display. This would require careful and systematic training to ensure the individual using the new technology was able to use it as intended.

Preparing for a training session

Effective training is an important part of managing health and safety and careful preparation prior to delivering a training session is necessary to ensure its success. One of the first considerations is to identify the particular aspect of health and safety that the training relates to, so that the objectives and the content (breadth and depth) of the training session can be established. Other factors to be considered when preparing a training session include:

- The training style and methods to be used. For example, lecture, visual presentation, role play, group work, use of equipment or site visit.

- The target audience. For example, their existing knowledge and skills, relevance of the subject being taught to them, their motivation.

- The number of trainees. Considering the number of trainees the tutor can train at one time when related to the training method.

- The time available for the training. It may be necessary to break the training down into a number of small parts and provide the training over a period of days.

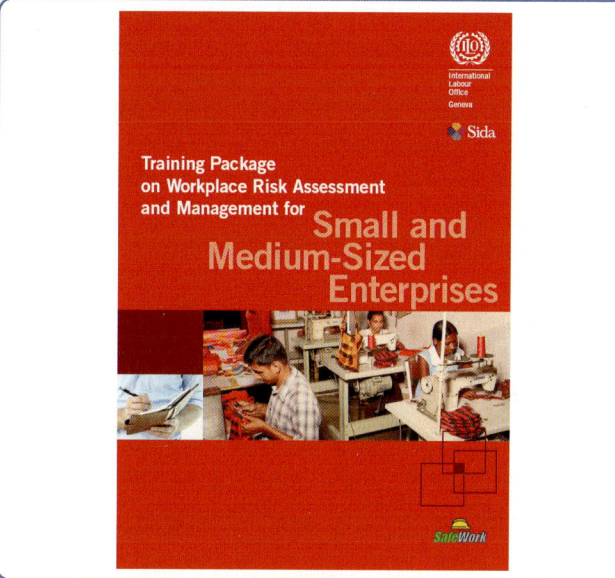

Figure 3-20: Health and Safety training package. Source: ILO.

- The knowledge, skills and experience required of the trainer; this will be particularly influenced by the training method to be used. For example, if a trainer is required to train a number of people in the safe use of a fork-lift truck, they will need to have sufficient skill in using a fork-lift truck to demonstrate the safe use. Assistance from trainers outside the organisation may be required to ensure the right expertise.

- Audio-visual and other training aids required. Effective training is often supported by training aids. It is important when preparing for training to ensure they work and that the trainer knows how to use them.

- The facilities needed to conduct the training. For example, their location (near work activities or off-site), room layout, size and lighting.

- Provision of refreshments, if necessary.

- How the effectiveness of the training is going to be evaluated. This can be done at the time of training (for example, by using course evaluation forms) and after the training (for example, by measuring the level of compliance with procedures or number of accidents/incident).

3.3 How human factors influence behaviour positively or negatively

As technical systems to prevent work-related harm have become more reliable, the focus of prevention strategies has turned to the human causes of accidents/incidents and ill-health. It is estimated that approximately 80% – 90% (Heinrich et al, 1980) of accidents/incidents are caused by human factors. This is because people are involved throughout the activities of an organisation, from design through to operation and maintenance.

The decisions and actions of managers and workers can prevent or cause harm. Many accidents/incidents and cases of ill-health are blamed on the individual who was directly involved in the work causing harm. This ignores the fundamental failures that lead to accident/incidents and ill-health. These are usually rooted deeper in the organisation's design, management and decision-making functions and are greatly influenced by human factors.

A considerable amount of work has been done in different parts of the world to study and describe the influence of human factors on behaviour in the workplace. One notable document that described the significance of human factors on health and safety was produced by the HSE in 1999, it is called 'Reducing error and influencing behaviour' (HSG48). In this document the HSE defined the term 'human factors', **see Figure 3-21.**

"Human factors refer to environmental, organisational and job factors, and human and individual characteristics which influence behaviour at work in a way which can affect health and safety."

Figure 3-21: Definition of human factors.
Source: UK, HSE, Reducing error and influencing behaviour, HSG48.

In HSG48 the HSE identifies the importance of three human factors that have a significant effect on the behaviour of individuals at work, these are the:

- Organisation, for example, culture, leadership, resources, work patterns, communications.

- Job, for example, task, workload, environment, display and controls, procedures.

- Individual, for example, competence, skills, personality, attitude, risk perception.

The HSE has shown that these three human factors are constantly active in the workplace and that they interact with each other, greatly influencing health and safety behaviour, see **Figure 3-22.**

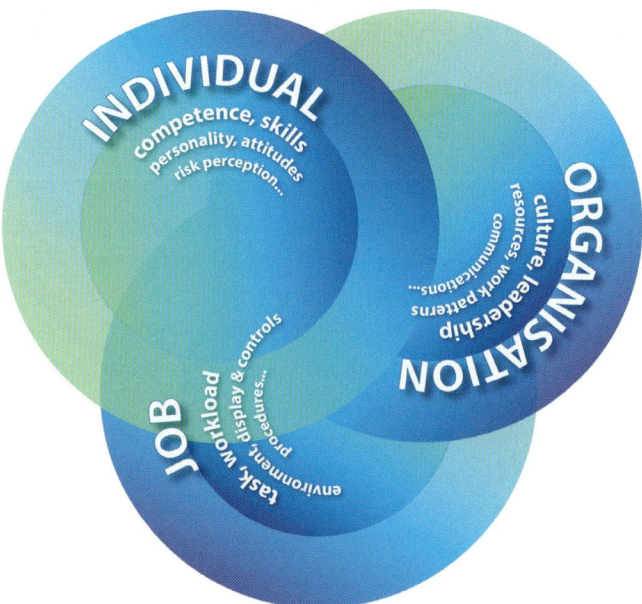

Figure 3-22: Three important human factors.
Source: UK, HSE, Reducing error and influencing behaviour, HSG48/RMS.

ORGANISATIONAL FACTORS

The organisational factors that influence health and safety behaviour include factors like:

- An organisation's culture.
- Its approach to responsibilities and leadership.
- The resources provided to achieve health and safety.
- How health and safety is communicated.
- The way the organisation organises work.
- The standards and procedures it sets.

These organisational factors have a major influence on individual and group behaviour, yet it is common for them to be overlooked during the design of work and when investigating accidents/incidents and ill-health.

As stated earlier in this element, the importance of a positive health and safety culture on behaviour at work is significant. If the culture is negative towards health and safety then the behaviour of many people in the organisation is likely to follow this and also be negative to health and safety. To minimise the effects of this organisational factor, organisations need to produce a culture that promotes manager and worker commitment to health and safety and emphasises that deviation from corporate health and safety aims and objectives, at whatever level, is not acceptable.

The leadership of those who have senior management responsibilities within an organisation may also have either a positive or a negative effect on the health and safety behaviour of others. Individuals will be influenced by the leadership of senior management and following their example. Individuals may also be reluctant to take actions necessary to ensure health and safety if their decisions are likely to be subject to criticism from those senior to them in the organisation or their peers. It is important that senior management leadership is positive and supports decisions that ensure health and safety.

Poor definition of responsibilities and poor coordination of health and safety can quickly lead to unacceptable health and safety behaviour at work. If managers and workers do not know their responsibilities for health and safety they are less likely to behave in the way the organisation wants them to and less likely to give health and safety proper regard. Where efforts to manage health and safety are poorly co-ordinated, they can become ineffective, causing workers to have to accept poor working conditions and leading to low motivation.

Where adequate resources are not provided it can cause work to be done with insufficient people or equipment to do it safely, this can lead to fatigue and error. In some cases this lack of resources may cause managers or workers to deliberately not follow the usual procedures. Inadequate resources can also lead to inadequate responses to accidents/incidents and ill-health when they occur.

This could mean that the work practices and conditions that caused the accident/incident or ill-health are not improved and the poor behaviour that created the work practices and conditions may continue. Only by allocating adequate resources to investigation and prevention can an organisation be confident that similar accidents/incidents and ill-health can be prevented in the future.

Poor communication can have a significant influence on health and safety behaviour. Where an organisation relies on one-way communication from its managers to workers this can prevent involvement, result in misunderstanding and lead to lower motivation. Where communication systems do not encourage two-way communication, managers and workers may not report situations that could lead to harm, which over time could result in acceptance of bad practices and other similar poor behaviour becoming common practice.

An important organisational factor is the overall approach to health and safety taken by the organisation. If the organisation does not take a systematic approach that manages risks and prevents human factor failures this can lead to poor health and safety behaviour, resulting in accidents/incidents and ill-health. It is important that organisations take opportunities to identify possible sources of failure and establish preventive measures to deal with them. This should include risk assessment processes as well as accident/incident and ill-health investigation processes.

Even when organisations establish and maintain clear procedures and standards for health and safety, individuals may not understand the relevance of the procedures or appreciate their significance in controlling risk. This could cause individuals to decide not to follow the procedures and to behave in a way that is not safe or healthy. When accidents/incidents happen, managers should not blame individuals for taking short cuts that seemed safe and were allowed to become routine if the managers have not explained the importance of and monitored the correct use of procedures. Procedures can also lapse from use because workers are discouraged from following the procedures by peer group pressure or other pressures, such as work patterns and production targets. Some of these pressures can be caused by poor planning of work activities. Where managers become aware of deficiencies in following health and safety procedures, but do not act to remedy them, the workers readily perceive that not following procedures is acceptable to managers. Health and safety procedures can therefore soon become unused if there is no system of ensuring that they are followed.

To manage the organisational aspects of human factors, organisations need to provide:

1) Clear and evident commitment which promotes a culture for health and safety, from everyone. This should include positive health and safety leadership from the most senior management.

2) Clear identification of responsibilities and effective coordination of health and safety effort.

3) Resources, including time and money, to ensure the best systems and work practices to minimise risk are established. In particular, adequate and effective management and supervision should be provided, with the power to remedy problems when found.

4) Effective communication of the objectives of management and the need for appropriate standards. This should be supported by arrangements for collaboration of managers and workers, which ensures consultation and involvement. Communication should positively encourage a constructive exchange of information at all levels.

5) A systematic approach to health and safety, usually by establishing a formal health and safety management system that identifies and manages risks. An analytical approach to identifying possible routes to human factor failure should be taken, including the use of risk assessment processes and the consideration of work patterns to ensure fatigue and boredom are minimised.

6) Procedures and standards for all aspects of work involving significant hazards and mechanisms for reviewing them. This should be supported by effective monitoring systems to check the implementation of the procedures and standards.

JOB FACTORS

Job factors are often caused by the presence of organisational factors and directly influence individual performance and the control of risks. There are a number of job factors that could influence health and safety behaviour in the workplace, including:

- The tasks individuals carry out.
- The effects of workload and the workplace environment.
- The design and maintenance of equipment and procedures.

The tasks that an individual is expected to do may be physically or mentally demanding, for example, a task may require a large amount of strength or have many complicated features to consider in order to come to a decision. The high demands of tasks compared with an individual's capabilities provide a potential for them to make errors.

Poor design of equipment can lead those using them to make errors, for example, information displays may be misread or the wrong controls operated in error. These errors can cause direct harm to the person using them or in some cases could lead to incorrect decisions that later result in harm. Equipment used by individuals should be designed in accordance with ergonomic principles to take into account any limitations in human performance of those using them, in particular, mental and physical ability. Ergonomics, in relation to equipment, is the study of the design and arrangement of equipment so that people will interact with the equipment in a safe, healthy, comfortable and efficient manner. One of the main principles of ergonomics is to match, as far as possible, the equipment and how it is used to the user, rather than match the user to the equipment. Following these ergonomic principles and, where possible, matching the job to the individual will help to minimise errors. For example, the use of ergonomic principles can greatly improve work equipment displays and the layout of controls. This can reduce the human errors made when reading displays and operating equipment controls.

Where equipment and tools are poorly maintained and they do not work correctly this can cause high levels of frustration and fatigue and could lead an individual to act in an unsafe or unhealthy way in order to get their work done, for example, if the drill bit fitted to a drilling machine frequently became loose and the worker had to keep removing the guard to re-fit the drill bit the worker may be tempted to leave the guard off.

The health and safety procedures and instructions an organisation uses are important job factors that should be considered.

Where procedures or instructions communicated to an individual are unclear or have missing information, the individual receiving them may make an error, for example, they may mix two chemicals that react to each other or might administer the wrong amount of medicine to a patient.

In addition, there are a number of other job factors that could cause mental and physical fatigue to an individual and cause them to make errors. For example, fatigue may result where the workload is unacceptably high, where the individual is affected by interruptions or where the work environment is unpleasant, for example, noisy or very hot.

To manage the job aspects of human factors, organisations therefore need to ensure the:

1) Identification and comprehensive analysis of the tasks expected of individuals and the assessment of likely errors related to the tasks. This should include the evaluation of manager and worker decision making.

2) Application of ergonomic principles to the design of person-equipment interfaces, including instrument displays of process information and the suitable positioning and labelling of control devices. Care should be taken to ensure the provision and maintenance of the correct tools and equipment for the job.

3) Design and consistency of presentation of procedures, supported by the provision of efficient and suitable communication of information related to the job.

4) Scheduling of work patterns, including shift organisation and workload, to control fatigue and stress. This should include procedures to cover for absence and for emergencies that might arise when doing the job. Where concentration is required to prevent errors, arrangements should be made to prevent interruptions. This should be supported by the organisation and control of the working environment so that it reduces fatigue and stress, including the workspace, lighting, noise and thermal conditions.

INDIVIDUAL FACTORS

Although organisational and job factors are important, individual factors can also greatly influence the health and safety behaviour of individuals in the workplace. These individual factors include the individual's personality, attitude, risk perception, skills and competence. All individuals have different physical and mental characteristics that can affect these factors and the way they behave.

Physical	Mental
Gender - for example, females of child bearing age should not be exposed to lead.	Attitude - for example, a person may have the attitude that all personal protective equipment is uncomfortable and refuse to wear it.
Size - for example, the large size of a person may restrict their movement in a confined space.	Motivation - for example, a person may feel that while doing their work in a particular way presents additional risks it is worth taking the risk to make the work easier or complete the work quicker.
Health - for example, colour blindness may prevent the correct identification of wiring by an electrician.	Perception - for example, a person may not recognise the risk from moving machinery and remove a guard or they may not recognise an emergency alarm and fail to respond to it.
Capability / strength - for example, an individual may not have sufficient capability to manually handle a load.	Mental capability - for example, a person may have limited mental ability, which leads them not to follow complex or lengthy health and safety instructions.

Figure 3-23: Examples of individial factors and their possible effect on health and safety. Source: RMS.

The individual differences related to mental characteristics come from a combination of the 'inherited characteristics' (passed on from the parents of the individual) and the various 'life experiences' through which the individual passes. Typical life experiences that influence the mental characteristics of an individual include:

- Experiences in the womb.
- Birth trauma.
- Family influences.
- Geographical location.
- Pre-school influences.
- Education - opportunities, quality, support.
- Occupational factors - training and retraining.
- Hobbies and interests.
- Own family influences - marriage, children.
- Ageing

These life experiences shape the mental characteristics (personality) of each person, such as their attitudes and perception. In addition, each individual develops physical characteristics, including their size, strength and any limitations arising from disability or illness. These characteristics combine to create a unique person, different from all other individuals. *Figure 3-23* shows some examples of individual factors and how they may influence the health and safety aspects of work.

It is important to know what a particular job involves so that the effects of individual factors can be minimised, especially for high hazard jobs. Jobs should be analysed to identify features of the job that may be affected by individual factors and the required characteristics of individuals to do the job safely and healthily.

Following the analysis of jobs, a review process should be used to assess if the job can be modified to enable a wide range of individuals to conduct it without harm to their health and safety. The job may also have to

be modified to take account of individuals with special needs, for example, disabilities. Where possible, the job an individual does should be adjusted by following good ergonomic principles to minimise the effects of individual factors on behaviour and performance. However, the effects of individual factors may not always be controlled by good ergonomic principles, other methods may also be required. The information gained from analysis of jobs can also be used to establish person specifications for the jobs. Where necessary, appropriate selection techniques may need to be used to match individuals to jobs that suit their characteristics, particularly where these characteristics directly influence health and safety. Also, some individual characteristics, such as skills and attitudes, can be modified by training and experience. Training should aim to give individuals the knowledge to allow them to understand their work processes, risks and control measures, and the skills that enable them to perform their job well and ensure health and safety. This training helps to improve human reliability because managers and workers become more aware of the process and task, their perception of risk increases to an appropriate level and an improvement in their attitude and motivation should take place. Further information on training and competence is provided later in this element.

Monitoring of individual performance should be carried out at all levels in an organisation. In particular, all workers should be monitored through direct supervision. This will ensure work activities are performed correctly and requirements for correct health and safety behaviour are met. An old maxim states: 'what gets measured, gets done.' This is particularly relevant for health and safety and if individuals are monitored they are more likely to behave safely and healthily when doing their work. Supervision also provides an opportunity to commend people and take action to improve behaviour. The degree of monitoring and supervision should be based on the level of health and safety risk related to the task and the level of competency of the worker carrying out the job.

For certain jobs there may be specified medical standards for which pre-employment/periodic health surveillance is necessary. These medical standards may relate to the practical requirements necessary to carry out the job and identify specific characteristics necessary to perform the job safely and healthily. For example, medical standards may be set for vehicle drivers and a medical examination required before an individual is allowed to become a driver. There may also be a need for routine health surveillance to determine the effects on individuals of their exposure to workplace hazards that influence behaviour, for example, fatigue due to an individual being exposed to high temperatures.

Similarly, it is useful to review the individual factors that could affect an individual's health and safety behaviour when they return to work following a significant period of sickness absence. This provides an opportunity to consider the reason for their absence, any arrangements that may enable their smooth return to normal duties and any underlying reasons for their absence. The underlying reasons for absence could include anxiety or stress related to their work and personal reasons like alcohol or drug abuse. Access to specialist assistance or counselling may be appropriate and temporary or permanent redeployment to other work may be necessary.

 **CASE STUDY**

Managing individual factors, the health of a support worker

A worker, who was employed as a support worker in a residential home for adults with learning difficulties, was diagnosed as having diabetes. The diabetes condition required the support worker to control diet, blood sugar levels and diabetes medication.

Support workers working at the residential home were required to provide assistance on a 24 hour basis, which meant one support worker remaining awake throughout the night shift and another asleep, but on call. Remaining awake during the night shift disrupted the support worker's sleep patterns and food intake, affecting the blood sugar levels, which made the diabetes condition worse.

There were a limited number of support workers available. If the support worker with diabetes did not do night shifts other members of the team would have to do more. The problem was

discussed as a group, involving the support worker with the diabetes condition, other support workers and their manager. It was agreed that the support worker with diabetes would not have to do shifts that required the support worker to remain awake, but would take a greater share of the weekend day shifts and the night shifts where the support worker slept and was on call. The support worker with the diabetes condition, other support workers and their manager were all happy with the arrangement.

In summary, to manage the individual aspects of human factors organisations should:

1) Write job and person specifications. Consider age (if this affects health and safety), physique, aptitude, personality, knowledge, skill, qualifications, experience.

2) Use ergonomic principles, where possible matching the job to the individual. Where necessary, match the aptitudes and skills of individuals to the requirements of jobs. Consider the needs of special groups of individuals, for example, those with disabilities.

3) Implement an effective training system.

4) Monitor individual performance, particularly supervision for individuals that do high risk jobs.

5) Provide pre-employment and periodic health surveillance where this is needed.

6) Review individual factors affecting an individual after periods of absence, including absence due to sickness or injury due to accidents/incidents.

7) Provide counselling and support for ill-health or stress.

Personality

The personality of an individual is a set of mental characteristics (pattern of thoughts, feelings, and behaviours) that influence an individual's values, attitudes, information processing (including perception), emotions and motivations in various situations. This makes the individual a unique person. Personality arises from within the individual and is influenced by their experiences in life and their interaction with other individuals. It is believed that an individual's personality usually remains fairly consistent throughout their life. The personality of an individual can therefore influence all aspects of their health and safety behaviour at work.

Attitude

Attitude may be defined as 'the tendency to respond in a particular way to a certain situation'. An individual's attitudes will influence the way in which situations are viewed by them and will create a response or pattern of behaviour.

Attitudes are one way in which individuals are different from each other. Attitudes guide how we react to other people, what things are important to us and our preferences. Attitudes are formed (not necessarily consciously) through a lifetime of experiences that shape the beliefs and values that establish our attitudes, for example, our attitude to working safely. Attitudes are used to provide the individual with a structure to their world, organising an individual's standardised response to situations, which saves the effort of thinking of a response that is acceptable to their values and beliefs each time they meet a situation they need to respond to. Attitudes are not directly observable, but can be assessed by observing an individual's thoughts, feelings and behaviours (physical or verbal). An individual's behaviour can therefore express their attitude towards something and reflect how they feel about that thing. Individuals often develop similar attitudes to those people they like and seek others with similar attitudes. This is very important when considering an individual's health and safety behaviour.

Individuals may develop attitudes about working safely, which can be formed over a long period of time and through interaction with other people and their attitudes. An individual may therefore develop a negative attitude to health and safety, which causes them to respond in a negative way to attempts to get them to work safely. This attitude may develop through the individual working in situations where other individuals hold a negative attitude and copying their behaviour and attitude. For example, an individual may develop a negative attitude to health and safety when working in situations where there is a negative health and safety culture in an organisation. The individual may express their negative attitude in their behaviour, verbally rejecting the need to work safely and physically refusing to work in a safe way. Because attitudes are formed over time, sometimes many years, they are not easily changed. Attempts to change such a fundamental part of an individual's personality will usually be resisted. The individual will often feel their values and beliefs are under threat. This is worth remembering in the context of health and safety propaganda campaigns that attempt to change attitudes. The difficulty in changing attitudes was summarised by Chairman Mao Tse Tung, *see Figure 3-24.*

"People's attitudes and opinions that have been formed over decades of life cannot be changed by holding a few meetings or giving a few lectures."

Figure 3-24: Observation made by Chairman Mao Tse Tung. Source: "Little Red Book".

Examples of verbal expression of negative attitudes affecting safe and healthy working include:

- 'It will never happen to me'.

- 'We have never had an accident/incident'.

- 'I am not planning to injure myself'.

- 'I know my limits'.

Actions to influence attitudes might include:

- Providing strong leadership that shows a positive attitude to health and safety.

- Providing training and provision of retraining when there is a need for reinforcement.

- Providing positive experiences (involvement), for example, improving an individual's attitude to wearing personal protective equipment by involving them in the selection of the equipment.

- Discussing the basis of the attitudes that individuals hold. For example, asking individuals to consider whether their values and beliefs are correct, providing them with other knowledge that could adjust their values and beliefs and helping them to reform their attitude.

- Providing individuals with an interaction with other people who have the correct attitude. For example, arranging for individuals to work with a group of people with the correct attitude to health and safety. The individuals may change their behaviour to make working with the other people in the group easier and become more positive to health and safety to get the group to be positive towards them.

- Providing positive reinforcement of the correct attitude and the health and safety behaviour that accompanies it. For example, recognition of an individual who is using personal protective equipment correctly.

Motivation

Motivation may be defined as 'the driving force behind the way a person acts in order to achieve a goal'. In the context of work situations there have been many attempts to identify what motivates people to work. The earliest theory (by F. W. Taylor) suggested that people worked for money and fear of losing their job. Financial reward was seen as the main motivator. This approach suggested that the more individuals were paid, the harder they worked. This led to a management philosophy that encouraged payment by results in the form of incentive schemes and motivation to do things that people did not like doing by provision of extra payments, for example, extra money to do dangerous or dirty jobs.

This philosophy was found to be unsuitable for the management of health and safety as most incentive schemes encouraged people to work unsafely, for example, rushing to get the job done. This often caused health and safety working practices to be ignored or compromised. Other motivational theories have been developed that establish a number of other themes related to motivation. Some of these themes emphasise the importance of involvement of managers and workers (team working and social interaction), setting goals/objectives, personal achievement and recognition of positive behaviour. Positive motivation (where safe and healthy working is encouraged and rewarded by recognition and praise) tends to be more effective than negative motivation (where workers fear disciplinary action for not ensuring health and safety) although both have a place and may be appropriate in different circumstances.

Actions to influence motivation might include:

- Establishing a positive health and safety culture, where risk taking is not accepted by managers or workers.

- Involvement of managers and workers in health and safety policy setting.

- Setting realistic goals/objectives with regard to accident/incident and work-related illness rates.

- Clarification of roles and responsibilities.

- Providing information and training to improving managers' and workers' knowledge of their responsibilities and the consequences of not working safely and healthily.

- Showing the commitment of the organisation to health and safety by providing resources and a safe and healthy working environment.

- Involving managers and workers in health and safety decisions, by consultation, team meetings and health and safety committees.

- Monitoring health and safety performance, including personal performance.

- Recognising and rewarding health and safety achievement, by developing a reward structure that recognises positive health and safety behaviour and performance.

Perception of risk

Perception may be defined as 'the way that a person views a situation'. Perception is the process of getting, selecting, organising and interpreting sensory information. The process of perception is a sequence of steps that begins with a stimulus in the environment and leads to our perception of the stimulus and an action in response to the stimulus. This process is continual and takes place without the individual thinking about the actual process at any given moment. The process of perception is illustrated in *Figure 3-25.*

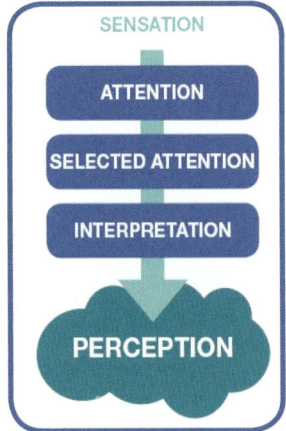

Figure 3-25: Model of perception. Source: RMS.

Within each individual, the attention and interpretation steps of the perception processes are closely interlinked and perception is influenced by a variety of factors, including:

- Attention factors, such as readiness of the individual to respond to a stimulus.
- The individual's past experience.
- The individual's motivation and emotional state.

The factors that influence the process and the way in which they operate are often referred to as the 'perceptual set' of the individual. This represents an accumulation of the individual's attention, experience, motivation and emotional state. These factors can cause an individual's perception to be distorted so that they perceive what they expect to perceive. For example, in the case of the two images in *Figure 3-26* the individual sees what they expect to see, influenced by what is common in their experience.

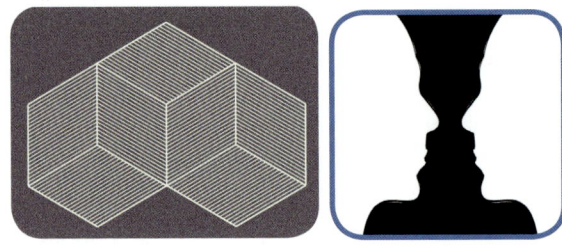

Figure 3-26a (Left) and 3-26(b): Examples of perception images. Source: Ambiguous.

In the case of image a), some people may see one box in the centre, sitting on a base with two sides or they may see two boxes to the side, hanging from a surface. Similarly, in image b), some people may see two people facing each other or a vase in the middle of the square. In work situations perception of health and safety hazards and risks is important. These examples illustrate that in regard to health and safety the subconscious process of perception takes place without the individual realising and may mean they genuinely may not see hazards in the workplace.

The influence an individual's perceptual set has on the perceptual process could be summarised by the phrases:

We do not see what is there.

We see what we expect to be there.

We do not see what we do not expect to be there.

Factors that influence the effectiveness of an individual's perception of hazards and risks may include:

- If an individual is doing a boring or repetitive job it may result in them thinking of other things than their work. This may result in a lowering of the impact of a stimulus and the stimulus may not get the individual's attention.

- Tiredness may reduce an individual's attention level and affect their perception of hazards. For example, they may not see a forklift truck that is approaching them and may step in front of it.

- Similarly, intense concentration on one task may make paying attention to another stimulus difficult or impossible. For example, an individual may fail to hear a warning of danger shouted by another person.

- An individual may not perceive a hazard they are aware of as a risk. For example, where an individual has no experience of the harm electricity can do they may not perceive it as a risk.

- Similarly, warnings of the hazard or risk may not be strong enough to get through the individual's perceptual set.

- Patterns of behaviour and habits that are learnt through experience can be carried from one situation to another where they are not appropriate. For example, a delivery driver may get used to driving at speed on a public highway and not perceive the need to change their speed when they enter a busy site.

- Some hazards may not be obvious or they could be hidden, making perception of their presence difficult, such as the hazard of electricity or certain gases like carbon monoxide.

- The presence of hazards may be masked by environmental issues such as poor lighting or ambient noise. Similarly, the use of personal protective equipment may interfere with the user's senses, limiting their perception.

- Individuals can get 'used to' a stimulus and, if it is not reinforced, it ceases to command their attention and is ignored. For example, they may get used to seeing obstructions in a walk route and after a time no longer perceive them.

Actions to influence perception might include:

- Making what we want people to perceive more obvious, for example, marking hazards so they are more visible.

- Providing information, instruction and training to influence an individual's interpretation of what should be perceived, for example, to improve recognition of hazards and risks.

- Providing experience to reinforce and organise the perception process so that the individual be-comes experienced in what the correct perception is, for example, providing practice fire drills so that when an individual hears a fire alarm it is promptly perceived as requiring action to leave the building.

Competence

If managers and workers are to be effective in contributing to the correct health and safety behaviour at work it is essential that they are competent. Ensuring competence improves health and safety behaviour because managers and workers become more aware of the health and safety aspects of work activities, their perception of risk and danger increases to an appropriate level and there should be an improvement in their attitude and motivation. In addition, competent managers and workers will have a better understanding of the rules and procedures and are more likely to follow them. The decisions made by competent managers are more likely to give proper regard to health and safety and lead to actions that support workers following rules and behaving appropriately.

Competence means more than just providing training and involves ensuring that the relevant ***knowledge, skill and experience*** are established. Studies have shown that human errors can occur through an absence of knowledge and skill. It is important that competencies held by managers and workers include the correct health and safety behaviour for their job. Where possible, correct health and safety behaviour should be learnt as part of the main job competencies, for example, when a manager is learning to manage contractors the correct health and safety behaviour expected of them and the contractors should be included. This will ensure that having the correct health and safety behaviour is seen as part of doing a job well and is more likely to become a normal part of doing the job.

Because individuals may not be capable of developing the competencies and behaving in the correct way needed for certain jobs, their suitability should be assessed. This will help to prevent situations where human errors and violations occur because a person is not capable of doing the job. For example, an individual may not have the aptitude, dexterity and physical ability needed to do certain jobs competently. Similarly, not everyone will be able to work at height or in confined spaces.

When new appointments of managers and workers are being made it should not be assumed that they have the competence that will provide the right health and safety behaviour, even if they are good at other aspects of their work, for example, quality. Managers and workers should be assessed for their range of competencies prior to appointment. Where competencies need to be developed it is important that managers and workers gain competence in a controlled way to ensure they learn the correct things and develop the right skills and behaviour. If this is not done they may not develop the correct health and safety behaviour, instead they may just become very skilled at their bad habits. In order to minimise human errors it may be necessary to arrange for individuals to gain experience under controlled conditions, for example, by close supervision. ILO-OSH 2001 emphasises the need to ensure competence in Clause 3.4.1, *see Figure 3-27.*

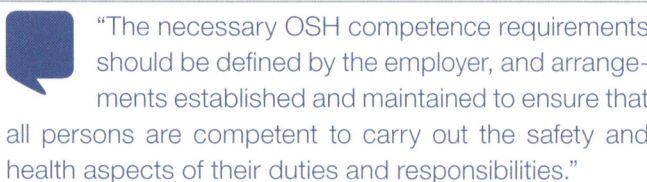

"The necessary OSH competence requirements should be defined by the employer, and arrangements established and maintained to ensure that all persons are competent to carry out the safety and health aspects of their duties and responsibilities."

Figure 3-27: Requirements for competence.
Source: ILO-OSH 2001, Clause 3.4.1.

Competence may be achieved and maintained by actions that include:

- Defining the competencies needed to carry out jobs and tasks safely and healthily.
- Assessing the competencies that managers and workers have and planning competence development.
- Providing the means to ensure that all managers and workers, including temporary workers, develop competencies they require in a controlled way, particularly for those involved in high risk work.
- Arranging for help, advice and supervision while they develop the correct competencies and health and safety behaviour.
- Ensuring individuals know the limits of their competency.
- Reviewing and refreshing competencies as needs change or where health and safety behaviour shows it is required.

 **REVIEW**

Explain the terms attitude, motivation and perception and give an example of each.

Human errors and violations

It is important to develop a good knowledge of how human factors influence health and safety behaviour at work, in particular with regard to understanding accident/incident causation. Human factors includes consideration of the individual and should include how human error and violations (human failures) are caused as this will help to develop a better understanding of accident/incident causation and provide wider opportunities for actions to reduce the likelihood of errors and violations.

It is estimated that approximately 80%–90% (Heinrich et al., 1980) of accidents/incidents are caused by human factors. Studies of major accidents/incidents have shown the importance of human factors as a cause in these events. They have also challenged the widely held belief that accidents/incidents are solely the result of 'human error' by the individual(s) directly involved in the activity related to the accident/incident. Attributing accidents/incidents to the 'human error' of these individuals alone has often been seen as sufficient explanation of the cause of accident/incidents and something that is beyond the control of managers. This ignores the failures of others, including managers, which can lead to accidents/incidents, for example, the failure of individuals involved in the organisation's management and decision-making processes. These failures are usually the result of 'human errors' by people other than those more immediately involved in the accidents/incidents and in many cases these failures can directly influence the behaviour of those individuals that are often accused of causing accidents/incidents. The table in *Figure 3-28* (see following page) illustrates the influence of human failure in some major accidents/incidents. Similar human failures have been identified in other major accidents/incidents that have taken place worldwide since these events.

As can be seen from the examples in *Figure 3-28* the prevention of human failure should be recognised as an important part of an accident/incident and ill-health prevention programme. There are two different types of human failure that need to be assessed and managed effectively - errors and violations. A human error is an action or decision that was not intended, which involved a deviation from an accepted standard, and which led to an undesirable outcome. A violation is a deliberate deviation from a rule or procedure.

Human errors

Human error is an important part of human factors and a particularly significant part of the health and safety related human failures that take place in the workplace. They are not the deliberate human failures, but are the unintended actions or decisions of individuals, often when trying to do their job well. Human errors fall into two categories,

Accident / incident	Consequences	Human contribution
Three Mile Island, Nuclear industry, USA, 1979	Serious damage to the core of the nuclear reactor.	Operators failed to recognise a valve that was stuck open due to poor design of the control panel. Maintenance failures had happened before, but no steps had been taken to prevent a recurrence.
Union Carbide Bhopal, Chemical processing, India, 1984.	The plant released a cloud of toxic methylisocynate. The death toll was 2,500 and over one quarter of the city's population was affected by the gas.	The leak was caused by a discharge of water into a storage tank. This was the result of a combination of operator error, poor maintenance, failed safety systems and poor health and safety management.
Space Shuttle 'Challenger', Aerospace, USA, 1986	Explosion killed all 7 astronauts on board.	There was an inadequate response to internal warnings about the faulty design of a seal. A decision was taken to go ahead with the launch in very cold temperatures despite the faulty seal. The decision made was a result of conflicting scheduling/safety goals, the mindset of individuals and the effects of fatigue.
Chernobyl, Nuclear industry, USSR, 1986.	A 1,000 MW nuclear reactor exploded releasing radioactivity over much of Europe. High environmental and human cost.	Causes were much debated, but the Soviet investigative team admitted "deliberate, systematic and numerous violations" of health and safety procedures by operators.
Herald of Free Enterprise, - Transport industry, UK, 1987.	Ferry sank killing 189 passengers and crew.	There was no system for checking that bow doors were shut. An inquiry reported that the company was "infected with the disease of sloppiness". The priority was to turn the ship around in record time.
Kings Cross fire, Transport industry, UK, 1987.	Major fire killed 31 people.	Organisational changes had led to poor escalator cleaning. The fire took hold because of inadequate firefighting equipment and poor staff training. There was a culture that viewed fires as inevitable.
Piper Alpha, Petrochemical industry, UK, 1988.	Major explosion on North Sea oil platform killed 167 workers.	The maintenance error that eventually led to the leak was the result of inexperience, poor procedures and poor learning. There was a breakdown in communications and the permit-to-work system at shift changeover and health and safety procedures were not properly practised.

Figure 3-28 (Above): Human failure in accidents/incidents. Source: UK, HSE, HSG48.

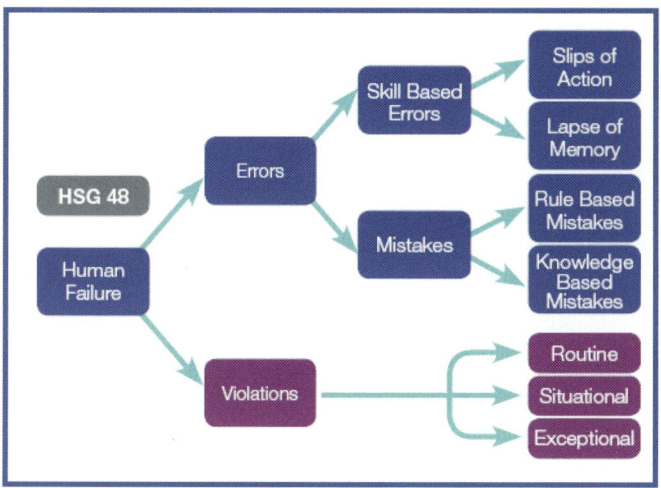

Figure 3-29: Human failures flow chart.
Source: UK HSE, HSG48.

skill-based errors and mistakes, ***see Figure 3-29*** (Left). Skill based errors are then divided into two types, slips of action and lapses of memory. Mistakes are also divided into two types, rule-based mistakes and knowledge-based mistakes.

Slips and lapses

Slips and lapses are the human errors that occur in the workplace during very familiar tasks, which can be carried out without much need for conscious attention. Once an individual has learned a skill, there is little need for conscious thought about what they are doing. They can carry out a task without having to think too much about the next step. An individual learns to ride a bicycle or drive a car in this way. They need to pay attention to the road and the traffic, but manipulate the pedals and change gear without thinking about it. If their attention is diverted by a distraction or interruption, they may fail to carry out the next action of a task, a slip, or forget the next action or lose their place, a lapse. Therefore, slips and lapses are the errors that are made by even the most experienced, well-trained and highly-motivated individuals whilst conducting routine tasks.

A slip is 'not doing what you are meant to do'. Examples of slips include:

- Performing an action too soon in a procedure or leaving it too late, for example, raising a load on the forks of a forklift truck while the truck is travelling along.
- Carrying out an action with too much or too little strength, for example, over-tightening a bolt.
- Performing an action in the wrong direction, for example, an operator of a crane pushing the directional movement control to the left instead of the right.

A lapse is 'forgetting to do something, or losing your place part of the way through a task'. Examples of lapses include:

- Forgetting to switch the local exhaust ventilation (LEV) system on when operating equipment that produces dust.
- Forgetting to nail down a wooden joist when building a floor.
- Taking a respiratory protection mask off to talk to someone and then forgetting to put it back on.
- Failing to secure a scaffold because the individual erecting it was interrupted during the task.
- Forgetting to empty the oil in a pump before moving it for maintenance, allowing it to leak on the floor.

Mistakes

Mistakes are also human errors, but are a little more complex than slips and lapses. Mistakes are decision making failures. Mistakes involve an individual doing the wrong thing, believing it to be right. The two main types of mistake are rule-based mistakes and knowledge-based mistakes.

Rule-based mistakes

There is a tendency for individuals at work to use familiar rules or procedures and sometimes they apply them in situations where they should not be applied. The wrong application of a rule to a situation can result in an error in the form of a rule-based mistake, for example, the use of a water-based fire extinguisher on a fire involving live electrical equipment. In this situation the individual applied the rule that required them to act promptly to extinguish the fire, but used the wrong type of fire extin-

guishing medium (water) to extinguish the fire. Using the wrong type of fire extinguisher could have caused them to have an electrical shock.

Knowledge-based mistakes

In some situations, such as unfamiliar circumstances, an individual may have to apply their knowledge and reason out what action is required. If the situation is misdiagnosed or the action is miscalculated, a knowledge-based mistake may occur.

For example, a driver may make a poor judgement when overtaking, leaving insufficient room to complete the manoeuvre before oncoming vehicles get dangerously close. Knowledge-based mistakes often occur with untrained and inexperienced people as they may base their decisions on misunderstandings and a lack of perception of risk. These errors also occur with trained, experienced people in situations where there are time pressures and where they are doing too many complex tasks at once.

An individual may be influenced by a number of job factors that can contribute to them making a mistake, for example, where there are too few people to do a task, distractions, confusing instructions, poor design of control panels or poor work environment. Individual factors like lack of sleep, fatigue, stress, medication, drugs and alcohol can also contribute to people making mistakes. When considering the various types of human error as a whole there are various actions that may be taken to influence the reduction of errors including:

1) Reducing work environment factors that can lead to increased errors, for example, extremes of heat, humidity, noise, vibration, poor lighting, and restricted workspace.

2) Reducing the effects of extremely demanding tasks, for example, tasks where there is a high workload, tasks demanding high levels of concentration, tasks that are very monotonous or repetitive and tasks where there are many distractions and interruptions.

3) Reducing organisational factors, for example, insufficient numbers of managers or workers, inflexible or over-demanding work schedules, peer pressure and conflicting attitudes to health and safety.

4) Reducing individual factors, for example, inadequate training and experience, high levels of fatigue, reduced alertness, family problems, ill-health, misuse of alcohol and drugs.

5) Reducing equipment factors, for example, poorly designed displays and controls, inaccurate and confusing operating instructions.

6) Simplifying complex tasks and planning for emergencies and other non-standard situations.

7) Ensuring procedures and instructions are clear, concise, available and up-to-date.

8) Ensuring proper supervision, particularly for inexperienced managers and workers.

9) Considering the possibility of human error when undertaking risk assessments.

10) Identifying causes of human errors during accident/incident and ill-health investigations.

11) Ensuring arrangements for the evaluation of health and safety include monitoring the effectiveness of measures taken to reduce error.

Violations

Violations are intentional human failures and involve an individual deliberately doing the wrong thing. Although violations are a deliberate failure to follow a rule or procedure, they are rarely acts of vandalism or sabotage, but are often carried out in order to get the job done, for example, using a convenient ladder of insufficient length. Many injuries and cases of ill-health are caused by the violation of health and safety rules or procedures. There are three types of violation - routine, situational and exceptional violations.

Routine violations

Routine violations are where breaking the health and safety rules or procedures has become the normal way of working. They can develop in a number of ways. The incorrect method may be developed and used because it is a quicker way to work or because the rules are seen as too restrictive. In some cases, new workers can join an organisation and learn the incorrect method from other workers, not realising they are wrong. Examples of routine violations include:

- Not removing dirty work clothes when taking refreshment breaks in a facility provided for eating food.
- Removing the guard on a dangerous machine so that the user of the machine can see the operation easier.
- Driving a fork-lift truck too fast.
- Not wearing hearing protection in a noisy workplace.

Actions to minimise routine violations include:

1) Increasing the chances of violations being detected, by monitoring and supervision.

2) Identifying and removing unnecessary rules.

3) Making rules and procedures remain relevant and practical.

4) Explaining the reasons for and relevance of the rules and procedures.

5) Improving design factors that affect the likelihood of individuals not following rules or procedures. For example, equipment that is excessively awkward, tiring or slow to use, frequent false alarms from instrumentation and uncomfortable personal protective equipment.

6) Involving managers and workers in writing rules and procedures to help to ensure their acceptance.

Situational violations

Situational violations may occur due to pressures related to the work being done. This can include time pressures, extreme weather conditions, provision of an inadequate number of people and provision of the wrong equipment/materials to do a task. These pressures may make it very difficult to comply with rules in a particular situation or workers may think that the rules are inappropriate under the circumstances. For example, rules for safe working at height may be ignored in a situation where work on a roof may need to be completed and the correct work at height equipment had not been provided.

Actions to minimise situational violations include:

1) Providing the correct equipment.

2) Improving the working environment.

3) Providing appropriate supervision.

4) Improving job design and planning.

5) Establishing a positive health and safety culture.

Exceptional violations

Exceptional violations occur when something has gone wrong and a decision is made to solve the problem in a way that involves breaking a rule and taking a risk.

For example, an individual may enter a confined space without suitable breathing apparatus if another person has collapsed in the confined space from lack of oxygen and needs to be rescued. Similarly, a hospital worker may carry out a one person lift of a patient that has fallen out of bed, putting their back at risk, because they considered it to be necessary in order to care for the patient. Though the individual carrying out the violation may perceive it to be necessary in the exceptional circumstances it is an incorrect belief that the benefits outweigh the risk.

Actions to minimise exceptional violations include:

1) Identify the possibility of violations in work activities as part of the risk assessment process.

2) Planning for exceptional situations.

3) Reducing the time pressure on managers and workers to act quickly in exceptional situations.

4) Providing safe and healthy procedures for exceptional situations.

5) Providing more training for abnormal and emergency situations.

LINK BETWEEN INDIVIDUAL, JOB AND ORGANISATIONAL FACTORS

The examination of human factors has shown that what may appear to be simple individual failings are often failings that have been shaped by the circumstances the individuals found themselves in. This will include the individual factors affecting them, the job factors and the organisational factors that cause these job factors to exist. The effects of a number of human factors may combine, causing individuals to fail to comply with health and safety rules or procedures and leading to them having poor health and safety behaviour.

An individual's attitude to health and safety (an individual factor) can be influenced by job factors that directly affect the individual. For example, if an individual is being affected by requirements to meet other priorities than health and safety (a job factor), this could lead to them developing an attitude that health and safety is not important. Further analysis might identify that this particular job factor was created by the presence of a poor health and safety culture (an organisational factor). This shows the link between individual, job and organisational factors.

This link is important to consider when analysing problems and devising solutions to improve health and safety behaviour. Attempts to improve the attitude of workers by the enforcement of health and safety rules are unlikely to be successful in a situation where organisational factors are creating job factors that strongly and negatively influence the attitude of workers. The enforcement of rules may change the behaviour of the individual for a short period, but it is unlikely to change their attitude or behaviour in the long term. Therefore, in this situation, it is important to ensure that the organisational factors that create the job factors are controlled in order for them to have a positive influence on the individual factors.

When the three human factors, organisational, job and individual, work together positively, they can help to ensure individuals adopt the correct behaviour. It is therefore important to ensure that all three factors are considered in order to adequately manage the human factors and ensure health and safety.

MEANING OF HAZARD, RISK, RISK PROFILING AND RISK ASSESSMENT

Hazard

A hazard is something that has the potential to cause harm. There are a number of definitions for the term hazard, including one used in the NEBOSH glossaries that are produced to assist international students.

 "A source of energy with the potential of causing immediate injury to personnel and damage to equipment, environment or structure; or the properties of toxic substances, such as chemicals, gases or radioactivity, that may cause health problems immediately, or in the short-term or longer-term, in people exposed to those substances."

Figure 3-30: Definition of hazard.
Source: NEBOSH, Technical Glossary.

Hazards can include:

- Articles, for example, tools such as chisels.
- Substances and chemicals such as pesticides or cement.
- Equipment, for example, mobile cranes or a fixed grinder for sharpening tools.
- Methods of work, for example, production line workers or workers at height.
- The working environment, for example, cold environments such as a frozen food storage warehouse, or hot and humid ones such as an industrial laundry.
- Other aspects of work organisation such as shift or lone working.

A practical example would be that of a substance, such as cement, which has a particular inherent hazard, i.e. chemical (alkaline) burn, which may occur if the cement or concrete comes into contact with the skin or eyes of a worker.

Risk

A risk relates to the likelihood of potential harm from a hazard. There are a number of definitions for the term risk, including one used in the NEBOSH glossaries that are produced to assist international students.

 "The likelihood or probability that the harm or loss will come about, taking into account the extent and severity of the outcome"

Figure 3-31: Definition of risk.
Source: NEBOSH, Technical Glossary.

When a worker drives a vehicle at work, such as a fork-lift truck, it has the potential to cause harm. The vehicle is therefore a hazard. When the vehicle is being driven, the potential exists to strike another worker. If there are a lot of workers in the area, this potential is higher or more likely to result in a worker getting struck by the vehicle. The impact may be very severe and break a bone or kill, or it may have less impact and push them over. If the vehicle driver checks by looking to see that no one is nearby then the vehicle is less likely to strike a worker. The probability of hitting a worker is reduced by adopting control measures.

Risk profiling

Risk profiling is a systematic and structured approach that helps to inform and guide risk management by providing an organisation with a detailed picture of the risks affecting its operation and the effectiveness of the control measures in place to mitigate the risks. It will also provide a framework for assurance and monitoring its higher risk priorities.

"The process of examining the nature and level of health and safety threats faced by an organisation."

Figure 3-32: Definition of health and safety risk profiling.
Source: RMS.

Risk assessment

Risk assessment relates to the process of evaluating the risks arising from the hazards, identifying preventive and protective measures, taking into account the adequacy of any existing controls, and deciding whether or not the risk is acceptable. There are a number of definitions for the term risk assessment, including one used in the NEBOSH glossaries that were produced to assist international students.

"The process of evaluating the risks to safety and health arising from hazards at work."

Figure 3-33: Definition of risk assessment.
Source: NEBOSH, Technical Glossary.

Risk assessment is an analytical process that identifies hazards, who may be harmed and in what way they may be harmed, and also takes into account factors that make the risk more likely and those that make it less likely.

RISK PROFILING

What is involved

Risk profiling involves examines the range of health and safety threats that could occur, considers the nature of the adverse effects arising from the threats and how likely they are to occur and have these effects.

The scope of the adverse effects are considered in terms of harm to people, disruption to the organisation and associated costs. When considering the likelihood and scope of these adverse effects the effectiveness of control measures known to be in place is determined.

The outcome of risk profiling will be that the right risks have been identified, prioritised for action and information provided to inform decisions about the risk control measures in place and those that are needed. Through this, the health and safety risk profile of an organisation informs all aspects of the approach to leading and managing its health and safety risks.

Who should be involved

Those who undertake risk profiling need to be competent to do so. In many cases this will mean involving those in the organisation that can provide a competent perspective on the threats and control measures. This may require the use of a team or the efforts of a competent individual to gather this wide perspective from various individuals in the organisation, for example by interviewing them. Some organisations may choose to use external expertise to help them develop their risk profile, anyone doing so must have a broad knowledge of the entire organisation and health and safety risk management.

The risk profiling process

The risk profile process examines the:

- Nature and level of the risks faced by the organisation.
- Likelihood of adverse effects occurring.
- Level of disruption, costs and consequences associated with the adverse effects.
- The nature of and effectiveness of the control measure in place to manage those risks.

Risk profiling involves gathering information about operations and process, including information generated from interviews, for example, with directors, senior managers, operational managers and other workers and any risk analysis activities that have been conducted. The accuracy of information gathered is confirmed and the profile generated from it is ranked to form the basis of an overall risk profile.

Every organisation will have its own risk profile, which will vary due to the nature and scale of the organisation and the context it operates in. In some organisations, the risk profile will consist of tangible and immediate safety risks, while in others the risks may involve longer term health-related effects, where it may be a long time before illness becomes apparent. Some organisations, such as those in the construction industry, may have a mixture of threats, spanning both immediate and longer-term hazards. In addition, the range of threats affecting an organisation can have adverse effects beyond health and safety and typically can also threaten quality, environmental and asset damage, because issues in one area could impact in another. For example, unsafe forklift truck driving may have a service or quality dimension as a result of damage to goods. The risk profile approach encourages the consideration of these wider adverse effects.

Health and safety risks can range from low hazard high frequency risks, for example, slips on floors, to high hazard low frequency risks, such as the risk of an oil refinery explosion, with the latter being at the top of the risk profile priority. As the profile is being developed care should be taken to ensure that minor risks have not been given too much priority and that major risks have not been overlooked.

Some organisations use risk registers to enable them to document the risk profile and monitor risks. A risk profile should contain:

- A summary of the main strategic and operational health and safety risks for the organisation.
- Quantification of these risks, in terms of likelihood and potential impact.
- Identification of the current control measures, their effectiveness and improvement potential.
- Identification of any controls not yet in place
- Identification of any new or emerging risks with plans on how to deal with them.
- A framework for the management of the risks, including a prioritised action plan with recommendations for improvements to address weaknesses or to further mitigate the risks.

The results of the risk profile should be used to target risks that require detailed risk assessment and to inform the establishment of the organisation's health and safety policy and objectives. It can also be used to determine the scope and emphasis of the organisation's health and safety management system, including the arrangements and processes to manage the risk. The risk profile should be reviewed to ensure it remains relevant and to provide information on progress with managing the risks.

PURPOSE OF RISK ASSESSMENT

The purpose of health and safety risk assessment is to provide information to enable organisations to manage risks and prevent workplace incidents, in particular accidents and ill-health. Risk assessment processes provide organisations with the means to identify hazards, sometimes before they result in harm, and to evaluate

the priority and resources needed to control them. This enables preventive and precautionary measures to be applied on a timely basis enabling risk to be reduced to an acceptable level or in some cases eliminated from work activities.

This is achieved through the:

- Identification of all the factors that may cause harm to workers and others (the hazards).
- The consideration of the chance of that harm being done to someone in the circumstances of a particular case, and the possible consequences that could come from it (the risks).
- Deciding what are suitable preventive and precautionary measures and the priority to take action to implement them.

Risk assessment involves:

1) The identification of hazards.

2) Identification of the population at risk, who might be harmed and how.

3) The evaluation of the risks from the hazards (likelihood and severity (consequence) and deciding on precautions (adequacy of current controls and need for additional controls).

4) Recording significant findings and the implementation of them.

5) Reviewing the risk assessment and updating it if necessary (periodically or when there is a significant change, for example, process or legislative change).

Most people undertake risk assessment as a normal part of their everyday lives. Routine activities, such as crossing the road and driving to work, call for a complex analysis of the hazards and risks involved in order to avoid injury. Therefore, most people are able to recognise hazards as they develop, and take corrective action. People do, however, have widely different perceptions regarding risk and may find it difficult to apply their experience to the formal workplace risk assessments required. The assessment of risks must be suitable and sufficient and should cover the whole of the organisation's activities. It should also be broad enough so that it remains valid for a reasonable period. Risks arising from routine activities associated with life in general can usually be ignored unless the work activity compounds these risks.

Requirements to conduct risk assessments

The International Labour Office (ILO) has established a number of health and safety conventions, recommendations and guidance. Conducting health and safety risk assessments is part of the principles and practices established by these requirements.

In particular, the ILO Occupational Safety and Health Convention C155 (Convention C155), Article 16, states:

> "1) Employers shall be required to ensure that, so far as is reasonably practicable, the workplaces, machinery, equipment and processes under their control are safe and without risk to health.
>
> 2) Employers shall be required to ensure that, so far as is reasonably practicable, the chemical, physical and biological substances and agents under their control are without risk to health when the appropriate measures of protection are taken."

Figure 3-34: Requirements of conducting health and safety risk assessments. Source: ILO, Occupational Safety and Health Convention C155, Article 16.

Although the ILO has not set out specific requirements for employers to conduct risk assessments, it does require, through Convention 187 Article 5 that member states protect workers by eliminating or minimising, so far as is reasonably practicable, work-related hazards and risks, in accordance with national law and practice, in order to prevent occupational injuries, diseases and deaths and promote safety and health in the workplace. Many member states will have national requirements for conducting risk assessments and the expectation is that these will be undertaken in accordance with that national law.

The requirement for employers to 'ensure that, so far as is reasonably practicable' in Article 16 of Convention C155 and Article 5 of Convention 187 means the employer should balance risks against costs (in terms of time, trouble and money). This requirement establishes the principle that employers should identify and assess the health and safety risks related to their work activities.

A growing number of countries have established specific legal requirements relating to risk assessments. Notable amongst these is Japan, which established a requirement to investigate danger of harm in the Industrial Safety and Health Act 1972, supplementing this with their Guidelines on Occupational Management Systems in 1999. The European Union established a requirement to conduct risk assessments in its Framework Directive in 1989; member countries of the European Union have accommodated this requirement in their national laws.

> "Article 6 - General obligations on employers
>
> 3) Without prejudice to the other provisions of this Directive, the employer shall, taking into account the nature of the activities of the enterprise and/ or establishment:
>
> (a) Evaluate the risks to the safety and health of workers, inter alia in the choice of work equipment, the chemical substances or preparations used, and the fitting-out of work places.
>
> Subsequent to this evaluation and as necessary, the preventive measures and the working and production methods implemented by the employer must:
>
> (a) Assure an improvement in the level of protection afforded to workers with regard to safety and health.
>
> (b) Be integrated into all the activities of the undertaking and/ or establishment and at all hierarchical levels.
>
> Article 9 - Various obligations on employers
>
> 1) The employer shall:
>
> (a) Be in possession of an assessment of the risks to safety and health at work, including those facing groups of workers exposed to particular risks.
>
> (b) Decide on the protective measures to be taken and, if necessary, the protective equipment to be used."

Figure 3-35: Risk assessment requirements.
Source: European Framework Directive on Safety and Health at Work (Directive 89/391 EEC).

In the UK, the requirement to conduct risk assessments was introduced by the Management of Health and Safety at Work Regulations (MHSWR) 1992 which was amended in 1999. Other countries, such as Abu Dhabi, have established requirements for risk assessment through decree and the introduction of health and safety management system frameworks.

A 'SUITABLE AND SUFFICIENT' RISK ASSESSMENT

The term 'suitable and sufficient' when used in relation to conducting risk assessments relates to how well they are done and the standard it needs to reach. The UK Health and Safety Executive's Managing for Health and Safety HSG65 considers that a risk assessment must be suitable and sufficient and says "it should show that:

- A proper check was made.
- You asked who might be affected.
- You dealt with all the obvious significant risks, taking into account the number of people who could be involved.
- The precautions were reasonable, and the remaining risk is low.
- You involved your workers or their representatives in the process.

Suitable is generally taken to relate to the type of process used to assess risk, and sufficient is usually taken to relate to the amount of process done. Where risk assessments are required to be suitable and sufficient, it means that the type of assessment and amount of assessment may vary in different circumstances. The level of detail in a risk assessment should be proportionate to the risk and appropriate to the nature of the work. This means that where risks are not complex, a relatively simple approach could be taken, but where risks are more complex, a more detailed approach would be necessary. Sufficient means just enough and is designed to avoid organisations having to waste resources.

Some countries may have specific legal parameters relating to what is a suitable and sufficient risk assessment; however, general considerations may include the following:

- For small organisations presenting few or simple hazards, a simple approach based on informed judgement and reference to appropriate guidance will be enough.
- In many intermediate cases the risk assessment will need to be more sophisticated. Some areas of the risk assessment may require specialist advice, for example, those relating to complex processes or where specialist analytical techniques are required to measure worker exposure to the hazard and to assess its impact.
- Large and hazardous sites will require the most developed and sophisticated risk assessments, particularly where there are complex or novel processes.
- Risk assessments must consider all those who might be affected, for example, contractors and members of the public.
- Employers are expected to take reasonable steps to help themselves identify risks, for example, by looking at appropriate sources of information or seeking advice from competent sources.
- The risk assessment should be appropriate to the nature of the work and should identify the period of time for which it is likely to remain valid.

The risk assessment should be conducted in a systematic way; adopting a step-by-step approach will assist in assuring that a suitable and sufficient risk assessment is conducted. The ILO OSH management system: a tool for continuous improvement; 5 steps to risk assessment is outlined in *Figure 3-36.*

Figure 3-36: 5 steps to risk assessment.
Source: ILO OSH management system: a tool for continuous improvement.

A GENERAL APPROACH TO RISK ASSESSMENT (5-STEPS)

Identifying hazards

Sources and forms of harm

The first step in the risk assessment process is to identify the work tasks and associated hazards. Hazards can be grouped under two main headings, health hazards and safety hazards.

The health hazard categories are:

CATEGORIES	EXAMPLES
Chemical	Paints, solvents, transport exhaust fumes
Biological	Bacteria, pathogens
Physical	Noise, vibration
Psychological	Occupational stress
Ergonomic	Repetitive strain injuries, manual handling

Safety hazard categories include:

Machinery.
Flying or falling objects.
Moving vehicle.
Struck against something fixed or stationary.
Mechanical handling, lifting or carrying.
Slips, trips and falls on the same level.
Working from a height.
Pressure equipment.
Dangerous substance.
Fire.
Confined space.
Electricity.
Animals.
Workplace violence.

The identification of hazards can be done in many ways. For complex activities it may be necessary to break the activity down into its component parts by using task analysis, for example. *Task analysis* identifies hazards before work starts. This is achieved by breaking the task down into its component steps and identifying the hazards associated with each step. Having established the hazards, safe and healthy working methods can be established that deal with them. For a large machine, task analysis could mean looking at:

- Installation.
- Normal operation.
- Breakdown.
- Cleaning.
- Adjustment.
- Dismantling.

The hazards associated with each part of the job or task could then be identified more easily and thoroughly. It is necessary to identify contingent hazards that could arise from failures of a system, component, or checking and maintenance, as well as continuing hazards, i.e. those that are present continuously.

Examples may include:

- Mechanical hazards.
- Electrical hazards.
- Thermal hazards.
- Noise and vibration.
- Radiation.
- Toxic materials.
- Ergonomic design.

Once the area/activity has been selected, the method(s) of hazard identification can be chosen. If a hazard is defined as being something with the potential to cause harm, then hazard identification can be carried out by observing the activity and noting the hazards as they occur in the actual work setting, for example by using task analysis. This has advantages over carrying out hazard identification as a desktop exercise using the health and safety manual, because the manual may not reflect how work is actually done and workers may have developed their own method of working, contrary to instructions and training.

In more complex organisations, it may be appropriate to use other, more advanced, hazard identification techniques:

- Hazard and operability studies (HAZOP), which are much used by the chemical industry at the design stage of processes and equipment.
- Reliability analysis and failure mode and effect analysis (FMEA), which are inductive techniques.
- Fault tree analysis (FTA) and event tree analysis (ETA), which are deductive techniques.

 **CONSIDER**

Step 1: Identify the hazards

Walk around your workplace and look at what could reasonably be expected to cause harm.

Check manufacturer's instructions or data sheets for chemicals and equipment as they can be very helpful in spelling out the hazards and putting them in their true perspective.

Remember to think about long-term hazards to health, for example, high levels of noise or exposure to harmful substances.

 **REVIEW**

What is meant by the terms 'hazard' and 'risk'? Provide an example for each.

SOURCES OF INFORMATION TO BE CONSULTED

Internal to the organisation

Incident and absence records

Incident, for example accident and ill-health, and absence records can provide information on the organisation's workers have actual experience of harm from hazards, this can be useful in identifying hazards that exist in the workplace, indicating how likely harm can result from them and the probable type and extent of harm that can result from exposure to the hazard.

Investigation reports

Investigation reports can provide information on the causes of accidents and ill health, which may identify hazards and the effectiveness of control measures. Immediate, underlying and root causes should be analysed and care taken to take this into account in the risk assessment process.

Results of audits/inspections

Audits and inspections can provide information on hazards in the workplace, including those that have not resulted in harm. They can also provide information on how effective the current control measures are, indicating where they are effective or not. The results of audits and inspections are particularly useful in identifying actual circumstances in the workplace, for example, whether procedures are followed, training takes place or guards sufficiently enclose hazards. In this way the circumstances can be taken into account when deciding the adequacy of current control measures and the need for further control measures.

Maintenance records

If a system of planned maintenance is implemented then, by definition, any breakdown of equipment is likely to be unintended. Therefore, breakdowns should be reported and investigated to determine the causes. Trends and patterns of failures can be analysed and taken into account during the risk assessment. The absence of failure of equipment used to control risks can support the effectiveness of the equipment and confirm its continued use as a control measure identified in the risk assessment documentation.

External to the organisation

Manufacturers' data

ILO conventions set out an expectation that manufacturers make adequate information available to those they supply with machinery, articles or substances. The information should be sufficient to enable the person to use the machine, article or substance safely and healthily. The information should cover reasonably foreseeable risks, including those arising from disposal and dismantling.

This information can be used in the risk assessment process to compare how the organisation uses the machinery, article or substance. This may show an exposure to a hazard that was not expected, confirm the adequacy of current control measures or indicate opportunities to improve the control measures.

If, as a result of testing or research, new hazards come to light, manufacturers may provide new information to those supplied. This should then be used to review the current risk assessment to see if the hazard had already been identified by other means and the adequacy of control measures compare with any advice provide by the manufacturer.

Legislation

These are prime sources of information that set out requirements to improve health and safety at international and national levels. Legislation can be difficult to read without some legal understanding, and it is also easy to miss changes and amendments unless an updating service is used. However, it is helpful to workplace organisations in setting out requirements that can encourage an organisation to put systems in place that ensure the identification and management of risks. Legislation may set specific technical limits to the exposure of workers to hazards, for example levels of noise. It may also specify minimum requirements for control measures. This information is important to take into account when conducting risk assessments as the organisation may have standards that are below the legal requirements.

European

The European Safety Agency (ESA) was set up by the European Union (EU) to bring together the vast reservoir of knowledge and information on OSH-related issues and preventive measures. This information is informative and can widen the perspective of those in the organisation conducting risk assessments and help them to understand the significance of hazards and what good practice control measures should be used.

The ESA makes good-practice information available covering a range of topics and sectors. These include good prevention practices, campaigns and an active publications programme producing everything from specialist information reports on hazards to fact sheets covering a wide variety of occupational health and safety problems. In addition, the ESA:

- Collects and disseminates technical, scientific and economic information.

- Promotes and supports co-operation and exchange of information and experience amongst the Member States.

- Organises conferences and seminars and exchanges of experts.

- Supplies the Community bodies and the Member States with the objective available technical, scientific and economic information they require to formulate and implement judicious and effective policies designed to protect the safety and health of workers.

- Provides the Commission in particular with the technical, scientific and economic information it re-quires to fulfil its tasks of identifying, preparing and evaluating legislation and measures in the area of the protection of the safety and health of workers, notably as regards the impact of legislation on enterprises, with particular reference to small and medium-sized enterprises.

HSE Publications

The UK Health and Safety Executive (HSE) publish both approved codes of practice (ACOPs) and guidance notes. The HSE provides a range of information in different forms. The range includes information related to different industries, including air transport, construction, food and drink, oil and gas, railways and textiles. Topics covered by this information include asbestos, confined spaces, falls from height, nanotechnology, stress and welding. Information is available directly from the website, in leaflets, practical guides, detailed codes, and DVDs and as electronic tools to aid health and safety risk assessment.

Electronic versions of HSE publications are available free via their website, http://www.hse.gov.uk.

Trade associations

Good practices are technical and/or practical documents prepared by a variety of interested parties, including labour inspectorates, social partners, accident insurance companies, trade/industry associations, and practitioners' associations. They often demonstrate solutions to occupational health and safety hazard control. As these solutions can be transferred to other employers, or other countries, information on good practice provides a useful source of reference. They contribute to the risk assessment process and the decisions regarding the suitability and effectiveness of control measures.

British, European and international standards

Of particular interest to health and safety is the ISO 45001:2018 'Occupational health and safety management systems – requirements with guidance'. ISO 45001 is intended to help organisations develop a framework for managing occupational health and safety so workers and others who may be affected by the organisation's activities are adequately protected. As part of the standard it includes requirements for risk assessment and defines a hierarchy of control measures, the guidance that is part of the document provides further advice on this.

There is also progressive harmonisation to European Standards (CEN) and the use of CE marking. Some British and European standards are being used as a basis for international standards and are being adopted as ISO standards, such as that for machinery health and safety BS EN ISO 12100-1. The British Standards Institute (BSI) provides some high-quality advice which is usually above the legal minimum standard.

Some of these standards provide detailed technical information that can inform more complex risk assessment process, providing good practice standards for control measures for a variety of hazards.

ILO and other authoritative sources

The International Labour Organization (ILO) deals with labour standards, fundamental principles and rights at work. They have a website and a database listing all types of work-related information. The ILO produces an Encyclopaedia of Occupational Health and Safety, a widely respected comprehensive publication which is now available free of charge on the ILO website. This provides comprehensive information on hazards, their risks and control measures, which is useful and supports the risk assessment process.

The World Health Organization is the United Nation's specialised agency for health and was established in 1948. It has information on health, where good health is defined as "a state of complete physical, mental and social well-being and not merely the absence of disease or infirmity".

The United States of America's Occupational Safety and Health Administration (OSHA) provides information related to job safety and health issues, as well as compliance resources. These cover topics as diverse as avian flu, black widow spiders, chainsaws, generator safety, lock out / tag out and winter driving. They provide information in a number of formats, including fact sheets, pocket guides and what they call 'quick cards' for easy reference; many are available in English and Spanish.

The government of Western Australia's Department of Commerce has established an organisation called WorkSafe which has created a web-based resource called SafeLine. The resource includes documents, images and videos related to workplace health and safety. The resource includes a library of health and safety material as well as many checklists on health and safety topics.

IT sources

There is an increasing amount of information available online, using a computer and an internet service provider, that can be used when conducting risk assessments. The World Wide Web provides access to many health and safety organisations, including the ILO: (http://www.ilo.org).

Care should be taken to verify information from an unknown web source through an alternative source; also, some material found may be out of date, applicable to another territory or incorrect.

There is also a range of electronic subscription services available that aim to keep the subscriber informed of relevant health and safety issues, providing access to information and an updating service.

REVIEW

Where might information about health and safety come from? Provide two external and two internal sources.

Identifying people at risk

In the risk assessment process it is necessary to identify the people at risk. In practice, this may mean noting groups or types of people at risk, for example, all nursing staff in the operating theatre or members of the public or patients. It is not necessary to list all individual workers in a category as this would be too much detail for most risk assessments.

A suitable and sufficient risk assessment will identify all groups of people at risk. When considering the people who might be affected, it is important to remember certain groups of workers who may work unusual hours, for example, security staff and cleaners. Similarly, maintenance workers need to be considered and, where relevant, the fact that they may be workers not employed by the organisation should be identified. Also, people who do not work at a fixed workplace, such as welfare workers, street cleaners, delivery workers, consultants and sales people, should be considered. For some workers in a shared workplace, it will be necessary to consult with another employer to determine any risks associated with the shared workplace. Members of the public, visitors, students, work experience people and even trespassers should be considered in the risk assessment process.

Specific groups at risk

Vulnerable people

Vulnerable people might be at risk from a particular hazard or may be affected by all hazards because of their lack of knowledge or understanding. For example, women of childbearing age (or more particularly any unborn foetus they may be carrying) may be deemed to be at risk from exposure to the hazards presented by lead.

The range of people that may be considered to be vulnerable includes new workers, those with disabilities, workers that have ill-health or are on medication, young people, pregnant women or nursing mothers, workers that have come from other countries where language or work practices may be different.

Operators

Typically, operators are individuals engaged in production-type activities where they have little control over their environment or work routine. Issues include repetitive strain, slips, trips and falls, together with a variety of equipment hazards. Consideration of the task and issues of fatigue and loss of concentration are usually significant.

Maintenance staff

Maintenance issues that may result in injury are usually associated with poor access and egress. Some maintenance work is carried out infrequently, resulting in a lack of familiarity, which can lead to serious mistakes. The fact that a maintenance worker is working alone may make the risk of harm and possible injuries worse.

Cleaners

Cleaners may be at risk from the materials that they use or that which they clean. Often the turnover of cleaners is high and their competency, in terms of such issues as to correct use and health effects of the materials they use or remove, is low.

Figure 3-37: Car washing. Source: BOSH Service.

Contractors

Contractors may be involved in work that has a particularly high risk due to its unusual nature or complexity. They are a particular group to identify as they may not be as familiar with the workplace as other workers. They may not understand the hazards of the workplace and may not be as equipped as other workers to deal with the hazards.

Visitors/public

Visitors and the public are particular groups of people that need to be identified because they may not perceive or understand hazards and may behave differently from workers. They are seen as a vulnerable group because of their lack of awareness and ability to protect themselves from hazards.

CONSIDER

Step 2: Decide who might be harmed

This does not mean listing everyone by name, but by general group, for example, welders, people working at height or the general public.

In each case, identify how they might be harmed and what type of injury or ill-health might occur, for example, falls of objects onto workers or the general public.

Some workers have particular requirements. New or young workers and people with disabilities may be at a higher risk, as may cleaners, visitors, contractors and maintenance workers etc., i.e. those who may not be in the workplace all the time.

If you share your workplace, you will need to think how your work affects others, as well as how their work affects your workers.

Evaluating risk and adequacy of current controls

Likelihood of harm and probable severity

After the hazards have been identified, it is necessary to evaluate the risk. The risk assessment process requires a judgement for each hazard to decide, realistically, what the most likely outcome is and how likely this is to occur. It may be a matter of simple subjective judgement or it may require a more complex technique depending on the complexity of the situation. In order to do this, at least two factors must be considered - the likelihood and the probable severity of harm.

Likelihood - when conducting a risk assessment, we take account of the circumstances in which the hazard may be encountered and the current controls in place, as these can greatly influence the likelihood of a person being harmed by a hazard. The circumstances may relate to environmental factors that can mask or make a hazard more obvious, for example, poor or good lighting; where the hazard is located, away from normal workers or in a busy walkway. The person encountering the hazard is another factor affecting likelihood. Someone who does not perceive the hazard because of lack of knowledge or reduced visual ability or other sensory impairment is more likely to come into contact with the hazard. Hazard controls in use may have only a limited effect, may fail, be defeated or become inactive at various times and these factors also influence the likelihood. Reliance on a control such as personal protective equipment would normally increase the likelihood compared to a control that put the hazard behind a protective barrier.

Other factors that might be considered include:

- Competence of workers.

- Levels and quality of supervision.

- Attitudes of workers and supervisors.

- Environmental conditions, for example, adverse weather.

- Frequency and duration of exposure.

- Work pressures.

Probable severity - this considers the probable outcome of contact with the hazard, which may include death, major injury, minor injury, damage to plant/equipment/product, or damage to the environment. It is important that this is the most probable outcome, and not a possible outcome, as it may be possible to think of some extreme circumstances in which all hazards may have the outcome of death. Again, it is important to take into account the nature of the hazard and the circumstances in which the hazard is encountered. The severity of contact with electricity is greatly reduced if the nature of the hazard is that it is operating at low voltage. Also, the circumstances may be that electricity is encountered in wet conditions, which could increase the probable severity due to the increased conductivity. Similarly, the person coming into contact with the hazard may have an influence on severity; for a large healthy adult the severity may be different than it would be for an unhealthy child.

Possible acute and chronic health effects

Risk assessment should consider both the acute and chronic effects. A risk, for example from exposure to a chemical, may present a harmful acute effect and if exposure to the chemical continues over time a worker may develop chronic effects. An acute effect may be irritation to the lungs when the chemical is breathed in, this may only sustain for a short period while being exposed. However, the repeated exposure over time may mean that sufficient dose of the chemical is absorbed into the lungs and was likely to lead to a chronic effect on the workers organs, for example, a form of cancer or

the lungs may be affected causing chronic obstructive pulmonary disease (COPD).

Risk rating and prioritisation

In order to manage risks, it is important to understand their scale: are they significant or not? One way to evaluate this is to describe the two components of risk, likelihood and probable severity, in terms of how large they are as a factor of the risk.

Different people may have a different perception of risk, as a result of training, life experiences and background. A method is therefore required in order to have a common approach and attempt to overcome individual differences. Risk can be rated according to the likelihood and probable severity of harm resulting from a hazard by using a guided approach to assigning a value to likelihood and severity. Words commonly associated with the scale of likelihood and severity are provided to guide the decision process. These words have a numeric value attached to them. The risk rating is a combination of the likelihood and severity value.

Risk rating = likelihood × severity

Where the:	Likelihood is how likely a loss will occur as a result of contact with the hazard.
	Severity is the probable amount of harm from contact with the hazard.
	Risk rating is the level of risk after current controls have been taken in to account.

Using the formula stated (Risk rating = Likelihood × Severity), the risk rating can be calculated. It will fall into the range of 1-25. This risk rating is used to prioritise the observed risks. This approach is known as a 'semi-quantitative' risk evaluation, and different organisations will use different numbers and descriptions to assign a value to risk. It is the general principle of evaluating and managing risk that is important, not the exact words used on a form. The risk rating is then classified.

There are many variations on this method and other variables can be used other than those proposed in the 5×5 matrix used in this example. The size of the matrix can be increased to 10×10 and many more subdivisions for severity and consequence can be added. The more complex and hazardous the process, the more sophisticated the approach to assessing risk must be. Ultimately, the numbers used can be scientifically developed, such as the hours before a pump fails in use, or the static electricity charge crude oil will gain when travelling down a pipeline. Assessing risk on this level would be known as a 'quantitative' risk assessment.

Likelihood categories

The likelihood can be assessed on a scale of 1 to 5.

5.	Almost certain	Absence of any management controls. If conditions remain unchanged there is almost a 100% certainty that an accident/incident will happen (for example, broken rung on a ladder, live exposed electrical conductor, and untrained personnel).
4.	High	Serious failures in management controls. The effects of human behaviour or other factors could cause an accident/incident, but it is unlikely without this additional factor (for example, ladder not secured properly, oil spilled on floor, poorly trained personnel).
3.	Medium	Insufficient or sub-standard controls in place. Loss is unlikely during normal operation; however, it may occur in emergencies or non-routine conditions (for example, keys left in fork-lift trucks, obstructed gangways, refresher training required).
2.	Low	The situation is generally well managed; however, occasional lapses could occur. This also applies to situations where people are required to behave safely in order to protect themselves but are well trained.
1.	Improbable	Loss, accident/incident or illness could only occur under exceptional conditions. The situation is well managed and all reasonable precautions have been taken.

Severity categories

The severity can be assessed on a scale of 1 to 5.

5.	Major	Causing death to one or more people or loss or damage is such that it could cause serious business disruption (for example, major fire, explosion or structural damage).
4.	High	Causing permanent disability (for example, loss of limb, sight or hearing, cancer, chronic obstructive pulmonary disease (COPD)).
3.	Medium	Causing temporary disability (for example, fractures, electrical burns, chemical burns, heat exhaustion).
2.	Low	Causing significant injuries (for example, sprains, bruises and lacerations, lung irritation, irritant dermatitis).
1.	Minor	Causing minor injuries (for example, cuts, scratches). No lost time likely other than for first-aid treatment, or repair of superficial damage, for example to interior decorations.

Principles to consider when controlling risk

The ILO document 'ILO-OSH 2001 – Guidelines on occupational safety and health management systems' establishes the following principles to consider when controlling risk, prevention and control procedures or arrangements should:

a) be adapted to the hazards and risks encountered by the organisation;

b) be reviewed and modified if necessary on a regular basis;

c) comply with national laws and regulations, and reflect good practice; and

d) consider the current state of knowledge, including information or reports from organisations, such as labour inspectorates, occupational safety and health services, and other services as appropriate.

This is further strengthened by the well accepted principles of prevention established by the European Union in the Framework Directive, see *Figure 3-38*.

The general principles of risk prevention are broad principles that are applied to risks in order to prevent them. The principles of prevention encourage that risks be avoided or combated at source, but they also establish further risk-prevention principles, such as adapting work to the individual. They are not hierarchical, but illustrate a good-practice approach to the prevention of risk. Therefore, for any given risk, all the principles of prevention should be considered and applied.

(a) "Avoiding risks.

(b) Evaluating the risks which cannot be avoided.

(c) Combating the risks at source.

(d) Adapting the work to the individual, especially as regards the design of work places, the choice of work equipment and the choice of working and production methods, with a view, in particular, to alleviating monotonous work and work at a predetermined work rate and to reducing their effect on health.

(e) Adapting to technical progress.

(f) Replacing the dangerous by the non-dangerous or the less dangerous.

(g) Developing a coherent overall prevention policy which covers technology, organization of work, working conditions, social relationships and the influence of factors related to the working environment.

(h) Giving collective protective measures priority over individual protective measures.

(i) Giving appropriate instructions to the workers."

Figure 3-38: Principles of prevention.
Source: Article 6 - European Framework Directive on Safety and Health at Work (Directive 89/391 EEC).

Avoiding risks

If risks are avoided completely, they do not have to be either controlled or monitored. For example, not using pesticides or not working at height.

Evaluating unavoidable risks

Carry out a suitable and sufficient assessment of risks.

Controlling (combating) hazards at source

Repairing a hole in the floor is safer than displaying a warning sign. Other examples are the use of local exhaust ventilation to remove a substance at source, the design of equipment so that mechanical movement is enclosed and does not create a hazard, and the replacement of a defective bearing to control the hazard of noise at source.

Adapting work to the individual

This emphasises the importance of human factors in modern control methods. If the well-being of the person is dealt with, there is less chance of the job causing ill-health and less chance of the person making mistakes which lead to accidents. Consideration should be given to the design of any equipment used. Frequently used controls should be close to the operator, start buttons should be positioned to avoid inadvertent use and stop buttons should be close to the operator and easy to operate in an emergency.

All equipment should be clearly labelled. Consideration should also be given to minimisation of fatigue.

Alleviating monotonous work by breaks or task rotation can help the individual to remain alert and pay attention to the task.

Adapting to technical progress

Technical progress can lead to improved, safer and healthier working conditions, for example, the provision of new non-slip floor surfaces or the bringing into use of less hazardous equipment such as new sound-proofed equipment to replace old noisy equipment. In recent years, this has included the use of waste chutes for removal of materials from scaffolds and the use of equipment which produces lower levels of vibration.

Replacing the dangerous by the non/less dangerous

For example, using a battery-operated drill rather than a mains-powered tool, providing compressed air at a lower pressure or providing water-based chemicals instead of solvent-based chemicals.

Developing an overall coherent prevention policy

Taking a holistic stance to the control of risk involves consideration of the organisation through the establishment of risk/control identification systems, consideration of the job, the use of task analysis and the selection of people, which includes consideration of human factors that affect an individual, such as mental and physical requirements.

Giving priority to collective protective measures over individual protective measures

Organisations with a less developed approach to health and safety may mistakenly see the solution to risks is to provide individuals with protective measures, such as warning people of hazards and provision of personal protective equipment - using the 'safe person' (and healthy person) strategy. A more developed and effective approach is to give priority, where possible, to collective measures that provide protection to all workers, such as provision of barriers around street works or cleaning up a slippery substance spill rather than putting signs to warn workers - using the 'safe place' (and healthy place) strategy.

These two strategies are supported by a third strategy, often called the 'safe system' (and healthy system) strategy. This strategy establishes the correct way to do things in the form of rules and procedures, for example, the rule that says 'clean up after you do work' in order that the place is left in a safe and healthy condition.

Reliance on only a safe/healthy person strategy is the weakest of controls. The preferred strategy is the safe/healthy place and priority must be given to using it where possible. By establishing a safe/healthy place, an organisation will ensure that all people that find themselves in it will gain protection. This approach is reflected in the hierarchy used for safeguarding dangerous parts of equipment - our first priority is to provide guards around the dangerous parts (making it a safe place), and for the remainder that cannot be enclosed in this way we provide information, instruction and training to those that use the equipment (making the safe person).

In practice, the most successful organisations use a combination of the three strategies, with the emphasis on making the place safe/healthy, supporting this with systems of work (procedures) and paying attention to the human element so that they support rather than undemine the other strategies. Many organisations that feel they have invested effort in getting the place and procedures right are taking a fresh look at the actions necessary to ensure the person is right also. Studies have shown that this too is a critical aspect of effective management of health and safety.

Providing appropriate instructions to workers

In order to meet the principle of providing appropriate instructions to workers it is essential to support the other principles by ensuring that workers know what is expected of the in relation to action taken to prevent risk.

General hierarchy of control

A hierarchy of control is a list of measures designed to control risks that are considered in their order of impor-

tance and effectiveness. The approach used in applying the hierarchy is to address each risk in the order of the priority expressed in the control hierarchy. The general hierarchy of control for the elimination of hazards and reduction of health and safety risks referred to in Clause 8.2 of ISO 45001:2018 is set out below:

a) Eliminate the hazard.

b) Substitute with less hazardous processes, operations, materials or equipment.

c) Use engineering controls and reorganization of work.

d) Use administrative controls, including training.

e) Use adequate personal protective equipment.

This general hierarchy of controls begins with the preferred measure to be used, elimination of the hazard, and goes through to the lowest level of protection, the use of personal protective equipment. The hierarchy of risk controls encourages the use of the highest level of control first. Only when we have found that it is not possible to use this control should we move to the next highest control. Often a combination of measures is used to control a risk adequately.

Elimination of hazards	Stopping using hazardous chemicals, applying ergonomic techniques when planning new workplaces, eliminating monotonous work or work that causes negative stress, removing mobile equipment like fork lift trucks from a work area.
Substitution with less hazardous processes, operations, materials or equipment	Replacing the hazardous with less hazardous, (for example, solvent based paint with water based paint), or change the physical form of materials used (for example, dust to pellets) or replace workplace materials (for example, slippery floor material with material that provides better grip for walking), changing processes (for example, requiring clients that present a risk of violence to visit controlled offices instead of seeing them in their home).
Engineering controls and reorganisation of work to limit exposure to the hazard	Providing collective preventive measures (for example, machine guarding, fume/dust extraction or isolation of noise with insulation, guard rails for workplaces at height, security glass screen between worker and visitors), isolating people from hazards (for example a 'glove box' for handling hazardous biological agents), segregation of workers from a hazard by distance (for example, locating observation points away from hot processes). Reorganisation of work to limit unhealthy work hours or workload or victimisation (for example, limiting hours worked without a break, job rotation, reassignment to a different team).
Administrative controls to remain alert to the hazard	Conducting periodic inspections (for example, of safety equipment), conducting training (for example, to prevent bullying and harassment, induction training), administrating competence (for example, forklift driving licences), coordination of activities (for example, managing contractors), control of work patterns (for example shifts or who does work at night), managing a health/medical surveillance programme (for example, related to hearing, hand-arm vibration, respiratory disorders, skin disorders), giving instructions to workers (for example, entry control processes like permit-to-work) and supervision.
Personal protective equipment to limit the effects of exposure to the hazard	Personal protective equipment, a physical barrier between the worker and the hazard (for example, protective glasses, hearing protection, gloves, protective helmet, anti-slip shoes).

Figure 3-39: Hierarchy of controls. Source: Adapted from ISO 45001:2018.

This means that when the current control measures are considered during the risk assessment process, the extent they relate to the hierarchy of controls should be determined and whether the appropriate control measure has been used or if other, better, control measures should be used.

It is important that the risk assessment evaluate the adequacy of the current control measures in place.

Eliminate the hazard/risk

Risks can be eliminated by avoiding situations that create them. This is the best option for controlling risk as it means that everyone is protected and there is no residual risk to manage. If hazards/risks are avoided completely they do not have to be either controlled or monitored. For example, not using pesticides or not working at height.

In many cases this may be a difficult option to achieve. However, it is something that should be considered by designers at the conception stage of any work activity.

For example, in a construction project, in order to avoid the risk of falling it is critical to design out the need to work at height or when a new workplace is being designed good ergonomic practices should be used to eliminate ergonomic hazards.

Substitution of the hazard/risk

Replacing the dangerous with something less dangerous can be a very effective preventive measure and provides a means to control the hazard/risk at source. Risks can be reduced to an acceptable level by substituting what is causing the risk with something less hazardous, for example substituting a water-based chemical for a solvent-based chemicals or by using a hazardous substance in a dilute form. Similarly, risks can be controlled by substituting the energy causing the risk for one of a lower level, for example, using a battery-operated drill rather than a mains-powered tool or providing compressed air at a lower pressure or replacing equipment that has high vibration characteristics with equipment that has low vibration characteristics.

Engineering controls or organisational control measures

It is possible to combat many risks at source by the use of engineering controls or organisational control measures, particularly as they are collective protective measures rather than personal protective measures. Controlling the hazard/risk at source is preferable to the use of other control measures, for example, repairing a hole in the floor is better than displaying a warning sign.

A common engineering control involves the *isolation* of the hazard from people. For example, enclosing the hazard, so there is a controlled barrier between people and the hazard. This could include fitting fixed guards around dangerous parts of a machine, installing sound absorbing enclosures around noise sources, totally enclosing dusty processes or those involving toxic chemicals, providing guard rails on scaffolds or barriers round temporary work activities. Other measures could include providing machinery that is remotely operated and to which materials are fed automatically, therefore separating the machine operator from danger areas or using robots for tasks that have high risk, such as spraying of isocyanate paints.

Some engineering controls limit the chance or amount of exposure to hazards. For example, control of the amount of dust/fumes released into the atmosphere by an activity may be achieved by using local exhaust ventilation. Other examples of engineering controls include those that respond to work conditions that can present an increased risk. For example, provision of an overload or over-run device on a crane/hoist, pressure relief valves fitted to boilers or other pressure vessels, automatic water sprinklers linked to heat detectors and devices that prevent equipment running if the operator is not in the correct position to avoid hazards it creates or that stop the equipment if the operator approaches the hazards. Organisation controls could include:

- Organising high hazard tasks so they take place when workers are not nearby.
- Organising work to minimise the exposure of any one worker to the hazard, for example, by reducing the frequency and duration of exposure to radiation, noise, vibration, heat or cold.
- Controlling the hours worked to ensure rest and avoid fatigue.

Administrative controls

Administrative controls are those that rely on the administration of a procedure as the control measure. A safe system of work is an example of an administrative control; it is a formal procedure which results from systematic examination of a task in order to identify all the hazards and the controls necessary for health and safety. It defines safe and healthy methods of working, to ensure that hazards are eliminated or risks minimised.

In some cases, the system of work is controlled by the use of structured instructions that help to ensure the system of work is carried out as intended, this is often called a method statement or procedure. Where work involves a significant risk, structured checklists are often used to control that each step in the system of work is carried out, this is often in the form of what is often called a permit-to-work.

Other examples of administrative controls include provision of licences to control who can do higher haz-ard tasks, managing health surveillance programmes, information, instruction, training, co-ordination and supervision of activities. Supervision is used to reinforce the use

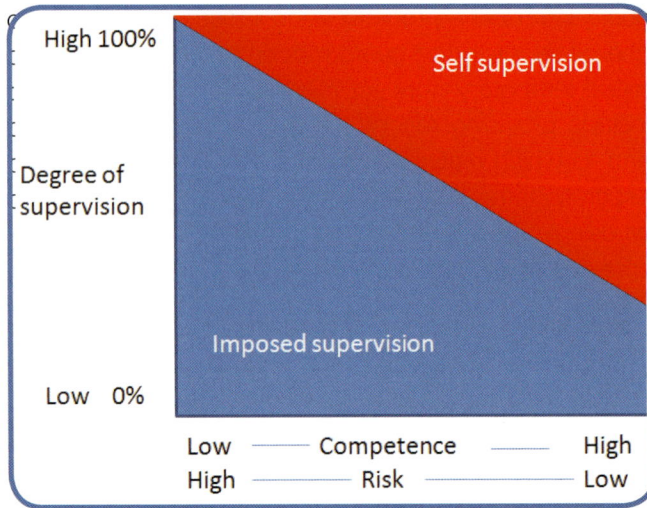

Figure 3-40: Competence v supervision. Source: RMS.

When considering supervision of individuals, it is important to also take account of their level of competence. In cases where the individual has qualifications, but no experience, and is therefore low in competence (for example, a young person straight from full-time education), supervision must increase accordingly.

Signs and warnings may accompany administrative controls to help to ensure people are aware of the hazards and respond as intended. Employers should provide and maintain specific health and safety signs whenever there is a significant risk that has not been avoided or controlled by other means, for example, by engineering controls and safe systems of work. Where a safety sign would not help to reduce that risk, there is no need to provide a sign. Signs which no longer apply should be removed to ensure the credibility of the signage elsewhere.

It is important to balance the amount and timing of supervision against the nature of the work being done. It is generally appropriate that the level of supervision necessary increases with the level of risk involved in the work.

1.	Prohibition	Circular signs. Red and white.	For example: • No smoking • No pedestrian access • No children • No unauthorized	
2.	Warning	Triangular signs Black on yellow	For example: • Caution - slippery floor surface • Toxic substance • Site traffic • Electrical hazard • Deep excavation	
3.	Safe conditions	Oblong/ square signs Green and white	For example: • Emergency assembly point • Fire exit • First-aid • Eye-wash station	
4.	Mandatory	Circular signs Blue and white	For example: • Hearing protection must be worn. • Safety helmets must be worn. • Safety boots must be worn. • All visitors must report to site office.	

Figure 3-41: Categories and examples of typical health and safety signs, ISO 7010:2011 Source: RMS.

Element 3

Figure 3-42: Hazard identification - obstruction.
Souce: RMS.

Figure 3-43: Hazard identification - restricted width.
Source: RMS.

Figure 3-44: Hazard identification - obstruction.
Source: RMS.

Figure 3-45: Hazard identification - restricted area.
Source: RMS.

In order to avoid possible confusion due to language ability and reading skills, the shape, colour and symbols used on health and safety signs should be to a standardised format. Most countries use four (4) categories of health and safety signs. Prohibition (must not do, for example, no naked lights); hazards (for example, electricity); safe conditions (for example, exit route, first-aid, and assembly point) and mandatory (instruct safe actions to take before proceeding, for example, fit hearing protection).

Although the format of the signs may vary throughout the world, they tend to follow similar principles. ISO 7010 'Graphic symbols - Safety colours and safety signs - Registered safety signs' sets out an internationally agreed approach to safety signs, as described in *Figure 3-41.*

Supplementary signs provide additional information. For example, where a noise hazard is identified by a warning sign - 'hearing protection available on request' may be added as supplementary information. Supplementary safety signs can also be used to mark hazards, for example, the edge of a raised platform, dangerous locations where objects may fall, or an area where work is going on that the public should not access. These are in distinct colours to attract the attention of workers and others that may encounter them. The colours used include yellow and black or red and white - in each case they consist of alternate colour stripes, often set at 45° angles.

 REVIEW

What are the safety colours/shapes for the following safety signs: Prohibition, Warning, Safe Condition and Mandatory? Give examples of where they may be used.

 CONSIDER

Hierarchy of controls. Think of one practical example for each control measure in the hierarchy, preferably from your own workplace.

Personal Protective Equipment

Where residual hazards/risks cannot be controlled by collective measures, the employer should provide for appropriate personal protective equipment (PPE), including clothing, at no cost, and should implement measures to ensure its use and maintenance.

Personal protective equipment is equipment that will protect the user against health or safety risks at work.

PPE includes the following when worn for health and safety reasons at work:

- Aprons, for example, to protect from spills.

- Adverse weather gear, for example, thermal protection.

- High-visibility clothing, for example, work alongside a road.

- Gloves, for example, gauntlet hand-forearm protection.

- Safety footwear and helmets, for example, on a construction site.

- Eye protection, for example, goggles when dispensing acids.

- Life-jackets, for example, when working over or near water.

- Respiratory protection equipment (RPE), for example, dust mask.

- Safety harness, for example, protect from falls from a height.

- Breathing apparatus, for example, work in confined space.

PPE should only be used for low risk activities when all other controls have been considered or to provide additional protection when other control measure are used. Even when substitution or engineering controls and safe systems of work have been applied, some hazards might remain that may have to be controlled by the use of PPE. For example, a respiratory protector may be used to protect against limitations/failure of local exhaust ventilation used to control volatile toxic substances. Similarly, hearing protection may be used to further reduce noise exposure of workers where enclosure of a noise source has not reduced noise levels sufficiently. Other examples of residual hazards that may be controlled by the use of PPE could include:

- Contaminated air, which may be a hazard to lungs.

- Falling materials, which may be a hazard to a worker's head and feet.

- Flying particles or splashes of corrosive liquids, which may be a hazard to eyes.

- Corrosive materials, which may be a hazard to skin.

- Extremes of heat or cold, which may be a hazard to a worker's body temperature.

In general, PPE should only be used as a last option, after all other options have been considered.

The reasons for this include:

- PPE may not provide adequate protection because of such factors as poor selection, poor fit, incompatibility with other types of PPE, contamination, and misuse or non-use by workers.

- PPE is likely to be uncomfortable and relies for its effectiveness on a conscious action by the user.

- In certain circumstances, its use can actually create additional risks (for instance, warning sounds masked by hearing protection).

- PPE only protects the wearer and not others who may be in the area and also at risk.

- The introduction of PPE may bring another hazard such as impaired vision, impaired movement or fatigue.

- PPE may only be capable of minimising injury rather than preventing it.

- If PPE fails, it fails to danger and does not provide 100% protection, it may not be possible to measure actual exposure to hazards such as dust, vapours and noise.

 "Whatever PPE is chosen, it should be remembered that, although some types of equipment do provide very high levels of protection, none provides 100%."

Figure 3-46: PPE quote. Source: Guidance on the UK Personal Protection Equipment Regulations 1992.

In many cases, the effectiveness of protection provided by PPE will depend upon how well it fits, for example, with respirators, breathing apparatus and noise protection devices. Factors that can influence fit are whether adequate training and instruction have been given, and certain other individual conditions such as:

- Long hair.
- Wearing of spectacles.
- Facial hair. Where a person has shaved prior to work, hair growth during a work period may necessitate the progressive adjustment of respiratory protection. Beards may prohibit the wearing of respiratory protection.

PPE will not be suitable unless:

- It is appropriate to the risk involved and the prevailing conditions of the workplace.
- It is ergonomic (user-friendly) in design and takes account of the state of health of the user.
- It fits the wearer correctly, perhaps after adjustment, and is comfortable to wear.
- It is effective in preventing or controlling the risk(s).

- It does not increase the overall risk to the user to an unacceptable level.

In addition:

- It must comply with relevant national/international directives or other specific standards.
- It must be compatible with other equipment where there is more than one risk to protect against.

The main benefits of using PPE as a control measure include their relatively low cost and portability when compared with engineering strategies, which can be difficult to implement or be very costly.

Good-practice principles encourage employers to:

- Ensure PPE is suitable for hazard and person.
- Provide adequate training/instruction.
- Issue, obtain signature and record.
- Set up monitoring systems.
- Organise routine exchange systems.
- Implement cleaning/sterilisation.
- Issue written/verbal instructions, define when and where to use.
- Provide suitable storage.

Training and instruction in the use of PPE includes both:

- Theoretical - provision of a clear understanding of the reasons for wearing the PPE, factors which may affect the performance, the cleaning, maintenance and identification of defects.
- Practical - practising the wearing, adjusting, removing, cleaning, maintenance and testing of PPE.

Application of control measures based on prioritisation of risk

Further or different control measures should be applied to risks that are determined to be inadequately dealt with by current controls. It is essential that the application of control measures be made based on the priority of the risk. The higher the risk, the greater the priority to apply control measures.

 "It is essential that the work to be done to eliminate or prevent risk is prioritised. The prioritisation should take account of the severity of the risk, the likely outcome of an incident, the numbers who might be affected and the time necessary for prevention."

Figure 3-47: Prioritisation of control measures.
Source: European Commission Guidance on Risk Assessment at Work.

Where risk control measures will take a period of time to set up, it may be necessary to put in place temporary control measures that may not be so effective but give sufficient short-term risk control while better controls are being established.

Where risks are particularly high and only low-level temporary control measures are available, it may be necessary to suspend the work activity until the necessary effective controls are in place.

Use of guidance

When making a judgement about risks and whether controls are adequate, care has to be taken to consider relevant guidance.

This may be in the form of guidance to legislation, official government guidance documents, industry-standard guidance and relevant international/local standards. If this is not done, then the risk assessment may be determined to be not suitable and sufficient.

Legislation requiring controls for specified hazards
Where specific duties set out control measures that must be taken, the risk assessment must consider whether these are in place. Duties to apply specific controls may be found in international and national legislation that is set out to control specific risks. This may be in the form of International Labour Organization (ILO) Conventions, for example, the 'Chemicals Convention C170', or national legislation, for example, the 'Occupational Safety and Health (Use and Standards of Exposure of Chemicals Hazardous to Health) Regulations 2000', or in the form of national legal codes, for example, the United Arab Emirates 'Abu Dhabi OSHAD-SF Code of Practice on Working at Heights', which includes requirements for preventing injuries from falling objects, *see Figure 3-48.*

When work is conducted at height, employers shall implement the following:

(i) Establish exclusion zones and enforce them under work at height areas to prevent unauthorized access to the area.

(ii) work is to stop while people traverse the exclusion zone.

(iii) place warning signs to warn people of hazards, all signage shall be in accordance with OSHAD-SF CoP 17.0 – Safety Signs and Signals.

(iv) Employees are to use bolt bags and tool carriers to carry small items and tools – these are not to impede the employee.

(v) Make sure that employees required to be in the exclusion zone including persons holding ladders and banksmen, wear hard hats.

(vi) Implement safe working platforms with appropriate toe boards to prevent falling objects.

(vii) Prevent tools and equipment used at height from falling by securing them with a lanyard.

Figure 3-48: Code of practice for working at height. Source: Abu Dhabi, OSHAD-SF CoP 23 Working at Heights.

Applying controls to specific hazards

Controls need to be identified and applied to specific hazards, for example, electrical, chemical, manual handling. Methods for the reduction in the levels of risk for a particular hazard will vary in accordance with the hazard.

For example, reduction of risk for a chemical substance may involve changing to a less harmful substance or using it in a dilute form, whereas reduction of risk in a manual handling activity may involve redesign of containers so each one is lighter to carry or a reduction in the distance that they are carried.

The controls applied should follow the general hierarchy of controls, see earlier in the element.

Residual risk

Residual risk is concerned with the risk which remains when controls have been decided. For example, whilst a slip which could result in a fall from height may be prevented by a guard rail, the potential to slip on the level would remain, and this would be the residual risk after the guard rail was provided. Similarly, if local exhaust ventilation equipment removes a substance from a workplace it will only work to a degree of efficiency. It may reduce the level of substance to just below the occupational exposure limit (OEL) set by national or international standards, but the amount of substance remaining in the atmosphere will have a residual risk.

If the residual risk is low after the control measures have been applied then it may be considered an **acceptable risk level**. The degree of acceptance will be influenced by what society considers as acceptable or tolerable. This may be influenced by how much knowledge there is on a given hazard or the level of risk to which people expect to be exposed. Societal standards change and risk acceptability reduces each year. For example, the noise action levels applicable in European Union (EU) countries have been reduced in over time, illustrating this change in acceptability.

If the residual risk is high a decision has to be made about whether the risk is tolerable or unacceptable. **Tolerable** implies that the risk is acceptable for the short term and further control measures should be developed to reduce the risk further. Unacceptable risk clearly implies that the level of risk is too high for the work to be allowed to continue. Much of the health and safety legislation created by countries places a general duty to reduce the level of risk so far as is reasonably practicable (assessment of the balance between the consequence of the risk against the time, trouble and cost to mitigate it to an acceptable level). When applying and meeting the term 'practicable', employers need to use any new improvements in technology.

This will also assist in keeping pace with acceptability, as society may change its view on acceptability based on what level of risk reduction can easily be achieved.

Distinction between priorities and timescales

Reducing a high risk may have a high priority; however, the action needed might take a very long time to carry out. It is therefore not always helpful to class all high risks as requiring a solution within an immediate timescale. It is important not to ignore the risk and often it is possible to carry out some aspects of lesser control on an interim basis while the longer-term solution is being established. A low risk may be low priority, but it may also be a very simple low-cost solution to risk reduction. A law court may find it difficult to understand why a low-cost simple solution was not speedily implemented. Interim and simple precautions may reduce risk from 'tolerable' to 'acceptable'. They may not be long-term solutions to the nature of the problem but help manage short-term risk while longer-term strategies are being developed.

Recording significant findings

Format

Employers should record the data considered in the risk assessment process and the significant findings of the risk assessments. This may be in writing or electronically, so long as it may be retrieved. The record should confirm when the assessment took place, the scope of the assessment (workplace, activity or specific risk) and who was involved. It should be noted that there are many forms and systems designed for recording assessments; while these may differ in design, the methodology broadly remains the same. The record will assist in providing proof that a suitable and sufficient risk assessment has been made.

Information to be recorded

The record should represent an effective statement of hazards and risks and should lead management to take action to eliminate and reduce risk. In its simplest form, the record will include details of the task/plant/process/activity, together with the hazards involved, the associated risks, persons affected by the hazard and the existing control measures. The necessary actions required to further reduce the risk are also recorded, sometimes separately. Some hazards, particularly those with a high-risk rating, may require a more detailed explanation or there may be a series of alternative actions. Information on risk assessments and any controls must be brought to the attention of those assigned the task of work. Risk assessment information should be included in lesson plans to ensure items are not missed when staff are trained or retrained.

No.	Hazard identification	Associated risks	Persons at risk	Existing controls	Likelihood	Severity	Current risk rating	Comments and actions

Figure 3-49: Risk assessment form. Source: RMS.

 **CONSIDER**

Step 4: Record your findings and implement them.

Putting the results of your risk assessment in to practice will make a difference.

Writing down the results and sharing them with your workers will encourage you all to make a difference.

Remember, prioritise and tackle the most important things first. As you complete each action, tick it off your plan.

Reasons for review

The risk assessment should be periodically reviewed and updated. In addition, a review of risk assessments should be carried out following any significant changes to a workplace. Examples of circumstances that would require the review of the validity of a risk assessment are:

Internal to the organisation

- When the results of monitoring, for example, routine analysis of accident and ill-health reports, damage accident reports, and 'near-miss' reports are adverse and not as expected.
- Following the findings of an accident investigation.
- Following consultation with workers.
- Following internal health and safety audits.

Internal changes

- A change in process, work methods (introduction of shifts) or materials.
- Changes in personnel.
- The introduction of new plant or technology.

External to the organisation

- Changes in legislation, for example, work exposure limits.
- New information becoming available from research, for example, identification of new asthma sensitisers.
- Following enforcement action that called into question the health and safety of a process.
- Third party health and safety audits.

Periodic review

As time passes the risk assessment should be periodically reviewed and updated.

The validity of risk assessment should be monitored through a combination of monitoring techniques such as:

- Preventive maintenance inspections.
- Health and safety representative/committee inspections.
- Statutory and maintenance scheme inspections, tests and examinations.
- Health and safety tours and inspections.
- Occupational health surveys.
- Air monitoring.

 CONSIDER

Step 5: Review and update your assessment

Few workplaces stay the same. New equipment, substances and procedures could lead to new hazards. Review should be carried out at least annually.

Look at your risk assessment again. Have there been any changes? Are there any improvements you need to make?

Have your workers spotted a problem?

Have you taken account of any accidents or near misses?

APPLICATION OF RISK ASSESSMENT FOR SPECIFIC TYPES OF RISK AND SPECIAL CASES

Examples of when specific risk assessment methodologies are required

The general approach to risk assessment described previously is likely to be adequate for the identification and assessment of general risk in a workplace. However, some specific types of risk and special cases may require an enhanced approach where by specific factors are considered. For example, it is generally accepted that specific risks arising from fire, noise and chemical substances require specific factors to be considered during the risk assessment process.

Similarly risks arising from display screen and manual handling activates benefit from an approach where specific risk factors are considered in the assessment process.

Why specific risk assessment methods are required

Specific risk methods that take account of factors specific to the particular risk are required to enable the proper, systematic consideration of all relevant issues that contribute to the risk.

Special case applications of risk assessment

Young persons

Another specific group of people who must be considered in risk assessments is young people (for most countries this is a person under 18 years of age) and children (generally taken to be under compulsory school-leaving age).

Young person risk assessments should consider the following risks:

- Work which is beyond their physical or psychological capacity.
- Work involving harmful exposure to agents that are toxic, carcinogenic or cause heritable genetic damage, harm to the unborn child or other chronic health effects.
- Work involving harmful exposure to radiation.
- Work involving risk of accidents that it may reasonably be assumed cannot be recognised or avoided by young people owing to their insufficient attention to safety or lack of experience or training.
- Work in which there is a risk to health from extreme cold or heat, noise or vibration.

Where children are involved, the significance of assessments should be communicated to the people who have parental responsibility.

Factors such as lack of knowledge, experience or training, see *Figure 3-50*, the tendency of young persons to take risks because of over-enthusiasm or poor perception and to respond more readily to peer pressure will make young people more vulnerable. Young people may also demonstrate an eagerness or willingness for work and agree to undertaking tasks they are not skilled enough to undertake. Young people may also have less well developed communication skills and fail to understand warnings they are given about hazards. All these factors

Figure 3-50: Young person craft training.
Source: www.theguardian.com.

should be considered when carrying out risk assessments. It may be necessary, therefore, to restrict young persons from carrying out certain high-risk activities (for example, machine operation). It may also be necessary to restrict hours of work and to conduct a thorough induction into the workplace.

Expectant and nursing mothers

There are many factors that may increase risks to pregnant workers in the workplace. These include: exposure to chemicals such as pesticides, lead and those causing intracellular changes (mutagens) or affecting the embryo (teratogenic); biological exposures (for example, rubella or hepatitis); exposure to physical agents such as ionising radiation and extremes of temperature; manual handling (especially later in pregnancy); ergonomic issues relating to prolonged standing or the adoption of awkward body movements; stress; whole body vibration and issues associated with the use and wearing of personal protective equipment.

Disabled workers

When a general risk assessment is conducted, it is common that the assessor's main focus is the majority of the working population being considered. For most situations, this will be people who are not disabled or otherwise vulnerable. It is essential that if disabled workers are, or are going to be, involved in the work being considered, the risk assessment take full account of them. This may mean conducting a separate risk assessment for the disabled workers. Because disabilities vary greatly, it may be necessary to conduct separate risk assessments related to the specific disabilities that the workers have. In this way they are more likely to be both suitable and sufficient. Common worker impairments include difficulty with hearing, vision, mobility, and learning. Care should be taken to consider the particular disabilities in relation to the job when it is functioning normally, when problems arise from it and where emergencies may exist.

CASE STUDY

A UK care worker notified her employer of her pregnancy. The employer looked back at the outcome of the initial risk assessment, which had identified that a possible risk for pregnant women was exposure to acts of violence (for example, difficult patients). The employer then conducted a specific risk assessment for the pregnant worker, who dealt with patients who were difficult and on occasion violent. As a result, the employer offered the care worker suitable alternative work at the same salary and reviewed the assessment at regular intervals. The employee accepted the alternative work and had a risk-free pregnancy. Following her maternity leave, the employee returned to work.

Lone workers

Risk assessments that relate to people working alone are a special case in that special risks can arise that are not present in other work. It should be remembered that work that is risk assessed on the basis of it normally being conducted in the main part of the day, whilst other people are around, can quickly become lone working when people work late, come in early or work weekends.

Essentially, the hazards from the work are often the same as they would be if the person was working with others, but the additional risk control of having others at hand to provide routine assistance, for example, to move something or hold something, are not available. Furthermore, lone working leads to the removal of another risk control - the ability of a co-worker to observe if another worker is suffering harm from the risks to which they are exposed or to respond or obtain assistance. When assessing the risks, it is therefore essential to identify how the risks are controlled to take account of the lone working and whether this is adequate.

3.5 Management of change

The objective of a management of change process is to minimise the introduction of new hazards and risks into the workplace and to identify opportunities for improvements as changes occur. The efforts and resources put in to the process should be proportionate to the complexity of the change, the scale of the hazards concerned and the degree to which the change may impact on the management of major hazards.

Typical changes

Typical types of change likely to be encountered in the workplace include:

- Organisational changes - for example, new workers, managers promoted, temporary contracted seasonal workers introduced.

- Activity changes – for example, changes to processes, equipment, infrastructure (construction work) and work practices.

- Material changes – for example, new chemicals, raw material changed from liquid form to powder, packaging changed from solid containers to plastic sacks, buying in bulk making the size of items larger.

Possible impact of change

Ineffective management of change is one of the leading causes of serious incidents. Some of the possible impacts of ineffective management of change could include:

- Changing a chemical to speed up a chemical reaction could lead to increased heat that the equipment is not designed to withstand and a fire may result.
- Changing equipment that works at a faster rate may lead to ergonomic hazards and fatigue of workers.
- Changing from manual handling of goods to the use of a battery powered fork lift truck would introduce the hazard of moving vehicles in the workplace, electrical hazards and explosive gas hazards from the battery charging process.
- Where construction activities are to take place in a workplace this represents a change to what is normal in that workplace. This could introduce hazards that would not normally be there, for example flammable liquids or sources of ignition, closure of access routes and obstruction of emergency equipment/exits.
- In some cases construction changes to a building could reveal hidden hazards, for example the presence of asbestos in a building, which could lead to uncontrolled exposure of workers.

Overall approach to the management of change

The overall approach to management of change includes consideration of health and safety hazards and risks prior to the introduction of the "change". This involves the identification of the hazards associated with the "change", assessment of the risks and implementation of the controls needed to mitigate the hazards and reduce the risks. In addition, the approach should identify and implement opportunities for improving risk by elimination of hazards or provision of control measures that can be applied as part of the change.

The main steps involved in management of change are as follows:

1) Identification of the change.

2) Preliminary screening of the health and safety impacts, for example design review.

3) Preliminary plans for the implementation of the change.

4) Risk assessment.

5) Development / approval of final plans, including control measures.

6) Implementation of the change.

7) Final assessment and close out.

Managing the impact of change

The main control measures to mitigate the impact of change include:

- Communication and co-operation.
- Risk assessment.
- Appointment of competent people.
- Segregation of work areas.
- Amendment of emergency procedures.
- Welfare provision.

Review of change

As the change progresses and when the change has been introduced, health and safety performance should be reviewed to ensure there are no adverse impacts due to hazards not identified or control measures not being effective. Adverse impacts identified at this stage may require further analysis and development of additional action items. The review should also identify what went well in the change process in order to learn from this to benefit future changes.

3.6 Safe systems of work for general work activities

WHY WORKERS SHOULD BE INVOLVED IN THEIR DEVELOPMENT

Worker involvement in the development of safe systems of work is essential to ensure safe behaviour when working to the system. The worker knows all the practical difficulties in working to theoretical systems of work, such as the absence of equipment specified or requirements that mean more workers are needed to do the task than are available. They have frequently encountered circumstances that make the work difficult and know how they get around them, such as poor access/egress.

By discussing proposed systems of work with workers, it is possible to learn these practicalities and ensure the system of work takes account of them. Involvement in this way often ensures a deeper understanding of hazards and risks and how safe systems of work will reduce those risks. This helps with ownership of the safe system of work and makes it more likely that the agreed system of work will be followed.

WHY PROCEDURES SHOULD BE DOCUMENTED

Where the safe system of work is complex or the risks of not working to it are high, it is important to define what must be done in written procedures. This does not mean that written procedures have to be complicated

or wordy; some of the most effective are those that are clear, simple and easily understood. The importance of creating written procedures is that they serve as a clear setting of standards of work, including how the risks of the work are to be combated. This is particularly relevant when communicating to a client how a contractor is going to conduct work safely. Written procedures can contribute to tender documents used to win work and will enable supervisors and others that monitor health and safety to do so against clear standards of performance.

Written procedures may relate to specific work to be done such as carrying out a cutting operation, or to general health and safety matters such as reporting accidents. The construction industry makes particular use of written procedures in the form of method statements. As the name suggests, these are documents that express how work is to be conducted, the order of events, the health and safety considerations and the measures that will make it safe and healthy for those that might be affected.

It is important that procedures express how the work is done, not just how it should be done in theory. If there is a poor match between the procedure and what can be done in practice it will devalue that procedure and others and cause confusion. If a good match is achieved, in a manner that is clear to workers and supervisors, there is a higher chance that it will be complied with and the supervisor will be more able to enforce it. Job/task analysis is not only necessary to identify the hazards and controls, but is essential for preparing written procedures. These can then be translated into comprehensive training plans to ensure the correct skill and knowledge content for the work to be carried out safely and healthily.

TECHNICAL, PROCEDURAL AND BEHAVIOURAL CONTROLS

In the workplace we use a variety of controls as part of safe systems of work. In order to give emphasis to them and to help identify how complete our approach is to health and safety we group controls into a number of categories. The categories tend to reflect the main strategies used to improve health and safety and the general principles of prevention referred to earlier in **Section 3.4,** but use slightly different words.

It is important to note that, in order to deal effectively with a hazard, we will need to use all three forms of control, to a greater or lesser degree, for example, the use of a hard hat for the control of a hazard of falling materials.

A hard hat is a technical control that will afford a degree of protection - the technical aspects of this come into play when we consider how long a hard hat lasts and what type is suitable for which work/person. Purchasing hard hats will not be effective if procedures are not in place to enable issue and replacement. These controls will be immediately undermined if we do not provide education and training to those that wear and enforce the wearing of hard hats in order to ensure the right behaviour of the wearer.

Risk control measures:

Technical	(place)	(job)
Procedural	(system)	(organisation)
Behavioural	(person)	(person)

Technical controls include:
- Equipment - design (for example, guarding) and maintenance.
- Access/egress - provision of wide aisles, access kept clear of storage items.
- Materials (substances and articles) - choice of packaging to make handling easier.
- Environment (temperature, light, dust, noise) - local exhaust ventilation (LEV).
- Issue of the correct personal protective equipment (PPE) for the task.

Procedural controls include:
- Policy and standards.
- Rules.
- Procedures.
- Permit-to-work.
- Authorisation and coordination of actions.
- Purchasing controls.
- Accident investigation and analysis.
- Emergency preparedness.
- Procedures in the issue, use and maintenance of PPE.

Behavioural controls include:
- Awareness, knowledge, skill, competence.
- Attitude, perception, motivation, communication.
- Supervision.
- Health surveillance.
- Training in the issue and use of PPE.

In practice, the most successful organisations use a combination of the three strategies, with the emphasis on making the place safe/healthy (providing technical controls) and supporting this with procedures and paying attention to the person. Many organisations that feel they have done all that is necessary to get the place and procedures correct are taking a fresh look at the actions necessary to ensure the person's behaviour is also correct. Studies have shown that this too is a critical aspect of effective management of health and safety.

DEVELOPING A SAFE SYSTEM OF WORK

Developing a safe system of work requires a systematic approach and generally requires the involvement of a number of people. The development process requires a number of stages:

- Analysing the task.
- Identification of hazards and risk assessment.
- Introducing controls and formulating procedures - including the definition of the safe and healthy method and the implementation of the system.
- Instruction and training in the operation of the system.
- Monitoring the system.

It is important that the development of a safe system of work involves relevant people; this could include managers, workers who will work to the system, maintenance workers, competent health and safety practitioners and specialists.

Analysing the task

Analysis of the task must consider not only the work to be done, but the environment where it is to be done, to allow a full consideration of the hazards to be made and the establishment of an effective safe system of work. Simple work activities that involve low risk may be analysed in a structured way, but without the complexity of a detailed review using documentation. Though a simpler approach may be taken, the principles of analysis are similar to those where a detailed review is involved. More complex or higher risk work activities may involve a detailed review known as job or task analysis. Job/task analysis consists of a formal step-by-step examination of the work to be carried out. Analysis can be:

1) Job based - considering all the work of a specific group of people, for example, all the work of operators of a process plant or fork-lift truck drivers.

2) Task based - considering specific work, for example, a manual handling activity that is part of the operation of process plant or the fork-lift truck task of stacking loads on the trailer of a vehicle.

The main difference between job and task analysis is that the steps of the work activity are smaller in task analysis. The steps in job analysis are made up of the tasks that comprise the job and the steps in task analysis are smaller parts of the task. There is a tendency for people to use the terms job and task to mean the same thing when discussing this process.

Job/task analysis for health and safety reasons is the identification of all hazards relating to the job/task, the accident/incident and ill-health prevention measures appropriate to a particular job/task and the behavioural factors that most significantly influence whether or not these measures are taken. The approach is formal, structured and is diagnostic as well as descriptive.

The process of job/task analysis involves a number of stages, typically this will include:

1) Select the job/task to be analysed.

2) Break the job/task down into a sequence of steps.

3) Identify potential hazards.

4) Identify current control measures.

5) Consider the effectiveness of the control measure and the factors that could affect this, for example, worker behaviour.

6) Determine what additional control measures are required.

The process is usually conducted by direct observation of the job/task, though a generic analysis may be conducted without direct observation by using the experience of a number of people to consider a job/task before it is introduced into the workplace. This can be useful in establishing initial safe systems of work that are adjusted as the job/task is actually introduced into the workplace. When conducting a job/task analysis, all aspects of the job/task should be considered and recorded in writing to ensure that nothing is overlooked.

Typical considerations in any analysis for development of a safe system of work are:

- The work to be done.
- Where the work is done.
- Equipment, materials and people used.
- The current controls.
- Adequacy of controls.
- Correct use by the workers of the controls.
- Behavioural factors relating to workers/supervisors/managers - consideration of errors and violations.

The objective of the analysis is therefore to establish the hazards and controls at each step of the work to ensure a safe and healthy result. How the progress of work, in particular health and safety conditions, will be monitored should also be considered, for example, gas testing, temperature or pressure levels, and measurement of emissions.

Select the job/task to be analysed

Ideally, all jobs/tasks should be systematically analysed, but it is usually necessary to decide which ones are done first. Factors to consider could include the frequency and severity of accidents/incidents and ill-health related to the job/task. Consideration should also be given to the existence of newly established or infrequent jobs/tasks, as these could present higher risks due to the lack of experience of workers.

Break the job/task down into a sequence of steps

When breaking the job/task down into steps, it is important to ensure that it enables sufficient depth of analysis for the purpose of the process. The steps should be recorded in the sequence that they occur. If the job/task is easily repeated it is often useful to concentrate on breaking down and recording the steps only and recording the other factors separately when the job/task is repeated. If this is not easy, it will be necessary to consider and record the other factors as they occur, as well. The steps should be recorded by starting each step with an action verb, for example, 'remove...' 'insert...'. The job/task should be observed in normal work conditions, which will mean that if the job/task is usually done at night the analysis should be done at night.

Identify potential hazards and current control measures

The hazards that may occur and the control measures in place are identified and recorded for each step. Hazards may be identified by observation, supported by asking the workers and others questions.

Consider such things as:

- Slips, trips and falls.
- Falling objects.
- Hazards presented by tools, machines or other equipment.
- Body parts getting caught in or between objects.
- Making harmful contact with moving objects.
- Strain from lifting, pushing, or pulling.
- Hot, toxic, or caustic substances.
- Dusts, fumes, mists or vapours in the air.
- Excessive noise or vibration.
- Exposure to extreme heat or cold.
- Lighting problems.
- Weather conditions that affect health and safety.

A range of control measures may be in place and include technical control measures like temperature limitation, procedural control measures like control of access to keys or behavioural control measures in the form of supervision. The control measures will also include the actions the worker takes to prevent harm from hazards, such as checking a lid is firmly in place on a drum of chemicals after use or holding a tool firmly.

Consider the effectiveness of the control measures

It is important to consider the effectiveness of the control measures and the factors that could mean the controls might not be used. This could include worker behaviour. The job/task may be observed 'as the work should be done', because the worker is being observed and taking care to do the work as expected. However, this may not be the way that workers usually do the job/task. Workers may do the work in a way they find easier, or external factors may force the worker's behaviour to change from the correct way of working. The job/task analysis should try to determine these influences on behaviour. The analysis may determine that the current control measures are not adequate and that improvements are necessary. This should be identified and recorded.

The results of a job/task analysis can be used to establish safe systems of work that set out the correct sequence for the job/task and the control measures that are to be used at each step to control the hazards. They may also make a specific contribution to the risk assessment process and improve such things as training, emergency procedures, reporting of information or the layout of work areas.

Example of part of a task analysis - changing a wheel of a vehicle:

	Step	Hazards	Control measures	Additional control measures
1	Park vehicle.	Vehicle too close to passing traffic. Vehicle on uneven, soft ground. Vehicle may roll.	Drive to area well clear of traffic. Turn on hazard warning lights. Choose a firm level parking area. Apply the parking brake.	Place blocks in front and back of the wheel diagonally opposite the wheel to be removed.
2	Loosen wheel nuts of wheel to be removed.	Injury to hands. Muscle strain.	Correct design of spanner that contacts all sides of the wheel nut, to reduce slipping. Gloves for grip and to protect. Long-handled spanner to enable suitable force.	Apply steady pressure, slowly.
3	Raise vehicle with jacking device	Difficulty in access under vehicle. Jacking device could slip. Vehicle could over-balance.	Suitable clothing or sheet to enable you to get down to level to fit jacking device. Locate jacking device in jacking point. Raise vehicle only high enough so wheel to be removed clears the ground and can be removed. If tyre of wheel being removed is flat raise sufficiently high to enable spare wheel to be fitted.	Use hand light to help see position of jacking point.
4	Continued...	Continued...	Continued...	Continued...

Figure 3-51: Example of part of a task analysis. Source: RMS.

Hazard identification and risk assessment

The identification of hazards and the assessment of risks are important factors arising from the analysis stage of developing safe systems of work. The analysis will provide data on the steps that comprise the work and will also usually provide information on the hazards related to the work and current control measures and indicate the need for further controls. This naturally leads into the stage in system of work development that considers these hazards and assesses risks related to them. Where a significant risk is identified, this may require further, more detailed analysis of the work activity, involving job/task analysis if this has not already been done. This in turn will lead to a more detailed risk assessment.

This is an important reflective stage of the development of the safe system of work. The analysis stage may only confirm what hazards exist and the current controls, it may not strongly challenge whether the risk remaining is acceptable. This stage reflects on the current situation and proposed additional controls determined in the analysis stage and considers whether the controls relate appropriately to the risk. For example, exposure to a hazardous substance may present a very high risk of death. This may have been identified in the analysis stage and the current control of workers using a nuisance dust mask determined as acceptable as it follows current practice. This risk assessment stage has to consider the scale of the risk and determine if the current or proposed control measures are adequate to manage the risk to an acceptable level. In the example given, this would mean challenging the reliance on a simple nuisance dust mask for such a high risk. From the reflection and decision making conducted at this risk assessment stage, appropriate control measures can be introduced into the system of work and workers can be trained how to minimise risks through use of the system of work.

For further information on hazard identification and risk assessment see '3.4 - Assessing risk' earlier in this element.

Introducing controls and formulating procedures

Define the safety and health methods

The first phase in introducing controls and formulating procedures is to define the safe and healthy methods. The earlier stages in the development of a safe system of work required the analysis of the job/task, identification of hazards and assessment of risk. These stages will have considered the control measures that are in place or are required to deal with hazards related to the work. This will enable a safe and healthy method of working to be developed and defined that incorporates these controls. In the example given in *Figure 3-51*, one of the controls would be to park the vehicle where there is minimal or

no traffic. A further control is to ensure the parking brake has been applied.

The method can include setting out the correct order of the steps in the work, if this is applicable. This safe and healthy method should be defined to establish it as a formal decision by the organisation on the way work must be done. In situations where the risk related to the work is high, this must be done in writing, often in a procedure. In situations where risks are low, this defined method may not be written down but agreed between the people involved in the work.

The following factors should be considered when deciding and defining safe and healthy methods:

- It should take account of the analysis, identification of hazards and risk assessment conducted.
- The method should adequately control hazards associated with the work. Where possible, hazards should be eliminated at source. The need for protective or special equipment should be identified, as should the need for the provision of temporary protection, guards or barriers.
- It should comply with the organisation's standards.
- It should comply with relevant legal or international best-practice standards.
- It should take account of manufacturer and supplier instructions.
- It should be as simple as possible and defined in a way that is understandable to workers.
- The defined method should identify specific responsibilities at various stages in the work, and the person in control of work should be clearly identified.
- Consideration should be given to the definition of who does what, where, when and how.
- The method should take account of likely emergencies, for example, fire or spillage. If there is a possibility that harm could result during the work, needs for emergency and rescue methods should be identified.
- It should be as simple as possible and defined in a way that is understandable to workers and in multiple languages where necessary.

Where the defined safe and healthy method is complicated or is a significant change from current practice, it may be tested using a simulated work situation and/or introduced with a pilot group of workers to validate that it is effective. This would take place before implementation and include the introduction and use of the control measures defined in the safe and healthy method of working.

Implementation

Once the system of work has been developed and agreed, preparation for implementation can proceed. Implementation is far harder than developing the safe system of work because the workers need to adopt the new working practice and use the control measures that have been developed. It is important that care be taken to ensure that controls referred to when the system of work was being developed and defined are available to the workers when the system of work is implemented. If they are not, this will devalue the system of work and cause the workers to have to work outside the defined method, possibly in an unsafe way. In the example given in *Figure 3-51*, if the correct spanner is not available then workers may use an oversize spanner that might slip when being tightened. The process of implementation will involve:

- The formal acceptance of the defined safe and healthy method, by the simple agreement to work in the proposed way by the people affected by it or by formal acceptance of the agreed written procedure by managers and workers. This will involve agreement on the date and, in some cases, the time the system of work becomes effective.
- Instruction and training in the operation of the system.
- A period of evaluation of the effectiveness of the system.
- If problems arise that necessitate modification to the system, formal approval should be gained, documentation of the modification should be made and the modification communicated to those involved with the system of work.

Instruction and training in the operation of the system

Provision for the communication of relevant information to all involved or affected by the system of work is an essential element of the development of a system of work. Many organisations may be tempted to issue procedures relating to safe systems of work without instruction and training. This can lead to a great deal of misunderstanding about what the system of work is and will often lead to people not being motivated to work to the system. This is particularly the case when the system is a change from the usual way that workers may have worked. Workers, and supervisors, can have a naturally high resistance to change.

Simply providing information will not usually be sufficient to change behaviour. A more assured way to introduce the system would be to utilise such things as 'tool box' talks, where team leaders brief their work group at the start of shift regarding any relevant day-to-day issues or the introduction of new technology or equipment. Tool box talks provide a practical opportunity to develop the workers' understanding of the operation of the system, and if these are done as part of a cascade training method the supervisor or similar person will have

received training from someone committed to the new system and will be better placed to get over the resistance of workers to change.

Article 24

The employer shall take necessary measures for preventing industrial accidents arising from the work actions or behaviour of workers.

Figure 3-52: Duty to ensure safety of work actions.
Source: Japan, Industrial Safety and Health Act 1972.

In addition to worker training, it is important to ensure that training of those that manage workers takes place. For example, supervisors should be adequately trained to enable them to identify the hazards and control strategies associated with work equipment under their control. They need to have sufficient training to identify that the defined method of working is being followed, when workers deviate from it and the consequences. Job/task analysis is not only useful to identify the hazards and controls, but is essential for preparing written procedures and specifying the skill and knowledge content of the work to be carried out. The analysis will not only identify the sequence of work but often the work rate expected - sometimes timeliness is an important consideration, particularly in relation to certain chemical manufacturing processes.

The analysis can then be incorporated into job training programmes. For certain high-risk tasks, this will often involve a training course to develop knowledge and understanding. This is then followed by practical application, either utilising a work simulator (for example, train driver / aircraft pilot) or close one-to-one supervision while on the job (for example, fork-lift truck driver). Where training requirements have been identified, a record of those affected should be made. The required training should be conducted, confirmed as successful and recorded.

MONITORING THE SYSTEM

All safe systems of work should be formally monitored and records kept of compliance and effectiveness. This can be done by direct observation or by discussion at team meetings or safety committee meetings. The safe system of work should not simply be imposed upon the people responsible for their operation. A system of monitoring and feedback should be implemented to ensure it is effective. Audits and investigation of accidents/incidents and ill-health that may occur can provide a valuable insight into whether systems of work are effective. Any permanent record of monitoring of the system of work must be kept and regularly reviewed by a member of the management team responsible for the system of work in order to identify the need for improvements or modification. If improvements or modification to the system are required formal approval should be gained, documentation of the modification should be made and the modification communicated to those involved with the system of work. If the change to the system of work is significant, workers and managers may require retraining.

3.7 Permit-to-work systems

MEANING OF A PERMIT-TO-WORK SYSTEM

A permit-to-work system is a formal written (administrative) system used to control certain types of jobs that have high hazard potential.

A permit-to-work system (PTW) should not be applied to every task (except in certain permanent high-risk work, for example, processing highly flammable liquids) as this may detract from its importance and lead to it being treated as nothing more than a bureaucratic, form-completion exercise.

The term 'permit-to-work' refers to the paper or electronic 'permit' document that is used as part of the overall system of work which has been devised to identify and control the risks involved in the work.

WHY PERMIT-TO-WORK SYSTEMS ARE USED

A permit-to-work system is an integral part of a safe system of work and can help to properly manage work activities. The purpose of a permit-to-work system is to ensure that full and proper consideration is given to the risks of the particular work that it relates to and to control the way of working in order to ensure a safe system of work is followed.

A permit-to-work system will:

- Ensure the proper authorisation of specified work.
- Confirm the identity, nature, timing, extent and limitations of the work.
- Establish criteria to be considered when identifying hazards and what they are.
- Confirm that hazards have been removed, where possible.
- Confirm that control measures are in place to deal with residual hazards.
- Confirm work is started, suspended, conducted, and finished safely.
- Control and confirm who has control of the location and equipment relating to the work when it passes between parties.
- Control change and consider other work activities that might interact with specified work.
- Provide a record of the steps in the process.

HOW PERMIT-TO-WORK SYSTEMS WORK AND ARE USED

Requirements of the system

The requirements of an effective permit-to-work system are that it must:

- Be formal and documented.
- Simple to operate.
- Have the commitment of those who operate and are affected by it.
- Provide concise and accurate information.
- Ensure liaison with controllers of other plant or work areas whose activities may be affected by the permit-to-work.
- Require that boundary or limits of work area are clearly marked or otherwise defined.
- Include contractors undertaking specific tasks in the permit-to-work system, including any briefing prior to commencement.
- Be supported by training in the system for all working under it and affected by it.

Permit-to-work document

The important features of a permit-to-work document are:

- A description of the task to be performed.
- An indication of the duration of the validity of the permit.
- Details and signature of the person authorising the work.
- Identification of the hazards affecting the work and the precautions to be taken.
- The isolations that have been made and the additional precautions required.
- A clear record that:
  - All foreseeable hazards have been considered.
  - All precautions are defined and taken in the correct sequence.
- Acceptor assumes responsibility for safe conduct of work.
- Establishes a safe system of work.
- An acknowledgement of acceptance by the workers carrying out the task, who would then need to indicate on the permit that the work had been completed and the area made safe in order for the permit to be cancelled.
- Overrides any other instructions until cancelled.
- Excludes work not described within the permit.
- In the event of a change to the programme of work, it must be amended or cancelled and a new per-mit issued.
- Only the originator may amend or cancel.

Hot work permit - applies only to area specified below

Part 1

Site...................................... Floor......................................

Nature of the job (including exact location)

....

....

....

The above location has been examined and the precautions listed on the reverse side have been taken.

Date..............................

Time of issue:........................

Time of expiry:

....

NB. This permit is only valid on the day of the issue.

Signature of person issuing permit:

....

Part 2

Signature of person receiving permit:

....

Time work started:

....

Time work finished and cleared up:

....

Part 3 | Final check up

Work areas and all adjacent areas to which sparks and heat might spread (such as floors above and below and opposite side of walls) were inspected one hour after the work finished and were found fire safe.

Signature of person carrying out final check:

....

After signing return permit to person who issued it,

Figure 3-53: Hot work permit - front of form.
Source: Lincsafe.

WHEN TO USE A PERMIT-TO-WORK SYSTEM

A permit-to-work system is a formal health and safety control system designed to prevent accidental injury/illness of workers, damage to plant, premises and product, particularly when work with a foreseeable high hazard content is undertaken and the precautions required are numerous and complex.

Good practice suggests that permit-to-work systems are normally considered most appropriate to:

- Non-production work (for example, maintenance,

Hot work permit - precautions

Hot Work Area

- Loose combustible material cleared.

- Non-moveable combustible material covered.

- Suitable extinguishers to hand.

- Gas cylinders fitted with a regulator and flashback arrester.

- Other personnel who may be affected by the work removed from the area.

Work on walls, ceilings or partitions

- Opposite side checked and combustibles moved away.

Welding, cutting or grinding work

- Work area screened to contain sparks.

Bitumen boilers, lead heaters, etc.

- Gas cylinders at least 3m from burner.

- If sited on roof, heat-insulating base provided.

Figure 3-54: Hot work permit - reverse of form.
Source: Lincsafe.

repair, inspection, testing, alteration, construction, dismantling, adaptation, modification, cleaning).
- Non-routine operations.
- Jobs where two or more individuals or groups need to coordinate activities to complete the job safe-ly.
- Jobs where there is a transfer of work and responsi-bilities from one group to another.

Hot work

This typically involves welding operations, such as pipework where the risk of sparks may ignite nearby flammable materials. Elimination or protection of such combustible items will need to be considered. The provision of firefighting equipment close at hand and trained personnel to deal with inadvertent ignition is also important.

It is common for national legislation relating to dangerous substances and explosive atmospheres to specify the application of permits to work for hazardous places or work involving hazardous activities that could lead to ignition/explosion, for example, hot work.

Work on non-live electrical systems

The high risks associated with work on or near electrical systems often justifies the use of a permit-to-work system. Work on electrical equipment, such as a trans-former, requires safe isolation, access and egress, work at a height and heavy lifting to be considered. The permit will help to ensure that adequate and effective isolation is in place, the electrical system is confirmed to be 'dead' and that all precautions are in place, including the competency of the workers involved.

Machinery/plant maintenance

Machinery/plant maintenance sometimes requires workers with different disciplines to work on large complex plant at the same time. This is aggravated by the fact that the plant may be spread over a number of floors in a building, such as power-generation plant, flour mills or lift systems in an office block.

This sort of work can involve many risks - related to a variety of services and energy sources, dangerous parts of the equipment, problems with access, risk of falling or being trapped inside the plant. These risks are best controlled by a well-established system of work, supported by a permit-to-work. The permit can ensure that the work is carefully planned and that power sources (electric or hydraulic) are appropriately isolated and locked off. The permit can also take into account the release of any stored energy, for example, electrical, hydraulic or kinetic energy.

Figure 3-55: Confined space - chamber. Source: RMS.

Confined spaces

A confined space is a place which is substantially enclosed (though not always entirely), and where serious injury can occur from hazardous substances or conditions within the space or nearby (for example, lack of oxygen). Failure to appreciate the dangers associated with confined spaces has led not only to the deaths of many workers but also to the demise of some of those who have attempted to rescue them.

Figure 3-56 (below): Example of a permit-to-work for entry in to confined spaces. Source: UK, HSE Guidance note on permit-to-work.

Permit-to-work Certificate				
PLANT DETAILS (Location, identifying number, etc.)		ACCEPTANCE OF CERTIFICATE Accepts all conditions of certificate	Signed Date Time	
WORK TO BE DONE				
WITHDRAWAL FROM SERVICE	Signed Date Time	COMPLETION OF WORK All work completed equipment returned for use.	Signed Date Time	
ISOLATION Dangerous fumes Electrical supply Sources of heat	Signed Date Time			
CLEANING AND PURGING of all dangerous materials	Signed Date Time	EXTENSION	Signed Date Time	
TESTING for contamination	Contaminations tested Results Signed Date Time			
I certify that I have personally examined the plant detailed above and satisfied myself that the above particulars are correct. (1) The plant is safe for entry without breathing apparatus (2) Breathing apparatus must be worn Other precautions necessary: Time of expiry certificate: Delete (1) or (2) Signed Date Time		This permit-to-work is now cancelled. A new permit will be required if work is to continue Signed Date Time		
		RETURN TO SERVICE	I accept the above plant back into service Signed Date Time	

A confined space may be any place, such as a chamber, tank, vat, silo, pit, well, pipe, sewer, flue, or similar, in which, by virtue of the enclosed nature, there is a foreseeable risk. Work in a confined space may present a risk of:

- Injury arising from fire or explosion.
- Loss of consciousness from increased body temperature.
- Loss of consciousness from asphyxiation from gas, fume, vapour or lack of oxygen.
- Drowning of any person from an increase in the level of liquid.
- Asphyxiation of a person as a result of being trapped by a free-flowing solid.

Work at height

Work at height (which may mean working near an excavation or cellar where a fall below ground level can cause serious injury) has been identified as presenting a significant risk to workers who may fall. The permit system used will ensure that avoiding work at height is considered and, if this does not prove possible, that fall-prevention strategies (for example, provision of a safe place) are used. If a safe place cannot be provided, then the permit must be used to consider minimisation of the distance of the fall by the use of safety devices (lanyards, nets, etc.). The permit is also likely to consider the influence of weather conditions on the safety of the task being undertaken, for example, high winds or ice.

Working at height includes:

- Ladders.
- Stepladders.
- MEWP (mobile elevated working platforms).
- Roof working.

CONSIDER

A maintenance worker is required to enter a chemical process tank containing a residual amount of a substance that gives off significant quantities of an inert gas that would replace oxygen in the tank when disturbed. The inert gas is heavier than air. The maintenance worker will have to disturb the substance in order to remove it. As much of the substance as possible has been drained away.

Entry is via an opening at the top of the tank. The worker will have to gain access by a ladder and will need to take temporary lighting. The maintenance worker has been provided with a respirator mask for use while doing the cleaning. The cleaning task is defined as a one-person activity on the job card.

Using the risk rating matrix, determine:	
The severity of risk related to this work	
Likelihood of harm	
Overall risk rating	

3.8 Emergency procedures

WHY EMERGENCY PROCEDURES NEED TO BE DEVELOPED

It is important to develop and implement emergency procedures for potentially major loss-causing events in order to bring the event under control promptly, reduce the effects of the event (on premises, equipment, materials, environment and people that might be affected) and enable a fast return to normal operations. In the absence of emergency procedures, there may not be a timely or suitable response to the emergency, allowing it to get out of control and cause more serious effects than if procedures were in place. Depending on the emergency, this could result in major loss of life, long term ill-health, environmental effects, significant damage to buildings and equipment and long delays in the organisation returning to normal operations. Some organisations or locations where emergencies take place that are not effectively controlled never recover from the effects.

Adequate emergency procedures should be in place, or developed, to control likely emergencies, for example, a fire, terrorist threat, spillage or release of hazardous substances, explosion, vehicle collision, severe weather, major injury, poisoning or exposure to pathogens, or release of radioactivity.

Requirements for an emergency procedure

Article 18 of the ILO Occupational Safety and Health Convention C155 requires employers to provide, where necessary, measures to deal with emergencies and accidents/incidents, including adequate first-aid arrangements.

In addition, the ILO's 'Guidelines on Occupational Safety and Health Management Systems - ILO-OSH 2001' include a specific clause requiring organisations to establish and maintain emergency prevention, preparedness and response arrangements as part of their management system. ISO 45001:2018 'Occupational health and safety management systems' also has a specific clause, Clause 8.2, which sets out requirements.

"8.2 Emergency preparedness and response

The organization shall establish, implement and maintain a process(es) needed to prepare for and respond to potential emergency situations, as identified in 6.1.2.1, including:

a) establishing a planned response to emergency situations, including the provision of first aid;

b) providing training for the planned response;

c) periodically testing and exercising the planned response capability;

d) evaluating performance and, as necessary, revising the planned response, including after testing and, in particular, after the occurrence of emergency situations;

e) communicating and providing relevant information to all workers on their duties and responsibilities;

f) communicating relevant information to contractors, visitors, emergency response services, government authorities and, as appropriate, the local community;

g) taking into account the needs and capabilities of all relevant interested parties and ensuring their involvement, as appropriate, in the development of the planned response.

The organization shall maintain and retain documented information on the process(es) and on the plans for responding to potential emergency situations."

Therefore, in some countries developing emergency procedures is a National legal requirement and emergency procedures are a common part of health and safety management systems.

WHAT TO INCLUDE IN EMERGENCY PROCEDURES

The employer should identify and assess the nature of any injury likely to be caused by the emergency and consider the distance to emergency hospital facilities to decide if it is near enough to provide medical services. It may be necessary to provide a first-aid room locally and to appoint specific people who are competent in emergency medical techniques, for example, people competent in resuscitation techniques.

Where organisations are dependent on the assistance of local emergency services, they should formalise the arrangement, for example, if it is necessary for the local emergency services to have available special anti-toxins or isolation facilities. Similarly, if the organisation's activities could lead to a major fire it is important to formalise arrangements with the local fire service to ensure an adequate and timely response from them.

This will usually mean agreeing the amount of support the emergency and rescue services can provide and what must be provided by the employer.

Typical scope of emergency procedures

Emergency procedures should cover:

- What might happen and how the alarm will be raised.
- How emergencies are communicated within the organisation, i.e. night and shift working, weekends, presence of temporary workers, contractors, members of the public and times when the premises are closed, for example, holidays.
- The planned responses to the emergency - for example, obtaining assistance from external emergency and rescue services and the first action of internal emergency response teams to limit the effects of the emergency.
- The people who will be involved in implementing the procedure and the training that they will need. This will mean identifying, where necessary, first-aiders, fire marshals, confined-space rescue teams, etc.
- Any equipment needed to deal with the emergency, for example, spill kits, firefighting equipment or rescue equipment.
- Nomination of competent people to take control (a competent person is someone with the necessary skills, knowledge and experience to manage health and safety).
- What other key people are needed, such as a nominated incident controller, someone who is able to provide technical and other site-specific information if necessary, or first-aiders.
- Plan of essential actions, such as emergency plant shutdown, isolation or making processes safe. Clearly identify important items such as shut-off valves and electrical isolators, fire dry and wet risers and hydrant points.
- Evacuation of people that might be affected - including methods to evacuate disabled and other vulnerable people to agreed safe assembly points. Consider distance to reach a place of safety or to get rescue equipment, and the need for suitable forms of emergency lighting.
- Ensuring that there are enough emergency exits for everyone to escape quickly, and that emergency doors and escape routes are kept unobstructed and clearly marked.
- Actions to reduce the effects of the emergency - for example, the provision of first-aid, prevention of environmental contamination and shutting down plant/equipment that could make the emergency worse.
- Actions to clean up after the emergency and return to normal operations - for example, checking the atmosphere of the area affected by the emergency to see if people can be allowed to return to work.

Arrangements for contacting emergency services

The employer should involve the emergency and rescue services as part of the process of establishing emergency arrangements for contacting them. This will allow the employer to discuss potential emergencies with the emergency and rescue services and agree arrangements to contact them at the time of an emergency, for example, in the event of a fire. Depending on the emergency arrangements, this could mean contacting internal emergency and rescue services initially and when the internal emergency team decide external services will be required, promptly contacting them. In some cases the emergency arrangements require internal and external emergency and rescue services to be contacted at the same time.

This early contact of external emergency services assists them in providing an effective response to locations that may not be near them or where time-delays in responding may be caused by traffic.

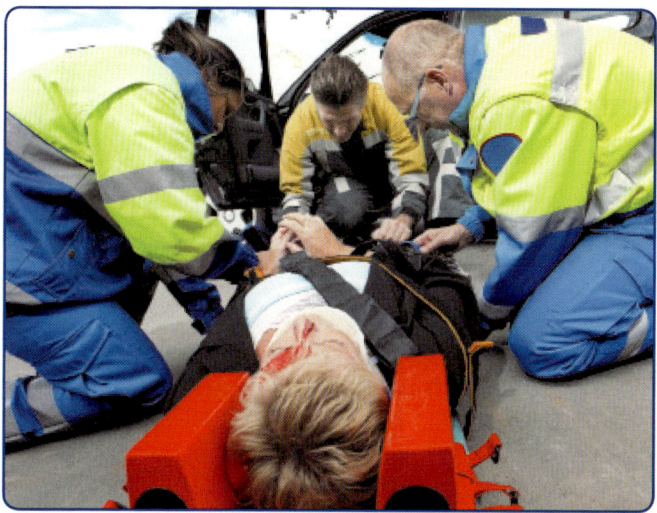

Figure 3-57: Emergency services.
Source: Shutterstock.

In many situations, external emergency and rescue services prefer to be contacted immediately when the emergency is identified and to respond to emergencies as a precaution, even if they are not subsequently required due to the successful efforts of the employer in controlling the emergency.

Consideration should be given to identifying those people who have a specific role and responsibility for contacting the external emergency services and who will be on site to receive the emergency services when they arrive so that they can be informed of the nature of the emergency and directed to its precise location.

To guide managers and workers in the correct response, emergency instructions may be posted throughout the organisation, for example, it is common to post instructions for responding to a fire. Contact telephone numbers should be well known and posted in a sufficient number of locations to ensure they are available when needed. The means of contacting the emergency and rescue services should be established.

Often this will be by using a standard telephone system, however arrangements may need to be made for circumstances when there is a failure of the telephone system at the time of the emergency. This might include contact by radio, mobile/cell telephone or on a special protected telephone line. Similarly, if workers are working away from the main workplace, consideration would have to be given to the availability of a mobile/cell telephone signal and contact may have to be made by other means, for example, a satellite-phone or a special telephone line provided for emergencies.

Figure 3-58: Emergency telephone symbol.
Source: www.sksigns.co.uk

In some cases the emergency and rescue services may be contacted by a remote monitoring organisation that calls the services in response to an automatic alarm from the organisation affected. It has been found that using remote monitoring organisations can greatly shorten response times in situations where the emergency takes place outside normal working hours.

As a precaution, it may be necessary to contact the external emergency and rescue services before complex hazardous work is to be carried out to alert them that they may be needed. This could include work in a confined space or other work where there is a significant risk that people may need to be rescued. This precautionary contact would enable the emergency and rescue services to be ready for the type of emergency that may relate to this work and in some cases they may position a suitable team nearer to where the work is taking place.

FIRST AID REQUIREMENTS

The purpose of first-aid is to:

- Preserve life.
- Prevent the condition requiring first-aid getting worse/minimise its consequences until medical help arrives.
- Promote recovery of the person requiring first-aid.
- Provide treatment where medical attention is not required.

To ensure adequate first-aid provision, an employer should make an assessment to determine the first-aid needs of the organisation. This will determine the organisation's requirements for first-aiders (quantity and competence) and first-aid equipment/facilities (quantity and type). When making the assessment, consideration should be given to a number of factors, including:

- The size of the organisation - in particular, the number and distribution of workers that need to be covered by the first-aid provision.

- The type of workforce - including the special needs of trainees, younger and older workers, those with known medical conditions and the disabled.

- The need to provide first-aid at all times when workers are on site - considering shift patterns and shift change-over periods, any work done outside normal working hours and cover for first-aiders due to their sickness or other absences.

- Different work activities - first-aid needs will depend on the type of work being done. Some activities, such as office work, have relatively few hazards and low levels of risk and will need less first-aid provisions than other activities that have more hazards or more specific hazards (construction or chemical sites).

- Past experience - previous accidents and ill-health should be considered, including their type, location and the type of harm caused.

- Ease of access to medical treatment - where external emergency medical services are not easily available, more first-aid provision may be required, for example, a first-aid room.

- Workers working away from the employer's premises - for some work activities first-aid arrangements may have to be made for workers working in locations away from the employer's fixed work places. For example, mobile teams of workers, particularly those working in remote locations.

- Workers of more than one employer working together - first-aid facilities may be shared, for example, when a number of contractors work together on a construction site they could share first-aid facilities or when individual workers go to an employer's site to do maintenance work they could use the employer's first-aid facilities.

- Provisions for people who are not the employer's workers - employers do not usually make first-aid provisions for any person other than their own workers, though this may be required by national legislation. Where organisations provide services to members of the public, for example, a theatre or shop, it is recommended that they are included in the first-aid arrangements that are provided.

- National legal requirements - the assessment should consider legal requirements and ensure minimum requirements are met or exceeded.

The first-aid assessment should be used to guide the employer in providing first-aid arrangements. A sufficient number of competent first-aiders should be readily available. Sometimes first-aiders will need specific competencies related to the risks of the work activities, for example, competencies in the use of a defibrillator or the treatment of chemical burns. There should be a sufficient number and size of first-aid boxes/containers to provide ready access to facilities.

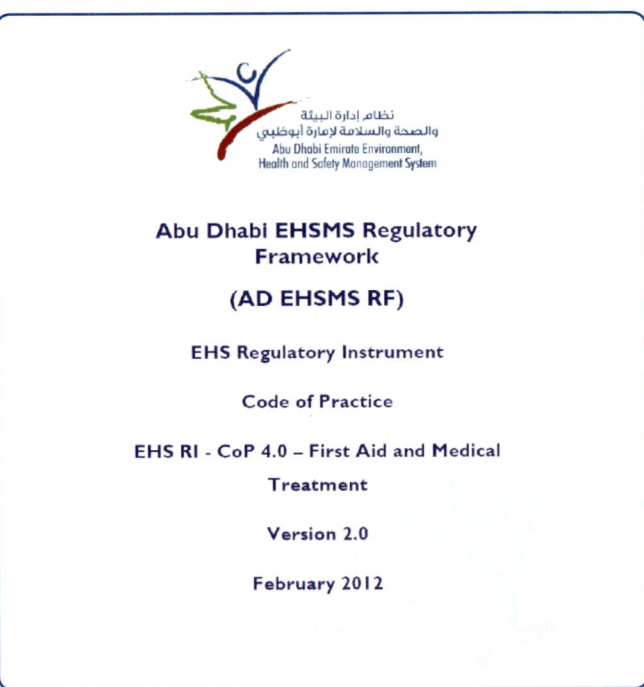

Figure 3-59: National first-aid requirements.
Source: Abu Dhabi.

Additional equipment and facilities may also be appropriate, for example, a stretcher, defibrillator or first-aid room. When first-aid arrangements have been established, the employer should inform workers of where first-aid equipment is and who is responsible for performing first-aid duties.

©RMS

The employer should ensure that first-aid treatment remains available throughout the time the employer's work activities take place. Additional first-aiders may be necessary to cover out of hours, shift or overtime working. In some organisations, security workers are available throughout the day and night, including at weekends, and it may be possible to train and appoint them as first-aiders to ensure first-aid can be provided when work is taking place out of normal working hours. When high hazard work is being done outside normal working hours it may be necessary to arrange for specific first-aiders to be available near to the high hazard work activity, rather than relying on general first-aid provision.

Where workers work away from their employer's premises or if the geographical area in which the work takes place is large, all workers may need to be trained and provided with first-aid equipment. For example, this may be appropriate for gas, oil, electrical or telecommunication workers, particularly where they are working in remote locations. A full first-aid box/container may be provided in vehicles and/or each individual could be provided with a personal first-aid pouch. A first-aid pouch is a first-aid container that has fewer contents than a full first-aid box/container and is usually provided for individuals to administer first-aid to themselves or for small teams to assist each other in case of minor injuries.

Where a site is shared by a number of employers it may be useful for them to make an agreement so that each employer does not have to make separate first-aid arrangements and to enable cover for the absence of each other's first-aiders. For example, if main contractors on construction projects make their first-aid facilities available to sub-contractors this would avoid the need for them to provide their own first-aid facilities. Similarly, where a contractor's workers work in another employer's premises, for example, doing maintenance on the employer's equipment, the contractor could make an agreement to access the employer's first-aid facilities. It is important that these agreements are made formally, preferably as a written agreement, as this will enable those making the agreement to be confident about the arrangements. The employer should inform workers of the arrangements made to provide first-aid.

REVIEW

What are the four main purposes of first-aid?

WHY PEOPLE NEED TRAINING AND EMERGENCY PROCEDURES NEED TESTING

People need to be training related to emergencies and emergency procedures so that in the event of an emergency the people who:

- May be in danger act in a calm, orderly and efficient way.
- Are designated with specific duties carry them out in an organised and effective manner.

Training will provide people with essential knowledge so they are better able to identify promptly when an emergency is taking place and take action to raise the alarm at an early time, enabling an opportunity to respond to the emergency and limit the effects. Training will provide those people who need to respond to the alarm with an understanding of what the alarm means to them and the action they must take. It also increases the chance that they will take the alarm seriously and take action without delay.

Those people designated with specific duties would need specific training in what these duties are and how to carry them out. In most cases this will require practical training, for example as a fire marshal checking an area or a person using equipment to minimise the effects of the emergency.

Training should include practice drills to reinforce the knowledge and establish the actions actions as a practiced routine, making this familiar and more likely to be repeated without hesitation or doubt.

Emergency procedures should be in writing and regularly tested through drills and exercises. Quick and effective action may help to ease an emergency situation and reduce the consequences. In an emergency people are more likely to respond reliably if they:

- Have clearly agreed responsibilities.
- Are well trained and competent.
- Take part in regular and realistic practice to rehearse actions.

The results of the drills and exercises should be recorded and the procedures amended as necessary to make them effective.

The procedures should also be subject to regular review to determine if any new factors are affecting them. For example, fire emergency procedures may need regular review throughout a major construction project as construction work activities can create changes that could affect exit routes and assembly points, requiring them to be redefined and those individuals affected by the changes to be re-trained.

For further information on arrangements for fire emergencies see *'Element 10.4 - Fire evacuation'*.

CONSIDER

What types of emergency could affect your organisation?

Who would need to be contacted if these emergencies occurred?

Sources of reference

Reference information provided, in particular web links, was correct at time of publication, but may have changed.

Ambient factors in the workplace, International Labour Organisation (ILO) Code of Practice (CoP), ISBN 92-2-11628-, http://www.ilo.org/safework/info/standards-and-instruments/WCMS_107729/lang--en/index.htm

Benefits of a Participatory Safety and Hazard Management Program in a British Colombia Forestry and Logging Organisation, 1986, Painter and Smith, http://www.moderntimesworkplace.com/good_reading/GRHealth/Safety_all.pdf

Code of Practice for Ambient Factors in the Workplace, 2001, ILO, http://www.ilo.org/safework/info/standards-and-instruments/codes/WCMS_107729/lang--en/index.htm

Common topic 4: Safety culture, HSE, http://www.hse.gov.uk/humanfactors/topics/common4.pdf

Emergency procedures, HSE website guidance, http://www.hse.gov.uk/toolbox/managing/emergency.htm

Ergonomics and human factors at work, INDG90, HSE Books, http://www.hse.gov.uk/pubns/indg90.pdf

European Framework Directive on Safety and Health at Work, European Union, Directive 89/391 EEC, https://osha.europa.eu/en/safety-and-health-legislation/european-directives, (Please select from list)

Example risk assessments, HSE, http://www.hse.gov.uk/risk/casestudies/index.htm

First aid in work, HSE, https://www.hse.gov.uk/simple-health-safety/firstaid/index.htm

Five Steps to Risk Assessment (INDG163), HSE Books, ISBN: 978-0-7176-6440-5, http://www.hse.gov.uk/pubns/indg163.pdf

Fundamental Principles of Occupational Health and Safety, ILO, ISBN 92-2-120454-5, http://www.ilo.org/global/publications/books/WCMS_093550/lang--en/index.htm

Global Database on Occupational Safety and Health Legislation (LEGOSH), http://www.ilo.org/dyn/legosh/en/f?p=14100:1:1900763574781::NO

Guidance on permit-to-work systems. A guide for the petroleum, chemical and allied industries, HSE Books, ISBN: 978-0-7176-2943-5, http://www.hse.gov.uk/pubns/books/hsg250.htm

Guidelines on Occupational Safety and Health Management Systems (ILO-OSH 2001), ILO, ISBN 92-2-111634-4, can be downloaded free from the ILO website - http://www.ilo.org/safework/info/standards-and-instruments/WCMS_107727/lang--en/index.htm

The health and safety toolbox, How to control risks at work, HSG268, HSE Books, ISBN: 978-0-7176-6587-7, http://www.hse.gov.uk/pUbns/priced/hsg268.pdf

Health and Safety at Work etc Act 1974, UK, www.legislation.gov.uk (please enter the name of the Act)

Human factors and ergonomics, HSE, http://www.hse.gov.uk/humanfactors/

Human factors: Organisational change, HSE, http://www.hse.gov.uk/humanfactors/topics/orgchange.htm

Human factors: Permit to work systems, HSE, http://www.hse.gov.uk/humanfactors/topics/ptw.htm

ILO OSH 2001: Guidelines on occupational safety and health management systems, ILO, http://www.ilo.org/global/publications/ilo-bookstore/order-online/books/WCMS_PUBL_9221116344_EN/lang--en/index.htm

Industrial Accident Prevention, Heinrich et al, 1980, 5th Edition, McGraw-Hill, New York.

Industrial Safety and Health Act, 1972, Japan, https://www.ilo.org/dyn/natlex/natlex4.detail?p_lang=en&p_isn=27779

International Labour Organisation (ILO) www.ilo.org

Investigating Incidents and Accidents at Work, HSG245, HSE Books, ISBN: 978-0-7176-2827-8, http://www.hse.gov.uk/pubns/books/hsg245.htm

Involving your workforce in health and safety: Good practice for all workplaces (HSG263), UK, HSE Books, ISBN: 978-0-717-66227-2, http://www.hse.gov.uk/pubns/books/hsg263.htm

ISO 45001:2018 Occupational health and safety management systems – requirements with guidance for use, International Organization for Standardization

ISO 7010 'Graphical symbols - Safety colours and safety signs - Registered safety signs', International

Organisation for Standardisation (ISO), www.iso.org (please enter the number of the standard)

Managing for health and safety, HSG65, HSE Books, ISBN: 978-0-7176-6456-6, http://www.hse.gov.uk/pUbns/priced/hsg65.pdf

Management of Health and Safety at Work Regulations (MHSWR) 1999, UK, http://www.legislation.gov.uk/uksi/1999/3242/contents/made

NORMLEX, ILO, (ILO database of international labour standards, including ILO Conventions, and National laws) http://www.ilo.org/dyn/normlex/en/f?p=NORMLEXPUB:1:0::NO:::

Organisational culture, HSE, http://www.hse.gov.uk/humanfactors/topics/culture.htm

Occupational Health and Safety Act, R.S.O, 1990, Ontario Canada, https://www.ontario.ca/laws/statute/90o01?_ga=2.4040304.1546202488.1562759892-1571920508.1562759892

Occupational health and safety management systems – requirements with guidance for use: ISO 45001: 2018, ISBN: 978-0-580-86393-6

Occupational Safety and Health Act 1994, Malaysia, www.dosh.gov.my (please follow the links via legis-lation to Acts)

Occupational Safety and Health Act 1993, South Africa, https://www.ilo.org/dyn/natlex/natlex4.detail?p_lang=en&p_isn=34695

Occupational Safety and Health Convention C155, 1981, ILO, http://www.ilo.org/dyn/normlex/en/f?p=1000:12000:0::NO::: (please select C155 from the Technical list).

Occupational Safety and Health Management System, 1999, Japan, JICOSH, https://www.jniosh.johas.go.jp/icpro/jicosh-old/english/guideline/oshms.html

Occupational Safety and Health Recommendation R164, 1981, ILO, http://www.ilo.org/dyn/normlex/en/f?p=1000:12010:0::NO::: (please select R164 from the Technical list).

Organisational culture, HSE, http://www.hse.gov.uk/humanfactors/topics/culture.htm

Personal Protective Equipment at Work Regulations 1992, UK, http://www.legislation.gov.uk/uksi/1992/2966/contents/made

Reducing Error and Influencing Behaviour (HSG48), UK, HSE Books, ISBN: 978-0-717-62452-2, http://www.hse.gov.uk/pubns/books/hsg48.htm

Risk assessment, a brief guide to controlling risks in the workplace, INDG163, http://www.hse.gov.uk/pubns/indg163.pdf

Safety and Health in Construction Convention C167, 1988, ILO, http://www.ilo.org/dyn/normlex/en/f?p=NORMLEXPUB:12100:0::NO::P12100_INSTRUMENT_ID:312312

Safety and Health in Construction Recommendation R175, 1988, ILO, http://www.ilo.org/dyn/normlex/en/f?p=NORMLEXPUB:12100:0::NO::P12100_INSTRUMENT_ID:312513

Safety and health in construction, ILO Code of Practice, ILO Geneva, ISBN: 92-2-107104-9, http://www.ilo.org/wcmsp5/groups/public/---ed_protect/---protrav/---safework/documents/normativeinstrument/wcms_164653.pdf

Safety management; What it means to us, Lauriski and Guymon, 1989, Mining Engineering (Littleton, Col-orado); (USA), 41:10. Journal ID: ISSN 0026-5187, https://www.osti.gov/biblio/5338922

Safe Work in Confined Spaces (ACoP, Regulations and guidance), L101, HSE Books, safe-work/documents/instructionalmaterial/wcms_215344.pdf, ISBN: 978-0-7176-6233-3, http://www.hse.gov.uk/pubns/priced/l101.pdf

Specific topic 3: Organisational change and transition management, HSE, http://www.hse.gov.uk/humanfactors/topics/specific3.pdf

Technical Glossary, NEBOSH, 2018, Version 7, https://www.nebosh.org.uk/policies-and-procedures/glossary-of-nebosh-policy-terms/

Training package on workplace risk assessment and management for small and medium size enterprises, 2013, ILO Geneva, ISBN: 978-92-2-127065-2 (web pdf) http://www.ilo.org/wcmsp5/groups/public/---ed_protect/---protrav/---safework/documents/instructionalmaterial/wcms_215344.pdf

Workplace first-aid kits. Specification for the contents of workplace first-aid kits BS 8599-1:2011, UK BSI, ISBN 978-0-580-72884-6, http://shop.bsigroup.com/ (please enter the number of the standard)

Web links to these references are provided on the RMS Publishing website for ease of use - www.rmspublishing.co.uk

STUDY QUESTIONS

1) What are the additional arrangements employers need to make to meet their responsibility to protect visitors, neighbours or members of the public from their work activities? (8)

2) What is meant by 'health and safety culture of an organisation'? (8)

3) Explain why an understanding of individual factors is important in the workplace. (8)

4) (a) Explain the role of worker participation in health and safety (4)

   (b) What are the benefits of participation for the employer? (4)

5) What are the key stages that need to be followed in the risk assessment process? (8)

6) (a) What are two reasons why visitors to a workplace might be at a greater risk of injury than a worker? (2)

   (b) What are the measures to be taken to ensure the health and safety of visitors in the workplace? (6)

7) What should you consider when developing and implementing a safe system of work for general activities? (8)

8) What is the function of a permit-to-work system? (8)

9) Why is it important to develop emergency procedures for the workplace? (8)

10) What are eight items that should be included in a hot work permit-to-work? (8)

11) (a) What is the purpose of first-aid? (2)

    (b) Explain the role of first-aiders. (6)

For guidance on how to answer NEBOSH questions please refer to the 'study question answer guidance' section located at the back of this guide.

Element 4

Health and safety monitoring and measuring

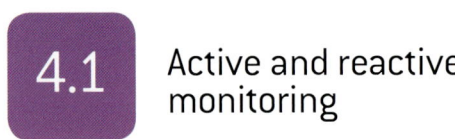

4.1 Active and reactive monitoring

THE DIFFERENCE BETWEEN ACTIVE AND REACTIVE MONITORING

Monitoring of health and safety performance

Organisations need to monitor their health and safety performance objectives to assess how effective they are, in the same way that they would measure finance, production, service or sales objectives. It is essential for organisations to learn from their experiences and take the opportunity to decide how to improve performance.

Monitoring provides the opportunity and information to enable management to:

- Assess the suitability and effectiveness of health and safety objectives and arrangements, including control measures.

- Make recommendations for review of the current management systems.

It is common practice in most countries that legal requirements mandate organisations to monitor and report on a range of accidents/incidents that may have occurred. A successful management system will include the requirement for the regular review of the results of monitoring to ensure the organisation achieves ongoing health and safety success. ILO Occupational Safety and Health recommendation R164, describes requirements for active and reactive monitoring:

"Recommendation 15

1) Employers should be required to verify the implementation of applicable standards on occupational safety and health regularly, for instance by environmental monitoring, and to undertake systematic safety audits from time to time.

2) Employers should be required to keep such records relevant to occupational safety and health and the working environment as are considered necessary by the competent authority or authorities; these might include records of all notifiable occupational accidents and injuries to health which arise in the course of or in connection with work, records of authorisation and exemptions under laws or regulations to supervision of the health of workers in the undertaking, and data concerning exposure to specified substances and agents."

Figure 4-1: Requirements for active and reactive monitoring.
Source: ILO Occupational Safety and Health recommendation R164.

Differences between active and reactive monitoring

An effective approach to the management of health and safety risks seeks to learn from all available monitoring sources. Therefore, there is a need for an organisation to use a range of both active and reactive monitoring to determine whether objectives have been met and are effective. The differences between these two categories:

1) Active monitoring, before the incident, involves identification through regular, planned observations of workplace conditions, systems and the actions of people, to ensure that performance standards are being implemented and management controls are working, for example, workplace and plant inspections.

2) Reactive monitoring, after the incident, involves learning from mistakes, whether they result in injuries, illness, property damage or near-misses, for example, accident/incident investigation.

Both monitoring methods require an understanding of the immediate and the underlying causes of accidents/incidents and ill health. Organisations need to ensure that information from both active and reactive monitoring is used to identify situations that create risks, and to do something about them. Priority should be given where risks are greatest.

 CONSIDER

Two key components of a monitoring system.

Active monitoring (checking standards before things go wrong). Are you achieving the objectives and standards you set yourself? Are they effective?

Reactive monitoring (after things go wrong). Investigating injuries, cases of illness, property damage and near-misses, identifying in each case why performance was substandard.

ACTIVE MONITORING METHODS

Objectives and usefulness of active monitoring

The primary objectives of active monitoring are to:

- Check that health and safety objectives, plans and arrangements have been implemented.
- Monitor the extent of compliance with the organisation's arrangements/procedures, and with its legislative/technical standards.

Active monitoring will tell the organisation about the reliability and effectiveness of its policy, objectives, arrangements and procedures, before their limitations are made obvious through accidents/incidents or ill health. This provides a good basis from which decisions and recommendations for maintenance and improvement may be made.

Active monitoring provides an opportunity for management to confirm commitment to health and safety objectives. It also reinforces a positive health and safety culture by recognising success and positive actions, instead of 'punishing' failure after an undesired event.

Health and safety performance in organisations that manage health and safety effectively is measured against established standards. This enables confirmation of compliance with standards and where improvement is required. By establishing standards of expectation, it enables deficiencies to be quickly translated into improvement actions. Organisations should make active monitoring an integral and normal part of the management function. As such it must take place at all levels and at all opportunities in the organisation's operation.

Managers should be given responsibility for the monitoring of objectives and compliance with performance standards for which they and their subordinates are responsible. The actual method of monitoring will depend on the situation and the position held by the person monitoring.

The success of action to manage risks is assessed through active monitoring involving a range of techniques. This includes techniques which examine technical measures (equipment, premises and substances), procedural measures (systems of work, method statements, safety cases, and permits-to-work), and behavioural measures (motivation, attitudes, and competencies).

Active monitoring methods

Inspections

There are two broad categories of inspections, those carried out to ensure the standard of workplace conditions are appropriate and maintained, for example, heating lighting, welfare, correct storage, and those which are usually specific to work equipment safety, for example, regular inspection of lifting devices, pressure systems and electrical equipment, that are in use in the workplace.

Workplace inspections

The role of health and safety inspections is to identify the health and safety status of what is being inspected and what improvements are needed. They are particularly well suited to identifying workplace hazards and inappropriate work activities to determine if they are under satisfactory control. Workplace inspections will improve the organisational health and safety culture, particularly

where workers' views are sought as part of the inspection. In this way, the employer's commitment to health and safety can be demonstrated, ownership of health and safety can be shared and worker morale can be increased by simple improvements being implemented at the time of the inspection. Typically, workplace inspections are carried out by a local supervisor, often accompanied by a worker health and safety representative, and involve walking through the area for which they are responsible. Inspection findings should be recorded and any corrective actions required should be carried out.

Work equipment inspections

In addition to general workplace inspections, inspections are used to confirm the safe condition of work equipment. National legislation usually sets out specific requirements for the inspection of work equipment to ensure that health and safety conditions are maintained and that any deterioration can be detected and remedied in good time. Work equipment inspections should be carried out in compliance with a schedule of work and frequency appropriate to the equipment. Inspections should be recorded and any corrective actions identified should be recorded together with the completion date for any identified repairs.

Limitations of inspections

Inspections may not identify an unsafe activity if the associated work was not taking place at the time of the inspection; some hazards are not obvious and may not be observed or identified by the inspector.

The failure of an inspector to identify a hazard or unsafe practice may be because of their lack of knowledge and competence for the work area under consideration.

Conversely the unsafe condition might have been observed, but was not followed through and actioned by a responsible person. There could be a situation where the responsibility for taking the corrective action was unclear, for example, if there had been a number of different employers on site.

The inspection may only be as good as the quality of the checklist. If a checklist is used that does not specifically cover the hazard observed it may not be recorded, the checklist may therefore, be regarded as inadequate for the risks being undertaken. It may also be a wilful act on behalf of the inspector to ignore the hazard or to not record the observation.

National legislation may also require the regular and timely inspection of such things as excavations, scaffolds and local exhaust ventilation systems.

Approach to inspections

When establishing an approach to inspections some of the factors to consider during the planning stage are:

- What needs inspecting?
- Who is to conduct the inspections and are they competent?
- When inspections should be conducted: changing circumstances or regular frequency?
- What standards are to be used?
- Is a checklist required?
- What equipment is to be used and does it need to be calibrated?
- Is any personal protective equipment (PPE) required?
- Where are findings recorded?
- Who will prepare the inspection report and develop the action plan?
- Who will be responsible for ensuring that any remedial action is carried out?

Frequency and type of inspection

There are different types of inspections for different purposes, they include:

- General workplace inspections - carried out by local first-line managers and worker health and safety representatives. An example of a routine inspection would be an inspection of an office location undertaken every three months where standards of housekeeping were monitored.

- Statutory thorough examination of equipment, for example, boilers, lifting equipment - carried out by specialist competent persons.

- Statutory inspections of equipment, for example, excavations and equipment used in work at height - carried out by a competent person. The ILO convention 152 Occupational Safety and Health (Dock Work) Convention 1979 on the thorough examination of lifting equipment falls into the category of statutory inspection.

- Preventive maintenance inspections of specific (critical) items - carried out by maintenance staff.

- Pre-use 'checks' of equipment, for example, vehicles, fork-lift trucks, access equipment - carried out by the user.

The frequency of inspections should be planned to take place at regular intervals. The time between most inspections is often at the employer's discretion.

Factors that will influence the timescales are the changing nature of the workplace, manufacturer's recommendations, type and frequency of use, environmental conditions, severity and previous history of failure, how well established the process is, findings from previous inspections, the presence of vulnerable workers, recommendation from enforcing agencies and workers voicing concerns. Principally this establishes a frequency on the basis of risk. Occasionally, maximum intervals are set by legislation such as that referring to local exhaust ventilation, lifting equipment, scaffolds and excavations.

General workplace inspections may be carried out in lower risk organisations that are not affected by change on a frequency so that every workplace is inspected once in a three-month period; this may suit an office environment. In organisations that have higher risks, such as manufacturing, this may be increased to once in a month. Where risks are higher and changes to the workplace are more likely, such as in maintenance or construction operations, this may be increased to weekly or daily.

Equipment inspections are usually made on a periodic basis and after changes are made to the equipment. National legislation may detail the period and circumstances for inspection of specific equipment.

Requirements are:

- Where safety depends on how it is installed or assembled in any position - before being used in that position.
- Where exposed to conditions causing deterioration which are liable to result in dangerous situations, to ensure that health and safety conditions are maintained and that any deterioration can be detected and remedied in good time - at suitable intervals and each time that exceptional circumstances which are liable to jeopardise the safety of the work equipment have occurred.
- Work platforms used for construction work in which a person could fall - inspected in position, or if a mobile work platform is inspected on the site, within the previous 7 days.

Competence and objectivity of inspector

Inspections normally involve physical examination of the workplace or equipment with a view to identifying hazards and determining if they are effectively controlled. They are usually carried out by a manager, worker health and safety representative, equipment user or technical specialist.

The inspector must be competent to inspect the specific area of work selected. The qualifications, knowledge, skill and experience may be set out in national legislation. This will usually include techniques of conducting the inspection, recognising conditions that are of a good standard, recognising conditions that are or will become substandard and the acceptable response to what they find. One of the most important competencies an inspector must have, is the ability to know their own limitations of competency, and what action they should take when they identify something that falls within their limitations. The term 'objectivity' in relation to inspections is usually associated with ideas such as reality, truth and reliability. The person conducting the inspection must be impartial in their approach and motivated to record actual events reliably and accurately.

Use of checklists

In many cases inspections are based on checklists. If this is the case, then managers relying on inspections must be mindful that the system of using a checklist has advantages and limitations.

An inspection checklist is typically a list of 'the way things should be'. When a work area or item of equipment fails this test, it is considered substandard and represents a hazard. Each substandard condition should be assessed and corrective action identified and details recorded.

Place of inspection: Date:			
Observation			Action required
	Yes	No	Action
Fire			
Extinguishers in place			
Clearly marked for type of fire			
Recently serviced			
Electrical			
Portable equipment in date tested			
Broken sockets, connectors			
General lighting			
Adequate illumination			
Emergency lighting operable			
Chemicals			
MDS for all chemicals			
Containers clearly labelled			
First aid			
Easy access to cabinets			
Cabinets correctly stocked			

Figure 4-2: Workplace Inspection checklist.
Source: RMS.

Checklists relating to work equipment tend to reflect the specific, critical health and safety conditions of the equipment.

Factors to be considered when creating a checklist for a general workplace inspection should include: substances or materials being used, condition of traffic routes and means of access and egress, work equipment, work practices (manual handling, etc.), work environment, electricity, fire precautions, welfare provision including first-aid arrangements and workstation ergonomics *see Figure 4-2* for a sample of a workplace inspection checklist.

Advantages of using a checklist include:

- Enables prior preparation and planning.
- Quick and easy to arrange.
- Brings a consistent approach.
- Clearly identifies standards.
- Thorough.
- Provides ready-made basis for inspection report.
- Provides evidence for audits.

Disadvantages of using a checklist include:

- Does not encourage the inspector to think beyond the scope of the checklist.
- Items not on checklist are not inspected.
- May tempt people who are not authorised/competent to carry out the inspection.
- Can be out of date if standards change.
- Inspectors might be tempted to fill in the checklist without checking the work area/equipment.

Sampling

The role of sampling is to select a representative, partial amount of a group of items, people or area; which is examined to establish facts about it and used to indicate the status of the whole.

When a representative sample is taken this is therefore considered to reasonably represent the situation for the whole group. A very small sample, such as the examination of three pieces of lifting tackle, may only give a simple, but acceptable indication of the situation relating to lifting tackle as a whole.

Various health and safety sampling activities may be conducted, including sampling specific hazards, good practice and general workplace hazards.

OBSERVATIONS OF PHYSICAL CONDITIONS

Category	Number Checked (C)	Number Standard (S)	% Meeting Standard $\frac{S \times 100}{C}$	Comments
EQUIPMENT *For Guidance Notes consider Audit 123 Level 1 - Sections 4 and 7*				
Guarding				
Hand tools				
Power tools				
Electrics (visual)				
Electrics (technical inspection)				
Pressure systems				
Ladders and mobile towers				
Personal protective equipment				
Total Score			Total % Compliance $\frac{\text{Total S} \times 100}{\text{Total C}}$	

Figure 4-3: Sampling - extract from Audit 123 Level 3 Section 1 Workbook [ISBN 978-1-900420-50-1]. Source: RMS.

Specific hazards

Specific hazards - such as noise or dust - because specific measuring equipment or techniques are often used those conducting this form of sampling will typically be those trained in appropriate hygiene techniques. Measurements are conducted on a planned basis to periodically confirm the status of levels of the hazard in specific work locations or after changes have been made to determine the effectiveness of the changes. The results of this type of sampling are formally recorded and used to compare with other measurements over time.

Good practice

Good practice - such as the wearing of personal protective equipment - typically conducted by first-line managers or health and safety practitioners. This type of sampling may be conducted periodically and after risk control measures have been introduced to enable positive recognition and reinforcement of good practice. Any forms or checklists should be designed to encourage the recording of good practice not just poor practice.

General workplace hazards

General workplace hazards - such as those identified during a defined walk through a work area - typically conducted by first-line managers, worker health and safety representatives and workers. In order to obtain information on a timely basis and to compare performance over a period regular samples can be taken, for example monthly and the frequency increased to weekly where it is important to gain specific information on how hazards are being managed where action is being taken to make improvements. It is not necessary for all hazards to be identified, simply that a selected amount of the work area is observed and this is used to represent the whole work area.

Tours

The role of tours is to provide an opportunity for management to explore the effectiveness of risk control measures through planned visits to the workplace. This ensures they can observe and discuss the controls in use by the workers carrying out the tasks.

It is important, when developing a positive health and safety culture, that management commitment is visible. The conducting of planned tours to workplaces to meet work groups is one effective way of achieving this. As such it is a monitoring method that senior and middle managers would find useful.

It is not necessary for the manager conducting a health and safety tour to be the line manager of the workers involved in the tour, it is also very useful for tours to be conducted by those managers who provide support functions, such as finance, sales, marketing and human resources as it provides them with a better insight into health and safety matters, enables the organisation to gain a perspective from someone other than a line manager and can improve their involvement in management meetings.

One of the advantages of conducting tours as a monitoring method is that it enables direct contact and communication between workers and senior management. This gives the manager a personal and accurate picture of work conditions and the understanding of workers. This may be in contrast to the understanding gained through line management communication channels. It also provides managers a chance to demonstrate their personal commitment to health and safety, show leadership when discussing health and safety matters and an opportunity to demonstrate a good health and safety example.

As well as listening to workers and observing the workplace a health and safety tour provides an opportunity to communicate important health and safety principles, find barriers to good health and safety standards and to motivate workers, including supervisors.

The frequency tours are carried out can depend on the amount of change in health and safety that the organisation is experiencing, for example a change in culture may mean that managers want to conduct tours to show their commitment and ownership of health and safety. The frequency may also be affected by the fact that major changes in health and safety process are affecting workers and tours can provide information on how they are reacting or whether measures to minimise the effects of change are working. It is not necessary for every workplace to receive a tour frequently, but that the manager is active in various places where workers are on a sufficiently frequent basis. A common approach would be that senior managers each conduct health and safety tours four times per year.

Although some managers have a natural ability to conduct tours in a positive and meaningful way most would benefit from some explanation of the approach and coaching in how to get the best out of conducting a tour. Without this guidance managers tend towards inspecting the workplace and fault finding, which can lead to a very negative outcome.

In order to be planned there must be an intended outcome from conducting a tour, for example, to find something out, communicate something and review a topic, even if this includes some free time for other less structured discussion. Checklists are not normally used to structure a tour, but those managers unfamiliar with conducting a tour and in the process of developing the skill would find a list of things to remember while carrying one out useful guidance.

Tours can indicate deficiencies or success in managers carrying the organisation's objectives through to action. It also provides a forum for gaining the viewpoint of workers directly, without the translation that takes place through formal management channels.

Details of the tour and outcomes, including improvement actions taken, must be recorded to be effective.

Allocation of responsibilities and priorities for action

Monitoring will not improve health and safety performance unless, where a deficiency is identified, corrective and preventive actions are identified, responsibilities are allocated and they are fully implemented. Standards will continue to deteriorate where workers/managers find their efforts to identify where improvements are needed are not followed through and actions are not implemented. They may quickly regard their efforts as 'a waste of time'.

It is, however, equally important to avoid putting in an inappropriate solution as this is bound to result in a waste of the organisation's resources (time, equipment, money etc.). To get the right solution implemented will usually take a quantity of management time and effort, and if the risk is significant it is clearly warranted. Whenever a substandard (at risk) situation is identified two important steps must be taken:

1)   Each situation must be evaluated to determine its risk potential.

2)   The underlying cause(s) of the situation must be identified.

The first step ensures that situations identified are prioritised and allocated for action on a worst first basis, and the second step ensures that the appropriate corrective action is taken to remedy causes, reduce the risk and prevent its return to the same level.

When situations are identified and corrective actions have been determined, the maximum time period by which the action is to be completed should be agreed, the time allocated should be dependent on the risk potential of the situation.

A simple approach to risk potential levels and allocation of priority/time for actions may follow the method shown in the following section:

- High risk (likely to cause a major loss) - complete within 24 hours.
- Moderate risk (likely to cause a serious loss) - complete within 7 days (1 month if preferred).
- Low risk (will possibly cause a minor loss) - complete within 30 days (3 months if preferred).

A structured approach is necessary to ensure that the actions needed are initiated and followed up (monitored) in such a way that they are not simply forgotten.

Responsibilities for action should be clearly allocated to those that have responsibility to ensure the action takes place.

When compiling an action plan the following issues have to be taken into consideration:

- The clear identification of any problem or defect.
- Prioritisation of actions.
- Details of corrective and preventive actions.
- The identification of who is responsible for carrying out the corrective and preventive action.
- The setting of realistic timescales for completion and dates for reviewing progress.
- The allocation of resources.
- Communicating details of the monitoring activity to interested parties.

REACTIVE MONITORING MEASURES

Objectives and usefulness of reactive monitoring

The objective of reactive monitoring is to measure the negative outcomes from the organisation's efforts to ensure health and safety, in order to identify the significance of these outcomes and opportunities for improvement. These negative outcomes are usually the result of undesired events, incidents. In order to carry out reactive monitoring effectively, systems must be in place to identify the incident, record it and report it. Without this nothing may be learnt. Indeed, what little data that is communicated might serve to reinforce that there is no need to put in a great deal of health and safety effort. If reporting etc. is planned and encouraged it is not uncommon to find a large increase in reported incidents. This does not necessarily mean an increase in incidents, merely an increase in reporting.

Information from these incidents contributes to the 'corporate memory' of the organisation, helping to prevent a repeat in another part of the organisation or at a later time. Though it should be remembered that the 'corporate memory' is said to be short, in the average organisation (one undergoing some change) it is likely to be in the order of four years. Data may also be gained from other organisations to reinforce or extend the organisation's experience of incidents and the hazards involved.

Distinction between different types of incident

Incident

An incident can be defined as an unplanned, uncontrolled event that may result in injury, ill-health, dangerous occurrence, property damage or near-miss.

"Incident

Occurrence arising out of, or in the course of, work that could or does result in injury and ill health."

Figure 4-4: Definition of incident. Source: ISO 45001: 2018 Occupational health and safety management systems requirements with guidance for use.

Injury

The term injury refers to physical harm to an individual and incidents that result in this type of harm are often called accidents.

"(a) The term occupational accident covers an occurrence arising out of, or in the course of, work which results in fatal or non-fatal injury."

Figure 4-5: Definition of accident. Source: ILO Protocol to Occupational Safety and Health Convention C155.

Accidents (injury incidents) may be further categorised by the scale of injury experienced by the worker, for example, fatal injury, major injury or injury requiring a set amount of absence from work or normal duties.

Ill-health

The term ill-health refers to harm to a person's health caused by their work and will include harm to health in a physiological or psychological way. Ill-health of this type will relate to the wide range of occupational diseases that workers may experience, for example, asbestosis, pneumoconiosis and silicosis, where the ill-health effects may take several years to develop. More recently, ill-health effects have been related to work load and stress.

National legislation may set out minimum requirements with regard to the identification, recording and reporting of specific types of disease.

"(b) The term occupational disease covers any disease contracted as a result of an exposure to hazards arising from work activity."

Figure 4-6: Definition of occupational disease. Source: ILO Protocol to Occupational Safety and Health Convention C155.

Dangerous occurrence

This term is used to describe significantly hazardous accidents/incidents, such as the failure of any load-bearing part of a lift or hoist; collapse of a mobile powered access platform; overturning of a mobile crane or a significant collapse of a scaffold. What is taken to be a dangerous occurrence may be different in each country depending on factors like the country's enforcement reporting requirements for accidents/incidents.

"(c) The term dangerous occurrence covers a readily identifiable event as defined under national laws and regulations, with potential to cause an injury or disease to persons at work or to the public."

Figure 4-7: Definition of dangerous occurrence. Source: ILO P155 Protocol of 2002 to the Occupational Safety and Health Convention, 1981.

Damage-only

Accidents/incidents at work cause substantial damage to equipment, property and materials annually. The study of the incidence of damage-only losses may be a useful predictive tool to identify situations that might result in injury to people or other significant loss to an organisation. For example, a series of collisions into a scaffold, requiring minor repair by replacement of a scaffold tube, may be predictive of a later collision involving total scaffold collapse and major personal injury. Considering damage-only accidents/incidents can contribute greatly to the risk assessment process, indicating potential for major financial and human loss.

Near-miss

The term 'near-miss' refers to an event (incident) which did not result in personal injury, equipment damage or some other loss, but under slightly different circumstances could have done (for example, a building block falling off a scaffold and landing on the floor). The difference between a near-miss and a fatal accident in terms of time and distance can be very small. Apart from being unpleasant and perhaps very costly, every incident constitutes an opportunity to correct some problem. For this purpose, a near-miss is just as valuable as a serious injury/damage, in fact even more valuable in its role of providing preventative analysis.

"An incident where no injury and ill health occurs, but has the potential to do so, may be referred to as a "near-miss", "near-hit" or "close call"."

Figure 4-8: Definition of near-miss. Source: ISO 45001:2018 Occupational health and safety management systems Requirements with guidance for use.

The investigation of 'near-miss' accidents/incidents and the identification of their underlying causes might allow preventive action to be taken before something more serious occurs. It also gives the right message that all failures are taken seriously by the employer and not just those that lead to injury.

Figure 4-9 on the following page demonstrates the difference between an accident, near-miss and undesired circumstances or unsafe condition.

Figure 4-9 (a) Accident. Source: UK HSE, HSG45

Figure 4-9 (b) Near-miss. Source: UK HSE, HSG45

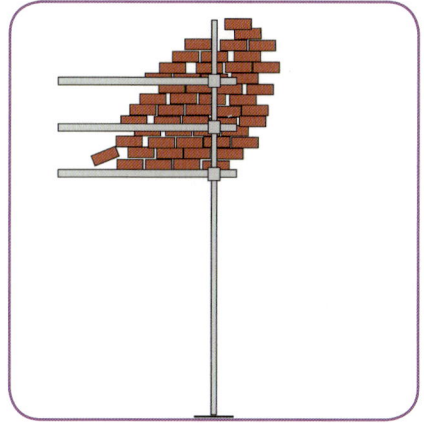

Figure 4-9 (c) Unsafe condition Source: UK HSE, HSG45

REACTIVE MONITORING METHODS

These methods are deemed to be after the incident and are therefore reactive monitoring measures:

- Identification.
- Reporting.
- Investigation.
- Collation of data and statistics, on the unwanted events.

These reactive monitoring methods are nearly always limited to measurement of the extent of failure of health and safety management arrangements and procedures (if we ignore unforeseeable events which may affect the business, for example, severe weather, flooding, fire spread from a third party). For example, historical records to show:

- Accidents/incidents, for example, resulting in lost time, physical injury.

- Dangerous occurrences, for example, significant damage to plant, equipment or facilities.

- Near-misses, for example, accidents/incidents with no measurable loss.

- Ill-health, for example, resulting from exposure to substances or repetitive actions.

- Complaints by workforce, or contractors, for example, headaches, acne, blanched fingers.

- Worker absence statistics.

- Worker accident/incident and injury statistics compared with national averages for the same sector of employment

- The extent of lost profits arising from damaged goods, lost production time and reduced output following a health and safety failure.

- Enforcement action, for example, issue of verbal instructions or written notices by an enforcement agency.

It is important to identify, in each case, why performance was substandard. Trends and common features may be identified, such as when, where and how these events occur. This provides an opportunity to learn and put into place improvements to the overall management system and to specific risk controls.

Incident statistics

Many organisations spend considerable time developing data on their health and safety performance based on a variety of incidents, including accidents, dangerous occurrences, near-misses and ill-health information. Whilst there is value in doing so it has the limitation of being after the incident. Incidents must occur to get the data, thus tending to influence effort to prevent a recurrence rather than action to prevent the event.

A low incident rate is not a guarantee that risks are being effectively controlled. In some cases, this might be a matter of good fortune, or the fact that incidents are not being reported, rather than effective management.

If organisations wait until an incident occurs to determine where health and safety effort is required then some sort of loss will usually have occurred. In order to gain sufficient management attention this could be an incident resulting in personal injury to someone.

Clearly this is an undesirable way of learning, particularly as, with an amount of effort, planning and thought, the incident could have been foreseen and prevented. The more mature organisation seeks to learn most from activities (for example, risk assessment) 'before the incident' or, at the very least, learn from those incidents that result in no personal injury, for example, near-misses.

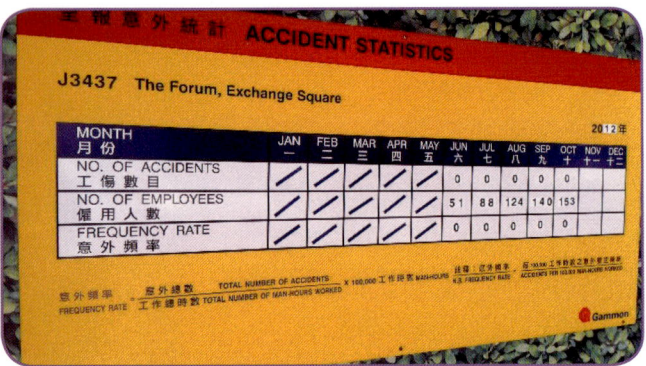

Figure 4-10: Accident statistics.
Source: Wikimedia.org

The obvious use of incident data is to identify specific problem areas by recording instances where control measures have failed. However, analysis of the data allows general trends to be shown in order perhaps to identify common root causes, as well as comparisons to be made with others in order to learn from successes elsewhere. Incident data can also help to raise awareness in the minds of both managers and workers of health and safety in general, and of specific problems in particular. In addition, collection of data allows costs to be calculated, which can increase the likelihood of resources being allocated.

There are several methods of presenting incident data for analysis, some of the more common indices used are incidence and frequency rates. Methods of presenting data should not be mixed and figures for indices should only be used to compare like with like, for example, the same organisation over time or similar organisations using the same indices. The multipliers used in calculating the indices vary depending on whether the rates are derived at international, national or workplace organisation level. There is no formal structure as to what multipliers should be used, but the main purpose of the multiplier is to bring the numbers to a manageable size.

Accident/ill-health incidence rates

The typical national competent authority formula for calculating an accident/ill-health incidence rate is:

Number of accidents/ill-health in the period x 100,000
Average number of people that worked during the period.

This is the rate per 100,000 workers and gives an incident rate for the organisation on the number of accidents/ill health in the workplace. It may not allow for part-time workers or overtime unless the organisation has converted these hours to full time equivalent workers. The calculated number should only be used for a comparison between one organisation and another or between different parts of the same organisation because risk levels may be different in other industries.

The UK government, as a competent authority, calculates national incidence rates using the 100,000 multiplier. The Japanese government uses a multiplier of 1,000 for incidence rates. When workplace organisations calculate their own incident rate, the number of people that worked in the period and accidents/ill-health will be far less than nationally and it is common practice to use a smaller multiplier, for example, 1,000.

Accident/ill-health frequency rates

The accident/ill-health frequency rate illustrates how often accidents/injuries are occurring during the period measured. Some national competent authorities prefer to calculate injury frequency rates per million hours worked. Considering the hours worked rather than the number of workers avoids the problem of part-time workers and overtime causing a distortion as it does in the incidence rate calculation. It should be noted that the International Labour Organization and many national competent authorities, for example, Japan and the UK, use 1,000,000 for the frequency rate, the OSHA/Bureau of Labour Statistics in the USA uses 200,000.

As with incidence rate, many workplace organisations tend to use a smaller multiplier, for example, 100,000.

Number of accident/ill-health in the period x 1,000,000
Total hours worked during the period

Accident/ill-health severity rates

Total number of days lost x 1,000
Total hours worked

The accident/ill-health severity rate does not necessarily correlate well with the seriousness of the injury. The data may be affected by the propensity and ability of people in different parts of the world to take time off after a particular injury.

Mean duration rate

Total number of days lost
Total number of accidents/ill-health

Duration rate

Number of hours worked
Total number of accidents/ill-health

Absence data

Analysing absence data is important for two reasons:

- The process enables an organisation to determine whether it has a work-related absence problem.
- It can help the organisation to understand the causal effects of absenteeism.

Assessing the magnitude of absenteeism

Days lost per worker and percentage lost time are the two measures used to gain an overall understanding of the magnitude of the problem.

Days lost per worker =	Total days lost
	Number of workers

Percentage lost time =	Number of days lost through absence x 100
	(Numbers of workers) x (Number working days)

Technically, percentage lost time is the better overall measure of absenteeism. Days lost per worker, however, has an advantage in terms of simplicity of calculation. For this reason, it is the measure most commonly used by organisations to assess absenteeism.

For incident statistics to be of value their limitations have to be understood. Variables in work methods, hours of work, hazard controls and management systems make it difficult to make comparisons outside the organisation deriving the data.

Indices such as these are best suited to comparison of performance of the same organisation over similar periods of time, for example, yearly. In this way trends may be observed and conclusions drawn. If comparisons are to be made outside the organisation, it should be remembered that other organisations might have a different understanding of the following:

- Definition of an accident (lost time or reportable).
- Hours worked may not be actual (contracted minimum hours may be used as it is easier to work out).
- The workers included (contractors may not be included).
- The multiplier used (International Labour Organization and a number of countries use 1,000,000 for the frequency rate, USA use 200,000).

Ratios of incident outcomes

Frank Bird developed the International Safety Rating System (ISRS), a safety management system in 1978, following his research into the causation of 1.75 million incidents in 21 industries. Bird showed that there is a fixed ratio between losses of different severity and incidents where no loss occurred, i.e. near-misses. This is illustrated in Bird's pyramid model of incident outcomes.

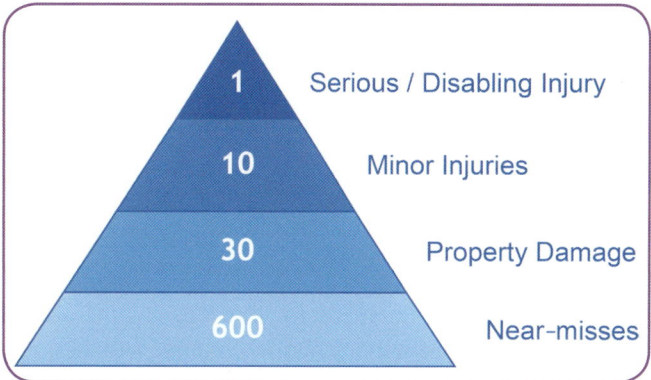

Figure 4-11 Incident ratio study. Source: Frank Bird.

As can be seen in *Figure 4-11*. It is generally accepted that 'near-misses' far outnumber incidents that result in injury. These incidents can therefore produce more data from which a greater understanding of the deficiencies in existing management systems, such as risk assessments and safe systems of work, can be identified and rectified before an incident resulting in injury occurs.

The incident ratio study described by Frank Bird illustrates that the organisation, site or department that has most incidents may, over time, be likely to have the most serious injuries/losses. This can usefully convey to managers and workers the importance of reporting incidents that do not involve injury to people, as they are more likely to occur and they represent an opportunity to establish their causes and the possibility that this will prevent later injury to people. It follows that if we thoroughly address the causes of near misses and minor loss incidents, we will automatically reduce the serious loss incident causes as well.

The incident ratio pyramid therefore has wide application. However, it is important that managers and workers do not misunderstand the incident ratio pyramid. The numbers shown in *Figure 4-11* only represent an average of the organisations evaluated by Frank Bird; some higher risk organisations may have a much narrower pyramid, illustrating that when incidents do happen, they have a higher likelihood of injury. In lower risk organisations the pyramid may be wider, as injury is less likely.

It is important that managers and workers do not see the incident pyramid as giving them a quota of near-misses that they can have before they have an injury incident; the first incident that happens may have circumstances that cause injury.

The principle that the incident ratio pyramid helps with is not to waste the learning opportunities presented by incidents that do not result in injury.

Other reactive monitoring measures

Complaints from the workforce

Legislation requires employers to provide a safe place of work. When workers are dissatisfied with working conditions or the safety management of risk, they are likely to complain about it. Therefore, the number of complaints from workers can be used as a reactive measure of how well (or not) the organisation is doing.

Enforcement action

The number of enforcement actions taken against the organisation is an indication of the health and safety performance of the organisation. Enforcement actions taken against an organisation is often one of the questions asked on pre tender qualification questionnaires. Where the organisation has to report that action has been taken it will be seen as a negative against the organisations' bid for the work.

Incident investigation

Incident investigation is a reactive monitoring measure that can provide a large amount of data on the incident being investigated and the procedures and arrangements in place to prevent it. It can also provide data on how effective the policy and objectives are in managing risk, often identifying barriers in the organisation that could have an effect on wider issues than the specific one being investigated, for example lack of management motivation and commitment or habitual health and safety deviations from procedures. For more information on incident investigation see *Section 4.2.*

LESSONS LEARNT

It is important that lessons learned from *adverse events* like incidents are shared with as many people who would benefit from it as is possible. As a minimum this must include different departments within an organisation. It would be regrettable if an incident happening in one department was seen as of no relevance to others, without considering the causes. The more that the root causes are examined the more likely the lessons are to be relevant to other departments. For example, examination of the causes of an accident/incident in one department may reveal a need to improve the job induction processes, which may affect all departments of an organisation.

In the same way, it is important to share lessons learned with other locations of the same organisation. Some industries share lessons learned across the whole of the organisations in an industry, for example, public transport. This is very important as they may have similar practices or equipment.

Trade associations (sector groups, for example, construction, paper manufacture, which are often formed over time to address common issues and to lobby government departments through collective representation) often take the lead in collating data on a non-attributable basis and sharing the lessons learned with participating member organisations. The collective experience of what is learned enables common problem activities and hazards to be identified and underlying causes, human, organisational and technical, to be examined. In some cases, the lessons are shared as an immediate alert, in other cases quarterly reports provide analysis of

---

BN0901A1649 - Transport of Zip-Up (Mobile Tower) Scaffolding

Background

- A task had been carried out in the HAST 18 area of B215 which required the use of a zip-up scaffold.
- The type of scaffold used was a Planet Platforms Protec scaffold (GRP tower).
- The scaffold had been dismantled and the base frames (1.35m wide / 15-20 kg weight) were being carried up the HAST 15 stair case which is of an open nature.
- One of the adjustable legs with affixed castor (approx 5-6 kg) became detached from the bottom of one of the frames and fell approximately 9m, glanced off some lagging and came to rest on the landing of a lower floor.

Details

- The adjustable leg / castor assembly is inserted into the base frame which has an adjustment collar that is designed to be used to adjust the legs to level up a scaffold on an uneven surface.
- The adjustment collar has only a small range of movement between 'lock' and 'unlock' (approx one eight of a turn). When in the unlock position, the adjustable leg can be fully released in an uncontrolled manner if the castor is not resting on something.

Key Learning

- <u>Do not use this type of scaffold with the legs extended until investigations have been concluded</u>.
  - If the adjustable legs are used to raise the height of a scaffold there is the potential that if not properly locked or if the collar is in a poor state of repair, if knocked the collar could release resulting in the scaffold tipping.
  - Remove the leg / castor assemblies from the base frame before transporting zip-up scaffolding.
  - Do not carry zip-up scaffold on stair cases if possible – look for alternative routes where a lift can be used instead.
  - If equipment has to be carried around Plant / on stair cases, ensure areas where personnel are at risk of falling items have barriers and signs are erected.

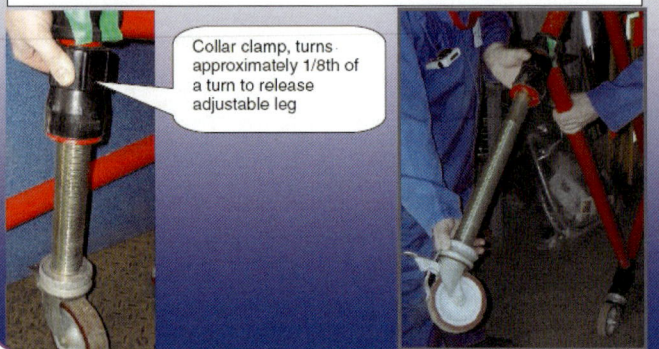

Collar clamp, turns approximately 1/8th of a turn to release adjustable leg

Figure 4-12: Incident lessons learned alert. Source: Jamie Lockie Sellafield Fellside site.

---

recent experiences and trends. See *Figure 4-12* which shows an example of a document which was circulated to all interested parties following investigations into an incident.

Learning should not be restricted to analysing what has gone wrong, it is essential that organisations commit to spending time and using processes to learn from what has gone well. It is important to learn from **beneficial events**, for example where the introduction of preventive measures has had a beneficial effect on risk. Similarly, if a health and safety behaviour change event is run and is successful the organisation should learn what made it such a success. These beneficial events may be considered as examples of good practice that can be discussed and shared between departments and other organisations. They can be discussed in management meetings, end of project meetings, health and safety committees and in periodic reviews of the management system.

It is important that the lessons learned from both beneficial and adverse events are considered as part of the regular review of an organisation's health and safety performance. What is learned may influence the organisation's policy, objectives and management system, therefore they should be considered by senior management in the organisation.

THE DIFFERENCES BETWEEN LEADING AND LAGGING INDICATORS

When considering the management of performance of an organisation we often refer to "leading indicators" and "lagging indicators". Lagging indicators are typically indicators of the outputs from the organisation and in the case of health and safety usually the outputs from its health and safety management system. These outputs can have a beneficial or adverse effect. For example, an output with a beneficial effect might be the reduction in the noise level in a workplace leading to reduced risk of noise induced deafness in workers and an output with an adverse effect might a new workplace layout leading to an increase in impact of moving vehicles with workers.

Leading indicators typically relate to the inputs that are made to the health and safety management system to achieve a given performance output and are the indicators that describe or measure this input. For example, in order to achieve a performance of a reduction in injuries from manual handling activities it may be decided to provide workers with manual handling training. A leading indicator would be the number of workers receiving manual handling training and a lagging indicator would be the quantity of manual handling injuries that occurred over a given period.

4.2 Investigating incidents

ROLE OF INCIDENT INVESTIGATION

The role of incident investigation includes to:

- Prevent a recurrence.

- Discover underlying and root causes.

- Establish legal liability, prepare for legal action and ensure legal obligations are complied with.

- Provide enforcement agency/insurance/worker compensation data.

- Determine compliance with the organisation's policy, processes and procedures.

- Identify weaknesses in health and safety systems, arrangements and procedures so that standards can be improved.

- Determine the economic loss caused by the incident.

- Gather data to produce statistics.

- Identify trends.

- Update risk assessments.

- Demonstrate the commitment of management to provide a safe and healthy place of work.

- Establish if internal disciplinary procedures are necessary.

- Support worker morale, not carrying out an incident investigation will have a negative effect on the health and safety culture as workers will assume the organisation does not value their health and safety.

Employers have a responsibility to prevent harm to workers and therefore, where an incident causes harm there is a need to investigate the causes and prevent a recurrence. In addition, the study by Frank Bird confirmed that in general organisations have more incidents that do not result in harm to people than those that do, this is illustrated by the 'accident triangle' shown earlier in *Figure 4-11.* Many of the incidents that do not result in harm have causes that in slightly different circumstances could result in harm to people, which emphasises the importance of investigation of all incidents. The findings of the investigation of incidents should be used to prevent a recurrence by improving workplace standards, processes, procedures and training requirements.

In addition, in some countries there is a specific legal requirement to report accidents, dangerous occurrences and ill-health to external organisations such as competent authorities, enforcement agencies, insurance institutions and worker compensation organisations. The information needed for the report will cause the employer to investigate the circumstances leading up to an incident.

The role and directive of any investigation of this nature should never seek to blame any individual or group of individuals. If human error is believed to be a significant cause, the reasons for this must be investigated. Lack of knowledge, training or unsuitability for the job may be the causes of this error. These are management and not worker failings. Only when these have been evaluated can the conclusion of wilful and intentional acts or omissions be considered.

The ILO Code of Practice for Recording and Notification of Occupational Accidents and Diseases sets out requirements at 'enterprise level' that each organisation should follow. This begins with the requirement for employers to investigate reported occupational accidents, commuting accidents, occupational diseases, dangerous occurrences and accidents/incidents.

The Code of Practice proposes that if the employer does not have the expertise to conduct an investigation, they should call upon the services of a person that does, even if this is from outside the enterprise. It further requires that arrangements be in place to enable the immediate investigation of occupational accidents, occupational diseases, dangerous occurrences and accidents/incidents.

DIFFERENT LEVELS OF INVESTIGATION

Ideally all incidents should be investigated, but the level of investigation may vary depending on the circumstances of the incident. The level of investigation should depend on the severity of actual or potential loss resulting from the incident, whichever is the greater. For example, a scaffold collapse may not have caused any injuries, but had the potential to cause major or fatal injuries. Therefore the investigation should, in part, be based on the worst possible case of harm which is reasonably foreseeable as a result of the incident in question. It should also consider how likely a similar incident is to occur. In this way the risk related to the incident is considered. The table below will assist you in determining the level of investigation appropriate for the incident.

RISK	LEVEL OF INVESTIGATION	
Minimal	Minimal level	In a minimal-level investigation, the relevant supervisor will look into the circumstances of the incident and try to learn any lessons which will prevent future occurrences.
Low	Low level	A low-level investigation will involve a short investigation by the relevant supervisor or line manager into the circumstances and immediate, underlying and root causes of the incident, to try to prevent a recurrence and to learn any general lessons.
Medium	Medium level	A medium-level investigation will involve a more detailed investigation by the relevant line manager, with the support of the health and safety practitioner and, where appropriate, worker representative and will look for the immediate, underlying and root causes.
High	High level	A high-level investigation will involve a team-based investigation, typically involving a senior manager, line manager, health and safety practitioner, technical specialist and, where appropriate, worker representative. It will be carried out under the supervision of top management/ directors and will look for the immediate, underlying, root causes and lessons that can be learned that can be applied across the organisation and other locations, if applicable.

Minimal and low-level investigations by the Supervisor

The supervisor as the person in immediate operating control of an area or activity is a logical person to gather information on all accidents that happen in their area of responsibility. It often will result in swift identification of the causes and remedial actions being implemented, and underlines the supervisor's responsibility for health and safety on a day to day basis. This investigation is normally all that is necessary for the majority of incidents.

Medium-level investigations by the Manager

Organisations that encourage management involvement identify incidents that warrant further investigation by an appropriate manager. Where there is a medium level risk a manager, independent of the line managers where the incident occurred, may be required to conduct a more detailed investigation. They would take advice, including from the health and safety practitioner, as appropriate.

Managers who receive reports of the investigation of minimal and low risk incidents conducted by supervisors may decide that a more detailed investigation is required. This may be in order to gain a better understanding of the root causes of the incident, something that may be outside the perspective of the supervisor.

High-level investigation

Where an incident has caused serious harm, or there has been the potential for serious harm to have occurred, for example, a near miss with high potential for harm, it will be necessary to carry out a high-level investigation using a team. Any person whose responsibilities or actions may have been involved in the incident being investigated, for example, a supervisor, should be excluded from the investigation team, but would be valuable as a witness. A typical high-level investigation team would usually include the following people:

- A senior manager from another department to act as an independent chairperson.
- A health and safety practitioner to advise on specific health and safety issues.
- A health and safety practitioner to provide management experience and perspective.
- An engineer or technical expert to provide any technical information required.
- A worker health and safety representative to provide worker experience and perspective.

Health and safety practitioner monitoring and investigation of incidents

The organisation's health and safety practitioner should monitor and evaluate the incidents reported and help to decide the level of investigation required. They will support medium and high-level investigations and monitor lower level investigations to determine if the investigation has been effective. Where necessary lower level incidents may be investigated by the health and safety practitioner to a greater depth than the initial investigation. Ill-health events may require involvement of other specialists with a medical or environmental workplace assessment background.

BASIC INCIDENT INVESTIGATION

Basic incident investigation steps

When an incident occurs prompt emergency action should be taken, for example provision of first aid, making the area where the incident happened safe and preserving the scene of the incident to enable it to be investigated. If it is necessary to disturb the site, the employer should arrange for a competent person to make a record of the site of the incident, including photographs, plans and the identification of witnesses, before intervention.

The ILO Code of Practice for Recording and Notification of Occupational Accidents and Diseases 10.2 requires the employer to arrange that the site of an accident be left undisturbed before the start of an investigation, unless first-aid provision or a further risk is presented. It further requires the employer to, as far as possible:

- Establish what happened.
- Determine the causes of what happened.
- Identify measures to prevent a recurrence.

Throughout the process of investigation, it must be clearly borne in mind that the objective is to prevent a recurrence of the incident, not to apportion blame. It is important to identify the true causes of the incident, not just the superficial ones. This cannot be achieved without the full commitment and assistance of witnesses and other persons who work in the area in which the incident happened. It follows that recommendations must be put into action, even though they may take a considerable amount of time, trouble and money.

An approach reflecting good practice would follow a step-by-step approach to investigations, for example:

Step 1: Gathering the information - the where, when, who, what and how of the incident. The information gathered will include results of interviews, photographs of the equipment involved and the area in which it was positioned at the time, sketches of the workplace layout, weather conditions, etc.

Step 2: Analysing the information - determine what has happened and analyse the information to identify the immediate, underlying and root causes. At this stage it should be considered if human error is a contributory factor. Job factors, human factors and organisational factors can all influence human behaviour and will all need to be considered in the analysis.

Step 3: Identifying suitable risk control measures. Possible solutions can be identified. This will involve looking at the technical, procedural and behavioural controls, with the technical or engineering risk control measures being more reliable than those that rely on human behaviour.

Step 4: The action plan and its implementation to identify which risk control measures should be implemented in the short and long-term. The risk control action plan should have SMART objectives, i.e. Specific, Measurable, Achievable and Reasonable, with Timescales. This will also state which risk assessments need to be reviewed and which procedures need to be updated; any trends that need further investigation; and the cost of the incident.

Step 1: Gathering information

The step to gather information should explore all reasonable lines of enquiry, be timely and structured – determining clearly what is known, what is not known and starting a record of the investigation process. This step focuses on determining where and when the incident took place, what the result of the incident was (for example, injury or ill health), who was involved (including witnesses) and how it occurred (the events that led up to the incident).

It is important to gather information as soon as possible after the incident. This helps to maintain its integrity and prevents confusion, for example, items at the scene of the incident might get moved, control measures not used might be put in place (guards replaced, personal protective equipment worn) and witnesses may discuss what they believe they saw and cause confusion. If necessary, work must stop and unauthorised access to the site, witnesses and other potential evidence prevented.

Gathering information includes:

- Physical information - from the scene of the incident, for example photographs, closed circuit television (CCTV) footage, making a plan of the area and noting details of any equipment involved.

- Verbal information – from witness statements, available recordings of instructions and conversations.

- Documented information – for example, risk assessments, policies, procedures, work instructions and training records.

The amount of time and effort spent on information gathering should be proportionate to the level of investigation. Collect all available and relevant information. That includes opinions, experiences, observations, sketches, measurements, photographs, check sheets, permits-to-work and details of the environmental conditions at the time etc. This information can be recorded initially in note form, with a formal report being completed later. These notes should be kept at least until the investigation is complete.

Interviews

Interviews are an important part of the investigation process and can quickly provide information on the incident and what lead up to it, providing further investigation trails. However, the witnesses may be distressed, very defensive and could feel that they might be blamed, so it is important to put the person being interviewed at ease, emphasising that the purpose of the interview is to help determine the facts to prevent a recurrence.

Good interview techniques will include:

- Interview of witnesses promptly after the incident, to avoid lapse of memory or distortion through witnesses debating what occurred.

- Conducting the interview in private one person at a time. Free from distractions and interruptions.

- Explanation of the purpose of the interview i.e. to prevent recurrence. The interviewer should explain what is to happen during and after the interview process.

- An explanation that notes will be taken of their responses during the interview.

- The use of open questions, such as, 'what were you doing before the incident occurred'. Open broad questions aid interviewee reflection, generally put them at ease and make them more responsive to the interview process.

- The use of closed questions, ones that can be answered with a single word or short phrase can be used to confirm a specific point, such as 'did the incident happen as soon as soon as you let go of the ladder?'

- The use of a sketch plan and photographs by the interviewer to assist in developing a good understanding of what happened and assist in determining root causes of the incident. Sketch plans of the workplace and the site of the incident can be used during interviews to provide a clear indication of the incident scene including the position of any injured person, witnesses, plant and equipment. Cameras can be used to record and preserve images of incident scenes or resultant injuries. This can be especially useful if the situation changes, through such things as corrective actions being put in place or changes in environmental conditions, while the investigation is continuing.

- Care to keep the person being interviewed at ease. Avoid complex language or jargon. It is especially important to observe the witnesses' body language to ensure that they are listening, giving full attention to the questions and remain at ease with the interview. The interviewer should maintain an even voice tone and manage their responses to answers in a calm and neutral manner.

- Recording the details: names of the interviewers, interviewee and anyone accompanying interviewees; place, data and time of the interview; and any significant comments or actions during the interview.

- Summarising the interviewer's understanding of the information provided by the witness to ensure there is a mutual understanding gained from the interview. If necessary and appropriate, ask the witness to write and sign a statement of truth to create a record of their testimony.

- Expressing appreciation for the information given by the witnesses. Thank them for their help, explain may need to discuss further at a later stage.

 **REVIEW**

List five good practice techniques for conducting an effective interview of witnesses following an accident.

What are the main reasons for investigating accidents?

What are the categories of staff who may be team members in an accident investigation team for a major incident?

Step 2: Analysing the information

The step of analysing information should be objective, unbiased and evidence based. The analysis should identify the sequence of events and adverse conditions leading to the incident, identify immediate, underlying and root causes, determine common themes from interviews, record findings.

An analysis of the information gathered involves examining all the facts, determining what happened and why. All the detailed information gathered should be assembled and analysed to identify what information is relevant and what information is missing. The information gathering and analysis are usually carried out together, as a progressive process when further lines of enquiry requiring additional information develop.

To be thorough and free from bias, the analysis must be carried out in a systematic way, so that all the possible causes and consequences of the incident are fully considered.

Identifying immediate causes

Immediate cause: the most obvious reason why an incident happened, this is usually due to workplace conditions and the actions of people, for example:

Workplace condition – for example, an electrical cable on the floor (trip hazard), a guard missing from a machine, poor light level or a slippery floor.

Actions – for example, overreaching on a ladder, driving too fast, not looking in the direction of walking or not lifting a load properly.

There may be several immediate causes identified in any one incident. These workplace conditions and actions result in workers being exposed to uncontrolled hazards that present a risk of injury or ill health.

Underlying causes

Underlying cause: these are the causes that led to the immediate causes and relate to the 'organisational', 'job' and 'individual' reason for an incident happening, for example; the hazard has not been adequately considered via a suitable and sufficient risk assessment; inadequate time allocated for a task, poor maintenance of equipment, wrong or no equipment/materials, absence of training, unclear procedures/instructions, unclear responsibilities, absence of supervision, poor co-ordination of work activities, no inspection of equipment, poor motivation of worker.

For example, the underlying causes of a worker overreaching on a ladder could be that the ladder was placed too far away from where it was needed, the ladder was too short, an obstruction at the base of the ladder prevented it being placed nearer, the worker did not understand the risk of overreaching, the work activity forced the worker to overreach, the worker did not have time to reposition the ladder, the worker was rushing to complete a job, the worker decided it was easier to overreach than reposition the ladder.

Root causes

Root cause: these are the causes that enabled or led to the underlying causes and relate to management failings for example, failure to identify training needs and assess competence, low priority given to risk assessment, poor selection of equipment, inadequate provision of equipment, poor planning of work activities, pressure to complete work quickly, other priorities put above health and safety, poor leadership, poor prioritisation, inadequate resources, failure to maintain the workplace and equipment, poor management of change, ill-defined or inadequate standards, poor communication, inadequate worker engagement, failure to implement risk control measures, failure to monitor health and safety.

These management system failings occur when the organisation does not establish an adequate health and safety policy, organisation, planning, support, operation, performance evaluation, improvement where an appropriate approach to health and safety risk identification and control for the organisation's activities is established.

The analysis of incident information often show that an incident is can be the result of many causes.

 CONSIDER

An incident involving an operator coming into contact with dangerous machinery could be the result of any of the following:

Inadequate or non-existent safety devices

Poor housekeeping

Loose clothing

Machine malfunction

Operator error

Think of an underlying cause for each of the immediate causes above, then for each of the underlying causes think of a root cause.

Examples

A worker slipped on a patch of oil on a warehouse floor, was admitted to hospital with an injury and remained there for several days. The oil was found close to a stack of pallets that had been left abandoned on the pedestrian walkway.

Immediate cause:

- Oil that had leaked onto the floor from equipment.

- The floor remaining in a slippery condition because the spillage was not cleaned up.

- The abandoned pallets blocking the walkway causing the injured worker to make a detour.

- Inadequate lighting at the scene of the incident.

- The worker who slipped was wearing unsuitable footwear and not looking where they were walking.

Underlying cause:

- No materials available to clean up oil leak.

- Leak left for someone else to clean up.

- Oil on floor not considered to be a hazard by workers in the area.

- No time available for workers to clean up oil.

- No clear storage area for pallets.

- Lighting units not working.

- Worker who slipped was late and rushing.

- Lack of supervision.

- Walkway poorly identified.

Root cause

- The absence of adequate risk assessments and safe systems of work.

- Failure to introduce procedures for routine maintenance of equipment and cleaning up leaks/spillages.

- Poor warehouse design with inadequate walkways.

- Failure by management to monitor working conditions in the warehouse to ensure workers are not exposed to risks to their health and safety.

- Failure to define responsibilities.

- Failure to make workers aware of hazards and consequences of not taking action to control them.

- Inadequate training or instruction of workers in those procedures that might have been introduced.

- A range of personal factors, for example, stress, fatigue, influence of drugs and alcohol.

Step 3: Identifying the risk control measures

After identifying the causes, risk control measures that effectively remove the root causes need to be identified. The information analysis stage may have identified a number of risk control measures that either failed or could have prevented the incident, if they had been in place. This should indicate possible solutions that should prevent a recurrence of the incident. This could include the implementation of additional risk control measures to prevent risk controls failing and measures that were identified to be missing.

These possible solutions need to be systematically evaluated and best solutions should be considered for implementation. In deciding which risk control measures to recommend and their priority, you should choose measures based on the risk control hierarchy. This means, where possible, measures that eliminate risk should be given priority followed by measures that reduce the risk at source. Generally, measures that rely on human intervention are less effective than engineering controls that require no human input. Each risk control measure has to be evaluated in its own right to assess its effectiveness in preventing a recurrence and if they can be successfully implemented. The fact that some measures might be more difficult and/or expensive to implement should not be a barrier to them being considered.

When identifying if the risk control measures in place to prevent incidents are appropriate ensure:

- Missing/inadequate/unused controls are identified.

- Compare conditions/practices as they were with that required by current legal requirements, codes of practice and guidance.

- Risk assessments are reviewed.

- Identify additional control measures (application of hierarchy of control).

- Provide meaningful and realistic recommendations based on the outcomes of the investigation.

Recommended risk control measures should include both short and long-term controls measures. Short-term control measures may be necessary to help to prevent a recurrence until longer-term control measures can be implemented to remove the root causes.

Ensure recommendations are meaningful and reflect the causes identified. If the only identified risk control measure recommended is 'workers need to take more care' this will indicate that the investigation has not been thorough enough and failed to identify root causes.

Reports

The report should include a summarised version of the facts and recommendations for risk control measures, together with discussion of controversial points and if necessary, appendices containing specialist reports (medical and technical), photographs and diagrams. This virtually finishes the work of the investigator, but management is still responsible for seeing that the necessary corrective actions are implemented and monitored to ensure that the causes are satisfactorily controlled. The line manager, health and safety practitioner and health and safety committee/members should monitor the actions regarding risk control measures to ensure they are completed and effective. The information included in the investigation report should cover:

- Date, time and location of the incident.
- Personal details of the injured parties (name, role, work history, injuries sustained).

- Description of the activity being carried out at the time.
- Immediate, underlying and root causes.
- Drawings and photographs used to aid understanding of the scene.
- Identification of any breaches of law.
- Details of witnesses and witness statements.
- Recommended risk control measures that are prioritised and with estimated costs.
- The cost of the incident to the organisation.

 CONSIDER

Consider your organisations' procedure, does the accident report contain a section which requires the identification of immediate, underlying and root causes?

Step 4: Action plan and its implementation

If several risk control measures are required, they should be carefully prioritised and put in the form of a risk control action plan, which sets out what needs to be done, when and by whom. In deciding priorities, the level of risk and the ability of each risk control measure to limit the risk should be considered. It may be necessary to implement some less effective short-term risk control measures while better longer term risk control measures are being developed and implemented.

It is important to ensure that the action plan deals effectively with the immediate, underlying and root causes of the incident investigated. It should also include actions related to any more general lessons that may be applied to prevent other incidents, for example assessment of competencies may be needed for other activities.

Senior managers who have the authority to influence performance must be responsible for creating an action plan based on the SMART (specific, measurable, achievable, reasonable, and time bound) principles. Responsibility for specific actions should be assigned to ensure the timetable for implementation is carried out. A senior manager should be appointed to be overall responsible for implementation of the action plan.

The results of the investigation, the action plan and arrangements to ensure the action plan is implemented and progress monitored should be communicate to everyone who needs to know.

Progress on the action plan should be regularly reviewed. Any significant departures from the plan should be explained and risk control measure rescheduled, if appropriate. Workers and their representatives should be kept fully informed of the contents of the risk control action plan and progress with its implementation.

TYPICAL EXAMPLE OF WHAT MIGHT BE REPORTED TO EXTERNAL AGENCIES

The following definitions are provided in the 'ILO Protocol 2002 to the Occupational Safety a Health Convention C155' and 'Code of Practice for Recording and Notification of Occupational Accidents and Diseases'.

Reporting requirements may include reporting to a competent authority, an enforcement agency, insurance institutions, worker compensation organisations or a doctor, depending on the event being reported and local requirements. The 'ILO Code of Practice for Recording and Notification of Occupational Accidents and Diseases' sets out requirements that occupational accidents resulting in death should be reported immediately and other accidents within a prescribed period of time set out by national legislation.

Major injuries

The Code of Practice for Recording and Notification of Occupational Accidents and Diseases does not specify types of major injury resulting from accidents that should be reported. This is left to national legislation.

In the UK, major injuries are classified as 'specified injuries to workers' which must be reported to the enforcement agency.

For example, the list of specified injuries are:

- Fractures, other than the fingers, thumbs or toes.

- Amputations.

- Any injury likely to lead to permanent loss of sight or reduction in sight.

- Any crush injury to the head or torso causing damage to the brain or internal organs.

- Serious burns (including scalding) which covers more than 10% of the body; causes significant damage to the eyes, respiratory system or other vital organs.

- Any scalping requiring hospital treatment.

- Any loss of consciousness caused by head injury or asphyxia.

- Any other injury arising from working in an enclosed space which leads to hypothermia or heat-induced illness or requires resuscitation or admittance to hospital for more than 24 hours.

Diseases

The Employment Injury Benefits Convention, 1964 (No. 121), established by the International Labour Organization (ILO), provides for the competent authority of a country to define occupational accidents and diseases for which certain compensation benefits will be provided. Schedule I of the Convention lists those diseases that are common and well recognised.

This list has subsequently been supported by the 'ILO List of Occupational Diseases Recommendation R194', which was updated in 2010. This ILO list plays a key role in harmonising the development of policy on occupational diseases and in promoting their prevention and it provides a clear statement of diseases or disorders that can and should be prevented. This includes the employer recording and reporting such diseases to external agencies. The ILO list of occupational diseases includes the following examples:

Diseases by agent

Physical agents include:

- Hearing Impairment caused by noise.
- Diseases caused by work in compressed air.
- Diseases caused by vibration (disorders of muscles, tendons, bones, joints, peripheral blood vessels or peripheral nerves).
- Diseases caused by ionising radiation.
- Diseases due to extremes of temperature.
- Diseases caused by optical (ultraviolet, visible light, infrared) radiations including laser.

Chemical agents, this includes diseases caused by:

- Lead or its compounds.
- Asphyxiants: for example, carbon monoxide.
- Vanadium or its toxic compounds.
- Carbon disulphide.
- Organic solvents.
- Mercury or its compounds.
- Thallium or its compounds.
- Phosphorus.
- Isocyanates.
- Pesticides

Biological agents and infections or parasitic diseases, conditions include:

- Anthrax.
- Toxic or inflammatory syndromes associated with bacterial or fungal contaminants.
- Hepatitis viruses.
- Leptospirosis.
- Tetanus.

Diseases by target organ systems

Respiratory diseases include:

- Chronic obstructive pulmonary diseases, for example, caused by coal, textile or wood dust.
- Pneumoconiosis, for example, mineral dust, asbestosis, silicosis.
- Bronchopulmonary diseases caused by dusts, for example, cotton, sugar cane and hard-metal.
- Asthma caused by recognised sensitizing agents.
- Aluminium.

Skin diseases include:

- Irritant contact dermatitis.
- Physical, chemical and biological agents.
- Allergic contact dermatitis.
- Other skin diseases caused by physical, chemical or biological agents at work.

Musculoskeletal disorders, includes diseases caused by specific work activities or the environment:

- Tenosynovitis due to rapid repetitive motion, forceful exertion and awkward postures of the wrist.
- Epicondylitis due to repetitive forceful work.
- Carpal tunnel syndrome.
- Bursitis, prolonged pressure of the elbow, shoulder, hip or knee region.

Mental and behavioural disorders include:

- Post-traumatic stress disorder.
- Other work-related mental disorders.

Occupational cancer, cancer caused by agents include:
- Asbestos.
- Coal tars, coal tar pitches or soot.
- Benzene.
- Wood dust.

Dangerous occurrences

Dangerous occurrences are accidents/incidents that have the potential to cause death or serious injury and so should be reported, even though no one is injured.

The ILO Code of Practice for Recording and Notification of Occupational Accidents and Diseases does not specify classes of dangerous occurrences that should

be reported. Examples of dangerous occurrences that may be required to be reported under national laws are:

- Collapse or failure of lifting equipment.
- Failure of a pressurised closed vessel.
- Collapse of a building.
- Explosion.
- Fire.

REPORTING OF EVENTS TO EXTERNAL AGENCIES

National practices vary considerably regarding the terms and definitions used for accidents/incidents. Some countries do not have a definition and rely on a simple reference in legislation to 'accidents occurring in the workplace', for example, Botswana and the United Kingdom, or to 'injuries occurring during the performance of work', for example, Norway. Others, such as the United States, have a definition that includes explicit reference to a sudden or unexpected event, as well as violent acts.

The definitions of what is to be recorded and notified have huge implications on the data to be collected and analysed.

 Occupational accident

"An occurrence arising out of, or in the course of, work which results in:

(a) Fatal occupational injury.

(b) Non-fatal occupational injury."

Commuting accident

"An accident occurring on the direct way between the place of work and:

(a) The worker's principal or secondary residence.

(b) The place where the worker usually takes their meals.

(c) The place where the worker usually receives their remuneration, which results in death or personal injury involving loss of working time."

Occupational disease

"A disease contracted as a result of an exposure to risk factors arising from work activity."

Dangerous occurrence

"Readily identifiable event as defined under national laws and regulations, with potential to cause an injury or disease to persons at work or the public."

Figure 4-13: Definitions of what should be recorded and notified. Source: ILO Protocol 2002, C155 and Code of Practice for Recording and Notification of Occupational Accidents and Diseases.

CONSIDER

Consider the requirements of the ILO Code of Practice for Recording and Notification of Occupational Accidents and Diseases. Does your organisation have a procedure which would ensure these requirements are fulfilled?

CONSIDER

Consider how accidents/incidents are recorded in your workplace.

Who is responsible for initially reporting accidents?

4.3   Health and safety auditing

MEANING OF THE TERM 'HEALTH AND SAFETY AUDIT'

An audit is a monitoring and measurement activity that determines the level of conformity/compliance of something to agreed standards. In the field of health and safety, the standards may be derived from relevant legislation. Therefore, an audit may determine the extent to which an organisation is compliant with legislation. In addition, audits provide a level of assurance and an opportunity for the organisation to learn what is working well, where conformance and compliance has been achieved, where risks are managed effectively and positive outcomes have been achieved.

Audits of an organisation's occupational health and safety management systems can assist the organisation to managing risk and reducing harm. An audit will systematically gather information and evidence on the appropriateness and effectiveness of the management system. The information and evidence would be drawn from sources that indicate how the system is constructed and how it functions. The audit process will identify strengths, weaknesses and areas of vulnerability. Through this process audits should identify any failings of the organisation's management system to achieve its intended outcome, in the form of nonconformity with standards, noncompliance with legal requirements, ineffectiveness in its control of risk and undesired outcomes.

The auditor would analyse the information and establish findings, conclusions and recommendations to enable the organisation to take any corrective or improvement action necessary. The outcome from an audit is a report to management, which will allow health and safety to be managed successfully and provide opportunities for continual improvement.

Figure 4-14: Auditing.
Source: Shutterstock.

Audit techniques have been developed through practical experience and conventions, but there is increasing formalisation through the publication of standards such as ISO 19011 – "Guidelines for auditing management systems". It is from this international standard that the formal definition in **Figure 4-15** is taken for use on ISO 45001:2018.

"...systematic, independent and documented process for obtaining audit evidence and evaluating it objectively to determine the extent to which the audit criteria are fulfilled."

Figure 4-15: Definition of term audit.
Source: Clause 3.32, ISO 45001:2018 Occupational health and safety management systems requirements with guidance for use.

WHY HEALTH AND SAFETY MANAGEMENT SYSTEMS SHOULD BE AUDITED

By auditing the health and safety management system current performance can be evaluated, informed decisions can be made, actions prioritised and resources allocated. Audits can confirm the parts of the management system that are working well and those parts that are not working well can be identified. The audit process enables the organisation to learn what works well, for example worker and management co-operation may have led to good outcomes. By identifying the positive aspects of the management system that are working well it provides an opportunity to commend the good practices and efforts of individuals that enabled success. By identifying the negative aspects of the management system it enables the identification of failings/non-conformities, enables the organisation to learn what led to these failings/non-conformities and provides an opportunity to take action that leads to an improvement in the management system.

If the audit includes comparison of the parts of the management system with good practice it can also determine if there are gaps in the structure of the management system.

Carrying out audits of the health and safety management system can also:

- Demonstrate to workers, management, clients, contractors, other stakeholders and other interested parties (including regulatory authority) a commitment to the management of health and safety.

- Provides a proactive opportunity to learn – an alternative to incident investigation.

- Provide verification and assurance of the effectiveness of the health and safety management system.

- Improve health and safety performance and worker morale – by helping to reduce rates of incidents, ill-health and associated costs.

- Lead to a positive health and safety culture - through being prepared to learn and take action where needed.

- Provide a structured route toward continual improvement and best practice.

- Enhance the organisation's reputation - a good health and safety record is a competitive advantage and a reflection of management strength.

DIFFERENCE BETWEEN AUDITS AND INSPECTIONS

Health and safety audits assess the health and safety system, or parts of it, to determine if the system is ensuring health and safety. One of the parts of the system that may be examined by an audit is active monitoring methods, like inspections. In this way, the audit would identify if the correct people were conducting them, using the right methods, at the right frequency and how effective they were.

Inspections usually involve the examination of the workplace, work equipment or work activities; with the purpose of identification of hazards, or conditions that can lead to hazards, and to put in controls to mitigate the hazards. It can therefore be said that inspections are concerned with hazard identification in the workplace, whereas auditing relates to the systems that manage the prevention and control of hazards. In summary *see Figure 4-16.*

Requirements	Audits	Inspections
Time to carry out	High	Low
Cost	High	Low
Focus	System compliance	Hazard control
Experience	Significant and range	Low and narrow
Number of people	Two	One

Figure 4-16: Comparison of audit and inspection requirements. Source: RMS.

TYPES OF AUDIT

When planning audits, it is useful to consider the scope of the intended audit. Questions to be asked could include:

- Will the audit cover health or safety or welfare or a combination of them all?

- Will the audit cover a single department or a process that crosses many departments?

- How comprehensive will the audit be?

- What audit criteria will be used to conduct the audit against?

These questions will help to define the type of audit, its scope and objectives.

The entire health and safety management system of an organisation should be subjected to a comprehensive audit from time to time.

Individual aspects of the health and safety system, arrangements, processes and procedures can also be subjected to individual audits, for example:

- Reporting and management of incident data.
- Occupational stress.
- Work at height.
- Fire prevention and control.
- Review of health and safety as part of the management system.

In addition, depending on the organisation, specific services or products provided to others can be audited to determine if health and safety risks are adequately managed.

EXTERNAL AND INTERNAL AUDITS

See Figure 4-17 for a summary of the advantages and disadvantages of external and internal audits.

THE AUDIT STAGES

Stage 1: Notification of audit and timetable for auditing

In general, an organisation should plan a programme of internal audits to help the organisation manage its health and safety risks. This should set out what is going to be audited, when it is planned to take place, who is the responsible person for what is being audited (the auditee) and who will carry out the audit. By planning audits in advance it enables prior notification and planning by those carrying out the audit and those being audited.

Where audits are not planned as part of a programme suitable start and completion dates for the audits should be agreed with the auditors and those to be audited. Similarly, the date by which the audit report is to be made available should be agreed.

Prior to each audit formal notification should take place, confirming the arrangements. It is usual to notify the auditee of the scope of the audit and the criteria that will be audited against, for example, a specific procedure, legislation, ISO standard or technical specification. In order to prepare for an audit, it is necessary to decide who needs to be interviewed and organise a timetable in order to meet them on a planned and organised basis. The timetable for carrying out the audit should be agreed with the auditee, including locations and people involved in the audit.

It is important that the people involved in the audit are notified of arrangements for interview and informed what type of documented information they may need to make available. In this way, there is a better chance that they will be available to participate in the audit and documented information provided at the time of audit. If specific workplaces or activities will be audited then the auditee should be notified of this.

	Advantages	Disadvantages
Internal audits	• Internal audits ensure local acceptance to implement recommendations and actions improving ownership of issues found. • The auditor often has intimate knowledge of the hazards and existing work practices. • An awareness of what might be appropriate for the industry. • Familiarity with the workforce including their strengths and weaknesses. • Relatively low cost and easier to arrange. • Builds internal competence.	• May not possess auditing skills. • May not be up to date with current legislation and best practice. • The auditor may also be responsible for implementation of any proposed changes and this might inhibit recommendations because of the effect on workload. • May be subject to pressure from management and time constraints
External audits	• External audits are usually impartial; auditors will have a range of experience of different types of work practices. • May be able to offer solutions to what might be considered unsolvable problems within. • Not inhibited by criticism. • Will assess the organisation's performance without prior bias. • Will more readily identify norms to working which are no longer appropriate (someone whose opinion is not prejudiced by past knowledge of the company).	• Need to plan well to identify nature and scope of the organisation. • Individuals may not be forthcoming, be nervous or resistant to discussing their workplace with an outsider. • May seek unrealistic targets. • May be expensive

Figure 4-17: Advantages and disadvantages of external and internal audits. Source: RMS.

Stage 2: Pre-audit preparations

Competence and independence

Audits should be conducted by people who are both independent of what is being audited and competent to carry out the audits. In deciding the necessary competence for an audit, an auditor's knowledge and skills related to the following should be considered:

a) Scope of the audit - objectives and extent of the audit.

b) Audit criteria – management system requirements, legal requirements, technical requirements.

c) Size, nature and complexity – the organisation, its products, services and processes.

d) Types and levels of risks.

e) Methods for auditing.

Health and safety practitioners that have received specific training in health and safety auditing techniques would usually be able to carry out health and safety audits as they are generally sufficiently independent of the function being audited. Audits can also be carried out by managers of the organisation, provided that the managers do not audit their own efforts directly (bias must be eliminated) and that the managers concerned have been trained in auditing techniques. Often a small team will be commissioned to conduct the audit, in order to widen the experience base and establish some degree of independence. An internal audit team may comprise three main groups of people:

1) A manager.

2) A representative from the workforce.

3) A health and safety practitioner.

Extra individuals with specific knowledge and skills may join the team when specific topics are being audited. Where an audit is to be conducted of the whole health and safety management system, an approach that establishes independence would be to conduct the audit using auditors from outside the organisation or location being audited (external auditors).

Auditors must be familiar with audit techniques, familiar with work practices, have the ability to interpret standards and be in a position to be able to keep up-to-date with new information and standards. Consideration should be given to auditors receiving formal, generic audit skills training to enable them to conduct audits efficiently and effectively.

Auditor characteristics and behaviour

When selecting the people to carry out health and safety audits, employers should be sure that the auditor is suitable and competent to deal with likely situations they could encounter during audit, for example ISO 19011 Guidelines for auditing management systems says they should be:

1) Ethical - fair, truthful, sincere, honest and discreet.

2) Open-minded - willing to consider alternative ideas or points of view.

3) Diplomatic - tactful in dealing with individuals.

4) Observant - actively observing physical surroundings and activities.

5) Perceptive - aware of and able to understand situations.

6) Versatile - able to readily adapt to different situations.

7) Tenacious - persistent and focused on achieving objectives.

8) Decisive - able to reach timely conclusions based on logical reasoning and analysis.

9) Self-reliant - able to act and function independently while interacting effectively with others.

10) Able to act with fortitude - able to act responsibly and ethically, even though these actions may not always be popular and may sometimes result in disagreement or confrontation.

It should be remembered that the auditors are responsible for reporting on their findings and line management are responsible for the implementation of any corrective or improvement actions arising from the audit.

Time and resources

Audits are an in-depth analysis of conformity with standards or compliance with legislation and must not be treated lightly. The planning of the audit can be very time consuming. Evidence gathering and verification can also take a long time, depending on the scope of the audit. When arranging audits care should be taken to plan to allow enough time to evaluate findings and compile an audit report. Employers should not apply pressure on the auditor to complete audits in less time than is appropriate and must be prepared to allocate sufficient time to the task.

Similarly, the auditor might need other resources than time, such as access to documentation, measuring equipment, electronic storage facilities and research facilities (internet, library, etc.), in order to do a thorough job.

Stage 3: Information gathering

During the audit, information relevant to the audit objectives, scope and criteria, should be collected by means of appropriate sampling. The audit process must be structured and coordinated in its assessment of what is being audited. This is best achieved by utilising audit check-lists developed or obtained before the audit. The audit process involves gathering information by interviewing people, observations in the workplace and assessment of documented information.

Prior to conducting the audit, information should be obtained on the results of prior audits, the organisational structure, policy and objectives of the organisation and other important information that will enable the audit to be effective, such as shift patterns, where work is taking place off site and critical activities that may be of interest to the audit.

Whilst some of this information may be assessed in the workplace, it is useful to gather information through analysis of documented information that set out the expectation of the organisation before conducting the audit, such as:

- Health and safety policy.

- Risk assessments.

- Procedures for method statements and permits to work.

- Maintenance procedures and records.

- Training records, etc.

- Records of inspections, such as lifting equipment and portable electrical appliances.

- Health surveillance records.

- Incident reports, including accidents and ill health records.

Interviews should be structured and provide the interviewee with opportunity to express what they are doing to meet the requirements being audited. It may be necessary to utilise pro-forma questionnaires to ensure that interviews are carried out in a structured manner and that all the information required is obtained efficiently and with the minimum of inconvenience to the parties involved.

The information gathered should be sufficient to assist the analysis process that determines the extent to which the organisation, processes, procedures, products and services being audited meet the audit criteria and the how effective this is in managing health and safety risks.

Stage 4: Information analysis

The information gathered should be analysed to determine the extent that audit criteria were being met and the effective management of health and safety risks to determine audit findings.

For example, an audit of an organisation's health and safety management system will determine if:

- A comprehensive system exists in relation to the audit criteria.
- Workers at all levels are fully aware of the requirements set out in the audit criteria and their responsibilities with respect to health and safety.
- Documented system reflects the practices.
- Procedures and processes should provide satisfactory health and safety to those they are intended to protect.
- Procedures, processes are being worked to.
- There are areas that are deficient and if there are non-conformities.
- There are areas where improvements can be made.

Assessment decisions

Only information that can be subject to some degree of verification should be accepted as audit evidence. Where the degree of verification is low the auditor should use their professional judgement to determine the degree of reliance that can be placed on it as evidence. Audit evidence leading to audit findings should be recorded.

 "Records, statements of fact or other information which are relevant to the audit criteria and verifiable."

Figure 4-18: Definition of term audit evidence.
Source: ISO 19011:2018

Audit findings should include identified conformity/non-conformity, good practices and opportunities for improvement, along with their supporting evidence. Non-conformities can be graded depending on the context of the organisation and its risks. This grading can be quantitative (for example, 1 to 5) and qualitative (for example, minor, major). Non-conformities and their supporting audit evidence should be recorded.Incident investigation as a reactive monitoring measure

Stage 5: Completion of audit report

Initial feedback

At the end of the information gathering and analysis stages the auditor or audit team should summarise and provide feedback on their initial findings to local senior management and, in particular, draw attention to any issues that are of such significance as to necessitate immediate attention. In addition to the provision of a detailed written report, a verbal presentation of the report may be provided soon after the close of the audit, in order to give an early opportunity for management to learn and take action.

Written report

The outcomes from the audit process (audit findings and conclusions) should be communicated in a written report, this would usually include findings and recommendations to improve or maintain the health and safety management system. The structure, style and approach to the written report should be agreed when planning the audit.

The written report should give a clear assessment of the overall performance of the organisation against the audit criteria. It should identify deficiencies, conformity/non-conformity and make recommendations for improvement.It should also identify the observed strengths and suggest how they can be built upon. Recommendations should flow logically from the main body of the report and should be clearly connected to the findings set out in the report. They should include justification for the recommendation by referring to information summarised from findings.

All audit reports need to be accurately and clearly communicated. The written audit report should be submitted to local senior management in draft form to enable factual accuracy to be checked and to ensure that the report is understood.

Correcting non-conformities

Responsibility for carrying out actions for correcting non-conformities or making improvements should be assigned to those in line management responsible for them, together with target completion dates. It is essential that management take ownership of both the audit and the subsequent action plans.

Progress on correcting non-conformities should be monitored; this can be through reports or feedback at meetings.

4.4 Review of health and safety performance

PURPOSE OF REVIEWING HEALTH AND SAFETY PERFORMANCE

The purpose of the health and safety performance review is to evaluate the effectiveness of what is being done when compared with experience and foreseeable factors that might affect future performance. A thorough review will identify substandard health and safety practices and conditions.

Performance review will also identify trends in relation to different types of incident, by analysis of relevant incident data and enable comparison of actual performance with previously set targets. Ongoing review will enable a 'benchmark' (used to measure performance using a specific indicator, for example, accident injury frequency rate) of the organisation's performance against that of similar organisations or an industry norm to facilitate continuous improvement.

The analysis would typically identify whether control measures are in use, assess their legal compliance and effectiveness and enable decisions on appropriate remedial measures for any deficiencies identified. The process will identify any new or changed risks; assess compliance with legal requirements and accepted national/international standards. The information should be provided to the Company's most senior management (Directors), to enable the allocation of resources and where appropriate the safety committee to improve communication, thereby improving morale and motivation of the workforce. Monitoring and review is a vital component of any safety management system and is essential if the system is externally accredited and audited by a specific body.

> "Reviewing is the process of making judgements about the adequacy of performance and taking decisions about the nature and timing of the actions necessary to remedy deficiencies. [The purpose is that] the organisation learns from all relevant experience and applies the lessons."

Figure 4-19: Reviewing performance.
Source: UK, Health and Safety Executive.

Learning from all relevant experience (including that of the organisation and of other organisations) needs to be done systematically, through regular reviews of performance. The review draws on sources like data from monitoring activities and from independent audits. These form the basis of continuous improvement, necessary to maintain compliance and effectiveness. This helps to maintain a management system that is fresh, dynamic, appropriate and effective.

During a review of an organisation's health and safety management system documents that may be examined include the health and safety policy together with completed risk assessments and safe systems of work; health and safety monitoring records such as inspections, audits and surveys that have been carried out; accident/incident data including accident investigation reports and reports on near misses; health surveillance records; any communications received following visits from the enforcement authorities; insurance company reports; results and measurements from environmental surveys; records of the maintenance of equipment together with information on any failures that have occurred; details of the emergency procedures in place and records of any complaints made by workers.

In some countries conducting a management review is not only good practice or part of a recognised management system it is a legal requirement. For example, in the UK Regulation 5 of the Management of Health and Safety at Work (MHSWR) 1999 requires:

"(1) Every employer shall make and give effect to such arrangements as are appropriate, having regard to the nature of his activities and the size of his undertaking, for the effective planning, organisation, control, monitoring and review of the preventive and protective measures."

Figure 4-20: Regulation 5 of MHSWR 1999 in the UK.
Source: Management of Health and Safety at Work (MHSWR) 1999.

Legislation in Singapore specifies that any worksite with a contract sum of $30 million should conduct an internal review of their Safety and Health Management System at least once in every 6 months.

WHAT THE REVIEW SHOULD CONSIDER

The ILO Guidelines on occupational safety and health management systems, summarises the approach which should be taken to carry out a systematic management of health and safety systems review:

"3.14 - Management review

3.14.1 - Management reviews should:

(a) Evaluate the overall strategy of the OSH management system to determine whether it meets planned performance objectives.

(b) Evaluate the OSH management system's ability to meet the overall needs of the organization and its stakeholders, including its workers and the regulatory authorities.

(c) Evaluate the need for changes to the OSH management system, including OSH policy and objectives.

(d) Identify what action is necessary to remedy any deficiencies in a timely manner, including adaptations of other aspects of the organization's management structure and performance measurement.

(e) Provide the feedback direction, including the determination of priorities, for meaningful planning and continual improvement.

(f) Evaluate progress towards the organization's OSH objectives and corrective action activities.

(g) Evaluate the effectiveness of follow-up actions from earlier management reviews."

3.14.2. The frequency and scope of periodic reviews of the OSH management system by the employer or the most senior accountable person should be defined according to the organization's needs and conditions.

3.14.3. The management review should consider:

(a) The results of work-related injuries, ill-health, diseases and incident investigations; performance moni-toring and measurement; and audit activities; and

(b) Additional internal and external inputs as well as changes, including organizational changes that could affect the OSH management system.

Figure 4-21: Scope of management review.
Source: ILO Guidelines on occupational safety and health management systems.

Evaluation of compliance with legal and organisational requirements

It is a well-established principle that health and safety policy and risk assessments are reviewed. This is done when something it is believed, may affect them has changed. Legislation may change and cause a review to determine if the arrangements and controls in place are relevant to the proposed change, for example, the European Regulation on classification, labelling and packaging of substances and mixtures, which came into force in all EU Member States in January 2009, and is known by its abbreviated form, 'the CLP Regulation' or just plain 'CLP'. The CLP Regulation adopts the United Nations' Globally Harmonised System on the classification and labelling of chemicals (GHS). The legislation on classification, applies to all EU member states and the CLP system has been adopted by many countries worldwide.

Accident and incident data, corrective and preventative actions

Whilst review of accident/incident data is important it should be remembered it is after the event. Any analysis should consider the potential for each occurrence to have resulted in a more significant outcome. Two elements to consider when reviewing accident/incident data are:

- The quality of planning before the event.
- Any failure of controls. This will often include the review of current risk assessments.

When carrying out the review, it is important to consider the effectiveness of corrective and preventative actions taken following the analysis of accident/incident data. The review should also consider the timeliness of fulfilment of the corrective and preventive actions and if there was delay what caused it.

The review of corrective and preventive actions applied following incidents may indicate a need for further strategic level action.

For example, a number of accidents/incidents may have led to action to provide refresher training for the specific risks associated with the hazards causing the accidents/incidents and this may indicate a need for a more organised, strategic approach to refresher training.

Inspections, surveys, tours and sampling

The primary purpose of workplace planned general inspections is to identify general workplace hazards that are out of control before they result in any harmful outcome. Similarly, other inspections are necessary to prevent harmful outcomes for specific hazards, for example, inspections of scaffolds, excavations, lifting equipment and pressure systems. The presence of hazards or substandard conditions of equipment identified at inspection might illustrate a need to review maintenance or usage.

Surveys, tours and sampling are active methods of measuring performance. The review of the outcome of these activities can provide information on the effectiveness of objectives in improving health and safety. For example, a survey of attitudes of workers may confirm that action taken to improve communications and confidence of workers has had an effect. It is important to review what is working, as well as what is not, the approach taken to introducing the improvements may indicate how future improvements may be made with maximum acceptance of all people involved.

In some cases, organisations may pilot improvement strategies in one location or department, the outcome of this may be reviewed before extending it across the organisation. Surveys, tours and sampling may provide the data for the review which will determine if it is to go ahead on the same basis or following adjustments.

Absences and sickness

Whilst physical injury will be recorded in the accident book and may have resulted in some investigation, it should be remembered that some workplace hazards or working environments may result in absence, sickness or ill-health. Sickness or general absence data should be collected and reviewed to determine if it has resulted from work or working environment issues. Such review may determine causes such as shift working, excessive work or poor environmental conditions, including inadequate ventilation or extremes of temperature or humidity.

 **CASE STUDY**

Jahi was 47 years old and an electrician. He had been experiencing lower back pain and was finding some aspects of his job difficult. His GP had referred him for physiotherapy. Jahi discussed the issues with his manager who was supportive and had temporarily re-arranged Jahi's work to enable him to manage lighter tasks and attend physiotherapy. Jahi has a positive outlook and enjoys his job. He continues to undertake gentle exercise and his manager maintains contact with him.

Quality assurance reports

A quality assurance program is a system of policies and procedures designed to continually improve specific business processes. Whilst most popular in manufacturing and service work sectors, quality assurance reports can be used to improve the efficiency and quality of any workflow.

Quality assurance reports are useful because they will identify the *'normal or abnormal'* features of a process and as such may be a useful indicator of the level of work performance in an area. The sort of problems that affect quality can also affect health and safety and the quality assurance reports may be the first indication of problems, particularly where quality has a stronger reporting and control ethic than health and safety.

Quality assurance performance information, in its various forms, can provide a useful confirmation on the effectiveness of health and safety systems that work in parallel with quality. This is particularly useful in contributing to the process of health and safety performance review.

Audits

Audits are designed to determine the effectiveness of the management system and in particular the degree of management control.

Audits will examine all types of data, documents and records to determine the degree of system compli-ance and their suitability with the passage of time. Quality assurance audits may identify system non-compliance that can have an effect on health and safety, particularly where the organisation provides a service or product to an end user outside the organisation, for example, the defective manufacture of an electric drill intended for sale.

The review should consider information from audits because the audit process should provide a strong source of independent, verified data on the management system, implementation of objectives or specific risk being audited. This may provide a different perspective to that delivered by data from managers as it should be free from bias.

Monitoring data/records/reports

Health surveillance monitoring data may provide the review with information that supports the effectiveness of health and safety performance or, that objectives have not been met. Because monitoring is focussed on specific aspects like the presence of a substance in the atmosphere or levels of blanching of a worker's fingers due to exposure to vibration, they are providing a very tangible perspective on the success or failure of effort applied to improve health and safety. The review process should consider the results of the monitoring and the action taken following them.

Monitoring data considered at review may also include reports of hazards identified in the workplace, particularly by workers. This can provide valuable feedback on how systems affect workers in practice. The reporting of hazards should be documented and actioned within an agreed time frame with appropriate feedback, in much the same way as customers complaints might be dealt with. The review should evaluate if this is being done effectively.

Figure 4-22: Analysis of data.
Source: Shutterstock.

External communication and complaints

Feedback on levels of performance should be encouraged from customers and neighbours; this will ensure that any health and safety issues are addressed before an incident occurs. Similarly, any communication or complaints from the enforcement authorities will need to be taken into account. In addition, suppliers may provide information on the performance of things they supply, gathered from other users. All of this data informs the review process of the wider perspective of others that might be affected by the organisation or may have an opinion on how it is performing.

With the worldwide increase in social media, feedback from customers may be made through channels such as forum comments, blogs, Facebook and Twitter. Communication made through these channels will need to be monitored and taken into account with its increasing use.

Figure 4-23: Social media.
Source: Unsplash.

Results of participation and consultation

At review, it is important to involve employee representatives in the outcomes of monitoring reports to ensure good worker and employer commitment to health and safety objectives. This will enable the status of outcomes achieved against planned objectives to be considered, and appropriate actions to be decided with worker representatives to address any objectives outstanding or planned over the next period to meet legal/good practice developments or proposed changes.

Objectives met

It is essential that an organisation review its progress against its objectives in order that they can be successfully fulfilled. It could be that objectives set when implemented meet resource problems or are not as effective as planned. By identifying this through a review process as they are implemented, they are more likely to be met and be effective.

Actions from previous management reviews

In a similar way to objectives, the actions identified from previous management reviews must be reviewed to determine their progress and effectiveness. This must be planned at intervals through the year to ensure they are completed effectively and on a timely basis. It is important that the lessons learned from the review process are acted on and progress made to improve performance through this.

Legal/good practice developments

Innovation in health and safety is happening all the time and even well intending organisations can find themselves out of step with legislation and good practice. It is essential that formal reviews of changes take place and improvement plans are put into place to take account of them. Information on legislative changes is often communicated well in advance of the actual change being made. This provides an early opportunity for the possible impacts to be reviewed and plans put in place to deal with it on a timely basis.

Assessing opportunities for improvement and the need for change

Regular reviews of health and safety performance allow managers to identify success and deficiencies in performance. The need for improvement or change may not be apparent from individual incidents, but a review of accident/incident statistics over time might show that certain control measures are not working as well as intended. For example, an organisation may have introduced a series of measures to reduce the risk of manual handling, but the statistics do not show an improvement in incidents to the level expected.

The review provides an opportunity to consider why this is and allows the organisation to make improvements on a timely basis. This may involve revising practical measures, for example, review of the way work is carried out including any unreasonable time constraints placed on workers; improved training and supervision. The information gathered from the review might suggest that changes in policy and procedures are required if improvements are to be maintained.

Reviews conducted by senior management are important in helping to shape health and safety objectives for the following period, and are essential for assessing opportunities for improvement and factors that may drive the need for change, such as changes in legislation, best practice and technology.

REPORTING ON HEALTH AND SAFETY PERFORMANCE

The results of the review of health and safety performance should be reported at senior management level. This is particularly important in situations where the review has been conducted by a work group drawn from the senior management team, as this will enable all of senior management to understand and accept the implications of the review.

The results of the review should be communicated widely in the organisation and in particular to those managers that have responsibility for responding to the actions arising from the review.

It is customary to include a statement of health and safety performance, along with other risks, within the annual report. Such reports should be available to all workers and other stakeholders.

"3.14.4 - The findings of the management review should be recorded and formally communicated to:

(a) The persons responsible for the relevant element(s) of the OSH management system so that they may take appropriate action.

(b) The safety and health committee, workers and their representatives."

Figure 4-24: Recording and reporting management review.
Source: ILO Guidelines on occupational safety and health management systems.

FEEDING INTO PLANS AS PART OF CONTINUOUS IMPROVEMENT

It is important that health and safety reviews take place in an analytical way, questioning if actions taken to date have been appropriate, effective and completed. From this review process objectives and actions to improve health and safety may be identified and fed into development and improvement plans.

The strategic level plans enable the production of local level plans through information cascade. In this way health and safety in an organisation is maintained dynamically, leading to continuous improvement.

Health and safety objectives should be established for all development/improvement plans and be subject to key performance indicators (KPI's) in the same way as KPI's are established for the other key business objectives, such as production or quality. For example, active reporting at meetings should be established for health and safety items such as the status of inspections and risk assessments.

Sources of reference

Reference information provided, in particular web links, was correct at time of publication, but may have changed.

Employee Injury Benefits Convention (No. 121), ILO, 1964,

https://www.ilo.org/dyn/normlex/en/f?p=NORMLEXPUB:12100:0::NO::P12100_ILO_CODE:C121

Guidelines for auditing management systems, ISO 19011:2018,

https://www.iso.org/standard/70017.html

ILO Guidelines on Occupational Safety and Health Management Systems (ILO-OSH 2001), ISBN: 92-2-111634-4 (2nd edition), can be downloaded free from ILO web site.

http://www.ilo.org/safework/info/standards-and-instruments/WCMS_107727/lang--en/index.htm

Investigating Incidents and Accidents at Work, HSG245, HSE Books, 2004, ISBN: 978-0-7176-2827-8
http://

www.hse.gov.uk/pubns/books/hsg245.htm

List of Occupational Diseases Recommendation R194, ILO, 2010,

https://www.ilo.org/dyn/normlex/en/f?p=NORMLEXPUB:12100:0::NO::P12100_ILO_CODE:R194

Management of Health and Safety at Work (MHSWR), 1999,

http://www.legislation.gov.uk/uksi/1999/3242/contents/made

Occupational Safety and Health Convention C155, ILO Protocol of 2002, 1989,

https://www.ilo.org/dyn/normlex/en/f?p=NORMLEXPUB:12100:0::NO::P12100_ILO_CODE:P155

Occupational Safety and Health (Dock Work) Convention, ILO convention 152, 1979,

https://www.ilo.org/dyn/normlex/en/f?p=NORMLEXPUB:12100:0::NO::P12100_ILO_CODE:C152

Occupational Safety and Health recommendation R164, ILO, 1981,

https://www.ilo.org/dyn/normlex/en/f?p=NORMLEXPUB:12100:0::NO::P12100_INSTRUMENT_ID:312502

Recording and Notification of Occupational accidents and Diseases, ILO Code of Practice, Geneva, 1996, ISBN: 92-2-109451-0

http://www.ilo.org/safework/info/standards-and-instruments/codes/WCMS_107800/lang--en/index.htm

Web links to these references are provided on the RMS Publishing website for ease of use – www.rmspublishing.co.uk

STUDY QUESTIONS

1) What are three different types of inspections that might be used in any workplace and give an example of each one? (8)

2) What are the advantages and disadvantages of the use of a checklist when carrying out inspections? (8)

3) What are the functions of an accident investigation? (8)

4) What is the purpose of active and reactive monitoring? (8)

5) What are the steps to be taken when conducting an accident investigation? (8)

6) What actions should be taken following an audit? (8)

7) What are the differences between audits and inspections? (8)

8) (a) Who in the organisation should receive reports on health and safety performance from managers? (2)

   (b) Why should others in the organisation also receive the reports? (6)

9) What are the advantages and disadvantages of internally conducted audits of an organisation? (8)

10) What information should you consider when carrying out a review of health and safety performance? (8)

For guidance on how to answer NEBOSH questions please refer to the 'study question answer guidance' section located at the back of this guide.

Element 5

Physical and psychological health

Contents

INTRODUCTION

In addition to chemical and biological health hazards there are also physical and psychological hazards which can affect the health of workers. These hazards are covered by the International Labour Organization (ILO) Code of Practice 'Ambient Factors in the Workplace' and the Code should be used as a basis for controlling the hazards posed by noise, vibration, ionising and non-ionising radiation, and the effects of work-related stress.

5.1 Noise

PHYSICAL AND PSYCHOLOGICAL EFFECTS ON HEARING OF EXPOSURE TO NOISE

The ear senses **sound**, which is transmitted in the form of pressure waves travelling through a substance, for example, air, water, metals. Any audible sound is noise. The ear has three basic regions **(see Figure 5-1).**

1) The **outer** ear channels the sound pressure waves along the ear canal where they impact on the eardrum, causing it to vibrate.

2) In the **middle** ear, the vibrations of the eardrum are transmitted through the three smallest bones in the body (known as the hammer, anvil and stirrup) to the inner ear.

3) In the **inner ear** the vibrations are transferred to the cochlea (the 'hearing' organ). The cochlea is filled with fluid and contains tiny hair cells (nerves), which respond to the vibrations by bending. Movement of these tiny hair cells causes signals to be sent to the brain via the acoustic/auditory nerve where it is interpreted as recognisable sound.

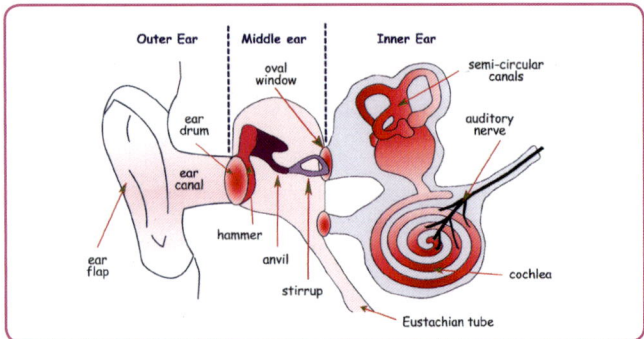

Figure 5-1: Diagram of the ear.
Source: eChalk Ltd.

Physical effects of noise

Exposure to high levels of noise can make the hair cells in the cochlea collapse and flatten. Short-term exposure usually results in the worker experiencing temporary effects, whereas long-term exposure often over a number of years, can result in permanent effects.

Typical hearing effects can include:

- Tinnitus or ringing in the ears, which may be either acute or chronic

- A threshold shift in hearing which can result in a temporary (acute) or permanent (chronic) inability to hear certain sounds. These two conditions are known as Temporary Threshold Shift (TTS) and Permanent Threshold Shift (PTS).

- **Noise-induced hearing loss (NIHL),** which can be of a temporary or permanent nature and can affect one ear or both. NIHL can be caused by a single exposure to an intense impulse sound, such as an explosion, or by continuous exposure to loud sounds over a period of time. Most NIHL is caused by permanent damage to the hair cells in the cochlea.

The Department of Occupational Safety and Health in Malaysia and United States Department of Labor have both reported statistically on the prevalence of Noise Induced Hearing Loss (NIHL) compared with other diseases:

 "Every year, approximately 30 million people in the United States are occupationally exposed to hazardous noise. Noise-related hearing loss has been listed as one of the most prevalent occupational health concerns in the United States for more than 25 years."

Figure 5-2: Exposure to noise and noise-induced hearing loss.
Source: USA, OSHA.

 "A total of 467 cases of occupational disease that were investigated were of noise induced hearing loss (NIHL) and the disease is still the most common occupational disease experienced by workers (70.4%) as compared with other diseases."

Figure 5-3: Prevalence of noise-induced hearing loss.
Source: Malaysia, Department of Occupational Safety and Health.

Permanent noise-induced hearing loss is cumulative, occurring gradually over a long period of time, and when established, the worker's hearing cannot recover.

The first sign of NIHL is often indicated by a difficulty in hearing high-pitched sounds, such as consonants (for example, 't', 'd', 's') and the voices of women and children. When more than one person is speaking or there is a background noise, the problem becomes worse.

In addition to the more constant noise exposure discussed earlier, a single very intense or explosive noise can damage the ear by dislocation of a bone or rupturing the ear drum. This is known as acoustic trauma. Background noise can also cause those with normal hearing ability to fail to hear warnings such as alarms, moving-vehicle warning horns, shouted warnings or instructions and other alarms that may sound.

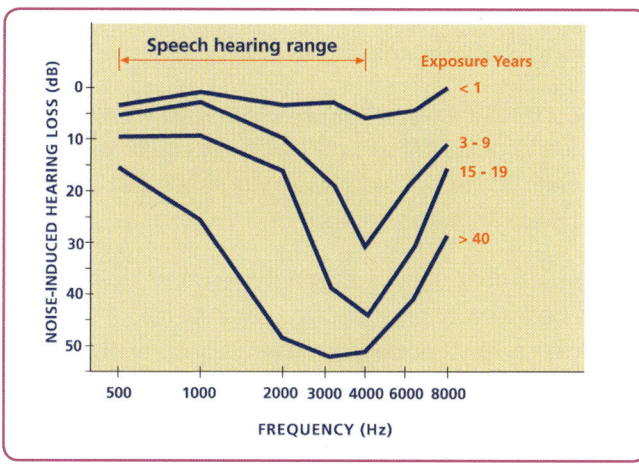

Figure 5-4: Effects of high levels of noise
Source: Australia, SafeWork SA.

Psychological effects of noise

Noise is often linked with adverse psychological effects such as stress, sleep disturbance or aggressive behaviour. It is frequently cited as the cause of conflict between workers, particularly in a noisy office environment where some individuals may need to concentrate on complex issues, but they find this difficult or impossible because of background noise levels.

In addition, the loss of hearing in the speech hearing range leads to a feeling of isolation as the person affected cannot contribute so easily to conversations, and pastimes may become affected, for example, listening to music, radio and television.

 **REVIEW**

What are the possible effects on hearing from exposure to noise?

What are the causes of noise-induced hearing loss?

THE MEANING OF COMMON SOUND MEASUREMENT TERMS

Sound power and pressure

For noise to occur power must be available. It is the sound power of a source (measured in watts) that causes the sound pressure (measured in pascals, Pa) to occur at a specific point.

Intensity and frequency

The *amplitude* of a sound wave represents the *intensity* of the sound pressure. When measuring the *amplitude* of sound there are two main parameters of interest *(as shown in Figure 5-5)*. One is related to the energy in the sound pressure wave and is known as the 'root mean square' (rms) value, and the other is the 'peak' level.

We use the 'rms' sound pressure for the majority of noise measurements, apart from some impulsive types of noise when the peak value is also measured.

A sound can have a *'frequency'* or *'pitch',* which is measured in cycles per second (Hz). Frequency in this context represents the number of times in a given time period that the sound wave repeats itself.

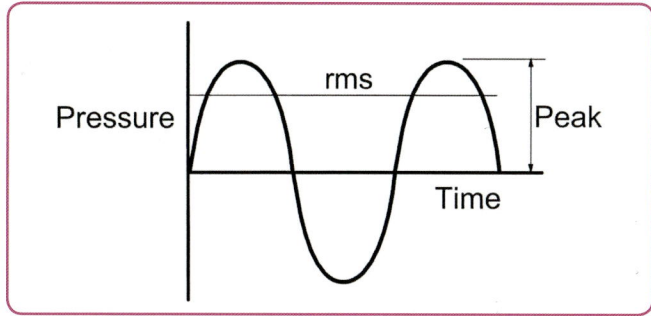

Figure 5-5: Rms and peak levels of a sound wave.
Source: RMS.

The decibel scale

Sound intensity or pressure is measured in a unit known as a pascal (Pa). The ear can detect pressures over a very wide range, from 20 µPa - 20 Pa. To measure in pascals therefore requires a very large range to exist and this is often inconvenient. A more helpful way of measuring sound is to use the decibel range. A decibel (dB) is a unit of sound pressure (intensity) measured on a logarithmic scale from a base level taken to be the threshold of hearing (0 dB). Typical noise levels are listed in *Figure 5-6*.

Source	DB
CHAINSAW	120
SMOKE DETECTOR AT 1 METRE	105
MACHINE SHOP	90
RADIO IN AVERAGE ROOM	70
LIBRARY	30
THRESHOLD OF HEARING	0

Some typical noise levels for equipment used on construction sites are:

DISC CUTTER	99-115 db
HAMMER DRILL	102-111 db
BREAKERS	103-113 db
EARTHMOVER	87-94 db

Some typical manufacturing noise levels:

CONVEYOR	115 db
LATHE	96 db
PACKING MACHINE	106 db

Figure 5-6: Typical noise levels.
Source: RMS.

A problem with decibels is that they are based on a logarithmic scale and cannot be added together in the conventional way. The basic way to interpret decibels is that whenever noise/sound doubles in intensity (loudness) the decibel reading will increase by 3; and when it halves the decibel reading will decrease by 3; for example:

- 2 dB + 2 dB = 5 dB (deciBel arithmetic).
- 85 dB + 85 dB = 88 dB (deciBel arithmetic).
- A 3 dB increase in sound pressure level corresponds to a doubling of the sound energy.
- A 10 dB increase in sound pressure level corresponds to a 10 times increase of the sound energy.
- A 20 dB increase in sound pressure level corresponds to a 100 times increase of the sound energy.

Weighting scales - the terms dB(A) and dB(C)

The human ear can hear sound over a range of frequencies, from 20 Hz up to approximately 20,000 Hz (20 kHz). However, the ear does not hear the same at all frequencies; it naturally reduces (attenuates) low frequencies and very high frequencies, respectively, to the range of speech. To take account of the response of the human ear, sound level meters use weighting scales or filters. The most widely used sound level filter is the A scale. Using this filter, the sound level meter is thus less sensitive to very high and very low frequencies. Measurements made on this scale are expressed as dB(A) or referred to as 'A-weighted'.

The majority of measurements are made in terms of dB(A), although other weightings are used in some circumstances. One of these is the C scale, which is used to assess very high sound pressure levels such as the acoustic emissions of machines. The C scale is used to determine peak sound pressure levels and is particularly useful for impact or explosive noises. It has a broader spectrum than the A-weighted scale and is more accurate at higher levels of noise. Measurements made on this scale are expressed as dB(C).

THE NEED FOR ASSESSMENT OF EXPOSURE TO NOISE

Employers have a responsibility to assess the risks from noise in order to reduce the risk of hearing damage to their workers by controlling exposure to noise. The employer should make a suitable and sufficient assessment of the risk created by work that is liable to expose workers to risk from noise.

Several countries set action levels and limits relating to exposure to noise in the workplace, and such levels and limits need to be considered when assessing the risks faced by workers and others.

The assessment should establish which workers are at risk of hearing damage and the level of risk. It will also identify sources of noise that particularly contribute to the noise level workers are exposed to, for example, equipment and specific activities. This will enable analysis of the options available to control the noise at source or by other means. The assessment will also help the employer when providing suitable hearing protection, marking out hearing protection zones and giving information, instruction and training to workers. The results of the risk assessment and controls implemented should be recorded to provide evidence of a structured risk assessment process.

Risk assessment

When conducting the risk assessment, consideration should be given to workers at particular risk. The assessment will need to identify the type, duration, effects of exposure including any additional exposure at the workplace (for example, in rest facilities) to determine whether the exposure limit/action values have been exceeded. Information from health surveillance records and manufacturer's information on noise levels should also be reviewed. The level of noise workers are exposed to should be assessed by:

- Observation.
- Reference to information on expected levels for work conditions and equipment.
- If necessary by measurement of the level of noise to which their workers may be exposed.

It is not always necessary to carry out measurements of noise exposure as part of the assessment; an estimate of noise levels may be enough to decide that controls are required. Estimation by observation may be sufficient to indicate there is a noise problem.

One example involves the assessor determining how easy it is for two people to hold a conversation at a distance of one or two metres from each other.

Test	Probable noise level	A risk assessment will be needed if the noise is at the level for more than:
The noise is intrusive but normal conversation is possible	80 dB	6 hours
You have to shout to talk to someone 2 m away	85 dB	2 hours
You have to shout to talk to someone 1 m away	90 dB	45 minutes

Figure 5-7: Simple tests to get a rough estimate on whether a risk assessment is required. Source: RMS.

Two metre rule: If conversation is difficult (need to raise the voice or repeat words) at a distance of two metres apart the noise level is likely to be above 85 dB.

One metre rule: If conversation is difficult (need to raise the voice or repeat words) at a distance of one metre apart the noise level is likely to be above 90 dB.

If it is necessary to be more certain whether noise exposure levels exceed acceptable values, **measurements** of actual noise levels may be required. A detailed assessment should include consideration of:

- Level, type and duration of exposure, including any exposure to peak sound pressure.
- Effects of exposure to **noise** on workers or groups of workers whose health is at particular risk from such exposure.
- Indirect effects on the health and safety of workers resulting from the interaction between noise and audible warning signals.
- Information provided by the manufacturers of **work** equipment.
- Availability of alternative equipment designed to reduce the emission of **noise**.
- Any extension of exposure to noise at the workplace beyond normal **working** hours, including exposure in rest facilities supervised by the employer.
- Appropriate information obtained following health surveillance, including, where possible, published information.
- Availability of personal hearing protectors with adequate noise-reduction (attenuation) characteristics.

Section 9.2.3 of the ILO Code of Conduct for 'Ambient Factors in the Workplace' specifies that noise measurements conducted in relation to noise assessments should be used to:

"(a) Quantify the level and duration of exposure of workers and compare it with exposure limits as established by the competent authority or internationally recognised standards.

(b) Identify and characterise the sources of noise and the exposed workers.

(c) Create a noise map for the determination of risk areas.

(d) Assess the need both for engineering noise prevention and control and for other appropriate measures and for their effective implementation.

(e) Evaluate the effectiveness of existing noise prevention and control measures."

Noise is measured using a sound pressure level meter, which works, in simple terms, by converting pressure variations into an electrical signal.

Figure 5-8: Noise measurement.
Source: Pulsar Instruments Plc.

This is achieved by capturing the sound with a microphone, pre-amplification of the resultant voltage signal and then processing the signal into the information required dependent on the type of meter. For example, 'A' weighting, the A-weighting filter covers the full audio range - 20 Hz to 20,000 Hz and represents the audible response of the human ear at these levels.

The microphone is the most critical component within the meter as its sensitivity and accuracy will determine the accuracy of the final reading. Meters can be set to fast or slow response depending on the characteristics of the noise level. Where levels are rapidly fluctuating, rapid measurements are required and the meter should be set to fast time weighting.

Measurement

In most situations, workers are exposed to variations in noise level over a period of time. When measuring noise we need to determine the average, or 'equivalent continuous level', over a period of time. This is often known as the L_{eq}. Sound pressure level meters are used to measure the:

- Equivalent continuous sound level (L_{eq}) - an average measure of intensity of sound over a reference period, usually the period of time over which the measurement was taken. Measured in dB(**A**).
- Daily personal exposure level, dB(**A**), $L_{EP,d}$ - this is equivalent to the L_{eq} over an 8-hour working day. The $L_{EP,d}$ is directly related to the risk of hearing damage.
- Peak pressure level, L_{peak} - this is the peak level of the sound pressure wave with no time constant applied. This is the loudest noise experienced during the measuring process. For noise at work measurements, the peak level should be measured in dB(**C**).

Personal dosimeters are available for situations where the task of the worker involves movement around the workplace and exposure is likely to vary. Sound pressure level meters should be calibrated using a portable acoustic calibrator and batteries checked before, during and after each measurement session. Laboratory calibration should be carried out annually or according to the manufacturer's instructions.

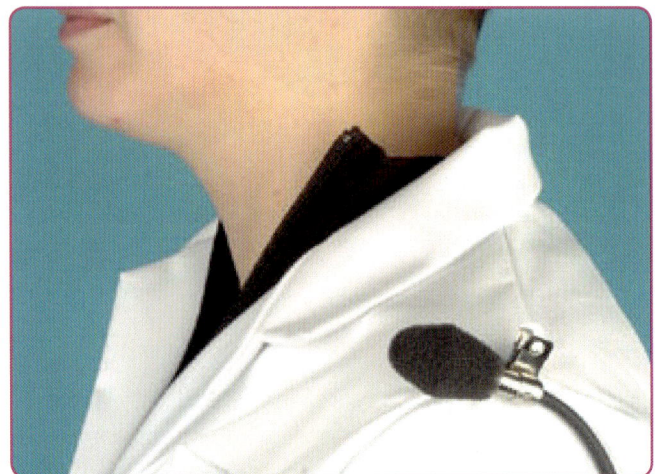

Figure 5-9: Correct position of a personal noise dosimeter.
Source: UK, HSE L108.

Workers or their representatives should be consulted, and significant findings, measures taken or planned to control the risks should be recorded. Reassessment should be carried out after action is taken to control exposure, or after a reasonable time, to establish the effectiveness of the controls.

Comparison of noise exposure levels with recognised exposure limit standards

Article 16 of the ILO Convention C155 'Occupational Safety and Health' requires employers to:

"1) Ensure that, so far as is reasonably practicable, the workplaces, machinery, equipment and processes under their control are safe and without risk to health.

2) Ensure that, so far as is reasonably practicable, the chemical, physical and biological substances and agents under their control are without risk to health when the appropriate measures of protection are taken."

Noise in the workplace is a likely hazard related to machinery and work processes and is an example of a physical agent (health) risk that should be controlled. As people respond differently to noise, the level at which noise will start to cause harm to workers' hearing varies. Research has established that long periods of repeated exposure to workplace noise levels between 75 dB(A) and 80 dB(A) present a small risk of the average worker developing a hearing disability. As noise levels increase, the risk becomes greater. For example, exposure to noise levels of 90 dB(A)-95 dB(A) presents a considerably greater risk of a worker developing hearing disability.

Information from research has been used to develop international noise exposure standards in the form of International Standardization Organization (ISO) 1999: 2013 'Acoustics - estimation of noise-induced hearing loss', which have been adopted in a number of countries.

Standards concerning acceptable levels of noise, and therefore exposure limits, are usually based on an 8-hour work period. They may also provide exposure limits for shorter and longer working periods. ISO 1999: 2013 recommends an acceptable level of noise is 85 dB(A) and many national competent authorities specify an acceptable 8-hour time-weighted average (TWA) noise level somewhere between 85 dB(A) and 90 dB(A), a common maximum acceptable noise exposure limit value being 87 dB(A) for an 8-hour work day. For example, 87 dB(A) is specified in Canadian federal legislation and in European legislation, though the European legislation also specifies action to control exposure at levels of 80 and 85 dB(A). 87 dB(A) is the level at which these countries have decided that a worker may be exposed to noise, taking into account personal hearing protection, each working day, with an acceptable level of risk that they may suffer from NIHL.

In Australia, the national standard for occupational noise follows the ISO 1999: 2013 recommendation and sets an acceptable noise exposure level of 85 dB(A), which may be achieved by the use of personal hearing protection. In the USA, OSHA specifies a legally acceptable noise level (permissible exposure limit - PEL) of 90 dB(A), but expects hearing conservation programmes to be introduced at levels of 85 dB(A).

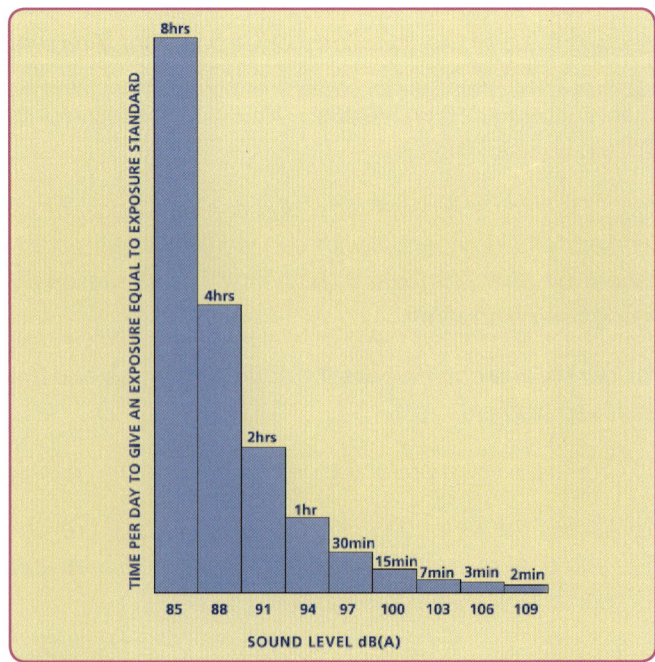

Figure 5-10: Exposure times.
Source: Australia, SafeWork SA.

It should be remembered that exposure to noise below the acceptable noise level does not mean a safe condition exists; it means that an 8-hour exposure to the acceptable noise level is considered to represent an acceptable level of risk to workers' hearing in the workplace. Some workers may be affected by noise exposures below this level, and therefore most national legislation will include a requirement to reduce exposure as far as is reasonably practicable, and at least below the legally acceptable noise level.

The amount of damage caused by noise depends on the total amount received by the worker's hearing over time. This means, as noise becomes more intense it causes damage to hearing in less time. A 3 dB(A) increase in noise level will produce twice the sound energy and cause the same damage in half the time. Therefore, the acceptable duration of exposure at this noise level is halved. For example, if a worker is exposed to a noise level of 88 dB(A) the acceptable duration would be 4 hours, half that of a worker exposed to 85 dB(A), 8 hours.

Similarly, 15 minutes of working in noise levels of 100 dB(A) may cause the same damage as 8 hours working in 85 dB(A).

Action and limit values approach

The European graduated approach to occupational exposure values has established a maximum acceptable noise exposure limit (exposure limit value) and two action levels. The exposure limit value is fixed at 87 dB(A) and the exposure action levels are fixed at 80 dB(A) (known as the lower level); and 85 dB(A) (upper level), taking into account the attenuation provided by any personal hearing protection worn by the workers.

These are levels of 'daily average noise exposure', as a time-weighted average over an 8-hour working day. If a worker's exposure to noise in the workplace varies greatly, the employer can choose to calculate weekly instead of daily noise exposure in determining whether levels or limits are exceeded. Weekly noise exposure means the average of daily noise exposures over a week and normalised to 5 working days.

The European approach also recognises the potential harm that may be done by particularly high-intensity noise experienced over a short period and sets limit and action levels for this situation in the form of peak sound pressures.

Lower exposure action values

Where a worker is likely to be exposed to noise at or above the lower exposure action level values, the employer must make hearing protection available if requested and provide the workers and their representatives with suitable and sufficient information, instruction and training. This should include:

- The nature of risks from exposure to noise.

- Organisational and technical measures taken in order to comply.

- Exposure limit values and upper and lower exposure action values.

- Significant findings of the risk assessment, including any measurements taken, with an explanation of those findings.

- Availability and provision of personal hearing protectors and their correct use.

- Why and how to detect and report signs of hearing damage.

- Entitlement to health surveillance.

- Safe working practices to minimise exposure to noise.

- The collective results of any health surveillance in a form calculated to prevent those results from being identified as relating to a particular person.

Figure 5-12: Noise hazard sign.
Source: RMS.

	Lower exposure action level values	Upper exposure action level values	Exposure limit values
Daily or weekly personal noise exposure (A-weighted)	80 dB	85 dB	87 dB
Peak sound pressure (C-weighted)	135 dB	137 dB	140 dB

Figure 5-11: Noise exposure values.
Source: UK, Control of Noise at Work Regulations.

Figure 5-13: Mandatory hearing protection sign.
Source: ISO 1710, Graphical symbols: safety colours and safety signs.

Upper exposure action values

Where the noise exposure of a worker is likely to be at or above the upper exposure action level values, the employer must:

- Provide workers with hearing protection, and ensure that it is worn.
- Ensure that the area is designated a hearing protection zone, fitted with mandatory hearing protection signs.
- Ensure access to the area is restricted where practicable.
- So far as reasonably practicable, ensure those workers entering the area wear hearing protection.

Exposure limit values

The employer must ensure that workers are not exposed to noise above an exposure limit value. If an exposure limit value is exceeded the employer must immediately:

- Reduce exposure to below the limit level.
- Identify the reason for the limit value being exceeded.
- Modify the organisational and technical measures to prevent a recurrence.

ILO CODE OF PRACTICE FOR WORKERS

The ILO Code requires that:

- For noise levels above the lower action level, workers must use any control equipment supplied by the employer and report any defects.
- For noise levels above the upper action level, workers must wear the hearing protection provided.
- Workers must notify their employer and seek medical assistance if their hearing deteriorates.
- Workers must present themselves for health surveillance.

BASIC NOISE CONTROL MEASURES

Employers should ensure that risk to their workers from exposure to noise is either eliminated at source or, where this is not reasonably practicable, reduced to as low as is reasonably practicable. Consideration should be given to reducing noise at source, blocking the transmission path and preventing worker exposure.

The International Labour Organization (ILO) Code of Practice 'Ambient Factors in the Workplace' and the Code should be used as a basis for controlling the hazards posed by noise.

The ILO Code requirements for noise can be summarised as requiring employers to:

- Assess noise levels.
- Reduce the risks from noise.
- Provide health surveillance for workers.
- Provide information, instruction, training and supervision as necessary.

Enclosure:	Total enclosure to contain noise at source.
Isolation:	A form of separation between the noise and the worker by distance or use of a barrier of an absorbent character (for example, an acoustic absorbent wall) in the path of noise transmission or relocation of workers into another room remote from the noise source.
Absorption:	When noise passes through porous materials (for example, foam, mineral or wool) some of its energy is absorbed.
Insulation:	Imposing a barrier (for example, a brick wall or lead sheet) between the noise source and the workers.
Lagging:	Insulation of pipework and fluid containers to reduce noise levels.
Damping:	Mechanical vibration can be reduced through conversion into heat by damping materials, for example, use of plastic gears, rubber coating on conveyors or rollers, reinforcement of metal panels.
Silencing:	Pipes/boxes can be designed to reduce air/gas noise (for example, engine exhaust silencers or duct silencers).
Work practice:	Modify material-handling processes to reduce the noise from shock and impact, for example, reducing the distance where objects fall onto hard surfaces or fixing damping material to surfaces or containers. Review frequency of maintenance programmes, for example, equipment lubrication and replacement of worn bearings.

Figure 5-14: Ways of reducing noise at source.
Source: RMS.

Reducing noise at the source

A company policy should be established which purchases only the quietest equipment and replaces outdated, noisy machinery for existing workplaces. This can be done in a number of ways, **see Figure 5-14.**

Blocking the noise transmission path

This can be done by carrying out the following:

- Relocating noisy machines or processes to remote areas of the workplace.
- Fitting sound-absorbent materials to ceilings and walls.
- Enclosing noisy machinery within sound-absorbent materials.
- Mounting noisy floor-standing machinery on rubber pads to reduce vibration.
- Fitting flexible or fixed screens or curtains of sound-absorbent material.

Noise can be controlled at different points in the transmission path that lead to exposure:

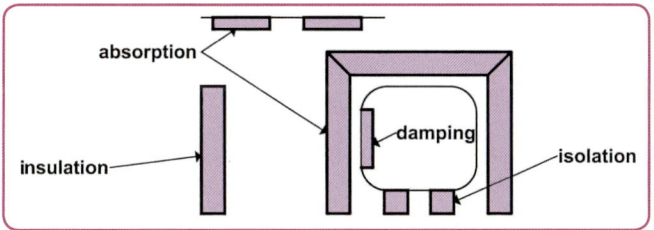

Figure 5-15: Main transmission control methods.
Source: RMS.

Preventing worker exposure to noise (Administrative controls)

This can be done by implementing the following:

- Scheduling noisy work for times when as few workers as possible are present.
- Building sound-proof booths for operators of noisy equipment.
- Minimising the amount of **time** a worker is exposed to noise, for example, by job rotation.

The employer has to take care that noise levels in rest facilities are suitable for their purpose. The employer should adapt measures provided to suit a group of workers or individual worker whose health is likely to be at particular risk from exposure to noise at work. Workers or their representatives should be consulted on the measures used.

 **REVIEW**

What control measures could be used to help reduce noise levels in a workplace?
Explain the 'two metre' rule in relation to noise in the workplace.

PERSONAL HEARING PROTECTION

Purpose

The purpose of personal hearing protection is to protect the user from the adverse effects on hearing caused by exposure to high levels of noise. All hearing protection must be capable of reducing exposure to below the acceptable noise level set nationally by the competent authority, for example, 85 dB(A) averaged over 8 hours.

Figure 5-16: Helmet mounted ear muffs.
Source: RPA.

The provision of hearing protection should only be considered after all attempts to reduce the exposure to noise by other means have proved ineffective in reducing noise levels satisfactorily, or where exposure above any nationally imposed action levels requiring hearing protection to be provided exists.

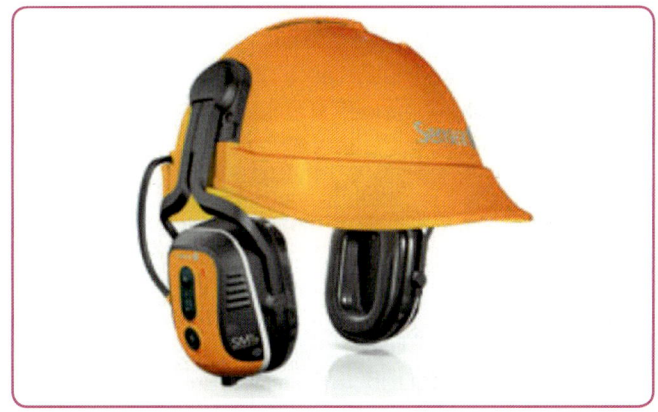

Figure 5-17: Ear muff with two-way communication.
Source: Davro Online Safety.

Custom-fit earplugs

A wide range of custom-fit earplugs is now available for people who work in noisy environments such as working with machinery, maintenance activities or musicians.

Custom-made earplugs ensure a perfect seal to the ear canal. The fit is designed to replicate the exact shape of each individual ear canal and this ensures a comfortable fit, greatly reducing workers' reluctance to wear them.

Custom-fit earplugs require impressions of the ear to be taken first by a trained audiologist so that a proper seal can be created. Normally it takes four to six weeks to make the earplugs after the impressions are taken.

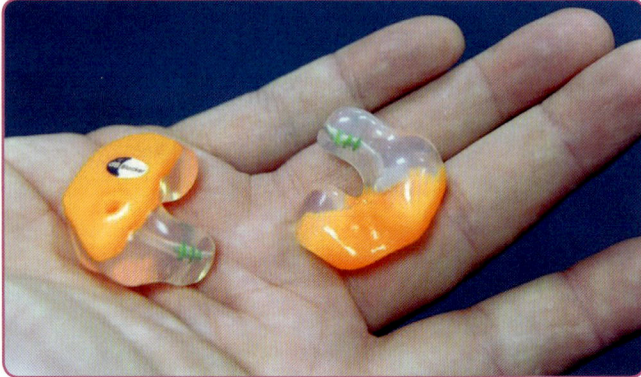

Figure 5-18: Custom-fit earplugs.
Source: US Air Force.

Application and limitations of earmuffs and earplugs

See Figure 5-19

Selection

When selecting personal hearing protectors employers must take into consideration several factors, including:

- Noise attenuation (reduction) capability.
- Compatibility with other PPE.
- Suitability for the work environment.
- Readily available.
- Comfort and personal choice.
- Issue to visitors.
- Provision of information and training.
- Care and maintenance.

Use

All personal hearing protectors should be used in accordance with employer's instructions, which should be based on manufacturer's instructions for use. Personal hearing protectors should only be used after adequate training has been given. Also, adequate supervision must be provided to ensure that training and instructions are being followed. Personal hearing protectors may not always provide adequate protection, due to any of the following reasons:

- Inadequate or lack of training
- Not fitted properly
- Long hair, spectacles or earrings may cause a poor seal to occur with banded ear muffs.
- Not wearing personal hearing protectors all of the time.
- Personal hearing protectors may become damaged, for example, ear muffs cracked.
- Specification of personal hearing protectors does not provide sufficient attenuation.

Maintenance

Employers should ensure that any personal hearing protection used are maintained in efficient working order. Simple maintenance can be carried out by the trained wearer, but more intricate repairs should only be carried out by specialist personnel. Maintenance in this context includes actions to keep personal hearing protection in use in good order. This will include the timely disposal of personal hearing protection that no longer affords adequate protection due to the effects of use.

Earmuffs:	Application:	Limitations:
These completely cover the ear and can be: - Banded. - Helmet mounted. - Communication muffs.	- Worn on the outside of the ear so less chance of infection. - Clearly visible therefore easy to monitor. - Can be integrated into other forms of personal protective equipment (PPE), for example, head protection.	- Can be uncomfortable when worn for long periods. - Incompatibility with other forms of PPE. - Effectiveness may be compromised by, for example, long hair, spectacles. - Requires correct storage facilities and regular maintenance.
Earplugs:	**Application:**	**Limitations:**
These are inserted in the ear canal and can be: - Pre-moulded. - User formable. - Custom moulded. - Banded plugs.	- Easy to use and store - but must be inserted correctly. - Available in many materials and designs, disposable. - Relatively lightweight and comfortable. Can be worn for long periods.	- They are subject to hygiene problems unless care is taken to keep them clean. - Correct size may be required. Should be determined by a competent person. - Interferes with communication. - Worn inside the ear, difficult to monitor.

Figure 5-19: Application and limitations of various types of hearing protection. Source: RMS.

Attenuation factors

It is essential to match the attenuation (noise reduction) provided by a personal hearing protector to the noise level and the desired attenuation performance. In addition, the frequency of the noise may need to be taken into account to ensure the personal hearing protector performs effectively at the frequency of the noise to which the worker is exposed. The attenuation data associated with personal hearing protectors should be supplied by the manufacturer with the product. It may come in a number of forms, using rating numbers that are applicable in different parts of the world. The rating numbers can provide a quick guide to the effectiveness of personal hearing protectors.

United States of America (USA)

NRR (Noise Reduction Rating) - this rating is used in the USA and is accepted for use in a variety of other countries. The range of NRRs available in the USA market extends from approximately 0 to 33 decibels. Data on the packaging of the personal hearing protector includes a chart showing mean attenuation values and standard deviations at each of the seven test frequencies (from 125 Hz through to 8,000 Hz).

Europe

SNR (Single Number Rating) - this rating number is used by the European Union and affiliated countries. In addition to an overall rating, the SNR further rates personal hearing protectors in terms of the particular noise environments in which they will be used: H for high-frequency noise environments, M for mid-frequency, and L for low-frequency.

The HML designation does not refer to noise level, but to the spectrum of noise. The noise we hear is made up of a wide range of different frequencies. Frequencies that are close together form a 'bandwidth' which, within the field of noise, is termed an 'octave band'. An octave band is defined where the upper frequency is twice the lower frequency thus a frequency range of 22Hz to 44 Hz would form one band width or Octave. The octave band is given by citing the centre frequency (31.5 Hz in the bandwidth given).

Common octave bands are 31.5 Hz, 63Hz, 125Hz, 250Hz, 500Hz, 1kHz, 2kHz, 4kHz and 8KHz. The standard test frequencies are 63Hz to 8kHz, therefore a personal hearing protector might be rated with an SNR of 26, H=32, M=23, and L=14.

The SNR value is the result of a lengthy mathematical calculation; it gives a single-number rating of a personal hearing protector's attenuation for a specified percentage of the population. The SNR is significantly lower than the average attenuation across all of its test frequencies as the calculation contains correction factors to make it applicable to the broader population. While it is not the perfect measure of attenuation, SNR is a very useful standardised method for describing a personal hearing protector's attenuation in a single number.

For example, if an environment has a weighted noise measurement of 100 dB(A) then by using an earplug with a SNR rating of 26 dB it will reduce the noise to 74 dB(A).

To determine the predicted noise attenuation (PNA) it is necessary that a noise reading be taken in both A-weighting and C-weighting mode, at the worker's ear. If the difference between the A-weighted reading and the C-weighted reading is greater than 2 a formula using the medium and low values is used; if it is less than or equal to 2 a formula using the high and medium value is used. This method enables the effects of the frequency of the noise to be taken into account by using a sound level meter with A and C weighting.

If readings have been taken using an octave band sound level meter, the octave band method can be used to produce a more accurate reflection of the effectiveness of hearing protectors. This can help to ensure a close match between the hearing protector and the noise concerned. It involves quite complicated mathematics, so to assist employers, the UK Health and Safety Executive (HSE) has produced a calculator, which can be found at http://www.hse.gov.uk/noise/calculator.htm. Calculators are also available for SNR and HML methods.

Whichever method is used to determine the PNA of hearing protection, in the UK, the HSE recommends reducing the value by 4 dB to reflect what they see as 'real-world factors'. This takes account of variances in fit or other user factors that may limit the effectiveness of the hearing protection.

Australia and New Zealand

SLC_{80} (Sound Level Conversion) - the SLC_{80} is a rating number used in Australia and New Zealand. It is an estimate of the amount of protection attained by 80% of users, based upon laboratory testing across a range of frequencies (from 125 Hz through to 8,000 Hz). Depending on the level of attenuation in the SLC_{80} rating, a classification is assigned to a personal hearing protector.

A Class 1 protector (SLC_{80} 10 to 13) may be used in noise up to 90 dB, a Class 2 protector (SLC_{80} 14 to 17) to 95 dB, a Class 3 protector (SLC_{80} 18 to 21) to 100 dB, and so on in 5 dB increments. Packaging for the personal hearing protector will often show the SLC_{80} number followed by the classification, for example, (SLC_{80} 27, Class 5).

THE ROLE OF HEALTH SURVEILLANCE

The role of health surveillance is to provide early detection of work-related ill-health. Audiometry is a medical testing procedure that establishes hearing sensitivity across a range of sound frequencies, which can then be monitored over time. It will assist with the identification of noise hazards and the evaluation of noise control measures.

By conducting health surveillance from the start of a worker's assignment to work it is possible to detect early signs of hearing loss and provide early intervention to limit the continuing effects.

Health surveillance can assist with confirming the success of noise controls at source and support the promotion of personal hearing protection. If a risk assessment indicates a risk to the health and safety of workers who are, or are liable to be, exposed to noise they should be put under suitable health surveillance (including testing of their hearing). The employer must keep and maintain a suitable health record. The employer should, providing reasonable notice is given, allow the worker access to their health record.

Where, as a result of health surveillance, a worker is found to have identifiable hearing damage, the employer should ensure that the worker is examined by a doctor. If the doctor, or any specialist to whom the doctor considers it necessary to refer the worker, considers that the damage is likely to be the result of exposure to work-related noise, the employer should:

- Ensure that a suitably qualified person informs the worker accordingly.
- Review the risk assessment.
- Review any measure taken to control the noise risk.
- Consider assigning the worker to alternative work.
- Ensure continued health surveillance.
- Provide for a review of the health of any other worker who has been similarly exposed.

Workers must, when required by the employer and at the cost of the employer, present themselves during working hours for health surveillance procedures.

REVIEW

List four items of personal hearing protection.
What control measures could be used to ensure a new worker's hearing protection in a noisy work environment?
Explain why it is important to determine noise action levels in some workplaces.

5.2 Vibration

INTRODUCTION

Hand-held vibrating equipment and machinery can produce risks to the health of workers through hand-arm vibration - often known as HAVS (hand-arm vibration syndrome). Standing or sitting on vibrating plant or machinery can result in WBVS (whole-body vibration syndrome).

Employers should seek to eliminate work which exposes workers to vibration risk or, where this is not reasonably practicable, assess the risk and seek to reduce or control the exposure to the lowest practicable level.

Advice for eliminating, limiting and controlling exposure exists within the ILO Code of Practice 'Ambient Factors in the Workplace'.

THE EFFECTS ON THE BODY OF EXPOSURE TO VIBRATION

Occupational exposure to vibration may arise in a number of ways, often reaching workers at intensity levels disturbing to comfort, efficiency, health and safety. Long-term, regular exposure to vibration is known to lead to permanent and debilitating health effects such as vibration white finger, loss of sensation, pain, and numbness in the hands, arms, spine and joints. These effects are collectively known as hand-arm or whole-body vibration syndrome. In the case of whole-body vibration it is transmitted to the worker through a contacting or supporting structure, which is itself vibrating, for example, a ship's deck, the seat or floor of a vehicle (tractor or dumper truck, - used for transporting loose material on a construction site), or where a whole structure is shaken by machinery (for example, in the processing of coal, iron ore or concrete), where the vibration is intentionally generated for impacting.

By far the most common route of harm to the human body is through the hands, wrists and arms of the subject - so-called segmental vibration, where there is actual contact with the vibrating source.

Measurement

The assessment of vibration exposure (acceleration) is measured in meters per second squared (m/s^2). The risk to health from vibration is affected by the frequency content of the vibration. When vibration is measured in accordance with BS EN ISO 5349-1:2001, vibration frequencies between 8 and 16 Hz are most important, and frequencies above and below this range make a smaller contribution to the measured vibration magnitude.

This process is called frequency weighting. Vibration meters intended for HAV and WBV measurement are equipped with a frequency weighting filter, to modify their sensitivity at different frequencies of vibration. In order to establish a complete measurement of vibration exposure it is necessary to measure the vibration on three axes, 'x', 'y' and 'z'; as shown in *Figure 5-20*, for the hand which relates to the axes in which the vibration is entering the hand and *Figure 5-21*, for the whole body. Where 'x' is taken as before and after vibration, 'y' as side-to-side vibration and 'z' as vertical vibration.

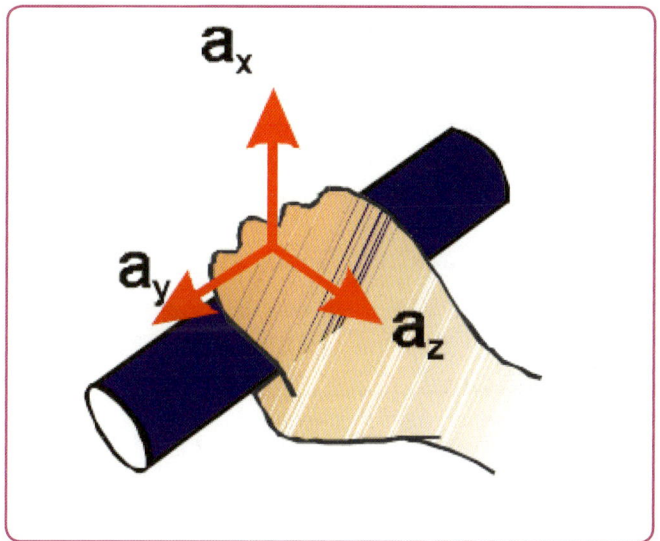

Figure 5-20: Frequency weighting filter.
Source: UK, HSE, RR795.

The vibration is measured in one axis at a time and once all three axes have been individually measured the combined vibration exposure of the worker can be determined.

The amount of exposure is determined by measuring acceleration in the units of m/s². Several types of instruments are available for measuring acceleration, the rate of change of velocity in speed or direction per unit time (for example, per second). Measuring acceleration can also give information about velocity and amplitude of vibration. The degree of harm is related to the magnitude of acceleration.

A typical vibration measurement system includes a device to sense the vibration (accelerometer), and an instrument to measure the level of vibration. The accelerometer produces an electrical signal and the size of this signal is proportional to the acceleration applied to it. A weighting system provides a single number as a measure of vibration exposure which is expressed as the frequency-weighted vibration exposure in metres per second squared (m/s²), units of acceleration.

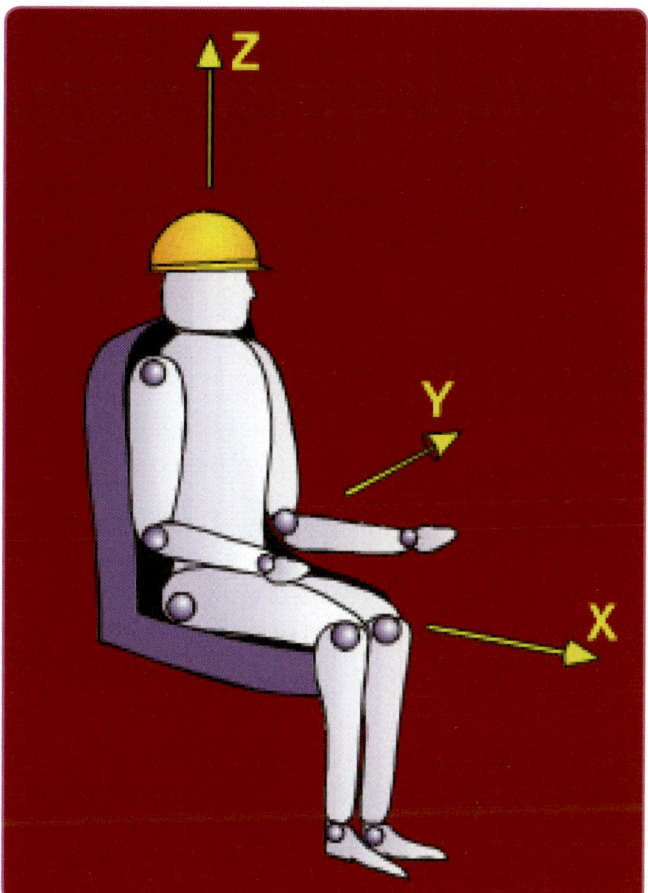

Figure 5-21: Triaxial measurements.
Source: UK, HSE.

The exposure values are expressed in terms of m/s² A(8), which expresses a person's exposure as an average over an eight-hour period. It does not fully represent the risks which are generated by vibration when it includes severe shocks and jolts (for instance, when driving over potholes or large rocks), which are considered to be an important risk factor in back pain. It is possible to make a basic assessment of the severity and frequency of shocks and jolts by observing the working vehicle and the movement of the driver in the seat, or by asking the driver about them.

Hand-arm vibration

The European Physical Agents (Vibration) Directive (2002/44/EC) deals with risks from vibration at work and distinguishes between vibration affecting the hand-arm system and the whole body. *See Figure 5-22* for the Directive's definition of the term 'hand-arm vibration'.

"'Hand-arm vibration': the mechanical vibration that, when transmitted to the human hand-arm system, entails risks to the health and safety of workers, in particular vascular, bone or joint, neurological or muscular disorders."

Figure 5-22: Definition of hand-arm vibration.
Source: EU, Directive 2002/44/EC - on physical agents - vibration.

Effects on the body

Prolonged intense vibration transmitted to the hands and arms by vibrating tools and equipment can lead to a condition known as *hand-arm vibration syndrome (HAVs)*. These are a range of conditions relating to long term damage to the circulatory system, nerves, soft tissues, bones and joints. The medical effects of sustained exposure to hand-arm vibration can be serious and permanent and are summarised in the following points:

- Vascular changes in the blood vessels of the fingers.
- Neurological changes in the peripheral nerves.
- Muscle and tendon damage in the fingers, hands, wrists and forearms.
- Suspected bone and joint changes.

Vibration generated by tools and equipment held in the hand can result in a significant reduction in blood flow to the hand. Prolonged exposure can cause the fingers go white and numb.

Leading to probably the best known of the conditions arising from the medical effects of exposure to vibration - vibration white finger (VWF).

This condition is also known as *Raynaud's phenomenon* (it has other causes in addition to exposure to vibration). Other symptoms of the condition are sharp tingling pains in the affected area and possible change in skin colour as blood vessels dilate when exposure to the vibration stops.

Figure 5-23: Use of circular saw - vibration.
Source: US Air Forces Central Command.

Contributory factors

As with all work-related ill-health there are a number of factors that when combined result in the problem occurring. These include:

- Vibration frequency - frequencies ranging from 2-1,500 Hz are potentially damaging, but the most serious is the 5-20 Hz range.
- Duration of exposure - this is the length of time the individual is exposed to the vibration.

- Contact force - this is the amount of grip or push used to guide or apply the tools or work piece. The tighter the grip, the greater the vibration to the hand.
- Factors affecting circulation - including medication and smoking.
- Individual susceptibility.

Examples of risk activities

- The use of hand-held chainsaws in forestry.
- The use of hand-held rotary tools in grinding, or in the sanding or polishing of metal, or the holding of material being ground, or metal being sanded or polished by rotary tools.
- The use of hand-held percussive metal-working tools, or the holding of metal being worked upon by percussive tools in riveting, caulking, chipping, hammering, fettling or swaging.
- The use of hand-held powered percussive drills or hand-held powered percussive hammers in demolition, or on roads or footpaths, including road construction.

Whole-body vibration

The European Physical Agents (Vibration) Directive (2002/44/EC) defines the term 'whole-body vibration', *see Figure 5-24*.

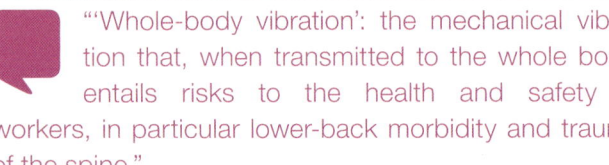

"'Whole-body vibration': the mechanical vibration that, when transmitted to the whole body, entails risks to the health and safety of workers, in particular lower-back morbidity and trauma of the spine."

Figure 5-24: Definition of whole-body vibration.
Source: EU, Directive 2002/44/EC - on physical agents - vibration.

Whole-body vibration (WBV) is vibration transmitted to the entire body via the seat or the feet, or both, often through driving or riding in motor vehicles (including fork-lift trucks and off-road vehicles) or through standing on vibrating floors (for example, near power presses in a stamping plant or near shakeout equipment in a foundry).

Whole-body vibration can cause lower-back and spine pain, fatigue, insomnia, stomach problems, and headaches shortly after or during exposure.

Studies show that WBV can increase heart rate, respiration and oxygen uptake and can cause changes in blood and urine. Eastern European research has noted workers having an overall ill feeling, which they call vibration sickness.

Prolonged exposure can lead to considerable back pain and time off work and may result in permanent injury and having to give up work.

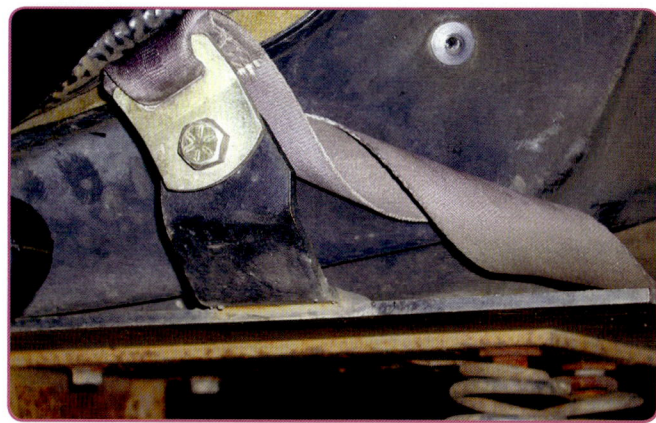

Figure 5-25: Dumper truck seat.
Source: RMS.

High-risk activities include driving rough-terrain vehicles (for example, dumper trucks) and the prolonged use of compactors.

The examples in **Figure 5-26**, are from agricultural four-wheel-drive tractors in the 90 - 125 kW engine power range, incorporating examples of WBV reduction technology. The data was gathered over a sampling period of about half a day. Available machinery on a range of farms for ploughing, spraying, cultivating and trailer work was put to normal use. Exposures during trailer transport work are very close to the exposure limit value and exposures during cultivation work can exceed the exposure limit value.

Exposures are above the UK exposure action value but below the exposure limit value. The highest reading for a machine is shown in bold. Higher levels of WBV exposure should be expected for use of older designs and smaller tractors.

ASSESSMENT OF EXPOSURE

The need for assessment of exposure to vibration

Employers have a responsibility to assess the risks from vibration in order to reduce the risk of damage to their workers' health by controlling exposure to vibration.

Figure 5-27: Whole-body vibration.
Source: UK, HSE Guidance L141.

The employer should make a suitable and sufficient assessment of the risk created by work that is liable to expose workers to vibration. Several countries set action levels and limits relating to the exposure of workers to vibration in the workplace, and such levels and limits need to be considered when assessing the risks faced by workers.

The assessment should establish which workers are at risk of health effects from vibration, and the level of risk. It will also identify sources of vibration that particularly contribute to the vibration level workers are exposed to, for example, equipment and specific activities. This will enable analysis of the options to control the vibration at source or by other means. The assessment will also help the employer when marking equipment or zones that present significant risk, giving information, instruction and training to workers and organising health

Machine	Daily exposure ranges in the three directions of vibration		
	X (fore and aft vibration)	Y (side-to-side vibration)	Z (vertical vibration)
Agricultural tractor - ploughing	0.3 - 0.6 m/s^2 A(8)	0.5 - 0.9 m/s^2 A(8)	0.3 - 0.5 m/s^2 A(8)
Agricultural tractor - cultivating	0.3 - 0.9 m/s^2 A(8)	0.5 - 1.4 m/s^2 A(8)	0.3 - 0.7 m/s^2 A(8)
Agricultural tractor - spraying	0.3 - 0.5 m/s^2 A(8)	0.4 - 0.8 m/s^2 A(8)	0.3 - 0.4 m/s^2 A(8)
Agricultural tractor - trailer work	0.3 - 1.1 m/s^2 A(8)	0.4 - 0.9 m/s^2 A(8)	0.3 - 0.5 m/s^2 A(8)

Figure 5-26: Example exposure in agriculture.
Source: UK, HSE L141.

surveillance for those particularly at risk. If recorded, it will provide evidence of a structured risk assessment process.

Risk assessment

The assessment should observe work practices, make reference to information regarding the magnitude of vibration from equipment and, if necessary, measurement of the magnitude of the vibration.

When conducting the risk assessment, consideration should also be given to the type, duration, effects of exposure, exposure limit/action values, effects on workers at particular risk, the effects of vibration on equipment and the ability to use it, manufacturers' information, availability of replacement equipment, the extension of exposure at the workplace (for example, in rest facilities), temperature and information from health surveillance. The risk assessment should be recorded as soon as is practicable after it is made and be reviewed regularly.

Comparison of vibration exposure levels with recognised exposure limit standards

The International Standardization Organization (ISO) 5349-1:2001 'Mechanical vibration - Measurement and evaluation of human exposure to hand-transmitted vibration - Part 1: General requirements sets out international recommendations for vibration exposure limits. These have been widely adopted globally, sometimes as guidance and at other times as part of a legal framework. For example, a European Union (EU) Directive ratified in 2005 reflected the recommendations in ISO 5349 and gave rise to national legislative vibration limits in European member states.

These specify the personal daily exposure limits and daily exposure action level values, normalised over an 8-hour reference period. Because the harm from vibration is considered to be related to a combination of the energy from the acceleration and the duration of exposure, it is possible to be exposed to higher levels than the action and limit values provided they are for proportionally shorter periods of time. For example, exposure to a magnitude of 5 m/s^2 for 4 hours would be equivalent to 2.5 m/s^2 for 8 hours.

Exposure action and limit values

	Daily exposure action level	Daily exposure limit values
Hand-arm vibration	2.5 m/s^2	5 m/s^2
Whole-body vibration	0.5 m/s^2	1.15 m/s^2

Figure 5-28: Vibration action and limit values.
Source: EU, Directive 2002/44/EC - on physical agents - vibration.

In the UK, the HSE have developed a points system, known as the 'Exposure points system and ready reckoner' in the form of a table, *see Figure 5-29*, for calculating daily vibration exposures. All that is needed is the vibration magnitude (level) and exposure time. The ready reckoner covers a range of vibration magnitudes up to 40 m/s^2 and a range of exposure times up to 10 hours.

The exposures for different combinations of vibration magnitude and exposure time are given in exposure points instead of values in m/s^2 A(8). The user may find the exposure points easier to work with than the A(8) values:

- Exposure points change simply with time: twice the exposure time, twice the number of points.
- Exposure points can be added together, for example, where a worker is exposed to two or more different sources of vibration in a day.
- The exposure action value (2.5 m/s² A(8)) is equal to 100 points.
- The exposure limit value (5 m/s² A(8)) is equal to 400 points.

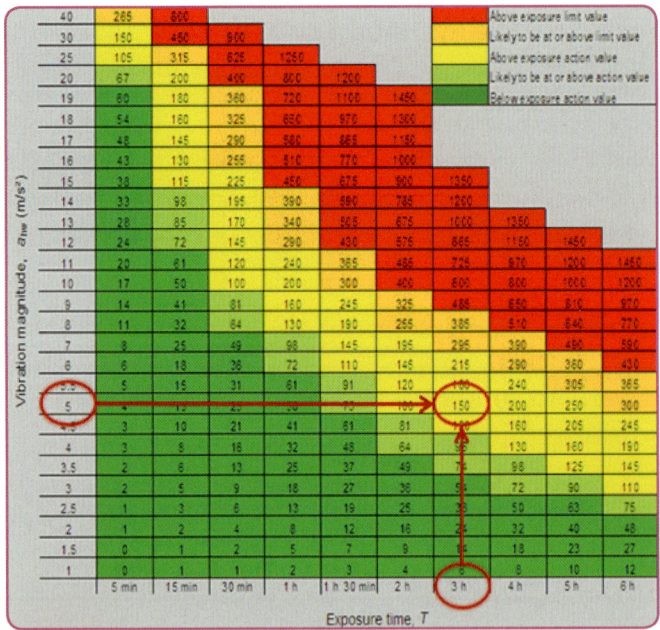

	Above limit value
	Likely to be above limit value
	Above action value
	Likely to be above action value
	Below action value

Figure 5-29: Vibration ready reckoner.
Source: UK, HSE L140.

Using the ready reckoner:

1) Find the vibration magnitude (level) for the tool or process (or the nearest value) on the grey scale on the left of the table.

2) Find the exposure time (or the nearest value) on the grey scale across the bottom of the table.

3) Find the value in the table where the magnitude and time intersect. The illustration shows how it works for a magnitude of 5 m/s^2 and an exposure time of 3 hours: in this case, the exposure corresponds to 150 points.

4) Compare the point's value with the exposure action and limit values (100 and 400 points respectively). In this example the score of 150 points lies above the exposure action value.

5) The colour of the square containing the exposure points value tells you whether the exposure exceeds, or is likely to exceed, the exposure action or limit value.

If a worker is exposed to more than one tool or process during the day, repeat steps 1-3 for each one, add the points, and compare the total with the exposure action value (100) and the exposure limit value (400).

Sourced and adapted from UK, HSE

Exposure action values

Where workers are likely to be exposed to vibration at or above an exposure action level value, the employer should place them under suitable health surveillance and provide them and their representatives with suitable and sufficient information, instruction and training.

Exposure limit values

In general, the employer must ensure that workers are not exposed to vibration above an exposure limit value. If an exposure limit value is exceeded the employer must immediately:

- Reduce exposure to below the limit value.
- Identify the reason for the limit value being exceeded.
- Modify the measures taken to prevent a recurrence.

Where exposure to vibration is usually below the exposure action value, but varies markedly from time to time, the exposure limit value may be occasionally exceeded, providing that:

- Any exposure to vibration averaged over one week is less than the exposure limit value.
- There is evidence to show that the risk from the actual pattern of exposure is less than the corresponding risk from constant exposure at the exposure limit value.
- Risk is reduced to as low a level as is reasonably practicable, taking into account the special circumstances.
- Workers concerned are subject to increased health surveillance.

BASIC VIBRATION CONTROL MEASURES

Preventive and precautionary measures

The employer should seek to eliminate the risk of vibration at source or, if not reasonably practicable, reduce it to as low a level as is reasonably practicable. Where it is not reasonably practicable to take preventive measures that eliminate the risk at source and the personal daily exposure action value is likely to be reached or exceeded, the employer should reduce exposure by implementing a programme of organisational and technical measures. These precautionary measures include the use of other methods of work (for example, by mechanisation), improved ergonomics, maintenance of equipment, design and layout, rest facilities, information, instruction and training, limitation by work rotation and breaks and the provision of personal protective equipment to protect from cold and damp. Measures should be adapted to take account of any group or individual worker whose health may be at particular risk from exposure to vibration.

The following precautionary measures should be considered when protecting people who work with vibrating equipment.

Choice of equipment

It is important to consider vibration characteristics when purchasing new equipment or selecting equipment for a task. Some equipment will provide better control of vibration at source; others will have damping measures provided to limit vibration transmission to the user.

Manufacturers should provide details of the 'vibration magnitude' of their equipment to enable employers to determine the extent of any vibration risks in use. Preference should be given to the purchase and use of low-vibration emission equipment where this is possible without it impeding work processes so that it takes more time to do the work and leads to an equivalent exposure.

Many manufacturers claim to use composite materials that, when moulded into hand-grips and fitted onto vibrating power tools, reduce vibration by up to 45%. Some equipment may provide beneficial design that can limit the effects of vibration, such as the routing of exhaust gases of portable petrol driven equipment through the operating handles, to keep the user's hands warm in cold weather conditions.

Maintenance

Equipment should be maintained at its optimum performance level, thereby reducing vibration to a minimum. For example, tools such as powered chisels should be kept sharp, and bearings of machinery such as grinders should be regularly maintained. Even a relatively small imbalance in set-up can be sufficient to produce high levels of vibration. It is important to maintain the suspension and engine mountings on rough-terrain vehicles in order to minimise the effects of WBV.

Limiting exposure

The work schedule for workers who may conduct work that exposes them to vibration should be examined to identify opportunities to limit the time they are exposed and the magnitude of the vibration to below the action levels set by competent national authorities, for example, by avoiding continuous vibration exposure through alternating this work with non-vibration work or lower vibration magnitude work or scheduling rest breaks. Where equipment creates a high magnitude of vibration this must be clearly identified and its use by a single worker limited to short periods. Care should be taken to organise work schedules so that rest periods from this high-risk work happen naturally in the process. Where they do not, it may be necessary to use reminders in the form of timed alarms or supervision.

Suitable personal protective equipment

Wearing gloves is recommended for safety of the hands and **protection against the cold** through the retention of heat. Maintaining a good temperature in the hands will help circulation of blood to the fingers. Care needs to be taken to select appropriate gloves, as the absorbent material in some gloves for thermal insulation may introduce a resonance frequency that may increase the total energy input to the hands.

Research indicates that anti-vibration gloves are generally unreliable as devices for controlling exposure to vi-bration that is transmitted to the worker's hands from equipment they are using. This is because they tend to have most effect on frequencies that are higher than the main vibration frequency of most power tools. To improve their reduction of lower vibration frequencies the gloves have to contain a thick layer of resilient material. Thick gloves greatly reduce dexterity and could cause the worker to grip the equipment tighter, increasing the risk of vibration injury. Other means of vibration control are more likely to deliver effective vibration reductions and should be used where possible.

Warm clothing can help workers exposed to vibration maintain a good core body temperature, which will assist circulation of blood to the hands. It may be necessary for workers to be provided with a warm location for rest breaks or periods when cold is affecting their circulation. Workers with established HAVS should avoid exposure to cold and thus minimise the number of blanching attacks *(see Figure 5-30)*. Workers with advanced HAVS, which **health surveillance** has determined as deteriorating, should be removed from further exposure.

Other precautionary measures

- The vibration characteristics of hand tools should be assessed and reference made to the ISO or national standard organisation guidelines.
- Development of a **purchasing policy** to include consideration of vibration and, where necessary, vibration isolating devices.
- **Training** of all exposed workers on the proper use of tools and the minimisation of exposure. The stronger the grip on the tool, the more energy enters the hand. There are working techniques for all tools and the expertise developed over time justifies an initial training period for new starters. Operators of vibrating equipment should be trained to recognise the early symptoms of HAVS and WBV and how to report them.
- A continuous review should be conducted with regard to redesigning tools, rescheduling work methods, or automating the process until such time as the risks associated with vibration are under control.
- As with any management system, the controls in place for vibration should be subject to inspection.

Employers should provide information, instruction and training to all workers (and their representatives) who are exposed to risk from vibration. This includes any organisational and technical measures taken, exposure limits and values, risk assessment findings, why and how to detect signs of injury, how to report signs of injury, entitlement to health surveillance and any personal or collective results of health surveillance and safe working practices. Information, instruction and training should be updated to take account of changes in the employer's work or methods. The employer should ensure all workers, whether employees or not, who carry out work in connection with the employer's duties have been provided with information, instruction and training.

ROLE OF HEALTH SURVEILLANCE

The role of health surveillance is to provide early detection of work-related ill-health, as it will assist with the identification of symptoms of the effects of vibration on health. Health surveillance can therefore prevent or diagnose any health effect linked with exposure to vibration. Health surveillance should be carried out for all workers where there is a risk to their health due to being exposed to vibration.

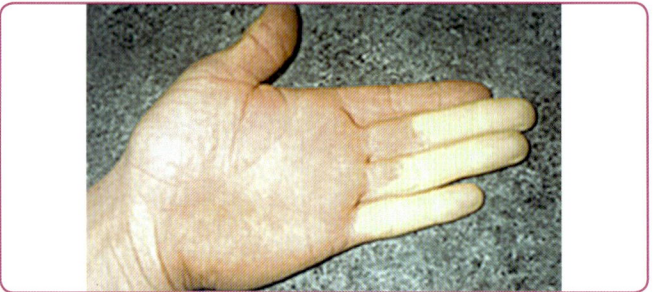

Figure 5-30: Blanched (white) finger typical of hand vibration syndrome (Caucasian race people). Source: UK, HSE L140.

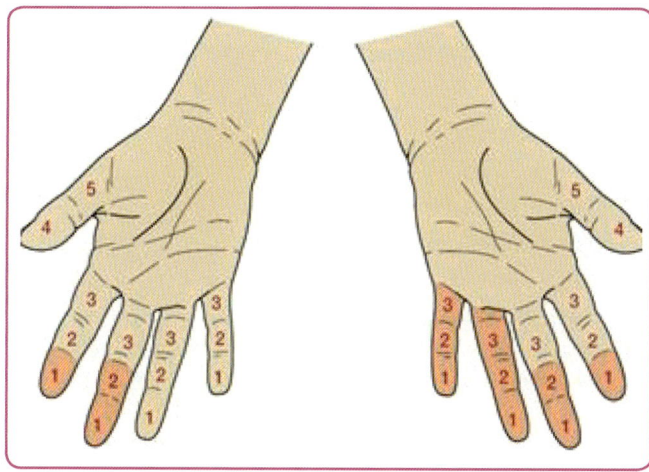

Figure 5-31: Finger zones used to identify degree of hand vibration damage (pin prick test - loss of touch sensitisation).
Source: UK, HSE L140.

By conducting health surveillance from the start of a worker's exposure to vibration risks, it is possible to detect early signs of the effects of vibration and provide early intervention to limit the continuing effects. Health surveillance can also assist with confirming the success of vibration control measures.

Surveillance for HAVS usually involves the worker or an occupational health specialist examining the hands to identify early signs of tingling or blanching, **see Figures 5-30 and 5-31**. For WBV, surveillance may be a simple reporting method or questionnaire relating to experience of lower-back discomfort or pain.

A record of health should be kept of any worker who undergoes health surveillance. The employer should, providing reasonable notice is given, provide the worker with access to their health records and provide copies to a relevant competent authority on request.

If health surveillance identifies a disease or adverse health effect, considered by a doctor or other occupational health professional to be a result of exposure to vibration, the employer should ensure that a qualified person informs the worker and provides information and advice. The employer should ensure they are kept informed of any significant findings from health surveillance, taking into account any medical confidentiality.

In addition, the employer should also:

- Review risk assessments.
- Review the measures taken to comply.
- Consider assigning the worker to other work.
- Review the health of any other worker who has been similarly exposed and consider alternative work.

ILO Recommendations

ILO Code of Practice 'Ambient Factors in the Workplace' requires that consideration should be given to providing a pre-employment medical examination for any worker who may have been exposed to HAVS previously and who may be employed to work with vibrating equipment in the future. Where symptoms are identified, the worker should not be offered work unless the risks from exposure are satisfactorily controlled. Furthermore, if the worker is exposed he or she should be examined periodically as prescribed under any relevant national legislation for HAVS.

REVIEW

What control measures can be taken to reduce the risks from WBV?
What are the symptoms of HAVs?

5.3 Radiation

DIFFERENCES BETWEEN NON-IONISING AND IONISING RADIATION

All matter is composed of **atoms**. Different atomic structures give rise to unique **elements**. Examples of common elements, which form the basic structure of life, are hydrogen, oxygen and carbon.

Atoms form the building blocks of nature and cannot be further sub-divided by chemical means. The centre of the atom is called the **nucleus**, which consists of **protons** and neutrons. Electrons take up orbit around the nucleus.

Protons:	Have a unit of mass and carry a positive electrical charge.
Neutrons:	These also have mass but no charge.
Electrons:	Have a mass about 2,000 times less than that of protons and carry a negative charge.

In an electrically neutral atom the number of electrons equals the number of protons (the positive and negative charges cancel each other out). If the atom loses an electron then a positively charged atom is created. The process of losing or gaining electrons is called **ionisation**.

If the matter that is ionised is a human cell, the cell chemistry will change and this will lead to functional changes in the body tissue. Some cells can repair radiation damage, others cannot. The cell's sensitivity to radiation is directly proportional to its reproductive function; bone marrow and reproductive organs are the most vulnerable, while muscle and central nervous system tissue are affected to a lesser extent.

Figure 5-32: Radiation symbol. Source: ISO 7010.

Ionising radiation is emitted from radioactive materials either directly or indirectly. It has an energy potential capable of changing the cellular composition of matter by penetrating, ionising and damaging body tissue and organs.

Non-ionising radiation has a relatively long wavelength and does not possess the energy needed to ionise matter. Instead the effect tends to be one of heating up cells rather than changing their composition.

Differences between non-ionising and ionising radiation

Ionising radiation is radiation, typically alpha and beta particles and gamma and X-rays, that has sufficient energy to produce ions by interacting with matter, whereas non-ionising radiation does not possess sufficient energy to cause the ionisation of matter.

TYPES, OCCUPATIONAL SOURCES AND HEALTH EFFECTS OF NON-IONISING RADIATION

Non-ionising radiation exists as optical radiation (ultraviolet, visible light, lasers and infrared) and electro-magnetic fields (electrical power transfer line frequencies, microwaves and radio frequencies).

Ultraviolet

Possible occupational sources

There are many possible sources of ultraviolet (UV) radiation to which people may be exposed at work. Sunlight is a natural source of UV radiation. Sunlight UV is filtered by the earth's ozone layer and the shorter high-energy wavelengths, which are more harmful to life forms, are absorbed before they reach ground level.

UV is produced by high temperature sources created by, for example, electrical welding *(see Figure 5-33)* or from fluorescent and tungsten lights and mercury vapour lamps. UV light has many applications; UV will attract insects and is used in food-preparation areas in the design of insect killers.

UV is used to sterilise contaminated water in water treatment, to cure adhesives and inks in printing processes, for metal surface inspection and crack detection, forgery detection (paper currency and works of art), for

leisure sun beds and tanning lamps. These sources are summarised in the following bullet points:

- The sun.
- Tungsten halogen lamps.
- Mercury vapour lamps.
- Electric arc welding and cutting.
- Insect killers, to attract the insects.
- Crack-detection equipment.
- Water treatment.
- Some lasers.
- Adhesive curing equipment.
- Forgery detectors.
- Sunbeds and sunlamps.

Potential health effects

Over-exposure to sunlight can cause sunburn and even blindness. Its effect is thermal and photochemical, producing burns and skin thickening, and eventually skin cancer.

Electric arcs used in welding operations and ultraviolet lamps can produce a harmful effect on the eyes resulting in inflammation (sometimes called 'arc-eye'), and cataract formation.

Natural and artificial light is used in most workplaces and many processes emit radiation in the visible spectrum range.

Figure 5-33: UV - from welding.
Source: Speedy Hire Plc.

Visible light

Possible occupational sources

Visible light can cause hazards in the workplace due to direct or reflected radiation in the range of visible light.

Possible sources of visible light at work include:

- Natural daylight.
- Lasers operating in the visible wavelength, for example, in surveying or level alignment equipment.
- Furnaces or fires.
- Molten metal, ceramics or glass.
- Gas or electrical welding or cutting.
- High-intensity light beams and light bulbs.
- Other high-intensity lights, such as in photocopiers and printers.
- The sun.

Potential health effects

Light in the visible frequency range can cause damage if it is present in sufficiently intense form. The eyes are particularly vulnerable, but skin tissue may also be damaged. Retinal damage may occur if chronic exposure to high levels of light takes place. If the visible light is focused on to the skin it may cause skin burning. Indirect danger may also be created by workers being temporarily dazzled.

Directive 2006/25/EC of the European Parliament and of the Council of 5 April 2006 on the minimum health and safety requirements regarding the exposure of workers to risks arising from physical agents (artificial optical radiation) places a requirement on Member States to assess the risks from artificial optical radiation. This includes ultraviolet, infrared and visible sources (luminaires) in the workplace. The directive does not include sunlight.

Infrared

Possible occupational sources

Any hot source which visibly glows is likely to be a source of infrared radiation, for example:

- Furnaces or fires.
- Molten metal or glass.
- Burning or welding.
- Heat lamps.
- Some lasers.
- The sun.

Potential health effects

Exposure results in a thermal effect such as skin burning and loss of body fluids (heat exhaustion and dehydration). The eyes can be damaged in the cornea and lens, which may become opaque (cataract). Retinal damage may also occur if the radiation is focused on the eye.

Radio frequency and microwaves

Possible occupational sources

Radio frequency radiation is produced by radio/television transmitters, intruder detectors and high voltage electricity cables. Microwave radiation is used in communication systems and in cooking equipment.

Potential health effects

Burns can be caused if workers using this type of equipment allow parts of the body that carry jewellery to enter the radio frequency field. The jewellery will absorb the energy of this form of radiation, get hot and may cause burns.

Intense fields at the source of transmitters will damage the body directly through this absorption/heating process, and particular precautions need to be taken to isolate radio/television transmitters to protect maintenance workers. Microwaves can produce the same deep heating effect in live tissue as they can in cooking.

Figure 5-34: Radio mast. Source: RMS.

TYPES, OCCUPATIONAL SOURCES AND HEALTH EFFECTS OF IONISING RADIATION

Types of radiation

Ionising radiation occurs as either electromagnetic rays, for example, gamma rays or X-rays, or in particles, for example, alpha and beta particles.

Radiation is emitted by a wide range of sources and appliances used throughout industry, medicine and research. It is also a naturally occurring part of the environment.

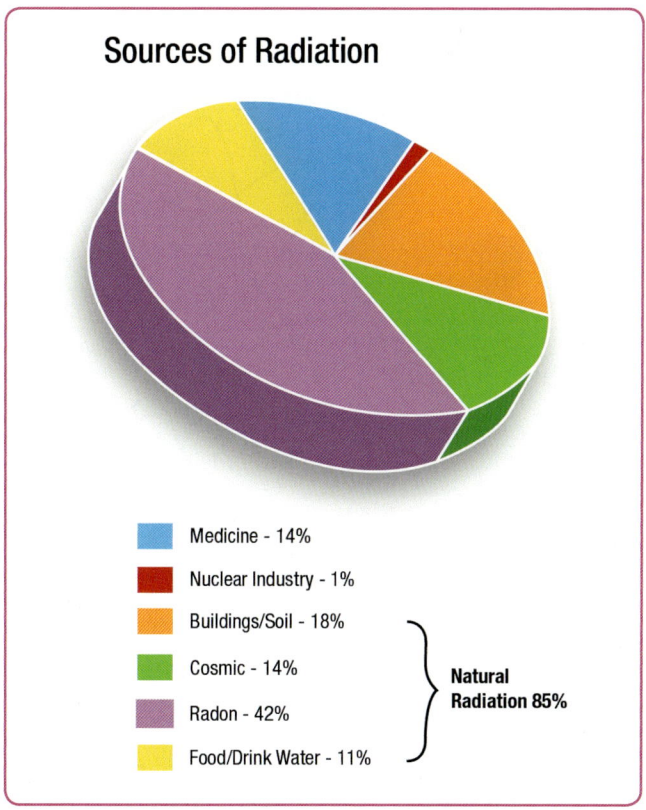

Figure 5-35: Sources of radiation.
Source: World Nuclear Association.

Ionising radiation is found naturally or in the workplace in the form of alpha, beta, gamma, and X-rays. The human body absorbs radiation readily from a wide variety of sources, mostly with adverse effects.

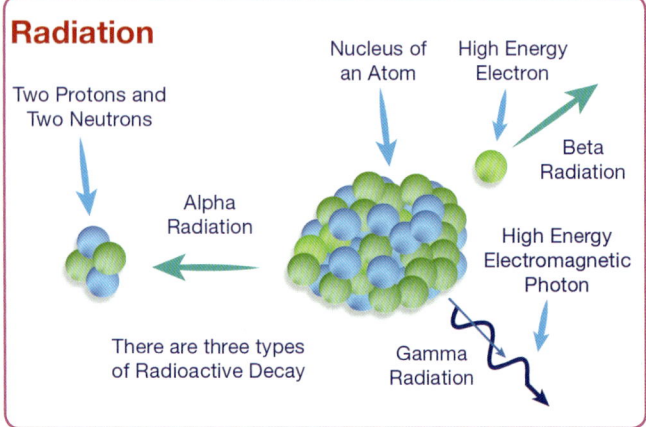

Figure 5-36: Ionising atomic decay.
Source: RMS.

There are a number of different types of ionising radiation, each with its different powers of penetration and effects on the body. Therefore, the type of radiation will determine the type and level of protection.

Alpha particles

Alpha particles are comparatively large. Alpha particles travel short distances in dense materials, and are unlikely to penetrate living skin tissue.

The principal risk is through ingestion or inhalation of a source, for example, radon gas emits alpha particles. This might place the material close to vulnerable tissue such as the lungs; when this happens the highly localised energy effect will destroy associated tissue of the organs affected.

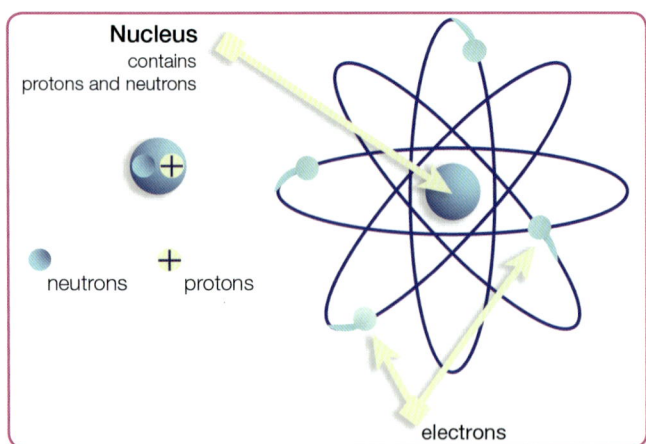

Figure 5-37: Structure of an atom.
Source: RMS.

Beta particles

Beta particles are much smaller and faster moving than alpha particles, and have a longer range, so they can damage and penetrate the skin. While they have greater penetrating power than alpha particles, beta particles are less ionising and take longer to effect the same degree of damage.

Gamma rays

These have great penetrating power. Gamma radiation passing through a normal atom will sometimes force the loss of an electron, leaving the atom positively charged; this is called an ion.

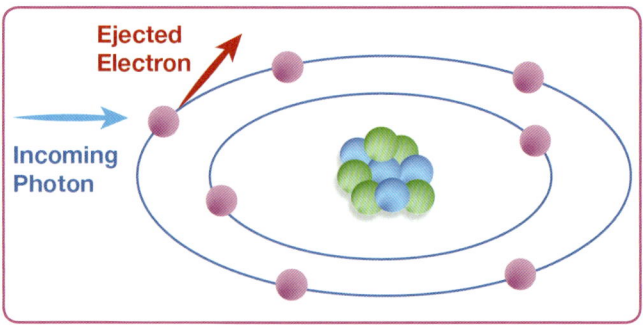

Figure 5-38: Ionisation of a normal atom.
Source: RMS.

X-rays

X-rays are very similar in their effects to gamma rays. Both X-rays and gamma rays have **high energy**, and **high penetration** power through fairly dense material. In low-density substances, including air, they may travel long distances.

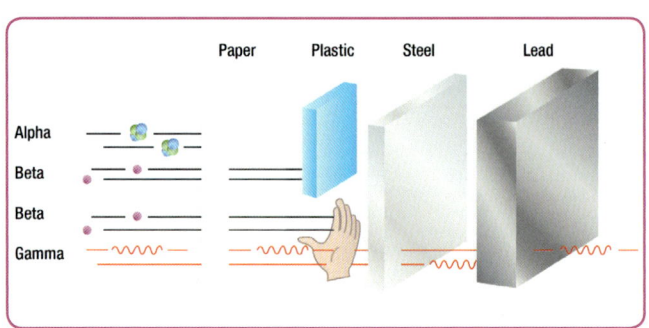

Figure 5-39: Relative penetration power of several types of ionising radiation (note gamma and X-ray have similar penetrating power).
Source: RMS.

Radon

Radon is a naturally occurring colourless, odourless radioactive gas that can seep out of the ground and enter buildings. It is a particularly common source of exposure to radiation and for many people easily exceeds exposure from nuclear power stations or hospital scans and X-rays.

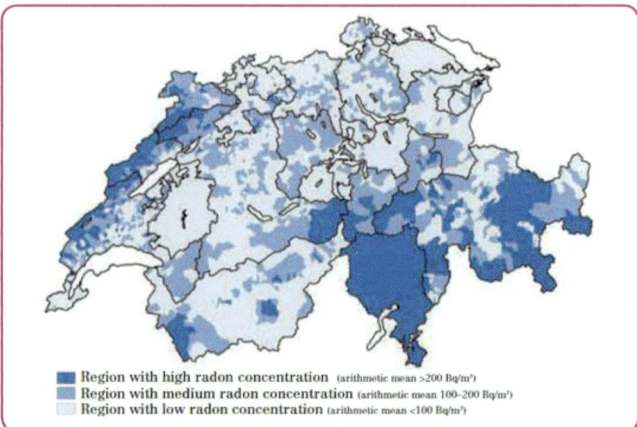

Figure 5-40: Map of radon distribution Switzerland 2009.
Source: World Health Organization.

Radon occurs naturally deep within the earth's core from decaying uranium, and it is particularly abundant in regions with granite bedrock. However, the gas disperses outdoors so levels are generally very low. The gas decays readily into other radioactive isotopes of lead, bismuth and polonium, including polonium 210, the dust from which can be inhaled and expose local tissue to intense alpha particle ionisation, resulting in cell death or malignancy. Radon has been identified as a common cause of lung cancer.

Radon causes over 1,100 deaths in the UK each year from lung cancer, *Figure 5-40* illustrates the radon map of Switzerland in 2009, showing that distribution is greatest in granite mountains. Many countries have similar radon maps used to identify at-risk areas.

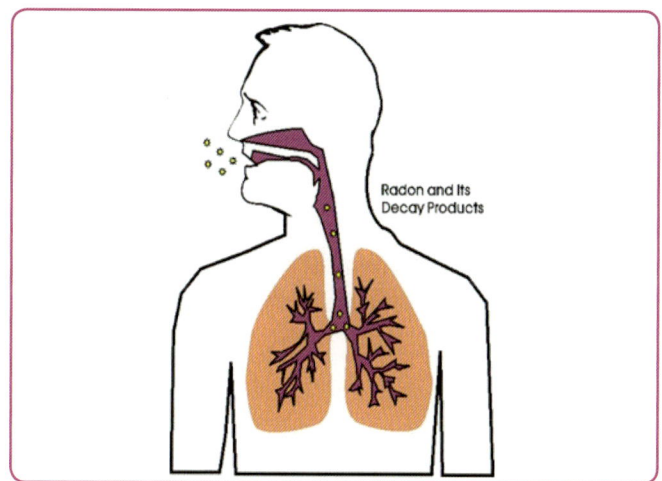

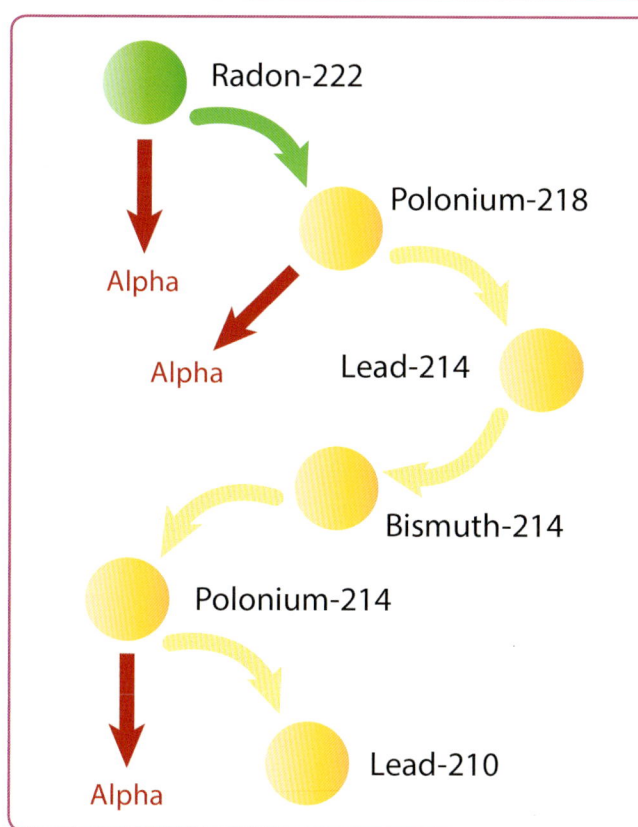

Figure 5-41: Radon decay a) and b).
Source: Administration and Business Portal.

Occupational sources of ionising radiation

The most familiar examples of ionising radiation in the workplace are in hospitals, dentists' surgeries and veterinary surgeries where X-rays are used extensively. X-ray machines are also used for security purposes at baggage-handling points in airports.

In addition, X-rays and gamma rays are used in non-destructive testing of metals, for example, site radiography of welds in pipelines. In other industries, ionising radiation is used for measurement, for example, in the paper industry for the thickness of paper, and in the food-processing industry for measuring the contents of sealed tins. While all workplaces can be at risk from radon, workplaces at higher risk tend to be those located in specific geological areas. The highest levels of radon are usually found in underground spaces such as basements, cellars, caves and mines, particularly where groundwater is present.

Examples include:

- Underground workers such as miners (uranium miners are exposed to the highest levels).
- Railway workers - tunnelling, especially under rivers.
- Utility company workers (when entering service ducts).
- Oil and gas drilling (drilling 'muds' or slurries and the processing of oil).
- Quarry workers.

"Radiation is permanently present throughout the environment, in the air, water, food, soil and in all living organisms. Large proportion of the average annual radiation dose received by people results from natural environmental sources. Each member of the world population is exposed, on average, to 2.4 mSv/yr of ionising radiation from natural sources. In some areas (in different countries of the world) the natural radiation dose may be 5 to 10-times higher to large number of people."

Figure 5-42: Sources of ionising radiation.
Source: World Health Organization.

Potential health effects of ionising radiation

Radiation damage to tissue and/or organs depends on the dose of radiation received. The potential damage from an absorbed dose depends on the type of radiation and the sensitivity of different tissues and organs.

The sievert (Sv) is a unit of radiation weighted dose, also called the effective dose. It is a way to measure ionising radiation in terms of the potential for causing harm. The Sv takes into account the type of radiation and sensitivity of tissues and organs. The Sv is a very large unit so it is more practical to use smaller units such as millisieverts (mSv) or microsieverts (µSv). There are one thousand µSv in one mSv, and one thousand mSv in one Sv. In addition to the amount of radiation (dose), it is often useful to express the rate at which this dose is delivered (dose rate), for example, µSv/hour or mSv/year.

Beyond certain thresholds, radiation can impair the functioning of tissues and/or organs and can produce acute effects such as skin redness, hair loss, radiation burns, or acute radiation syndrome. These effects are more severe at higher doses and higher dose rates. For instance, the dose threshold for acute radiation syndrome is about 1 Sv (1000 mSv). Typical radiation dose limits would be around 20 mSv/year.

If the dose is low or delivered over a long period of time (low dose rate), the damaged cells are more likely to repair themselves successfully. However, long-term effects may still occur if the cell damage is repaired but incorporates errors in the DNA, transforming an irradiated cell that still retains its capacity for cell division. This transformation may lead to cancer after years or even decades have passed.

Effects of this type will not always occur, but their likelihood is proportional to the radiation dose. This risk is higher for children and adolescents, as they are significantly more sensitive to radiation exposure than adults. For example, a typical limit for a young person would be around 6 mSv/year.

THE BASIC MEANS OF CONTROLLING EXPOSURE TO IONISING AND NON-IONISING RADIATION

Controls for non-ionising radiation

Employers should conduct a specific risk assessment for risks relating to non-ionising radiation and use it to eliminate or reduce risks. Exposure limit levels should be set for this form of radiation as part of the control measures. Information and training should be provided to those that may be affected, which includes workers and others carrying out work on behalf of the employer. Medical examination and health surveillance should be provided for those workers that receive over-exposure.

Ultraviolet

Protection from natural sources of UV is relatively simple and includes providing outdoor workers with barrier creams, suitable lightweight UV-rated clothing and head protection, and eye protection, as appropriate. In addition, workers should be encouraged to take breaks in the shade where possible and consideration should be given to adjusting work schedules so outside tasks may be conducted at times of day that are less affected by UV. Control of artificial sources of UV includes segregation of UV-emitting processes and the use of warning signs. UV radiation emitted from industrial processes

can be isolated by physical shielding such as partitions. Users of UV-emitting equipment, such as welders, can protect themselves by the use of goggles and protective clothing - the latter to avoid 'sunburn'.

Assistants in welding processes often fail to appreciate the extent of their own exposure, and they require similar protection. Workers should check skin exposed to UV frequently to identify possible effects that might lead to skin cancer.

Visible light

The eye detects visible light. It has two protective control mechanisms of its own, the eyelids ('blinking' or closing the eyelids) and the iris (dilates). These are normally sufficient to provide general protection, as the eyelid has a reaction time of 150 milliseconds. However, where this is not adequate because of the intensity of the light or sustained exposure of the eye to it, other precautions should be considered, including confinement of high-intensity sources, matt finishes to nearby paintwork, and provision of protective glasses for outdoor workers in snow, sand or near large bodies of water. Where the high-intensity visible light is artificially created, those not involved in the process must also be protected. Warning signs should be posted and access restricted to the process area.

Infrared

Controls to limit exposure include engineered measures, such as remote controls, screening, interlocks and clamps to hold material to enable the worker to be outside the exposure area. This can be supplemented by forms of PPE where exposure cannot be prevented, for example, face shields, goggles or other protective eyewear, coveralls and gloves, and limiting exposure time through rotation and job-sharing. It is important to protect others not directly involved in the process by using screens, curtains and restricted access.

Radio frequency and microwaves

Radio frequency and microwave radiation can usually be shielded at point of generation to protect the users. If size and function prohibits this, restrictions on entry and working near an energised microwave device will be needed. Metals, tools, flammable and explosive materials should not be left in the electromagnetic field generated by microwave equipment. Appropriate warning devices should be part of the controls for such appliances.

Controls for ionising radiation

The standard approach to controlling ionising radiation involves, time, distance and shielding.

Reduced time

Exposure to ionising radiation can be controlled by reducing the duration of exposure through redesigning work patterns, giving consideration to shift working, job rotation, etc. The dose received will also depend upon the time of the exposure. These factors must be taken into account when devising suitable operator controls.

Increased distance

Radiation intensity reduces the further away a person is from the source. This is subject to the inverse-square law, which means that energy received (dose) is inversely proportional to the square of the distance from the source.

Shielding

The type of shielding required to give adequate protection will depend on the penetration power of the radiation involved. For example, it may vary from thin sheets of silver paper to protect from beta particles through to several centimetres of concrete and lead for protection against gamma or X-rays.

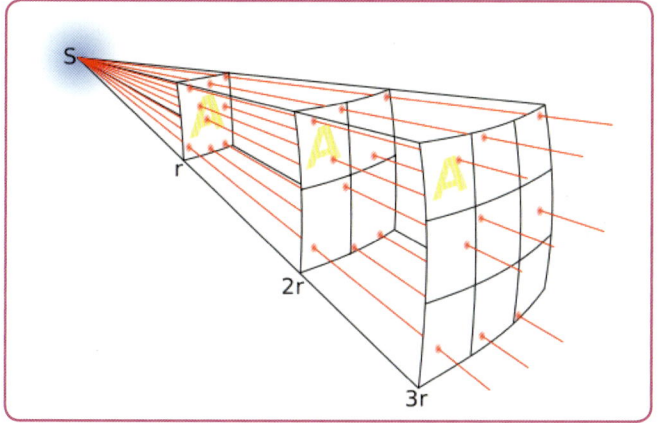

Figure 5-43: The inverse-square law. Source: Wikipedia.

In addition to the previous specific controls, the following general principles must be observed:

- Radiation should only be introduced to the workplace if there is a positive benefit.
- Safety information must be obtained from suppliers about the type(s) of radiation emitted or likely to be emitted by their equipment.
- Safety procedures must be reviewed regularly.
- Protective equipment provided must be suitable and appropriate, as required by relevant regulations. It must be checked and maintained regularly.
- Emergency plans must cover the potential radiation emergency.
- Written authorisation by permit should be used to account for all purchase/use, storage, transport and disposal of radioactive substances.

THE BASIC MEANS OF CONTROLLING EXPOSURES TO RADON

Above-ground workplaces

For the vast majority of above-ground workplaces the risk assessment should include radon measurements in appropriate ground-floor rooms where the building is located in a radon-affected area.

Below-ground workplaces

For occupied below-ground workplaces (for example, occupied greater than an average of an hour per week/ 52 hours per year), or those containing an open water source, the risk assessment should include radon measurements.

Control of radon exposure in new buildings can be done by installing a 'radon-proof membrane' within the floor structure. In areas more seriously affected by radon it may be necessary to install a 'radon sump' to vent the gas into the atmosphere. A radon sump has a pipe connecting a space under a solid floor to the outside.

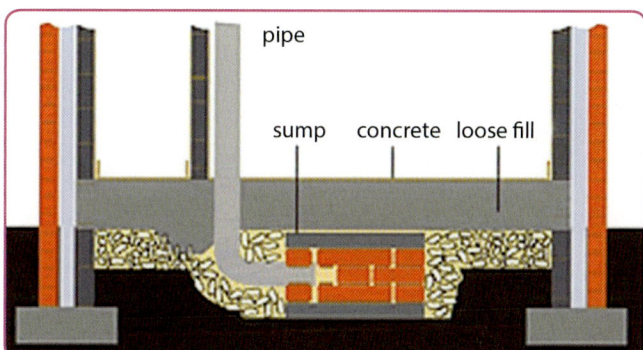

Figure 5-44: Radon sump. Source: UK, HSE.

A small electric fan in the pipe continually sucks the radon from under the house and expels it harmlessly to the atmosphere. Modern sumps are often constructed from outside the building so there is no disruption inside. In existing buildings, it is not usually possible to provide a radon-proof barrier, so alternative measures are used to control the increasing concentration of radon in the building and subsequent exposure to it.

Such measures include a mixture of active and passive systems such as improved under-floor and indoor ventilation in the area, positive pressure ventilation of occupied areas, installation of radon sumps and extraction pipework and sealing large gaps in floors and walls in contact with the ground.

There are relatively simple tests for radon gas and detectors are safe and simple to use. In some countries these tests are routinely done in areas of known systematic hazards.

Figure 5-45: Radon measurement.
Source: theradonshop.com

Detection involves a collector (a hollow plastic shell containing a piece of clear plastic that records damage caused by radon) that the user hangs in the lowest occupied floor of the premises for 2 to 7 days. The user then sends the collector to a laboratory for analysis. The detectors do not emit anything and do not collect anything dangerous. However, they can be damaged by heat or submersion in water and should not be opened.

BASIC RADIATION PROTECTION STRATEGIES

Protection strategies for non-ionising radiation

Basic radiation protection strategies relating to non-ionising radiation include:

- Identification of the sources of non-ionising radiation and consideration of the risks related to exposure.
- Identification of classes of worker exposed to non-ionising radiation.
- Where the radiation is emitted from a controlled source, a Laser Protection Supervisor should be appointed to give specialist advice, ensure the energy from the source is reduced as far as reasonably practicable, using engineering measures (i.e. screening), protective eyewear and ensuring information and training for users while monitoring and enforcing control measures.
- Shielding the emission of energy from sources, for example, closing furnace openings when not in use, using glass filters and other forms of enclosure around lighting sources, curtains around welding activities, and enclosure of work areas near radio wave transmitters or provision of UV-protective cream.
- Increasing the distance of workers from the source, for example, the positioning of microwave transmitters so they are high enough above work areas, positioning welding activities away from other workers.
- Limiting the time that a worker is exposed to the source, for example, ensuring workers that work outside in the sun take regular breaks inside or in the shade, or job rotation for those involved in hot processes such as furnace work.
- Provision of information and training on the risks of exposure to non-ionising radiation and the measures to limit exposure.
- Provision of monitoring and health surveillance.

Protection strategies for ionising radiation

Radiation protection strategies for ionising radiation are a little more specific and reflect the principles and recommendations made by the International Commission on Radiological Protection (ICRP).

The three basic principles of radiological protection are:

1) Justification of activities that could cause or affect radiation exposure.

2) Optimisation of protection in order to keep doses as low as reasonably achievable (this includes shielding, minimising time exposed and maximising distance from the radiation source).

3) The use of dose limits.

The recommendations of the ICRP include establishing responsibilities, assessing risk, establishing prevention/control measures and conducting health surveillance.

Responsibilities

Every employer should, in relation to any work with ionising radiation, take all necessary steps to restrict so far as is reasonably practicable the extent to which workers are exposed to ionising radiation. Those with responsibilities under the radiation protection programme should be assigned responsibilities in writing. A radiation protection adviser/officer should be appointed to assist the employer with the radiation protection programme.

The ILO Code 'Radiation Protection Workers (Ionising Radiation)' sets out the following requirements:

"2.2. Duties and responsibilities of employers.

2.2.1. The responsibility for providing adequate protection of workers against radiations rests with the employer, even if the employer is a subcontractor.

2.2.2. (1) When two or more employers undertake activities simultaneously at one workplace, they should collaborate in order to ensure compliance with national regulations. This collaboration does not relieve the employer of the duty to secure the health and safety of his employees."

Radiation protection advisers

At least one radiation protection adviser (RPA) should be appointed in writing by employers using ionising radiation. The numbers of RPAs appointed should be appropriate to the risk and the area where advice is needed. Employers should consult the RPA on:

- The implementation of controlled and supervised areas.
- The prior examination of plans for installations and the acceptance into service of new or modified sources of ionising radiation in relation to any engineering controls, design features, safety features and warning devices provided to restrict exposure to ionising radiation.
- The regular calibration of equipment provided for monitoring levels of ionising radiation and the regular

checking that such equipment is serviceable and correctly used.

- The periodic examination and testing of engineering controls, design features, safety features and warning devices and regular checking of systems of work provided to restrict exposure to ionising radiation.

The employer should provide the RPAs with adequate information and facilities to allow them to fulfil their functions.

Assessment

The employer should make an assessment of work involving exposure to radiation and use the assessment to contribute to the design of a radiation protection programme. The assessment should include consideration of the:

- Limits and technical conditions for the operation of the radiation source.
- Ways in which technical, procedural and behavioural measures related to radiation protection might fail and lead to potential exposures, and the possible consequences of such failures.
- Nature and magnitude of potential exposures and the likelihood of their occurrence.

Arrangements

The employer should follow the procedures for notification, registration or licensing that are laid down by the competent authority.

The requirements of the ILO Code 'Radiation Protection Workers (Ionising Radiation)' for arrangements include:

"2.2.4.   The employer should make the administrative and organisational arrangements necessary for controlling the exposure of workers to radiations and radioactive materials. He should therefore appoint the appropriate staff, provide the necessary protective equipment, including radiation-measuring systems, maintain buildings, installations and workplaces, and organise work in such a way as to ensure that the radiation exposure of each worker, including his internal exposure, is controlled and complies with the provisions of this code.

2.2.5.   The employer should structure the administrative and organisational arrangements in such a way that they operate in a smooth manner and that an effective safety programme consistent with the requirements of this code is implemented.

2.2.6.   The employer should establish a policy for the protection of the health and safety of workers, comprising appropriate measures, during planning and operation, to prevent any unnecessary exposure in the installation under his control."

Prevention and control

Basic radiation protection strategies to prevent and control exposure include the application of the principles of time, distance and shielding. Exposure should be limited to as few people as possible and those individuals should be monitored and exposure levels maintained within limits. In the case of ionising radiation, where possible, only sealed sources should be used and a system developed to minimise dose levels to individuals. This work should be under the control of an RPA (officer) Prevention and control of radiation exposure will include:

- Categorisation/classification of workers.
- Dose limitation, including arrangements for monitoring workers' exposure and radiation levels in the workplace. A system for recording and reporting information related to the control of exposures.
- Designation of controlled or supervised areas.
- Local rules for classified workers and the supervision of work.
- Information and training programme.

Categorisation/classification of workers

The ILO Code of Practice 'Radiation protection of workers (ionising radiations)' sets out the following requirements:

"4.1.1.   For the purpose of this code there are two categories of workers:

(a)   Workers engaged in radiation work.

(b)   Workers not engaged in radiation work, but who might be exposed to radiations because of their work.

4.1.2.   Workers engaged in radiation work are workers to whom the dose limits given in paragraph 5.4.3 apply.

4.1.3.   Workers not engaged in radiation work should be treated, as far as restricting radiation exposure is concerned, as if they were members of the public.

4.1.4.   No person under the age of 16 should be considered to be a worker engaged in radiation work for the purpose of this code.

4.1.5.   No worker, student, apprentice or trainee under the age of 18 should be allowed to be engaged in radiation work in radiation Working Condition A (see paragraph 4.3.1); such persons may only, therefore, work in Working Condition B."

Workers engaged in radiation work are usually called **'classified workers'**, relating to their classification under work conditions A or B. The ILO Code sets out the two working conditions expressed for classified workers, working condition A and working condition B.

"4.3.1. For the purpose of this code there are two classes of working conditions for workers engaged in radiation work:

(a)   Working Condition A - where the annual exposures might exceed three-tenths of the dose limits (given in paragraph 5.4.3).

(b)   Working Condition B - where it is most unlikely that the annual exposures will exceed three-tenths of the dose limits (given in paragraph 5.4.3)."

Training and information

As part of the radiation protection programme, the employer should establish a training and information programme.

The programme should ensure that all workers receive adequate information on the:

- Health risks due to their occupational exposure to radiation.
- Significance of their actions and how they may affect radiation protection measures.

All radiation workers should receive adequate training on radiation protection. Those assigned responsibilities in the radiation protection programme, for example, the RPA or radiation protection supervisor, should receive appropriate information and training. Appropriate management should receive training on the basic principles of radiological protection, their main responsibility regarding radiation risk management and the principal elements of the radiation protection programme.

Specific information should be provided to female workers who are likely to enter controlled or supervised areas on the risk to an embryo or foetus due to exposure to radiation and on the importance of notifying the employer as soon as she suspects she is pregnant.

THE ROLE OF MONITORING AND HEALTH SURVEILLANCE

The role of radiation monitoring is to ensure that working conditions in workplaces exposed to radiation are kept under review and that the current level of radiation exposure of workers is known. The main functions of monitoring are to:

- Check that areas have been correctly designated for the hazards that exist.
- Identify any changes to radiation exposure levels so that appropriate control measures for restricting exposure can be proposed.
- Detect breakdowns in controls or systems, so as to indicate whether conditions are satisfactory for continuing work in that area.

- Ensure workers use the controls provided and report any defects.
- Ensure workers use personal protection where its use is designated as mandatory.
- Provide information on those who may be at risk and in need of health surveillance.

The role of health surveillance is to provide the identification of symptoms and early detection of ill-health arising from exposure to occupational sources of radiation. Health surveillance can be part of a prevention strategy, allowing interventions to be made that limit further exposure beyond acceptable limits.

The health of the employee should also be established before the individual becomes a classified worker and when the individual ceases to be a classified worker. Special consideration will be required for classified workers who are new or expectant mothers. The health surveillance undertaken will include examination of the skin and respiratory system, taking into account exposure records and sickness absence records.

Figure 5-46: Personal dose monitor.
Source: RMS.

Non-ionising radiation

Monitoring of workplace levels of non-ionising radiation is important for both radio frequency and optical forms of radiation as they are not a visible hazard and levels may increase and not be obvious until harm is done to workers. Where workers are exposed to significant levels of non-ionising radiation, employers should arrange for appropriate health surveillance by occupational health specialists, who should assess the possible need for medical examination of workers, including ophthalmic and skin examination.

Figure 5-47: Contamination monitoring. Source: www.anythingradioactive.com.

In particular, employers should arrange for workers using Class 3 or 4 lasers to receive ophthalmic examinations:

a) Pre- and post-assignment to laser work.

b) After an apparent or suspected harmful exposure of the worker's eyes.

Ionising radiation

Where workers are exposed to ionising radiation, the employer who designates an area as a controlled or supervised area should ensure that levels of ionising radiation are adequately monitored and that working conditions in those areas are kept under review.

Employers should ensure that classified workers working with ionising radiation are under adequate health surveillance by an appointed doctor or medical specialist to determine their fitness for the work being undertaken.

Fitness in this sense is not restricted to possible health effects from exposure to ionising radiation. Those conducting the health surveillance will need to take account of specific features of the work with ionising radiation in relation to the fitness of the individual.

This will include such things as:

- Their ability to wear any personal protective equipment (including respiratory protective equipment) required to restrict exposure.
- Whether they have a skin disease that could affect their ability to undertake work involving unsealed radioactive materials.
- Whether they have serious psychological disorders that could affect their ability to undertake work with radiation sources that involves a special level of responsibility for safety.

 **REVIEW**

Define the terms ionising and non-ionising radiation and identify one industrial source for each.

What are the health risks associated with visible radiation?

What are the monitoring activities which should be undertaken to protect a classified worker?

 **5.4** Work-related mental ill-health

FREQUENCY AND EXTENT OF MENTAL ILL-HEALTH AT WORK

Common mental ill-health and work-related stress can exist independently – people can experience work-related stress without having anxiety, depression or other mental ill-health problems. They can also have anxiety and depression without experiencing work-related stress.

An existing mental ill-health problem can be aggravated by work-related stress, making it more difficult to control. If work-related stress reaches a point where it has triggered an existing mental ill-health problem, it becomes hard to separate one from the other. Work-related stress and other mental ill-health problems often go together and the symptoms can be very similar.

Mental ill-health is a significant issue in the general working age population, which is reflected in the statistics compiled by the UK Mental Health Foundation who determined the following global perspective on mental ill-health:

- A recent index of 301 diseases compiled by the World Health Organisation (WHO) in 2016 found mental health problems to be one of the main causes of the overall disease burden worldwide. They were shown to account for 21.2% of years lived with disability worldwide.
- According to the 2013 Global Burden of Disease study, the predominant mental health problem world-wide is depression, followed by anxiety, schizophrenia and bipolar disorder. In 2013, depression was the second leading cause of years lived with disability worldwide, behind lower back pain. In 26 countries, depression was the primary driver of disability. Depressive disorders also contribute to the burden of suicide and heart disease on mortality and disability; they have both a direct and an indirect impact on the length and quality of life.

- The World Health Organization (WHO) estimates that between 35% and 50% of people with severe mental health problems in developed countries, and 76 – 85% in developing countries, receive no treatment.

From this information it can be seen that there is a range of mental ill-health that workers could have, but it is not clear their relationship to work, however the main mental ill-health problems were confirmed to be depression and anxiety.

Statistics compiled by the UK Mental Health Foundation, in relation to the UK, determined that according to the UK National Institute for Health and Care Excellence (NICE), common mental health problems include depression, generalised anxiety disorder (GAD), social anxiety disorder, panic disorder, obsessive compulsive disorder (OCD) and post-traumatic stress disorder (PTSD). A study in 2014 confirmed that 17.5% of working-age adults had symptoms of common mental ill-health. Comparing data from 2007 with 2014 the Mental Health Foundation confirmed that the prevalence of depression and general anxiety disorder in adults over 16 years of age had increased.

Mental health problem	2007 (%)	2014 (%)
Depression	2.3	3.3
General anxiety disorder	4.4	5.9

When considering post-traumatic stress disorder (PTSD) a survey conducted in 2014 found that about one third of adults in England reported having at least one traumatic event in their life and 4.4% of adults screened posi-tively for having PTSD in the previous month, whereas a lower number indicated that they believed they had PTSD and only 13% of those screened positive had been diagnosed by a health professional.

The 2014 UK Adult Psychiatric Morbidity Survey concluded that every week approximately 1 in 6 people experi-ence common mental health problems in the workplace and that women in full-time employment are nearly twice as likely to have a common mental health problem as full-time employed men (19.8% vs 10.9%).

Mental ill-health affects the productivity of those in work by impairing their ability to function at full capacity and, in the UK, it is reported that it causes about 40% of all days lost through sickness absence (Sainsbury Centre for Mental Health, 2007). Compared with many common physical conditions, mental health problems are often gradual in onset and long lasting. Reports by the Confederation for British Industry in 2007 identified that on average each mental ill-health sickness absence spell lasts 21 days and for this reason, although mental ill-health causes just 25% of absences of less than seven days, they account for 47% of long-term absences. The International Labour Organisation (ILO) reports in their document 'Mental health and work: Impacts, issues and good practices' that the Association of Canadian insurance companies estimates that 230-50% of worker disability allowances are paid in relation to mental ill-health problems.

The Sainsbury Centre for Mental Health has estimated (in 2007) that impaired work efficiency (sometimes called 'presenteeism') due to mental ill-health costs the UK £15.1 billion, or £605 for every employee in the UK, which is almost twice the estimated £8.4 billion annual cost of absenteeism. Some US studies have put the cost of presenteeism at four or five times the cost of absenteeism.

The overall cost of depression in England in 2000 was estimated to be £9 billion. More than £8 billion of this cost was due to lost productivity as a result of work days lost, which is 20 times larger than the estimated £0.4 billion costs of the National Health Service (NHS) for dealing with the mental ill-health.

The effect of stress-related conditions can be difficult to calculate, but it has been estimated that the harmful effects of stress cost the UK economy around £4 billion a year, with almost 13 million working days lost and over half a million people reporting that they were suffering from work-related stress.

In the period 2017/2018 there were 0.6 million new or longstanding cases of depression, anxiety and stress identified through the self-reporting UK Labour Force Survey.

	Cost per average employee (£)	Total cost to UK employers (£billion)	% of total
Absenteeism	335	8.4	32.4
Presenteeism	605	15.1	58.4
Replacing worker	95	2.4	9.2
Total	1035	25.9	100

Figure 5-48: Estimated annual costs to UK employers of mental ill-health. Source: Sainsbury Center for Mental Health, 2007.

A study in 2005 found that workers in particular jobs are more likely to develop common mental ill-health problems. This includes teachers, nurses, social workers, probation officers and fire fighters, police officers, the armed forces and medical practitioners.

Mental ill-health problems can cause fatigue, poor memory, impaired attention and difficulty concentrating. These problems can be made worse by the effects of medication. The psychological effects of an individual's mental ill-health can result in the following workplace issues:

- Lack of motivation.
- Lack of commitment.
- Poor timekeeping.
- Increase in mistakes.
- Increase in sickness absence.
- Poor decision making.
- Poor planning.
- Tension between colleagues/supervisors.
- Poor service to clients.
- Deterioration in work relations.
- Increase in disciplinary problems.

COMMON SYMPTOMS OF WORKERS WITH MENTAL ILL-HEALTH

Depression

Most people experience short-term feelings of sadness or anxiety during difficult times. This is not usually a sign of depression. Depression is a low mood that lasts for a long time, and affects everyday life.

Depression can affect people in different ways and can cause a wide variety of symptoms. The severity of the symptoms may vary. At its mildest individuals may have a persistently low temperament and at its most severe depression can make those affected feel life is no longer worth living and suicidal.

Common symptoms of depression individuals might have include:

How they feel

- Down, sad, upset or tearful.
- Empty and numb.
- Restless, agitated or irritable.
- Guilty, worthless and negative towards themselves.
- No self-confidence or self-esteem.
- Isolated and unable to relate to other people.
- Finding no pleasure in life or things they used to enjoy.
- A sense of unreality.
- Hopeless and despairing.
- Suicidal.

How they behave

- Difficulty concentrating or remembering things.
- Difficulty speaking, thinking clearly or making decisions.
- Feeling tired all the time.
- Difficulty sleeping, or sleeping too much.
- Moving very slowly, or being restless and agitated.
- Physical aches and pains with no obvious physical cause.
- Avoiding social events and activities they used to enjoy.
- Loss of appetite and weight or eating too much and gaining weight.
- Losing interest in sex.
- Using more tobacco, alcohol or other drugs than usual.
- Self-harming or suicidal behavior.

Depression can lead an individual to feel unable to cope with everyday tasks. Their reduced functional capability means they are no longer able to apply themselves to everyday tasks that they once could do with little difficulty. It may be that they are no longer able to function properly in their workplace or, if they are a student, they may find that they are no longer able to concentrate on studying. In the home this may result in an inability to gather enough energy to perform everyday activities, for example, those related to personal hygiene. This leads the depressed individual to feel incapable of completing basic and important tasks. This then makes the possibility of completing those tasks even more unlikely and this cycle leads to lower self-esteem and further incapability.

Depressed individuals therefore often develop low self-esteem and do not value their own ideas, views and opinions. They tend to constantly worry that what they think and do is not 'good enough'. Eventually, it is not uncommon for depressed individuals to find it impossible to concentrate or gather enough mental energy to hold a conversation.

Treatment for depression involves either medication or talking therapy treatments, or usually a combination of the two. The kind of treatment that a physician recommends will be based on the type and seriousness of the depression diagnosed.

Anxiety/panic attacks

Anxiety and panic disorders are common, but can cause extreme distress to individuals. Anxiety is a feeling of unease, such as worry or fear, which may be mild or severe. Panic attacks are a type of fear response. They are an exaggeration of your body's normal response to danger, stress or excitement.

Anxiety

Everyone has feelings of anxiety at some point in their life. For example, they may feel worried about sitting an examination or having a medical test or job interview. During times like these, feeling anxious can be perfectly normal. However, some people find it hard to control their anxiety, their feelings of anxiety are more extreme, often constant and affect their ability to cope with everyday tasks. There are a range of anxiety disorders and some of the more common ones are:

Generalised anxiety disorder (GAD) – this means having regular or uncontrollable worries about many different things in everyday life, which could include work.

Social anxiety disorder – means the individual affected experiences extreme anxiety triggered by social situations (such as weddings, workplaces or any situation in which you have to talk to another person). It is also known as social phobia. A phobia is an extreme anxiety triggered by a particular situation (such as social situations or going outside the house) or a particular object (such as spiders).

Obsessive-compulsive disorder (OCD) – this means the individual has anxiety involving repetitive thoughts, behaviours or compulsions.

Anxiety feels different for everyone. An individual might experience some of the symptoms listed below:

How they feel

- Tense, nervous or unable to relax.
- The world is speeding up or slowing down.
- They cannot stop worrying and have a sense of dread or fear the worst.
- Thinking a lot about bad past experiences, often thinking over a situation again and again.
- Worrying a lot about things that might hap-pen in the future.
- Worrying about anxiety itself, for example worrying about when panic attacks might happen.
- Other people can see they are anxious and are looking at them.
- A need for reassurance from other people or worrying that people are angry or upset with them.
- Worrying that they are losing touch with reali-ty.
- Disconnected from their mind or body, like they are watching someone else or the world around them is not real.

How they behave

- An agitated feeling in their stomach.
- Feeling faint, light-headed or dizzy.
- Breathing faster.
- A fast, very strong or irregular heartbeat.
- Sweating, hot flushes or very cold.
- Trembling or shaking.

- Nausea (feeling sick).
- Needing the toilet more or less often.
- Headaches, backache or other aches and pains.
- Feeling restless or unable to sit still.
- Problems sleeping.
- Grinding their teeth, especially at night.
- Changes in their sex drive.
- Having panic attacks.

Symptoms that may particularly show themselves at work may include loss of interest, poor concentration, low mood and irritability.

Panic

Panic disorder is where a person has regular or frequent panic attacks without a clear cause or trigger. Experiencing panic disorder can mean that the person affected feels constantly afraid of having another panic attack, to the point that this anxiety itself can trigger panic attacks.

During a panic attack an individual might feel very afraid that they are losing control, going to faint, having a heart attack or going to die. The physical symptoms of a panic attack can build up very quickly and are similar to the symptoms for extreme anxiety.

Work-related stress

Stress is the feeling of being under too much mental or emotional pressure. Pressure turns into stress when an individual feels unable to cope with the pressure. Many of life's pressures can cause stress, particularly work, home life, relationships and financial problems. People have different ways of reacting to pressure, therefore a situation that feels stressful to one person may be motivating to someone else. When an individual feels stressed, it can make it difficult to manage these pressures and can affect everything else in their life. For example, stress caused by work-related pressures could affect a worker's home life.

The World Health Organization (WHO) defines the meaning of work-related stress as follows:

"Work-related stress is the response people may have when presented with work demands and pressures that are not matched to their knowledge and abilities and which challenge their ability to cope."

Figure 5-49: Definition of stress.
Source: WHO, Work Organization and Stress.

In the UK, the Health and Safety Executive defines stress as:

"Stress is the adverse reaction people have to excessive pressures or other demands placed on them."

Figure 5-50: Definition of stress.
Source: UK, HSE.

Being exposed to situations of excessive pressure can bring about changes in an individual's behaviour, as well as their physical and psychological well-being. In particular, it can lead to worker's developing common mental ill-health problems like depression or anxiety.

Figure 5-51: Stress words.
Source: Venture Galleries.

Physical effects of stress include:

- Increased heart rate/sweating.
- Headache and dizziness.
- Blurred vision.
- Aching neck and shoulders.
- Skin rashes.
- Lowered resistance to infection.

Behavioural effects of stress include:

- Poor concentration.
- Irritability.
- Erratic sleep patterns.
- Increased smoking.
- Increased alcohol or drug use.

Psychological effects of stress include:

- Increased anxiety.
- Feeling unable to cope with everyday tasks.
- Low self-esteem.
- Depression.

Post-traumatic stress disorder

Post-traumatic Stress Disorder (PTSD) is developed after being involved in or witnessing traumatic events, such as a natural disaster, serious accident, terrorist act, war/combat or other violent assault. The events are therefore usually very frightening, upsetting or distressing. The person may not be exposed to the traumatic event directly, for example and individual may learn about the accidental death of a work colleague or violent death of a close member of their family. It can also occur as a result of repeated exposure to horrible details of trauma, such as what police officers may experience when investigating cases. People working in a range of roles in the emergency services may therefore experience PTSD.

PTSD can affect how an individual feels, thinks, behaves and their body functions. Common symptoms of PTSD include restless sleep, sweating, loss of appetite and difficulty concentrating. This could be because when an individual feels stressed their body release hormones called cortisol and adrenaline. This is the body's automatic way of preparing to respond to a threat (sometimes called the 'fight, flight or freeze' response). Studies have shown that someone with PTSD will continue producing these hormones when they are no longer in danger, which is thought to explain some symptoms, such as extreme alertness and being easily startled.

The symptoms of post-traumatic stress disorder (PTSD) can have a significant impact on an individual's usual life. In most cases, the symptoms develop during the first month after a traumatic event, but in a small number of cases there may be a delay of months or even years before symptoms start to appear. Some people with PTSD experience long periods when their symptoms are less noticeable, followed by periods where they get worse. Other people have constant severe symptoms.

The specific symptoms of PTSD can vary widely between individuals, but generally fall into the following three categories – re-experiencing, avoidance/emotional numbing and hyperarousal.

Re-experiencing

Re-experiencing is the most common symptom of PTSD. This is when an individual involuntarily and vividly relives the traumatic event in the form of:

- Vivid visual recollections or sensations from the traumatic event ('flashbacks'), feeling like the trauma is happening at the time.
- Vivid traumatic dreams ('nightmares').
- Physical sensations, such as pain, sweating, nausea (feeling sick) or trembling.

Some people have constant negative thoughts about their experience, repeatedly asking themselves questions that prevent them coming to terms with the traumatic event. For example, they may wonder why the event hap-pened to them and if they could have done anything to prevent it, which can lead to feelings of self-doubt, blaming themselves, guilt or shame.

Avoidance and emotional numbing

Trying to avoid being reminded of the traumatic event is another common symptom of PTSD. This usually means the individual avoiding certain people or places that remind them of the trauma, or avoiding talking to anyone about their experience. Many people with PTSD try to push memories of the trauma out of their mind, often distracting themselves with work or hobbies. Some people attempt to deal with their feelings by trying not to feel anything at all, known as emotional numbing. This can lead to the person becoming isolated and withdrawn, and they may also give up activities they used to enjoy.

Avoidance and emotional numbing symptoms can include:

- Having to keep busy.
- Avoiding anything that reminds them of the trauma.
- Being unable to remember details of what happened in the traumatic event.
- Feeling emotionally numb or cut off from their feelings.
- Feeling physically numb or detached from their body.
- Being unable to express affection.
- Using alcohol or drugs to avoid memories.

Hyperarousal (feeling 'on edge')

Someone with PTSD may be very anxious and find it difficult to relax. They may be constantly aware of threats, feel nowhere is safe, not trust anyone and be easily startled. This is known as hyperarousal. Hyperarousal often leads to:

- Irritability.
- Angry outbursts.
- Sleeping problems (insomnia).
- Difficulty concentrating

Because of these, potentially extreme effects on the individual suffering PTSD they sometimes lead to work-related problems and the breakdown of personal relationships.

CAUSES OF MENTAL ILL-HEALTH AND CONTROL MEASURES

Work-related mental ill-health, like other risks, should be subject to a risk assessment in order to identify the causes, evaluate the current controls and decide on actions to manage the risk.

There are six main factors that should be managed to minimise the risk of mental ill-health in the workplace: demands, control, support, working relationships, role and change. These are explained below in the form of causes, standards to be achieved in the workplace that will minimise mental ill-health and the control measures to eliminate or minimise the causes and achieve the standard.

Demand

Causes
Job content issues, for example:

- Monotonous, under-stimulating, meaningless tasks.
- Lack of variety.

- Unpleasant or high risk tasks.
- Environmental aspects of the job, for example, noise, vibration, extremes of temperature and lighting.
- Lone working.
- The job involves making unwelcome interventions, for example, law enforcement activities.
- Involves working with substandard resources, for example equipment that frequently breaks down or fellow workers that absent themselves at short notice.
- Workers are asked to do things they are not capable of doing, for example, they require specific competencies the worker does not have, or are too complex or emotionally demanding.

Workload and work pace, for example:

- Having too much or too little to do.
- Working under time pressures.
- Working to unrealistic deadlines.
- Workload is reactive and depends on external factors, such as demand from customer, national authorities, 'headquarters'.

Working hours, for example:

- Strict and inflexible working schedules.
- Long and unsocial hours.
- Unpredictable working hours.
- Badly designed shift systems.
- Inadequate rest periods.
- Communication equipment is issued to workers that makes them contactable out of normal hours.

Control measures
The standard to be achieved is that the demands of the job are within the capability of the workers and there are systems in place locally to respond to any individual concerns. Control measures to eliminate or minimise the causes and meet this standard include:

- Balancing the demands of the work to the agreed hours of work; consider shift working, the amount of additional hours worked and unsocial hours.
- Provision of regular and suitable breaks from work and rest periods.
- Matching worker skills and abilities to the job demands.
- Designing jobs so they are within the capability of workers.
- Minimising the work environment risks, such as noise and temperature.
- Provision of suitable facilities for work breaks, refreshment and rest.

Control

Causes
Participation and personal control, for example:

- Lack of participation in decision making.
- Lack of control, for example, over work methods, work pace, working hours, rest periods and the work environment.
- Workload and work activities are imposed by external organisations, such as national authorities or 'headquarters'.
- Lack of opportunity for workers to develop their competence.
- Lack of opportunity for workers to use their competence and initiative.

Control measures

The standard required is that workers are able to bring to the attention of the organisation issues to do with their work and that there are systems in place locally to respond to any individual concerns.

Control measures to eliminate or minimise the causes and meet this standard include:

- Providing workers with control over their pace and manner of work.
- Reducing the effects of repetitive and monotonous work by job rotation.
- Encouraging workers to use their skills and initiative to do the work.
- Encouraging workers to develop new skills and so enabling them to do more challenging or new work.
- Providing workers with opportunity to influence when breaks are taken.
- Consulting workers regarding work patterns.
- Encouraging the participation of workers in improving work.
- Providing mechanisms where workplace problems can be discussed.

Figure 5-52: Managing your stress.
Source: RMS/Positive Productive.

Support

Causes

Support from the organisation, management and other workers, for example:

- Poor leadership and culture.

- Conflicting priorities, for example, health and safety versus productivity.
- Lack of clarity about organisational objectives and structure.
- Lack of or unclear policies and procedures.
- Poor communication.
- Lack of support from line managers or co-workers or a feeling of isolation.
- Lack of constructive feedback from line managers or co-workers.
- Unclear how to access resources or lack of provision of requested resources to do the job.

Control measures

The standard required is that workers should receive adequate information and support from their colleagues and superiors and there are systems in place locally to respond to any individual concerns. Control measures to eliminate or minimise the causes and meet this standard include:

- Providing systems to inform workers of important issues and decisions.
- Establishing policies and procedures that provide support, particularly where workers may feel other factors are putting them under pressure.
- Providing systems that enable and encourage managers to identify where workers need support; consider where workers deal with the public in demanding environments, where new workers are introduced and times of high demand.
- Providing systems that enable and encourage managers to provide support to workers; consider particularly those working remotely by virtue of their location or time of working and new workers.
- Encouraging co-workers to support each other.
- Ensuring workers understand what resources and support are available and how they can get it.
- Providing regular constructive feedback to workers.

Work Relationships

Causes

Interpersonal relationships, for example:

- Inadequate, inconsiderate or unsupportive supervision/management.
- Poor relationships with co-workers.
- Bullying, harassment, abuse or violence (see Figure 5-53).
- Isolated or solitary work.
- No agreed procedures for dealing with problems or complaints.
- Unacceptable behavior tolerated.
- Workers (at all levels) treated unfairly.
- Poor communication between individuals and management.

"Persistent, offensive, abusive, intimidating, malicious or insulting behaviour, abuse of power or unfair penal sanctions which make the recipient feel upset, threatened, humiliated, or vulnerable, which undermines their self-confidence and which may cause them to suffer stress. Harassment, in general terms, is unwanted conduct affecting the dignity of men and women in the workplace. It may be related to age, sex, race, disability, religion, sexual orientation, nationality or any personal characteristic of the individual, and may be persistent or an isolated incident. The key is that the actions or comments are viewed as demeaning and unacceptable to the recipient."

Figure 5-53: Definition of bullying or harassment.
Source: UK, CIPD.

Control measures

The standard required is that workers should feel that they are not subjected to unacceptable behaviours, for example, bullying at work, and there are systems in place locally to respond to any individual concerns. Improving work relationships and attempting to modify people's attitudes and behaviour is a difficult and time-consuming process.

Effective strategies include regular communication with workers, provision of accurate and honest information on the effect of organisational changes on them, adopting partnership approaches to problems and provision of support. Employers should promote a culture that respects the dignity of others. Specific control measures to eliminate or minimise the causes and meet this standard include:

- Promoting positive behaviour that avoids conflict and leads to fairness; consider co-workers, customers and suppliers.
- Establishing policies and procedures that resolve unacceptable behaviour that leads to conflict.
- Encouraging managers to deal with unacceptable behaviour, such as harassment, discrimination or bullying.
- Encouraging workers to report unacceptable behaviour.
- Establishing policies that treat workers equally, for example, male and female, young and old, or disabled and able workers.
- Establishing systems that ensure communication with managers and workers; consider timeliness of communication to those that work isolated by location or time.
- Establishing systems that ensure involvement and consultation, such as regarding the process of conducting risk assessments.

Role

Causes

Role in the organisation, for example:

- Unclear role and responsibilities.
- Conflicting roles within the same job.
- Unclear or conflicting work objectives.
- Responsibility for people.
- Continuously dealing with other people and their problems.
- Organisational change affecting roles.
- Job insecurity.
- Lack of promotion prospects.
- Under-promotion or over-promotion.
- Work of 'low social value'.
- Demanding productivity payment schemes.
- Unclear or unfair performance evaluation systems.
- Being over-skilled or under-skilled for the job.

Control measures

The standard required is that workers must clearly understand their role and responsibilities and there are systems in place locally to respond to any individual concerns.

Control measures to eliminate or minimise the causes and meet this standard include:

- Ensuring workers have the knowledge, skill and experience to conduct their role or are being supported appropriately; consider workers undertaking new or difficult work.
- Ensuring role requirements are compatible, for example, that the need to manage costs does not conflict with health and safety.
- Ensuring role requirements are clear.
- Ensuring workers and their managers understand the roles and responsibilities.
- Providing systems to enable workers to raise concerns about role uncertainty or conflict; particularly consider the work-life balance for those workers who provide care for others outside their normal work routine.

Change

Causes

Organisational change, for example:

- Unclear reason for change.
- No prior warning of the change.
- No consultation on change or opportunity to influence the proposed change.
- Extent and impact of the change is unclear.
- Frequency of changes.
- Change is very large.
- Absence of or unclear timetable for changes.
- Absence of support during changes.
- Absence of information and training to enable transition to new circumstances.

Control measures

The standard required is that workers feel the organisation engages workers frequently when undergoing organisational change, and there are systems in place locally to respond to any individual concerns.

Control measures to eliminate or minimise the causes and meet this standard include:

- Providing workers with timely information to help them understand the change, reasons for it and timing of effects.
- Ensuring worker consultation on proposed changes.
- Providing workers with information on likely impacts of change on their jobs.
- Providing training and support through the period of change.

In addition to the six main factors that are managed in order to minimise the risk of mental ill-health, there are a number of general control measures that should form part of a mental health management strategy. These include:

- Introducing a mental health policy and procedures to demonstrate to managers, workers, worker representatives and enforcing agencies that the organisation recognises mental health as a serious issue worthy of a commitment to manage the risk.
- Providing training and support for workers and all levels of management in the form of mental health awareness and mental health management training, as appropriate.
- Recognising good work performance and the positive contribution of workers and responding to poor performance to ensure equality of treatment.
- Addressing the issue of work-life balance, which may include consideration of job-sharing, part-time work, voluntary reduced hours, home-working, flexible-time, etc.
- Listening to the concerns of workers and managers and providing practical and emotional support.
- Promoting general health and well-being awareness initiatives within the organisation, such as diet, exercise and fitness programmes.

HOME-WORK INTERFACE

Mental ill-health can be a reaction to events or experiences in someone's home life, work life or a combination of both. The pressure of an increasingly demanding work culture and more complex personal life is perhaps the biggest and most pressing challenge to the mental health of the general population.

A worker could have common mental ill-health problems with no obvious causes, which could affect them in their home life and at work. Mental ill-health can also be caused by specific problems outside work, in a work-er's home life, for example bereavement, divorce, postnatal depression or a new medical condition. In addition, work pressures could lead to a worker developing mental ill-health that affects them at work and in their home life.

The cumulative effect of increased working hours and higher work demands for some workers is likely to have an important effect on their home life, which may prove damaging to their mental health. In addition, home life issues, such as commuting, childcare and care of relatives, put workers under pressure that may, on its own or in conjunction with work pressures, lead to mental ill-health. Both of these issues, work and home life, can lead to individuals neglecting the factors in their lives that help them to be more resilient and resist the effects of pressures and less likely to have mental health problems, for example taking exercise and socialising.

Commuting

Commuting to and from work can put workers under great pressure, often as a result of factors outside their control, for example delays in their journey caused by weather, increase in the number of commuters, unexpected failure/repairs to a transport system. In addition, this can be made worse by issues in their home life causing delays that affect them and prevent them leaving in good time to begin their commute to work. These delays could lead to sanctions form the employer or, if they occur on their journey home, poor family relations or consequences to arrangements for childcare or care of dependent relatives.

In addition, the duration and sometimes arduous nature of commuting can lead to fatigue, which could lower an individual's resilience and resistance to mental ill-health. For some workers these effects could add sufficient additional mental burden, leading to stress and possible depression or anxiety. Those workers who suffer from an anxiety disorder could develop a panic attack when exposed to challenging commuting issues.

Childcare

Childcare issues can arise through unexpected loss of childcare provision, for example a relative who provides the childcare could become ill or the child may become ill and the childcare service refuse to provide care until the child gets better. Where the childcare is in the form of schooling the child may be prevented from going to school because the school has closed due to staff shortages, perhaps due to poor weather.

Whatever the reason, the change in circumstances that are heavily relied on by a family will put a great deal of pressure on the family and may mean one of the members of the family not being able to work. This additional burden is likely to add pressure and be difficult to cope with. There may also be possible sanctions from their employer and loss of income. As with the pressure of commuting, the pressure that can comes from problems with childcare may cause mental ill-health or make a mental ill-health condition become intolerable and require the individual to seek medical intervention. The experience or chance of childcare problems could lead to the development of an anxiety disorder, where the individual affected constantly anticipates and fears the possibility of childcare arrangements failing.

Relocation

Relocation of an individual's job or home can be one of the most difficult situations that they may encounter as it causes most of the structure of their life to change, for example, new relationships to be made at work or with neighbours at home or different arrangements for childcare and commuting. Many of these new situations will have aspects of deep uncertainty. These new situations and other unknown factors can be very disturbing for those involved. This could be a sufficiently strong and sustained pressure to cause mental ill-health in any or all of those involved. The pressure may fracture the usual support that those involved give each other, sometimes because they may be separated for a period of time due to the relocation. This separation and loss of support could cause a feeling of isolation and may lead to depression or social anxieties.

Care of relatives

Care of frail or vulnerable relatives, for example elderly parents, may feel like a duty that the individual must meet at all costs and those being cared for could put unrealistic demands on the individual. Both of these fac-tors could lead the individual to strive to have high expectations of themselves with regard to meeting the de-mands. Where these prove to be more difficult than expected or a sustained challenge, the individual may feel that they are failing and may begin to develop depression or

become anxious about not meeting expectations. As with the other home life issues discussed it is likely that these additional demands of caring for relatives could also lead to fatigue and may limit the opportunities of the individual to benefit from things that maintain their resilience, for example exercise, socialising, time for relaxation and personal space.

PEOPLE WITH MENTAL ILL-HEALTH CONTINUING TO WORK

There is an increasing recognition that most people with mental ill-health can continue to work effectively and one of the best things for their wellbeing is for them to be in appropriate work, which takes regard for their mental ill-health.

It is important to remember that studies have shown the longer a worker is off work with mental ill-health the more difficult it becomes for them to return to work and the less likely it is that they will return to work at all. Therefore, where possible, workers should remain in work while receiving treatment or return to work at the earliest opportunity.

For some workers affected by mental ill-health this may mean making some adjustments to their work. These may be of a temporary nature as they readjust to work after developing the mental ill-health or become unnecessary as their mental ill-health improves and symptoms diminish. It is therefore important to support their return by making appropriate adjustments.

It is also important to remember that for people with mental ill-health, for example depression, improvement in ability to function well at work may be reduced for a period. In order to improve wellbeing and help recovery, a doctor might confirm a worker to be fit to return to work before they have the full capability to attain previous levels of work performance. This creates the risk that

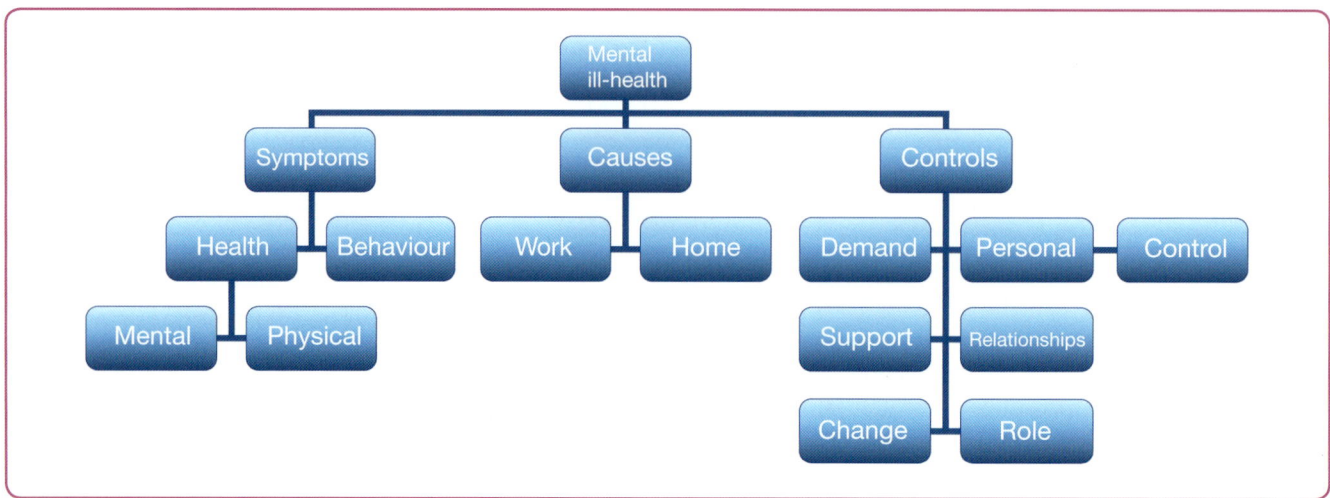

Figure 5-54: Mental ill-health in summary.
Source: RMS.

employers interpret poor performance by the worker as lack of effort or motivation or competence and so create the conditions in which it is more likely that the worker believes that they are failing and starting to become ill again. The worker is therefore likely to need an encouraging work environment in order to adjust to work, improve and work well.

There are a range of treatments available for common mental ill-health, including medication and psychological therapies, for example, cognitive behavioural therapy (CBT) and self-help. These can support the individual affected and enable them to continue to work. With this opportunity and support it is likely that most people with mental ill-health can continue in work.

Good work gives the worker a social identity, purpose, social contacts, a means of structuring/occupying time, activity, involvement and a sense of personal achievement. This can improve the resilience of the worker and reduce the likelihood of them being affected by mental ill-health.

Figure 5-55: Stress prevention at work checkpoints.
Source: ILO.

 **REVIEW**

Other than those associated with the environment, list four possible work-related causes of increased stress levels amongst workers.

5.5 Violence at work

TYPES OF VIOLENCE AT WORK

Violence at work is not limited to physical violence, it also includes verbal abuse and threats.

When physical violence is involved, the injuries to those workers affected are obvious. However, those subjected to constant and repeated verbal abuse and threats may suffer less visible psychological effects, such as stress, anxiety and depression.

 "Any action, incident or behaviour that departs from reasonable conduct in which a person is assaulted, threatened, harmed, injured in the course of, or as a direct result of, his or her work."

Figure 5-56: Definition of work-related violence.
Source: ILO Code of practice on workplace violence.

 "Any incident, in which a person is abused, threatened or assaulted in circumstances relating to their work."

Figure 5-57: Definition of work-related violence.
Source: UK, HSE.

Workers may be sworn at, threatened or even attacked. Verbal abuse and threats are the most common types of violence at work, physically attacks are comparatively rare. It is more common in those jobs where workers have face-to-face contact with the public. Physical attacks are obviously dangerous, but serious or persistent verbal abuse or threats can cause psychologically damage to a worker's health.

JOBS AND ACTIVITIES WHICH INCREASE THE RISK OF VIOLENCE AT WORK

The most common job and activity risk factors relating to violence are:

- The position a person holds - a worker may hold a position of authority over another person; how they use this authority and the effects on the person can cause disagreement, resentment and could cause the person to be more aggressive.
- The nature of the work - if a worker is working alone this may lead them to be more at risk of violence. The nature of the work may be dealing with people that are emotionally charged, and this may cause them to react to normal things in an unpredictable way.
- The location of the work - if the worker is working isolated from others this may leave them at risk. Some locations may have a tendency for violence, for example, certain parts of a city or housing estates in deprived areas.

- The time of working - working during late evening and early morning may mean there are fewer people around and those people that are around may be more likely to have violent tendencies.
- Alcohol and drugs - can make some people more aggressive. Because their perception and behaviour is more unpredictable, it may lead to misunderstandings and violence.
- Visible appearance - violence may arise simply because someone does not like a worker's visible appearance and what this may represent. This can be accentuated if the appearance is one of privilege and wealth.
- The availability of weapons - if weapons, improvised or normal, are available to be used a sudden outburst that may have been verbal could escalate to involve major personal injury because a weapon was available, for example, knives in a restaurant or home, shotgun on a farm or drinking glass in a bar.

Those jobs where workers have contact with the public and are most at risk are where workers are engaged in caring, education, cash handling, and representing authority. This include police, fire, medical, social workers and those in customer services. Lone workers, those working with people under the influence of drugs and alcohol and those who handle money or valuables are particularly at risk.

In addition, it is necessary to consider possible violence between co-workers. Maintenance and construction activities often need to be completed within a fixed timescale and can result in a lot of pressure on workers to work quickly and without error. This can lead to a good deal of tension which might result in violent outbursts between co-workers including contractors.

Determine the size of the problem

Employers should determine the extent to which the risk factors affect their organisation. In order to determine the scale of the problem employers should:

- Ask workers informally, through managers and health and safety representatives.
- Encourage workers to report all incidents and keep detailed records.
- Classify all incidents according to their actual or potential severity of outcome.
- Try to predict what might happen and how violence may arise.

CONTROL MEASURES TO REDUCE RISKS FROM VIOLENCE AT WORK

Policy

The policy for dealing with violence should be written into the health and safety policy statement so that all workers are aware of it.

This will encourage workers to cooperate with the policy and report further incidents.

Determine controls relevant to the risk

Employers should establish appropriate control measures to eliminate or minimise the effect of the risk factors that affect their organisation.

In order to determine controls relevant to the risk, employers should:

- Provide a means of raising the alarm and a plan for dealing with violent situations.
- Brief workers on problems that may foreseeably occur.
- Provide information, for example, case histories, and incidents.
- Set up policy for alcohol and substance misuse.
- Set up procedures for reporting any occurrence.
- Identify who might be harmed and how.
- Identify who is most vulnerable.
- Train workers to recognise early signs of violence and in de-escalation techniques (modification of their own behaviour), for example, case histories.
- Providing counselling for both perpetrators and victims.
- Consider relocation of perpetrators elsewhere.
- Consider the ambience of the environment in terms of colour, lighting, seating, space and even the type of background music.
- Put up signage to warn violence will not be tolerated.
- Remove any items that could be used as a weapon. Even pens and pencils can be used as weapons.
- Improve security - for example, closed circuit television (CCTV) with recording facility, coded locks, wider and higher counters for those who deal with members of the public, panic alarms.
- Redesign the job - for example, use credit cards rather than cash, bank money more frequently. Check arrangements for lone workers and consider two workers where risks are high.
- Vary the routine when handling cash and other items that some might consider to be valuable such as drugs and medication.
- Arrange safe transport or secure car parking for people who work late at night.
- Make sure that crowds, particularly queues, are managed properly to prevent violence as a result of frustration caused by waiting for attention. Consider signing and updating current waiting time; introduce pre-appointments whenever possible.
- Place screens between workers and the public.

Monitor

Check regularly to see if the arrangements are working by consulting workers and worker health and safety representatives. If violence is still a problem, review work practices and risk controls.

Dealing with incidents

If there is a violent incident in the workplace it will be necessary to consider the following:

- Debriefing - victims might need to talk through their experience as soon as possible.
- Providing time off work - individuals may need differing times to recover.
- Support - in some cases victims might need counselling. Consider a phased return to work.
- Legal help - legal assistance may be appropriate in serious cases.
- Other workers - may need guidance or counselling to help them react appropriately.

 **REVIEW**

What control measures might an employer consider to minimise the risk of violence against workers?
List five examples of occupations that are at high risk from violence.

5.6 Substance abuse at work

TYPES OF SUBSTANCES ABUSED AT WORK

Many substances may affect a worker's ability to work safely and they do not have to be taken whilst at work to have an effect. Substances can have a residual effect even when they are taken outside working hours. This can have a significant effect on the health and safety of the worker and those that may be affected by their work. It is imperative that organisations take the issue of substance misuse seriously in order to protect health and safety, especially in high risk activities such as chemical, nuclear, construction and transport etc.

Alcohol

Even at low levels of consumption the body is affected by alcohol, reducing reaction time and impeding coordination. It also affects thinking, judgement and mood and can have a significant effect on behaviour when conducting routine work. Whilst large amounts of alcohol in one session can put a strain on the body's functions, in particular the liver, it can also affect muscle function and stamina.

Drinking alcohol raises the drinker's blood pressure and this can increase the risk of coronary heart disease and some kinds of stroke. Regularly drinking smaller amounts also increases the risk of cirrhosis of the liver. The effect of alcohol is not limited to physical effects - people who drink very heavily may develop psychological and emotional problems, including depression.

 **CASE STUDY**

Brandon was a 25 year old male with alcohol related problems. He had been having an increasing number of absences from work, was often late on duty and with the smell of alcohol on his breath. The manager discussed the issues with Brandon, who agreed that he required help to overcome his excessive drinking. Advice from occupational health resulted in Brandon being referred for alcohol counselling during the working day. With the full support of the manager Brandon responded well, stopped drinking and his attendance improved. Further follow up by occupational health demonstrated that the early intervention and support provided for Brandon prevented his excessive drinking.

Legal/illegal drugs

The effects on the workplace of some drugs can be catastrophic if a worker is a regular user. Even the casual user may not be fully fit for work after taking a drug the night before or the weekend before work.

Legal drugs

Legal drugs include those that are prescribed by a medical practitioner and those that may be obtained without prescription from a pharmacy. Access to drugs with or without prescription varies in different countries. In some countries opium based drugs, such as codeine or morphine, may be prescribed for pain relief. The depressant effects of this drug when used for short-term use under medical supervision would not normally be a concern related to work, however in larger doses they may cause a worker to have reduced mental functions that could affect their use of machinery, tools or driving. Antihistamines, used to treat allergic reactions, may cause drowsiness and impaired thinking that can create risks for workers in a number of jobs. Drugs prescribed to treat depression, antidepressants, can also slow reaction speeds and affect activities that require concentration.

Benzodiazepines are a range of drugs that are in wide use to treat insomnia, anxiety and epilepsy. They have a number of common effects that could affect the health and safety of workers, the main ones being drowsiness, dizziness, confusion, light-headedness and memory loss.

Qualaquin (quinine sulphate) is an antimalarial drug used to treat malaria, a disease caused by parasites. Parasites that cause malaria typically enter the body through the bite of a mosquito. Malaria is common in areas such as Africa, South America, and Southern Asia. Common side effects of Qualaquin include decreased hearing, dizziness and blurred vision.

The use of legal drugs should be considered where the work being done may be particularly influenced by their effects, for example, work at height, driving or operating mechanical handling equipment or safety critical work like control of a permit-to-work system.

Illegal drugs

Illegal drugs are those that have been classified by the State as requiring control because of the seriousness of their effects. Controls are usually imposed to limit their supply and use, including criminal penalties. There are many types and variants of illegal drugs, some of the common ones include cannabis, depressants and stimulants. These drugs could be taken both in the workplace and away from it. In either case their use will have detrimental effects on the user of them, resulting in higher risks when driving or operating machinery and also putting others at risk. The effects on the individual of using cannabis include:

- Euphoria.
- Slow reactions.
- Poor coordination.
- Short term memory affected.
- Distortion in perception of space and time.
- Drowsiness.
- Inability to think clearly.
- Anxiety and depression.

The effects of cannabis can continue for up to three weeks in the frequent user.

Depressants include opiates like heroin, opium, morphine and codeine. Opiates begin to affect the central nervous system almost immediately after use. Because the drug suppresses the central nervous system the user experiences 'cloudy' mental function, breathing slows and may reach a point of respiratory failure. They are extremely physically and psychologically addictive drugs.

Stimulants like amphetamines and cocaine and MDMA (3,4-methylenedioxy-N-methylamphetamine) make people feel more lively, awake, energetic, confident and, in the case of MDMA, give a sense of well-being. The effects on the individual will vary depending on the drug, but generally as the drug wears off the person may become anxious, irritable and restless, however even when they feel desperate for sleep the drug may continue to keep them awake. Finally, exhaustion and often intense mood swings affect the user. The range of adverse health effects include increased heart rate, high blood pressure, high temperature, nausea, chills, sweating, involuntary teeth clenching, muscle cramps and blurred vision.

In the case of MDMA some users report undesired effects like anxiety, agitation and recklessness. In the hours after taking MDMA it causes significant reduction in mental abilities that can last for a week, particularly those affecting memory.

Solvents

Solvents are readily available in the form of paints, varnishes, inks, glues, aerosols, and degreasing agents, particularly in the workplace. Different solvents can affect an individual in different ways, but in particular the central nervous system is readily affected by breathing in solvent vapours. When an individual is exposed to solvent vapours in low concentrations they tend to cause nausea, headache, dizziness and light-headedness. When higher concentrations of vapours are involved they can cause unconsciousness or death. Long term use can affect the brain, kidneys and liver. The loss of coordination and disorientation from the effects on the central nervous system can lead to workers making errors at work that may affect themselves or others. As with other drugs that affect the central nervous system in a similar way workers that rely on high levels of coordination and concentration, like driving and operating equipment are particularly at risk.

RISK TO HEALTH AND SAFETY FROM SUBSTANCE ABUSE AT WORK

The regular use of substances like alcohol and drugs is becoming increasingly common in some societies. People who start consuming alcohol or drugs usually have a nil or low dependency and do so for recreational reasons. This however can escalate quickly into abuse when larger quantities of alcohol or drugs are consumed more frequently and become habitual. Drugs may be prescribed for a worker with a medical condition as treatment or drugs that are controlled, where use is illegal, such as cocaine and heroin. The effects of alcohol or drugs can vary dependent upon the individual's state of health and fitness, and resilience to the chemicals. Alcohol and drugs can remain in the body for a considerable time after consumption and continue to affect the individual when at work.

Apart from the general effects on the individual's health (for example, cirrhosis of the liver from alcohol abuse), the risks related to substance misuse and work is derived from the effects the substances have on the individual's biological and psychological systems. These effects create risks to the health and safety of the worker and others.

Effects on health and safety include:

- Poor coordination and balance.
- Perception ability reduced.
- Overall state of poor health including fatigue, poor concentration and stress.
- Poor attitude, lack of adherence to rules.
- Increased risk of violence.
- Increased likelihood of transport incidents.
- Reduction in work rate which may encourage risk taking when catching up.

The range of sensory impairments that the misuse of substances can have means that a worker's or manager's ability to carry out their work safely may be affected because their coordination, concentration, memory, judgement or balance may be affected. This would make work at height, with equipment, tools, transport and many other activities that rely on care and accuracy to be affected. It is important to remember that substance misuse affects managers as well as workers and that although middle and senior managers may not work with hazardous equipment they do make decisions that can affect the health and safety of workers and others. These decisions may be seriously influenced by the effects of the substances. In addition, misuse of sub-stances may change the behaviour of the worker or manager such that they do not follow normal patterns of behaviour or rules. This could involve taking unnecessary risks, not bothering to use health and safety precautions or communicating important information poorly. In some cases they may become aggressive, particularly if challenged regarding their not following rules, poor behaviour or misuse of substances.

Risk factors to consider when assessing whether workplace risks exist or may arise from alcohol and other drugs include:

- Availability of substances - in some workplaces individuals are more likely to be exposed to the presence of alcohol, drugs or solvents, perhaps because they are legitimately involved in their manufacture, storage, transport and distribution, for example, the production and sale of beer, pharmaceuticals or glues. Substance availability may tempt workers to use it.

- Type of workplace culture - for example, there may be a culture at work that encourages or accepts consumption of alcohol and/or other drugs at the workplace.

- Usage of substances in relevant social groups - if this increases, decreases or is a known problem, it may have an impact at the workplace. This relates to groups outside work, but also includes groups that may consume drugs and/or alcohol when they meet before work or at breaks during the working day. In some situations the social group may be a work group, such as those that get involved with sales and promotion and spend time in situations where clients are entertained.

- Patterns of substance consumption - different patterns of use create different risks. For example, people who use large amounts on single occasions may create different risks compared to people who are regular users. Individuals may use alcohol, illegal drugs or solvents only on rare occasions, weekends or occasionally outside work or they may regularly take breaks from work in an environment where substances are available. An individual may use a prescribed drug to treat a medical condition on a single occasion, however the condition may become persistent and the use of the prescription drug and dose levels may increase.

- Inadequate job design and stress - unrealistic performance targets and deadlines, excessive responsibility, monotonous work or low job satisfaction may, in some instances, be risk factors. For example, symptoms of stress are sometimes associated with poor health, including substance related problems. Drugs prescribed to moderate the effects of stress may have side effects that create risks to health and safety.

- Excessive working hours, shift work and the absence of rest - illegal drugs, such as amphetamines or cocaine, or prescription medication may be taken in order to keep awake or maintain attention.

- Poor working conditions - for example, such as hot or dangerous environments may contribute to substance misuse. Sometimes people drink beer because they feel it is justified by the hot work they do and that they need to replenish lost liquid. In situations where work is perceived to be dangerous or undesirable the substances may inhibit thought processes or distort perception.

- Interpersonal factors - for example, bullying at work may increase risks.

- Isolation from family and friends - workers in isolated areas who are separated from family and friends may be more likely to consume substances due to boredom, loneliness or lack of social activities.

- Inadequate supervision - jobs where there is inadequate supervision may increase the risk of substance related problems because there is an absence of guidance and control of behaviour, which over time can reinforce the acceptability of the misuse of substances.

The level of risk in a workplace may partly be determined by analysis of absence from work, accidents/incidents and disciplinary records.

This can be supported by observation of changes in behaviour, including:

- Sudden mood changes.
- Unusual irritability or aggression.
- Tendency towards confusion.
- Fluctuations in concentration and energy.
- Impaired job performance.
- Poor time keeping.
- Increased sickness.
- Deterioration in relationships - colleagues, managers, customers, family.
- Dishonesty and theft.

In addition, the presence of items related to substance misuse could be identified in the workplace and in an individual's possessions, for example, alcohol containers, syringes and ligatures, scorched tin foils or spoons with their base burnt or small empty packages in which the drugs may have been contained. Though some of these items may have a use not related to substance abuse they can indicate a potential risk. Some organisations may also supplement this analysis with an interview of workers and managers, self-declaration opportunities or an independent drug and alcohol test.

CONTROL MEASURES TO REDUCE RISKS FROM SUBSTANCE ABUSE AT WORK

The use of substances like alcohol and drugs is a personal choice over which employers usually have little or no control when it is conducted in the worker's own time. However, employers should establish a policy to deal with drugs and alcohol should it start to impact on the worker's performance at work.

Control measures should be introduced to eliminate or reduce the risks of people being injured or harmed at work due to substance misuse. This can be achieved through adopting a number of control measures as appropriate and the approach taken should be tailored to meet the needs of the particular workplace. Deciding which strategies to adopt will depend on a number of different 'risk factors' (see earlier), the extent of substance misuse, the nature of the organisation and the size and resources available.

The control measures may include:

- Developing a substance misuse policy and supporting procedures for all levels of individual in the work-place, managers and workers. The procedures should include how to deal with impaired persons in the workplace.

- Communicating the policies and procedures on substance misuse and the general expectations for health and safety related to this.

- Encouraging those in management positions to support the policies and procedures.

- Providing information, education and training on the risks, symptoms and expectations to everyone that has an influence on substance misuse, managers, workers, contractors, suppliers, neighbouring establishments and families.

- Implementing control measures related to tasks, processes and equipment that require an individual to have a high level of concentration or coordination and where a risk assessment has identified a high level of risk if they are impaired by substances. This may include pre-employment and periodic testing for substance misuse for those that are involved in safety critical work. It is usual not to presume use or non-use of alcohol or drugs for those that conduct safety critical work, treating all those involved in such work equally.

"3.2 - Contents of an alcohol and drug policy 3.2.1 - A policy for the management of alcohol and drugs in the workplace should include information and procedures on:

a) Measures to reduce alcohol and drug-related problems in the workplace through proper personnel manage-ment, good employment practices, improved working conditions, proper arrangement of work, and consul-tation between management and workers and their representatives.

b) Measures to prohibit or restrict the availability of alcohol and drugs in the workplace.

c) Prevention of alcohol and drug-related problems in the workplace through information, education, training and any other relevant programmes.

d) Identification, assessment and referral of those who have alcohol or drug-related problems.

e) Measures relating to intervention and treatment and rehabilitation of individuals with alcohol or drug-related problems.

f) Rules governing conduct in the workplace relating to alcohol and drugs, the violation of which could result in the invoking of disciplinary procedures up to and including dismissal.

g) Equal opportunities for persons who have, or who have previously had, alcohol and drug-related problems, in accordance with national laws and Regulations."

Figure 5-58: Content of an alcohol and drugs policy.
Source: ILO Code of practice on alcohol and drugs in the workplace.

- In a simple approach, all workers that come on to a site, such as a hospital, construction or power generation site, might be considered to be in safety critical work. All workers would work to the same rules, which included banning workers being under the influence of substances while at work, offering them opportunities to talk to someone about how this affects them and carrying out random drugs and alcohol tests.

- Reducing the influence of factors that may contribute to stress or fatigue, for example, redesigning jobs, controlling hours of work and providing regular breaks.

- Providing suitable arrangements for workers and managers to take breaks where the environment is more conducive to avoiding substance misuse, for example, making non-alcoholic drinks available or providing recreational options where boredom might be a risk factor or where workers are isolated from family and friends.

- Increased supervision of those groups that may be more likely to engage in substance misuse, for example, those with opportunity because of its availability in the workplace, those in dangerous, undesirable or hot work and those that have poor interpersonal relations.

- Providing access to rehabilitation advice, counselling and support groups early in the apparent development of substance misuse problems.

- Disciplinary procedures for those workers who refuse to be tested or who fail a test, refuse to attend rehabilitation or are found to have alcohol or drugs on their person when at work.

All control measures applied should be monitored and reviewed on an ongoing basis. When accidents/incidents occur it is common to consider alcohol and drugs as factors. It may not always be possible to test an injured party to identify if they caused their own injuries by being under the influence of alcohol or drugs, but, for example, it may be possible to test a driver of a fork-lift truck that drove into someone on site.

 **REVIEW**

How might substance abuse cause health and safety problems in the workplace?

What measures can employers put in place to reduce such problems?

List three drugs that may be classified as stimulants and three drugs that may be classified as depressants.

Sources of reference

Reference information provided, in particular web links, was correct at time of publication, but may have changed.

Ambient factors in the Workplace, International Labour Organization (ILO) Code of Practice (CoP), ISBN 92-2-11628-X http://www.ilo.org/safework/info/standards-and-instruments/WCMS_107729/lang--en/index.htm

Controlling Noise at Work, The Control of Noise at Work Regulations, Guidance on Regulations, second edition 2005, L108, HSE Books, ISBN: 978-0-7176-6164-4, http://www.hse.gov.uk/pubns/priced/l108.pdf

Drug misuse at work a guide for employers HSE Books INDG91 www.hse.gov.uk/pubns/indg91.pdf

Electromagnetic fields at work, A guide to the Control of Electromagnetic Fields at Work Regulations, HSG281, HSE Books, http://www.hse.gov.uk/pubns/priced/hsg281.pdf

Essentials of Health and Safety at Work, fourth edition 2006, HSE Books, ISBN: 978-0-7176-6179-4

Hand-arm vibration, Control of Vibration at Work Regulations 2005, Guidance on Regulations, L140, HSE Books, ISBN: 978-0-7176-6125-1, http://www.hse.gov.uk/pubns/priced/l140.pdf

The health and safety toolbox, How to control risks at work, HSG268, HSE Books, ISBN: 978-0-7176-6587-7 , http://www.hse.gov.uk/pUbns/priced/hsg268.pdf

How to tackle work-related stress. A guide for employers on making the Management Standards work, INDG430, HSE Books, http://www.hse.gov.uk/pubns/indg430.pdf

HSE Stress Management Standards www.hse.gov.uk/stress/standards

ILO Radiation Protection Convention, 1960 (No. 115) – C115, 1960, R114, http://www.ilo.org/dyn/normlex/en/f?p=NORMLEXPUB:12100:0::NO:12100:P12100_INSTRUMENT_ID:312260:NO

ILO Radiation Protection Recommendation, 1960, R114, http://www.ilo.org/dyn/normlex/

en/f?p=1000:12100:0::NO::P12100_ILO_CODE:R114

ILO Working Environment (Air, Pollution, Noise and Vibration) Convention, 1977 (No 148) - C148, http://www.ilo.org/dyn/normlex/en/f?p=1000:12100:0::NO::P12100_ILO_CODE:C148

ILO Working Environment (Air, Pollution, Noise and Vibration) Recommendation, 1977, R156, http://www.ilo.org/dyn/normlex/en/f?p=1000:12100:0::NO::P12100_ILO_CODE:R156

Management of alcohol and drug related issues in the workplace, ILO CoP, 1999 ISBN: 92-2-109455-3 www.ilo.org/wcmsp5/groups/public/---ed_protect/---protrav/---safework/documents/normativeinstrument/wcms_107799.pdf

Personal Protective Equipment at Work (second edition), Personal Protective Equipment at Work Regulations 1992 (as amended), Guidance on Regulations, HSE Books, ISBN: 978-0-7176-6139-3, http://www.hse.gov.uk/pubns/priced/l25.pdf

Protection of workers against noise and vibration in the working environment, ILO Code of Practice, https://www.ilo.org/global/topics/safety-and-health-at-work/normative-instruments/code-of-practice/WCMS_107878/lang--en/index.htm

Radiation Protection C115 and R114, 1960, http://www.ilo.org/dyn/normlex/en/f?p=1000:12100:0::NO::P12100_ILO_CODE:C115

http://www.ilo.org/dyn/normlex/en/f?p=1000:12100:0::NO::P12100_ILO_CODE:R114

Radiation protection of workers, ionising radiations, ILO CoP, 1987, ISBN: 9-22-105996-0, http://www.ilo.org/wcmsp5/groups/public/---ed_protect/---protrav/---safework/documents/normativeinstrument/wcms_107833.pdf

Radon in the workplace: http://www.hse.gov.uk/radiation/ionising/radon.htm#testingradon

Tackling work-related stress using the Management Standards approach, A step-by-step workbook, WBK1, HSE Books, ISBN 978-0-7176-6715-4, http://www.hse.gov.uk/pubns/wbk01.pdf

The Control of Noise at Work Regulations 2005, UK, www.legislation.gov.uk (please enter the name of the Act)

The Control of Vibration at Work Regulations 2005, UK, www.legislation.gov.uk (please enter the name of the Act)

The Ionising Radiations Regulations 2017, UK, www.legislation.gov.uk (please enter the name of the Act)

The Personal Protective Equipment at Work Regulations 1992, UK, www.legislation.gov.uk (please enter the name of the Act)

Upper Limb Disorders in the Workplace - A Guide, second edition 2002, HSG60, HSE Books, ISBN: 978-0-7176-1978-8, http:// www.hse.gov.uk/pubns/priced/hsg60.pdf

Violence at work: A guide for employers, HSE Books, INDG69, www.hse.gov.uk/pubns/indg69.pdf

What are the Stress management standards, HSE, http://www.hse.gov.uk/stress/standards/

Whole-body vibration; The Control of Vibration at Work Regulations 2005, Guidance on Regulations, L141, HSE Books, ISBN: 978-0-7176-6126-8, http://www.hse.gov.uk/pubns/priced/l141.pdf

Workplace violence in services sectors and measures to combat this phenomenon, ILO Code of Practice, https://www.ilo.org/wcmsp5/groups/public/---ed_protect/---protrav/---safework/documents/normativeinstrument/wcms_107705.pdf

Working Environment (Air, Pollution, Noise and Vibration) C148 and R156, 1977;

http://www.ilo.org/dyn/normlex/en/f?p=1000:12100:0::NO::P12100_ILO_CODE:C148

http://www.ilo.org/dyn/normlex/en/f?p=1000:12100:0::NO::P12100_ILO_CODE:R156

Web links to these references are provided on the RMS Publishing website for ease of use - www.rmspublishing.co.uk

STUDY QUESTIONS

1) What circumstances may lead to occupational stress amongst workers? (8)

2) What are the health effects that may be caused by exposure to ionising radiation? (8)

3) Explain four control methods to reduce noise at source. (8)

4) Explain the technique of audiometry used in health surveillance and say why it is necessary. (8)

5) What are the typical health effects that might be shown by those who use vibrating hand-held tools regularly at work? (8)

6) What are the control measures which should be taken when a noise survey has determined an average noise level of 83dbA in a factory workplace? (8)

7) What are risks to workers from the misuse of substances at work? (8)

8) How can an employer determine the size of the problem of violence at work? (8)

For guidance on how to answer NEBOSH questions please refer to the 'study question answer guidance' section located at the back of this guide.

Element 6

Musculoskeletal health

Content

MEANING OF MUSCULOSKELETAL DISORDERS AND WORK-RELATED UPPER LIMB DISORDERS

Work-related musculoskeletal disorders (MSDs) are disorders of various parts of the body. These are caused by work and working conditions. Disorders can affect the muscles, joints, tendons, ligaments, nerves, bones and the localised blood circulation system. Most work-related MSDs are cumulative, resulting from repeated exposure to high or low intensity loads over a long period of time. The main areas of the body affected are the back, neck, shoulders and upper limbs, but the lower limbs may also be affected. Some MSDs, such as carpal tunnel syndrome in the wrist, are specific because of their well-defined signs and symptoms. Others are classed as non-specific because only pain or discomfort exists without clear evidence of a specific disorder or occupational disease.

Work-related upper limb disorder (WRULD) is a generic term for that group of disorders that affect the neck or any part of the arm from the fingers to the shoulders. Recognised WRULDs conditions include carpal tunnel syndrome and tenosynovitis.

The European Agency for Safety and Health in the Workplace recognises musculoskeletal disorders as a significant problem:

> "There is substantial evidence to suggest that neck and upper limb musculoskeletal disorders are a significant problem within the European Union. Some member states have identified a major ill-health and financial burden associated with these problems."

Figure 6-1: Prevalence of musculoskeletal disorders.
Source: European Agency for Safety and Health in the Workplace.

POSSIBLE ILL-HEALTH CONDITIONS FROM POORLY DESIGNED TASKS AND WORKSTATIONS

A worker's body will be affected to a varying degree by poorly designed tasks and workstations that involve bending, reaching, twisting, repetitive movements and poor posture. Frequent repetition of work movements and high force demands on the hand are significant risk factors, especially when they occur together. Bent postures of the wrist at work and low environmental temperature have also been identified as risk factors. These poorly designed tasks and workstations can lead to workers developing ill-health conditions in the form of work-related upper limb disorders.

WRULDs were first defined in medical literature as long ago as the 19th century as conditions caused by *forceful, frequent, twisting and repetitive movements.*

Figure 6-2: WRULDs caused by conveyor work (repetitive twisting). Source: Bremerhaven.

WRULDs are musculoskeletal disorders caused by exposure in the workplace affecting the soft tissues, muscles, tendons, ligaments, nerves, blood vessels, joints and bursae (a small fluid-filled sac which provides a cushion from injury between bones and tendons and/or muscles around a joint) of the fingers, hand, wrist, arm, shoulder and neck.

Common WRULDs include carpal tunnel syndrome, tenosynovitis, tendinitis, peritendinitis, epicondylitis and Trigger finger (stenosing tenosynovitis). The conditions are usually caused by repetitive movements and aggravated by excessive workloads, inadequate rest periods and sustained or constrained postures.

This can result in pain, soreness or inflammatory conditions of soft tissues, such as muscles and the synovial lining of the tendon sheath. Clinical signs and symptoms are stiffness or pain in joints and inability to straighten or bend the joints or crepitus (a grating sensation in the joint). In addition there can be aches, pain, tenderness, stiffness, weakness, tingling, numbness, cramp, swelling in the muscles of the arms and neck.

Carpal tunnel syndrome

Carpal tunnel syndrome is a condition where the median nerve that passes through the carpal tunnel in the wrist become trapped when tendons or ligaments in the wrist become enlarged, often from inflammation. The narrowed tunnel of bones, ligaments and tendons in the wrist pinch the nerves that reach the fingers and the muscles at the base of the thumb.

The first symptoms usually appear at night due to the wrists being flexed during sleep. Symptoms range from a burning, tingling numbness in the fingers, especially the thumb and the index and middle fingers, to difficulty gripping or making a fist, to dropping things. It is a common condition often caused by using display screen equipment.

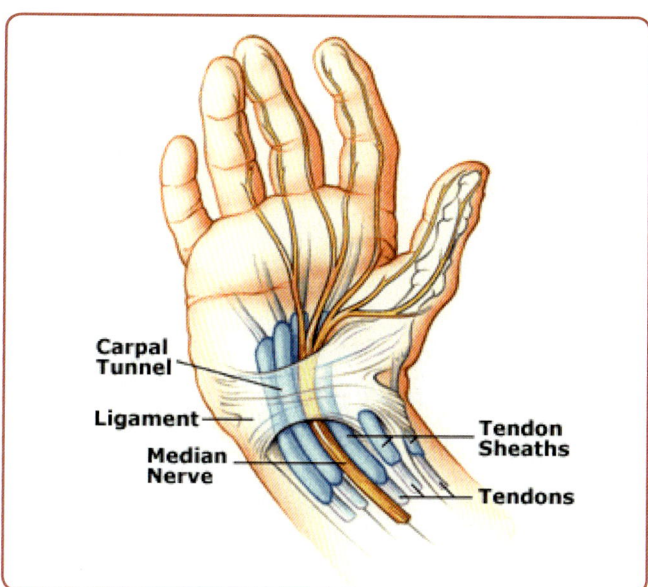

Figure 6-3: Carpal tunnel syndrome.
Source: www.sodahead.com.

Tenosynovitis

Tenosynovitis is inflammation of the tendon sheath at the wrist. The tendon sheath makes a tiny amount of oily fluid that lies between the tendon and the tendon sheath, which provides lubrication and allowing the tendon to move easily through the sheath. Tenosynovitis occurs when prolonged and repetitive use of the tendon be-comes excessive and the tendon sheath can no longer provide sufficient lubricant to the tendon. As a result, the tendon sheath thickens and becomes inflamed. Symptoms include pain and swelling, usually near where the tendon attaches to the bone. It is a common condition often caused by repetitive finger movements, for example, when typing while using display screen equipment.

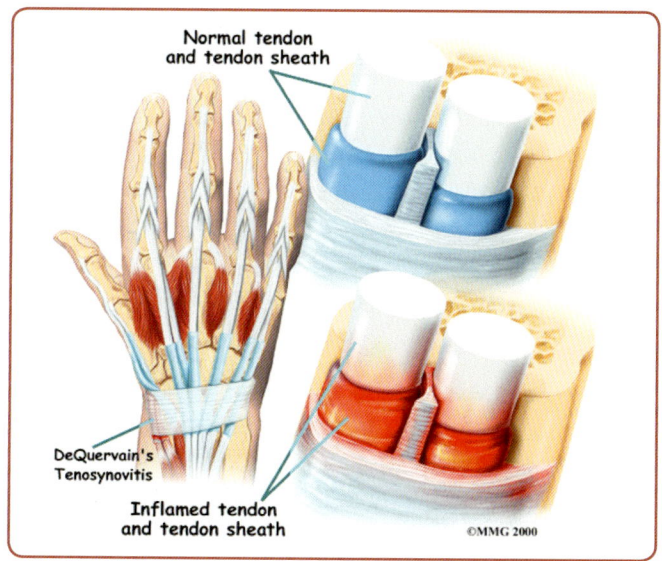

Figure 6-4: Tenosynovitis.
Source: www.rayur.com.

Tendinitis

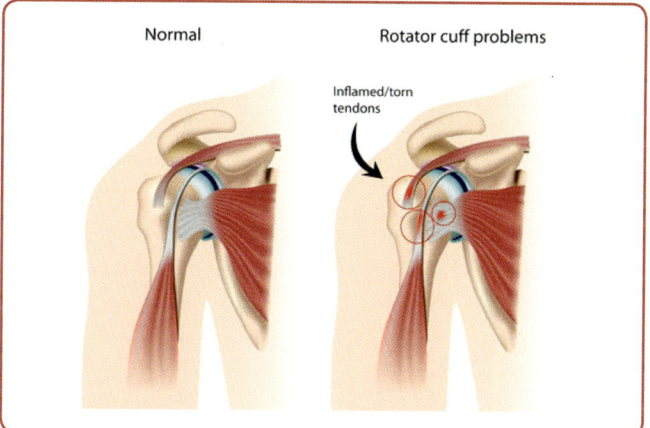

Figure 6-5: Tendinitis.
Source: Beth Israel Deaconess Medical Center.

Tendinitis involves inflammation of a tendon, the fibrous cord that attaches muscle to bone. It usually affects only one part of the body at a time and lasts a short time. However, in some cases it involves tissues that are continuously irritated. It may result from an injury, activity or exercise that repeats the same movement. Rotator cuff tendinitis is one of the most common causes of shoulder pain.

Peritendinitis

Peritendinitis is an inflammation of the sheath that surrounds a tendon. It can be associated with tendinitis, an inflammation of the tendon itself. This condition is most commonly seen in overuse injuries, such as those involving continually repeated tasks (for example, data input into a computer) or for those people who do not get enough rest between tasks or following injury. The primary treatment is resting in order to allow the inflamed tissue to heal.

Epicondylitis

Also referred to as 'tennis elbow' or lateral epicondylitis, this is a condition that occurs when the outer part of the elbow becomes painful and tender, usually because of a specific strain, overuse, or a direct impact (for example, when playing sports, such as tennis). Sometimes no specific cause is found, but it can be caused in the workplace by excessive lifting or twisting at the wrist, for example by using a screwdriver. Tennis elbow is similar to golfer's elbow (medial epicondylitis), which affects the other side of the elbow.

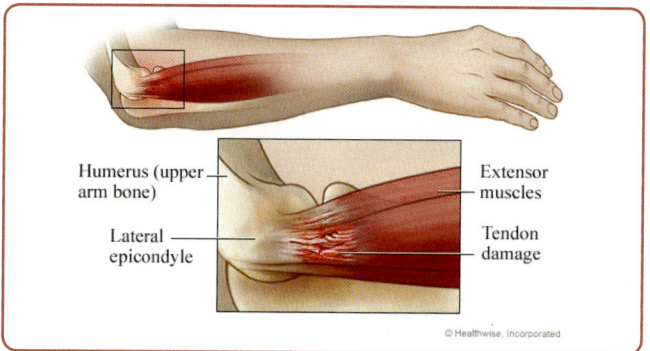

Figure 6-6: Epicondylitis.
Source: Myhealth.alberta.ca.

Trigger finger (stenosing tenosynovitis)

A painful condition that affects the tendons in the hand, which may involve one or more fingers, usually in the dominant hand. It may occur through repetitive movements at work. Trigger finger occurs if there is a problem with the tendon or sheath, such as swelling, which means the tendon can no longer slide easily through the sheath and it can become bunched up to form the nodule. This makes it harder to bend the affected finger or thumb. If the tendon gets caught in the opening of the sheath, the finger can click painfully as it is straightened.

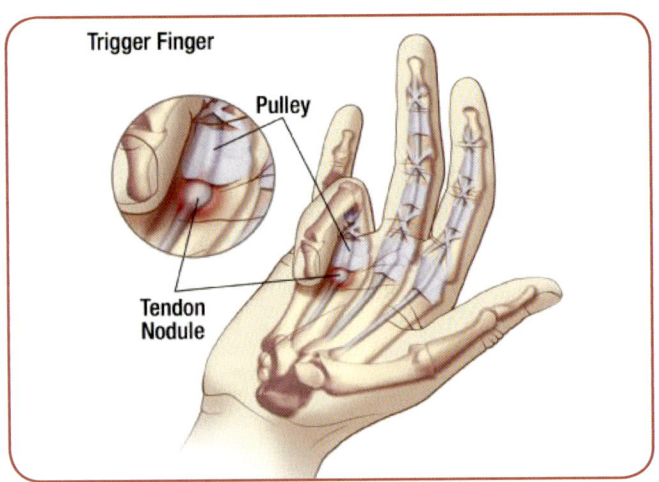

Figure 6-7: Trigger finger. Source: 3pointproducts.

The occurrence of WRLUDs, varies widely according to the type of work done by workers. High incidences of WRLUD, such as tenosynovitis or peritendinitis, have been reported among manufacturing workers, such as food-processing workers, butchers, packers and assemblers. Some recent studies show that high incidence rates exist even in modern industries.

Study population	Rate per 100 person-years
700 Muscovite tea packers	40.5
12,000 car factory workers	0.3
7,600 workers of diverse trades	0.4
102 male meat cutters	12.5
107 female sausage makers	16.8
118 female packers	25.3
141 men in non-strenuous jobs	0.9
197 women in non-strenuous jobs	0.7

Figure 6-8: Incidence of tenosynovitis/peritendinitis in various populations. Source: International Labour Office Encyclopaedia.

Workers who are unaccustomed to hand-intensive work, either as a new worker or after an absence from work, have been found to be at an increased risk.

The UK National Health Service reports in their guidance leaflet for employers on upper limb disorders that:

* 1 in 20 adults gets carpal tunnel syndrome.
* 1 in 100 men and 1 in 50 women get tenosynovitis.
* 1 in 75 men and 1 in 90 women get lateral epicondylitis (tennis elbow).

An estimated minimum of 6.6 million working days were lost in Britain due to WRULDs in 2018, with each affected worker taking, on average, 14 days off work. Costs to employers of WRULDs were estimated to be at least £15.0 billion.

The European Agency for Safety and Health at Work reported that statistics from Holland, accumulated over a fifteen-year period, suggest that each year approximately a third of construction workers have suffered from WRULDs and the trend over the last few years in the period was that this was increasing. The aches, pains and fatigue suffered doing tasks that cause these conditions will eventually impair the worker's ability to do tasks and lead to poor performance. Although present approaches to treatment are largely effective, provided the condition is treated in its early stages, it is essential to consider control measures to avoid or minimise the risk of WRLUDs occurring.

CONTROL MEASURES TO AVOID OR MINIMISE RISKS

Avoiding or minimising risks from poorly designed tasks and workstations should consider:

* Task - including repetitive or strenuous tasks, the duration of tasks and those that create poor posture.
* Environment - including lighting level or glare and temperature.
* Equipment - including user requirements, adjustability, matching the workplace to individual needs of workers.

Tasks

Anyone who repeatedly uses their hands and arms to perform tasks may be at risk of developing a WRULD. Keyboard operators, workers on factory assembly lines, agricultural workers, musicians, dressmakers, packing workers, bricklayers, supermarket checkout operators and cleaners are examples of workers at particular risk.

Repetitive

Tasks should be assessed to determine the WRULD risk factors. For example, a task is repetitive in nature if the same series of operations are repeated in a short period of time, such as ten or more times per minute. Repetitive tasks require the same movements to be made using the same muscle groups. If this repetitive movement is carried out over a prolonged period injury may occur to the muscles and ligaments affected.

Many tasks have repetitive movements, for example, stitching pieces of cloth, manufacturing individual parts, packaging individual items, checkout tasks in a supermarket, sorting parcels and trimming vegetables. This means that many workers may be exposed to the risks from repetitive movements.

Examples of repetitive tasks found in construction activities include:

- Hammering
- Drilling
- Driving screws
- Sawing
- Laying bricks
- Painting with brushes
- Plastering
- Digging
- Loading and unloading small items – for example, tiles or bricks.

Control measures to avoid or minimise risks of WRULDs from *repetitive* tasks include:

- Breaking up work periods involving a lot of repetition with several short breaks instead of one break in the middle of the work period.
- Allow for short, frequent pauses for very intensive work.

Strenuous

Similarly, tasks that are of a *strenuous* nature and require a worker to use significant *force* to carry them out should be considered. For example, building blocks (typically weighing around 10kg) are often used in the construction of walls. The blocks are picked up from a storage stack and moved to their position on the wall, before placing them in position in a controlled manner. In order to do this the worker uses a power grip to hold the block between their thumb, palm and fingers. Tasks such as this require high muscle forces to be applied and can lead to fatigue or injury if there is insufficient time for recovery. The risk is increased if the wrist cannot be held straight when the force is applied, for example resulting from inappropriate working height for the task.

Reduce the amount of *force* required to do the task by such control measures as:

- Reduce the weight of items, or the distance they are moved or slide the item instead of lifting it.
- Provide some means of increasing mechanical advantage, such as levers, or other means of mechanical assistance.
- Distribute force requirements over several fingers rather than one.

- Allow workers to use alternate hands to carry out tasks.
- Enable the stronger muscle groups to be used where possible, for example can a foot operated pedal be used to provide force.
- Ensure any handles and/or controls that are used are well maintained and easy to manipulate without requiring the application of unnecessary force.
- Provide lightweight tools where possible or provide a support, jig or counterbalance to help reduce the force need to use the tool.
- Ensure all tools are well maintained so they operate efficiently and require as little force as is possible to perform the task.
- Ensure the right tools are used for the job, which can reduce the amount of force required to perform tasks

Duration

The risk of experiencing a WRULD increases with the *duration* that a task is carried out. Duration refers to the length of time in each shift and the number of working days the task is performed. Duration is an important factor that affects the risk of developing a WRULD. Whilst injury may occur conducting repetitive and strenuous tasks over a short duration if the task requires a lot of effort, work requiring less effort for a long duration is very likely to lead to WRULDs.

Tasks conducted over a long duration could include, holding a polishing tool for a long time while cleaning, drivers keeping their head in a set posture while driving a vehicle for long distances, grasping small tools to carry out medical or dental tasks or holding construction material above a worker's head while another worker fixes it in place.

It is generally accepted that many types of WRULD are cumulative in nature, in that they develop progressively as tasks are carried out over time in situations where the parts of the body doing the work have insufficient time for recovery. Therefore, when the duration of a task is increased the risk of developing a WRULD is increased.

Tasks conducted for a short duration are unlikely to create a significant risk of developing WRULDs, except where the task is exceptionally demanding and/or workers have not been allowed to get accustomed to the demands of the task. This can occur when a worker returns to work after a holiday or if there is a significant increase in the speed of doing the task.

Reduce the effects of the *duration* of the task to help to avoid or minimise WRULDs by such things as:

- Develop a work/rest regime that provides sufficient time for recovery.
- Share a high-risk task among a team by rotating workers between tasks (each task needs to be suffi-

ciently different to benefit the worker).

- Allow workers to carry out more than one different step of a process, this could reduce the duration of exposure to a WRULD provided the steps are sufficiently different from each other.
- Introduce short frequent breaks for the task, but not necessarily a rest, for example a chance to change posture for a short period.
- Allow workers a short period to get accustomed to the duration of work when starting work for the first time and after return from a holiday or illness.
- Monitor and manage overtime working.

Posture

Tasks that require workers to adopt a particularly *awkward posture* or to hold a *fixed posture* for a period of time present a high risk of causing WRULDs.

An *awkward posture* is where a part of the body, for example, an arm joint, is used significantly beyond its neutral position. For example, when a worker's arm is hanging straight down with the elbow by the side of the body, the shoulder is in a neutral position. However, when a worker is performing overhead work, such as repairing equipment or accessing objects from a high shelf, their shoulders are moved significantly from the neutral position. When awkward postures are adopted, because muscles are less efficient at the extremes of the joint range, additional muscle effort is needed to do the task. The awkward postures can also lead to compression of soft tissue structures and muscle damage.

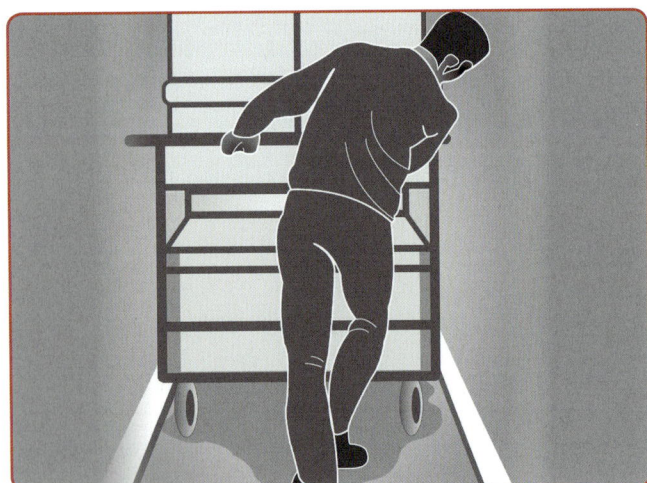

Figure 6-9: Awkward posture.
Source: Australia, Comcare.

Awkward postures include:

- Posture that is unbalanced or asymmetrical.
- Postures that require extreme joint angles or bending and twisting.
- Squatting while servicing plant or a vehicle.
- Working with arms overhead.
- Bending over a desk, table or workbench.
- Using a hand tool that causes the wrist to be bent

to the side.

- Kneeling while working concrete to a smooth finish or laying carpet.
- Bending the neck or back to the side to see around bulky items pushed on a trolley.

Fixed postures occur when a part of a worker's body is held in a fixed position for extended periods without the muscles and other soft tissues being allowed to relax. Fixed postures require the continuous contraction of the same muscles to maintain the posture of a part of the body or to hold the amount of force being exerted constant. For example, when holding and carrying a container or box it is likely that the hands and arms are in a fixed posture. A fixed posture restricts blood flow to the muscles and tendons resulting in less opportunity for recovery and metabolic waste removal. Therefore, muscles held in fixed postures fatigue very quickly. When tasks involve fixed postures and high muscle effort, fatigue will usually force the worker to take a rest. With lower muscle effort the level of fatigue is not so evident, which can lead the worker to spend too long in the same posture without taking a rest, resulting in the development of WRULDs.

Figure 6-10: Fixed posture.
Source: Australia, Comcare.

Working with the arms extended above the shoulder or extended above the head is common in maintenance and construction work, for example when painting a ceiling or installing light fittings. When working with the arms stretched upwards the small shoulder muscles have to do more in order to hold the weight of the arms. The force on the shoulder muscles is extremely high if the worker also holds a tool or materials in their hand at a distance from the shoulder. In some situations, workers also have to bend their neck backwards to see the work being done, which also strains the neck. These situations significantly increase the risk of shoulder and neck disorders.

Reduce the **awkward** and **fixed postures** of tasks to help to avoid or minimise WRULDs by such things as:

- Introduce short frequent pauses in the task to provide a chance to change posture for a short period.
- Reduce the size/weight of items that need to be held in a fixed position.
- Provide mechanical devices to hold items raised at height ready for fixing.
- Provide power tools to enable the fixing task to be done quickly.
- Revise the task to enable variation of work to avoid long periods in awkward or fixed postures.
- Reduce the distance items are held away from the body by better layout of workstations.
- Provide equipment, for example a platform, to enable workers to work at the same height as equipment they are maintaining.
- Enable equipment to be easily removed from where it is located to enable maintenance to be conducted without awkward or fixed postures.

Figure 6-11: Poor posture.
Source: Speedy Hire Plc.

Environment

Working in extremes of **temperature** will make doing simple tasks more fatiguing. In cold environments the circulation of blood to muscles is impaired and maximum strength capacity is diminished, which makes it harder to carry out tasks that are normally relatively easy to do. Cold hands can also result in reduced sensation and more muscle force being needed to hold tools and materials, especially if the worker has to wear gloves. This can lead to the more rapid onset of fatigue and to the development of disorders.

When conducting tasks in high temperatures it can cause workers to sweat, which may cause their hands become slippery with sweat and make it difficult for them to handle things. This can lead to poor grip that requires more muscle force or could cause a sudden movement when their hands slip and lose control of what they are holding, resulting in damage to muscles or other soft tissue.

The **lighting** requirement of tasks are important to consider as the posture a worker adopts to do a task can be influence by their need to see the work. Their visual requirements and posture may be affected by inadequate level of light, shadow, glare, reflections or flickering light. They can cause workers to adopt a bent neck and poor shoulder postures to enable them to see their work, leading to an increased risk of developing WRULDs.

The work environment may include workstations that have sharp edges or hard surfaces. If the work requires the worker to be in contact with the workstation it can put local pressure on the area of the workers body in contact with the edges and surfaces. If the contact causes significant pressure and/or lasts too long, it can restrict circulation and causes compression of underlying nerves and soft tissue.

Reduce the **environmental** issues affecting tasks to help to avoid or minimise WRULDs by such things as:

- Make sure that the temperature is comfortable,
- Avoid positioning workstations too near to air vents.
- Ensure task illumination is at a level that allows the worker to comfortably view the work without altering their posture. Avoid extreme levels of lighting – low levels (typically below 100-200 lux) or high levels (typically greater than 800 lux).
- Make sure general lighting is good or provide a specific light source for the task.
- Avoid reflected light, for example sunlight on computer monitors.
- Avoid glare by moving light sources, providing blinds on windows or moving workstations.
- Avoid shadows or glare.
- Organise the layout and positioning of items on workstations to minimise the worker having to reach, twist, bend or stoop, for example when collecting components and assembling them.
- Avoid temporary arrangements where materials and equipment are placed on the floor requiring a worker to kneel and bend to conduct tasks, for example where a welder might weld material during part of a maintenance task.
- Avoid tasks that require a worker to be contact with sharp edges and hard surfaces.
- Improve the layout of the workplace to avoid tasks requiring high muscle force in awkward or fixed postures.

Equipment

Equipment may be unduly heavy or not designed to take account of the different size, shape or strength of workers. Objects and attachments may act as obstacles causing the worker to reach over or round them. Where a workstation has confined space for the worker's legs this can cause poor posture. Seats provided may not be adjustable or where they are adjustable there may not be enough space to make effective use of the adjustable features. Equipment and materials may be positioned so they are not in easy reach of the worker.

Matching the workplace to the individual needs of workers

In managing risks from WRULDs a number of changes may need to be made and ergonomic solutions should be given first consideration. This means matching the workplace, work equipment and the work to the individual needs of the worker, rather than making the worker adapt to fit the workplace, work equipment and work undertaken. Equipment design should therefore take into account the *ergonomic requirements of the worker* and, where possible, allow the worker to *adjust* any settings to suit their needs, for example, setting a workbench height, positioning storage trays for materials used to assemble parts and adjusting the height of their computer screen.

Ergonomics (derived from the Greek *ergon*, meaning work, and *nomos*, meaning natural law) is the scientific study of human work. It is a broad area of study that includes the disciplines of psychology, physiology, anatomy and design engineering. Ergonomics places the human being at the centre of the study, where individual capabilities and the human potential for making mistakes are considered. The fundamental principle of ergonomics is that good job design, by reducing worker error, fatigue and discomfort, is likely to lead to maximum effectiveness.

When matching the workplace to the individual needs of workers it is necessary to take into account that individuals have different physical capabilities due to gender, height, weight, age and level of fitness. This will usually meant making adjustments to limit the risk of developing WRULDs.

New workers, particularly young workers, and those returning to work from a holiday, sickness or injury, may need to be introduced to a slower rate of work than other workers. This enables them to get their body used doing tasks that use their upper limbs, assimilate training more effectively and develop good work practices before having to concentrate on working fast.

Reduce the *equipment* issues affecting tasks to help to avoid or minimise WRULDs by such things as:

- Design workplaces and equipment (including tools) for workers of different sizes, shape, strength and for left-handed workers.
- Tools should have comfortable handles and make the work easier – avoid rigid hard surfaced handles, sharp edges or narrow handles that put pressure on the hand when they are gripped.
- Tool handles should enable a straight wrist posture (like when a handshake is given) and avoid awkward hand and wrist postures.
- Provide tools with a suitable size of grip, consider the effect of wearing gloves.
- Tools like pliers should not require a wide hand span to use them.
- Use powered tools to reduce the force a worker has to apply to do a task.
- Arrange the position, height and layout of the workstation so that it is appropriate for the workers to do the work- enable work to be done with the joints at about the mid-points of their range of motion.
- Use work fixtures and jigs to hold the work in more accessible positions and to make it easier to see the work to be done.
- Improve the location of objects or attachments to prevent them acting as obstacles and causing poor posture.

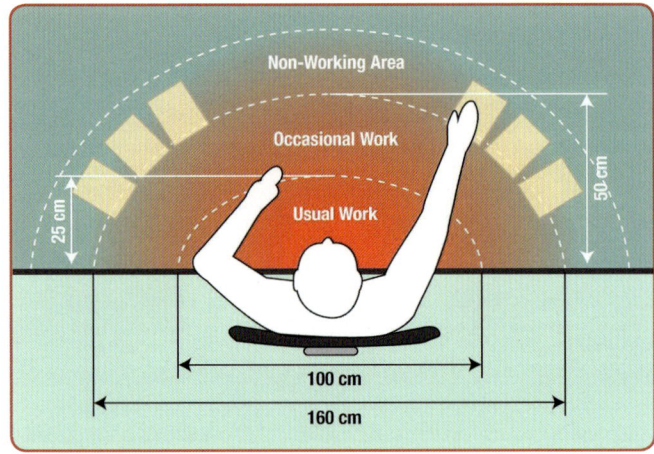

Figure 6-12: Ergonomic workstation assessment.
Source: RMS.

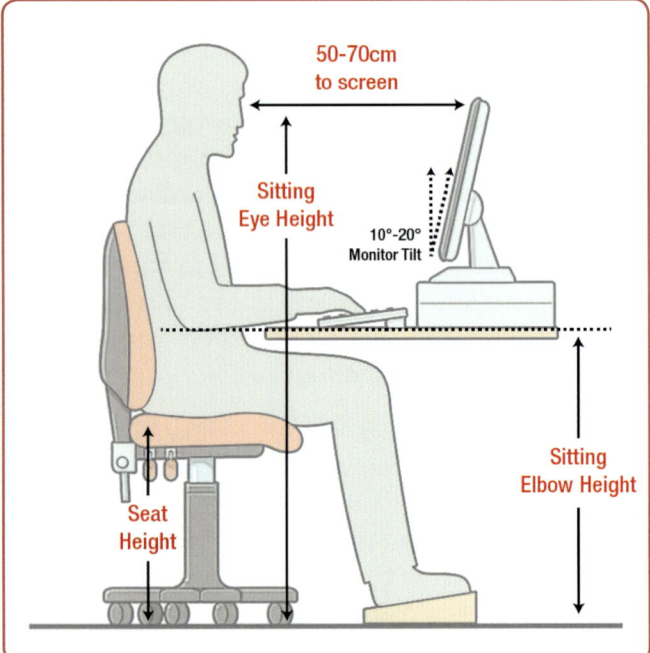

Figure 6-13: DSE user/workstation assessment.
Source: RMS.

- Place equipment and materials that are needed more frequently nearest to the worker.
- Ensure seats are adjustable.
- Ensure that there is sufficient space to enable workers to stretch, make changes in leg and foot posture and make effective use of the adjustable features of their chairs.
- Provide platforms, adjustable chairs and footrests for smaller worker to achieve optimal working height to do the work.
- Standing workstations should be used for jobs that require a lot of body movement and greater force.

 CASE STUDY

A call centre company identified the need to display larger characters on a call centre operator's screen. The company acquired and installed suitable software to enable this. Following the change, the operator achieved the same efficiency levels as the rest of the other centre operators and commented that *"it enabled me to feel I was making a real contribution to the overall team effort."*

 REVIEW

What typical work tasks may result in poor posture being adopted by the worker?

What is meant by the term ergonomics?

 REVIEW

What work equipment should be considered when carrying out a display screen equipment workstation assessment?

Give an example of a typical musculoskeletal injury that might affect a display screen user.

What control measure could be used to prevent or reduce the risk of carpal tunnel syndrome when carrying out assembly of a component at a workstation?

Control measures for a display screen equipment workstation

Display screen equipment (DSE) is a device or equipment that has a display screen for graphics, words or numbers. DSE includes both conventional display screens and those used in the latest technologies such as laptops, touch-screens and other similar devices. The main risks that may arise in work with DSE are musculo-skeletal disorders such as back pain or upper limb disorders, visual fatigue, and mental stress. While the risks to individual users are often low, they can still be significant if good practice is not followed. Control measures should include the following good practices.

DISPLAY SCREENS

- Display screen is in good working condition, including swivel and tilt functions and screen quality.
- Display screen is set at the correct height for the user.
- Screen size is suitable for all tasks and characters are easy to read from the working distance.
- The screen is free from distracting reflections and glare.
- The user can adjust the screen's brightness and contrast.
- Multi-screen arrangements are configured for efficiency and good working postures.

KEYBOARD

- Keyboard is separate from the screen.
- Type of keyboard is appropriate for the user.
- Keyboard is in good working condition, including tilt and the keys are easy to read.
- Keyboard is comfortable to use.
- A wrist rest is provided if needed.

MOUSE

- Mouse is in good working condition.
- Mouse is comfortable to use, including pointer speed of movement.
- Type of mouse provided is appropriate to the user.

SOFTWARE

- Software works reliably.
- Software is suitable for the tasks.
- The user has received sufficient training to use the software.

DOCUMENT WORK

- A suitable document holder is provided, if needed.

DESK

- Work surface free is from clutter and well organised.
- Work surface is large enough for all items to be correctly positioned.
- Sufficient off-desk storage is provided for files and folders.
- The area under the desk is free from obstructions.
- The desk height is adjustable.

CHAIR

- Seat height is adjustable.
- Backrest height is adjustable.
- Backrest is tilt adjustable.
- Chair has a working swivel mechanism.
- Stable 5-star base with suitable castors.
- Upholstered with suitable materials for the environment.

WORKING POSTURE

- User can sit comfortably in their chair and is supported in an upright position with good lumbar support.
- User can sit at a safe working height and distance relative to the desk, with their feet fully supported.
- The DSE is organized to encourage good working postures.
- Supplementary items, such as the telephone, can be accessed easily.

ENVIRONMENT

- Lighting is suitable and sufficient, without visual concerns, including adjustable window coverings where necessary.
- Local, task lighting has been provided, where necessary.
- User is not affected by cold draughts and the air temperature is acceptable.
- No distracting noise caused by equipment.
- User can safely enter and exit the workstation.

6.2 Manual handling hazards and control measures

COMMON TYPES OF MANUAL HANDLING HAZARD AND INJURY

In their booklet 'Protecting Workers' Health Series No 5 'Preventing Musculoskeletal Disorders in the Workplace' the World Health Organization reported that back injuries generally constitute 60% of musculoskeletal disorders.

In the UK, around 25% of all injuries reported to the competent authority, the Health and Safety Executive, have been attributed to manual handling operations,

Manual handling injuries arise from **hazardous events** such as stooping while lifting, holding the load away from the body, and undertaking twisting movements of the trunk of the body or frequent or prolonged effort. Injuries can also arise from manually handling heavy/bulky/unwieldy/unstable loads, or loads which have sharp/hot/slippery surfaces. Other injuries can be caused by

workplace space constraints and the lack of capability of the individual. Manual handling operations can cause many types of **injury**.

The most common injuries are:

- Prolapsed (herniated) spinal disc.
- Muscle strain and sprain.
- Torn or overstretched tendons and ligaments.
- Rupture of a section of the abdominal wall can cause a hernia.
- Loads with sharp edges can cause cuts.
- Dropped loads can result in bruises, fractures and crushing injuries.

Figure 6-14: Manual handling.
Source: Shutterstock.

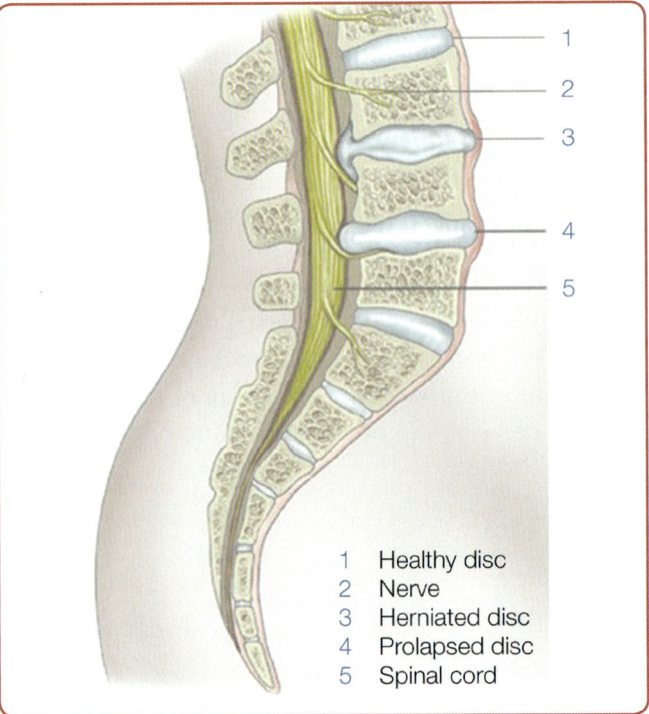

Figure 6-15: Lumber spine illustration.
Source: UK, NHS.

1 Healthy disc
2 Nerve
3 Herniated disc
4 Prolapsed disc
5 Spinal cord

Element 6

Damage to intervertebral disc

The spine is made up of individual bones (vertebrae) separated by intervertebral discs. Strain on the disc can cause it to bulge and protrude out between the vertebrae. This is known as a prolapsed or herniated disc. The disc can then press on nearby nerves or the spinal cord. If the fibrous case of the disc collapses the disc can lose its shape and compress – this can cause bones in the spinal column to touch or it can increase pressure on the spinal cord.

The above conditions can cause pain and inflammation in the area of the spinal cord where pressure occurs. Pain may also be felt in the area of the body that the part of the spinal cord affected relates to, such as in the lumbar region of the spine which affects the sciatic nerve (this condition is called sciatica). The damage to the disc is usually caused by too much pressure being applied to it during lifting and handling operations that involve 'top-heavy' bending. This is where the knees are not bent sufficiently and the head and upper body are bent over, causing the spine to bend in a curved manner (sometimes called stooping). This bending creates a high degree of leverage force on the base of the spine, which leads to the extreme pressure exerted on the disc. The force and therefore pressure on the disc is accentuated by the carrying of heavy loads.

Though top-heavy bending is a particular cause of this damage, poor posture due to leaning over for a sustained period could lead to similar damage.

Sprains/strains, fractures and lacerations

The most common injuries that arise from manual handling operations are sprains, strains, fractures and lacerations.

Sprains and strains

Sprains and strains often occur in the back or in the arm and wrists, though injuries in the legs can also occur where the leg has been hyper flexed and where the soft tissues (ligament, tendons and muscles) are overstretched. This can occur through kneeling or handling loads beyond the worker's limit.

- A *sprain* is an injury to a ligament. The ligament is tough fibrous tissue that connects a bone to another bone. Ligament injuries typically involve overstretching or a tearing of this fibrous tissue, i.e. a 'torn ligament'.
- A *strain* is an injury to either a muscle or a tendon that connects muscles to bones. Depending on the severity of the injury, a strain may be a simple overstretch of the muscle or tendon or it may result in a partial or complete tear. Rupture of the muscles in a section of the abdominal wall can cause a hernia.

Sprains and strains tend to occur when a person's body has been overloaded due to a large steady load being applied, a sudden smaller load being applied without the opportunity of the ligament or tendon to stretch, or a part of the body being forced to move in an unusual way. This may be because a person is reaching over to an extreme level (sometimes even without picking up a load, as their own body weight may be sufficient to cause damage) or two people picking up a load together in an uncoordinated way. It could also be due to a load slipping that causes someone to move into an awkward position to prevent it falling completely.

Fractures and lacerations

A fracture is a break in a bone, which is usually the result of trauma where the physical force exerted on the bone is stronger than the bone itself. Fractures of the hands and feet are the most likely type of fractures to arise from manual handling work and may be due to the load that is being lifted accidentally being dropped. Lacerations are often caused by the unprotected handling of loads with sharp corners or edges, or a person's grip slipping when trying to prevent dropping a load.

REVIEW

What factors that should be considered before attempting to lift a load manually?

CONSIDER

Select a manual handling activity that you or other people carry out at work.
What are the risk factors associated with the task, the load, the individual and the environment?

GOOD HANDLING TECHNIQUE FOR MANUALLY LIFTING LOADS

Lifting techniques using kinetic handling principles

Kinetic lifting is a method of lifting that takes into account the body's characteristics and how it works most effectively. It ensures that lifting is carried out without placing unnecessary strain and pressure on vulnerable body parts.

The principles of good lifting and moving techniques have been established for some considerable time.

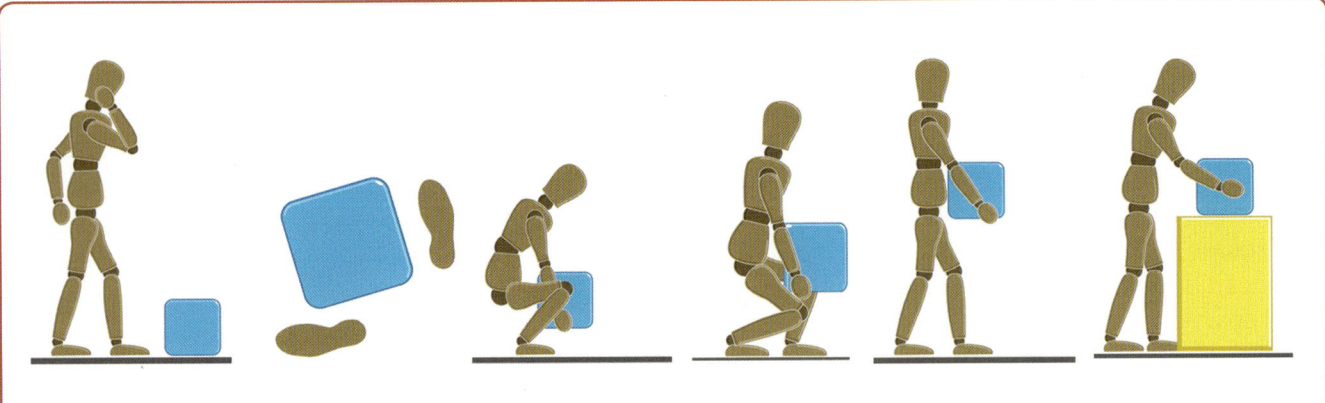

Figure 6-16: Basic lifting principles.
Source: UK HSE, Guidance L23.

The basic principles are:

- **Think before lifting/handling.** Plan the lift. Can handling aids be used? Where is the load going to be placed? Will help be needed with the load? Remove obstructions such as discarded wrapping materials. When a load is to be lifted or lowered a significant distance, consider resting the load midway on a table or bench. This will provide an opportunity for the worker's body to recover and to renew their grip of the load

- **Adopt a stable position.** The feet should be apart with one leg slightly forward to maintain balance (alongside the load, if it is on the ground). The worker should be prepared to move their feet during the lift to maintain their stability. Avoid tight clothing or unsuitable footwear, which may make this difficult.

- **Get a good hold of the load.** Where possible, the load should be hugged as close as possible to the body. This may be better than gripping it tightly with hands only.

- **Start in a good posture.** At the start of the lift, slight bending of the back, hips and knees is preferable to fully flexing the back (stooping) or fully flexing the hips and knees (squatting).

- **Do not flex the back any further while lifting.** This can happen if the legs begin to straighten before starting to raise the load.

- **Keep the load close to the waist.** Keep the load close to the body for as long as possible while lifting. Keep the heaviest side of the load next to the body. If a close approach to the load is not possible before starting to lift it, try to slide the load towards the body before attempting to lift it.

- **Avoid twisting the back or leaning sideways,** especially while the back is bent. Shoulders should be kept level and facing in the same direction as the hips. Turning by moving the feet is better than twisting and lifting at the same time.

- **Keep the head up when handling.** Look ahead once the load has been held securely, not down at the load.

- **Move smoothly.** The load should not be jerked or snatched as this can make it harder to keep control and may increase the risk of injury.

- **Do not lift or handle more than can be easily managed.** There is a difference between what people can lift and what they can safely lift. If in doubt, seek advice or get help.

- **Put the load down, then adjust its position.** If precise positioning of the load is necessary, put it down first, then slide it into the desired position.

AVOIDING OR MINIMISING MANUAL HANDLING RISKS

Avoiding manual handling risks

Wherever reasonably practicable, manual handling should be avoided, for example, by redesigning the task to avoid moving the load, by automating the process or mechanising it by using mechanical handling aids. Decisions on whether to use automation or mechanisation are best taken at the design stage, or when processes are being updated, as they can involve complex arrangements and capital expenditure, for example, where the pro-cess changes from handling raw materials in sacks or bags to bulk delivery, or where storage changes to using gravity or pneumatic transfer systems.

Avoiding moving the load

In some situations, manual handling can be avoided by not handling the load at all. For example, if a heavy load needs to be wrapped in stretch/shrink plastic film, it might be possible to bring the wrapping process to the load.

A similar approach is often adopted to avoid moving a patient in hospital, where equipment is brought to the patient rather than taking the patient to the equipment.

Automation

Automation may be particularly appropriate where a high-volume of items are moved and the unit costs of the items justify the large initial capital expense of automation. This is likely in food processing or where high-volume component dispatch is required. Examples of manual handling processes that may be suitable for automation include bottle or can filling, sorting items like letters and parcels, transferring materials from production areas into warehousing (**see Figure 6-17**) and order picking for dispatch (for example, electrical or automotive components). Automated systems will often combine a variety of techniques, but they often depend on the movement of goods and material from one point to another using conveyor systems and/or remotely operated material transfer trucks.

Figure 6-17: Conveyor track.
Source: TGW-Group.

Mechanisation

This involves the use of handling aids. Although mechanisation may retain some elements of manual handling, mechanisation means bodily forces are applied more efficiently. Examples are:

- Hoists – Can raise a load and support the weight of a load while allowing a worker to position the load.
- Trolley, sack truck, roller truck or hoist – Removes need to carry load and reduces effort to move loads horizontally.
- Chutes and roller conveyors – A way of using gravity to move loads from one place to another.

Figure 6-18: Brick/tile conveyor.
Source: Mace Industries Ltd.

Figure 6-18 shows a method of moving roofing tiles from ground level to roof height using a hoist system, which eliminates the need to carry the tile(s) up the ladder.

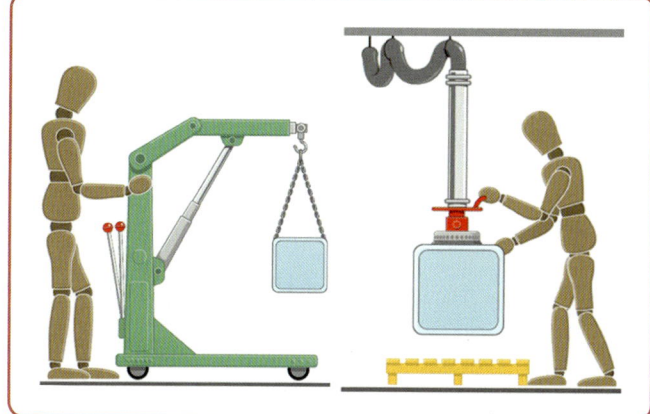

Figure 6-19: Mechanical aids.
Source: UK HSE, Guidance L23.

Minimising manual handling risks

Where risks cannot be avoided, control measures should be put in place to minimise the risk of manual handling injuries to the lowest level reasonably practicable. Practical measures that may be taken to reduce the risk of injury can be decided by considering the task, individual, load, and environment (TILE). This should be carried out with a view to following ergonomic principles and matching the manual handling operation to the individual needs of workers rather than the other way round.

Task

Many tasks have repetitive movements, for example, stacking boxes on a pallet, sorting parcels, moving patients, digging an excavation or laying bricks. This means that workers carrying out these tasks may be exposed to the risk of harm from **repetitive manual handling movements.** Repetitive movements take place when manual handling tasks are made up of a sequence of actions, of fairly short duration, which are repeated a significant number of times and are almost always the same. This requires the same muscle groups to be used repeatedly. Rapid or prolonged repetitive movements may not allow sufficient time for recovery and can cause muscle fatigue or inflammation of soft tissues.

Awkward manual handling movements are where a part of the body is moved well beyond its neutral position, for example, bending at the waist and twisting. Awkward bending movements often takes place where workers do not adopt good lifting techniques, particularly where features of the workplace encourage the worker to use a top heavy bending action, which involves bending at the waist to move a load, for example removing a load from a low storage place. This type of bending action puts severe strain on the lower back and should be avoided.

A twisting motion of the spine of 45 degrees will result in a 10% reduction in lifting capability, while a twisting motion of 90 degrees will result in a 20% reduction. Twisting motions can also put severe strain on the lower back and workers should not be expected to be able to lift loads of the same weight as they might be capable of if lifted directly in front of them.

Figure 6-20: Awkward movements.
Source: UK, HSE, HSG60.

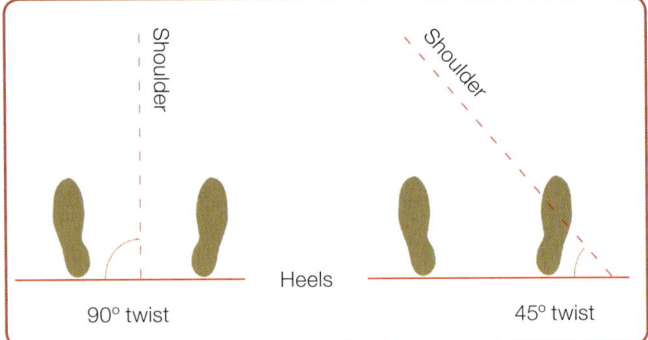

Figure 6-21: Twisting.
Source: UK HSE, Guidance L23.

When team handling is used for tasks that are beyond the capability of one person it should be remembered that the proportion of the load carried by each member of the team will vary. Therefore, the load that can be handled by a team will be less than the sum of the loads which each individual could cope with. This should be taken into account when planning to lift loads.

It is important that where it is reasonably practicable risks arising from manual handling tasks are minimised. Consideration should be given to control measures that:

- Avoid large vertical movements - provide somewhere for the load to be placed/rested at about the mid-point in the vertical movement, this will enable recovery of muscles and a change of grip before completing the vertical movement.

- Hold loads close to the body - the use of protective clothing can reduce the risk of manual handling injury as it helps workers to hold loads close to their body without affecting their own clothes.

- Avoid awkward postures like twisting, stooping or reaching upwards – improve the workplace layout and how it is organised, for example avoid storing items at a low level and ensure workers have room to move their feet to avoid twisting.

- Avoid lifting from floor level or above shoulder height, especially heavy loads - organise where loads are placed, put them on a shelf or place on a device that allows the load to be lowered, if necessary, and then raised before lifting it again.

- Reduce carrying distances – restrict carrying

distance to less than 10 metres before a rest is taken, use a manual handling aid when moving a load long distances, for example a trolley or wheeled cage.

- Avoid repetitive handling or minimise the effects - vary the work to enable one set of muscles to rest while others are used, reduce the speed required for the repetitive movements to reduce the effort required over a period of time and allow pauses in the work to enable recovery.

- Allow sufficient rest for demanding tasks – strenuous tasks that are carried out for too long or too frequently need rest time to enable muscles affected to recover.

- Allow the work rate to be controlled by the worker – to enable them to identify the need for and take sufficient rest.

- Ensure loads moved while seated remain within the weight capacity of the worker and regard is taken of good practice guidance - for example, limiting loads to 5kg for men and 3kg for women when moved close to their body.

- Push or pull loads instead of lifting and carrying them - generally pushing and pulling loads imposes less loading on the body.

- Avoid strenuous pushing or pulling - push rather than pull and make sure trollies and containers are not overloaded

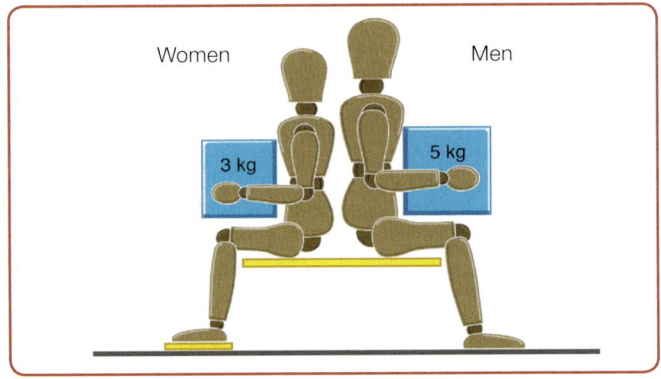

Figure 6-22: Handling while seated.
Source: UK HSE, Guidance L23.

Individual

An individual's physical capability to manually handle loads could significantly affect their ability to perform a manual handling task safely. Their capability can be influenced by a number of factors, including their gender, age, medical condition (general health, previous injuries, pregnancy). In addition, the training and information they are provided with will influence how they carry out tasks. Also, the type of clothing workers wear may inhibit good movement or, if the worker does not have suitable clothing to protect their own clothes, they may not want to hold the load close to their body because it might make their clothes dirty or damage them. Holding a load away from the body greatly influence the forces on the body, for example on the lower back.

Figure 6-23: Manual handling.
Source: Shutterstock.

To minimise the manual handling risk factors related to the individual consideration should be given to control measures that:

- Ensure workers' clothing and footwear is suitable for their work - provide protective clothing or PPE that does not restrict movement and allows the worker to hold loads close to their body.
- Consider the worker's capability before assigning manual handling tasks to them - a worker's state of health can significantly affect their capability to perform manual handling tasks safely, also, some age groups tend to have reduced physical capability (particularly those under 20 years of age and those above 45 years of age).
- Prevent workers with a health problem or physical disability carrying out manual handling work that is likely to endanger them and assign them to other work that suits their capability.
- Prevent pregnant workers carrying out manual handling work that is likely to endanger them or their unborn child and assign them to other work that suits their capability.
- Provide workers with relevant manual handling information - information on the load and the task, for example the weight of the load and specific precautions the worker must take before and during manual handling tasks are carried out.
- Provide manual handling training to workers and their managers – provide general training (including risks, injuries and control measures, including good manual handling technique) and specific training for more unusual tasks.

Load

The nature of a load can contribute to increased manual handling risks due to its size, weight, centre of gravity, stability, how easy it is to grip and any hazardous surfaces (sharp, hot, substance contamination) or contents it may have. Manual handling risks related to the load can be minimised by considering the design of the load, for example, designing it to be smaller by concentrating the substances contained in it or breaking the load up into suitable, smaller sized containers. It might also mean designing in handles or features that make it easier to grip the load, such as 'sticky grip' areas on plastic sacks.

To minimise manual handling risks related to the load, consideration should therefore be given to control measures that take account of the load's weight, size, grip and whether it is sharp or hot and its stability. Each of these risk factors and control measures to minimise the risks are outlined below.

Weight

Article 11 of ILO Convention C184 'Safety and Health in Agriculture' addresses 'Handling and transport of materials' and states:

 2) Workers shall not be required or permitted to engage in the manual handling or transport of a load which by reason of its weight or nature is likely to jeopardize their safety or health.

Figure 6-24: Requirement to prevent harm from manual handling.
Source: ILO Safety and Health in Agriculture Convention C184.

Generally, the heavier the load the higher the risk of injury. Therefore, risks can be minimised by ensuring workers only manually handle loads of a reasonable weight. Many countries have used simple maximum weight limits for the lifting of loads. However, this has been found to be an over-simplistic approach. Research has identified that it is not possible to set a simple, specific weight limit for all manual handling tasks because the individual capability of workers and the task being conducted greatly influence what is an acceptable weight. Many of the weight limits set by countries have tended to be too high for average workers and work practices. The ILO indicates that load weight limits to reduce risk should not be so high and should be more like those adopted by the USA in 1991, which is 23kg.

Work in the UK has confirmed this to be an effective weight limit to reduce risk and the maximum weight limit approach was adapted to reflect the fact that workers may lift from different positions and that there are differences in the lifting capacity of women and men. This work has been published as the UK Health and Safety Executive's (HSE) lifting guidelines. In the UK, therefore, a single maximum weight limit has been removed from legislation and replaced by the HSE lifting guidelines. The UK guidelines set out an *approximate* boundary within which manual handling operations are unlikely to create a significant risk of injury.

The UK guideline figures are not absolute weight limits. They may be exceeded where a more detailed assessment shows it is safe to do so. However, the guideline figures should not normally be exceeded by more than a factor of about two. The guideline figures for weight will give reasonable protection to nearly all men and between

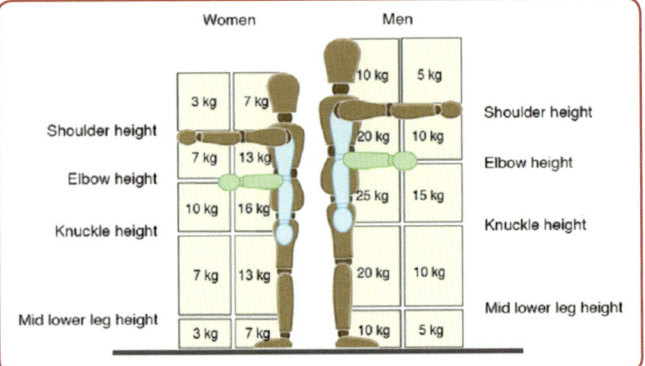

Figure 6-25: Lifting and lowering.
Source: UK HSE, Guidance L23.

It should be remembered that one of the main methods used to reduce the risk of manual handling is to change the load by repackaging it into smaller weights or breaking a bulk load down into smaller batches. This solution avoids the need to lift heavy loads. However, it will increase the frequency of manual handling movements and may lead to a significant repetitive movement risk.

Where the weight of a load is beyond the capability of one person the worker should get assistance from another worker or use a manual handling aid.

Size

Large (wide, tall or bulky) loads could make it difficult for a workers to see where they are going when moving a load. Also, in order to get a wide or bulky load close to the body, the worker has to open their arms to reach around the load and hold it. Therefore the size of the load may make it hard for a worker to hold the load when it is being moved. Also, the worker's arm muscles cannot produce as much holding force with the arms held open reaching round a load as when the arms are held closer together. In this situation, the muscles will get tired more rapidly. To enable the worker to keep the load as close to the body when lifting and carrying it the size of loads should be limited to enable the worker to hold it comfortably. This could include designing the load to be smaller through concentration of substances contained in it or breaking the load up into suitable-sized containers. Where this is not possible the worker should get assistance or use a manual handling aid.

Grip

Loads that are difficult to grip can result in the load slipping, causing sudden movement of the load. Ordinary gloves usually make gripping a load more difficult than gripping it with bare hands. Special manual handling gloves that have additional surfaces designed to provide grip will help to minimise this risk. Risks can also be minimised by providing the load with handles or using aids for gripping, for example, suction pads when carrying plate material or hand–held hooks for soft material. Where the load is contaminated with substances that make griping

the load more difficult it may be necessary to clean the surface to remove slippery substances on it before moving it. Where the packaging material makes it difficult to grip the load, it may be possible to change the material or add a slip resistant material to the surface, such as 'sticky grip' areas on plastic sacks.

Sharp, hot, cold or harmful

Any loads to be handled that have hazardous substances on their surface, sharp corners, jagged edges or rough surfaces increase the risk of workers' injuring their hands when they handle them and may cause the worker to drop the load suddenly. Loads containing hazardous substances or loads that are hot/cold can also injure workers, especially in the event of a collision. Personal protective equipment should be provided to protect the worker or the load should be covered in suitable protective material before moving it. In some cases, it may be appropriate to use handling aids to hold the load or use alternative means to move the load.

Stability

Unstable loads or loads with moving contents, such as a liquid, can cause uneven loading of the muscles and sudden movements of the load can make workers lose their balance and fall. Where possible, the moving parts of a load, for example components in machinery, should be prevented from moving before lifting. Where a load contains a liquid, consider whether the liquid can be emptied out before moving the load or if this is not possible consider the design of the container and whether baffles can be added to restrict movement of the liquid.

Some loads may be unbalanced because the centre of gravity of the load is not where expected and a worker may try to move the load by holding it with the centre of gravity too far away from their body or too much to one side. This can lead to uneven loading of muscles, damage to soft tissues and fatigue. It is important that where the centre of gravity of a load is in a position that is not obvious, sufficient information is given to the worker before they try to lift it so they can take this into account, for example by providing markings on the load. The worker can then reposition the side of the load nearest the centre of gravity close to their body before lifting the load or get assistance.

Environment

The design or layout of the work environment may cause a worker to take on an awkward posture while moving a load, the temperature of the workplace may lead to fatigue or difficulty in gripping the load and poor lighting level may cause a worker to trip while moving a load. In addition, movement of a load between levels, for example up or down stairs, can present risks. The risks from manual handling may be minimised by the use of good design and layout of the workplace and workstations. This involves placing items where they can be conveniently handled to prevent workers needing to stoop, improving work layouts so that travel distances are minimised and arranging that items can be picked up or put down at a suitable height. In general, working levels for workstations should be waist high, with tools, materials and equipment to be handled placed in front of the worker and within easy reach. The layout should suit the worker and there should be adequate room to perform the task safely.

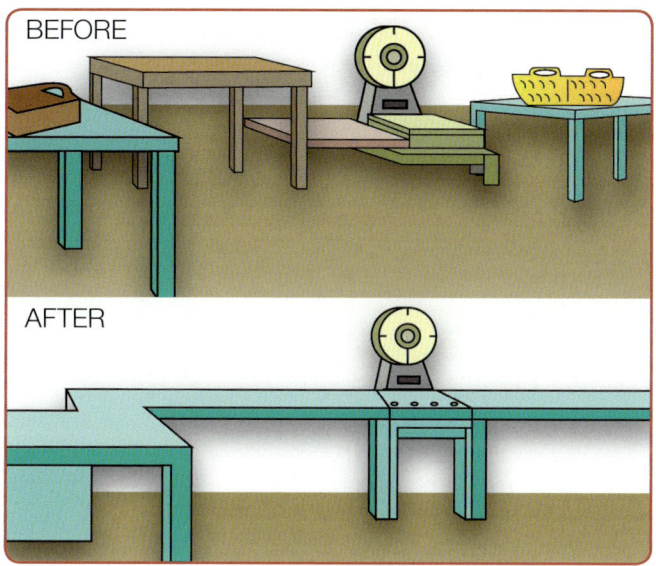

Figure 6-26: Eliminate height differences of work surfaces. Source: RMS/ILO, Ergonomic checkpoints.

To minimise manual handling risks related to the environment consideration should be given to control measures that:

- Design work layouts to reduce risk – minimise travel distance, reduce reaching (upwards, down and horizontally) and to enable loads to be picked up and put down at a suitable height.
- Remove restrictions on posture - ensure the working environment has sufficient headroom, so that workers do not have to stoop.
- Remove obstructions to free movement - ensure gangways, work space and working areas are adequate to allow room to manoeuvre during manual handling tasks.
- Provide a suitable floor for good movement – replace uneven floors and ensure good housekeeping, for example spilt liquids are cleaned up. This should help to ensure the worker is well balanced when carrying a load. If manual handling aids are used, for example a sack truck ensure the floor is suitable for it and that the equipment moves easily.
- Avoid variations in floor level, for example stairs, steps, ladders and steep ramps – by taking routes that do not have these variations and using lifts or hoists.
- Ensure workstations are of a uniform height - this

would reduce the need for raising or lowering loads when transferring them from one workstation to another.

- Avoid strong air movement when lifting large loads, for example gusts of wind - provide protective barriers or restrict work in strong winds.
- Ensure the general working environment is comfortable and suits manual handling activities - for example, humidity, heating/cooling, ventilation and lighting.

 CONSIDER

In a workplace of your choice, review how an understanding and application of ergonomic principles might improve a work task you have identified.

6.3 Load-handling equipment

HAZARDS AND CONTROLS FOR COMMON LOAD HANDLING EQUIPMENT

Sack truck

A sack truck is a simple device fitted with two wheels on which the load is pivoted and supported when the truck is manually tilted back and pushed.

A risk assessment must be made of manual handling operations associated with using equipment of this type. There is still a need to manually handle materials when using a sack truck.

Figure 6-27: Sack truck. Source: RMS.

Hazards and controls

Sack trucks are typically manually powered and the hazards arising from their use are generally of an ergonomic nature relating to posture and over-exertion through manual handling. If large heavy loads are placed on the sack truck this can exert strong downward forces when the truck is tipped back, due to the cantilever effect, the worker then has to apply additional force to overcome that of the load and enable it to be tipped back. The worker may find this strenuous and have difficulty holding it tipped back while moving. This could lead to the truck tipping forward or backwards suddenly.

Mechanical hazards are restricted to the wheels of the truck, which only move when the truck is pushed by the user. Other associated hazards are tripping and falling while using the equipment and manual handling back and strain injuries.

Control measures include indicating a safe working load for the equipment and the provision of information/instruction in the safe loading and use of the equipment for the operator. A manual handling assessment may be required when using this equipment.

Trolleys

A trolley is a platform fitted with four wheels, on which the load is placed and it is pushed manually, usually using a handle. Some trolleys have an additional cage fitted to the platform to contain loose loads or to allow the load to be stacked to a height without them falling off when the trolley is moved, for example, trolleys that are used to move goods from a delivery vehicle into supermarket stores.

Figure 6-28: Trolley with cage.
Source: Spacepac Industries Pty Ltd.

Hazards and controls

Trolleys are moved manually, therefore there are hazards related to the positioning of handles used to push the trolley, which could lead to poor posture. If the trolley is overloaded it could lead to over-exertion when pushing it.

The trolley could impact someone when it is moving or a wheel may run over someone's foot. If the load on the trolley is stacked too high or becomes unstable due to the trolley hitting an obstacle on the floor the load may fall from the trolley. Although a trolley fitted with a cage

might prevent the load falling in these circumstances, if the cage is tall compared to the width of the trolley, the trolley can become unstable on slopes or rough ground and could fall over.

Control measures include marking on the trolley the safe working load and safe load sacking height. Information/instruction should be provided on the safe loading and use of the trolley. Ensure the size of wheels are suitable for the surface of the floor it is used on, larger wheels are less affected by a floor with a rough surface, and the wheels are maintained and lubricated to enable it to move easily. The trolley should be pushed rather than pulled where possible. This will enable the worker to use their body weight and the strongest muscles in their legs to move the trolley efficiently. A manual handling assessment may be required when using this equipment.

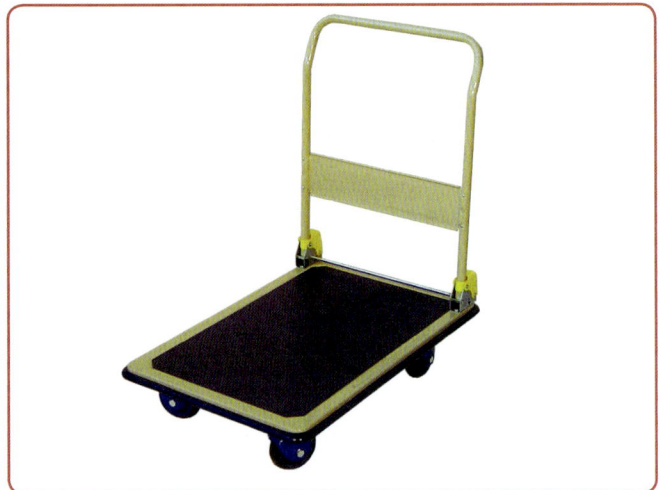

Figure 6-29: Platform trolley.
Source: Spacepac Industries Pty Ltd.

The pallet truck

This truck has two elevating forks for insertion below the top deck of a pallet. When the forks are raised the load is moved clear of the ground to allow movement.

This truck may be designed for pedestrian or rider control. They usually have no mast and cannot be used for stacking. Pallet trucks may be powered or non-powered.

Figure 6-30: Battery-powered pallet truck.
Source: RMS.

Hazards and controls

Pallet trucks can be driven both manually or by a quiet-running electric motor. Hazards include crushing from moving loads or from the momentum of the equipment when stopping. Other hazards include trapping in the forks of the equipment, manual handling and electrical hazards from battery power points.

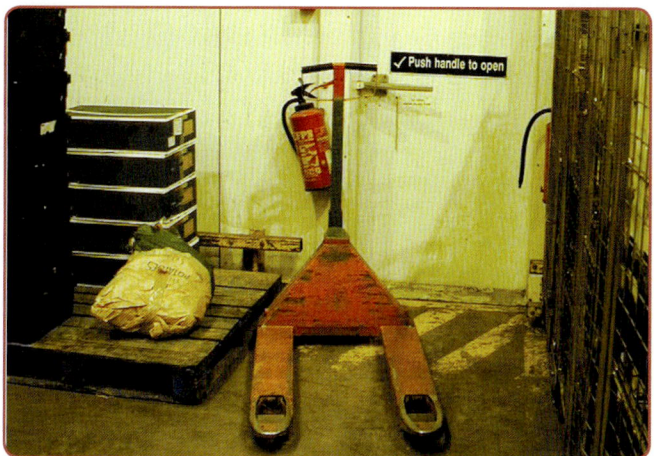

Figure 6-31: Manually operated pallet truck.
Source: RMS.

Some pallet trucks have a lifting mechanism to raise and lower the load. Control measures should include making sure that users are trained and authorised; identification of safe working loads; inspection and maintenance; and designated areas for parking the equipment.

People-handling hoists

To reduce the risk of manual handling injuries to carers, a patient hoist, located on firm level floors, is often used where there is a need to lift and transfer people, for example, to transfer a patient from a chair to a bed, into a bathtub or onto a toilet.

There are various types of people-handling devices, some types use manual systems (where the hoist is lifted by operating a manual crank handle), while others use a combination of manual and electrically driven mechanisms.

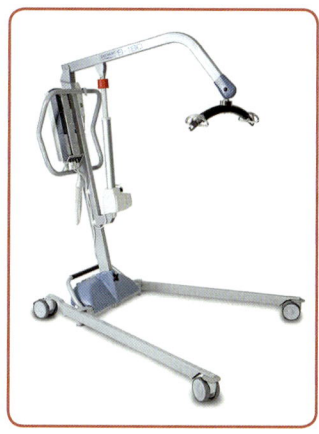 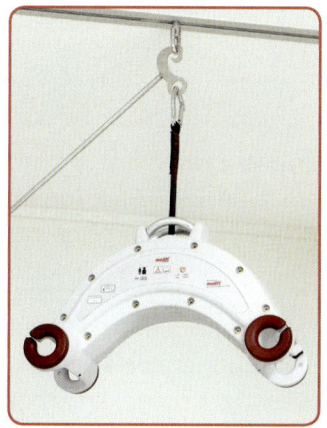

Figure 6-32: Mobile hoist.
Source: Prism Medical UK.

Figure 6-33: Ceiling hoist.
Source: Dolphin Mobility.

Hazards and controls

Mobile people-handling hoists require some effort to move the device and patient, presenting a risk of manual handling injury to the carer's back. Care should be taken to steer and locate the patient carefully, particularly when navigating around bends in corridors or doorways.

The route to be travelled should be inspected to ensure there is no risk of the patient colliding with obstructions when being moved. The hoist wheels should always be locked before loading or unloading patients to prevent uncontrolled movement of the lifting device.

Consideration should be given to the carer's footwear, which needs to be appropriate to the task, with particular emphasis on non-slip soles and low heels to prevent slips and trips when transferring patients. There is a possible additional risk of crush injury to the toes or feet from the wheels of the hoist **(see Figure 6-32)**.

Mobile hoists are not designed for movement of the patient over long distances. If the hoist is battery operated, the battery condition should be checked before use to avoid the risk of loss of power when transferring a patient. When not in use, hoists need to be stored safely out of the general work area. Hoists fitted with batteries will require a charging point for the battery. The charging should be carried out in a well-ventilated designated area to avoid trip hazards with the charging leads and risk of electric discharge or shock.

Ceiling hoists run on permanently fixed tracks **(see Figure 6-33)**. They provide less flexibility than a mobile system but are stable and secure. They can only be fitted where there are robust joists available in the ceiling or the ceiling is made of concrete. However, they take up less floor space than mobile hoists and may be operated by a remote control. They are better suited to moving the patient over long distances.

Other safety factors need to be considered for both types of hoist:

- They should only be used by workers trained and authorised in their correct use and storage.
- The safe working load must be clearly marked and care should be taken to assess whether this is adequate for the patient to be lifted.
- The equipment (including patient slings) should be subject to regular maintenance and, because these devices are used to lift people, they should be thoroughly examined by a competent person every six months.
- Care should be taken to inspect lifting slings for damage or tears. Only cleaning agents recommended by the equipment manufacturer should be used to avoid the risk of chemical degradation of the carrying harness.

People-handling aids

Hazards and controls

Lateral movement of patients is a common procedure in hospitals and care homes. Although manual handling is minimised where possible, performing this procedure many times during a working day can amount to considerable risk of workers experiencing musculoskeletal disorders. Risks to workers are increased where the transfer surfaces are not the same height or there are obstacles or gaps between them.

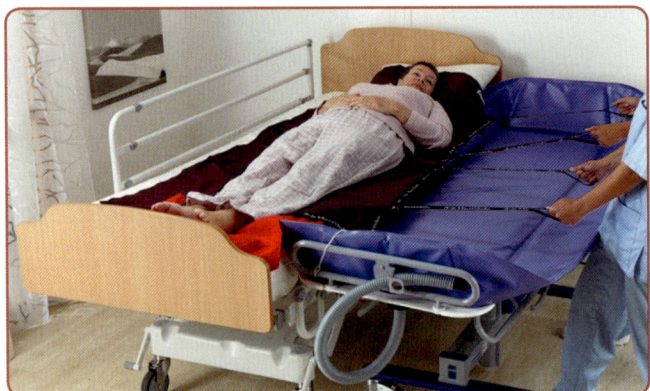

Figure 6-34: Slide sheets.
Source: ArjoHuntleigh.

Slide sheets allow basic handling without the need to lift the patient. They are invaluable (they reduce the friction significantly over conventional bedding sheets) when moving a patient on or off a bed, for turning the patient in bed and as an aid to sitting the patient up in bed.

Slide sheets reduce manual handling effort and strain, minimising the risk of back and other injuries to carers. They reduce skin tears and bruising to patients and promote patient comfort and dignity.

The **walk belt** allows carers to assist those patients who can bear their own weight but may need extra support when walking. The belt encourages patients to walk more often, with the comfort of knowing that a carer is there to assist.

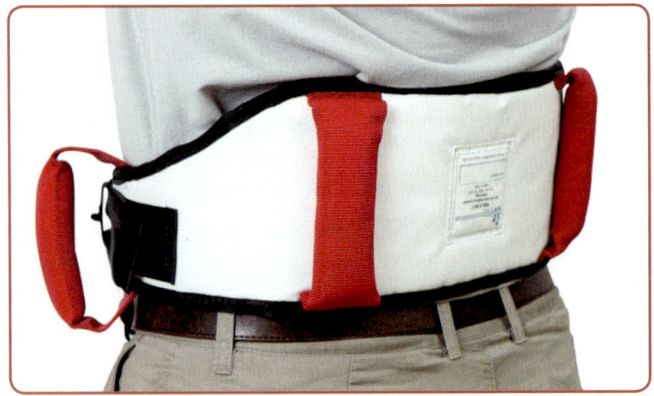

Figure 6-35: Walk belt.
Source: Patient Handling™.

The use of a walk belt increases the risk to carers since they still have to support some of the patient's weight. Normally the belt should only be used when it is impracticable to use a stretcher or wheelchair. The task should involve two carers working together. When fitted properly, the belt should not ride up as this can cause discomfort for the patient and increase the risk of instability. Fastenings (often hook-and-loop) should be checked for security and positioned so that the patient cannot unfasten them unexpectedly during transfer.

 REVIEW

Give examples of manual methods of transporting a load that would eliminate or reduce manual handling.

 CONSIDER

Are appropriate manual handling aids available to workers in your workplace?

FORK-LIFT TRUCKS

Hazards

Although the fork-lift truck is a very useful machine for moving materials in many industries, it features prominently in workplace accidents/incidents.

In the UK alone, there are about 20 deaths and 5,000 injuries each year that can be attributed to fork-lift trucks.

These can be analysed as follows:

- Injuries to driver   40%.
- Injuries to assistant   20%.
- Injuries to pedestrians   40%, of which 80% were fractures with some 60% resulting in injuries to ankles and feet.

Unless preventive action is taken, these accidents/incidents are likely to increase as fork-lift trucks are increasingly used in the workplace.

From accident investigations it can be seen that about 45% of accidents/incidents can be wholly or partly attributed to operator error, thus proper operator training is essential. There are, however, many other causes, including inadequate premises and gangways, poor truck maintenance and lighting.

The hazards related to the use of fork-lift trucks include:

- Overloading – exceeding maximum rated capacity.
- Driving too fast.
- Sudden braking.
- Driving on slopes.
- Driving with the load elevated.
- Driving over debris.
- Under-inflated (pneumatic) tyres or badly cut (solid) tyres.
- Driving with the load incorrectly positioned on the forks.
- Overturning.
- Poor floor surface, for example, holes, such as drains or potholes.
- Failure of load-bearing parts (for example, lifting chain).
- Collision with:
  - Vehicles
  - Buildings
  - Pedestrians
- Loss of load.
- Insecure load.

How some of these hazards can occur when using a fork-lift truck is explained below. It should be noted that fork-lift trucks, unless specifically designed to do so, are not intended for carrying passengers. Many of the hazards outlined for fork-lift trucks are equally applicable to other equipment used for load handling and may be seen as generic hazards of mobile powered load handling equipment.

Overturning

The stability of fork-lift trucks is particularly affected by the forces generated when turning, especially at speed, or if the equipment is tilted sideways, for example, by travelling across an incline or by the wheels running into a pothole or over an obstruction. The danger of a fork-lift truck being turned on its side is greater with the load in the raised position **(see Figure 6-36a)**, than in the lowered position **(see Figure 6-36b)**.

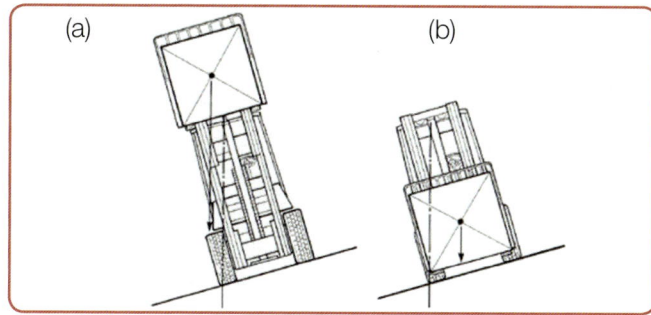

Figure 6-36: Overturning of lift truck.
Source: UK HSE, HSG6 (no longer available).

Overbalancing

The mass of a counterbalance fork-lift truck acts as a counterweight so that the load can be lifted and moved without the fork-lift truck overbalancing and tipping forward **(see Figure 6-37a)**. However, the fork-lift truck can be overbalanced and tip forward if the fork-lift truck is overloaded.

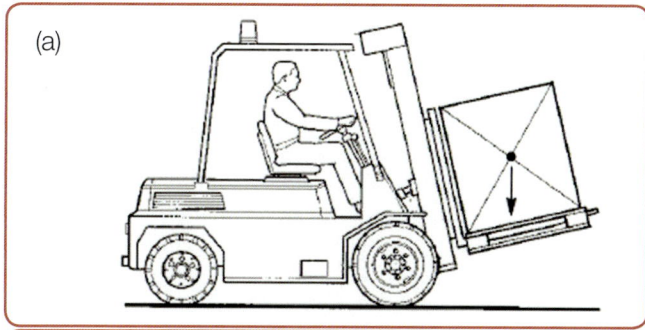

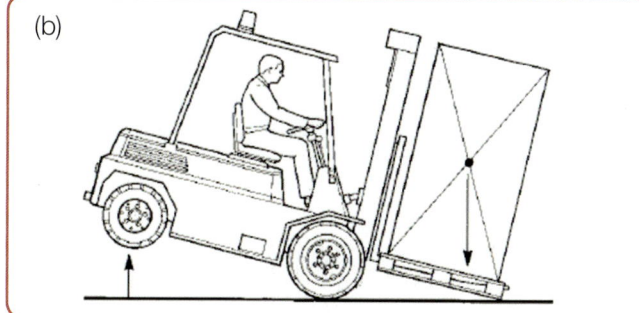

Figure 6-37: Overbalancing of fork-lift truck.
Source: UK HSE, HSG6 (no longer available).

Overloading may be caused if the load is too heavy **(see Figure 6-37b)**, if the load is incorrectly placed on the forks so that it is too far forward **(see Figure 6-37c)** or if the fork-lift truck accelerates or brakes sharply while carrying a heavy load.

This may not cause the equipment to overturn, but the overbalancing can injure the operator and lead to loss of control of steering. Control of steering may be lost because counterbalance fork-lift trucks usually have rear-wheel steering and overbalancing lifts the rear of a counterbalance fork-lift truck from the ground, preventing the steering wheels from contacting the ground properly.

Collisions with other vehicles, pedestrians or fixed objects

People may appear unexpectedly from a part of a building structure or workers intent on their work may step away from where they are working to collect materials or tools. Often the space in workplaces such as construction sites is restricted. Material may be stored at a height because it is large or to maximise the available space. This in turn leads to restricted visibility, especially at busy junctions where vehicles come together. Loads transported by fork-lift trucks may obstruct the driver's vision and cause them to look to alternate sides of the load in order to see past it. Any of these may lead to collisions with other vehicles and pedestrians or, in the avoidance of these, fixed objects.

Figure 6-38: Poor visibility loading.
Source: RMS.

When travelling with a load a fork-lift truck should have the forks tilted to cradle the load. If this is not done there is a risk that the load may slide from the forks and hit someone or something.

Fork-lift trucks often have a limited or no suspension system to reduce the effects of the wheels going over rough surfaces. This means that the shock of impact with ruts and small bumps can be transmitted readily to the forks and the load, or part of it, may be dislodged and caused to fall.

A fork-lift truck has an open framework that can allow a person to put a part of their body, typically their head or hand, outside the protective structure. Many injuries have occurred to fork-lift truck drivers who, while driving forward with a load obstructing their view, lean out to look round the load and their head strikes a fixed object.

Loading and unloading

There is a hazard of objects falling on a vehicle driver during loading and unloading operations, particularly where the fork-lift truck is used to load materials at a height or to remove materials from delivery vehicles.

Precautions and procedures

Traffic routes

The precautions related to traffic routes include:

- Separate routes, designated crossing places and suitable barriers at recognised danger spots.
- Roads, gangways and aisles should be wide enough and have sufficient overhead clearance for the largest fork-lift truck.
- Clear direction signs.
- Sharp bends and overhead obstructions should be avoided.
- The floor surface should be in good condition.
- Any gradient in a fork-lift truck operating area should be kept as small as possible. Trucks should reverse down gradients and not drive across them.
- Consideration to ensure vehicles are able to negotiate speed-retarding humps which are put in place.

Sufficient and suitable parking areas should be provided away from the main work area. Designated parking areas must be clearly identified. They should be on firm, level ground and located such that they do not obstruct fire points or other traffic routes, especially those for emergency access and exit.

Figure 6-39: Width of traffic route and barriers.
Source: RMS.

When the fork-lift truck is parked or left unattended the driver should ensure that the mast is tilted slightly forward with the forks resting on the floor, the power is switched off, the brake is applied, and the key removed and returned to a responsible person to prevent unauthorised use.

Protection of people

There is a need to alert people to the hazard when working in or near a vehicle operating area. Signs may be used to indicate the general operation of vehicles in the area. These can be supplemented by **visual and audible warning systems** that confirm the presence of a vehicle, such as a flashing light on the top of a fork-lift truck. The fork-lift truck should be fitted with a horn which can be sounded when approaching blind bends.

Driver protection

In many work vehicle accidents/incidents the driver is injured because the vehicle does not offer protection when it rolls over or it does not restrain the driver to prevent them falling out of the vehicle and being injured by the fall or the vehicle falling on them. Vehicles with open cabs, such as fork-lift trucks, may present this risk. Where there is a risk of the fork-lift truck overturning, a restraint system should be fitted and would typically involve some sort of adjustable seat belt.

Selection of equipment

There are many types of fork-lift truck available for a range of activities. Some situations require the use of specialist trucks such as reach trucks, overhead telescopic or rough terrain trucks. Many accidents/incidents happen due to the incorrect selection and/or use of fork-lift trucks.

Figure 6-40: Rider-operated fork-lift truck (reach type).
Source: RMS.

Figure 6-41: Rider-operated pallet truck.
Source: RMS.

Rider-operated counterbalance fork-lift truck

This type of fork-lift truck has a counterweight to balance the load on the forks. The forks and load project out from the front of the truck. Loads can be raised or lowered vertically and the mast may be tilted forward or backward.

This type of fork-lift truck is only suitable for use on substantially firm, smooth, level and prepared surfaces. A wide range of attachments is available.

Figure 6-42: Rider-operated fork-lift truck (counterbalance type).
Source: RMS.

Rider-operated reach fork-lift truck

This type of fork-lift truck is called a reach truck because the mast is capable of being moved forward (reached out) to pick up the load. For travelling, the load is retracted and carried within the wheelbase of the fork-lift truck. This allows greater manoeuvrability in areas where space is restricted.

This type of fork-lift truck is only suitable for use on substantially firm, smooth, level and prepared surfaces and is used particularly in warehouses.

Rough terrain counterbalance fork-lift truck

The rough terrain counterbalance fork-lift truck is similar in design to the standard counterbalanced fork-lift truck but is equipped with larger wheels and pneumatic tyres, giving it greater ground clearance.

It has greater ability to operate on uneven and soft ground and is mainly used in the construction industry and in agriculture. It may be used with a range of attachments.

Figure 6-43: Rough terrain counterbalance fork-lift truck.
Source: UK HSE, HSG6.

Telescopic materials handler

This form of fork-lift truck is fitted with a boom that is pivoted at the rear of the machine. The boom is raised and lowered by hydraulic rams. In addition, the boom can be extended or retracted to give extra reach or height. These machines may be two or four-wheel drive, and have two-wheel, four-wheel or crab steering. They are used mainly in agriculture and the construction industry. A range of attachments may be used with them.

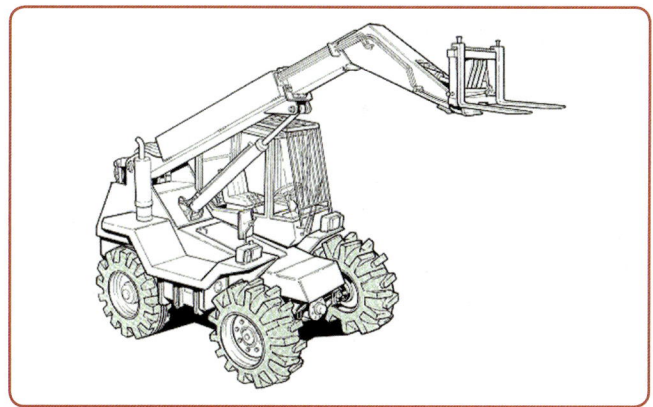

Figure 6-44: Telescopic materials handler.
Source: UK HSE, HSG6.

Side-loading fork-lift truck

When using a side-loading fork-lift truck the driver is positioned at the front and to one side of the load-bearing deck. The load is carried on the deck of the truck, the mast being traversed out sideways to pick up or set down the load.

This type of fork-lift truck is used for stacking and moving long loads such as bales of timber and pipes, and may be fitted with stabilisers for use when picking up or setting down loads.

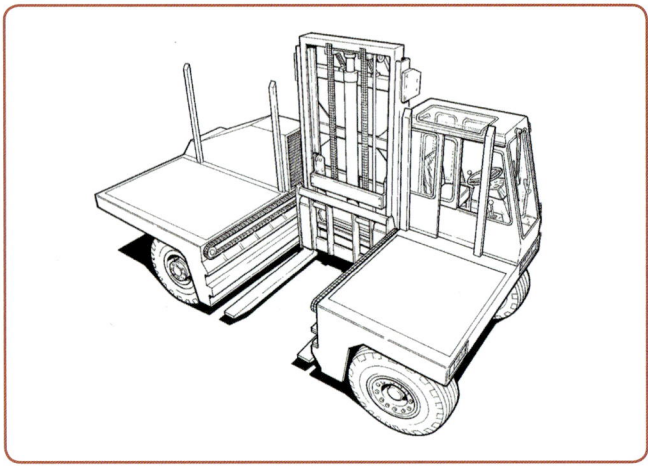

Figure 6-45: Side-loading fork-lift truck.
Source: UK HSE, HSG6.

Pedestrian-controlled fork-lift truck

This type of fork-lift truck has a limited lift height, usually not greater than two metres. It may be electrically or manually powered for lifting and for movement. The operator walks with the truck and controls it with a handle.

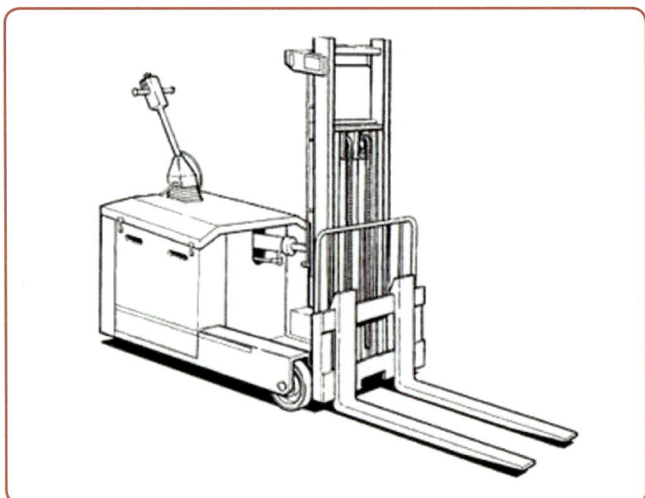

Figure 6-46: Pedestrian-controlled fork-lift truck
Source: UK HSE, HSG6.

Large fork-lift truck

The large fork-lift truck may be either masted or telescopic and is often fitted with a spreader for lifting freight containers. The spreader may be attached to the side or top of the container. These are specialist fork-lift trucks used mainly in container terminals. When choosing the right truck for the job the following factors should be taken into account:

- Power source – the choice of battery, LPG or diesel will depend on whether the truck is to be used indoors or outdoors. Factors to consider include, combustion fumes from LPG and diesel and manual handling with battery and LPG.
- Tyres – solid or pneumatic, depending on the terrain.
- Size and capacity – dependent on the size and nature of the loads to be moved.
- Height of the mast.

- Audible and/or visual warning systems fitted according to the proximity of pedestrians.
- Protection provided for the driver against overturning or the possibility of falling objects.
- Training given to operators must be related specifically to the type of truck.
- Provision of a suitable mechanism to prevent unauthorised use, for example, key or electronic pad.

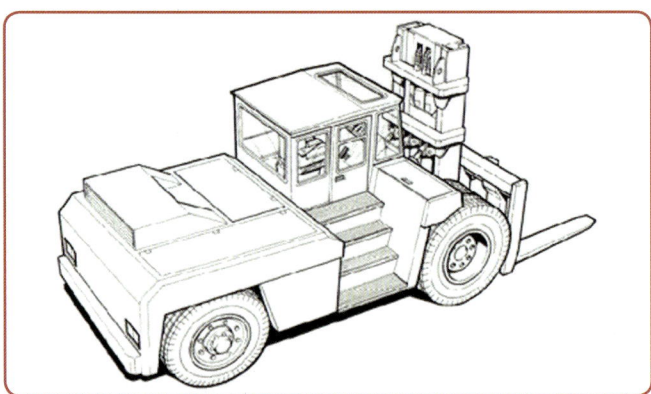

Figure 6-47: Large fork-lift truck.
Source: UK HSE, HSG6.

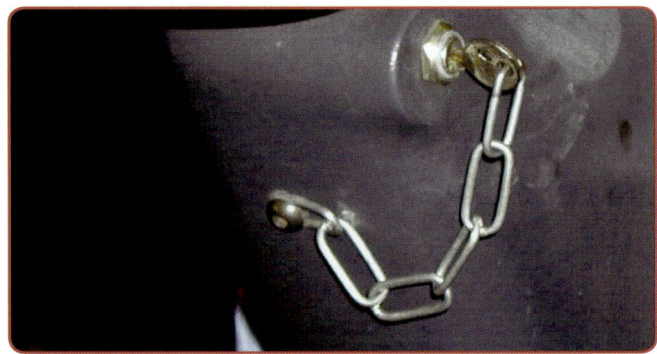

Figure 6-48: Keys – unauthorised use not controlled.
Source: RMS.

Figure 6-49: PIN pad to prevent unauthorised use.
Source: RMS.

Provision of information

Drivers and supervisors of fork-lift trucks should be familiar with the following information, which should be shown on the truck:

- Name of the manufacturer (or authorised supplier) of the lift truck.
- Model designation.
- Serial number.
- Unladen weight.
- Rated weight carrying capacity.
- Load centre distance.
- Maximum lift height.
- Inflation pressures of pneumatic tyres.

In addition, the functions of all the controls should be clearly marked so that they can be seen from the driver's position.

Drivers

No person should be permitted to drive a fork-lift truck unless they have been selected, trained and authorised to do so, or are undergoing properly organised formal training.

Selection of drivers

The safe use of fork-lift trucks calls for a reasonable degree of both physical and mental fitness and intelligence. The selection procedure should be devised to identify people who have shown themselves reliable and mature.

Training

Training should consist of three stages, the last being the one in which drivers are introduced to their future work environment.

Stage 1 - should contain the basic skills and knowledge required to operate the fork-lift truck safely. This will include:

- Understanding the basic mechanics and balance of the machine.

- Carrying out routine daily checks.

Stage 2 - under strict training conditions, in an area which workers not involved in the training and other people that may be harmed are excluded. This stage should include:

- Knowledge of the operating principles and controls.

- Use of the fork-lift truck in gangways, slopes, cold-stores, confined spaces and bad weather conditions, as appropriate.

- The work to be undertaken, for example, loading and unloading vehicles.

- On completion of training, the drivers should be examined and tested to ensure that they have achieved the required standard.

Stage 3 - after successfully completing the first two stages, the driver should be given further familiarisation and where necessary instruction in the usual place of work for the fork-lift truck.

Refresher training - if high standards are to be maintained, periodic refresher training and testing is essential good practice. Some countries have legislation that specifies refresher training is given within a specified time period. Some countries require periodic driver health surveillance as well.

 **CASE STUDY**

A site worker suffered severe injuries when he was trapped against a doorframe by a fork-lift truck driven by an untrained driver.

When a delivery arrived earlier than expected, there was not a trained fork-lift truck driver available on site. The delivery driver decided to operate the site fork-lift truck himself to unload. He reversed into pallets, over-corrected and reversed into the site worker.

The site manager should have made sure that only authorised people could use the fork-lift truck and that the keys were not available. The site manager and the delivery driver's employer should have liaised and agreed procedures for unloading deliveries. These procedures should have included fixing a time for vehicles to arrive with deliveries. The delivery driver should not have tried to operate a site vehicle without authorisation.

Daily checks/pre-use inspection

A checklist for daily checks should include items such as the condition and pressure of tyres, the integrity and proper functioning of lights, horns, brakes, steering and mirrors, the absence of oil leaks and that the seat is securely fixed (with properly functioning and intact restraints where fitted). The fork-lift truck should also be checked for obvious signs of damage to its bodywork and lifting forks mechanism and for the security of any equipment fitted such as liquid petroleum gas (LPG) cylinders.

If the fork-lift truck is to use attachment accessories, for example, forks or drum grabs, the driver should ensure they are correctly fitted and secured.

Thorough examination of load-bearing parts

The load-bearing parts of fork-lift trucks, for example, the lifting chain, should be thoroughly examined periodically to determine any wear or damage that could affect the integrity of the lifting equipment that forms part of the fork-lift truck. Some countries have legislation that specifies thorough examinations are made within a specified time period, for example, every 12 months.

Summary of precautions and procedures for fork-lift trucks

- Make someone responsible for fork-lift truck operation.
- Select and train drivers thoroughly.
- Daily fork-lift checks.
- Keep keys secure. *Do not leave in the ignition*.
- Maintain and light gangways.
- Separate fork-lift trucks and pedestrians.

A vigorous management policy covering driver training, vehicle maintenance and sound systems of work, supported by good supervision, will reduce personal injury and damage to equipment and materials. This in turn will lead to better utilisation of plant and increased materials handling efficiency.

LIFTS AND HOISTS

Hazards

Lifts and hoists are used for transporting people and goods between different levels. They can be found in a variety of buildings and structures, including temporary structures on constructions sites. Small, mobile devices are also used to lift and hoist materials or people. In general, the hazards associated with lifts and hoists are the same as with any other lifting equipment.

- The lift/hoist may overturn or collapse.

- The lift/hoist may strike people who are near or under the platform or cage.

- The lift/hoist may fail to stop in a safe position at the top, bottom or landing level.

- As the platform or cage moves, the load, including people, may come into contact with fixed or moving structures and objects, for example, landing level structures.

- The supporting ropes may fail and the platform/cage fall to the ground.

- The load, including people, or part of the load may fall.

- The lift/hoist may fail in a high position.

- People in and around the lift/hoist machinery may get trapped or entangled in moving parts.

Precautions and procedures

All lifts and hoists

If lifts and hoist are properly designed, installed and maintained there is relatively little risk from their operation.

All lifts and hoists require:

- Adequate design, including safety devices that are required by national legislation.
- Robust construction.
- Correct selection and installation.
- Holdback mechanisms that operate where the lifting mechanism fails, for example rope failure.
- Overrun trip devices to prevent the platform/cage being lifted or lowered beyond a safe limit. For example, overrun trip devices that prevent the platform/cage contacting the lifting gear at the top or being winched over the top of a drum.
- Guards on lift/hoist machinery, for example, to prevent entanglement with the moving parts of the machinery.
- Guards to prevent contact with the lift cage or platform when it is moving.
- Landing gates (securely closed during operation).
- Suitable lighting on all lift/hoist landings to reduce the risk of people tripping or falling.
- The safe working load of the lift/hoist clearly marked and visible to the user/operator.
- Operation by competent people.
- If the lift/hoist is for materials only, a prominent warning notice on the platform or cage to stop people riding on it.
- Regular inspection to ensure the lift/hoist is in working order. This may be as a daily, pre-use operator check and/or weekly inspection by an engineer.
- Regular servicing by a reputable maintenance company (approximately every three months). The maintenance contract should include removing rubbish as it may cause obstructions and contribute to the risk of a fire. The service report provided should relate to the efficient working of the lift/hoist and is not a substitute for a thorough examination.
- Periodic thorough examination. For example, passenger lifts/hoists should be thoroughly examined every six months by a competent person and goods lifts every 12 months. Alternatively, they should be examined at intervals detailed in an examination scheme drawn up by a competent person based on an assessment of risks.
- The results of inspections and thorough examinations to be recorded. Any remedial work identified should receive prompt attention.
- Unauthorised access to lift/hoist machinery to be prevented by keeping access under key control - which is kept in a secure position, controlled by a responsible person and available at all times to authorised people.

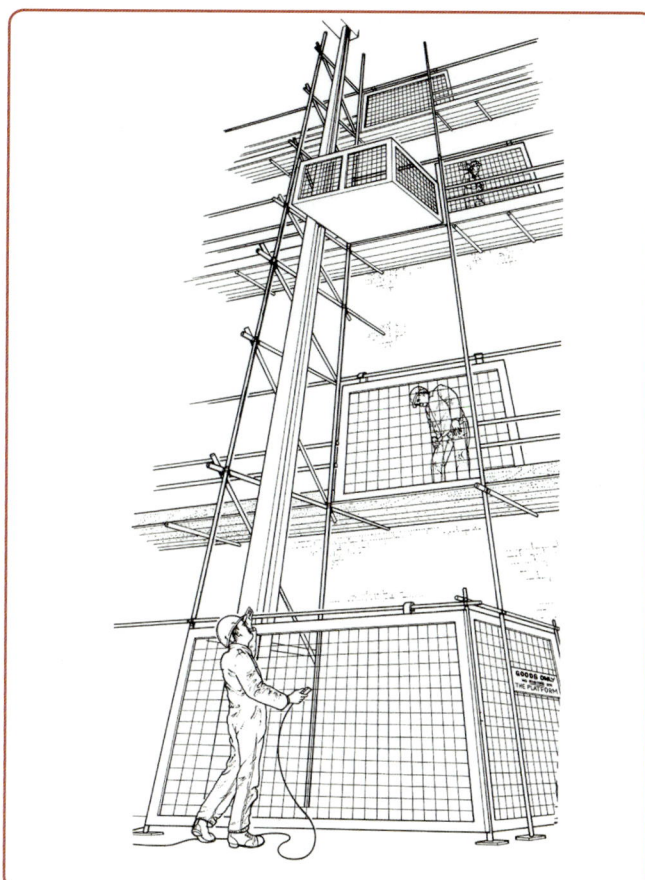

Figure 6-50: Lift/hoist.
Source: UK HSE, HSG150.

Passenger lifts and hoists

In addition, passenger lifts/hoists require more sophisticated controls measures:

- Operating controls inside the cage.
- Electromagnetic interlocks on the cage doors.
- If within a building, the enclosing shaft must be of fire-proof construction.
- The lift/hoist should be protected by a substantial enclosure to prevent anyone from being struck by any moving part of the hoist or material falling down the hoist way.
- Gates must be provided at all access landings, including at ground level. The gates must be kept shut, except when the platform is at the landing. The controls should be arranged so that the lift/hoist can be operated from one position only, which may be from within the lift/hoist.
- The safe working load of the lift/hoist must be clearly marked and visible to the operator, and cage-controlled lifts/hoists must be equipped with effective overload warning devices. Passenger lifts must be fitted with an alarm that can be activated by users from within the cage, if a fault condition arises, trained workers should be available to effect any rescue which may be required.

Procedures should be in place to ensure that any lift/hoist malfunction is reported to the supervisor immediately. A system should be developed for rescuing people trapped in passenger lifts/hoists and where this is to be carried out by workers they should be provide with adequate training on this procedure. Written rescue procedures should be displayed at appropriate locations and it should be ensured that the alarm bell used by passengers to warn that they are trapped can be activated.

Additional precautions and procedures for small mobile hoists controlled by an operator

Procedures and precautions for the safe operation of a small, mobile hoist controlled by an operator, such as that shown in **Figure 6-51**, include ensuring that only workers who have been trained in the proper use of hoists should be allowed to operate them. The lifting capacity of the hoist must be clearly marked and be visible to the operator. Before operation, operators should check that the hoist chains or ropes are of sufficient strength and length to safely lift the load. On a chain hoist, they should make sure the hook has a safety catch so that if the chain becomes slack the hook will not come loose.

Figure 6-51: Lift/hoist.
Source: Northern Tool + Equipment.

The oil level on hydraulic hoists should also be periodically checked.

Procedures should require that the hoist is positioned on firm level ground, with wheels locked during the lifting operation.

The load must be firmly secured to the hook of the hoist with appropriate lifting accessories and the weight of the load must be within the safe working load of the equipment. There should be a method of signalling that the lift is about to take place and the lift should be carried out at a steady rate. The load should be transported over only a short distance before being lowered and steps taken to prohibit work under the suspended load.

CONVEYORS

Conveyors use belts, rollers or screws (augers) *(see Figure 6-54)* to move goods or materials and are common in manufacturing and distribution workplaces.

Hazards

The mechanical hazards relating to conveyors include:

- **Drawing-in**   Clothing or limbs being drawn into in-running nips caused by moving parts.
- **Contact**   With moving parts (cut and abrasion).
- **Entanglement**   With rollers and drive mechanism.

Non- mechanical hazards include:

- **Striking**   Falling objects, especially from overhead conveyors.
- **Manual handling**   Loading and unloading components/packages.
- **Noise**   From mechanical movement.

Types of conveyor

The three basic types of conveyor are belt, roller and screw.

Belt

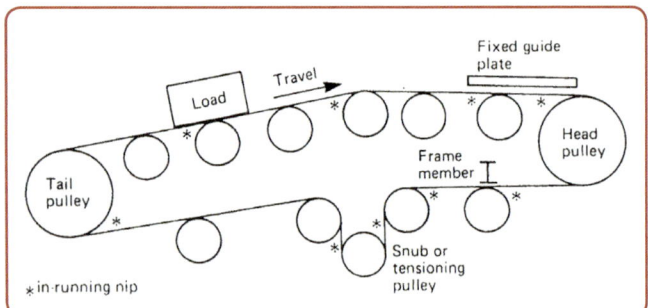

Figure 6-52: Diagrammatic layout of belt conveyor showing in-running nips. Source: J. Ridley; Safety at Work; Fourth Edition - Courtesy HSE, UK.

Belt conveyors transport materials on a moving belt. Drawing-in trapping points are created between the belt and the rotating conveyor drum (pulley) and any interme-diate rollers. The 'head and tail' drums (pulleys) create the main risks. Guards can be fitted to enclose the whole of the sides of the conveyor or at each drawing-in point (sometimes referred to as an 'in-running nip').

Roller

The hazards of roller conveyors are similar to belt conveyors, in that drawing-in points are created as the rollers rotate and move past fixed objects. Power-driven rollers create drawing-in points that exert a lot of energy on anything that gets trapped in them.

Figure 6-53: Roller conveyor.
Source: RMS.

It is essential to have guards on power drives and drawing-in points (in-running nips). Free-running rollers also create drawing-in points as they move past fixed objects or other rollers that are stationary or moving in the opposite direction. However, because they are free running, the energy exerted at the drawing-in point will be limited to that created by the momentum of the free-running roller.

Screw

Screw or worm conveyors push material forward by a rotating screw action. Screw conveyors can cause severe injuries and should be guarded or covered at all times. A locking-off system is required for maintenance and repairs.

Figure 6-54: Screw conveyor with a section of guarding removed showing screw (auger) mechanism. Source: Locom Engineering Ltd.

Precautions and procedures

The precautions relating to conveyors include:

- Warning alarms to alert people that the conveyor is about to start moving.
- Fixed guards at drawing-in points, including on drums (pulleys).
- Enclosure of conveyed items by side guards.
- Emergency stop buttons near the operator and along the length of the conveyor approximately 30 metres (100 feet) apart (or closer).
- Emergency stop buttons can be supplemented by trip wires (emergency stops operated by pulling a wire), if necessary, along the full length of the conveyor.

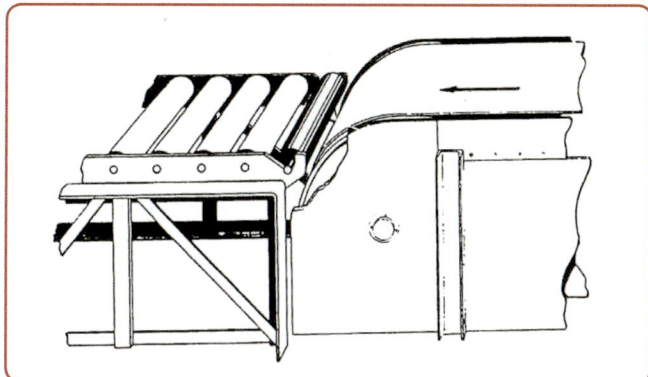

Figure 6-55: Preventing free-running roller trap.
Source: J. Ridley; Safety at Work; Fourth Edition - Courtesy UK HSE.

- Keep all stopping/starting control devices free from obstructions.
- Safe access at regular intervals.
- Do not climb, step, sit or ride on the conveyor at any time.
- Do not load conveyor outside of its design limits.
- Do not remove or alter conveyor guards or safety divides.
- Identify and train operators in the location and function of all stop/start controls.
- Avoid loose clothing and keep clothing, fingers, hair, and other parts of the body away from the conveyor.
- Provision of a system for workers to report defects.

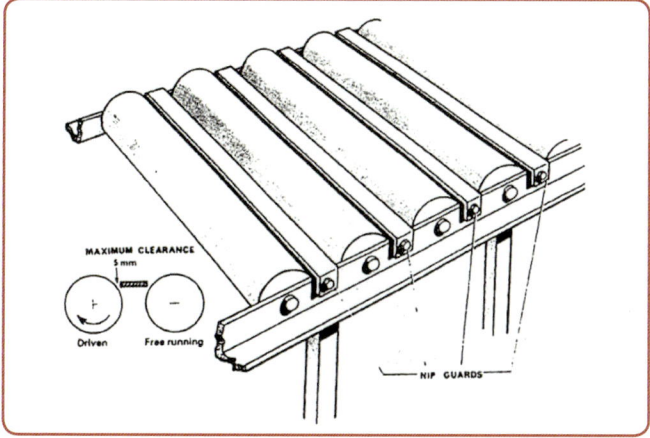

Figure 6-56: Guards between alternate drive rollers.
Source: J. Ridley; Safety at Work; Fourth Edition - Courtesy UK HSE.

Figure 6-57: Belt conveyor guard.
Source: RMS.

The safe procedures relating to conveyors include:

- Only trained personnel should operate the conveyor.
- All workers must be clear of the conveyor before starting.
- Wear bump caps.
- Keep the area around conveyors clear of obstructions.
- Restrict access to the area around the conveyor.
- Establish safe procedures to deal with items that fall from or become jammed in conveyors.
- Regular maintenance by competent people.
- Do not perform maintenance on a conveyor until the motor is disconnected and locked off.

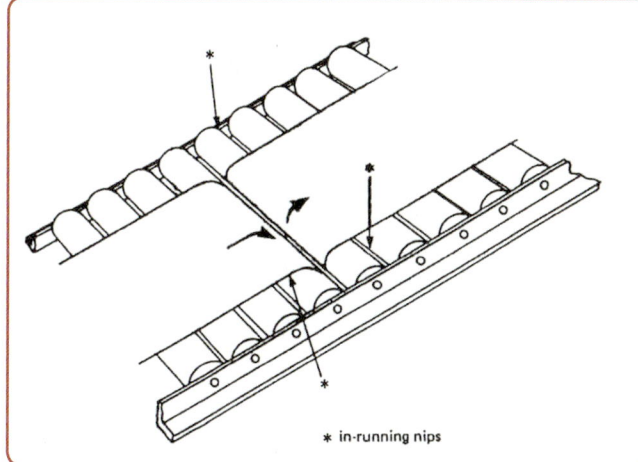

Figure 6-58: Drawing-in points on roller conveyor with belts.
Source: J. Ridley; Safety at Work; Fourth Edition - Courtesy UK HSE.

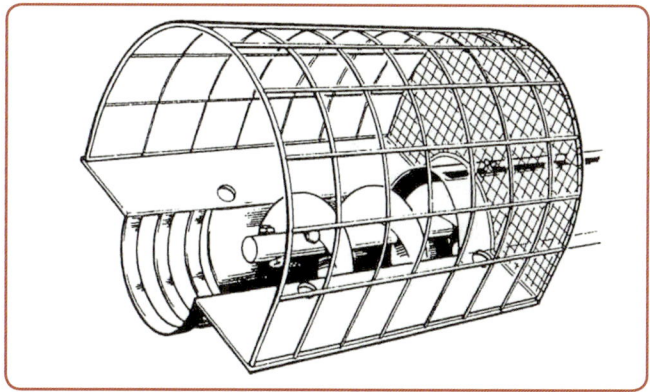

Figure 6-59: Screw conveyor guarding.
Source: J. Ridley; Safety at Work; Fourth Edition - Courtesy UK HSE.

CRANES

Hazards

The principal hazards associated with any crane lifting operation are:

- **Overturning** which can be caused by operating outside the capabilities of the machine, uneven or weak ground (cellars or drains), outriggers not extended, insufficient counterweight, and adverse weather and by striking obstructions.
- **Overloading** by exceeding the operating capacity or operating radii, or by failure of safety devices.
- **Collision** with other cranes, overhead cables or structures.

- **Failure of load-bearing part** from structural components of the crane itself or an accessory fitted to it. This may be due to overloading or degradation of the load-bearing part due to damage, use (wear) or faults (corrosion).
- **Loss of load** from failure of lifting tackle, incorrect hook fittings or slinging procedure.
- **People in and around the crane** may get entangled in or trapped by moving parts.

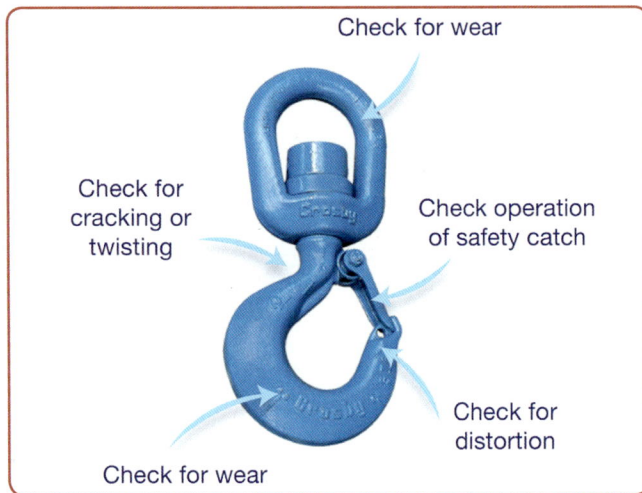

Figure 3-60: Hook-inspection.
Source: A. Noble & Son Ltd.

Factors that will affect all cranes

The factors that can increase the risks from using cranes, which affect all cranes, include:

- Soft or uneven ground conditions.
- Underground voids or cellars that are not capable of bearing the weight of the crane and its load.
- Load-bearing capacity of the crane must be sufficient for the task.
- Adverse weather conditions such as heavy rain, high wind and extremes of temperature.
- Workers or members of the public nearby.
- Insufficient room for the lift.
- Proximity to overhead power lines, buildings or other cranes.
- Tall cranes, for example, construction tower cranes, located near an airport or in a flight path.

Precautions and procedures

General requirements for cranes

The main control measures associated with any crane and lifting operation include ensuring that:

- Lifting operations are properly planned by a competent person, appropriately supervised and carried out in a safe manner.
- The ground the crane stands on is capable of bearing the load; check for underground services and cellars.
- The ground is level; if not, select a crane with hydraulic level-adjustment stabilisers.

Figure 6-61: Tower crane.
Source: Domson.

- The load-bearing capacity of the crane is sufficient for the task.
- The correct procedure is followed when erecting or dismantling any crane.
- The crane is positioned so that there is enough room for the lift and to avoid collision hazards.
- Non-essential people are kept clear of the work area.
- The crane is not operated in adverse weather conditions.
- The structural integrity of the crane is maintained; check for any signs of corrosion.

Figure 6-62: Ballast at base of tower crane.
Source: Wolffkran.

- A pre-use check by operator is carried out.
- Lifting equipment is of adequate strength and stability for the load. Stresses induced at mounting or fixing points must be taken into account. Similarly, every part of a load, and anything attached to it and used in lifting must be of adequate strength.
- The safe working load (SWL) is clearly marked on lifting machinery, equipment and accessories in order to ensure safe use. Where the SWL depends on the configuration of the machinery, it must be clearly marked for each configuration used and kept with the machinery.
- Passengers are not carried without authorisation, and never on lifting tackle.

Figure 6-63: Lifting operation.
Source: RMS.

- Equipment that is not designed for lifting persons, but which might be used as such, must have appropriate markings to the effect that it is not to be used for passengers.
- Load indicators are fitted. It is preferable that there are two types. This is a requirement with jib cranes, but beneficial in all cranes.
- Load/radius indicator is fitted. This shows the radius the crane is working at and the safe load for that radius. It must be visible to the operator.
- Automatic safe load indicator is fitted. This provides a visible warning when the SWL is approached and audible warning when the SWL is exceeded.
- Controls are clearly identified and are of the 'hold to run' type.
- Over-travel switches are fitted. These are limit switches to prevent the hook or sheave block being wound up to the cable drum.
- Access is provided. Safe access should be provided for the operator and for use during inspection and maintenance/emergency.
- Operating position provides clear visibility of hook and load, with the controls easily reached.
- Lifting tackle, for example, chains, slings, wire ropes, eyebolts and shackles, should be tested before use and thoroughly examined periodically.

Figure 6-64: Lifting points on load.
Source: RMS.

Accessories

Lifting accessories include slings, hooks, chains, eyebolts, shackles, lifting beams and cradles. This equipment is designed with the aim of assisting in lifting items without the need for manual force. Because these accessories are in a constantly changing environment and are in and out of use, they need to be protected from damage; a failure of any one item could result in a fatality. For example, lifting eyes need to be correctly fitted, slings have to be used with the correct technique and all equipment must be properly stored when not in use to prevent damage.

Figure 6-65: Safety latch on hook.
Source: RMS.

Figure 6-66: Lifting accessory - sling.
Source: RMS.

Accessories must be attached correctly and safely to the load by a competent person, and then the lifting equipment takes over the task of providing the necessary required power to perform the lift. As with all lifting equipment, accessories must be regularly inspected and certificated and only used by trained and authorised persons.

Operator training and practices

The effective management of lifting operations should involve selecting competent persons including the crane operator and appointed person who will supervise the lifting operation. Crane operators and slingers should be fit and strong enough for the work. A safe system of work, including safety rules, should be developed and communicated to all those involved. Circumstances differ from site to site and additional rules should be added to cover different circumstances, equipment and conditions. Training should be provided for the safe operation of the particular equipment. In particular, crane operators should receive systematic training similar to the approach used to train fork-lift truck operators, discussed earlier in this element. Typically this would include 3 stages:

Stage 1 - should contain the basic skills and knowledge required to operate the crane safely, to understand the basic mechanics and balance of the machine, and to carry out routine daily checks.

Stage 2 - under strict training conditions in an area which workers not involved in the training and other people that may be harmed are excluded. This stage should include:

- Knowledge of the crane operating principles and controls.
- Use of the crane in likely work conditions and doing the type of work to be undertaken, for example, loading and unloading vehicles.
- On completion of training, the crane operators should be examined and tested to ensure that they have achieved the required standard.

Stage 3 - after successfully completing the first two stages, crane operators should be given further familiarisation and where necessary instruction in the usual place of work for the crane operators.

General rules for safe operation of a crane

Always Ensure the lift is planned and supervised by a competent person and that the crane operator and slingers are competent.

Always Select the right appliance and lifting accessories for the job.

Always Site the crane on solid stable ground away-from structures or overhead cables.

Always Check the crane has been maintained and certified in accordance with local laws.

Always Ensure the appliance is stable when lifting, for example, not outside lifting radius, and that outriggers are correctly positioned.

Always Check the weather conditions and follow manufacturer's instructions about maximum wind speeds.

Always Use correct slinging methods.

Always Protect sling from sharp edges – pack out and lower onto spacers.

Always Ensure the sling is securely attached to the hook.

Always Use standard signals that are understood by those involved in the crane operation.

Always Ensure the load is lifted to correct height and moved at an appropriate speed.

Figure 6-67: Crane operation.
Source: RMS.

Never Use equipment if damaged (check before use) – for example, stretched or not free movement, worn or corroded, outside inspection date.

Never Exceed the safe working load.

Never Lift with sling angles greater than 120 degrees.

Never Lift a load over people.

Never Drag a load or allow sudden shock loading.

Rules should be established for when a crane operator takes over a crane and before use of the crane. For example, before taking over a mobile crane, the driver must always check around it, and check the pressure of tyres, the engine for fuel, lubrication oil, water and the compressed air system. All controls, such as lifting equipment clutches and brakes, should be tested to see that all lifting ropes run smoothly and safe load indicators function correctly.

The following are examples of specific safe operating rules for a **mobile crane**.

The operator must:

- Always lower the crane jib onto its rest (if fitted) or to the lowest operating position before travelling unladen, and point in the direction of travel, taking care of steep slopes.

- Always understand the signalling system and observe the signals of the appointed signaller/'banksman'.

- Never permit unauthorised persons to travel on the crane.

- Never use the crane to replace normal means of transport, or as a towing tractor.

- Always check that the crane is on firm and level ground before lifting and that spring locks/outriggers are properly in position.

- Never overload the crane. Always keep a constant watch on the load radius indicator.

- Always ensure movements are made with caution. Violent handling produces excess loading on the crane structure and machinery.

- Always make allowances for adverse weather conditions.

- Never cause loads to swing. Always position the crane so that the pull on the hoist rope is vertical.

- Always ensure that the load is properly slung. A load considered unsafe should not be lifted.

- Always ensure that all persons are in a safe position before any movement is carried out.

- Always make certain before hoisting that the hook is not attached to any anchored load or fixed object.

- Never drag slings when travelling.

- Always ensure that the jib, hook or load is in a position to clear any obstruction when the crane is slewing (swinging in a sideways or circular motion), but the load must not be lifted unnecessarily high.

- Always drive smoothly - drive safely. Remember that cranes are safe only when they are used as recommended by the makers. This applies in particular to speciality cranes.

- Always be on the lookout for overhead obstructions, particularly electric cables.

- Never tamper with or disconnect safe load indicators.

- Always stop the crane if the hoist or jib ropes become slack or out of their grooves, and report the condition.

- Always report all defects to the supervisor and never attempt to use a crane with a suspected serious defect until it has been rectified and certified by a competent person that it is not dangerous.

- Always make sure the crane is safe when leaving it unattended: ensure that the power is off, the engine stopped, the ignition key removed, the load unhooked, and the hook is raised up to a safe position.

- Always ensure special devices (for example, magnets and grabs) are used only for the purpose intended and in accordance with the instruction given.

- Always keep the crane clean and tidy.

- Always park the crane safely after use, remember to apply all brakes, slew locks, and secure rail clamps when fitted. Some cranes, however, particularly tower cranes, must be left to weather vane and the manufacturer's instructions must be clearly adhered to. Park the crane where the weather vaning jib will not strike any object. Lock the cabin before leaving the crane.

Figure 6-68: Poorly planned lifting operation - the rigger is at risk of falling due to lack of fall protection; inappropriate ladder used for access.
Source: RMS.

REQUIREMENTS FOR LIFTING OPERATIONS

Control of lifting operations

Employers should ensure that every operation involving lifting equipment for the purposes of lifting or lowering of a load is organised safely.

This will include ensuring the following:

- Lifting operations are properly planned by a competent person.
- Appropriate supervision must be provided.
- Work is carried out in a safe manner.

Figure 6-69: Lifting operations.
Source: RMS.

Strong, stable and suitable equipment

Strength

Employers should ensure that:

- Lifting equipment is of adequate strength and stability for each load, having regard in particular to the stress induced at its mounting or fixing point.
- Every part of a load and anything attached to it and used in lifting it is of adequate strength.

When assessing the strength of the lifting equipment for its proposed use, the combined weight of the load and lifting accessories should be taken into account. It is important to consider the load, task and environment in order to match the strength of the lifting equipment to the circumstances of use. For example, if the environment is hot or cold this can affect the lifting capacity of the equipment. In order to counteract this effect, equipment with a higher rated safe working load may be needed.

If the load to be lifted is a person, equipment with a generous capacity above the person's weight should be selected in order to provide an increased factor for safety. If the load is likely to move unexpectedly, because of the movement of an animal or liquids in a container, this sudden movement can put additional forces on the equipment and may necessitate equipment with higher strength to be selected.

When lifting a load that is submerged in water, the initial lifting weight will be misleading because the load will be supported by the water. When the load emerges from the water its support will no longer be available and this sudden increase in weight can put additional stress on the crane and its lifting accessories.

When conducting the lifting task, the lifting accessories may be used in such a way that its lifting capacity may be reduced below its stated safe working load; sharp corners on a load and 'back hooking' can have this effect. In these circumstances, accessories with a higher rated safe working load may be required. It is essential to remember that in a lifting operation the equipment only has an overall lifting capacity equivalent to the item with the lowest strength.

For example, in a situation where a crane with a lifting capacity of 50 tonnes is used with a hook of 10 tonnes capacity and a wire rope sling of 5 tonnes capacity, this would give an overall maximum lifting strength/capacity of 5 tonnes. Any alterations or repairs to the crane must be in accordance with the manufacturer's instructions to maintain the strength of the crane.

Stability

A number of factors can affect the stability of lifting equipment, for example, wind conditions, slopes/ cambers, stability of ground conditions and how the load is to be lifted. Lifting equipment must be positioned and installed so that it does not tip over when in use.

Anchoring can be achieved by securing with guy ropes, bolting the structure to a foundation, using ballast as counterweights or using outriggers to bring the centre of gravity down to the base area.

Mobile lifting equipment should be sited on firm ground with the wheels or outrigger feet having their weight distributed over a large surface area. Care should be taken that the equipment is not positioned over cellars, drains or underground cavities, or positioned near excavations.

Sloping ground should be avoided as this can shift the load radius out or in, away from the safe working position. In the uphill position, the greatest danger occurs when the load is set down. This can cause the mobile lifting equipment to tip over. In the downhill position, the load moves out of the radius and may cause the equipment to tip forward. If it is necessary to use a crane to lift a load on a slope it is important to select a crane with sufficient load lifting capability and stability to counter-act the effects of the slope on the load and the crane.

Figure 6-70: Siting and stability.
Source: RMS.

Suitability

The employer should ensure that work equipment is used only for operations for which it has been designed, including consideration of the working conditions for each lifting operation. In the UK, Regulation 4 of the Provision and Use of Work Equipment Regulations 1998 states:

 "Every employer shall ensure that work equipment is used only for operations for which, and under conditions for which, it is suitable in order to avoid any reasonably foreseeable risk to the health and safety of any person."

Figure 6-71: Suitability of work equipment.
Source: UK, Provision and Use of Work Equipment Regulations (PUWER) 1998.

In order for lifting equipment to be suitable it must be of the correct type for the task, have a safe working load limit in excess of the load being lifted, and have the correct type and combination of lifting accessories attached.

Lifting equipment used within industry varies and includes mobile cranes, static tower cranes and overhead travelling cranes. The type of lifting equipment selected will depend on a number of factors, including the weight of the load to be lifted, the radius of operation, the height of the lift, the time available, and the frequency of the lifting activities. This equipment is often very heavy, which means its weight can cause the ground underneath to sink or collapse.

Other factors like height and size may have to be considered as there may be limitations on site roads between structures or there may be overhead restrictions. Careful consideration of these factors must be made when selecting the correct crane. Selecting lifting equipment to carry out a lifting activity should be done at the planning stage, where the most suitable equipment can be identified that is able to meet all of the lifting requirements and the limitations of the location.

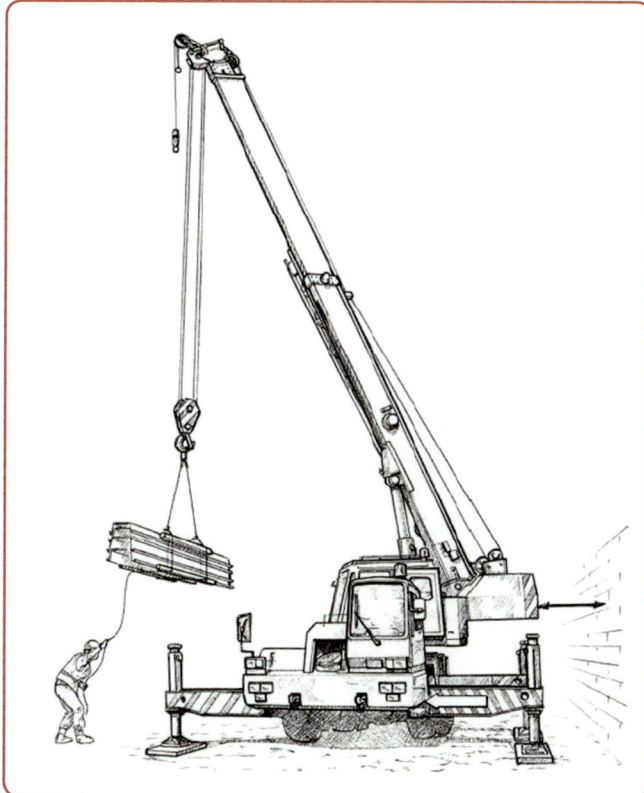

Figure 6-72: Danger zone – crane and fixed item.
Source: UK, HSE.

Positioned and installed correctly

Lifting equipment must be positioned or installed so that the risk of the equipment striking a person is as low as is reasonably practicable. Similarly, the risk of a load drifting, falling freely or being unintentionally released must also be considered and equipment positioned to take account of this.

All nearby hazards, including overhead cables and uninsulated power supply conductors, should be identified and removed or covered by safe working procedures such as locking-off and permit systems. The possibility of striking other lifting equipment or structures should also be examined.

Detailed consideration must be given to the location of any heavy piece of lifting equipment due to the fact that additional weight is distributed to the ground through the loading of the equipment when performing a lift. Surveys must be carried out to determine the nature of the ground, whether soft or firm, and what underground hazards are present, such as buried services or hollow voids. If the ground proves to be soft, it can be covered using timber, digger mats or hard core to prevent the equipment or its outriggers sinking when under load.

The surrounding environment must also be taken into consideration and factors may include highways, railways, electricity cables, areas of public interest. The area around where lifting equipment is sited should be securely fenced, including the extremes of the lift radius, with an additional factor of safety to allow for emergency arrangements such as emergency vehicle access or safety in the event of a collapse or fall.

Where practicable, lifting equipment should be positioned and installed such that loads are not carried or suspended over areas occupied by people. Where this is necessary, appropriate systems of work should be used to ensure it is done safely.

If the operator cannot observe the full path of the load, an appointed person (and assistants as appropriate) should be used to communicate the position of the load and provide directions to avoid striking anything or anyone.

Visibly marked

The safe working load (SWL) must be clearly marked on lifting machinery, equipment and accessories in order to ensure safe use. Where the SWL depends on the configuration of the machinery, it must be clearly marked for each configuration used and kept with the machinery. Accessories must be marked with supplementary information that indicates the characteristics for its safe use, for example, safe angles of lift.

Equipment designed for lifting people must be clearly marked as such and equipment which is not designed for lifting persons, but which might be used as such, must have appropriate markings to the effect that it is not to be used for lifting people.

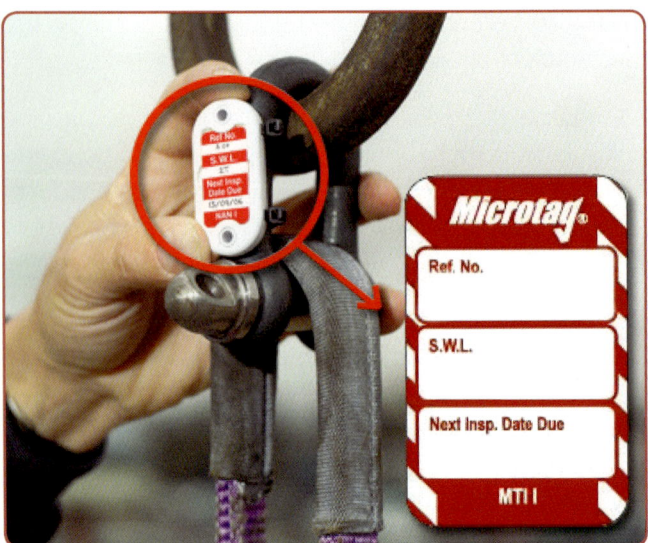

Figure 6-73: Marking of accessories.
Source: Scafftag.

Planned, supervised and carried out in a safe manner by competent people

The employer should ensure that every operation involving lifting equipment is:

- Properly planned by a competent person.
- Appropriately supervised.
- Carried out in a safe manner.

The person planning the operation should have adequate practical and theoretical knowledge and experience of planning lifting operations.

The plan should address the risks identified by the risk assessment and identify the resources required, the procedures and the responsibilities so that any lifting operation is carried out safely. The plan should ensure that the lifting equipment remains safe for the range of lifting operations for which the equipment might be used.

The type of lifting equipment that is to be used and the complexity of the lifting operations will dictate the degree of planning required for the lifting operation. Planning combines two parts:

1) Initial planning to ensure that lifting equipment is provided which is suitable for the range of tasks that it will have to carry out.
2) Planning of individual lifting operations so that they can be carried out safely with the lifting equipment provided.

Factors that should be considered when formulating a plan include:

- The load that is being lifted - weight, shape, centres of gravity, surface condition, lifting points.
- The equipment and accessories being used for the operation and suitability - certification validity.
- The proposed route that the load will take, including

the destination and checks for obstructions.
- The team required to carry out the lift - competencies and numbers required.
- Production of a safe system of work, risk assessments, permits to work.
- The environment in which the lift will take place - ground conditions, weather, local population.
- Securing areas below the lift - information, restrictions, demarcation and barriers.
- A suitable trial to determine the reaction of the lifting equipment prior to full lift.
- Completion of the operation and any dismantling required.

It is important that someone takes supervisory control of lifting operations at the time they are being conducted. Although the operator may be skilled in lifting techniques, this may not be enough to ensure safety, because other factors may influence whether the overall operation is conducted safely, for example, people may stray into the area. The level of supervision necessary is influenced by the nature of the work and the competence of those involved in using the equipment and assisting with the lifting operation. The supervisor of the lifting operation must remain in control and stop the operation if it is not carried out satisfactorily.

Lifting equipment and accessories should be subject to a pre-use check in order to determine their condition and suitability. In addition, care should be taken to ensure the lifting accessories used are compatible with the task and that the load is protected or supported such that it does not disintegrate when lifted.

Lifting operations should not be carried out where adverse weather conditions occur, such as fog, sand/dust storm, poor lighting, strong wind or where heavy rainfall makes ground conditions unstable. It is important that measures are taken to prevent lifting equipment overturning and that there is sufficient room for it to operate without contacting other objects.

Lifting equipment should not be used to drag loads and should not be overloaded. Special arrangements need to be in place when lifting equipment not normally used for lifting people is used for that purpose, for example, de-rating the working load limit, ensuring communication is in place between the people being lifted and the operator, and ensuring the operation controls are manned at all times.

Employers must ensure that any person who uses a piece of work equipment has received adequate training for purposes of health and safety, including training in the methods that may be adopted when using work equipment, any risks which such use may entail and precautions to be taken. Various appointments, with specified responsibilities, may be made in order to ensure the safety of lifting operations on site. The various appointments are as follows:

Competent person	Appointed to plan the operation.
Load handler	Attaches and detaches the load.
Authorised person	Ensures the load safely attached.
Operator	Appointed to operate the equipment.
Responsible person	Appointed to communicate the position of the load (banksman).
Assistants	Appointed to relay communications.

Each person given responsibilities must be competent to carry them out, usually this will mean that they must be adequately trained and experienced. The only exception is when they are under the direct supervision of a competent person for training requirements.

Special requirements for lifting equipment for lifting persons

The general use of people-handling hoists and people-handling aids has been covered earlier in this element in 'Manually operated load handling equipment'. This section looks at the special requirements for lifting people. The employer should ensure that lifting equipment for lifting persons:

- Is such as to prevent a person using it being crushed, trapped or struck or falling from the carrier.
- Is such as to prevent, so far as is reasonably practicable, a person using it being crushed, trapped or struck or falling from the carrier while carrying out activities from the carrier.
- Has suitable devices to prevent the risk of a carrier falling.
- Is such that a person trapped in any carrier is not thereby exposed to danger and can be freed.

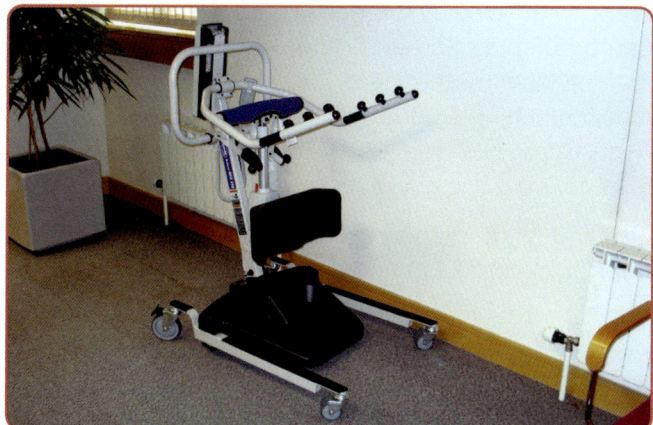

Figure 6-74: Mobile hoist for lifting people.
Source: RMS.

In addition, the employer should ensure that if the risks described earlier cannot be prevented for reasons inherent in the site and height differences:

- The carrier has an enhanced safety coefficient suspension rope or chain.
- The rope or chain is inspected by a competent person every working day.

Special arrangements need to be in place when lifting equipment not normally used for people is used for that purpose, for example, de-rating the working load limit, ensuring communication is in place between the people and operator, and ensuring the operation controls are manned at all times. Lifting equipment for lifting people may be subject to specific requirements for statutory examination prescribed by national legislation.

PERIODIC INSPECTION AND EXAMINATION/ TESTING OF LIFTING EQUIPMENT

To ensure that damaged or dangerously worn equipment does not remain in service, all items of lifting equipment must be periodically examined by a competent person.

Where the safety of lifting equipment depends on the installation conditions it must be thoroughly examined prior to first use, after assembly and on change of location in order to ensure that it has been installed correctly and is safe to operate. Lifting equipment exposed to conditions causing deterioration that is liable to result in dangerous situations should be thoroughly examined by a competent person:

- At least every 6 months - lifting equipment for lifting persons and lifting accessories.
- At least every 12 months - other lifting equipment.
- In either case, in accordance with an examination scheme.
- On each occurrence of exceptional circumstances liable to jeopardise the safety of the lifting equipment.

The person who carries out the thorough examination should, as soon as is practicable, make a written report of the results. This should be signed by the competent person who carried out the task and should be kept available for inspection for the period of validity of the report.

Where appropriate to ensure health and safety, inspections should be carried out at suitable intervals between thorough examinations. Examinations and inspections should ensure that the good condition of equipment is maintained and that any deterioration can be detected and remedied in good time.

The term 'competent' is generally taken to mean someone who is qualified and experienced in carrying out such examinations. The code of practice applicable to all workplaces in Western Australia defines a competent person as:

"**'Competent person'**, in relation to the doing of anything, means a person who has acquired, through training, qualification or experience or a combination of those things, the knowledge and skills required to do that thing competently."

Figure 6-75: Competent person definition.
Source: Western Australia, Code of practice – Prevention of falls at workplaces.

 **REVIEW**

What factors should be considered during the planning stage to reduce the risk of injury to workers and others before using lifting equipment?

CONSIDER

Review work equipment in your workplace and identify any mechanical lifting equipment in use.

Sources of reference

Reference information provided, in particular web links, was correct at time of publication, but may have changed.

Directive 2009/104/EC - use of work equipment https://osha.europa.eu/en/legislation/directives/3

Ergonomics and human factors at work, INDG90, HSE Books, http://www.hse.gov.uk/pubns/indg90.pdf

Ergonomic Checkpoints: Practical and easy-to-implement solutions for improving safety, health and working conditions, second edition, ILO Geneva 2010, ISBN 978-9-221226-66-6 http://www.ilo.org/wcmsp5/groups/public/---dgreports/---dcomm/---publ/documents/publication/wcms_120133.pdf

Lift-truck training – advice for employers, INDG462, HSE, http://www.hse.gov.uk/pubns/indg462.pdf

Lifting equipment at work – a brief guide, INDG290(rev1), HSE, http://www.hse.gov.uk/pubns/indg290.pdf

Manual Handling, Manual Handling Operations Regulations (MHOR) 1992 (as amended), Guidance on **Regulations, L23, third edition 2004, HSE Books,** ISBN: 978-0-717628-23-0, http://www.hse.gov.uk/pubns/priced/l23.pdf

Provision and Use of Work Equipment Regulations (PUWER) 1998, UK, www.legislation.gov.uk (please enter the name of the Regulation)

Rider-operated lift trucks, L117, 2013, HSE Books, ISBN: 978-0-717664-41-2 http://www.hse.gov.uk/pUbns/priced/l117.pdf

Safe use of lifting equipment, Lifting Operations and Lifting Equipment Regulations (LOLER) 1998, ACoP and Guidance, L113, HSE Books, ISBN: 978-0-717616-28-2 http://www.hse.gov.uk/pubns/priced/l113.pdf

Safe Use of Work Equipment, ACoP and guidance (part III in particular), L22, third edition 2008, HSE Books, ISBN: 978-0-717662-95-1 http://www.hse.gov.uk/pubns/books/l22.htm

Safety and health in the use of machinery, ILO CoP http://www.ilo.org/wcmsp5/groups/public/---ed_protect/---protrav/---safework/documents/normativeinstrument/wcms_164653.pdf

Seating at Work, HSG57, HSE Books, ISBN: 978-0-7176-1231-4, http://www.hse.gov.uk/pubns/priced/hsg57.pdf

The Health and Safety (Display Screen Equipment) Regulations (DSE) 1992 (as amended) www.legislation.gov.uk (please enter the name of the Regulation)

The health and safety toolbox, How to control risks at work, HSG268, HSE Books, ISBN: 978-0-7176-6587-7 , http://www.hse.gov.uk/pUbns/priced/hsg268.pdf

The Lifting Operations and Lifting Equipment Regulations (LOLER) 1998 www.legislation.gov.uk (please enter the name of the Regulation)

The Manual Handling Operations Regulations (MHOR) 1992 www.legislation.gov.uk (please enter the name of the Regulation)

Understanding ergonomics at work, INDG90 (rev2), HSE Books http://www.hse.gov.uk/pubns/indg90.pdf

Upper Limb Disorders in the Workplace, HSG60 HSE Books, ISBN: 978-0-7176-1978-8, http://www.hse.gov.uk/pubns/priced/hsg60.pdf

Work with display screen equipment: Health and Safety (Display Screen Equipment) Regulations 1992 as amended by the Health and Safety (Miscellaneous Amendments) Regulations 2002: Guidance on Regulations, L26, HSE Books ISBN: 978-0-7176-2582-6, http://www.hse.gov.uk/pubns/priced/l26.pdf

Working with display screen equipment (DSE): A brief guide, INDG36 (rev4), HSE Books, ISBN: 978-0-717664-72-6 http://www.hse.gov.uk/pubns/indg36.htm

Web links to these references are provided on the RMS Publishing website for ease of use –www.rmspublishing.co.uk

STUDY QUESTIONS

1) Explain, with examples, how mechanisation may reduce fatigue or injury from manual handling operations. (8)

2) (a) Why should periodic inspection and examination of lifting equipment be carried out? (4)

   (b) Who should be selected to do the inspection and examination? (4)

3) What are the principal hazards associated with a mobile crane lifting operation? (8)

4) Explain the musculoskeletal disorders sprain and strain. (8)

5) What should be considered when carrying out a manual handling risk assessment? (8)

6) (a) What are the health effects associated with carpal tunnel syndrome? (6)

   (b) Identify two examples of work tasks which may result in work-related upper limb disorders. (2)

For guidance on how to answer NEBOSH questions please refer to the 'study question answer guidance' section located at the back of this guide.

This page is intentionally left blank

Element 7

Chemical and biological agents

Contents

7.1 Hazardous substances

FORMS OF CHEMICAL AGENTS

The form taken by a hazardous substance is a contributory factor in its potential for harm. The form affects how easily a substance gains entry to the body, how it is absorbed into the body and how it reaches a susceptible site. Chemical agents take many forms, the most common being the primary forms or states – solids, liquids, gases – and the derivative forms – dusts, fibres, fumes, mists and vapours.

Solids

Solids are materials that are solid at normal temperature and pressure. The atoms, molecules or ions that make up a solid may be arranged in an orderly repeating pattern or irregularly. Materials whose constituents are arranged in a regular pattern are known as crystals. Crystals that are large enough to see and handle are known as crystallites. Other materials are called polycrystalline, which simply means they are composed of many crystallites of varying size and orientation. Almost all common metals, silicon and many ceramics are polycrystalline. Unlike a liquid, a solid object does not flow to take on the shape of its container, nor does it expand to fill the entire volume available to it like a gas. The risk from hazardous solids increases with reduction in particle size, particularly when it becomes a dust that can become airborne.

Liquids

Liquids are substances that are liquid at normal temperature and pressure. Liquids have a definite volume but no fixed shape. Similar to a gas, a liquid is able to flow and take the shape of a container. Some liquids, such as water, resist compression, while others can be compressed. Unlike a gas, a liquid does not disperse to fill every space of a container and maintains a fairly constant density. The density of a liquid is usually close to that of a solid and much higher than that of a gas.

Gases

Gases are formless fluids usually produced by chemical processes involving combustion or by the interaction of chemical substances. A gas will normally seek to fill the space completely into which it is liberated – for example, chlorine gas, carbon monoxide gas and methane gas.

Dusts

Dusts are solid airborne particles, often created by operations such as grinding, crushing, milling, sanding or demolition – for example, silica dust, flour dust and cement dust.

Fibres

Fibres are solid airborne particles that are significantly longer in length than they are in width, for example, cotton fibres, mineral wool fibres and asbestos fibres.

Fumes

Fumes are solid airborne particles formed by condensation from the gaseous state – for example, lead fume, welding fume.

Mists

Mists are finely dispersed liquid droplets suspended in air. Mists are mainly created by spraying, foaming, pickling and electro-plating – for example, mists from a water pressure washer, paint spray, pesticide spray, sprayed oil-based cutting fluids.

Vapour

Vapour is the gaseous form of a material normally encountered in a liquid or solid state at normal room temperature and pressure. Typical examples of vapours are those released from liquid solvents – for example, trichloroethylene, which releases vapours when the container holding it is opened.

"The terms 'dust', 'mist' and 'vapours' are defined as follows:

Dust: Solid particles of a substance or mixture suspended in a gas (usually air).

Mist: Liquid droplets of a substance or mixture suspended in a gas (usually air).

Vapour: The gaseous form of a substance or mixture released from its liquid or solid state."

Figure 7-1: Definition of dust, mist and vapour.
Source: UN Globally Harmonised System of Classification and Labelling of Chemicals (GHS) - part 3.

FORMS OF BIOLOGICAL AGENTS

Fungi

Fungi are a variety of organisms that act in a parasitic manner, feeding on organic matter. Most are either harmless or positively beneficial to health; however, a number cause harm to humans and may be fatal.

An example of fungi is the mould from rotten hay called aspergilla, which causes aspergillosis ('farmer's lung'). Farmer's lung is an allergic reaction to the mould. This occurs deep in the lungs in the alveoli region (where the exchange of oxygen and carbon dioxide takes place). It leads to shortness of breath, which gets progressively worse at each exposure. The resulting attack is similar to asthma. Aspergilla can also cause short-term effects of irritation to the eyes and nose and coughing.

Figure 7-2: Mould - Aspergillus.
Source: National Geographic.

Figure 7-2 shows Aspergillus with its characteristic chains of spores emerging from the head. Moulds from the same family can cause ringworm and athlete's foot.

Bacteria

Bacteria are single-cell organisms. Most bacteria are harmless to humans and many are beneficial. The bacteria that can cause disease are called pathogens. Examples of harmful bacteria are leptospira (causing Weil's disease), bacillus anthracis (causing anthrax), and legionella pneumophila (causing Legionnaires' disease).

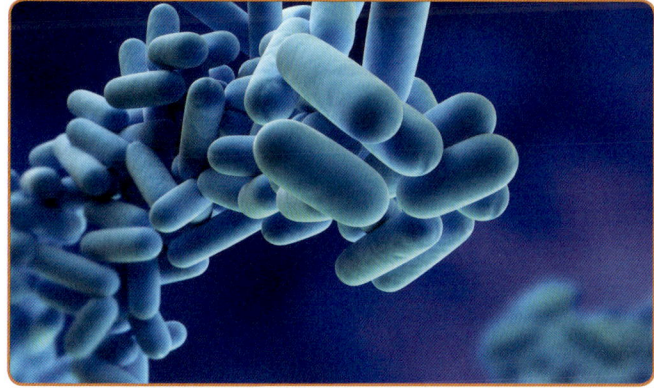

Figure 7-3: Bacteria - legionella pneumophila.
Source: European Hospital.

Viruses

Viruses are the smallest known type of infectious agent. They invade the cells of other organisms, which they take over and where they make copies of themselves, and while not all cause disease many of them do. Examples of viruses are hepatitis, which can cause liver damage, and the human immunodeficiency virus (HIV), which causes acquired immune deficiency syndrome (AIDS).

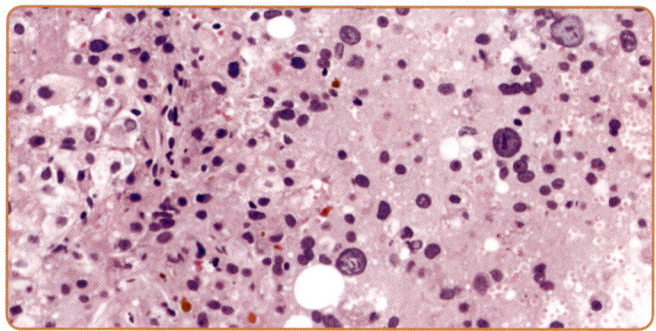

Figure 7-4: Virus - hepatitis C.
Source: National Public Radio.

DIFFERENCE BETWEEN ACUTE AND CHRONIC HEALTH EFFECTS

The effect of a substance on the body depends not only on the substance, but also on the dose and the susceptibility of the individual. No substance can be considered non-toxic; there are only differences in degree of effect.

Acute effect

An acute effect is an immediate or rapidly produced adverse effect following a single or short-term exposure to an offending agent, which is usually reversible (the obvious exception being death). Examples of acute effects are those from exposure to solvents, which affects the central nervous system, causing dizziness and lack of coordination, or carbon monoxide, which affects the level of oxygen in the blood, causing fainting.

Chronic effect

A chronic effect is an adverse health effect produced as a result of prolonged or repeated exposure to an agent. The gradual or latent effect develops over time and is often irreversible. The effect may go unrecognised for a number of years. Examples of chronic effects are lead or mercury poisoning, cancer and asthma.

Other common terms used in the context of occupational health are:

Toxicology

The study of the body's responses to substances. In order to interpret toxicological data and information, the meaning of the following terms should be understood.

Toxicity

The ability of a chemical substance to produce injury once it reaches a susceptible site in or on the body. A poisonous substance (for example, organic lead) causes harm to biological systems and interferes with the normal functions of the body. The effects may be acute or chronic, local or systemic.

Dose

The level of environmental contamination multiplied by the length of time (duration) of exposure to the contaminant.

Local effect

Usually confined to the initial point of contact. Possible sites affected include the skin, mucous membranes or the eyes, nose or throat. Examples are burns to the skin by corrosive substances (acids and alkalis), or dermatitis caused by solvents.

Systemic effect

Occurs in parts of the body other than at the point of initial contact. Frequently the circulatory system provides a means to distribute the substance round the body to a target organ/system.

Target organs

An organ of the human body on which a specified toxic material exerts its effects, for example, lungs, liver, brain, skin, bladder or eyes.

Target systems

Central nervous system, circulatory system, respiratory system, and reproductive system.

Examples of substances that have a systemic effect and their target organs/systems are:

- *Alcohol:* central nervous system, liver.
- *Lead:* bone marrow and brain.
- *Mercury:* central nervous system.

It must be noted that many chemicals in use today can have both an acute and chronic effect. A simple example is lead. Symptoms of acute lead toxicity include upset stomach (gastrointestinal problems), dullness, restlessness, irritability, poor attention span, headaches, kidney damage, hypertension and hallucinations. Health effects of chronic exposure to lead are blood disorder effects, such as anaemia, or neurological disturbances, including headache, irritability, lethargy, convulsions, muscle weakness, tremors and paralysis. Chronic lead exposure also causes cardiovascular and renal toxicity. In children, lead exposure may lead to cognitive deficits, such as a decrease in intelligence quotient (IQ). Chronic exposure to lead can cause adverse effects on both male and female reproductive functions.

HEALTH HAZARD CLASSIFICATIONS

UN Globally Harmonized System of Classification and Labelling of Chemicals

The UN has established a non-legally binding international agreement called 'Globally Harmonized System of Classification and Labelling of Chemicals (GHS)'. It has been widely accepted globally and is being established in national legislation of the countries that are adopting it. Within GHS is a classification of chemicals based on their effect on human health. Criteria for classifying chemicals have been developed for the following health hazard classes:

- Acute toxicity.
- Skin corrosion/irritation.
- Serious eye damage/eye irritation.
- Respiratory or skin sensitization.
- Germ cell mutagenicity.
- Carcinogenicity.
- Reproductive toxicity.
- Specific target organ toxicity - single exposure.
- Specific target organ toxicity - repeated exposure.
- Aspiration hazard.

A substance may be classified under more than one category, although those with specific target organ toxicity only apply if other classifications do not.

Acute toxicity

"Acute toxicity refers to those adverse effects occurring following oral or dermal administration of a single dose of a substance, or multiple doses given within 24 hours, or an inhalation exposure of 4 hours." Source: GHS.

The classification is divided into five categories related to the level of toxicity of the chemical. The subdivisions arise from the dose of the substance that would prove lethal to 50% of the population exposed to that substance (hence LD_{50}). The classification of categories and labelling for acute skin and inhalation toxicity are the same as for oral exposure, except for slightly different hazard statements, *(see Figure 7-5)*.

An example of a substance with acute oral toxicity is arsenic, which is a systemic poison. Another example is methanol, which is known to cause lethal intoxications in humans (mostly via ingestion) in relatively low doses and is a category 3 substance for acute oral toxicity.

Acute oral toxicity - Annex 1 of GHS					
	Category 1	Category 2	Category 3	Category 4	Category 5
LD_{50}	≤ 5 mg/kg	>5 <50 mg/kg	>50 <300 mg/kg	>300 <2,000 mg/kg	>2000 <5,000 mg/kg
Pictogram	 ☠	 ☠	 ☠	 !	No symbol
Signal word	Danger	Danger	Danger	Warning	Warning
Hazard statement	Fatal if swallowed	Fatal if swallowed	Toxic if swallowed	Harmful if swallowed	May be harmful if swallowed

Figure 7-5: Acute oral toxicity. Source: Annex 1 of GHS.

Skin corrosion/irritation - Annex 1 of GHS					
	Category 1A	Category 1B	Category 1C	Category 2	Category 3
Pictogram					No symbol
Signal word	Danger	Danger	Danger	Warning	Warning
Hazard statement	Causes severe skin burns and eye damage	Causes severe skin burns and eye damage	Causes severe skin burns and eye damage	Causes skin irritation	Causes mild skin irritation

Figure 7-6: Skin corrosion/irritation. Source: Annex 1 of GHS.

Skin corrosion/irritation

"Skin corrosion is the production of irreversible damage to the skin; namely, visible dead skin cells through the epidermis (outermost skin cells) and into dermis (the middle layer of skin cells), following the application of a test substance for 4 hours. Corrosive reactions are typified by ulcers, bleeding, bloody scabs, and, by the end of observation at 14 days, by discolouration due to blanching of the skin, complete areas of alopecia, and scars." Source: GHS.

The corrosion effect is divided into three categories related to duration of exposure necessary to create an effect, *(see Figure 7-6)*.

"Skin irritation is the production of reversible damage to the skin following the application of a test substance for up to 4 hours." Source: GHS.

Sulphuric (battery) acid and sodium hydroxide (caustic soda) are examples of substances classified under skin corrosion/irritation.

Serious eye damage/eye irritation

"Serious eye damage is the production of tissue damage in the eye, or serious physical decay of vision, following application of a test substance to the anterior (outer) surface of the eye, which is not fully reversible within 21 days of application."

"Eye irritation is the production of changes in the eye following the application of a test substance to the anterior surface of the eye, which are fully reversible within 21 days of application." Source: GHS.

This generally relates to effects on the cornea, iris or conjunctiva, *(see Figure 7-7)*.

The difference between category 1 and 2 for eye injury is whether the harm to the eye is fully reversible within the observation period. A category 2B substance is one where it is considered to be mildly irritating to eyes and fully reversible within 7 days of observation.

Serious eye injury/eye irritation - Annex 1 of GHS			
	Category 1	Category 2A	Category 2B
Pictogram			No symbol
Signal word	Danger	Warning	Warning
Hazard statement	Causes severe eye damage	Causes severe eye irritation	Causes eye irritation

Figure 7-7: Serious eye injury/eye irritation. Source: Annex 1 of GHS.

Respiratory sensitisation - Annex 1 of GHS			
	Category 1	**Category 1A**	**Category 1B**
Pictogram			
Signal word	Danger	Danger	Danger
Hazard statement	May cause allergy or asthma symptoms or breathing difficulty if inhaled	May cause allergy or asthma symptoms or breathing difficulty if inhaled	May cause allergy or asthma symptoms or breathing difficulty if inhaled

Figure 7-8: Respiratory sensitisation. Source: Annex 1 of GHS.

Respiratory or skin sensitisation

"A respiratory sensitiser is a substance that will lead to hypersensitivity of the airways following inhalation of the substance'. 'A skin sensitiser is a substance that will lead to an allergic response following skin contact."
Source: GHS.

Sensitisation takes place in two stages. The first is the recognition stage where, following contact with the airways or skin, the substance is recognised by the body as a pathogen (something that can cause harm to health). The second stage is the antibody generation and allergic response to further exposure to the substance. The three categories of respiratory sensitisers relate to the type and level of evidence that identifies the substance

as a sensitiser. Category 1 has been established by direct evidence that exposure will lead to specific hyper-sensitivity, for example, asthma, rhinitis/conjunctivitis and alveolitis (inflammation of the alveoli), whereas category 1B substances show a low to moderate frequency of occurrence of sensitisation. Flour dust and iso-cyanates are examples of respiratory sensitisers.

As with respiratory sensitisers, the three categories for skin sensitisers relate to the type and level of evidence that identifies the substance as a sensitiser. Category 1 substances are ones where there is strong documented evidence of causing allergic contact dermatitis. Nickel and epoxy resins are examples of skin sensitisers, **(see Figures 7-8 and 7-9)**.

Skin sensitisation - Annex 1 of GHS			
	Category 1	**Category 1A**	**Category 1B**
Pictogram			
Signal word	Warning	Warning	Warning
Hazard statement	May cause an allergic skin reaction	May cause an allergic skin reaction	May cause an allergic skin reaction

Figure 7-9: Skin sensitisation. Source: Annex 1 of GHS.

Germ cell mutagenicity - Annex 1 of GHS		
	Category 1A	Category 1B	Category 2
Pictogram			
Signal word	Danger	Danger	Warning
Hazard statement	May cause genetic defects	May cause genetic defects	Suspected of causing genetic defects

Figure 7-10: Germ cell mutagenicity. Source: Annex 1 of GHS.

Germ cell mutagenicity

"This hazard class is primarily concerned with chemicals that may cause mutations in germ cells of humans that can be transmitted to the descendants of a person." Source: GHS

The different categories of mutagenicity reflect the degree of knowledge about the chemical and the indications that it may cause mutagenicity. Category 1 substances are known or are presumed because of related evidence to induce heritable mutations, whereas category 2 substances are ones where there is concern that they may induce heritable mutations. If a specific route of exposure is proven to be the only route causing harm, this route must be stated in the hazard statement, *(see Figure 7-10)*.

Carcinogenicity

"The term carcinogen denotes a substance or mixture that induces cancer or increases its incidence. Substances and mixtures that have induced benign and malignant tumours in well-performed experimental studies on animals are considered also to be presumed or suspected human carcinogens unless there is strong evidence that the mechanism of tumour formation is not relevant to humans." Source: GHS.

As with the different categories of mutagenicity, the categories for carcinogenicity reflect the degree of knowledge about the chemical and the indications that it may cause cancer. Category 1 substances are known or are presumed because of related evidence to induce cancer, whereas category 2 substances are ones where there is concern that they may induce cancer. If a specific route of exposure is proven to be the only route causing harm, this route must be stated in the hazard statement. Benzyl chloride is an example of a carcinogenic substance. Another example is benzene, which affects bone marrow, causing leukaemia, *(see Figure 7-11)*.

Carcinogenicity - Annex 1 of GHS		
	Category 1A	Category 1B	Category 2
Pictogram			
Signal word	Danger	Danger	Warning
Hazard statement	May cause cancer	May cause cancer	Suspected of causing cancer

Figure 7-11: Carcinogenicity. Source: Annex 1 of GHS.

Reproductive toxicity - Annex 1 of GHS			
	Category 1A	Category 1B	Category 2	Additional category on effects on or via lactation
Pictogram				No pictogram
Signal word	Danger	Danger	Warning	No signal word
Hazard statement	May damage fertility or the unborn child	May damage fertility or the unborn child	Suspected of damaging fertility or the unborn child	May cause harm to breast-fed children

Figure 7-12: Reproductive toxicity. Source: Annex 1 of GHS.

Reproductive toxicity

"Reproductive toxicity includes effects on sexual function and fertility in adult males and females, as well as developmental toxicity in the offspring. The genetically based inheritable effects in offspring come under the classification 'germ cell mutagenicity'." Source: GHS.

The classification covers two main headings:

1) Adverse effects on sexual function and fertility - including alterations to the reproductive system, effects on the onset of puberty or the reproductive cycle, sexual behaviour, fertility and pregnancy outcomes.

2) Adverse effects on development of the offspring - including interference with the development of the foetus or child, before or after birth, resulting from exposure of either parent prior to conception or during development of the offspring.

The classification principally provides a warning for pregnant women and men and women of reproductive capacity. Category 1 substances are known or are presumed because of related evidence to be a human reproductive toxicant, whereas category 2 substances are ones where there is concern that they may be a human reproductive toxicant. If a specific route of exposure is proven to be the only route causing harm, this route must be stated in the hazard statement.

The UN GHS advises that 'for many substances there is no information on their potential to cause adverse effects on the offspring via lactation. However, substances that are absorbed by women and have been shown to interfere with lactation, or which may be present (including metabolites) in breast milk in amounts sufficient to cause concern for the health of a breast-fed child, should be classified to indicate this property hazardous to breast-

fed babies'. An example of a substance classified as reproductive toxic is 2-ethoxyethanol (a solvent used in commercial and industrial cleaning operations), which has been implicated in impairing fertility. In addition, the effect of lead on the development of the brain of an unborn foetus has long been established, *(see Figure 7-12)*.

Specific target organ toxicity - single exposure

This classification is for substances that produce specific, non-lethal target organ toxicity arising from a single exposure. The effects may be reversible or non-reversible, immediate or delayed, but are not covered in other classifications. Specific target organ toxicity can occur by any route and therefore includes oral, dermal and inhalation routes.

Category 1 substances are known to have produced significant toxicity in humans. Category 2 substances are presumed, because of related evidence, to produce significant toxicity in humans, whereas category 3 substances cause transient target organ effects that affect the respiratory tract or have a narcotic effect. Narcotic effects involve depression of the central nervous system, including drowsiness, loss of reflexes, lack of coordination, vertigo and reduced alertness.

The symptoms may include severe headache, nausea, dizziness, sleepiness, irritability, fatigue, impaired memory function, perception, coordination and reaction time. Some solvents causing narcosis or central nervous system failure may take effect after a single exposure, *(see Figure 7-13 on following page).*

Specific target organ toxicity - single exposure - Annex 1 of GHS				
	Category 1	Category 2	Category 3	Category 3
Pictogram				
Signal word	Danger	Warning	Warning	Warning
Hazard statement	Causes damage to organs	May cause damage to organs	(Respiratory tract irritation) May cause respiratory irritation	(Narcotic effects) May cause drowsiness or dizziness

Figure 7-13: Specific target organ toxicity - single exposure. Source: Annex 1 of GHS.

Specific target organ toxicity - repeated exposure

As with the similar classification for single exposure, this classification is for substances that produce specific, non-lethal target organ toxicity, but arising from repeated exposure. The effects may be reversible or non-reversible, immediate or delayed, but are not covered in other classifications. Specific target organ toxicity can occur by any route and therefore includes oral, dermal (via absorption or injection) and inhalation routes.

Category 1 substances are known to have produced significant toxicity in humans. Category 2 substances are presumed, because of related evidence, to produce significant toxicity in humans. If a specific route of exposure is proven to be the only route causing harm, this route must be stated in the hazard statement.

Similarly, if specific organs are known to be affected these organs must be stated, *(see Figure 7-14)*.

Aspiration hazard

"Aspiration means the entry of a liquid or solid chemical directly through the oral or nasal cavity, or indirectly from vomiting, into the trachea and lower respiratory system." Source: GHS.

Aspiration toxicity includes severe acute effects such as chemical pneumonia, varying degrees of pulmonary injury or death following aspiration.

Category 1 substances are known or are regarded to have caused aspiration toxicity in humans. Category 2 substances are those that cause concern of aspiration toxicity in humans, *(see Figure 7-15)*.

Specific target organ toxicity - repeat exposure - Annex 1 of GHS		
	Category 1	Category 2
Pictogram		
Signal word	Danger	Warning
Hazard statement	Causes damage to organs through prolonged or repeated exposure	May cause damage to organs through prolonged or repeated exposure

Figure 7-14: Specific target organ toxicity - repeat exposure. Source: Annex 1 of GHS.

Element 7

Aspiration hazard - Annex 1 of GHS		
	Category 1	Category 2
Pictogram	 	 
Signal word	Danger	Warning
Hazard statement	May be fatal if swallowed and enters airways	May be harmful if swallowed and enters airways

Figure 7-15: Aspiration hazard. Source: Annex 1 of GHS.

REVIEW

What is meant by the terms 'acute' and 'chronic' health effects?

Explain the term 'aspiration hazard'.

Explain the terms 'target organ' and 'target system' and give an example of each.

7.2 Assessment of health risks

ROUTES OF ENTRY OF HAZARDOUS SUBSTANCES INTO THE BODY

Inhalation

The most significant industrial entry route is inhalation. It has been estimated that at least 90% of industrial poisons are absorbed through the lungs.

Harmful substances can directly attack the lung tissue, causing a local effect, or may pass through to the blood system, to be carried round the body and affect target organs such as the liver. The effects of substances that enter the body through inhalation may be local or systemic.

Local effect

A local effect is where the hazardous substance has an effect where it first contacts the body. For example, silicosis, caused by inhalation of silica dust – where dust causes scarring of the lung, causing inelastic fibrous tissue to develop and reducing lung capacity

Systemic effect

A systemic effect is where the hazardous substance has an effect on a different site from where it first contacted the body. For example, anoxia, caused by the inhalation of carbon monoxide - the carbon dioxide replaces oxygen in the bloodstream, affecting the nervous system.

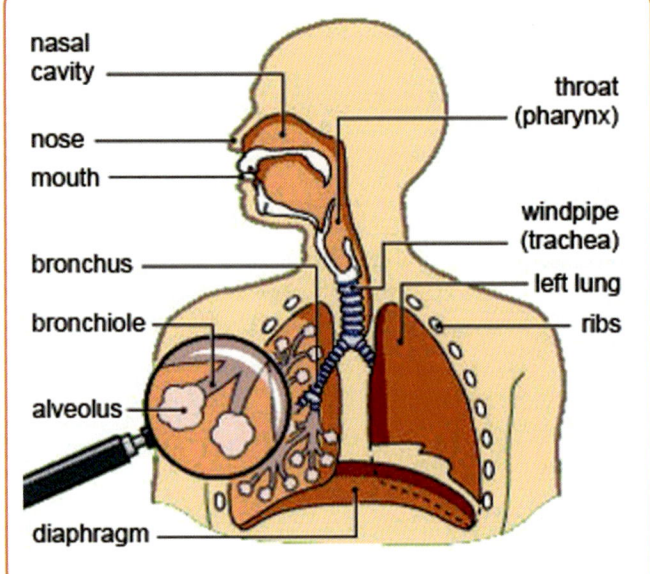

Figure 7-16: Respiratory system.
Source: BBC.

Ingestion

The ingestion route normally presents the least problem as it is unlikely that any significant quantity of harmful liquid or solid will be swallowed without deliberate intent.

However, accidents/incidents will occur where small amounts of contaminant are transferred from the fingers to the mouth if eating, drinking or smoking is allowed in chemical areas or where a substance has been decanted into a container normally used for drinking.

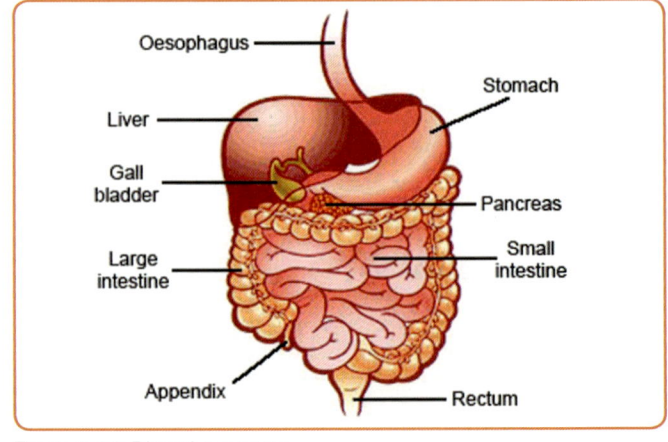

Figure 7-17: Digestive system.
Source: STEM.

The sense of taste will often be a defence if chemicals are taken in through this route, causing the person to spit it out.

If the substance is taken in, vomiting and/or excretion may mean the substance does not cause a systemic problem, though a direct effect, for example, ingestion of an acid, may destroy cells in the mouth, oesophagus or stomach.

Where hazardous substances are ingested, they may pass into the digestive system and be absorbed through the intestine to the blood system and may cause harm in another part of the body.

Absorption (skin contact)

Substances can enter through the skin, via cuts or abrasions and through the conjunctiva of the eye; this is called absorption. Solvents such as organic solvents, for example, toluene and trichloroethylene, can enter due to accidental exposure or if they are used for washing.

The substance may have a local effect, such as de-fatting of the skin, resulting in inflammation and cracking of the horny layer, or pass through into the blood system, causing damage to the brain, bone marrow and liver.

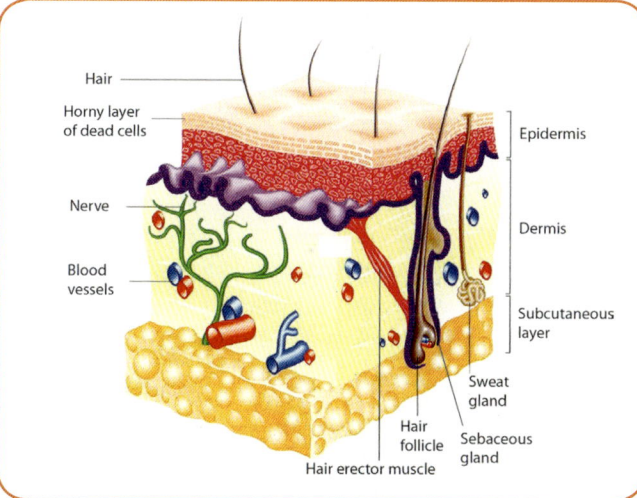

Figure 7-18: Skin layer.
Source: SHP.

Dermatitis

Dermatitis is caused by exposure to substances that interfere with normal skin physiology, leading to inflammation of the skin, usually on the hands, wrists and forearms. The skin turns red and, in some cases, may be itchy. Small blisters may occur and the condition may take the form of dry and cracked skin.

Contact dermatitis (irritant contact dermatitis)

If a person's skin is frequently in contact with some substances or is exposed for a long duration, such persistent contact can lead to irritation and then dermatitis. There are many chemicals used in the workplace that may irritate the skin, leading to this condition, including cement, soaps, detergents, epoxy resins and hardeners, acrylic sealants, bitumen and solvents used in paints or glues. Removal from contact with the substance usually allows normal cell repair. A similar level of repeat exposure results in the same response. This class of dermatitis is called contact dermatitis.

Sensitisation dermatitis

A second form of dermatitis is called sensitisation dermatitis. In cases of sensitisation dermatitis a person exposed to the substance develops dermatitis in the usual way. When removed from exposure to the substance, the dermatitis usually repairs, but the body gets ready for later exposures by preparing its defence mechanisms. A subsequent small exposure is enough to cause a major response from the immune system.

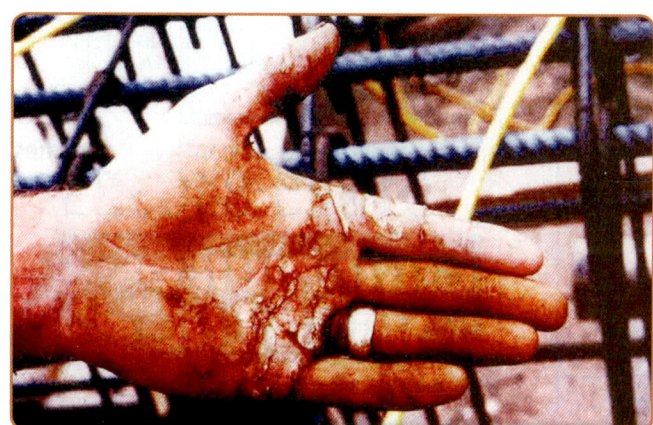

Figure 7-19: Dermatitis.
Source: SHP.

The person will have become sensitised and will no longer be able to tolerate even small exposures to the substance without a reaction occurring.

Although the range of substances an individual may become sensitised to varies, some substances have the tendency to cause sensitisation in a large number of people. For example, cement may contain two well-known sensitisers, chromate and cobalt. Epoxy resins also have a tendency to cause sensitisation.

Dermatitis can be prevented by:

- Pre-employment screening for sensitive individuals.
- Careful attention to skin hygiene principles.
- Clean working conditions and properly planned work systems.
- Use of protective equipment.
- Prompt attention to cuts, abrasions and spillages onto the skin.
- Avoid the overuse of hand cleaners containing abrasives such as ground pumice stone.
- Correct selection and use of skin barrier cream, used to repel water, oil or solvents from exposed hands.

CASE STUDY

Workers at a company premises in Bristol, UK, were exposed to hazardous chemicals over a four-year period, leading to the onset of a disease called 'allergic contact dermatitis'. One employee suffered four years of his skin blistering, cracking, splitting and weeping because of this allergic dermatitis.

Two other employees also suffered the symptoms of allergic dermatitis, including fingers and hands becoming so badly swollen and blistered that one could not do up his shirt buttons without his fingers splitting open. All three employees had been working with photographic chemicals.

The company was fined a total of £100,000 and ordered to pay £30,000 costs. It was fined £30,000 for breaching the UK Health and Safety at Work Act 1974, and £10,000 for 6 separate breaches of the UK Control of Substances Hazardous to Health (COSHH) Regulations for not making adequate risk assessments, not preventing or controlling exposure of employees to chemicals, and for not providing any 'health surveillance' of employees at risk. It was also fined £10,000 for not reporting a case of allergic contact dermatitis to the UK HSE.

Injection

Injection is a forceful breach of the skin, perhaps as a result of injury, which can carry harmful substances through the skin barrier. For example, handling broken glass that cuts the skin and transfers a biological or chemical agent. On construction sites there are many items that present a hazard of penetration, such as nails in broken-up timber structures that might be trodden on and penetrate the foot, presenting a risk of infection from tetanus.

In addition, some land or buildings being worked on may have been used by intravenous drug users and their needles may present a risk of injection of a virus such as hepatitis. The forced injection of an agent into the body provides an easy route past the skin, which usually acts as the body's defence mechanism and protects people from the effects of many agents that do not have the ability to penetrate.

When identifying possible routes of entry, it must be remembered that many substances have multiple possibilities. For example, trichloroethylene, which is a known carcinogen, will absorb through the skin. However, when in use, it also gives off very harmful vapours that may be inhaled and, because it is liquid, there is the possibility of accidental ingestion.

THE BODY'S DEFENCE MECHANISMS

The body's response against the invasion of substances likely to cause damage can be divided into external or superficial defences and internal or cellular defences.

Superficial defence mechanisms

Respiratory (inhalation):

NOSE

On inhalation, many substances and minor organisms are successfully trapped by nasal hairs, for example, larger wood and cement dust particles. Gases such as carbon monoxide are inhaled through the nose and easily defeat the defence mechanisms, the most superficial of which is smell. Carbon monoxide is highly poisonous to humans and has no smell. Vehicle exhausts are a common source of carbon monoxide in many countries.

RESPIRATORY TRACT

The next line of defence against inhalation or substances harmful to health begins here, where a series of reflexes activate the coughing and sneezing mechanisms to forcibly expel the triggering substances. Silica and cement dust generated during construction activities may be trapped in the respiratory tract and ejected by coughing and sneezing.

CILIARY ESCALATOR

The passages of the respiratory system are also lined with mucus and well supplied with fine hair cells which sweep rhythmically towards the outside and pass along large particles.

The respiratory system narrows as it enters the lungs where the ciliary escalator assumes more and more importance as the effective defence. Smaller particles of agents, such as some silica particles, are dealt with at this stage. The smallest particles, such as organic solvent vapours and asbestos, reach the alveoli and are either deposited or exhaled. Legionella bacteria that are inhaled can also defeat the ciliary escalator and reach the alveoli. Hardwood dusts (oak, mahogany, etc.), which are carcinogenic, if inhaled during cutting or machining operations, are usually trapped by the mucus in the escalator.

Gastrointestinal (ingestion):

MOUTH

The mouth is used for the ingestion of substances in general. Saliva in the mouth provides a useful defence against hazardous substances that are not excessively acid or alkaline or present in large quantities.

GASTROINTESTINAL TRACT

Acid in the stomach also provides a useful defence similar to saliva. Vomiting and diarrhoea are additional reflex mechanisms which act to remove substances or quantities that the body is not equipped to deal with.

Skin (absorption):

SKIN

The body's largest organ provides a useful barrier against the absorption of many foreign organisms and chemicals (but not against all of them). Its effect is, however, limited by its physical characteristics.

The outer part of the skin is covered in an oily layer and substances have to overcome this before they can damage the skin or enter the body. The outer part of the epidermis is made up of dead skin cells. These are readily sacrificed to substances without harm to the newer cells underneath. Repeated or prolonged exposure could defeat this. When attacked by substances, the skin may blister in order to protect the layers beneath. Openings in the skin such as sweat pores, hair follicles and cuts can allow entry, and the skin itself may be permeable to some chemicals, for example, toluene.

Workers in the sewage industry or agriculture (animals) can come into contact with Leptospira bacteria, which can also pass through breaks in the skin. Other blood-borne viruses, such as Hepatitis, may defeat the skin by being injected into the bloodstream. Occupations at risk include those in health-care.

Cellular mechanisms:

The cells of the body possess their own defence systems.

SCAVENGING ACTION

A type of white blood cell called a macrophage attacks invading particles in order to destroy them and remove them from the body. This process is known as phagocytosis.

SECRETION OF DEFENSIVE SUBSTANCES

This is done by some specialised cells.

Adrenaline is a hormone produced by the adrenal glands during high stress or exciting situations. This powerful hormone is part of the human body's acute stress response system, also called the 'fight or flight' response. It works by stimulating the heart rate, contracting blood vessels, and dilating air passages, all of which work to increase blood flow to the muscles and oxygen to the lungs.

Histamine is a chemical that is released when the body is exposed to an allergen. Allergens may include airborne allergens (such as pollen and dust mites), certain foods (such as peanuts and shellfish) or insect venom. Histamine is released in an effort to protect the body from an allergen; however, sometimes an overload of histamine can result in life-threatening symptoms.

PREVENTION OF EXCESSIVE BLOOD LOSS

Reduced circulation through blood clotting and coagulation prevents excessive bleeding and slows or prevents the entry of bacteria.

Heparin is an anticoagulant (blood thinner) that prevents the formation of blood clots and is produced naturally in the lungs and liver.

REPAIR OF DAMAGED TISSUES

This is a necessary defence mechanism that includes removal of dead cells, increased availability of defender cells and replacement of tissue strength, for example, scar tissue caused by silica.

THE LYMPHATIC SYSTEM

Acts as a form of 'drainage system' throughout the body for the removal of foreign substances. Lymphatic glands or nodes at specific points in the system act as selective filters, preventing infection from entering the blood system. In many cases, a localised inflammation occurs in the node at this time.

Other practical measures to complement the body's protection mechanisms

Practical measures include:

- Maintain good personal hygiene.
- Do not apply cosmetics in the workplace.
- Do not allow eating or drinking in the workplace.
- Provide proper containers/storage for food and drink.
- Provide and ensure use of appropriate personal protective equipment.
- Take care when removing contaminated protective clothing.

WHAT NEEDS TO BE TAKEN INTO ACCOUNT WHEN ASSESSING HEALTH RISKS

In order to assess the health risks, it is necessary to consider the following:

- The form of the substance, for example, solid, liquid, dust or gas. The form of the substance directly influences the way that the substance will enter the body. Solid objects, such as concrete or metal, cannot enter the body in their solid state. But if the concrete is cut with a disc cutter, dust is generated, which if not controlled may be inhaled. If the steel contains chromium, when it is machined or welded, particles which become airborne in the form of dust or fume can be inhaled and prove toxic.

- The classification of the hazard, for example, toxic, corrosive or sensitiser. If the substance is toxic, small quantities inhaled, ingested or absorbed through the skin may have a catastrophic effect on the well-being of the individual who is exposed. Sensitisers may cause allergic dermatitis when the employee comes into contact with the substance, or respiratory sensitisers may cause breathing difficulties when inhaled.

- How much of the substance will be present and its concentration. Ammonia occurs naturally in the environment and humans are regularly exposed to low levels of ammonia in air, soil and water. Ammonia exists naturally in the air at levels between 1 and 5 parts in a billion parts of air (ppb). Ingesting liquid ammonia at 35ppm, for example, from the use of a typical household cleaner, may cause burns to the mouth and throat. Exposure to higher concentrations of liquid ammonia in the eyes causes severe chemical eye burns and can lead to blindness.

- The routes of entry into the body, for example, inhalation, ingestion or skin absorption. Some substances cause harm immediately on contact with the skin, for example, acids. Other substances may be present as a gas, mist or fume and be in the air. Humans have to breathe so airborne substances are likely to be inhaled as a route of entry into the body.

- Whether the substance has an acute or chronic effect or both. Acute effects can often warn the person that they have come into contact with a hazardous substance and a reaction occurs almost instantly. This allows the person to take avoidance measures. Lower-level concentrations of substances, however, may not cause significant harm initially, but the damage builds up over time. Substances such as organic solvents can cause skin or eye irritation quickly (acute) and have longer-term exposure consequences (chronic), where damage to the liver and kidneys can occur.

- The extent to which the body's defences will deal with the substance. The body's defence systems try to prevent harm but also help the body to heal itself when it gets injured or sick. However, hazards arising from bacteria, viruses, chemicals, dusts, vapours, extreme temperatures or work processes can break down (weaken) the body's defence systems. In some instances, the body's defence mechanisms can be defeated and illness or death can result.

- The first signs of damage or ill-health. The body has nerve receptors located all over its surface, so often the first sign of damage will be the sensation of pain. Visual signs of damage may be blood from a cut or break in the skin. When a harmful substance has been ingested or inhaled the first signs are often vomiting or diarrhoea or restrictions in breathing.

- The vulnerability of the people involved in the process, for example young persons, pregnant workers, anyone who has existing health problems, such as skin problems or bronchitis. Anyone with a pre-existing medical condition or with underdeveloped defence systems will be at higher risk than workers who are fit and well.

- The effectiveness of existing control measures. Preventing exposure to hazardous substances is the most effective way of controlling health risks. Where exposure cannot be prevented, measures should be taken to eliminate or reduce the risk of harm. The effectiveness with which control measures have been applied will greatly affect the degree of exposure of the employee.

Considering all these issues will help the assessor decide whether the risks to health are tolerable or acceptable or whether further controls are needed.

SOURCES OF INFORMATION

Product labels

Hazardous chemicals have to be labelled so that workers are informed about their effects when they are exposed to them. The label should draw attention to the inherent hazards to those handling or using the chemical and provide information on precautions to prevent harm.

European regulation on classification, labelling and packaging of chemical substances

The European Union (EU) Regulation on Classification, Labelling and Packaging of Chemical Substances and Mixtures (CLP) introduced throughout the EU a system for classifying and labelling chemicals based on the United Nations' Globally Harmonised System (GHS). The CLP regulation is concerned with the hazards of chemical substances and mixtures and how to inform others about them. Its purpose is to protect people and the environment from the effects of those chemicals by requiring suppliers to classify and provide information about the hazards of the chemicals and to package them safely.

It is the responsibility of manufacturers to establish what the hazards of substances and mixtures are before they are placed on the market, and to classify them in line with the identified hazards. When a substance or a mixture is hazardous, it has to be labelled in accordance with CLP so that workers are informed about its effects if it is used. Note that 'mixture' is the same as the term 'preparation', which has been used previously.

All substances available for use in the workplace should be labelled in accordance with CLP, including the use of the appropriate CLP hazard pictogram. Information on the use of symbols to indicate the harm related to various dangerous substance hazard classifications is provided earlier in this element *(see also 'Health hazard classifications').*

The CLP regulation requires hazardous substances to carry a hazard label in a specified format, which is made up of specific symbols (known as 'pictograms'), warnings and precautions. These pictograms and the wording that supports them are defined in the CLP regulation, which requires chemical suppliers to use them where hazardous properties have been identified. Article 17 of the CLP Regulation requires that hazard labels include the following elements:

- Name, address and telephone number of the supplier(s).

- The nominal quantity of the substance or mixture in the package where this is being made available to the general public, unless this quantity is specified elsewhere on the package.

- Product identifiers.

- Hazard pictograms, where applicable.

- The relevant signal word, where applicable.

- Hazard statements, where applicable.

- Appropriate precautionary statements, where applicable.

- A section for supplemental information, where applicable.

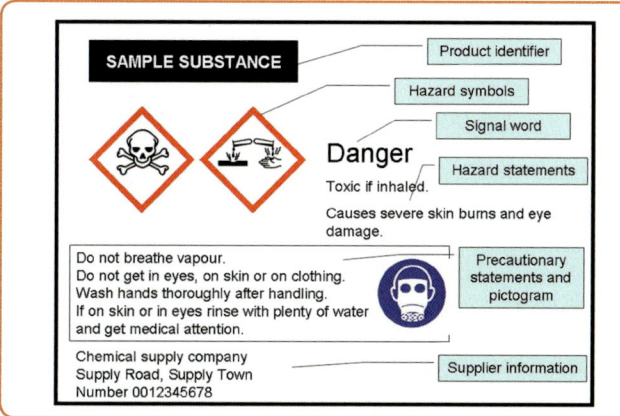

Figure 7-20: Sample GHS chemical labels.
Source: RMS.

A hazard *pictogram* is a graphical composition that includes a symbol plus other graphic elements, such as a border, background pattern or colour that is intended to convey specific information on the hazard concerned.

The pictogram forms an integral part of the label and gives an immediate idea of the types of hazards that the substance or mixture may cause.

A signal word indicates the relative level of severity of hazardous substances to alert the potential reader of the significance of the hazard. More severe hazards are identified by the signal word 'danger', while less severe hazards are identified by the signal word 'warning'.

A hazard statement is a phrase assigned to a hazard class and category that describes the nature and severity of the hazard.

A precautionary statement is a phrase and/or pictogram that describe recommended measure(s) to prevent or minimise adverse effects resulting from exposure to a hazardous substance or mixture due to its use. A maximum of six precautionary statements is allowed under CLP requirements. Precautionary statements cover statements for prevention, response to problems or contamination, storage and disposal.

HAZARD STATEMENTS	
H300	FATAL IF SWALLOWED
H320	CAUSES EYE IRRITATION
H336	MAY CAUSE DROWSINESS OR DIZZINESS
PRECAUTIONARY STATEMENTS	
P102	KEEP OUT OF REACH OF CHILDREN
P271	USE ONLY OUTDOORS OR IN WELL VENTILATED AREA
P331	DO **NOT** INDUCE VOMITING

Figure 7-21: Hazard and precautionary statements.
Source: EU, CLP Regulation, Annex III and IV.

Guidance documents

Guidance documents are used in many countries to set occupational exposure limit values *(see Section 7.3 - Occupational exposure limits)* for work activities. They are set by competent national authorities or other institutions acting nationally or internationally. Guidance documents set limits for concentrations of hazardous substances in workplace air.

Occupational exposure limit values for hazardous substances represent an important source of information for risk assessment and management of health risks. They provide valuable information for occupational health and safety activities involving hazardous substances and may carry legal status under national laws. As knowledge is gained about the health effects of substances, the occupational exposure limits are updated.

Therefore, it is important to obtain current data on occupational exposure limits. There are a number of guidance documents that are updated regularly and provide useful sources of information, including the EU list of indicative occupational exposure limit values. There is also guidance in the form of the UK Health and Safety Executive (HSE) list of workplace exposure limits and the USA American Conference of Governmental Industrial Hygienists' (ACGIH) list of threshold limit values.

Occupational exposure limits related to a large number of chemicals are set out in the UK HSE Guidance Note EH40, which is prepared and published annually. The Guidance Note EH40 contains lists of occupational exposure limits, called workplace exposure limits (WELs) in the document, for use with the UK Control of Substances Hazardous to Health Regulations (COSHH) 2002. It also contains a description of the limit-setting process, technical definitions and explanatory notes.

In the EU, the list of indicative occupational exposure limit values (IOELVs), also commonly referred to as the List of Indicative Limit Values, contains human exposure limits to hazardous substances specified by the Council of the European Union, based on expert research and advice. They are not binding on member states but must be taken into consideration in setting national occupational exposure limits. The EU IOELVs are taken into account when the UK WELs are established. EH40 may contain WELs for substances that do not have an EU indicative occupational exposure limit value, because data may be established in the UK that enables a limit to be set before there is agreement in the EU.

The USA ACGIH list of occupational exposure limits, known as threshold limit values (TLV), are established as advisory limit levels and have no direct link to legal requirements. These advisory limits are set using the knowledge and experience of this group of people, based solely on health factors. However, as they do not take account of economic or feasibility factors they may not be easily applied to all workplaces.

Safety data sheets

GHS establishes a requirement to prepare safety data sheets (SDS) for chemicals that constitute a health hazard. This requirement has been adopted globally by many countries. For example, in the EU, the regulation known as REACH (Registration, Evaluation, Authorisation and Restriction of Chemicals) is the system for controlling chemicals in EU member states, including the preparation of SDSs.

SDSs established in accordance with the GHS require the following sixteen headings to be included:

1) Identification of the substance/mixture and of the company/undertaking, including the product identifier shown on the label and its recommended use.

2) Hazards identification, including signal words, hazard statements and precautionary statements - the pictogram of hazard symbols may be provided.

3) Composition/information on ingredients.

4) First-aid measures, including the urgency with which it may be needed and symptoms of any effects.

5) Firefighting measures, including suitable media and hazards of combustion.

6) Accidental release measures, including spills containment and recovery.

7) Handling and storage, including any incompatibilities.

8) Exposure controls/personal protection, including exposure limit values and requirements for such things as local exhaust ventilation and specific personal protective equipment (PPE).

9) Physical and chemical properties, such as odour, flash point, vapour pressure.

10) Stability and reactivity, including conditions to avoid, for example, heat or incompatible materials.

11) Toxicological information, including the health hazards, for example, carcinogenicity, route of entry and symptoms.

12) Ecological information, in order to evaluate the environmental impact of dealing with spills and arrangements for waste.

13) Disposal considerations, including methods and properties that may affect landfill or incineration.

14) Transport information, including UN number and transport hazards.

15) Regulatory information, including relevant national information on the regulatory status of the chemical where it is being supplied.

16) Other information, including the date the SDS was prepared and changes from previous versions.

Section 5 of the ILO Code for 'Safety in the Use of Chemicals at Work' sets out a requirement that:

'The supplier should provide an employer with essential information about hazardous chemicals in the form of a chemical safety data sheet. The information should be given in the official language of the country in which the employer is located or in another language, agreed to in writing by the employer'.

The Code also requires that suppliers ensure that any revisions to chemical SDSs for hazardous chemicals are prepared and are provided to employers. The

GHS requirement for SDSs only relates to occupational situations.

In general, SDS information is not required or provided for:

- The offer or sale of dangerous substances or mixtures to the general public, provided sufficient information is given to enable users to take the necessary measures as regards safety, protection of human health and the environment.
- If the substances/mixtures are not classified as dangerous.
- Certain products intended for the final user, for example, medicinal products or cosmetics.

LIMITATIONS OF INFORMATION USED WHEN ASSESSING RISKS TO HEALTH

Information provided by manufacturers and suppliers of hazardous chemicals and that contained within occupational exposure limit sources may be very technical and require a specialist to explain its relevance to a given activity. Some substances have good toxicological information, usually gained from past experience of harm. Many others have a limited amount of useful toxicological information available to guide us as to the harm they may produce. This can lead to a reliance on data that is only the best understanding at the time and this may have to be revised as knowledge of the substance changes.

Individual susceptibility of workers differs by, for example, age, gender or ethnic origin, and this can limit the value of information sourced from manufacturers and guidance documents related to occupational exposure limits.

Exposure history varies over the working life of an individual and current exposure may not indicate that the individual may suffer due to a cumulative effect from earlier exposures. For example, an individual may have been or be engaged in a number of processes within a variety of workplaces or personal pastimes. These limitations are reflected in the use of occupational exposure limits, where the limits are considered to be a maximum and control measures may be applied to control assessed risks to well below this limit.

ROLE AND LIMITATIONS OF HAZARDOUS SUBSTANCE MONITORING

Role of hazardous substance monitoring

The role of hazardous substance monitoring is to determine the level of likely exposure of workers to substances in order to establish the likely effects on the worker. Hazardous substance monitoring can help to identify and assess health risks in the workplace. It can be used to determine compliance with national or other worker exposure limits and determine what controls are required

to remain within the limits. Monitoring is required to ensure that exposure has not exceeded a control limit. Monitoring also helps to identify deterioration in control measures that may result in harm if exposure is not adequately controlled. Monitoring can establish if current controls are adequate to limit exposure and assist in choosing appropriate PPE. Monitoring can provide information on patterns of exposure and levels of risk that managers can use and provide to workers. It can also indicate the need for health surveillance of groups and individual workers.

Limitations of hazardous substance monitoring

As described previously, the health effects of exposure to hazardous substances can be acute or chronic. It is therefore necessary to use appropriate methods of measurement to distinguish these effects. It is also important to understand the limitation of any hazardous substance monitoring method used, for example, the risk of cross-contamination of similar substances and the fact that general workplace monitoring may not represent specific worker exposure. Hazardous substance monitoring of the presence of a chemical in the air may not represent the worker's complete exposure as, for example, there may be additional exposure to the chemical by skin contact or orally through poor hygiene arrangements.

One of the main limitations of hazardous substance monitoring is the competence of the person conducting it. It relies on that person conducting monitoring at the time that represents real exposure of workers and using methods that will give reliable measurements. When carrying out monitoring related to dusts, it is important to discern the amount of dust that can penetrate the airway and cause harm. If the person conducting the monitoring does not understand the difference between 'total inhalable dust' and 'respirable dust' there is a risk that unsuitable measurements may be taken.

'Total inhalable dust' is the amount of airborne material that enters the nose and mouth during breathing and is available for deposition in the body. 'Respirable dust' is the amount of airborne material that penetrates to the gas exchange region of the lung.

General approach to hazardous substance monitoring

When embarking upon a monitoring campaign to assess the hazardous risk to the workforce, several factors need to be considered, the substance(s), who might be affected, the sample period and the method of sampling.

The substance

Considering the substance involves a review of the materials, processes and operating procedures being used within a process, coupled with discussions with management, workers and health and safety practitioners. A brief 'walk-through' survey can also be useful as a guide to the extent of monitoring that may be necessary. Health and safety data sheets are also of use. When the background work has been completed it can then be decided what is to be measured.

Who might be affected

Who might be affected depends on the size and diversity of the group that the hazardous substance monitoring relates to. From the group of workers being considered, the sample of workers to be monitored should be selected; this must be representative of the group and the work undertaken. Selecting the workers with the highest exposure can be a reasonable starting point. If the group is large then random sampling may have to be used, but care has to be exercised to ensure a representative sample is taken. The group should also be informed of the reason for the monitoring to ensure their full co-operation.

Sample period

Where monitoring is not continuous, the sample period selected should reflect the workers' exposure patterns. Factors to consider include the total exposure period and whether exposure is at the same level continuously. If so, a sample taken at any convenient time would be suitable. However, if exposure is occasional, the sample period must reflect this occasional exposure.

The duration of the sample period will also be influenced by whether the hazard presented by the substance is acute or chronic. Substances that present an acute hazard are usually sampled over a shorter sample period that reflects the short-term worker exposure limits set by international or national authorities, whereas substances that represent a chronic hazard may be sampled over a longer period to better reflect the long-term worker exposure limits.

The sample period may also be influenced by the type of equipment available, including whether it provides a continuous, long-term or a short-term measurement.

Method of monitoring

The particular measurement method will be based on the hazard presented by the substance, any limits in detection ability and what resources are available to measure the presence of the substance accurately. A typical approach is outlined in *Figure 7-22*.

Air samples should be collected in the worker's breathing zone. Personal dose measurement devices may conveniently be used for this purpose. Sampling should always be carried out during working hours when the activity involving hazardous substances is in operation. It may be necessary to repeat sampling at different times of the year to take account of seasonal effects on hazardous substances in the workplace.

In order to obtain indications of the distribution of contamination throughout the general atmosphere of the working area, samples should also be taken as close as possible to the sources of the hazardous substances, to indicate the standard of process control at source, and at various places in the working area, to indicate the overall effectiveness of control measures.

The minimum volume of the air sample to be taken for each analysis should be determined according to the sensitivity of the analytical method. The accuracy of the measuring method must be considered and noted with the results of the monitoring, as it can significantly affect results obtained from some monitoring equipment.

Measurements to determine	Suitable methods of measurement
Chronic hazard	Continuous personal dose measurement. Continuous measurements of average background levels. Short-term measurement of contaminant levels at selected positions and times.
Acute hazard	Continuous personal monitoring with rapid response. Continuous background monitoring with rapid response. Short-term measurement of background contaminant levels at selected positions and times.
Environmental control status	Continuous background monitoring. Short-term measurement of background contaminant levels at selected positions and times.
Whether area is safe to enter	Direct-reading instruments that provide an instantaneous measurement.

Figure 7-22: Sampling strategy. Source: RMS.

The ILO "Occupational Exposure to Airborne Substances Harmful to Health" suggests the time it should take to sample substances in the working environment, (see *Figure 7-23*, refer to **Section 7.3 - Occupational exposure limits)**.

> "The sampling of substances for which a ceiling limit exists, or the measurement of the concentration of such substances in the working environment, should normally last 15 minutes."

Figure 7-23: Sampling hazardous substances.
Source: ILO, Occupational exposure to airborne substances harmful to health.

BASIC MONITORING EQUIPMENT

Hazardous substance monitoring is often conducted by a specialist occupational hygienist, therefore detailed knowledge of monitoring equipment may not be required. However information on basic monitoring equipment is provided to give an understanding of how a hazardous monitoring strategy might be applied and outline the role and limitations of basic monitoring equipment.

Short-term samplers

Stain tube detectors (multi-gas/vapour)

A stain tube detector is a simple device for the measurement of contamination on a grab (short-term) sampling basis. It incorporates a glass detector tube filled with inert material. The material is impregnated with a chemical reagent that changes colour ('stains') in proportion to the quantity of contaminant as a known quantity of air is drawn through the tube.

There are several different manufacturers of detector tubes, including Dräger and Gastec. It is important that the literature provided with the pumps and tubes is followed. These provide a quick and easy way to detect the presence of a particular airborne contaminant. However, they possess inherent inaccuracies, and tube manufacturers claim a relative standard deviation of 20% or less (i.e. 1ppm in 5ppm). Types of tube construction include:

- The most common is the simple stain length tube, but it may contain filter layers, drying layers, or oxidation layers.
- Double tube or tube containing separate ampoules, avoids incompatibility or reaction during storage.
- Comparison tube.
- Narrow tube to achieve better resolution at low concentrations.

This list illustrates the main types of tube; however, there are many variations and the manufacturer's operating instructions must be read and fully understood before tubes are used.

There are four types of pump in general use:

1) Bellows pump.

2) Piston pump.

3) Ball pump.

4) Battery-operated pump.

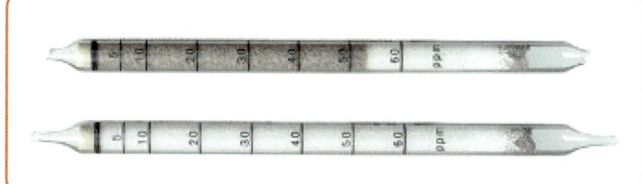

Figure 7-24: Dräger short-term detection tubes (before and after).
Source: Drager.

Pumps and tubes made by different manufacturers may not be compatible and they should not be mixed when taking measurements.

How to use tubes:

- Choose tube to measure material of interest and expected range.
- Check tubes are in date.
- Check leak tightness of pump.
- Read instructions to ensure there are no limitations due to temperature, pressure, humidity or interfering substances.
- Break off tips of tube, prepare tube if necessary and insert correctly into pump.
- Arrows normally indicate the direction of air flow.
- Draw the requisite number of strokes to cause the given quantity of air to pass through the tube.
- Immediately, unless operating instructions say otherwise, evaluate the amount of contaminant by examining the stain and comparing it against the graduations on the tube.
- Remove tube and discard according to instructions.
- Purge pump to remove any contaminants.

Figure 7-25: Gas detector pump.
Source: Drager.

Advantages

The advantages of short-term samplers are:

- Quick and easy to use.
- Instant reading without further analysis.
- Does not require much expertise to use.
- Relatively inexpensive.

Disadvantages

The disadvantages of short-term samples are:

- Tubes only measure a specific contaminant.
- Accuracy varies - some are only useful as an indication of the presence of contaminants.
- Is only a grab sample (taken at a single location point and may not represent the workplace as a whole).
- Relies on operator to accurately count pump strokes (manual versions).
- Only suitable for gases and vapours (not dusts).
- The tubes must be used within a short time period to prevent decay of the reagent and loss of reliability.
- The tubes are fragile and break easily.
- Used/expired tubes must be disposed of as chemical waste in accordance with local regulations.

Direct-reading dust sampler

A simple method of measuring dust is by direct observation of the effect of the dust on a strong beam of light, for example, using a dust lamp (Tyndall effect). High levels of small particles of dust show up under this strong beam of light. Other ways are by means of a direct-reading instrument. This establishes the level of dust by, for example, scattering of light. Some also collect the dust sample.

Advantages

The advantages of direct dust samplers are:

- Instant reading.
- Continuous monitoring.
- Can record electronically.
- Can be linked to an alarm.
- Suitable for clean-room environments.

Disadvantages

The disadvantages of direct dust samplers are:

- Some direct-reading instruments can be expensive.
- Do not usually differentiate between dusts of different types.
- Most effective on dusts of a spherical nature.

Long-term samplers

Personal long-term samplers

Passive personal samplers

Passive samplers are so described because they have no mechanism to draw in a sample of the contaminant but instead rely on passive means to sample. As such, they take a time to perform this function, for example, acting as an absorber taking in contaminant vapours over the period of a working day. Some passive samplers, like gas badges, are generally fitted to the lapel and change colour to indicate contamination.

Active personal samplers

Filtration devices are used for dusts, mists and fumes. A known volume of air is pumped through a sampling head and the contaminant filtered out. By comparing the quantity of air with the amount of contaminant a measurement is made. The filter is either weighed or an actual count of particles is done to establish the amount, as with asbestos. The type of dust can be determined by further laboratory analysis. Active samplers are used in two forms, for personal sampling and for static sampling.

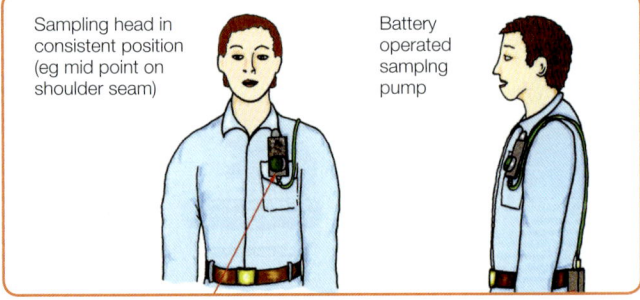

Figure 7-26: Personal sampling equipment.
Source: ROSPA OS&H.

Static long-term samplers

Static long-term sampling devices are placed in the working area. They sample continuously over the length of a shift, or a longer period if necessary. Mains or battery-operated pumps are used. Very small quantities of contaminant may be detected. The techniques employed include absorption, bubblers and filtration; they are similar in principle to personal samplers, but the equipment is designed for static use.

Advantages of static long-term samplers are:

- Will monitor the workplace over a long period of time.
- Will accurately identify 8 hour time weighted average.

Disadvantages of static long-term samplers are:

- Will not generally identify a specific type of contaminant.
- Will not identify multiple exposure i.e. more than one contaminant.
- Does not identify personal exposure.
- Unless very sophisticated, will not read variations in level of contamination during the measurement period.

Element 7

Smoke tubes

Smoke tubes are simple devices that generate a 'smoke' by means of a chemical reaction. A tube similar in type to those used in stain tube detectors is selected, its ends broken (which starts the chemical reaction) and it is inserted into a small hand bellows. By gently pumping the bellows smoke is emitted. Air flow can be studied by watching the smoke. This can be used to survey extraction and ventilation arrangements to determine their extent of influence.

REVIEW

Explain the term 'dermatitis' and how it may be prevented in the workplace.

What is the function of the ciliary escalator in the respiratory tract?

List eight headings that should be included on a safety data sheet supplied with a hazardous substance.

7.3 Occupational exposure limits

PURPOSE OF OCCUPATIONAL EXPOSURE LIMITS

The purpose of occupational exposure limits is to assist in the assessment of risk and control of exposure of workers to a variety of substances that can have harmful effects on health. Failure to control exposure to such substances can lead to many forms of ill-health. Therefore, it is important to know in advance what risks may arise from chemicals, what are acceptable occupational exposure limits and how to protect people at work. The occupational exposure limits set by competent national or international authorities will help to guide action and establish effective control by requiring organisations to put prevention and control measures in place to ensure occupational exposure limits are not exceeded. An occupational exposure limit is the maximum concentration of an airborne contaminant to which an employee may be exposed without causing damage to their health, measured within a reference time period.

The ILO Code of Practice for 'Safety in the Use of Chemicals at Work' requires that exposure to hazardous substances should be prevented, where reasonably practicable, for example, by changing the process. Where this cannot be done, exposure should be reduced by other methods, for example, by substituting it for something with less risk to health or by enclosing the process. Other methods may also include control of

worker exposure levels to within occupational exposure limits by limiting the concentration of the chemical in the atmosphere or inhaled by the worker and reducing the time workers are exposed to the chemical.

LONG-TERM AND SHORT-TERM LIMITS

The health effects of exposure to chemicals vary considerably depending on the nature of the chemical and the pattern of exposure. Some effects require prolonged or accumulated exposure, whereas others may arise from very short or a single exposure to a particular chemical.

Long-term limits

Long-term exposure limits (LTEL) are concerned with the total exposure of the worker to a concentration of the chemical, averaged over a reference period that represents a work period, usually 8 hours. They are therefore a time-weighted average (TWA) worker exposure.

This is appropriate for protecting against the effects of long-term exposure to a chemical that may have chronic health effects. If workers are exposed for longer periods than 8 hours, because they work longer work periods, the long-term exposure limits have to be adjusted to reflect this.

Examples of substances that have long-term occupational exposure limits established by a competent national or international authority include those set by the UK in HSE Guidance Note EH40; an example is shown in *Figure 7-27*.

Short-term exposure limits

Short-term exposure limits (STEL) are primarily aimed at avoiding the acute health effects of exposure to chemicals, for example eye irritation, or at least reducing the risk of their occurrence. Short-term exposure limits the total exposure of the worker to a concentration of the chemical averaged over a short reference period, usually 15 minutes. For those chemicals that do not have a short-term exposure limit, it is usually recommended that a figure of three times the long-term limit be used as a guideline for controlling short-term peaks in exposure.

Examples of substances that have short-term occupational exposure limits established by a competent national or international authority include those set by the UK in HSE Guidance Note EH40; this is shown in *Figure 7-28*. Note that some of these also have long-term exposure limits.

It can be seen, for example, that trichloroethylene has both a long- and short-term exposure limit to accommodate its acute effect (narcosis) and its chronic effect (possible cancer). It is also expressed as parts per million (ppm) to protect from its harmful vapours and, because processes that use trichloroethylene can create mists, it is also regulated by milligrams per cubic metre.

Substance	8 hour time-weighted average occupational exposure limit
Benzene	1ppm or 3.25 mg/m^3
Ferrous foundry particulate	Inhalable dust 10 mg/m^3 - Respirable dust 4 mg/m^3
Formaldehyde	2ppm or 2.5 mg/m^3

Figure 7-27: Examples of substances with long-term occupational exposure limits. Source: UK, HSE, EH40.

Substance	Time-weighted average occupational exposure limit	
	Short-term (15 minutes)	Long-term (8 hours)
Chlorine	0.5ppm or 1.5 mg/m^3	None
Phenol	4ppm or 16 mg/m^3	2ppm or 7.8 mg/m^3
Phosgene	0.06ppm or 0.25 mg/m^3	0.02ppm or 0.08 mg/m^3
Trichloroethylene	150ppm or 820 mg/m^3	100mg/m^3 or 550 mg/m^3

Figure 7-28: Examples of substances with short-term occupational exposure limits. Source: UK, HSE, EH40.

Ceiling exposure limits

Some workplace activities give rise to frequent short (less than 15 minutes) periods of high exposure that, if averaged over time, do not exceed either an 8-hour LTEL or a 15-minute STEL. These exposures have the potential to cause harm from special risks, such as carcinogenicity, skin absorption or allergic reaction, and may have a 'ceiling' exposure limit set for them. The ceiling exposure limit is a maximum allowable or acceptable concentration (MAC) that should not be exceeded at any time and ensures protection against acute effects, particularly those that may develop from a single exposure.

WHY TIME-WEIGHTED AVERAGES ARE USED

When considering the health effects of a hazardous substance, it is important to know what amount a worker has been exposed to over a period of time. This is known as the dose. The dose will influence the amount of the substance taken into the worker's body and any effects that may result from exposure.

In workplace situations, the concentration of a substance in the air a worker breathes may vary significantly over the reference period, depending on the activity being carried out. For example, the process of mixing powder in a vessel may give a high concentration of the substance in the air the worker breathes while the powder is added,

but while the mixing is taking place, the concentration in the air may reduce as water is added to the powder.

If a measurement was taken while filling the vessel with powder this would give the highest concentration, while if a measurement was taken during mixing this would give the lowest concentration. Neither measurement would represent the dose received by the worker over time as one would be too high and the other too low.

For this reason, the high and low concentration measurements taken are averaged over a reference time period. This is done by determining the amount of time that each of the concentrations is valid and averaging the combined concentrations/times over the reference period. This is the average of the concentrations of substance in the air the worker breathes over the reference period and is called a time weighted average.

Occupational exposure limits are set in such a way as to control the dose that a worker's body may receive by establishing a time-weighted average concentration of a substance that a worker may be exposed to over a reference period, for example, 8 hours or 15 minutes.

Because occupational exposure limits are set as time-weighted averages, a worker exposed to the occupational exposure limit may be exposed to concentrations of a hazardous substance that are higher than the specified time-weighted average for some of the time. This is acceptable provided the average concentration is not exceeded. As some work activities conducted in the period may create a concentration that is particularly low, this will offset those when it is high.

This effect is particularly important to consider where workers have a variety of tasks and take breaks away from the workplace where exposure is usually negligible.

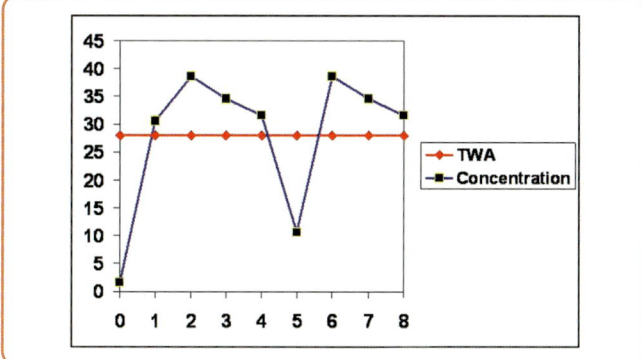

Figure 7-29: Time-weighted average compared with actual concentration. Source: RMS.

LIMITATIONS OF EXPOSURE LIMITS

There are many reasons why control of exposure should not be based solely on occupational exposure limits:

- **Compound routes of entry.** Many substances, for example trichloroethylene, have the ability to absorb through the skin. Occupational exposure limits only account for routes of entry by inhalation.

- **Individual susceptibility.** The majority of the work to establish occupational exposure limits has been based on the average male physiology from the countries in which studies were conducted. The occupational exposure limits relate to average males and it is possible for individuals to have a higher than average susceptibility to some substances. In addition, there may be a difference in susceptibility of females to some substances. Some work has been done where specific health-related effects have been noted amongst females, for example, exposure to lead compounds.

- **Origin of exposure limit data.** Work done to date has mainly been based upon exposure to individuals in the developed countries, for example, Europe and the USA.

- **Variations in control.** Local exhaust ventilation systems may not always work consistently because of lack of maintenance, overwhelming levels of contamination, etc. Changes in the environmental conditions, for example, increased humidity or air pressure, may increase the harmful potential of the substance. This may lead to occupational exposure limits being exceeded inadvertently.

- **Errors in monitoring.** Measuring microscopic amounts of contamination requires very accurate and sensitive equipment. Devices have a range of accuracies, and lack of maintenance and misuse can lead to inaccuracies in monitoring due to equipment becoming contaminated.

- **Synergistic effects.** The occupational exposure limits that are available relate to single substances; the effects of multiple substances in the workplace need to be considered.

COMPARISON OF MEASUREMENTS TO RECOGNISED STANDARDS

Comparison of workplace measurements with occupational limits

When identifying hazards and establishing risk controls, it is important to consider relevant occupational exposure limits set by competent national authorities or internationally recognised standards. When workplace exposure measurements are made, they should be compared with these limits or recognised standards. This will help to establish the presence of a workplace hazard and the level of risk, and will indicate the need to establish prevention and control measures. The comparison of measurements to occupational exposure limits will also show how effective the prevention and control measures will need to be if the occupational exposure limits are to be met.

Where occupational exposure limits are exceeded to a significant extent, this indicates the importance of taking immediate measures to reduce the level of occupational exposure to the substance. As measures are put in place, new workplace exposure measurements can be made and compared with the established occupational exposure limits. This will determine the level of progress towards meeting the occupational exposure limits and might indicate success or the need for further work.

Workplace exposure measurements should be made at intervals of time and after changes have been made in the workplace. The results of the measurements should be compared with occupational exposure limits to enable health risks to be monitored and action taken if the performance of control measures declines.

Occupational exposure limits are maximum allowable (acceptable) concentrations (MAC) of airborne substances under specified circumstances and are measured in either parts per million or per cubic metre of air. Airborne substances can be solid (dust), liquid (mist/aerosol), gas, vapours or fumes. Solids and liquids can be measured by weight (milligrams – mg), for example, an occupational exposure limit for cement dust may be expressed as 10 mg/m3. Gas, vapours and fumes are weightless.

Therefore, the occupational exposure limit is expressed as a concentration in the atmosphere – for example, an occupational exposure limit for trichloroethylene may be expressed as 100 parts per million (ppm). Occupational exposure limit values may be set relating to long-term exposure, short-term exposure or to establish a maximum allowable concentration for all circumstances, sometimes called a ceiling limit.

Occupational exposure limits may be known by different names, for example, 'indicative occupational exposure limit values' (IOELVs) in the EU, 'workplace exposure limits' (WELs) in the UK, 'occupational exposure standards' (OES) in Australia and 'threshold Limit values (TLVs) in the USA. They are all published to establish limits that employers refer to when controlling exposure of workers to substances that may harm their health. They do not address safety issues, such as flammable concentrations, and it should be noted that the occupational exposure limits do not generally include exposure to biological substances, as these are usually specified separately.

In most cases, other than the TLVs established by the American Congress of Government Industrial Hygienists, they are set to establish legal binding limits that employers must meet. Although occupational exposure limits are mostly connected to national law, their status can vary in some countries between obligation, indication and recommendation. Some countries prepare more than one list, one with an obligatory or binding character, another with more indicative or orientating character. The absence of a substance from established lists of occupational exposure limits does not mean that it will not cause adverse effects on health. Limits tend to be set where there is sufficient evidence to establish the limit. In some cases, this evidence may still be in the process of being established.

UK workplace exposure limits

In the UK, the HSE Guidance Note EH40 sets workplace exposure limits (WELs) in order to help protect the health of workers. It is a requirement of the UK Control of Substances Hazardous to Health Regulations (COSHH) 2002 that WELs must not be exceeded. WELs are concentrations of hazardous substances in the air, averaged over reference periods and are TWA concentrations. The two reference periods that are used are 8 hours and 15 minutes, which correspond with the EU IOELVs. The 8-hour WEL is known as a LTEL (long-term exposure limit) and is used to help protect against chronic ill-health effects. 15-minute STELs (short-term exposure limits) are set to protect against acute ill-health effects such as eye irritation, which may happen within minutes of exposure, or where there is the risk of high-level peak exposures over short time intervals during the working day.

European Union occupational exposure indicative limit values

IOELVs, previously known as indicative limit values (ILVs), are European legal limits of exposure to chemicals that are set to protect the health of workers in the EU from the ill-health effects of hazardous substances in the workplace. The EU Chemical Agents Directive requires member states to establish national laws to control the exposure of workers to chemicals. An EU directive establishes a requirement that member states of the EU take account of the IOELVs that are set out in the list that accompanies the directive, and subsequent lists as other substances are added, when they establish national limit values. Substances are added to the EU list of IOELVs as member states agree them. The list remains relatively short as it is limited to chemicals where agreement has been established. The IOELVs establish values for a reference period of an 8-hour time-weighted average, and for a short-term period of 15 minutes.

The EU has also established a number of binding occupational limit values (BOLVs) for certain chemicals with proven significant health risks, for example, benzene and vinyl chloride. Where a BOLV is established at EU level, Member States must establish a corresponding national binding occupational exposure limit value, which can be stricter, but cannot exceed the EU BOLV.

American Conference of Governmental Industrial Hygienists (ACGIH) threshold limit values (USA)

The ACGIH TLVs and most other worker exposure limits used in the USA and some other countries are limits that refer to airborne concentrations of substances and represent conditions under which "it is believed that nearly all workers may be repeatedly exposed day after day without adverse health effects" (ACGIH 1994). They are guidelines determined by the ACGIH, a voluntary body of knowledgeable people, for the use of those in the workplace who manage health risks.

The ACGIH emphasises that the limits are based solely on health factors, with no consideration to economic factors or technical feasibility. They represent the cumulative knowledge and experience of the individuals who decide them in the USA. They are not developed for use as legal standards and the ACGIH does not advocate this use. They remain guidelines set by knowledgeable people and should not be confused with exposure limits that have regulatory status, for example, permissible exposure limits (PELs) set by OSHA in the USA.

Ensuring the occupational exposure limit is not exceeded

In support of Article 16 of the ILO Occupational Safety and Health Convention C155, Article 12 of the ILO Chemicals Convention C170 establishes the following responsibility of employers regarding exposure of workers to chemicals, *(see Figure 7-30)*.

"(a) Ensure that workers are not exposed to chemicals to an extent which exceeds exposure limits or other exposure criteria for the evaluation and control of the working environment established by the competent authority, or by a body approved or recognised by the competent authority, in accordance with national or international standards."

Figure 7-30: Responsibility of employers.
Source: ILO Occupational Safety and Health Convention C155.

Though occupational exposure limits may be set for substances, employers retain a responsibility to reduce exposure levels to as low as is reasonably practicable. The limits should be considered to be a maximum allowable concentration and it is undesirable to see them as a normal working level if measures can easily be put in place to reduce them significantly below the workplace exposure limit. Work practices and control strategies should be constantly reviewed to ensure the lowest levels of exposure are achieved.

If the levels of exposure are to be maintained with confidence below the workplace exposure limit, it will be necessary to work far enough below them to account for changes in work situation. This is particularly important with those workplace exposure limits that are set for substances that are carcinogens or sensitisers. By working 'at the limit' employers do not allow for sensitive people who may be affected by relatively low exposures. Nor do they account for variations or inaccuracies in monitoring, sudden surges of contaminant or partial failures of control measures.

 **REVIEW**

Explain the significance of 'ceiling exposure' limits.

Give four reasons why control of exposure should not be based solely on workplace exposure limits.

7.4 Control measures

THE NEED TO PREVENT EXPOSURE OR ADEQUATELY CONTROL IT

Prevention

Article 16 of the ILO Occupational Safety and Health Convention C155 establishes the requirements of employers to prevent and control of risks from hazardous substances, *(see Figure 7-31)*.

"Ensure that, so far as is reasonably practicable, the workplaces, machinery, equipment and processes under their control are safe and without risk to health. Ensure that, so far as is reasonably practicable, the chemical, physical and biological substances and agents under their control are without risk to health when the appropriate measures of protection are taken."

Figure 7-31: Employer's responsibilities.
Source: ILO Occupational Safety and Health Convention C155.

This is further supported by the ILO Chemical Convention C170, Article 13 which requires employers to establish operational controls, *(see Figure 7-32)*.

"Make an assessment of the risks arising from the use of chemicals at work, and shall protect workers against such risks by appropriate means, such as:

The choice of chemicals that eliminate or minimise the risk.

The choice of technology that eliminates or minimises the risk.

The use of adequate engineering control measures.

The adoption of working systems and practices that eliminate or minimise the risk.

The adoption of adequate occupational hygiene measures.

Where recourse to the above measures does not suffice, the provision and proper maintenance of personal protective equipment and clothing, at no cost to the worker, and the implementation of measures to ensure their use'.

Figure 7-32: Operational controls.
Source: ILO Chemical Convention C170.

In the UK, regulation 7(1) of the Control of Substances Hazardous to Health Regulations 2002 states that:

> "Every employer shall ensure that the exposure of his employees to substances hazardous to health is either prevented or, where this is not reasonably practicable, adequately controlled."

Figure 7-33: Regulation 7(1).
Source: UK, Control of Substances Hazardous to Health Regulations (COSHH) 2002 (amended)

National and international protocols and good practice establish a clear need to prevent exposure where it is reasonably practicable. If exposure is prevented then no harm to workers can be done. This is the most effective control measure and, although it may involve substantial investment, change is usually the best long-term solution to health risks presented by substances.

In many workplaces, it will not be possible or reasonably practicable to prevent exposure to substances hazardous to health completely. If exposure cannot be prevented by eliminating the use of a hazardous substance employers should adequately control exposure.

The principle of 'so far as is reasonably practicable' allows employers to balance cost against the degree of risk. This principle has to be applied when selecting adequate control measures to protect workers from harmful exposures to hazardous substances.

Control measures have to be proportionate to the risk and take into account the nature of the hazard, the frequency and duration of exposure and the number and type of people exposed. If the risk is low and not likely to cause long-term harm, the control measures may be simple procedural issues, such as replacing the lid tightly on containers or a regime of regular cleaning. However, if the risk is higher and the consequences of exposure to hazardous substance is likely to lead to workplace diseases, such as dermatitis, asthma or cancer, then more robust and reliable control measures must be implemented.

COMMON MEASURES USED TO CONTROL EXPOSURE TO HAZARDOUS SUBSTANCES

ILO Code of Practice recommendations

The ILO "Code of Practice for Ambient Factors in the Workplace" recommends measures to prevent and control hazardous substances, *(see Figure 7-34)*. It firstly recommends elimination of hazardous substances or replacing hazardous substances or modifying processes to control exposure. It further recommends the implementation of a programme of control measures, which is clarified further in the code of practice and discussed specifically later in this element. The other control

measure recommended is to minimise the use of the more harmful substances, toxic substances, where this is possible. ILO "Code of Practice for Safety in the Use of Chemicals at Work" makes the same recommendations.

> "4.3. Prevention and control
>
> 4.3.1. Where the assessment of hazards or risks shows that control measures are inadequate or likely to become inadequate, risks should be:
>
> a. Eliminated by ceasing to use such hazardous substances or replacing them with less hazardous substances or modified processes;
>
> b. Minimised by designing and implementing a programme of action;
>
> c. Reduced by minimising the use of toxic substances, where feasible.

Figure 7-34: Requirement to eliminate risks.
Source: ILO, Code of Practice for Ambient Conditions in the Workplace

Elimination

This represents an extreme form of control and is appropriately used where high risk is present from such things as carcinogens (cancer forming substances). It is usually achieved through the prohibition of use of these substances. Care must be taken to ensure that all stock is safely disposed of and that controls are in place to prevent the substance being re-introduced to the workplace, even as a sample or for research. In a similar way, an employer may decide to cease activities that use hazardous substances that present a high risk. If this level of control is not achievable then another must be selected.

Replacing

The replacing of a highly toxic substance with a substance that is less toxic is a very useful control measure, for example, benzene might be replaced with toluene, asbestos might be replaced with glass fibre, or solvent-based adhesives might be replaced with water-based adhesives.

Process modification

An example of process modification is changing from a paint-spraying process, where large volumes of airborne vapours are released, to a process that uses a paint brush. Similarly, the emission of vapours from substances used in a chemical process may be minimised by changing the process to limit the temperature of the substance so that fewer vapours are given off.

Element 7

Figure 7-35: Reduced exposure - bulk supply. Source: RMS.

It is sometimes possible to change the composition or form of the substance used in a process, for example if a substance, such as cement, requires mixing of its constituent parts it may be ordered pre-mixed, thus avoiding the potential exposure to the substances when particles of the substance become airborne during the mixing process. Similarly, a substance may be used in pellet form, rather than powder form, in order to limit the possible emission of dust into the atmosphere.

Processes that usually create airborne dust may also be controlled by changing the process so that water is used to keep the material being worked on wet. For example, material being removed from a building in a demolition process may be watered down so that any small particles become a liquid slurry.

Programme of action

The ILO "Code of Practice for Ambient Factors in the Workplace" recommends a programme of action as further control measures for hazardous substances, **(see Figure 7-36)**. The code of practice recommends a number of control measures under the following headings:

a) Good design and installation practice

b) Work systems and practices

c) Personal protection

Each of the control measures recommended under these headings is discussed later in this element.

The ILO "Code of Practice for Safety in the Use of Chemicals at Work" makes the same recommendations.

"4. Hazardous substances, 4.3 Prevention and control

4.3.2. Control measures ....could include any combination of the following:

a) Good design and installation practice:

(i) Totally enclosed process and handling systems.

(ii) Segregation of the hazardous process from the operators or from other processes.

(iii) Plants processes or work systems which minimise generation of, or suppress or contain, hazardous dust, fumes, etc., and which limit the area of contamination in the event of spills and leaks.

(iv) Partial enclosure, with local exhaust ventilation.

(v) Local exhaust ventilation.

(vi) Sufficient general ventilation.

b) Work systems and practices:

(i) Reduction of the numbers of workers exposed and exclusion of non-essential access.

(ii) Reduction in the period of exposure of workers.

(iii) Regular cleaning of contaminated walls, surfaces, etc.

(iv) Use and proper maintenance of engineering control measures.

(v) Provision of means for safe storage and disposal of chemicals hazardous to health.

c) Personal protection:

(i) Where the above measures do not suffice, suitable personal protective equipment should be provided until such time as the risk is eliminated or minimised to a level that would not pose a threat to health.

(ii) Prohibition of eating, chewing, drinking and smoking in contaminated areas.

(iii) Provision of adequate facilities for washing, changing and storage of clothing, including arrangements for laundering contaminated clothing.

(iv) Use of signs and notices.

(v) Adequate arrangements in the event of an emergency."

Figure 7-36: Control measures for hazardous substances. Source: ILO, Code of Practice for Ambient Factors in the Workplace.

Minimising the use of toxic substances

A useful approach to controlling exposure to hazardous substances is to focus on minimising the actual quantity of toxic substance that are in use. This may be achieved by limiting the amount used in the process, for example by using a dilute form of the substance such as an acid, or the quantity stored in the workplace, for example by controlling the amount available to the minimum required for a work period.

Programme of action - Good design and installation practice

Totally enclosed processes and handling systems

Plant and equipment should be designed and installed to contain hazardous substances used at work and prevent their release into the workplace, including vapour and dust from the substances.

Containment of a hazardous substance is best achieved by totally enclosing processes involving the use of hazardous substances. Total enclosure of processes can be more easily achieved where plant and equipment are automated or operated remotely. The opportunity to do this should be considered during the design of plant and equipment and the process to be used.

The strategy of total enclosure of hazardous substances is based on the containment of substance to prevent its free movement in the working environment. It may take a number of forms, for example, enclosure of activities, laboratory 'glove box' for handling toxic substances, pipelines and enclosed conveyors. Total enclosure of the process is used in the removal of asbestos, where the work being done is enclosed in plastic sheeting in order to segregate the work from the surrounding areas. In a similar way, enclosure may be used for building cleaning processes that use impact abrasive spraying technique's, or when spray protection (insulation or paint) is applied to a structure or item during manufacture or maintenance.

Bulk storage of hazardous substances, with fixed pipework transfer, should be used in preference to small container storage, where appropriate.

Figure 7-37: Employees are segregated from the spray painting process. Source: Sima.

Segregation of process and people

In situations where the hazardous substances cannot be totally enclosed close to their source it may be preferable that the workforce be segregated from the process by providing physical separation in the form of a refuge, for example, a control room of a chemical process. Physical segregation can be a relatively simple method, such as distancing (segregating) them from the general workforce and other processes.

To further reduce risks from hazardous substances, plant, equipment and storage should be separated from other processes, from incompatible substances and other areas outside the control of the employer.

Minimising generation, suppression or containment of hazardous substances

It is important, where possible, to minimise emission, release and spread of hazardous substances by effective design and operation of processes and task activities. A useful approach to control exposure to hazardous substances is to reduce the actual quantity of the substance that is generated. For example, prevention of the generation of large volumes of airborne vapours may be achieved by use of a paintbrush rather than an aerosol or paint-spraying equipment. Generation of vapours from hazardous substances can also be minimised by process controls that limit the temperature of the substance so that fewer vapours are given off. In a similar way, processes that usually generate dust may be controlled by suppressing the dust by keeping the process wet.

Figure 7-38: Bunded storage.
Source: visualworkwear.au

In its simplest sense, suppression of hazardous substances can mean covering the containers of substances that generate volatile vapours, for example, putting the lid on tins of solvent-based products after use or using floating balls to cover the surface of a tank of hot acid used to dip components in.

Where particularly hazardous substances are used in enclosed plant and equipment, in order to reduce leaks of the substance the plant and equipment should be designed to ensure a slight negative pressure within it, where the process allows. This will ensure that air from the workplace is drawn into any gaps in the enclosure and prevent the release of the substance into the workplace. This will contain the hazardous substance within the plant or equipment, preventing worker exposure to the risk of harm. It will also help to ensure that cleaning of the workplace to remove hazardous substance contamination can be kept to a minimum.

All processes and equipment should be checked and maintained to ensure that containment of hazardous substances is effective, for example, seals on joints may need to be replaced or kept tight.

To prevent the spread of a hazardous substances in the event of its release, a secondary means of containment should be provided in accordance with established criteria, such as bund walls for hazardous liquids and containment areas for the evaporation of cryogenic liquids. A "bund wall" is a properly designed and constructed containment wall to contain the contents of a storage vessel that is enclosed by the wall in the event that the storage vessel leaks.

Figure 7-39: Water cascade used in wire manufacturing. Source: siriowire.com.

Partial enclosed processes, with local exhaust ventilation

Where total enclosure of a process involving hazardous chemicals is not reasonably practicable, it may be possible to partially enclose the process and supplement this with local exhaust ventilation equipment. This might be the case where it is necessary to place batches of components into a tank of a substance meaning that the hazardous substance cannot be totally enclosed. The tank might be fitted with partial enclosure and local exhaust ventilation over the top of the tank or around the edge of the tank. Similarly, an area in a laboratory used for mixing chemicals might have partial enclosure to three sides and the top, open front and local exhaust ventilation at the back extracting chemical contaminants.

Figure 7-40: Plastic enclosure for asbestos removal. Source: merryhillenvirotec.com.

Local exhaust ventilation

Where other control measures are not suitable local exhaust ventilation (LEV) equipment should be provided to ensure that workplace exposure limits are controlled so that they do not exceed good practice levels or occupational exposure limits specified by a competent authority are not exceeded.

Hazards that can be controlled by the effective use of LEV include: dust that could cause coughing, sneezing and various other respiratory diseases; substances that might cause sensitisation or other toxic effects; allergens that could aggravate asthmatic conditions; microorganisms that can cause diseases; asphyxiants that can lead to breathing difficulties, unconsciousness and, ultimately, death.

General principles and application

The LEV should be designed, constructed and installed to ensure either the effective removal of contaminated air from the workplace to a suitable place or the filtering/treatment of the contaminated air and its return to the workplace.

For efficient operation and to prevent exposure of the worker, the LEV intake points should be located as close as possible to the points where the hazardous substances are emitted. The length of ducting and the number of bends in the system should be kept to a minimum to enable efficient operation.

Various LEV systems are in use in the workplace and the main types are called receptors and captors. Receptor systems are designed to use low velocity extraction and are best suited to gases, vapours and lightweight dusts travelling at low velocity from the source. The work process creating the contaminant usually causes the contaminant to move towards the receptor system intake, for example where vapours are given off by a process and rise upwards to a receptor hood. Captor systems are designed to use high velocity extraction and are best suited to remove metal fume from welding operations or waste dust from high speed grinding or cutting equipment. The captor system uses its high velocity of air to capture volatile contaminant particles that may quickly move away from the source of where they are generated and have enough velocity of their own to escape a receptor system.

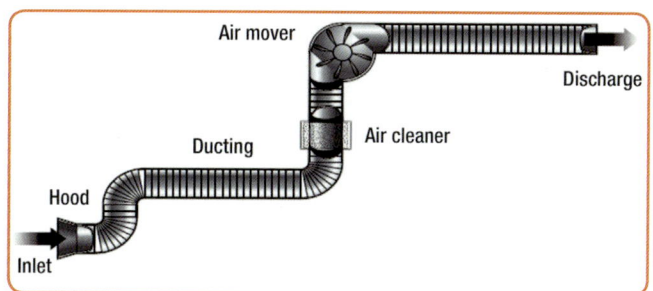

Figure 7-41: Common elements of a simple LEV system. Source: UK, HSE, INDG408.

Figure 7-42: Captor system on circular saw. Source: RMS.

Figure 7-43: LEV fan and motor. Source: Custom Dans Australia.

Figure 7-44: Flexible hose and captor hood. Source: RMS.

Figure 7-45: Portable self-contained unit.
Source: Industrial Air Solutions Inc.

Figure 7-46: Length of ducting with curves. Source: RMS.

The common elements of a simple LEV system are:

- **Hood(s)** to collect airborne contaminants at, or near, where they are created (the source).
- **Ducts** to carry the airborne contaminants away from the process, **(See Figure 7-46)**, which shows the length of ducting, with curves not corners.
- **Air cleaner** to filter and clean the extracted air.
- **Fan** must be the right size and type to deliver sufficient suction at the hood, **(See Figure 7-43)**, which shows the size of fan and motor required for industrial-scale LEV systems.
- **Discharge** the safe release of cleaned, extracted air into the atmosphere.

The efficiency of LEV systems can be affected by many factors including the following:

- Damaged ducting.
- Unauthorised alterations.
- Incorrect hood location, **(see Figure 7-44)**, which shows how a captor hood can be repositioned to suit the work activity by the use of a flexible hose.
- Too many bends or sharp bends in ducts.
- Blocked or defective filters.
- Leaving too many ports open.
- Process changes leading to overwhelming amounts of contamination.
- Fan strength or incorrect adjustment of fan.
- Excessive amounts of contamination.

The cost of heating/cooling air introduced to the workplace to replace that removed by the LEV system may encourage some employers to reduce extraction rates. When arranging installation of LEV it is vital that the pre- and post-ventilation contamination levels are specified and the required reduction in contamination levels should be part of the commissioning contract.

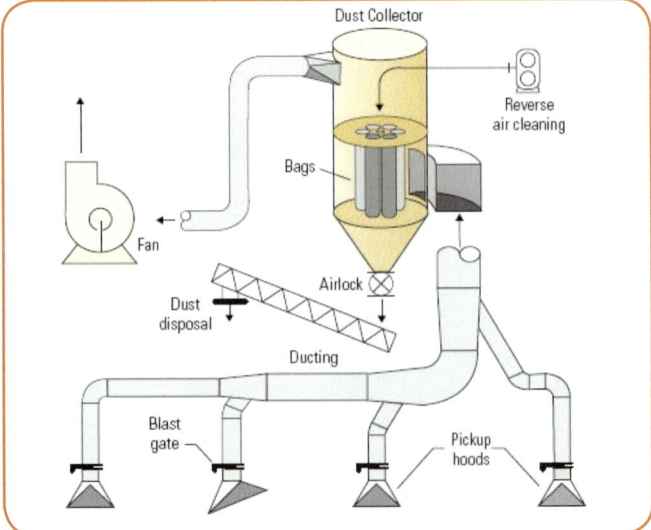
Figure 7-47: Multiple ports on one extraction duct.
Source: Wikipedia.org

©RMS

Requirements for thorough examination and inspection

LEV systems should be thoroughly examined at suitable specified intervals to ensure that they are continuing to perform as originally intended. The interval and content of the thorough examination is often specified by national legislation. For example, in the UK, the Control of Substances Hazardous to Health (COSHH) Regulation establishes a requirement that a thorough examination and test must take place at least once every 14 months, to ensure maintenance is carried out; for specified processes this is more frequent. For example, LEV systems provided for jute manufacture and blasting of metal castings should be thoroughly examined monthly.

Requirements for records of thorough examination may also be specified by national legislation; the UK requirement is that records must be kept available for at least 5 years from the date on which the inspection was carried out.

Regular inspection of LEV systems should be carried out in addition to the thorough examination. This should identify that the features for effectiveness are in place, for example, blanking covers for inlets not being used, and that no obvious defects exist.

The majority of ventilation systems, although effective in protecting workers' health from airborne contaminants, can create other hazards. One of the main hazards that needs to be considered when designing a LEV system is the noise the LEV system creates. Even if it has been considered as a design feature when establishing LEV systems, the noise level of the system should be monitored in the workplace on a periodic basis.

General ventilation

Where local exhaust ventilation is used, work areas should be supplied with clean air in the form of general ventilation to balance the volume of extracted air as it is exhausted through the extraction system. This ensures efficient extraction and the reduction of the concentration of hazardous substances in the workplace. Where local exhaust ventilation is not used it is still important to provide adequate general ventilation into the work-place to reduce the level of general contaminants and improve levels of fresh air.

General ventilation systems will only affect levels of general contamination in the air in a workplace and will not prevent specific hazardous substances entering a person's breathing zone. Because general ventilation does not target any specific source of contaminant and it relies on dispersal and dilution instead of specific removal, it can therefore only be used with nuisance contaminants that are themselves fairly mobile in air.

General ventilation may only be used as the only means of control in circumstances where there is:

- Non-toxic contaminant or vapour (not dusts).
- Contaminant that is uniformly produced in small, known quantities.
- No discrete point of release.
- No other practical means of reducing levels.

The limitations of general ventilation are:

- It is not suitable for substances of high toxicity.
- Large sudden releases of a contaminant are not coped with well.
- It tends to move dust around rather than remove it.
- Areas in the location may exist where natural airflow does not create air movement.
- Stagnant air or non-moving air is likely to hold high concentrations of the contaminant.

The flow rates of general ventilation should be sufficient to change the air of the work area according to any specified requirements of good practice or a competent authority. This will take into account the size of the workplace, the nature and quantity of any contaminant, the working conditions and numbers of workers.

General ventilation may be achieved by driving air into a work area, causing the air to flow around it, diluting contaminants in the work area and then allowing the air out of the work area through general leakage or through ventilation ducts to the open air. A variation on this is where air may be forcibly removed from the work area, but not associated with a particular contaminant source, and fresh air is allowed in through ventilation ducts to dilute the contaminated air in the work area. Sometimes a combination of these two approaches is used, for example, general air conditioning provided in an office environment.

Recirculation of all extracted air into the workplace should be avoided where possible as it can lead to a decline in the quality of the air being breathed by workers, where recirculation is allowed:

1) Effective methods should be used to decontaminate the air, which should be regularly checked and maintained.

2) Some air should be vented during recirculation and replaced by fresh air to avoid an accumulation of possible contamination.

3) The rate of replacement by fresh air should be designed to ensure that hazardous limits or criteria for the control of the working environment, established by good practice or a competent authority, are not exceeded.

4) Account should be taken in the design of the need to prevent any inadvertent release of hazardous substances into the workplace from causing a hazard and spreading it to other working areas, for example, the provision of emergency shut down of the ventilation system.

Programme of action - work systems and practices

Work systems and practices should be devised and followed for all uses of hazardous substances at work to protect workers against the risks which have been identified. Work systems and practices should be devised after other appropriate means to eliminate and minimise risks have been chosen (i.e. the appropriate chemicals, technology and engineering control measures).

The work systems and practices should incorporate the most effective use of the control measures provided. It should make it clear who is in charge of the work, specify the particular tasks included in that work (and which individuals have responsibilities where there is an overlap) and provide for the exchange of necessary information at shift changeover time.

Other than for straightforward tasks, work systems and practices should be described in writing. In particular, written work systems and practices should be devised and followed where they are of primary importance, for example, during routine maintenance, examination and repair of plant and equipment and the transfer of substances (including loading and unloading).

In some cases, the possible risks presented by hazardous substances are very high, for example during the maintenance of plant and equipment where entry is necessary. In such cases a formal written systems and practices, in the form of a "permit-to-work" system, is required.

Work systems and practices for emergency shut-down of processes should be established.

Reduction in the number of people exposed to the hazard

Reduction of the number of workers exposed to hazardous substances may be achieved by a number of means, such as keeping the number of workers working with hazardous substances to a minimum and segregating them from the general workforce. For example, maintenance work that releases hazardous substances could be conducted when other workers are not working in the area. Similarly, a person supervising the process may not need to do this from a position in close proximity to those doing the work and in some cases it may be able to carry out the supervision remotely, using closed circuit television (CCTV).

Exclusion of workers

The protection of young workers in certain work activities is important, for example work with some hazardous substances, as they may be more affected by them than the average worker because they are still developing physically. A good example is working with lead, which could lead to kidney, nerve and brain damage. Young workers are usually excluded from lead processes to protect them from the effects of lead. There also remains the possibility of gender-linked vulnerability of women of child-bearing capability as exposure to lead can affect the unborn child, particularly in the early weeks of pregnancy. Exclusion of pregnant workers from working with lead may therefore be appropriate. Exclusion of these workers from exposure to these substances provides a high level of control in these circumstances.

In addition, all non-essential workers and other people should be excluded from entering areas where hazardous substances are present, for example contractors and visitors could be excluded from entering this part of the workplace.

Reduction in the period of exposure of workers

The level of harm to a worker's health is dependent on the concentration of a hazardous substance that they are exposed to over time. In situations where the concentration of the substance cannot be reduced easily, reduction in the time workers are exposed to the substance will reduce the harm to their health. Individual worker time exposure reduction can be achieved by introducing worker rotation during the work period. For example, rotation of four workers over an 8-hour work period so that each worker spends an equal portion of time doing the same task that exposes them to a substance would reduce individual worker exposure by 75%.

Similarly, those workers only needed during a specific part of a process that exposes workers to hazardous substances could move out of the hazardous area to a safe distance when they are not needed and only return when needed.

Regular cleaning of contaminated walls, surfaces, etc.

Regular cleaning of surfaces, for example walls, floors and the surfaces of plant/equipment, will help to reduce incidental exposure to hazardous substances that have settled on the surfaces, preventing the contamination being transferred to the clothing or hands of workers. Cleaning contaminated floors would assist in preventing the spread of contamination by vehicles and worker's footwear to other workplaces that would otherwise be clean and therefore reduce the possibility of dust being disturbed and becoming airborne again.

Use and proper maintenance of engineering control measures

It is important that engineering controls provided to control exposure of workers to hazardous substances work correctly, for example, provision of enclosures and local exhaust ventilation. This will require workers to use them properly and for the employer to ensure they are maintained and effective. Enclosures to protect workers will only provide protection if workers ensure access doors are not prevented from closing by accidental or deliberate action and local exhaust ventilation may require a worker to take action to start it and adjust its position to ensure it is effective. The employer should have a planned maintenance programme for critical equipment used to control exposure to hazardous substances, for example, local exhaust ventilation equipment should be thoroughly examined and tested periodically and repairs, such as sealing leaks and cleaning filters, should be made to ensure it remains effective.

Provision of means for safe storage and disposal of hazardous chemicals

The safe storage of hazardous substances should take regard of the nature and quantity of the substance, including how it might harm workers and would typically include:

- Stock levels of hazardous substances in the workplace should be kept to the minimum, for example restricting the quantity stored to the maximum amount that would be used in a single work period. Larger quantities should be kept in a specific storage area designed for the purpose and located away from the workplace.

- All newly purchased chemicals should have a label on them identifying their hazard category (for example, corrosive, oxidising, toxic etc.). Storage areas for large quantity storage may need additional signs warning of the hazards related to the substances stored.

- Stored in a way that enables access to them to be controlled, for example a locked cupboard or store.

- Store like materials with like. It is essential to segregate incompatible substances to prevent dangerous interactions. For example, sulphuric acid should not be stored with substances containing chlorine.

- Store in suitable containers that prevent spill or loss of vapour. Additional spill containment may be required in the storage container or storage area. Some storage arrangements may require ventilation, at high or low level depending on whether vapours from the substances stored are lighter or heavier than air. Ensure containers put into or return to storage

are sealed properly to avoid unnecessary leakage of fumes or vapours.

- Store containers in a way that makes handling them easy and avoids them being dropped, which could lead to spills and exposure to the hazardous substance. For example, large breakable containers, particularly of liquids, should be stored below shoulder height.

- Arrangements should be made to deal with spills, unplanned exposure to the substances and emergencies, for example spill kits and suitable first aid.

- Hazardous substances that are no longer required should be disposed in controlled way that prevents harm to people and the environment.

Programme of action - personal protection

Personal protective equipment

Where other control measures, such as good design and work practices, do not sufficiently control the hazard suitable personal protective equipment (PPE) should be provided until such time as the risk is eliminated or minimised to a level at which personal protective equipment is not be required. This will provide more effective protection of workers in circumstances where the control measures provide limited protection or where there is a possibility that the control measure may fail. An example of this is where the control measure provided is local exhaust ventilation (LEV). In this situation, the LEV will remove an amount of the contaminant, but to increase the effectiveness of protection the worker may also wear respiratory protective equipment (RPE), such as a respirator.

In some cases it may be determined that the use of personal protective equipment on its own is the only reasonably practical control measure that can be used to protect workers from the hazardous substance. Personal protective equipment includes respiratory protective equipment, protective clothing and footwear, equipment to protect the hands, face and eyes.

Use of PPE for control of hazardous substances is considered in this element, below. The general use of PPE is considered in more detail in 'Element 3.4 – Assessing risk' and other PPE for use with other hazards are referred to in appropriate elements.

Respiratory protective equipment
Purpose, application and effectiveness

If there is a risk of harm to health that can be caused by breathing in hazardous substances, such as dusts, fumes, vapours, gases or even micro-organisms, then respiratory protective equipment (RPE) may be needed to supplement other control measures or in circumstances that other measures are not reasonably practicable. RPE includes a very wide range of devices, from simple respirators offering basic protection against low levels of nuisance dusts to self-contained breathing apparatus.

Types of respiratory protection equipment

There are two main categories of respiratory protection:

1) Respirators.

2) Breathing apparatus.

Respirators

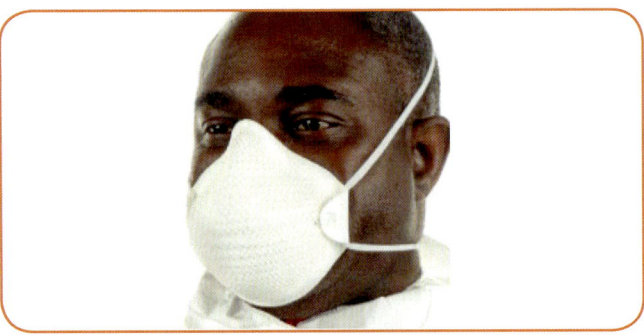

Figure 7-48: 3M Filtering face-piece respirator. Source: RMS.

Respirators filter the air breathed but do not provide additional oxygen. There are a number of types of respirator that provide a variety of degrees of protection, from dealing with nuisance dusts to high-efficiency respirators for solvents or asbestos.

Some respirators may be described as providing non-specific protection from contaminants, whereas others are designed to protect the user from a very specific contaminant such as solvent vapours.

There are five main types of respirators:

1) Filtering face-piece.

2) Half-mask respirator.

3) Full-face respirator.

4) Powered air purifying respirator.

5) Powered visor respirator.

Figure 7-49: Half-mask respirator. Source: Shutterstock.

The advantages of a respirator are:

- Unrestricted movement.
- Often lightweight and comfortable.
- Can be worn for long periods.

The limitations of a respirator are:

- Purifies the air by drawing it through a filter to remove contaminants. Therefore can only be used when there is sufficient oxygen in the atmosphere.
- Requires careful selection and confirmation of correct fit for the user by a competent person.
- Requires regular maintenance.
- Knowing when a filter or cartridge is at the end of its useful life.
- Requires correct storage facilities.
- Can give a 'closed in'/claustrophobic feeling.
- Relies on user for correct use, adjustment etc.
- Incompatible with some other forms of PPE.
- Performance can be affected by beards and long hair.
- Interferes with other senses, for example, sense of smell.

Figure 7-50: Full-face canister respirator. Source: Shutterstock.

Breathing apparatus

Breathing apparatus provides a separate supply of air (including oxygen) to that which surrounds the person. Because of the self-contained nature of breathing apparatus it may be used to provide a high degree of protection from a variety of toxic contaminants and may be used in situations where the actual contaminant is not known or there is more than one contaminant. There are three types of breathing apparatus:

1) Fresh-air hose apparatus - clean air from uncontaminated source.

2) Compressed air line apparatus from compressed air source.

3) Self-contained breathing apparatus - from cylinder.

Figure 7-51: Breathing apparatus - air from cylinder. Source: Haxton Safety.

The advantages of breathing apparatus are:

- Supplies clean air from an uncontaminated source. Therefore can be worn in oxygen-deficient atmospheres.
- Has a high assigned protection factor (APF). Therefore may be used in an atmosphere with high levels of toxic substance.
- Can be worn for long periods if connected to a permanent supply of air.

The limitations of breathing apparatus are:

- Can be heavy and cumbersome, which restricts movement.
- Requires careful selection and confirmation of correct fit for the user by a competent person.
- Requires special training.
- Requires arrangements to monitor/supervise user and for emergencies.
- Can give a 'closed in'/claustrophobic feeling.
- Relies on user for correct use, adjustment etc.
- Incompatible with some other forms of PPE.
- Performance can be affected by, for example, long hair.
- Interferes with other senses, for example, sense of smell.
- Requires correct storage facilities.

Selection, use and maintenance of RPE

There are a number of issues to consider in the selection of RPE, not least the advantages and limitations shown in the previous list. A general approach must not only take account of the needs derived from the work to be done and the contaminant to be protected from, but must include suitability for the person. This includes issues such as face fit and the ability of the person to use the equipment for a sustained period, if this is required. One of the important factors is to ensure that the equipment will provide the level of protection required.

This is indicated by the assigned protection factor (APF) given to the equipment by the manufacturer – the higher the factor, the more protection provided. With a little knowledge, it is possible to work out what APF is needed using the following formula:

$$APF= \frac{\text{Concentration of contaminent in the workplace}}{\text{Concentration of contaminent in the face-piece}}$$

It is important to understand that this factor is only an indication of what the equipment will provide. Actual protection may be different due to fit and the task being conducted.

Every worker must use any PPE provided in accordance with the training and instructions that they have received. It is important that the worker using the RPE conduct a face fit test to ensure the equipment is providing the protection needed. Where RPE (other than disposable respiratory protective equipment) is provided, the employer must ensure that thorough examination and, where appropriate, testing of that equipment is carried out at suitable intervals.

Other protective equipment and clothing

Hand/arm protection including gloves

There are numerous types of gloves and gauntlets available that offer protection from hazardous substances, such as acids, alkalis and solvents.

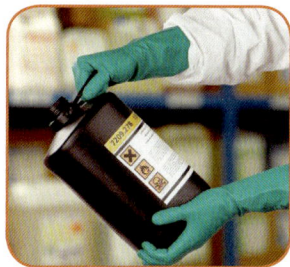

Figure 7-52: Gloves. Source: Speedy Hire plc.

The materials used in the manufacture of these products are an essential feature to consider when making the selection. There are several types of rubber (latex, nitrile, PVC, butyl), all giving different levels of protection against hazardous substances. Care should be taken to ensure that, where necessary, protection is provided for the arms, this may require gloves that extend far enough to provide this protection.

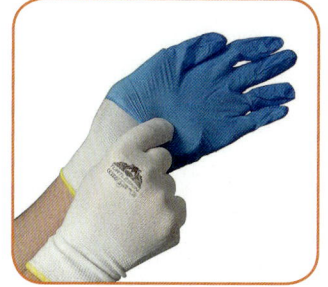

Figure 7-53: Protective clothing - gloves. Source: Haxton Safety.

Protective clothing

Protective clothing available that offer protection from hazards might include overalls, aprons, boots and full deluge suits.

Types	Advantages	Limitations
Spectacles	• Lightweight, easy to wear. • Can incorporate prescription lenses. • Do not 'mist up'.	• Do not give all-round protection. • Relies on the wearer for use.
Goggles	• Give all-round protection. • Can be worn over prescription lenses. • Capable of high impact protection. • Can protect against other hazards, for example, dust, molten metal.	• Tendency to 'mist up'. • Uncomfortable when worn for long periods. • Can affect peripheral vision.
Face shields (visors)	• Gives full-face protection against splashes. • Can incorporate a fan which creates air movement for comfort and protection against low-level contaminants. • Can be integrated into other PPE, for example, head protection.	• Require care in use, otherwise can become dirty and scratched. • Can affect peripheral vision. • Unless the visor is provided with extra sealing gusset around the visor, substances may go underneath the visor to the face.

Eye and face protection

When selecting suitable eye protection, some of the factors to be considered are: type and form of the hazard, for example protection from dusts, liquids, vapours, or gases, amount of protection, comfort and user acceptability, compatibility, maintenance requirements, training requirements and cost.

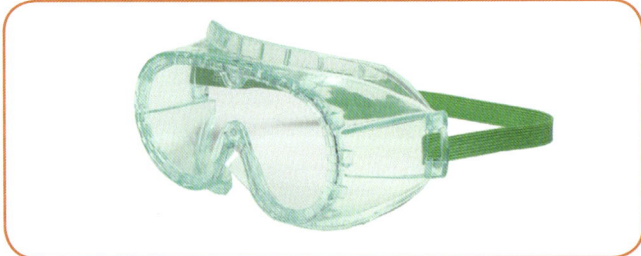

Figure 7-54: Goggles. Source: Gempler's.

Figure 7-55: Spectacles. Source: Pixabay.

Figure 7-56: Eye protection used with ear protection.   Source: Speedy Hire plc.

Prohibition of eating, chewing, drinking and smoking in contaminated areas

In workplaces that have a risk of contamination by hazardous substances it is important to prohibit eating, chewing, drinking and smoking in order to limit the likelihood of workers contacting their face around the mouth and ingesting the substance. In order to encourage workers to follow these prohibitions access to suitable facilities away from the contamination should be provided so they can be carried out. Workers should ensure they have removed contaminated clothing and wash any contamination from their hands before entering the facilities and carrying out any of the above actions.

The ILO "Code of Practice for Safety in the Use of Chemicals at Work" requires employers to provide adequate welfare facilities and makes particular reference to good personal hygiene practice, (**see Figure 7-57**).

 "To reduce the risk of ingesting chemicals hazardous to health, workers should not eat, chew, drink or smoke in a work area that is contaminated by such chemicals"

Figure 7-57: Control measures for chemicals, personal hygiene. Source: ILO, Code of Practice for Safety in the Use of Chemicals at Work.

Provision of adequate facilities for washing, changing and storage of clothing, including arrangements for laundering contaminated clothing

Personal hygiene has an important role in the protection of the health and safety of workers. Laid-down procedures and standards are necessary for preventing the spread of hazardous substance contamination.

The provision of adequate washing/showering facilities away from areas contaminated with hazardous substances is important to enable workers to remove contamination from their body. These facilities must be located conveniently close to where workers work, get changed, toilet facilities and where they eat to enable them to use them at rest periods and at the end of the day.

The use of barrier creams may also be additional personal protection considerations for hazardous substance risks. Where this is considered appropriate, facilities should be provided.

Facilities for changing and storage of clothing should enable workers to put clean work clothing on at the start of the work period and remove contaminated clothing at the end of the work period. Work clothing needs to be stored separate from non-work clothing.

The provision of laundry facilities for contaminated work clothing reduces workers from repeated exposure to contamination on the clothing. It is important that work clothing contaminated with hazardous substances is not taken home for laundering as this may expose members of the family to unnecessary contamination.

Where personal hygiene is critical, for example, when stripping asbestos, a 'three-room system' is employed. Workers enter the 'clean end' and take non-work clothes off and store them and put work clothes on, leaving by means of the 'dirty end'. When work has been completed they return by means of the 'dirty end', carry out personal hygiene, remove work clothing and put it ready to be cleaned and leave by means of the 'clean end' after dressing in their non-work clothing.

Use of signs and notices

Signs and notices should be provided to inform workers of the presence of hazardous substances, the type of hazard they present and the control measures necessary. They should also be used to inform non-essential workers not to enter the workplace. This should assist in ensuring that workers take the actions necessary to protect themselves.

Adequate arrangements in the event of an emergency

All of these control measures should be supported with arrangements for circumstances where there is an emergency involving hazardous substances, for example, where a worker gets a substance in their eye or where a substance is spilt in the workroom. These arrangements should include a means of raising the alarm and getting help, first aid facilities, containment/absorbent materials for spills and possible evacuation of the work-place to a safe area.

Health and medical surveillance

'Health surveillance' is concerned with collecting and using information about a worker's health related to their work and systematically looking for work-related ill-health in workers exposed to certain health risks. Health surveillance of workers may be useful for:

- Assessment of the health of workers in relation to risks caused by exposure to chemicals.
- Early diagnosis of work-related diseases and injuries caused by exposure to hazardous chemicals.
- Assessment of the worker's ability to wear or use required respiratory or other PPE.

Health surveillance may be appropriate where exposure to hazardous substances is such that an identifiable disease or adverse health effect may be linked to the exposure. There must be a reasonable likelihood that the disease or effect may occur under the particular conditions of work prevailing and valid techniques must exist to detect such conditions and effects.

The ILO advises in the Code of Practice 'Ambient Factors in the Workplace' that health surveillance may be appropriate where workers are exposed to the following types of hazardous substances:

"(a) Substances (dusts, fibres, liquids, fumes, gases) that have a recognized systemic toxicity (i.e. an insidious poisonous effect).

(b) Substances known to cause chronic effects (e.g. occupational asthma).

(c) Substances known to be sensitizers, irritants or allergens.

(d) Substances that are known or suspected carcinogens, teratogens, mutagens or harmful to reproductive health (reprotoxic substances).

(e) Other substances likely to have adverse health effects under particular conditions or in cases of fluctuations in ambient conditions."

The ILO Code of Practice 'Safety in the Use of Chemicals at Work' sets out a similar requirement. Health surveillance may include, where appropriate, investigations, examination and biological monitoring to detect exposure levels and early biological effects and responses. The types of checks that can be included in a health surveillance programme include:

- Clinical examinations.
- Diagnostic tests, such as by X-ray or ultrasonic scan.
- Function measurements, for example, a lung function test.
- Skin checks for signs of rashes.
- Biological monitoring by carrying out tests of blood or urine.

It may also include simple techniques for the early detection of effects on health, for example, a health assessment by questionnaire and self-checks once the symptoms have been explained.

Where a valid and generally accepted method of biological monitoring of workers' health exists for the early detection of the effects on health of exposure to specific occupational risks, it may be used to identify workers who need a detailed medical examination, subject to the individual worker's consent.

Biological monitoring may be particularly useful in circumstances where:

- There is likely to be significant skin absorption and/or gastrointestinal tract uptake following ingestion.

- There is a reasonably well-defined relationship between biological monitoring and effect.

- It gives information on accumulated dose and target organ burden related to toxicity.

Biological monitoring guidance values (BMGVs) or biological limit values (BLV) are usually set at national level for situations where they are likely to be of practical value, suitable monitoring methods exist and there are sufficient data available (**see Figure 7-58**).

BMGVs/BLVs may not be legally binding and often any biological monitoring undertaken in association with a guidance value needs to be conducted on a voluntary basis (i.e. with the fully informed consent of all concerned). Where a BMGV/BLV is exceeded, it does not necessarily mean that any corresponding airborne standard has been exceeded nor that ill-health will occur. Such limits are intended to indicate that investigation into current control measures and work practices is necessary. Similarly, a low BMGV/BLV should not suggest that there is no need to reduce workplace exposure further.

It is good practice for the employer to keep records of health surveillance, including biological monitoring,

in respect of each worker for an appropriate duration related to the health risk. Typically, this will be at least 30 years, but may be 40 or 50 years. The national competent authority should make arrangements in accordance with national practice to ensure that medical records are maintained for establishments that have closed down.

Provision of information and training to those working with hazardous substances

For control measures to be effective, people need to know their purpose and how to use them. Furthermore, workers should be consulted during the development of control measures. Workers who have been actively involved in the design of controls are more likely to appreciate the need for their use and, therefore, more likely to use them correctly.

The ILO Code of Practice for Ambient Factors in the Workplace establishes a recommendation regarding information and training that the employer should carry out, the employer should ensure sufficient, specific systematic training and information on **(see Figure 7-59).**

 "a) the nature and degree of hazards and risks from hazardous substances which may occur, particularly in the case of an emergency.

b) the protection of their safety and health and that o others from hazardous substances which may be present, in particular by using correct and prescribed methods of handling, storage and transport of hazardous substances, and waste disposal.

c) the correct and effective use of control and protection measures and of personal protective equipment."

Figure 7-59: Information and training by the employer.
Source: ILO Code of Practice - Ambient Factors in the Workplace.

Training should include elements of theory as well as practice. Training in the use and application of control measures and PPE should take account of recommendations and instructions supplied by the manufacturer. Employers should ensure that their relevant managers

Substance	Biological monitoring guidance value	Sampling time
Butan-2-one	70µmol butan-2-one/L in urine	Post shift
Carbon monoxide	30ppm carbon monoxide in end-tidal breath	Post shift
Lindane (Organ chlorine pesticide)	35nmol/L(10µg/L) of Lindane in whole blood (equivalent to 70nmol/Lidane in plasma)	Random
Xylene, o-, m-, p- or mixed isomers	650 mmol methyl hippuric acid/mol creatinine in urine	Post shift

Figure 7-58: Examples of biological monitoring values (BMGVs).
Source: : UK, EH40/2005 Workplace exposure limits.

are appropriately trained so that they can thoroughly instruct the workers regarding the precautions to be taken in their jobs and in the event of emergencies.

Employers should ensure special provisions for training and information are applied for newly recruited workers and for illiterate workers or foreign workers who may encounter language difficulties.

In accordance with the provisions of the Chemicals Convention, (C170), and the ILO Code of Practice Safety in the Use of Chemicals at Work the workers concerned and their representatives should be given information, in forms and languages which they easily understand, on the:

a) Nature of chemicals used at work, the hazardous properties and precautionary measures to be taken.

b) Form of labels and markings, and structure of chemical safety data sheets.

Employers should inform workers and their representatives, as appropriate, of the results of workplace assessments and of their health surveillance in relation to risks caused by exposure to hazardous substances, and in particular those workers who have specific needs for protection related to their health condition.

Figure 7-60: Working with hazardous substances.
Source: Shutterstock.

ADDITIONAL CONTROLS FOR SUBSTANCES WITH SPECIFIC EFFECTS

Carcinogens

Carcinogens are substances that have been identified as having the ability to cause cancer. Examples of these include arsenic, hardwood dusts and used engine oils.

Genetic damage

Substances have been identified that cause changes to DNA, increasing the number of genetic mutations above natural background levels. These are known as mutagens. The changes can lead to cancer in the individual affected or be passed to their offspring's genetic material, for example thalidomide and plutonium oxide.

Due to the serious and irreversible nature of cancer and genetic changes, an employer's first objective must be to prevent exposure to carcinogens and mutagens. These substances should not be used or processes carried out with them, if a safer alternative can be used instead. Where this is not feasible, suitable control measures should include:

- Totally enclosed systems.
- Where total enclosure is not possible, exposure to these substances must kept to as low a level as possible through the use of appropriate plant and process control measures, such as handling systems and LEV (these measures should not produce other risks in the workplace).
- Storage of carcinogens/mutagens must be kept to the minimum needed for the process, in closed, labelled containers with warning and hazard signs, including waste products until safe disposal.
- Areas where carcinogens/mutagens are present must be identified and segregated to prevent spread to other areas.
- The number of people exposed and the duration of exposure must be kept to the minimum necessary to do the work.
- PPE is considered a secondary protection measure used in combination with other control measures.
- Measures should be in place for monitoring of workplace exposure and health surveillance for work involving carcinogens and mutagens.

Occupational asthma

Occupational asthma is caused by substances in the workplace that trigger a state of specific airway hyper-responsiveness in an individual, resulting in breathlessness, chest tightness or wheezing. These substances are known as asthmagens or respiratory sensitisers and include such things as flour/grain dust, wood dust, isocyanates and solder. Not everyone who becomes sensitised goes on to get asthma, but once the lungs become hypersensitive, further exposure to the substance, even at quite low levels, may trigger an attack.

Exposure to these substances should be prevented, and where that is not possible, kept as low as reasonably practicable. Control measures used should take account of long-term time-weighted averages and short-term peak exposures to the substance. The additional control measures for carcinogens and mutagens would also be useful for substances that can lead to occupational asthma. It is also important that workers know what emergency action is required if someone has an asthma attack.

If an individual develops occupational asthma, their exposure must be controlled to prevent any further attacks. Because the worker may have become sensi-

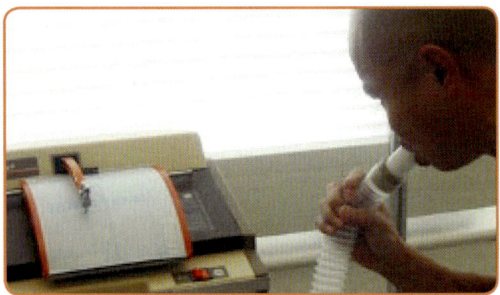

Figure 7-61: Person taking lung function test.
Source: www.elitecycling.co.uk

tised to the substance they may have to be reassigned to work that does not expose them to breathing the substance causing the asthma, for their own protection. Workers who work with substances that can cause asthma, asthmagens, must have regular health surveillance to detect any changes in respiratory function. The purpose of health surveillance is to monitor and protect the health of individual workers. Collecting simple information may enable early detection of ill health caused by work and identify the need for improved control measures.

CASE STUDY

An employee developed occupational asthma after working for a large multinational company in Gloucester, UK. He was employed between 1995 and 2004 as a solderer and was exposed to rosin-based (colophony) solder fume during his career.

His health was deteriorating from 1999 onwards, and he was taking time off work due to breathing difficulties

The company did not have adequate control measures in place and failed to install fume extraction equipment to remove rosin-based fumes from the workroom air or from the breathing zones of its solderers.

The company did not substitute the rosin-based solder with rosin-free solder until December 2003, despite an assessment having identified the need to in 1999.

Employees, including the asthma suffer, were not placed under a health surveillance scheme at any time.

As a result of action taken by HSE, the company was fined £100,000 with £30,000 costs. This attracted local and national media attention.

REVIEW

What factors reduce the effectiveness of LEV?

There are two main categories of respiratory protection, respirators and breathing apparatus. List two advantages and two disadvantages when using breathing apparatus.

7.5 Specific agents

ASBESTOS

Health risks

Asbestos is a general term used to describe a range of mineral fibres (commonly referred to by colour, for example, white, chrysotile; brown, amosite; and blue, crocidolite). Asbestos was mainly used as an insulating and fire-resisting material (and is still used in some countries). Asbestos fibres readily become airborne when disturbed and may enter the lungs, where they cause a number of fatal and serious diseases, the main ones being asbestosis, pleural thickening, mesothelioma and lung cancer.

Asbestosis is a serious scarring condition of the lungs that normally occurs after high levels of exposure to asbestos fibres. The scarring causes the lungs to lose their elasticity and can result in progressive breathlessness, and in severe cases can be fatal.

Asbestosis typically takes more than 10 years to develop. The primary symptom of asbestosis is generally the slow onset of shortness of breath on exertion.

In severe, advanced cases, this may lead to respiratory failure. Coughing is not usually a typical symptom, unless the patient has other respiratory tract diseases.

Figure 7-62. Asbestos label.
Source: Scaftag.

Pleural thickening a problem that generally happens after high levels of exposure to asbestos fibres. Scarring of the lining of the lung (pleura) causes it to thicken and swell. The pleura is a two-layered membrane that surrounds the lungs and lines the inside of the ribcage. If the thickening and swelling covers a large area of the pleura the lung is squeezed, which can cause breathlessness and discomfort in the chest.

Mesothelioma is a cancer which affects the 'mesothelium' (the lining of the lungs (pleura) and the lining surrounding the lower digestive tract (peritoneum)). Even low levels of exposure to asbestos fibres can cause mesothelioma, if exposure is repeated over time. Unfortunately, mesothelioma is incurable and, by the time it is diagnosed, it is almost always fatal. Symptoms of mesothelioma may not appear until 20 to 50 years after exposure to asbestos fibres. Shortness of breath, cough, and pain in the chest due to an accumulation of fluid in the pleural space are often symptoms of pleural mesothelioma.

Asbestos-related lung cancer develops in the tubes that carry air in and out of the lungs, for example, in the bronchus. The cancer can grow within the lung and can spread outside the lung through the rest of the body.

Controls

There may be national prohibition on the importation, supply and use of all forms of asbestos. This may extend to prohibition of second-hand use of asbestos products, such as asbestos cement sheets and asbestos tiles. If existing asbestos-containing materials are in good condition it may be acceptable to leave them in place and to monitor and manage their condition.

General controls related to asbestos include:

- The presence of asbestos should be identified, recorded and labelled.

- An assessment must be done of work that exposes workers to asbestos.

- Information must be provided on the location and condition of the asbestos to anyone who is liable to work on or disturb it.

- A written plan of work is required for work with asbestos.

- Work with asbestos may need to be notified to a national competent authority and a licence for the work obtained.

- Exposure must be prevented or reduced by controls.

- Controls must be used and maintained.

- The employer is responsible for the cleaning of personal protective clothing.

- Awareness training, should be provided for workers, such as cable installer's, decorators, that may inadvertently expose themselves to asbestos fibres during their work. This training should enable them to identify possible asbestos containing materials or residues, before or when carrying out cutting or drilling tasks.

National legislation may require that employers who work with asbestos in circumstances that represent a significant risk be licensed.

Managing asbestos in buildings

Asbestos has been widely used in building materials for a long time, though some countries have established programmes to phase out its use because of the risks to health. As long as the asbestos-containing material (ACM) is in good condition, and is not being or going to be disturbed or damaged, there is negligible risk. But if it is disturbed or damaged, it can become a danger to health, because people may breathe in any asbestos fibres released into the air.

Workers who may be particularly at risk of being exposed to asbestos when carrying out building maintenance and repair jobs include:

- Construction and demolition contractors, roofers, electricians, painters.

- Decorators, joiners, plumbers, gas fitters, plasterers, shop fitters, heating and ventilation engineers, and surveyors.

- Anyone dealing with electronics, for example, phone and information technology (IT) engineers, and alarm installers.

- General maintenance engineers and others who work on the fabric of a building.

If asbestos is present that can be readily disturbed, is in poor condition and not managed properly, all people in the building could be put at risk. Asbestos has been used in many parts of buildings, (*see Figure 7-63*) for examples of uses and locations where asbestos can be found.

Asbestos products	What it was used for
Sprayed asbestos (limpet).	Fire protection in ducts and to structural steel work, fire breaks in ceiling voids etc.
Lagging.	Thermal insulation of pipes and boilers.
Asbestos insulating boards (AIB).	Fire protection, thermal insulation, wall partitions, ducts, soffits, ceiling and wall panels.
Asbestos cement products, flat or corrugated sheets.	Roofing and wall cladding, gutters, rainwater pipes, water tanks.
Certain textured coatings.	Decorative plasters, paints.
Bitumen or vinyl materials.	Roofing felt, floor and ceiling tiles.
General uses.	Vehicle brake linings, woven fires, ropes used as high temperature gaskets for furnaces, jet engines, chemical pipelines. Electrical insulation for hotplate wiring, electrical fuse wire holders and in building insulation and sound absorption. Filters for cigarettes. Artificial (chrysotile) snow effects in Hollywood films made in the USA in the 1920's and 1930's.

Figure 7-63: Examples of uses and locations where asbestos can be found. Source: RMS.

High risk materials

Figure 7-64: Asbestos pipe lagging. Source: UK, HSE.

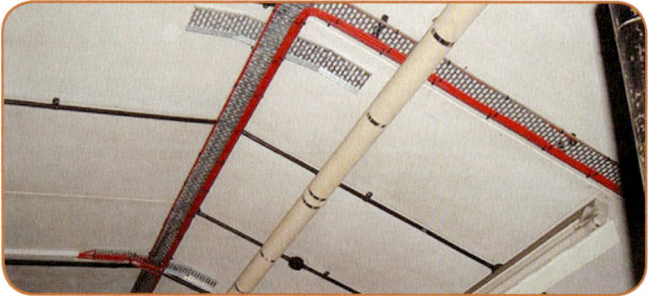

Figure 7-65: Asbestos insulating board (AIB). Source UK, HSE.

Figure 7-66: Textured decorative coating. Source: UK, HSE.

Figure 7-67: Window with AIB panel. Source: UK, HSE.

Normally lower-risk materials

A number of countries have established legal responsibilities that require those that control or occupy buildings to manage the asbestos risks. In the UK, the duty to manage asbestos is included in the Control of Asbestos Regulations (CAR) 2012. This sets out some useful principles that can be applied anywhere.

Figure 7-68: Asbestos cement roof sheeting. Source: UK, HSE.

Control measures to manage the risk from asbestos in buildings include:

- Carrying out a survey to determine if there is asbestos in the premises (or assessing if ACMs are liable to be present and making a presumption that materials contain asbestos, unless there is strong evidence that they do not), its location and its condition.

- Identify and record the location and condition of any ACMs or presumed ACMs in the premises.

- Assess the risks from any material identified.

- Prepare a plan that sets out in detail how risks identified are to be managed.

- Establish a system for providing information on the location and condition of ACMs for anyone who is liable to work on or disturb it.

- Record the roles and responsibilities of those who manage asbestos in the organisation.

- Review and monitor the plan and the arrangements made to prevent risk.

Figure 7-69: Asbestos-containing floor tiles.
Source: UK, HSE.

BLOOD-BORNE VIRUSES

Blood-borne viruses (BBVs) are mainly found in blood or bodily fluids and can transfer from one person to another in these fluids. The main BBVs of concern are human immunodeficiency virus (HIV), hepatitis B and hepatitis C.

Health risks

HIV

The earliest symptoms of HIV can resemble flu and they generally clear up within a month or two. These symptoms may include fever, headache, fatigue, and swelling in the lymph nodes, particularly those in the neck and groin. However, not everyone who acquires HIV will experience these symptoms. Similarly, for several years, perhaps as long as a decade, a person with HIV might not have any symptoms at all.

HIV progresses differently for each person affected. The course of the disease is determined by the specific infections or complications a person with HIV develops. HIV complications can affect different parts of the body. Some are localised to the mouth (white patches in the mouth or pharynx), others in the brain (dementia, difficulty in processing information, brain tumours) and others result in total body changes like losing body weight (loss of muscle as well as fat). Skin conditions in the form of sores or lesions are also common.

HIV is transmitted through contact with body fluids, in particular blood, semen, vaginal secretions and breast milk. It is not transmitted through casual contact. The main method of transfer at work is through laceration or puncture of the skin or contamination of the eyes.

Hepatitis

Hepatitis is an inflammation of the liver usually caused by a virus, although some forms of the disease can be caused by the effects of drugs, toxins, disease or alcohol abuse. There are three types of hepatitis: A, B and C. Each of these types is caused by a different virus.

Vaccines are available to prevent infection by the hepatitis A and B viruses, but no vaccines are currently available to prevent hepatitis C infection, and current treatments for it are not effective in all cases.

Symptoms of Hepatitis A

Hepatitis A is a type of viral liver infection that is widespread in parts of the world such as Africa and India. A person infected with hepatitis A virus, can shed the virus in their stool for a period of time during the illness. Food and water can be contaminated by food handlers who have hepatitis A and do not wash their hands adequately after using the toilet. Initial symptoms of hepatitis A are similar to flu and include experiencing a low-grade fever with body temperatures usually no higher than 39.5°C (103.1°F). In addition, there may be joint pain and feeling/being sick. This may then be followed by symptoms related to the liver becoming infected, such as:

- Yellowing of the skin and eyes (jaundice).
- Passing very dark coloured urine and pale faeces (stools).
- Abdominal pain.
- Itchy skin.

Symptoms usually clear up within two months, although occasionally last up to six months. Older adults tend to have more severe symptoms. In most cases, the liver will make a full recovery.

Symptoms of Hepatitis B

The hepatitis B virus spreads when blood, semen, or another bodily fluid from an infected person enters the body of another individual. Since the virus is extremely infectious - 50 to 100 times more so than HIV - even brief, direct contact could be enough to cause infection. Drug users who share syringes and drug equipment have an increased risk of getting infected. It's estimated that around 16% of new hepatitis B infections, in Europe, are from drug equipment use.

Until the virus has been cleared from those infected, they can pass it onto others. Most people remain healthy without any symptoms while they fight off the virus that causes hepatitis B. Some individuals will not even know they have been infected.

If there are any symptoms, they will develop on average 60-90 days after exposure to the virus.

Common symptoms of hepatitis B include:

- Flu-like symptoms, such as tiredness, general aches and pains, headaches and a high temperature of or above 38°C (100.4°F).
- Loss of appetite and weight loss.
- Feeling sick or being sick.
- Diarrhoea.
- Pain in your upper right-hand side.
- Yellowing of the skin and eyes (jaundice).

Symptoms will usually pass within one to three months. Hepatitis B is said to be chronic when the infection lasts for longer than six months. The symptoms are usually much milder and tend to come and go. In many cases, people with chronic hepatitis B infection will not experience any noticeable symptoms. Symptoms of chronic hepatitis B may include:

- Feeling tired all the time (fatigue).
- Loss of appetite.
- Feeling sick.
- Abdominal pain.
- Muscle and joint pains.
- Itchy skin.

Symptoms of Hepatitis C

Hepatitis C spreads when the blood from a person infected with hepatitis C virus gets into the body. This may occur, for example, through a shared contaminated syringe, or blood transfusion.

The earliest symptoms of infection by hepatitis C may start about six to seven weeks after exposure to the virus. This is considered the acute phase of the virus. Symptoms may include loss of appetite, fever, nausea, fatigue, vomiting, joint pain, jaundice, dark urine or abdominal pain. However, it is not unusual to be infected with hepatitis C and never experience any of these symptoms.

Some people are able to completely recover from hepatitis C after the acute phase. Others will develop the chronic form of the disease. Chronic cases of hepatitis C range from mild to severe and it is possible to have chronic hepatitis C with a mild degree of liver damage yet experience no serious health problems. Occasionally severe infection with hepatitis C may result in liver failure or cirrhosis (scarring) of the liver. Liver failure may occur more rapidly if an individual has HIV in addition to hepatitis C. Most people do not know if they are infected and may live for many years without symptoms. A proportion of people take 20 to 30 years to develop severe liver disease, some recover completely with treatment, and others recover without any treatment at all. A small proportion of infected people develop liver cancer.

Risk occupations and risk factors

Workers who come into contact with bodily fluids from other humans are at risk from BBVs, particularly if their work also involves sharp or abrasive implements or substances that may break the skin. Health-care workers are at an obvious risk. Less obvious, perhaps, are those who work for cleansing or recreation/parks departments and workers who conduct bodily searches or searches of personal effects. Workers who work in these circumstances may come into contact with used needles.

The level of risk will depend on:
- Frequency and scale of contact with bodily fluids.
- The type of fluid or material they come into contact with.
- The activity the person must conduct in relation to the infectious material.
- The nature of the infection contained in the material.

Risk controls

Where a risk of exposure to BBVs has been identified, simple, inexpensive measures to prevent or control risks can be taken:

- Ensure good personal hygiene practices are observed, in particular, hygienic hand-washing.
- Use procedures such as avoiding the use of sharps such as needles, blades, glass, etc.
- Use PPE such as gloves, eye protection, face masks, etc.
- Ensure contaminated waste is disposed of in a safe manner, for example, in a sharps disposal bin.
- Use disposable equipment where there is a risk of BBV contamination, otherwise decontamination procedures must be strictly complied with.
- Ensure workers are aware of immediate steps to be followed upon contamination with blood or other body fluids, for example, needle stick injuries procedure.
- Vaccination where appropriate.
- Decontamination and disinfection procedures.

CARBON MONOXIDE

Carbon monoxide is a colourless and odourless chemical asphyxiant produced as a by-product of incomplete combustion of carbon fuels, for example, gas water heaters, compressors, pumps, dumper trucks or generators. It is a particular risk when operated in poorly ventilated confined areas where workers are forced to inhale it. Carbon monoxide has a great affinity (200 times that of oxygen) with the haemoglobin red blood cells, which means it will inhibit oxygen uptake by red blood cells, resulting in chemical asphyxiation, leading to collapse and death.

Acute effects

The earliest symptoms, especially from low-level exposures, are often non-specific and readily confused with other illnesses, typically flu-like viral syndromes, depression, chronic fatigue syndrome, and migraine or other headaches. This often makes the diagnosis of carbon monoxide poisoning difficult. If suspected, the diagnosis can be confirmed by measurement of blood carboxyhaemoglobin. Common problems encountered are difficulty with higher intellectual functions and short-term memory, dementia, irritability, gait disturbance, speech disturbances, Parkinson-like syndromes, cortical blindness, depression and death.

Chronic effects

Long-term, repeat exposures present a greater risk to persons with coronary heart disease and in pregnant women.

Chronic exposure may increase the incidence of cardio-vascular symptoms in some workers, such as motor vehicle examiners, firefighters and welders. Patients often complain of persistent headaches, light-headedness, depression, confusion and nausea. Upon removal from exposure, the symptoms usually resolve themselves.

CEMENT

Cement is used extensively in the construction industry as part of the mix for mortar and concrete. It is mildly corrosive and can cause harm in the following ways:

- Skin contact - causing contact dermatitis. If it is trapped inside a worker's boot or glove then it can cause severe chemical burns.
- Eye contact - causing irritation and inflammation.
- Inhalation - causing irritation of the nose and throat and possible long-term respiratory problems.

Figure 7-70 shows severe burns sustained by a worker after kneeling in wet cement for three hours; his right leg had to be amputated. Where possible, exposure should be eliminated by pouring and vibrating the cement into position.

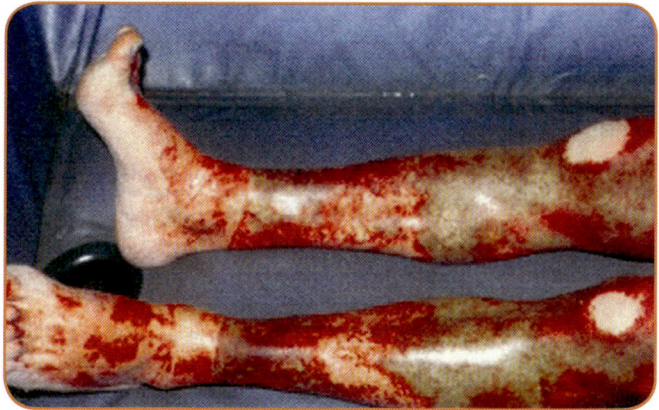

Figure 7-70: Cement burns.
Source: SHP.

Gloves, protective overalls, boots and masks should be worn and any contaminated clothing should be removed and not reused until thoroughly cleaned. Good hygiene standards should be maintained, for example, washing the skin and hands thoroughly after contact with cement.

LEGIONNELLA

Legionnellosis is a potentially fatal disease and is a type of pneumonia caused by legionella pneumophila, a bacterium. Legionnaires' disease is the most severe form of this disease. Legionella Pneumophila is widely present in water and frequently present in water-cooling systems and hot water systems. Large workplace buildings are therefore susceptible to infected water systems, especially hotels and hospitals.

The organism is widespread in the environment, but needs certain conditions to multiply, for example, the presence of sludge, scale, algae, rust and organic material plus a temperature of 20-50°C. Transmission is via inhalation of the organism in contaminated aerosols. Smoking, age and gender may increase susceptibility.

Symptoms are aching muscles, headaches and fever followed by a cough. Confusion, emotional disturbance and delirium may follow the acute phase. In some cases, death may occur. The fatality rate for those that contract Legionnaires' disease has ranged between 5% and 30% during outbreaks; for example, those occurring in the UK are reported to be about 12%. The death rate for older people and the infirm who contract the disease is often as high as 50%.

Greatest risk areas are showers used for bathing, air conditioning sprays, water-cooling towers and recirculating water-cooling systems. The hazard can be controlled by proper design of water systems, disinfection/chlorination of water or heating water to 60°C. Other forms of the illness that are not so severe are Lochgoilhead fever and Pontiac fever.

LEPTOSPIRA

The leptospira (spiral-shaped) bacteria enters the body through broken skin or mucous membrane, causing leptospirosis (Weil's disease). Rodents represent the most important reservoir of infections, especially rats (also gerbils, voles and field mice). **(See Figure 7-72**: Leptospira Scanning Electron Micrograph (SEM)).

Other sources of infection are dogs, hedgehogs, foxes, pigs and cattle. These animals are not necessarily ill, but carry leptospires in their kidneys and excrete it in their urine. Infection can be transmitted directly via direct contact with blood, tissues, organs or urine of one of the host animals or indirectly by a contaminated environment. Symptoms vary but include flu-like illness, conjunctivitis, liver damage (including jaundice), kidney failure and meningitis.

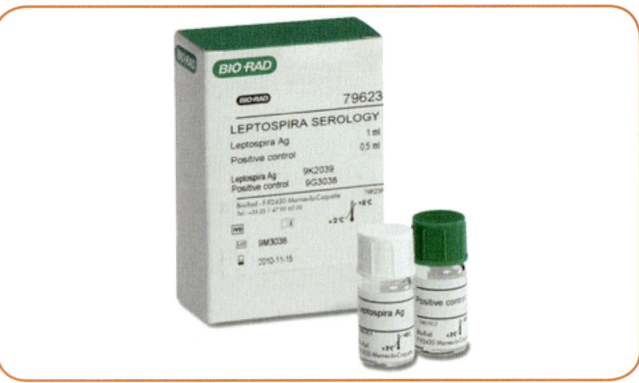

Figure 7-71: Treatment for Leptospira. Source: Bio-Rad.

If untreated, infection may be fatal. Those workers most at risk from leptospira bacteria are those who work where rats prevail, and include water and sewage work, demolition or refurbishment of old unoccupied buildings, and those working on sites adjoining rivers and other watercourses. The bacteria's survival depends on protection from direct sunlight, so it survives well in watercourses and ditches protected by vegetation. Typical control measures include good personal hygiene (for example, hand-washing), PPE (gloves), covering cuts and grazes, preventing animal (rat) infestations and issuing a worker with an 'At Risk' card to show a doctor if they develop flu-like symptoms after being exposed to contaminated water.

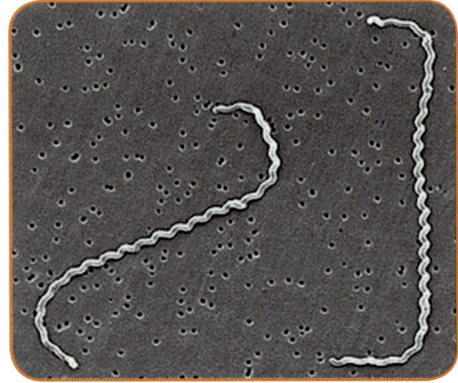

Figure 7-72: Leptospira magnified many times shown on SEM shows spiral shape.
Source: Wikipedia.

SILICA

Silica exists naturally as a crystalline mineral. A common variety is quartz (tridymite, cristobalite). Industrially, silica is used in the morphous (after heating) form, for example, fumed silica, silica gel. In construction activities, it may be encountered in stonework or work with quartz-based tiles. Inhalation of silica dust can result in silicosis, a fibrosis of the lung.

Silicosis (also known as Grinder's disease and Potter's rot) is a form of occupational lung disease caused by inhalation of crystalline silica dust, and is marked by inflammation and scarring in the form of nodular lesions in the upper lobes of the lungs.

Silicosis is the most common occupational lung disease worldwide. It occurs everywhere, but is especially common in developing countries. From 1991 to 1995, China reported more than 24,000 deaths due to silicosis each year. In the United States, it is estimated that 1 to 2 million workers have had occupational exposure to crystalline silica dust and 59,000 of these workers will develop silicosis at some point in their lives. Silicosis (especially the acute form) is characterised by shortness of breath, fever and cyanosis (bluish skin). It may often be misdiagnosed as pulmonary oedema (fluid in the lungs), pneumonia or tuberculosis. Typical control measures include preventing the dust from being inhaled by using alternative methods of processing, using dust suppression or dust collection (LEV). The employee should wear RPE and be subject to health surveillance (lung function testing, etc.).

WOOD DUST

Exposure to wood dust has long been associated with a variety of adverse health effects, including dermatitis, allergic respiratory effects, mucosal and non-allergic respiratory effects and cancer. Contact with the irritant compounds in wood sap can cause dermatitis and other allergic reactions. The respiratory effects of wood-dust exposure include asthma, hypersensitivity, pneumonitis and chronic bronchitis. Health problems arise from work with both softwoods and hardwoods; however, hardwoods are associated with causing nasal cancer. The main operations likely to produce high dust levels in the woodworking industry are:

- Machining operations, particularly sawing, routing and turning.
- Sanding by machine and by hand.
- During factory cleaning, especially when compressed air lines are used for blowing dust from walls, ledges and other surfaces.
- Using compressed air lines to blow dust off furniture and other articles before spraying.
- High airborne dust levels can also occur during the bagging of dust from dust-extraction systems.

 REVIEW

List two occupations that may expose workers to the risk of asbestosis.

Asbestos has been used in building construction and products for many years, list eight uses of asbestos.

What health effects may occur in workers from high levels of dust in the wood-working industry?

What two chronic health effects are associated with hepatitis B infection?

Sources of reference

Reference information provided, in particular web links, was correct at time of publication, but may have changed.

Ambient factors in the workplace, International Labour Organization (ILO) Code of Practice (CoP), ISBN 92-2-11628-X http://www.ilo.org/safework/info/standards-and-instruments/codes/WCMS_107729/lang--en/index.htm

Asbestos, convention and recommendation, C162 and R172, 1986 https://www.ilo.org/dyn/normlex/en/f?p=NORMLEXPUB:12100:0::NO::P12100_ILO_CODE:R172

https://www.ilo.org/dyn/normlex/en/f?p=NORMLEXPUB:12100:0::NO::P12100_ILO_CODE:C162

Asbestos: The Survey Guide HSG 264, HSE Books, ISBN: 978-0-7176-6502-0 http://www.hse.gov.uk/pubns/books/hsg264.htm

Chemicals, convention and recommendation, C170 and R177, 1990 https://www.ilo.org/dyn/normlex/en/f?p=NORMLEXPUB:12100:0::NO::P12100_ILO_CODE:C170

https://www.ilo.org/dyn/normlex/en/f?p=1000:12100:::NO:12100:P12100_INSTRUMENT_ID:312515

Classification, Labelling and Packaging of Substances and Mixtures Regulations EC No 1272/2008 – https://echa.europa.eu/documents/10162/23036412/clp_labelling_en.pdf/89628d94-573a-4024-86cc-0b4052a74d65

Control of exposure to silica dust, A guide for employees, INDG463, HSE Books, https://www.hse.gov.uk/pubns/indg463.pdf

Controlling Airborne Contaminants at Work: A Guide to Local Exhaust Ventilation, HSG258, second edition 2011, HSE Books, ISBN: 978-0-717664-15-3 http://www.hse.gov.uk/pubns/books/hsg258.htm

Factsheet about leptospirosis, European Centre for Disease Prevention and Control, https://www.ecdc.europa.eu/en/leptospirosis/factsheet

Health hazards, wood dust, Occupational Safety & Health Administration, United States Department of Labour, https://www.osha.gov/SLTC/etools/sawmills/dust.html

HIV/AIDS and the world of work, ILO Code of Practice, https://www.ilo.org/aids/Publications/WCMS_113783/lang--en/index.htm

ILO Asbestos, Convention, C162, ILO, 1986, https://www.ilo.org/dyn/normlex/en/f?p=NORMLEXPUB:12100:0::NO::P12100_ILO_CODE:C162%20

ILO Asbestos, Recommendation, C172, ILO, 1986, https://www.ilo.org/dyn/normlex/en/f?p=1000:12100:0::NO::P12100_ILO_CODE:R172

ILO Chemicals, Convention, C170, 1990, https://www.ilo.org/dyn/normlex/en/f?p=1000:12100:0::NO::P12100_ILO_CODE:C170

ILO Chemicals, Recommendation, R177, 1990, https://www.ilo.org/dyn/normlex/en/f?p=1000:12100:0::NO::P12100_ILO_CODE:R177

ILO Working Environment (Air, Pollution, Noise and Vibration) Convention, 1977 (No 148) - C148, https://www.ilo.org/dyn/normlex/en/f?p=1000:12100:0::NO::P12100_ILO_CODE:C148

ILO Working Environment (Air, Pollution, Noise and Vibration) Recommendation, 1977, R156, https://www.ilo.org/dyn/normlex/en/f?p=1000:12100:0::NO::P12100_ILO_CODE:R156

Information on the Control of Asbestos Regulations 2012 http://www.hse.gov.uk/asbestos/regulations.htm

Legionellosis, World Health Organization, https://www.who.int/ith/diseases/legionellosis/en/

Managing and working with asbestos, Control of Asbestos Regulations 2012. Approved Code of Practice and guidance ISBN: 978-0-717666-18-8 http://www.hse.gov.uk/pubns/books/l143.htm

Managing Asbestos in Buildings: A brief guide, INDG223(rev5) 2012, HSE Books http://www.hse.gov.uk/pubns/indg223.pdf

Managing skin exposure risks at work, HSG262, HSE Books, ISBN: 978-0-7176-6649-2, http://www.hse.gov.uk/pubns/priced/hsg262.pdf

Occupational cancer, convention and recommendation, C139 and R147, 1974 https://www.ilo.org/dyn/normlex/en/f?p=NORMLEXPUB:12100:0::NO::P12100_ILO_CODE:C139

https://www.ilo.org/dyn/normlex/en/f?p=NORMLEXPUB:12100:::NO:12100:P12100_ILO_CODE:R147:NO

Occupational Exposure Limits, EH40/2005, HSE Books, ISBN: 978-0-717664-46-7 http://www.hse.gov.uk/pubns/books/eh40.htm

Occupational exposure to airborne substances harmful to health, ILO CoP, ILO Geneva ISBN: 978-9-221024-42-3 https://www.ilo.org/wcmsp5/groups/public/---ed_protect/---protrav/---safework/documents/normativeinstrument/wcms_107851.pdf

Occupational health services, convention and recommendation, C161 and R171, 1985

https://www.ilo.org/dyn/normlex/en/f?p=NORMLEXPUB:12100:0::NO::P12100_ILO_CODE:C161

https://www.ilo.org/dyn/normlex/en/f?p=1000:12100:::NO:12100:P12100_INSTRUMENT_ID:312509

Registration, Evaluation, Authorisation and Restriction of Chemicals (REACH) http://www.hse.gov.uk/reach/

Regulation (EC) No 1272/2008 on classification, labelling and packaging of substances and mixtures (CLP), https://echa.europa.eu/guidance-documents/guidance-on-clp

Safety in the use of asbestos, ILO Code of Practice, https://www.ilo.org/wcmsp5/groups/public/---ed_protect/---protrav/---safework/documents/normativeinstrument/wcms_107843.pdf

Safety in the use of chemicals at work, International Labour Organization (ILO) Code of Practice (CoP), ILO, 1993. ISBN: 978-9-221080-06-4 http://www.ilo.org/wcmsp5/groups/public/@ed_protect/@protrav/@safework/documents/normativeinstrument/wcms_107823.pdf

   Section 6: Operational control measures and

   Section 7: Design and installation

Step by Step Guide to COSHH Assessment, HSG97, second edition 2004, HSE Books, ISBN: 978-0-717627-85-1 http://www.hse.gov.uk/pubns/books/hsg97.htm

The Control of Asbestos Regulations 2012, UK, www.legislation.gov.uk (please enter the name of the Regulation)

The Control of Substances Hazardous to Health Regulations 2002, UK, www.legislation.gov.uk (please enter the name of the Regulation)

The health and safety toolbox, How to control risks at work, HSG268, HSE Books, ISBN: 978-0-7176-6587-7 , http://www.hse.gov.uk/pUbns/priced/hsg268.pdf

The Personal Protective Equipment at Work Regulations 1992, UK, www.legislation.gov.uk (please enter the name of the Regulation)

Wood dust, Controlling the risks, Woodworking Sheet No 23, HSE, http://www.hse.gov.uk/pubns/wis23.pdf

Web links to these references are provided on the RMS Publishing website for ease of use - www.rmspublishing.co.uk.

STUDY QUESTIONS

1) What are two forms of biological agent and give a workplace example for each. (8)

2) What are the practical measures that complement the body's protection mechanisms against ill-health from agents in the workplace? (8)

3) What are the limitations of information provided by manufacturers and suppliers in assessing risks to health? (8)

4) What are the basic components of a local exhaust ventilation (LEV) system? (8)

5) What are the factors that reduce the efficiency of a local exhaust ventilation (LEV) system? (8)

6) What are the circumstances where breathing apparatus may be the preferred respiratory protection rather than the use of respirators? (8)

For guidance on how to answer NEBOSH questions please refer to the 'study question answer guidance' section located at the back of this guide.

Element 8

General workplace issues

8.1 Health, welfare and work environment

HEALTH AND WELFARE PROVISIONS

In this section welfare provision means access to facilities and services that will ensure the worker's health and welfare. Conventions, Recommendations and Codes of Practice will dictate, or give guidance on local standards to be achieved. The provision of first-aid is often considered as part of an employer's duties to provide health and welfare provision. The topic of first-aid is covered in **Element 3.8 – 'Emergency procedures'**

Supply of drinking water

An adequate supply of wholesome drinking water or some other wholesome drink must be provided in the workplace. The supply needs to be accessible to workers. Whenever the distribution of running drinking water is practicable, preference should be given to this system. If water is not provided in the form of a fountain, then drinking vessels must also be provided.

Figure 8-1: Safe Drinking water sign.
Source: Rivington signs.

The supply outlet from taps should be labelled, 'Suitable for drinking', or 'Unsuitable for drinking' as appropriate. ILO Hygiene (Commerce and Offices) Recommendation R120 sets out additional useful advice that should be considered when meeting the requirement for drinking water:

"*Recommendation 29:*

1) *Any containers used to distribute drinking water or any other authorised drink should:*

   a) *Be tightly closed and where appropriate fitted with a tap.*

   b) *Be clearly marked as to the nature of their contents.*

   c) *Not be buckets, tubs or other receptacles with a wide open top (with or without a lid) in which it is possible to dip an instrument to draw off liquid.*

   d) *Be kept clean at all times.*

2) *A sufficient number of drinking vessels should be provided and there should be facilities for washing them with clean water.*

3) *Cups, the use of which is shared by a number of workers should be forbidden.*

Figure 8-2: Drinking water. Source: Semcog.

Figure 8-3: Drinking water. Source: Pexels.

Recommendation 30:

1) *Water which does not come from an officially approved source for the distribution of drinking water should not be distributed as drinking water, unless the competent health authority expressly authorises such distribution and holds periodical inspections.*

2) *Any method of distribution other than that practised by the officially approved local supply service should be notified to the competent health authority for its approval.*

Recommendation 31:

1) *Any distribution of water not fit for drinking should be so labelled at the points where it can be drawn off.*

2) *There should be no inter-connection, open or potential, between drinking water systems and systems of water not fit for drinking."*

Figure 8-4: Do not drink sign.
Source: ISO 7010.

Washing facilities

Suitable and sufficient washing facilities must be provided at readily accessible places, including showers where necessary because of the nature of the work or for health reasons. It is recommended that there is one wash basin for every 25 employees and, if necessary, provision should be made for people with disabilities. There must be a supply of clean, hot and cold or warm water, running water so far as is practicable, soap or other means of cleaning, towels or other means of drying, sinks large enough to wash face, hands and fore-arms.

The rooms that contain these facilities must be kept clean, ventilated, lit and maintained.

ILO Hygiene (Commerce and Offices) Recommendation R120 sets out additional useful advice that should be considered when meeting the requirement for washing facilities:

"Recommendation 32: Sufficient and suitable washing facilities should be provided for the use of workers in suitable places and should be properly maintained.

Recommendation 33:

1) *These facilities should, to the greatest possible extent, include washstands, with hot water if necessary, and, where the nature of the work so requires, showers with hot water.*

2) *Soap should be made available to workers.*

3) *Appropriate products (such as detergents, special cleansing creams or powders) should be made available to workers wherever the nature of the work so requires; the use of products harmful to health for personal cleanliness should be forbidden.*

4) *Towels, preferably individual, or other suitable means of drying themselves should be made available to workers. Towels for common use which do not provide a fresh clean portion for each use should be forbidden.*

Recommendation 34:

1) *Water provided for washstands and showers should not present any health risks.*

2) *Where water used in washstands and showers is not fit for drinking, this should be clearly indicated.*

Figure 8-5: Washing facilities.
Source: Health.mil

Recommendation 35: Separate washing facilities should be provided for men and women, except in very small establishments where common facilities may be provided with the approval of the competent authority.

Recommendation 36: The number of washstands and showers should be fixed by the competent authority having regard to the number of workers and the nature of their work."

Sanitary conveniences

Readily accessible, suitable and sufficient sanitary conveniences must be provided. The conveniences must be adequate for the numbers and gender employed, lit, kept clean and maintained in an orderly fashion. It is recommended that there is at least one sanitary convenience for every 25 employees and, if necessary, provision should be made for people with disabilities. Where flushing water cannot be provided to aid cleaning a facility that uses chemicals or a dry (composting) toilet that uses aerobic processing to treat waste should be provided.

Separate conveniences for male and female workers must be provided except where the convenience is in a separate room and the door of which is capable of being locked from the inside.

ILO Hygiene (Commerce and Offices) Recommendation R120 sets out additional useful advice that should be considered when meeting the requirement for sanitary conveniences:

"Recommendation 38:

1) *Sanitary conveniences should be so partitioned as to ensure sufficient privacy.*

2) *As far as possible sanitary conveniences should be supplied with flushing systems and traps and with toilet paper or some other hygienic means of cleaning.*

3) *Appropriately designed receptacles with lids or other suitable disposal units such as incinerators should be provided in sanitary conveniences for women.*

4) *As far as possible, conveniently accessible washstands in sufficient number should be provided near conveniences.*

5) *Recommendation 41: Sanitary conveniences should be adequately ventilated and so located as to prevent nuisances. They should not communicate directly with workplaces, rest rooms or canteens, but should be separated there from by an antechamber or by an open space. Approaches to outdoor conveniences should be roofed."*

Accommodation for clothing

Suitable and sufficient accommodation must be provided in a suitable location for personal clothing where a worker is required to change clothing for work. Separate storage may be necessary for clothing worn at work to prevent possible cross contamination with the worker's clothes. Clothing accommodation must be secure when personal clothing not worn at work is being stored. Suitable drying facilities should be provided for clothing when necessary.

ILO Hygiene (Commerce and Offices) Recommendation R120 sets out additional useful advice that should be considered when meeting the requirement for clothing:

"Recommendation 47:

1) *Changing rooms should contain:*

   a) *Properly ventilated personal cupboards or other suitable receptacles of sufficient dimensions, which can be locked.*

   b) *A sufficient number of seats.*

2) *Separate compartments for street clothes and working attire should be provided whenever workers are engaged in operations necessitating the wearing of working attire that may be contaminated, heavily soiled, stained or impregnated."*

Facilities for changing clothing

Where special clothing must be worn at work for reasons of health or to conform to standards and a person cannot change in another room, suitable and sufficient changing facilities must be provided. Separate facilities or separate use of facilities for male and female workers must be taken into account.

Changing facilities should be readily accessible to workrooms (and eating facilities if provided). The facilities provided should be sufficiently large to enable the maximum number of workers to use them comfortably and quickly at any one time.

Rest and eating facilities

Readily accessible, suitable and sufficient rest facilities should be provided. Such rest facilities must be provided in one or more rest rooms.

Where meals are regularly eaten at work, suitable and sufficient facilities must be provided for their consumption. Where food eaten in the workplace is liable to become contaminated, suitable facilities for eating meals must be included in the rest facilities. Suitable rest facilities must be provided for pregnant women and nursing mothers. Rest facilities should include suitable and sufficient seats and tables for the number of workers likely to use them at any one time. Work seats in offices or other clean environments may be acceptable as rest facilities provided workers are not subjected to excessive disturbance during rest periods.

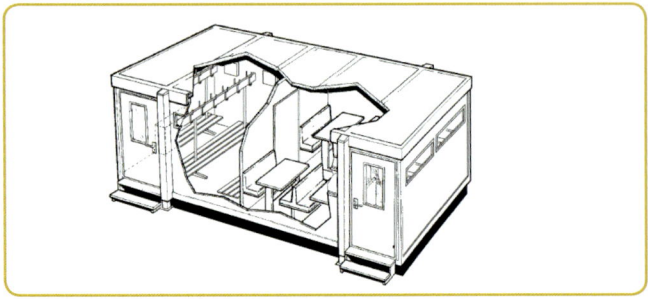

Figure 8-6: Welfare unit restroom/drying room etc.
Source: UK, HSE HSG150.

Eating facilities should include a facility for preparing or obtaining a hot drink, and where hot food cannot be readily obtained, means should be provided to enable workers to heat their own food. Canteens, etc. may be used as rest facilities providing there is no obligation to buy food.

Non-smoking facilities should be available for workers to avoid contamination with second hand smoke. Smoking (cigarettes/cigars/pipes etc.) in the workplace is strictly controlled by many statute laws in a wide range of countries. Provision does not have to be made for workers who smoke but if facilities are provided they should be at a distance away from other workers and buildings.

ILO Hygiene (Commerce and Offices) Recommendation R120 sets out additional useful advice that should be considered when meeting the requirement for rest and eating facilities:

"Rest Rooms - Recommendation 69. Where alternative facilities are not available for workers to take temporary rest during working hours, a rest room should be provided, where this is desirable, having regard to the nature of the work and any other relevant conditions and circumstances. In particular, rest rooms should be provided to meet the needs of women workers; of workers engaged on particularly arduous or special work requiring temporary rest during working hours; or of workers employed on broken shifts.

Recommendation 70: The facilities so provided should include at least:

   a) *A room in which provision suited to the climate is made for relieving discomfort from cold or heat.*

   b) *Adequate ventilation and lighting.*

   c) *Suitable seating facilities in sufficient numbers.*

Mess rooms - Recommendation 67:

   a) *Mess rooms should be provided with sufficient seats and tables.*

   b) *Within or in the immediate vicinity of mess rooms arrangements for heating meals, cool drinking water and hot water should be available.*

   c) *Covered waste bins should be provided.*

Figure 8-7: Rest and eating facilities.
Source: ILO, Ergonomic checkpoints.

Recommendation 68:

1) *Mess rooms should be separate from any place in which there is exposure to toxic substances.*

2) *The wearing of contaminated work clothing in mess rooms should be forbidden."*

Seating

Suitable and sufficient seats should be provided and maintained for workers who have in the course of their work reasonable opportunity for sitting without detriment to their work. Seats should be adequate in number and reasonably near the work position of the workers. The seat should be of adequate design, construction and dimension suitable for the worker and work. The seats should facilitate good posture and be comfortable; a foot-rest should be provided where necessary.

Seats should be provided for use during breaks. Rest rooms should have sufficient seats with backrests for the number of workers who may use them at any one time. Seats in the work area can be counted as eating facilities provided the place is clean and there is a suitable surface to place food. There should also be facilities for pregnant and nursing mothers to lie down if required. The Western Australia -Occupational Safety and Health Regulations 1996, 3.19 - Seating state

1) If an employee's work is done from a sitting position or is of a kind that can be satisfactorily done from a sitting position then the employer must provide and maintain seating:

a) That is designed having regard to the nature of the work to be performed and the characteristics of the workstation.

b) That is strongly constructed, stable, comfortable and of suitable size and height for the employee.

c) If practicable, has a backrest or is otherwise designed to provide back support.

2) If an employee's work is done from a standing position and the employee's work allows the employee to sit from time to time then, to the extent practicable, the employer must provide and maintain seating so that the employee may sit down for the periods when the employee is not working."

Figure 8-8: Requirements for seating.
Source: Western Australia Occupational Safety and Health Regulations 1996.

Ventilation

All premises used by workers must have sufficient and suitable ventilation, natural or artificial or both, supplying fresh or purified air. This includes all places in which work is carried on, or which contain sanitary or other facilities for the common use of workers. Steps should be taken to ensure that workers are not subject to unnecessary draughts.

All plant providing ventilation must be maintained regularly and be fitted with an audible and/or visual warning of failure of the plant.

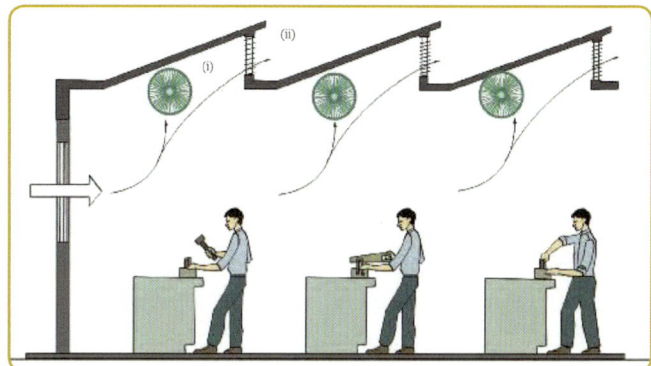

Figure 8-9: A combined ventilation system.
Source: ILO, Ergonomic checkpoints.

ILO Hygiene (Commerce and Offices) Recommendation R120 sets out additional useful advice that should be considered when meeting the requirement for ventilation:

"Recommendation 12 - In particular:

a) *Apparatus ensuring natural or artificial ventilation should be so designed as to introduce a sufficient quantity of fresh or purified air per person and per hour into an area, taking into account the nature and conditions of the work.*

b) *Arrangements should be made to remove or make harmless, as far as possible, fumes, dust and any other obnoxious or harmful impurities which may be generated in the course of work.*

c) *The normal speed of movement of air at fixed work-stations should not be harmful to the health or comfort of the persons working there.*

d) *As far as possible and in so far as conditions require, appropriate measures should be taken to ensure that in enclosed premises a suitable hygrometric (humidity) level in the air is maintained.*

Recommendation 3 - Where a workplace is wholly or substantially air conditioned, suitable means of emergency ventilation, natural or artificial, should be provided."

Heating

The level of heating/cooling should be appropriate to provide physical comfort. Whenever possible the individual should be able to adjust their workplace to achieve this objective. Sufficient thermometers should be provided to enable workers to check the temperature at the workplace. Any equipment used to heat or cool the workplace must not emit vapours or gases that cause injury or illness.

Workplaces can vary greatly:

- Very cold and exposed, such as an offshore oil rig platform or a frozen food storage warehouse.

- General warehousing at ambient temperature.

- General office where the nature of the work is sedentary.

- High temperature processing such as a laundry.

- High temperature manufacturing such as with glass, steel or ceramics production.

As comfortable and steady a temperature as circumstances permit should be maintained in all premises used by workers. Workers should be able to work in comfort and, if necessary, the workplace must be thermally insulated to give protection from extremes of temperature.

The nature of the work and the working environment will need to be assessed to achieve the correct level. Several factors should be considered, such as the work demands, personal capability of the worker, degree of heat or cold, wind speed and humidity.

ILO Hygiene (Commerce and Offices) Recommendation R120 sets out additional useful advice that should be considered when meeting the requirement for temperature:

"Recommendation 19 - In all places in which work is carried on, or through which workers may have to pass, or which contain sanitary or other facilities provided for the common use of workers, the best possible conditions of temperature, humidity and movement of air should be maintained, having regard to the nature of work and the climate.

Recommendation 20 - No worker should be required to work regularly in an extreme temperature.

Recommendation 21 - No worker should be required to work regularly in conditions involving sudden variations in temperature which are considered by the competent authority to be harmful to health.

Recommendation 22:

1) *No worker should be required to work regularly in the immediate neighbourhood of equipment radiating a large amount of heat or causing an intense cooling of the surrounding air, considered by the competent authority to be harmful to health, unless suitable control measures are taken, the time of the worker's exposure is reduced, or he is provided with suitable protective equipment or clothing.*

2) *Fixed or movable screens, deflectors or other suitable devices should be provided and used to protect workers against any large-scale intake of cold or heat, including the heat of the sun.*

Recommendation 23:

1) *No worker should be required to work at an outdoor sales counter in low temperatures likely to be harmful unless suitable means of warming himself are available.*

2) *No worker should be required to work at an outdoor sales counter in high temperatures likely to be harmful unless suitable means of protection against such high temperatures are available.*

Recommendation 24 - The use of methods of heating or cooling likely to cause harmful or obnoxious fumes in the atmosphere of premises should be forbidden.

Recommendation 25 - When work is carried out in a very low or a very high temperature, workers should be given a shortened working day or breaks included in the working hours, or other relevant measures taken."

The ILO recommends that national competent authorities establish maximum/minimum standards of temperature, having regard to the climate and to the nature of the undertaking and of the work.

Generally, countries have not set fixed standards in primary legislation preferring to set out general requirements to protect workers from extremes of heat and cold, and if working in premises to provide heating and cooling to enable worker comfort.

For example, the Singapore, Workplace Safety and Health (General Provisions) Regulations 1996 state under the heading of protection against excessive heat or cold and harmful radiations:

"10) It shall be the duty of the occupier of a workplace to take all reasonably practicable measures to ensure that persons at work in the workplace are protected from excessive heat or cold and harmful radiations."

Figure 8-10: Requirements for temperature.
Source: Singapore, Workplace Safety and Health (General Provisions) Regulations 1996.

The absence of temperature standards in primary legislation leaves any detail to be found in advisory codes of practice, for example, in the UK, the HSE code of practice to the Workplace (Health, Safety and Welfare) Regulations 1992 states that a temperature of 16°C should be maintained for sedentary work, for example, in offices, and a temperature of 13°C for work that requires physical effort; no maximum temperature is specified.

It has been found that setting temperature standards for high temperatures can be complex and different standards may be established for thermal comfort to those standards set to prevent heat stress illness.

The American Conference of Governmental Industrial Hygienists (ACGIH) recommends Threshold Limit

Values (TLVs) for working in hot environments. These limits are given in units of WBGT (wet bulb globe temperature) degrees Celsius (°C). The WBGT unit takes into account environmental factors namely, air temperature, humidity and air movement, which contribute to perception of hotness by people. In some workplace situations, solar load (heat from radiant sources) is also considered in determining the WBGT.

Some countries have adopted these TLVs as occupational exposure limits and others use them as guidelines to control heat stress in the workplace. The TLVs are adjusted to relate to the work/rest cycle, work rate and whether the worker is acclimatised. For example, light work for an acclimatised worker with good opportunities for rest would have a limit of 31°C, whereas very heavy work with good opportunities for rest would have a limit of 30°C.

See also - 'Exposure to extremes of temperature' later in this section.

Lighting

All premises used by workers must have sufficient and suitable lighting. Workplaces should, as far as possible, have natural lighting. Windows and skylights must be kept clean and additional lighting provided if necessary.

ILO Hygiene (Commerce and Offices) Recommendation R120 sets out additional useful advice that should be considered when meeting the requirement for lighting:

"Recommendation 14 - In all places in which work is carried on, or through which workers may have to pass or which contain sanitary or other facilities provided for the common use of workers, there should be, as long as the places are likely to be used, sufficient and suitable lighting, natural or artificial, or both.

Recommendation 15 - In particular, all practicable measures should be taken:

a) *To ensure visual comfort:*

  i) *By openings for natural lighting which are appropriately distributed and of sufficient size.*

  ii) *By a careful choice and appropriate distribution of artificial lighting.*

  iii) *By a careful choice of colours for the premises and their equipment.*

b) *To prevent discomfort or disorders caused by glare, excessive contrasts between light and shade, reflection of light and over-strong direct lighting.*

c) *To eliminate harmful flickering whenever artificial lighting is used."*

Lighting the task and workstation

In general, for each visual task a minimum quantity of light arriving on each unit area of the object in view (i.e. a minimum 'planar' illuminance) is required. The value of this minimum luminance depends primarily on the size of the detail which must be perceived, but will also depend on the visual contrast that the task makes with the background against which it is seen, the duration of the task, whether or not errors may have serious consequences and the presence or absence of daylight.

Ideally the workstation should be designed in such a way as to make the task the brightest part of the field of view. Research has shown that favourable conditions exist when the task has a luminance, which is about three times that of its immediate surrounds, and when the immediate surrounds, again, have about three times the luminance of the general surrounds to the workstation.

These conditions can be achieved by a combination of general and local lighting used to illuminate the work surfaces, which have appropriately chosen reflectance. For example, when a desk lamp provides local lighting for white paper seen against a grey blotting pad placed on a desktop served by a general installation of ceiling-mounted fluorescent lighting, an approximation to these desirable conditions is then easily obtained.

Not all problems caused by poor lighting can be overcome simply by the provision of more light. When tasks are visually demanding, the quality of the available light is at least as important as its quantity.

Daylight has qualities which are difficult to imitate artificially. Moreover, the provision of a distant view seen through a window will provide welcome relief to eyes which must focus at their 'near point' while work is in progress.

Aspects of lighting quality, which should receive attention when lighting the workstation are, the control of glare, the provision of adequate modelling *(see following section - 'Lighting the interior of the workplace')* and, where necessary, good colour rendering.

Figure 8-11: Reflected light from window.
Source: RMS.

Lighting the interior of the workplace

The basic need is to provide sufficient and suitable light in the circulation areas to allow movement of personnel, materials and equipment between workstations to take place conveniently and in safety. The aim should be to provide conditions that remain comfortable to the eye as it passes from one zone of the workplace to another, and to make available all information relevant to the well-being and safety of the workforce which can be received through the sense of sight. All the aspects of quantity and quality considered under task lighting would assume some significance in this larger scale application and so contribute to the achievement of these desirable conditions.

Attention should be drawn to hazards, such as changes in floor level or flights of stairs, with increased levels of illuminance served by luminaires that are carefully positioned to provide good three-dimensional modelling, while preventing any direct sight of the unshielded source, so as to avoid disabling glare. Machinery, which makes fast cyclic movements, should be carefully illuminated to prevent the occurrence of stroboscopic effects.

Sudden changes in lighting levels should not occur between neighbouring zones of the workplace. Levels should be graded to allow time for the eye to adapt. The processes of dark adaptation can take several minutes (more than half an hour in extreme cases) during which time the efficiency of the eye is severely reduced, making accidents/incidents more likely.

Differences in the colour characteristics of so-called 'white light' sources are sometimes apparent to the eye and although its ability to colour-adapt is considerable, the occurrence of frequent noticeable changes due to the use of various kinds of sources in the same interior may accelerate the onset of fatigue as well as making fine colour judgments impossible.

To summarise, some of the factors to be considered when determining appropriate lighting levels in indoor workplaces are:

- The tasks being carried out.
- Colour rendition.
- The type of equipment being used.
- The size of the workplace.
- Areas in shadow that might require additional, temporary lighting.
- The availability of natural light, bearing in mind that it changes with the seasons and time of day.
- The location of equipment being used.
- Levels of contrast between workplaces.
- The need for emergency lighting.
- Suitability for the environment (for example, intrinsically safe in flammable atmospheres).
- Lighting must not flicker or create a stroboscopic effect.

Cleanliness

All premises used by workers, and the equipment of such premises, must be properly maintained and kept clean. The accumulation of waste can be the source of several health and safety related problems. Poor storage of materials, including waste, creates a common source of fuel for unwanted fires. Waste materials such as off-cuts of wood and banding wire left in gangways and corridors is the cause of many trips and falls. Food waste attracts vermin and increases the likelihood of biological hazards. It is much easier to detect problems such as leaks in machinery that is normally maintained in a clean and tidy condition. ILO Hygiene (Commerce and Offices) Recommendation R120 sets out useful advice that should be considered when meeting the requirement for cleanliness:

"Recommendation 5 - All places in which work is carried on, or through which workers may have to pass, or which contain sanitary or other facilities provided for the common use of workers, and the equipment of such places, should be properly maintained.

Recommendation 6:

1) *All such places and equipment should be kept clean.*

2) *In particular the following should be regularly cleaned:*

   a) *Floors, stairs and passages.*

   b) *Windows used for lighting and sources of artificial lighting.*

   c) *Walls, ceilings and equipment.*

Recommendation 7 - Cleaning should be carried out:

   a) *By means raising the minimum amount of dust.*

   b) *Outside working hours, except in particular circumstances or where cleaning during working hours can be effected without disadvantage for the workers.*

Recommendation 8 - Cloakrooms, lavatories, washstands and, if necessary, other facilities for the common use of workers should be regularly cleaned and periodically disinfected.

Recommendation 9 - All refuse and waste likely to give off obnoxious, toxic or harmful substances, or be a source of infection, should be made harmless, removed or isolated at the earliest possible moment; disposal should be in accordance with standards approved by the competent authority.

Recommendation 10 - Removal and disposal arrangements for other refuse and waste should be made and sufficient receptacles for such refuse and waste should be provided in suitable places."

Workspace

All workplaces must be so laid out and workstations so arranged that there is no harmful effect on the health of the worker. Each worker should have sufficient unobstructed working space to perform their work without risk to their health. Consideration should be given to the:

- Floor area to be provided in enclosed premises for each worker regularly working there.
- Minimum unobstructed volume of space to be provided in enclosed premises for each worker regularly working there.
- Minimum height of new enclosed premises in which work is to be regularly performed.

EXPOSURE TO EXTREMES OF TEMPERATURE

Human body temperature

The human body maintains temperature by balancing heat loss and heat gain. The average 'normal' core body temperature is approximately 37°C (99°F). The body generates heat energy by the conversion of foodstuffs; energy is generated through muscle action. At rest, typically 80 watts of energy is produced, whereas during heavy physical exercise perhaps 500 watts is produced. The body loses (exchanges) heat energy by the process of perspiration (sweating) evaporating, conduction of heat through contact of the feet with the floor and by radiation (infra-red energy loss). The clothing worn and the effects of the environment around the person can greatly influence the amount of heat lost or gained.

Effects of exposure to extremes of temperature

Workers can be exposed to a varying degree of conditions and resultant temperatures. The effects of excessive cold or heat can have harmful effects on worker's health and accidents/incidents can result due to fatigue or thermal stress. People working outside their thermal comfort range can suffer a dramatic loss of concentration and efficiency. The precise effects will depend upon the type of work being carried out, the rate of air movement and temperature and humidity.

One of the fundamental mechanisms by which the body regulates its temperature is by perspiration. Important factors that aid or hinder this process are air movement and humidity. Humidity relates to the moisture content in the air. Air with a relatively high humidity has little capacity to cool the body by 'wicking' (drawing off moisture by capillary action) away sweat, whereas low relative humidity can cause dry skin and has been identified as a possible factor in facial dermatitis (occasional itching or reddened skin) reported by some display screen equipment users. High levels of air movement can lead to a cooling effect, particularly if the air is cool,

this is sometimes known as the wind chill effect. The air movement can have a significant effect on the body temperature, reducing an already cool body to dangerously low levels.

Industries and occupations particularly susceptible to extremes of temperature include hot metal process working (foundries and forges), refrigerated warehouses, those who carry out hot work (for example, burning and welding) or work in confined spaces where temperatures are often uncomfortably high. Similarly, those who work outdoors can be subject to extreme weather conditions that can lead to very hot or very cold temperatures.

Figure 8-12: Hot metal process working.
Source: RMS.

The main effects of working at high and low temperatures are outlined as follows:

General effects
Under cold conditions:

- Loss of concentration in mental work.
- Reduced manipulative powers in manual work.
- Discomfort caused by shivering.
- Frost burns caused by skin contact with very cold surfaces.
- Extreme cold can also cause problems with plant/equipment such as the brittle failure of racking.

Under hot conditions:

- Loss of concentration.
- Reduced activity rate.
- Discomfort caused by sweating.
- Dehydration as a result of sweating.
- Muscle cramping as a result of lost body salts during sweating.
- Heat exhaustion just prior to heat stroke.

Heat stress
- Heat rash or 'prickly heat'.
- Heat exhaustion.
- Anhidrotic heat exhaustion.
- Heat cramps.
- Heat stroke.

- Fainting due to vasodilatation (widening of the blood vessels, especially the arteries, leading to increased blood flow or reduced blood pressure).
- Skin disorder.
- Fatigue, nausea, headache, giddiness.
- Insufficient moisture to sweat.
- Painful spasms of muscles - insufficient salt.
- Breakdown of control mechanisms, body temperatures soar, immediate cooling of body temperature required otherwise death ensues.

Cold

- Hypothermia.
- Frost nip/bite.
- Trench foot, (also known as immersion foot) occurs when the feet are wet for long periods of time.
- Violent shivering.
- Chilblains.

Control measures

Where workplace temperatures cannot be maintained, for example, when working outside or where processes cause high or low temperatures, areas should be provided to enable workers to warm themselves or a refuge from heat may be necessary to reduce body temperature. Practical measures and adequate protection must be provided against adverse weather conditions, for example, workers may be provided with a sheeted area to carry out their work, in order to protect them from the rain, wind or radiant heat.

Personal controlling measures to minimise the effects of exposure to cold temperature include ensuring a high calorie diet, physical exercise and suitable protective clothing. The insulating quality of clothing, and therefore its ability to keep the wearer warm, is measured in Togs. 'Tog' is a word not an acronym, the individual letters do not stand for anything. A tog is a measure of thermal resistance of a unit area and is commonly used in the textile industry. Clothing for cold temperature conditions should have a high tog rating of 20–25 tog. The worker's feet should be insulated to avoid loss through conduction and their head should be covered as much as possible, to avoid excessive heat loss. Clothing should be light in weight (to allow easy movement), be dark in colour (to prevent heat loss through radiation), prevent absorption of water and be suitably fastened to reduce the effects of wind chill.

The effects of exposure to heat can be minimised by wearing suitable clothing, such as light and loose fitting clothing or reflective clothing if working with very hot sources of heat. Consideration should also be given to humidity levels and workload - high humidity and workload will increase the worker's temperature rapidly. It may be necessary to control the workload in situations

Figure 8-13: Clothing with a high tog rating.
Source: RMS.

where workers work in high temperatures, particularly where there is high humidity. This might involve providing frequent work breaks to allow the worker to recover, by replacing fluids that were lost through evaporation and reducing their body core temperature.

Most of the Arab states, including Bahrain, Kuwait, Qatar, United Arab Emirates and Saudi Arabia have implemented prohibition of work in open areas between set times during the summer months. The times each country has elected to specify varies, but typically this might cover the period June to August, within the hours of 12.00hrs to 15.00hrs.

Control measures that employers should consider to reduce the effects of extremes of temperature should include:

- Increasing the distance from the temperature source.
- Insulate heat sources, for example, cover hot pipes with thermal insulation.
- Shield heat sources - to reduce radiant heat.
- Shielding or insulation of cold sources
- Screening around the work area.
- Allow workers to become acclimatised to the conditions.
- Regular work breaks with fluid intake of isotonic drinks (isotonic drinks contain similar concentrations of salt and sugar as in the human body).
- Provide hot food and hot drinks for cold workplaces.
- Improved ventilation and humidity control.
- Provide suitable clothing.
- Medical screening of workers for individual susceptibility, for example, Raynaud's syndrome (individuals who suffer with poor circulation to hands and feet).
- Control of working in extremes of temperature, for example, banning work outside during midday in the summer.

8.2 Working at height

WHAT AFFECTS RISK FROM WORKING AT HEIGHT

Vertical distance

Where work is carried out above 2 metres (vertical distance) the risks of major injury or death are considered to be very significant, for example, work on a roof or scaffold. However, major injuries can still occur if a fall results when carrying out tasks at a height of less than 2 metres, for example, while fitting false ceilings or installing utilities inside buildings.

The ILO Safety and Health in Construction Convention C167 Article 18 states:

"Work at heights including roof work

1) Where necessary to guard against danger, or where the height of a structure or its slope exceeds that pre-scribed by national laws or Regulations, preventive measures shall be taken against the fall of workers and tools or other objects or materials.

2) Where workers are required to work on or near roofs or other places covered with fragile material, through which they are liable to fall, preventive measures shall be taken against their inadvertently stepping on or falling through the fragile material."

Figure 8-14: Requirement for roof work.
Source: ILO Safety and Health in Construction Convention C167.

Some national legislation reflects this risk and requires controls to be in place to manage the risk of falling, whatever the height; for example, in the UK this is required by the Work at Height Regulations (WAH) 2005. This legislation applies to workplaces in general, and therefore includes construction activities. Under these Regulations the interpretation of 'work at height' includes any place of work at ground level, above or below ground level that a person could fall a distance liable to cause personal injury and includes places for obtaining access or egress, except by staircase in a permanent workplace.

Roofs

Figure 8-15: Risk of fall from a height - roof used for storage.
Source: RMS.

Working at height and on roofs carries a high risk of accidents, unless proper procedures and precautions are taken. The danger of people or materials falling affects the safety of those working at height and those working beneath. Particular danger arises from two types of roof - fragile roofs and sloping roofs.

Fragile roofs

Any roof that has not been specifically designed to carry a load, other than that which may be imposed on the roof by weather, should be considered a fragile roof. Sheets used in roofing material may deteriorate over time leaving the remaining material in a particularly fragile state. Materials such as asbestos, cement, glass, corrugated metal, slates, tiles or plastic are likely to be unable to bear the weight of a person. A sheet of plastic or metal roofing material may look intact, but it might only have a small fraction of its original strength. In a similar way, plastic roof material used in roof lights will be affected by exposure to sunlight, leaving it brittle and weak. If a worker puts their weight on the fragile roof material it is likely to fail suddenly and cause the worker to fall through the roof.

It should not be assumed that it is safe to walk on newly installed roof material, even though it will have its original strength this may not be enough to bear the weight of a worker. All fragile roofs and/or access routes to them should be marked with an appropriate warning sign.

The employer must ensure that before anyone has access to a roof or begins work on a roof suitable and sufficient steps are taken to prevent workers falling through any fragile roof material. Measures should include ensuring that no worker passes across or works on (or near) fragile roof material when it is reasonably practicable to carry out work without doing so.

If workers pass or work near fragile roof material suitable and sufficient guardrails and other means of fall protection must be in place. If work has to be done from a fragile roof then suitable and sufficient means of support for the workers and materials must be provided.

Sloping roofs

Sloping roofs are those with a pitch greater than 10 degrees. The slope of the roof may cause workers working on top of the roof to slide off the edge of the roof. Falls from the edge of sloping roofs can cause serious injury even when the eaves are relatively low. The hazard of sloping roofs is less obvious when the pitch is small, causing people to underestimate the possibility of workers sliding off the edge.

The material and therefore the surface of the roof have a significant influence on the hazard, for example, a smooth sheet metal surface can present a significant hazard even when the pitch is small.

The chances of a worker falling are increased when working on roofs that are wet or covered in moss growth and in extreme weather conditions such as high winds. A build-up of dry particles or grit on a roof can also present a surface that leads to a high risk of slipping as the particles become free to move and form a mobile layer between the roof and the worker's foot. The worker's footwear can also significantly influence the risk, smooth flat soled footwear may seem suitable in dry conditions, but may not provide sufficient grip to deal with surface water in wet conditions.

Figure 8-16: Falls from a height - worker on a roof
Source: Pixabay

Deterioration of materials

The materials of a structure can deteriorate over time, which can reduce the strength or properties of the material. Materials that have deteriorated particularly present a hazard when people walk on them causing them to break and the workers to fall or when the material cannot hold its own weigh and pieces of the material fall on workers below. The rate of deterioration of materials will accelerate if the structure is exposed to adverse weather conditions (including extremes of temperature) or attack by chemicals, animals, insects etc. Plastic

materials may become brittle when exposed to sunlight and metal materials can become reduced in thickness and strength due to rusting. It may not always be obvious that deterioration has occurred and this should be a factor considered at the pre-work assessment.

Unprotected edges

Roofs, scaffolds, unfinished steel work and access platforms may sometimes have open sides. This increases the likelihood of someone or something falling, particularly if people have to approach them, work at them or pass by them repeatedly. It is very easy in these circumstances to lose perception of the hazard and forget that it is there. Errors, such as stepping back over an edge, overreaching, or being pushed over the edge whilst manoeuvring materials can easily lead to fatal falls.

Figure 8-17: Working at an unprotected edge.
Source: Richard Neale, ILO Theme summary 14 - Working at height.

Unstable/poorly maintained access equipment

If access equipment like mobile elevating work platforms (MEWPs), ladders and scaffolds is not stable it could move suddenly or fall over causing those using it to fall. The instability is often caused by unstable or uneven ground conditions where the equipment is positioned. The higher the equipment operates the risk of it becoming unstable usually increases, as the ratio of the height compared with the width of the base increases. High winds can also increase the instability of the equipment.

Poorly maintained access equipment can lead to sudden failure of the equipment and workers falling from height. The effective working of the hydraulics of a MEWP is critical and failure to maintain this could lead to a sudden and catastrophic failure of the MEWP whilst it is extended. *See also - Mobile elevating work platforms - later in this Element.*

Scaffolds rely on the strength and security of its load-bearing parts, failure of these components can lead to the collapse or overturning of the scaffold. Scaffolds need periodic inspection and maintenance to ensure load-bearing parts are still secured and critical items, such as brakes on mobile scaffolds, are in place and effective.

Cracks may occur in the sides of wooden ladders and loose rungs can lead to failure. They may warp or rot if left outside unprotected from the sun and adverse weather. Defects in ladders may be hidden if they are painted or covered in construction materials.

Weather

Adverse weather can have a significant effect on the safety of those working at height. Rain, snow and ice increase the risk of slips and falling from a roof. When handling large objects, such as roof panels, high wind can be a serious problem and may cause a worker to be blown off the roof. Extremely cold temperatures can increase the likelihood of brittle failure of materials and therefore increase the likelihood of failure of roof supports, scaffold components and plastic roof lights.

In addition, moisture can freeze; increasing the slipperiness of surfaces and on many occasions the presence of ice is not easily visible. Workers exposed to the cold can lose their dexterity and when hot, sweat may cause them to lose their grip.

Falling materials

The risk of falling materials causing injury is increased where workplaces at height are not kept clear of loose materials and no methods are provided to prevent materials rolling or being kicked off the edges. Workers and people nearby may be at risk of being injured when materials, such as old roofing material, are thrown from a workplace at height.

The risk of materials falling is increased by:
- Poor housekeeping of people working at height.
- Absence of toe boards, edge protection or nets.
- Incorrect hooking and slinging when using a crane.
- Incorrect assembly of gin wheels for raising materials.
- Surplus materials incorrectly stacked.
- Loose materials.
- Absence of safe means to get debris and other materials to the ground.
- Open, unprotected edges.
- Deterioration of materials, causing pieces of the material to fall.
- Gaps in platform surfaces.

HIERARCHY FOR SELECTING EQUIPMENT FOR WORKING AT HEIGHT

In summary, the hierarchy for selecting equipment for working safely at height is:

- Avoid working at height.
- Prevent a fall from occurring by using an existing workplace that is known to be safe or use suitable equipment.
- Minimise the distance and/or consequence of a fall.

Avoiding working at height

Where possible, work at height should be avoided. This could be achieved by using different equipment or methods of work to conduct work at ground level, conducting work from ground level or using other methods. For example, instead of manually filling equipment hoppers at height, bulk delivery or automatic feed systems could be used. Where workers have to lubricate or adjust equipment set at height by hand, options to automate or route the lubrication/adjustment mechanisms to ground level should be considered. Instead of assembling materials or equipment at a height pre-assembly before delivery or on the ground on site should be considered.

If holes required in materials are drilled at ground level, for example, holes to enable assembly of roof frames, this can greatly reduce the work needed to be done at height. In a similar way, materials can be treated or painted on the ground before assembly or fitting. Long reach or extendable tools can be used to allow cleaning or other tasks to be conducted from the ground. Where equipment, such as light units, requires maintenance an option may be to lower it sufficiently to enable bulbs to be changed and cleaning to be conducted from the ground.

Preventing a fall by using existing workplaces or suitable equipment

Use an existing safe workplace

Do not work at height unless it is absolutely unavoidable. If work must be performed at height, use an existing safe workplace that is:

- Stable and of sufficient strength and rigidity for the intended use - for example a solid roof or platform that is an integral part of a fixed structure.

- Sufficient dimensions to provide a safe working area, having regard to the work to be done – sufficient to permit the safe movement of workers and the safe use of any equipment or materials.

- Suitable and sufficient means for preventing a fall - for example, fixed guardrails.

- A surface which has no gaps that a person, materials or objects could fall through.

- Constructed, used and maintained in a safe condition - to prevent slipping or tripping, and, where it has moving parts, has appropriate devices to prevent its inadvertent movement during work at height or a worker being caught between it and any adjacent structure.

It is best to use an existing workplace that is known to be safe where possible. Erecting a temporary workplace at height is a less desirable option as it has its own added risks. By using more permanent workplaces, design features that provide collective protection of those at risk can be built in to the structure more easily, making the need to use of personal protective measures less likely.

Provide work equipment to prevent falls

Where no existing, safe workplace is available to enable work at height it will be necessary to provide work equipment that enables work at height and prevents workers falling. This could be in the form of scaffolding, barriers that provide edge protection (temporary or fixed), provision of a mobile elevating work platform (MEWP) or work restraint systems.

Figure 8-18: Use of a work restraint system to prevent worker going near the unprotected edge.
Source: UK, HSE, HSG150.

When planning the work at height it is important to select the most appropriate equipment for the work to be carried out, considering the activity and the work environment. Collective measures must be given priority over personal measures, for example, barriers or scaffolding would provide collective measures and would be preferable to using work restraint systems. Similarly, to maintain safe access and egress, temporary staircases should be considered before the use of ladders.

When using a work restraint system in a mobile elevating work platform (MEWP), it should always be anchored to the inside of the cradle and not the outside *(see Figure 8-75)*.

Minimise the distance and/or consequence of a fall

Collective measures

Collective fall-arrest measures are preferred to personal protective measures because they provide protection for all workers at risk of falling rather than protection for individuals. Collective measures include equipment designed to minimise the distance and consequence of falls by providing an energy absorbing 'soft landing', such as nets and air bags. Collective measures can simplify systems of work and can protect not only workers, but others such as supervisors.

Figure 8-19: Soft landing system - safety net.
Source: ProNet Safety Services.

Figure 8-20: Soft landing system - air bags.
Source: thxuk.com

Safety nets are installed underneath the structure being worked on so that there is an amount of free movement of the net to enable the energy of the person falling to be absorbed. It is essential that there are no obstructions underneath the net that might injure a person falling on the net, such as materials and equipment. Materials that have fallen onto the net should be removed promptly to ensure they do not injure someone who might fall onto them. It is important that measures are in place to rescue anyone who falls onto the net.

Other energy absorbing systems use a number of bags that are filled with an energy absorbing material and are connected together to fill the area where protection is required. They may be filled with air or small pieces of polystyrene or another energy absorbing material. The air filled bags are kept inflated by an air compressor which replaces any air that leaks from the bag. The bags must have sufficient depth and energy absorbent material to absorb the energy of the person falling. Other energy absorbing systems include thick, solid energy absorbing materials formed into mats.

Where energy absorbing systems, such as safety nets, are used they must be installed as close as possible beneath the surface that people can fall from, securely attached and able to withstand a person falling onto them. They must be installed and maintained by competent personnel. Although they do not prevent people falling, if they are installed correctly they will minimise the distance of the fall and minimise the consequences of the fall by absorbing the energy of the person falling. If the work height increases as work progresses, the energy absorbing system must be repositioned at each higher level to minimise the distance that a worker can fall.

Energy absorbing systems should:

- Provide sufficient clearance form solid surfaces when arresting a fall.
- Have anchors and other fixings that have sufficient strength and stability for the purpose of supporting the foreseeable loading in the event of a fall and during any subsequent rescue.
- Be securely attached to all the required anchor points.
- In the case of energy absorbing bags or mats, be stable.

Personal protective measures

If fall prevention measures (for example, working platforms, barriers or guard rails) or collective fall-arrest measures are not practical, alternative personal protective measures must be used. This could include work positioning systems, rope access systems or personal fall-arrest systems. Personal fall-arrest systems are widely used, and comprise a line, means of anchoring the line, means of absorbing the energy of the fall and a harness worn by the user. It should be remembered that they only provide protection for the user and the harness itself may cause injury when the worker's fall comes to a sudden stop.

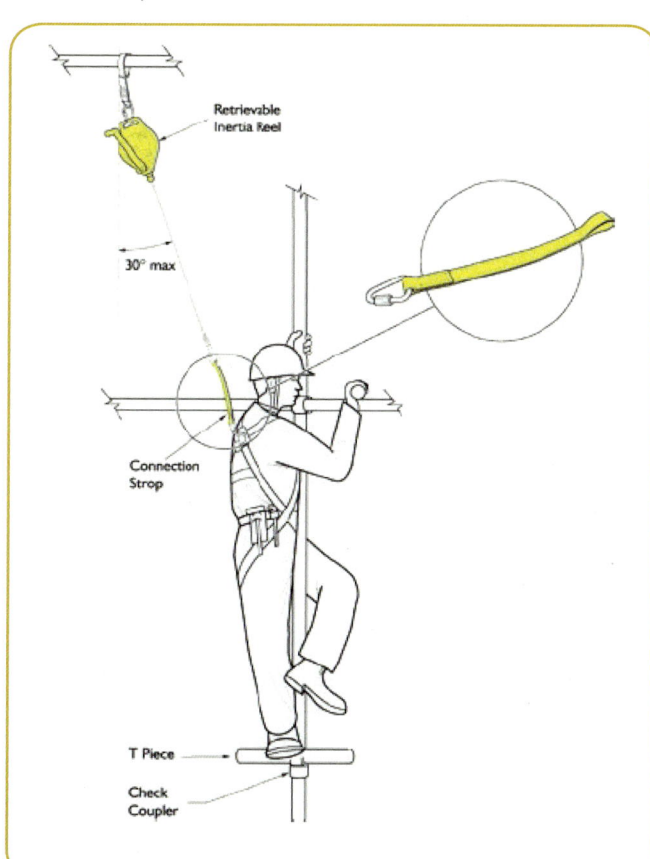

Figure 8-21: Personal fall-arrest system.
Source: NASC.

A fall-arrest system must incorporate means of absorbing the energy of a fall and limiting the forces applied to the user's body. Care should be taken to ensure that when used the fall arrest system it is not at risk of the line being cut and that it is free to operate as intended. Consideration should be made to ensuring at all times it is in use that there is a 'clear zone' around where the worker might fall and that this should take account of any pendulum effect. Arrangements must be made for the rescue of the worker who has fallen.

A personal fall-arrest system should only be used if it is:

- Suitable and of sufficient strength for the purposes for which it is being used - having regard to the work being carried out and any foreseeable loading.
- Correctly fitted.
- Designed to minimise injury to the user and adjusted to prevent the user slipping out of it if they fall - has a means of absorbing the energy of the fall.
- Securely attached to all the required anchor points - anchors must be capable of supporting the foreseeable loading when arresting the fall and during any subsequent rescue.
- Secured in a position that enables the worker to fall safely and for their fall to be arrested.

MAIN PRECAUTIONS NECESSARY TO PREVENT FALLS AND FALLING MATERIAL

Proper planning and supervision of work

Planning

Planning must include the selection of suitable equipment, take account of emergencies and give consideration to weather conditions impacting on safety.

Avoid, prevent, minimise

Work at height must only be carried out when it is not reasonably practicable to carry out the work at ground level or from ground level. If work at height does take place, suitable and sufficient measures must be taken to prevent a fall of any distance and, if this is not possible, to minimise the distance and the consequences of any fall liable to cause injury.

The ILO Safety and Health in Construction Convention C167 Article 14 states:

> "Scaffolds and ladders
>
> 1) Where work cannot safely be done on or from the ground or from part of a building or other permanent structure, a safe and suitable scaffold shall be provided and maintained, or other equally safe and suitable provision shall be made."

Figure 8-22: Requirement to use a work at height hierarchy.
Source: ILO Safety and Health in Construction Convention C167.

Selection of equipment

Careful consideration during the risk assessment phase of work planning should establish which equipment is best suited for the working environment and the work to be done. Consideration should be given to the working conditions, risks to people and the frequency and duration of use. Give collective measures priority over personal protection measures.

Fall of materials and objects

Employers should plan to take measures to prevent injury to any person from the fall of any material or object. Where possible this must be by the prevention of materials or objects falling. Where this is not reasonably practicable, measures must be taken to prevent any person being struck by falling material or object, for example, by provision of nets or other structures to catch materials.

In addition, employers should ensure that nothing is thrown or tipped from height in circumstances where it is liable to cause injury to any person. Where an area presents a risk of a person being struck by an item falling from height it must be equipped with measures that prevent unauthorised persons from entering the area and the area must be clearly indicated. This could involve fencing around the area and provision of warning signs.

Emergencies and rescue

Planning should consider what will need to be done when a worker falls from height and the nature of this emergency. For example, they may have fallen into a net or be attached to a personal fall-arrest system, which could involve injuries and the person being suspended for a period of time before they can be rescued.

Carried out safely

Supervision

It is essential that work at height is carried out safely, which will mean ensuring prevention and protection measures are put into place and used correctly. Although those workers that use the measures have a responsibility to ensure they use them correctly, adequate supervision must be provided to ensure they do.

Avoiding working in adverse weather conditions

Adverse weather can include wind, sand/dust storms, rain, sun, cold, snow and ice. Each of these conditions, particularly in extreme cases, can present a significant hazard to work at height. When long-term projects are planned methods are often adjusted to minimise the effects of these conditions. Work areas can be covered over at an early stage to enable work to be conducted in relative safety. In some cases, adverse weather must be considered formally and work may have to cease until conditions improve, for example, work on a roof in icy conditions or high winds. Similar approaches may have to be taken for operating a mobile elevating work platform (MEWP) in windy conditions.

Inspection

It is good practice, and often a legal requirement in many territories, to inspect access equipment routinely and to ensure that the person carrying out the inspection is competent to do so.

The Western Australia, Code of practice - Prevention of falls at workplaces defines competency as:

 "'Competent person', in relation to the doing of anything, means a person who has acquired, through training, qualification or experience or a combination of those things, the knowledge and skills required to do that thing competently."

Figure 8-23: Competent person.
Source: Western Australia, Code of practice - Prevention of falls at workplaces.

Typical inspection requirements for equipment used for working at height are:

- Where safety depends on how it is installed or assembled - before it is used in that position.

- Temporary work platforms, for example, a scaffold, or mobile work platform used for work in which a person could fall 2 metres or more (national legislation may specify a different fall distance) - regularly whilst in position, typically at intervals not exceeding 7 days.

- Other work equipment for work at height - at suitable intervals.

- Each time it has been affected by exceptional circumstances that are liable to affect the safety of the work equipment, for example, extremes of weather (for example, high winds or flood) or seismic conditions.

For example, The ILO, Code of practice - Safety and health in construction 4.4.2 states:

"- Inspection by the competent person should more particularly ascertain that:

a) The scaffold is of a suitable type and adequate for the job.

b) Materials used in its construction are sound and of sufficient strength.

c) It is of sound construction and stable.

d) That the required safeguards are in position."

Figure 8-24: Inspection criteria for scaffolds.
Source: ILO Code of practice - Safety and health in construction.

Results of an inspection should be recorded and kept until the next inspection. An inspection report containing the particulars set out in the following list should be prepared before the end of the working period within which the inspection is completed and, within 24 hours of completing the inspection, and be provided to the person it was carried out for.

The report should be kept at the site where the inspection was carried out until the work is completed and afterwards at an office of the person on whose behalf it was carried out, for typically 3 months. Reports on inspections should include the following particulars:

1) Name and address of person for whom the inspection is carried out.

2) Location of the access equipment.

3) Description of the access equipment.

4) Date and time of inspection.

5) Details of any matter identified that could give rise to a risk to the health and safety of any person.

6) Details of any action taken as a result of item 5 above.

7) Details of any further action considered necessary.

8) Name and position of the person making the report.

Competent

Those engaged in any activity in relation to work at height must be competent; and, if under training, be supervised by a competent person.

EMERGENCY RESCUE

When selecting equipment for work at height, it is important that the employer consider the additional risks that may arise from emergencies and the need for evacuation of or rescue from the equipment. This will include the need to establish clear emergency escape routes from major scaffold installations, rescue arrangements for workers where a mobile elevating work platform (MEWP) fails in its raised position and rescue for those on a fall arrest net or in a harnesses. Steps must be taken to minimise the risk of injury due to the fall or contact with the fall-arrest system. Even a short fall onto a net or other fall arresting system could cause minor injuries or fractures. This should be anticipated and workers taught how to minimise the likelihood of injury.

Where a worker has fallen from height, but has been protected by personal fall-arrest equipment, such as a harness, significant health effects, known as suspension trauma, may be experienced if they are not rescued quickly. This is mainly due to blood pooling in the legs, reducing the amount circulating through the rest of the body, which has consequential effects for vital organs, including the brain, heart and kidneys. Unless the individual is rescued quickly the lack of oxygenated blood to vital organs can be fatal. Therefore, a rescue procedure and equipment must be available and

rescue practised. The rescue procedure should take into account that the worker's sudden transition from a vertical to a horizontal position, when rescued and laid down, can lead to a massive amount of deoxygenated blood entering the heart, causing cardiac arrest.

INSTRUCTION, TRAINING AND OTHER MEASURES

Instruction and training

Employers should ensure that no person engages in any activity, including organisation, planning and supervision, in relation to work at height unless they are competent to do so. If they are being trained, they must be supervised by a competent person. Workers should receive full training and instruction on the use of work at height equipment to prevent falls and minimise the distance and consequences of a fall. Where personal fall-arrest systems are to be used, this should include how to wear the equipment, how to fit it to anchor points, what is a suitable anchor point and how to attach it at a height that minimises the fall (for example, above the worker's head where possible).

Training should also include the checks that need to be made on collective fall-arrest systems before they are used, this will include checks on the security of nets and the adequacy of airbags. Workers will also need to be trained in how to get off/out of this equipment safely when it has arrested a fall and on emergency arrangements for rescue, where workers cannot assist themselves.

> "Every employer shall ensure that no person engages in any activity, including organisation, planning and supervision, in relation to work at height or work equipment for use in such work unless he is competent to do so or, if being trained, is being supervised by a competent person."

Figure 8-25: Requirement for competence.
Source: UK, Working at Height Regulations (WAH) 2005.

Where a risk of falling remains, after fall protection systems have been provided, additional training and suitable and sufficient other measures should also be taken to prevent injury to any person falling a distance liable to cause injury. This may include training them in techniques that would limit the effect of the fall, controlling what equipment they use that might fall with them, ceasing processes in the area that may cause the worker to be disorientated/overcome leading to a fall, covering items that the person may fall onto and providing means to assist those that have fallen.

Requirements for head protection

Head protection, usually in the form of a hard hat, is required where there is a foreseeable risk of injury to a worker's head from being struck by falling materials.

Actions to be taken include:

- Decide on which areas of the site where hats have to be worn.
- Make site rules and tell everyone in the area.
- Provide workers with hard hats.
- Make sure hard hats are worn and worn correctly.

A wide range of hard hats are available. Let workers try a few and decide which is most suitable for the job and for them. Some hard hats have extra features including a sweatband for the forehead and a soft, or webbing harness. Although these hard hats are slightly more expensive, they are much more comfortable and therefore more likely to be worn.

GENERAL PRECAUTIONS WHEN USING COMMON FORMS OF WORK EQUIPMENT TO PREVENT FALLS

Ladders

Use of ladder

Ladders are primarily a means of vertical access to a workplace. However, they are often used to carry out short-term work at height and this frequently results in injuries. Many injuries involving ladders happen during work lasting 30 minutes or less. Before using a ladder to work from, consider whether it is the right equipment for the job.

Figure 8-26: Ladder as access and workplace.
Source: RMS.

Ladders are only suitable as a workplace for light work of short duration, and for a large majority of activities a mobile scaffold tower or a mobile elevating work platform (MEWP) is likely to be more suitable and safer. Generally, ladders should be considered as access equipment and use of a ladder as a work platform should be discouraged.

There are situations when working from a ladder would be inappropriate, for example:

- Where the equipment or materials used are large or awkward.
- When two hands are needed or the work area is large.
- Excessive height.
- Work of long duration.
- Where the ladder cannot be secured or made stable.
- Where the ladder cannot be protected from vehicles etc.
- Adverse weather conditions.

Safe working practices for the use of ladders include:

- Pre-use selection and inspection. The ladder to be used needs to be long enough to access the point where work will take place without the user overreaching or standing on the top three rungs. The ladder must not be overloaded and needs to be suitable for the worker and the equipment/materials that they may be carrying. Make sure the ladder is in good condition. Check the rungs and stiles for warping, cracking or splintering, the condition of the feet, and for any other defects. Do not use defective or painted ladders.

- Position the ladder properly for safe access and out of the way of vehicles. Do not rest ladders against fragile surfaces.

- Ladders must stand on a firm, level base, and be positioned approximately at an angle of 75° (1 unit horizontally to 4 units vertically), *(see Figure 8-27; note: a means of securing the ladder is omitted for clarity)*.

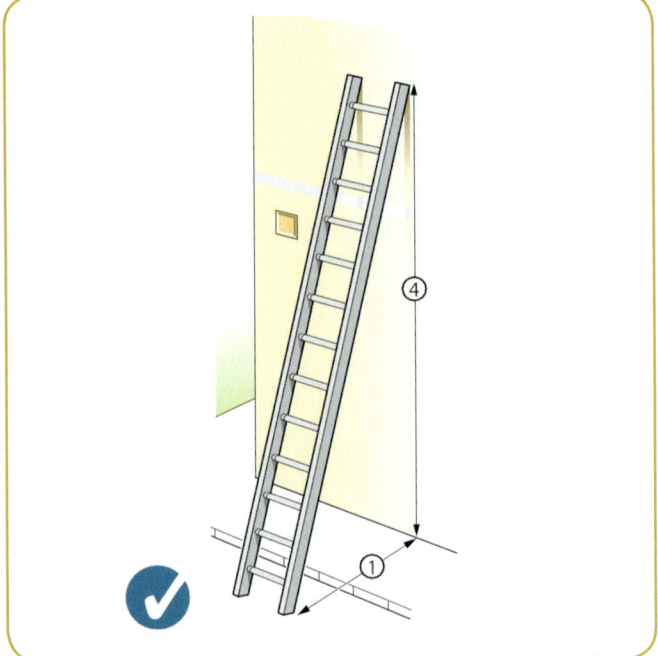

Figure 8-27: Correct 1 in 4 angle.
Source: UK, HSE, INDG402.

- Ladders must be properly tied near the top, even if only in use for a short time, while being tied a ladder must be footed. If not tied, ladders must be secured near the bottom, footed or weighted.

- If the ladder is being used to gain access to a landing place it should extend about 1 metre above the landing place.

Figure 8-28: Improper use of ladder.
Source: RMS.

- Both hands should be kept free to grip the ladder when climbing or descending, always maintain three points of contact on the ladder.

- Beware of wet, greasy or icy rungs and make sure soles of footwear are clean.

- Ensure the ladder is not overloaded.

- Only one person to climb at a time.

- Ladders should be used only for work of short duration and not in inclement weather.

- Use a holster or tool bag to carry tools securely.

Figure 8-29: Inappropriate storage.
Source: Lincsafe.

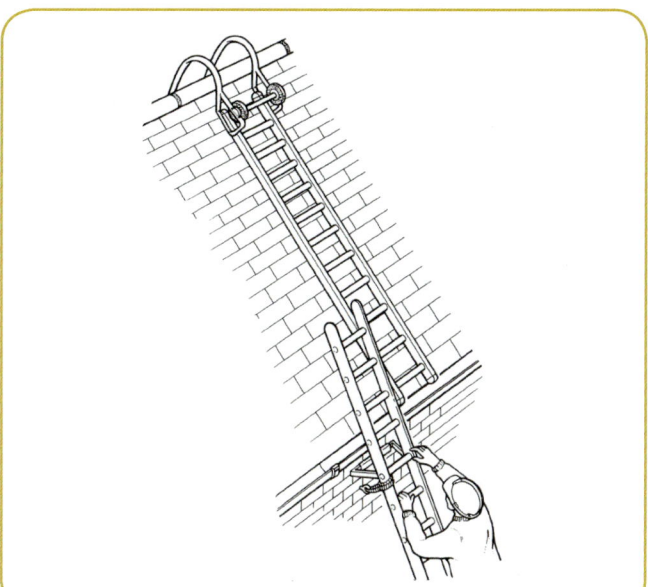
Figure 8-30: Use of roof ladders.
Source: UK, HSE, HSG150.

Step ladders

Step ladders require careful use. They are subject to the same general health and safety rules as ladders. However, in addition, they will not withstand any degree of side loading and overturn very easily. When using a step ladder overreaching should be avoided at all times and care should be taken to avoid side loading.

Figure 8-31: Incorrect use of a step ladder.
Source: UK, HSE, INDG402.

Ensure the step ladder is placed on a firm level surface to minimise the possibility of it overturning sideways.

Step ladder stays must be 'locked out' properly before use. The top step of a stepladder should not be used as a working platform unless it has been specifically designed for that purpose. Typically three clear steps should be left to ensure support and stability, depending on the size and design of the step ladder.

Figure 8-32: Correct use of a stepladder.
Source: UK, HSE, INDG402.

Figure 8-33: Correct use of a stepladder.
Source: UK, HSE, INDG402.

Vertical ladders and hoops

Maintenance workers may make use of vertical ladders to gain access to a roof or similar areas, they are often used where large process plant is installed over a number of floors and occasional access is required to parts of the plant.

There is a risk of fatigue when climbing long vertical ladders, because the climber needs to physically pull and step their body up the ladder. The ladder hoops are designed to create a 'tunnel' to prevent people falling away from the ladder through fatigue, but do not prevent sliding down the ladder *(see Figure 8-34)*. Hoops need to be used in conjunction with rest platforms, at intervals of 9 metres, to allow the climber to rest and recover from fatigue.

Figure 8-34: Ladder hoop.
Source: RMS.

Scaffolds

General Points

Scaffold terms

Base plate	A small square metal plate that distributes the load from a standard or a raker (scaffold standard used as an outrigger).
Brace	A tube fixed diagonally across two or more members in a scaffold for stability.
Fan	A structure of scaffold boards fixed to standards that is fixed to the outside of a scaffold at an angle inclined upwards and located below a work platform to catch debris that may fall.
Guardrail	A tube fixed to standards (uprights) to prevent personnel from falling.
Ledger	A tube spanning horizontally and tying the scaffold longitudinally. It may act as a support for put logs or transoms.
Putlog	A tube with a flattened end, spanning from a horizontal member to a bearing in or on a brick wall. It may support scaffold boards.
Raker	Scaffold standard used as an outrigger to prevent the scaffold falling away from a structure.
Reveal pin	A screw jack, fitting in the end of a tube.
Reveal tie	A tube wedged by means of a reveal screw between two opposite surfaces (for example, window reveals) to make a friction anchorage for tying a scaffold.
Sole board	Large pieces of timber put underneath the base plate to further spread the load over a wide surface area, especially where the ground is soft.
Standard/ column	A vertical or near vertical supporting member.
Tie	A member used for fixing the scaffold to the building or other structure for stability.
Transom	A tube spanning across ledgers to tie a scaffold transversely (normally at 90° to the building face). It may also support a working platform.

Scaffolding can be a very effective means of access to height and provide protection in preventing falls. However, there are specific safe working practices that apply to the design, construction, erecting and use of scaffolds.

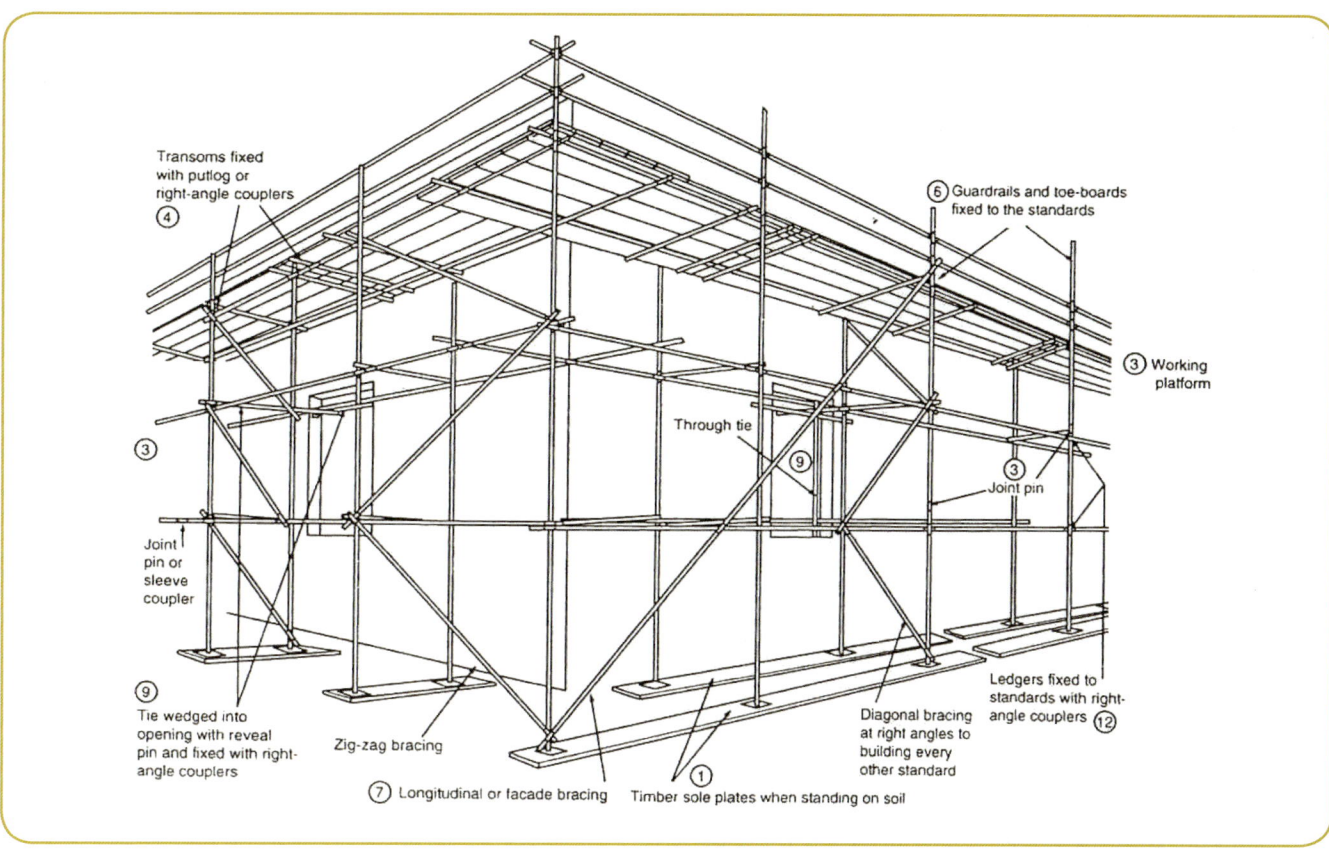

Figure 8-35: Independent tied scaffold (numbers relate to explanatory list, which is not included). Source: UK, HSE, HSG150.

Reasons why scaffolds collapse include:

- Incorrect erection.
- Overloading.
- Uneven distribution of loads.
- Poor ground conditions.
- Adverse weather.
- Insufficient or inappropriate ties.
- Interference with ties.
- Incompatible components.
- Unauthorised alteration.
- Being struck by a vehicle

Common misuse of scaffolding include:

- Removal of bracings and ties.
- Removal of scaffold boards.
- Removal of handrails or toe boards.
- Excavations near the scaffold.
- Use of the scaffold for propping shuttering when it was not designed for that purpose.
- Vehicle collision.

Independent tied scaffold

This type of scaffold typically uses two sets of standards; one near to the structure and the other set at the width of the work platform. It is erected so that it is independent from the structure and does not rely on it for its primary stability. However, as the name suggests, it is usual to tie the scaffold to the structure in order to prevent the scaffold falling towards or away from the structure.

Base plates and sole boards

Safe working practices for the use of base plates and sole boards include:

- A base plate must be used under every standard - it spreads the load and helps to keep the standard vertical.

- Sole boards are used to spread the weight of the scaffold and to provide a firm surface on which to erect a scaffold, particularly on soft ground. Sole boards must be sound and sufficient, and should run under at least two standards at a time.

Figure 8-36: Independent tied scaffold.
Source: RMS.

Figure 8-37: Base plates and sole boards.
Source: Lincsafe.

Figure 8-38: Base plate and protection.
Source: RMS.

Standards

Safe working practices for the use of standards include:

- All standards must be truly plumb, or leaning only a little towards the structure. One standard out of plumb will 'bow' and push the others.

- Any joints in standards must be staggered.

Ledgers

Safe working practices for the use of ledgers include:

- Ledgers must be truly horizontal, and not more than 2.7 metres above the ground or the ledgers below, on any scaffold.

- Any joints in ledgers must be made with sleeve couplers, and the joints must be staggered.

- Ledger bracings must be fixed to alternate standards on every platform. They must not be fixed to the handrail.

Boards

Safe working practices for the use of boards include:

- Boards supported by transoms or putlogs must be close fitting.

- They must be free from cracks or splits or large knots and must not be damaged in any way that could cause weakness.

Working platforms

Safe working practices for the use of working platforms include:

- Wide enough to allow workers to walk on it safely and to use any equipment or material necessary for their work - at least 600mm wide.

- Working platforms at height must be fitted with guardrails and toe boards. The height at which this requirement applies may be defined under national laws, regulations or codes of practice. A common safe working practice is to fit guardrails and toe boards when the working platform height is two metres or more above the ground or floor level.

- If the platform also carries materials, then the space between guardrails and toe boards must be reduced to a maximum 470mm (UK standard). This can be done with an intermediate rail, mesh or similar material.

- As well as toe boards, the platform itself should be constructed to prevent any object that may be used on the platform from falling through gaps or holes and causing injury to people working below. For scaffolds used on site a close-boarded platform would be enough, however for work over public areas a double-boarded platform sandwiching a polythene sheet may be needed.

- Where a ladder passes through a working platform, the access must be as small as practicable.

- Any access point through a working platform must be covered when it is not being used, and clearly marked to show its purpose.

- If a gap or opening is created in a working platform for work activities, then access to it must immediately be prevented by fitting guard rails etc.

- Trestles must not be put up on a working platform as workers using them would be working above the height of the guard rail.

- If a working platform becomes covered with ice, snow, grease or any other slippery material, then suitable action must be taken to reduce the hazard, by sprinkling sand, salt, sawdust, etc.

- Rubbish or unused materials must not be left on working platforms.

- Working platforms must not be used for "storing" materials - all materials placed on a working platform must be for immediate use only.

- Loading of materials and equipment must be spread evenly over working platforms to the fullest extent possible.

- Where loading of the platform cannot be distributed evenly (as may happen with bricklayers' materials) then the larger weights should be kept nearest to the standards.

- Any working platform near to fragile items, for example, windows, should have sheeting fitted.

- General access must not be allowed to any working platform until the erection of the scaffold has been completed - access must be blocked off to any section of scaffold that is not yet finished.

- All loose materials and equipment must be taken off working platforms before the scaffold and platform is dismantled.

Toe boards

These are usually scaffold boards placed against the standards at right angles to the surface of the working platform. They help prevent materials from falling from the scaffold and people slipping under guardrails. They must be suitable and sufficient to prevent the fall of any person or any material or object from a place of work.

Safe working practices for the use of toe boards include:

- The toe boards should be fixed to the inside of the standards with toe board clips. Joints must be as near as possible to a standard.

- Toe boards should be continuous around the platform where a guardrail is required.

- Any toe board that is removed temporarily for access or for any other reason must be replaced as soon as possible.

Figure 8-39: Scaffold boards and toe boards - some defective.
Source: RMS.

Guardrails

These are horizontal scaffold tubes that help to prevent people falling from a scaffold. Safe working practices for the use of guardrails include:

- They must be fixed to the inside of the standards, at least 950mm (UK standard) from the platform.

- An intermediate guardrail must be positioned such that the gap between it and the top guardrail and the toe boards is no more than 470mm (UK standard).

- Guardrails must always be fitted with load-bearing couplers.

- Guardrails are joined with sleeve couplers.

- Joints in guardrails must be near to a standard.

- Guardrail must go all round the work platform.

- Where there is a gap between the structures then a guardrail must be fitted to the inside of the working platform as well as the outside.

- Guardrails must always be carried around the end of a scaffold, to make a 'stop end'.

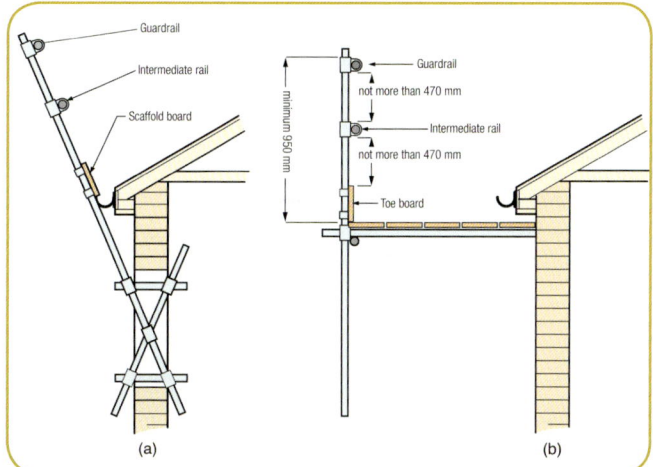

Figure 8-40: Toeboards and sloping roof edge protection.
Source: UK, HSE INDG28.

Mesh (brick) guards

These wire mesh infill panels may be used instead of or as well as a mid-rail. They sometimes have an integral toe board. When fitted they assist in preventing people and objects from falling between the rails and the toe board.

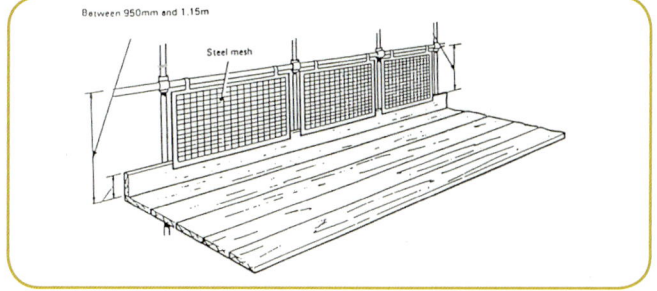

Figure 8-41: Brick guards. Source: UK, HSE, HSG150.

The Western Australian Occupational Safety and Health Regulations (WAOSHR) 1996 Regulation

3.55(5) states the following requirements for guard rails, toe boards and mesh guards:

Figure 8-42: Requirement for edge protection.
Source: Western Australian, Occupational Safety and Health Regulations 1996.

Ties, bracing and rakers

Ensuring stability of a scaffold is critical. In the case of an independent tied scaffold the scaffold is set a small distance away from the structure and ties connect it to the structure and prevent the scaffold falling away from or towards the structure. One of the ways to do this is to use a 'through tie' which is set into place through an opening in the structure such as a window.

Figure 8-43: Tie through window.
Source: RMS.

In addition to the use of ties, it is important that the scaffold is a rigid structure. A scaffold comprising standards, transoms and ledgers alone may not be rigid enough, particularly in the case of tall scaffolds. In order to improve rigidity a system of braces and rakers are used.

Braces (scaffold poles attached across the standards) are used diagonally in opposite directions, to the front and sides of the scaffold, to provide a rigid structure.

Figure 8-44: Ladder access.
Source: RMS.

Rakers (scaffold poles attached to the ledgers) may be set at a diagonal outwards from the front of the scaffold to the ground to prevent the scaffold falling away from the structure. They may be used in addition to ties or added where a tie has to be removed in order to do work on the opening where it is fixed.

Debris netting

Debris netting is often fixed to the sides of a scaffold to limit the amount of debris that may come from work being done on it escaping from the scaffold. It provides a tough, durable and inexpensive method of helping to provide protection from the danger of falling debris and windblown waste material. It allows good light transmission and reduces the effects of adverse weather. However, the netting may act as a sail in high winds, so the strength and rigidity of the scaffold needs to take this factor into account at the scaffold design stage.

Figure 8-45: Nets and sheets.
Source: RMS.

Debris netting may also be slung underneath steelwork or where roof work is being conducted to catch items that may fall. In this situation, it should not be assumed that the debris netting is sufficient to hold the weight of a person who might fall.

Figure 8-46: Nets.
Source: RMS.

Signs

Safety signs are used to provide people with information relating to the work being carried out, to control or divert people and, most importantly, of any dangers or hazards. Signs may be situated close to a location where a scaffold is being used in order to give prior warning of its presence.

Figure 8-47: Signs.
Source: RMS.

Figure 8-48: Protected scaffold with decorative wood.
Source: Book of wood.

Marking

Where there is a risk of collision with equipment used to gain access to a height, it should carry hazard marking. For example, the standards of a scaffold at ground level on a street may be marked with hazard tape or high-visibility foam and warning lights.

Figure 8-49: Marking.
Source: RMS.

Figure 8-50: Protected scaffold with hi-visibility protected foam.
Source: Alligata.

Lighting

Suitable and adequate lighting should be provided when light levels are poor to enable work on the working platform to be carried out. Lighting will also be required to enable any possible hazards the scaffold presents to be seen clearly and allow people to take the required actions to avoid a collision with it. Scaffolds located near roads may be fitted with suitably coloured lighting to warn traffic of its presence.

Fans

Fans are scaffold boards fixed on scaffold tubes set at an upward angle out from a scaffold in order to catch debris that may fall from the scaffold. They may be used at the entrances to buildings, to protect people entering and leaving the building that the scaffold is erected against, or on the face of the building if it runs along-side a public footpath.

Figure 8-51: Fans.
Source: RMS.

They are also used where a scaffold is erected along-side a pedestrian walkway where there is a need to have an increased confidence that materials that might fall cannot drop onto people below. In some cases, it may be necessary for horizontal barriers to be erected to direct pedestrians under the fan.

Mobile tower scaffolds

Mobile tower scaffolds are widely used as they are convenient for work which involves frequent access to a height over a short period of time in a number of locations that are spaced apart. However, they are often incorrectly erected or misused and incidents/accidents occur due to people/materials falling or the tower overturning/collapsing. They must be erected and dismantled by trained, competent workers, strictly in accordance with the supplier's instructions. All parts must be in good condition and from the same manufacturer.

Safe working practices for the use of mobile tower scaffolds include:

- Positioned on firm and level ground.
- The height of an untied, independent tower scaffold must never exceed the manufacturer's recommendations. A general guide to acceptable limits are:
- Outdoor use - 3 times the minimum base width.
- Indoor use - 3.5 times the minimum base width.
- If the height of the tower scaffold is to exceed these maximum figures then the scaffold must be secured (tied) to the structure or outriggers (rakers) used.
- Working platforms must only be accessed by safe means. Only use internal stairs or fixed ladders and never climb on the outside.
- Before accessing a mobile tower scaffold, the wheels must be turned outwards and the wheel brakes locked 'on'.
- Never move a mobile tower scaffold unless the working platform is clear of people, materials, tools etc.

Figure 8-52: Mobile tower, wheels with brakes.
Source: RMS.

- Mobile tower scaffolds must only be moved by pushing them at base level. They must not be pulled along by workers whilst people are standing on the working platform. Care must be taken to avoid obstructions at base level and overhead.
- Never use a mobile tower scaffold near live overhead power lines or cables.
- Working platforms must always be fully boarded out. Guardrails and toe boards must be fitted if there is a risk of a fall from a height that presents a significant risk, for example, more than two metres. Inspections must be carried out by a competent person - before first use, after substantial alteration and after any event likely to have affected its stability.

Figure 8-53: Mobile tower scaffold. Source: RMS.   Figure 8-54: Mobile tower scaffold. Source: RMS.

Mobile elevating work platforms

A mobile elevating work platform (MEWP) is, as the name suggests, a means of providing a work platform at height. The equipment is designed to be movable, under its own power or by being towed, so that it can easily be set up in a location where it is needed. Various mechanical and hydraulic means are used to elevate the work platform to the desired height, including telescopic arms and scissor lifts. The versatility of this equipment, which enables the easy placement of a platform at height, makes it a popular piece of access equipment. Often, to do similar work by other means would take a lot of time or be very difficult. They are now widely available and there is a tendency for people to oversimplify their use and allow people to operate them without prior training and experience. This places users and others at high risk of serious injury.

Figure 8-55: Mobile elevating work platform (MEWP).
Source: RMS.

Some MEWPs can be used on rough terrain. This usually means that they are safe to use on uneven ground, (**see Figure 8-59**). The MEWP's limitations should always be checked in the manufacturer's handbook before moving on to uneven or sloping ground. They should only be operated within their defined stability working area. A harness with a lanyard attached to the platform acts as a work restraint and provides extra protection against falls especially when the platform is being raised or lowered. MEWPs and similar equipment used in poor lighting conditions on or near roads and walkways must use standard vehicle lighting.

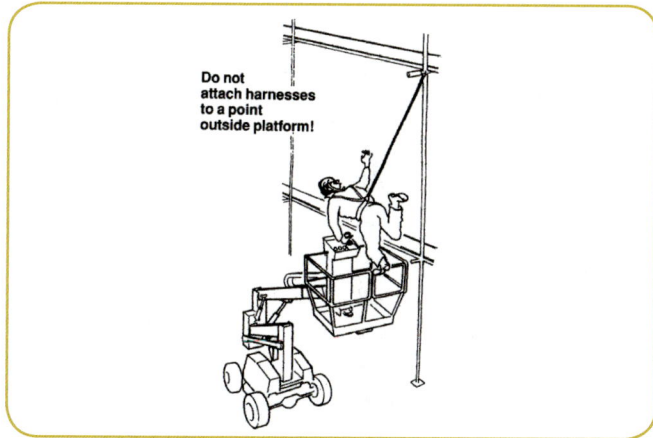

Figure 8-56: Use of harness with a MEWP.
Source: UK, HSE, HSG150.

Use of mobile elevating work platforms

Mobile elevating work platforms can provide excellent safe access to high level work. When using a MEWP the following safe working practices should be used:

Always ensure

- Whoever is operating it is fully trained and competent.
- Any statutory inspections or testing has been carried out.
- The work platform is fitted with guardrails and toe boards.
- It is used on suitable firm and level ground. The ground may have to be prepared in advance.
- Tyres are properly inflated.
- The work area is cordoned off to prevent access below the work platform, (**see Figure 8-54**).

- That it is well lit if being used on a public highway or in poor lighting.
- Outriggers are extended and chocked as necessary before raising the platform.
- Harnesses are used by workers.
- All involved know what to do if the machine fails with the platform in the raised position.

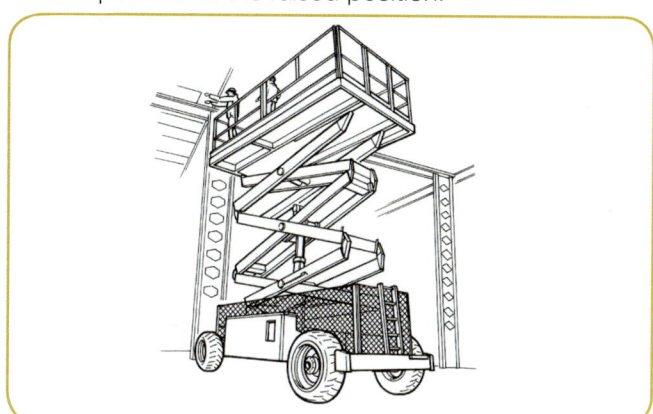

Figure 8-57: Scissor lift MEWP.
Source: UK, HSE, HSG150.

Never

- Operate MEWPs close to overhead cables or dangerous machinery.
- Allow a part of the hand or arm to protrude into a traffic route when working near vehicles.
- Move the equipment with the platform in the raised position unless the equipment is especially designed to allow this to be done safely (the manufacturer's instructions should be checked).
- Overload or overreach from the platform.

Figure 8-58: Scissor lift MEWP.
Source: RMS.

Figure 8-59: Rough terrain MEWP.
Source RMS.

Trestles

Trestles are pre-fabricated steel, aluminium or wood supports, of approximately 500mm to 1 metre width, that may be of fixed height or may be height adjustable by means of sliding struts with varying fixing points (pin method) or various cross bars to suit the height required. Scaffold boards are placed on top of them, stretching from one to the other, in order to make a working platform.

These should only be used where work cannot be carried out from the ground, but where a scaffold would be impracticable.

A good example of where they may be used would be where a plasterer is installing and plastering a new ceiling. Typical working heights when using a trestle system range from 300mm to 1 metre. Edge protection should be fitted wherever practical. As with any work carried out above ground level, a risk assessment must be carried out to determine if they are the appropriate equipment for the work to be conducted.

There are many configurations of locking and adjustable trestles. Platforms based on trestles should be fully boarded, adequately supported and provided with edge protection where appropriate. Safe means of access should be provided to trestle platforms, usually by stepladders.

When using trestles the following safe working practices should be used:

Always

- Set up the equipment on a firm, level, non-slip surface.

- On soft ground, stand the equipment on boards to stop it sinking in.

- Place each trestle at 1.5 metre intervals, which allows the scaffold boards to be adequately supported.

- Then open each up to the height required, ensure the locking pins are properly located.

Never

- Do anything that involves applying a lot of side force when using the trestle. The trestle could topple over.

- Exceed the maximum safe working load of a scaffold board (150kg evenly spaced). Remember to include the weight of the user, tools and materials in your calculations.

- Use steps, boxes, etc, placed on top of the trestle boards to gain extra height.

Staging platforms

Staging platforms can be made of metal alloy or wood and are often used for linking trestle systems or tower scaffolds together safely. They also provide a safe work platform for work on fragile roofs. They are produced in various lengths. The same rules apply for edge protection as with other scaffold platforms.

When using staging platforms the following safe working practices should be used. Ensure the staging platforms are:

- Of sufficient dimensions to allow safe passage and safe use of equipment and materials.

- Free from trip hazards or gaps through which persons or materials could fall.

- Fitted with toe boards and guardrails (if these requirements are not considered necessary for a specific platform, then this should be shown in the risk assessment, i.e. that not installing a toe board and/or a guardrail had been considered and why it was not necessary).

- Kept clean and tidy, for example, mortar and debris should not be allowed to build-up on platforms.

- Not loaded to the extent that there is a risk of collapse or deformation that could affect its safe use. This is particularly relevant in relation to block work loaded on staging platforms.

- Erected on firm level ground to ensure equipment remains stable during use.

Leading edge protection

Leading edges are created as roof sheets are laid or ones are removed. Falls from a leading edge need to be prevented; work at the leading edge requires careful planning to develop a safe system of work.

Staging platforms, fitted with guardrails or suitable barriers and toe boards, positioned in advance of the leading edge can provide some protection prevent falls during some leading edge work activities.

Figure 8-60: Staging platform with guard rail edge protection. Source RMS.

However, the staging platforms will need to be used in conjunction with work restraint systems (fall prevention) /fall-arrest systems (fall protection) attached to a suitable fixing. Close supervision of this system of work will be needed as it is often difficult for the fall prevention/protection equipment to remain safely clipped to the fixing at all times throughout the work activity.

Safety nets are the preferred protection for reducing the distance and consequences of falls at the leading edge, as they provide collective protection to everyone on the roof. Safety nets should be erected as close to the work surface of the leading edge as possible by trained riggers and be strong enough to take the weight of people who fall onto them.

In some countries, longer term fall protection is added to buildings that have roof sheets to provide protection during construction and at the time roof sheets are removed. This involves the installation of galvanised steel safety mesh on top of the roof steelwork prior to the laying of the roof sheets (**see Figure 8-61**). The safety mesh is rolled over the structure from work platforms set in position at the sides. It remains in position through-out the life of the building and provides leading edge protection at the time of laying or removal of the roof sheets

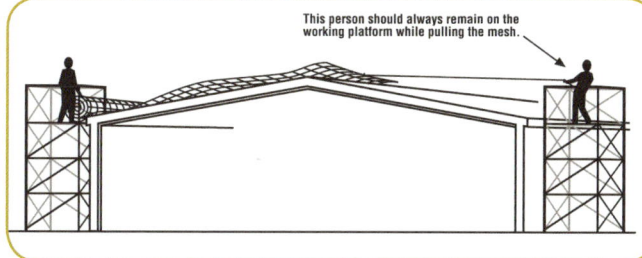

Figure 8-61: Safety mesh.
Source: Western Australia, Code of practice - Prevention of falls at workplaces.

 **REVIEW**

What are the features of a ladder that would ensure it is suitable for working at height?

Outline a safe method of working on a fragile roof.

When should scaffolding be inspected?

Outline safe methods of working on a leading edge on a roof.

PREVENTION OF FALLING MATERIALS THROUGH SAFE STACKING AND STORAGE

Materials and objects should be stacked and stored so they are not likely to fall and cause injury. Even small objects can cause injury if they are allowed to fall. The materials may be in short-term storage in the workplace in order to be available for use or be in longer-term storage. They may be raw materials; materials being worked on as part of a process or finished materials or objects.

It is important to organise stacking and storage to ensure stability and reduce the possibility of materials falling. In addition to prevention of materials falling, in some cases it is necessary for the stacking and storage arrangements to take account of specific hazards associated with them or properties they have that are incompatible with each other.

Stacking and storage of materials should be carefully planned as this will enable the arrangements to take account of the various characteristics of the materials and the need to access them for use or to transport them to another location. For example, the shape, size and weight of the materials to be stacked or stored will greatly influence the ease of handling and prevention of their falling. In an office or similar environment, this may mean providing shelving in the form of cupboards with doors to help contain the variety of items that need to be stacked and stored.

In manufacturing activities many items are brought together to make a finished product, this often involves specific stacking and storage arrangements to take account of the variety of sizes, shapes and weights of the materials. It also involves arrangements to enable the efficient work flow and use of the materials, which will mean that workers frequently interact with the materials and may therefore be at risk of injury if the materials should fall.

Warehouse and retail workplaces have to take account of the rapid stacking and storage of materials and their equally rapid removal. This constant change to stacking and storage means that there is an increased risk of materials falling. The many small items that are stacked or stored may need to be placed in bins, trays or other similar containers that can enable their safe removal and if required the selection of one or more items from the container. Where materials are displayed for the public to select an item for purchase it is important that the shelves used have a retaining edge or other mechanism that limits the likelihood of the material being dislodged and falling during or after selection.

The risk of falling materials related to chemical process activities is influenced by the need to stack or store large volumes of material as the process produces it and before it is despatched to customers. Though the materials are often in regular sized and shaped containers, they are often stacked to a significant height.

Agriculture and forestry work will often involve materials that are large and unusual in shape and may require particular care and arrangements to ensure materials do not fall when stacked and stored. In particular, the shape of the material may be round, for example, tree trunks that have been cut down or hay that has been rolled into a circular shape. When they are stacked on top of each other their round shape and the weight of the items at the top may make the lower items roll, causing the stack to become unstable and the material to fall.

A construction site is continually changing, with a variety of activities being conducted at once, as well as workers and materials being moved around to conduct construction activities. For this reason, it is important that the site is kept tidy. Safe stacking and storage of materials used in construction can help to keep the vehicle and pedestrian routes clear and keep the materials themselves from causing harm, for example, preventing knocking over bricks that could fall from height.

In vehicle maintenance operations the risk of falling materials can arise from such things as the stacking of scrap tyres or exhausts in an unstructured way, resulting in an unstable pile of material. To ensure safety and prevent falling materials, it is important to stack and store these items in as organised way as possible.

> "A fatally unsafe stacking system cost a dairy products manufacturer in Australia a $300,000 fine and a conviction. WorkSafe's investigation found the practice of stacking bulk salt bags at a dairy products manufacturer's site was unsafe as the tops of lower bags may not be sufficiently level to safely accommodate those bags on top. The danger was such that if the surface of a bag was not level, even a small disturbing force could cause the stack to topple; this led to a stack of material falling and the death of a worker. It was determined that the circumstances were in breach of Section 21 of the Occupational Health and Safety Act 2004."

Figure 8-62: Fatality due to poor stacking and storage.
Source: Australia, Victoria Worksafe.

Stacking

Stacking may be done by creating freestanding stacks or by using a racking system in which to stack the materials.

Free standing stacks

It is important that the stack be stable in order to prevent its collapse and material to fall. It is essential that the stack be placed on a firm and level base. To avoid the stack sinking in and leaning over, the ground must be firm and strong enough. This can be achieved by stacking on prepared concrete surfaces or if necessary by preparing soft ground by compacting it and laying material on it to spread the load of the stack. The location of the stack must be level, because if it is not level, when additional material is added to the stack the effects of being stacked at an angle, instead of being perpendicular, will cause the stack to overturn and material to fall. This means that stacking on slopes must be avoided and care taken to ensure the stack is not affected by loose material on the floor or dips and drains in the floor, all of which can make the stack lean at an angle and cause material to fall from the stack.

In order to create a stable base and to enable easy stacking and removal the material should be placed on a pallet or in a stackable cage, crate or similar stillage. This will provide improved stability as each stack is added and reduce the risk of materials falling.

It is important to ensure the material being stacked is as stable on the pallet as possible. Where loose items of a uniform size, like boxes or bricks, are placed on a pallet and are to be used in a stack, they can be stabilised by varying the pattern of cross-layering in a way that interlocks them and bonds them together.

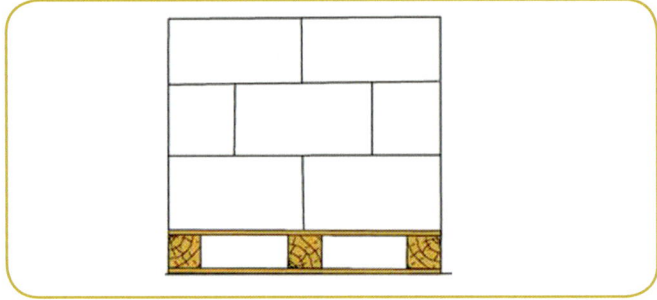

Figure 6-83: Bonding a stack by cross-laying.
Source: UK, HSE, HSG76.

In addition, the material in the stack can be held in place by the addition of straps or by applying stretch/shrink wrap plastic around each load. Care should be taken when using strapping as over tensioning of the straps can cause damage to the load or pallet and create a weak point that might lead to collapse of the stack.

Figure 8-64: Load secured by plastic film.
Source: UK, HSE, HSG76.

Different types of container should be stacked separately. Round materials, for example, pipes and drums that are stacked on top of each other should be blocked/wedged in the bottom tiers to prevent them rolling and the collapse or falling of items from the stack. They should be stacked in a symmetrical pattern to even out the forces arising from their weight. When round materials are stacked on end, planks, sheets of wood or pallets must be placed between each tier to make a firm, flat, stacking surface and to bind the material in the tier to each other.

Material in sacks or similar bags must be stacked by stepping back the layers and cross-laying them by laying them at 90 degrees to the previous layer. Similarly, large bags and bundles must be stacked in interlocking rows to remain secure.

"Cargo, pallets and other material stored in tiers shall be stacked in such a manner as to provide stability against sliding and collapse."

Figure 8-65: Safe stacking and storage.
Source: USA, OSHA Standard 1917.14.

Figure 8-66: Cage for loose materials.
Source: UK, HSE, HSG76.

Loose items can be contained in cages or similar stillages of mesh or sheet construction. This will enable them to be stacked in quantity without the risk of the materials falling. The stillages have feet and tops that interlock and provide a good stable structure. It is important to ensure the cages or stillages are not overfilled to prevent loose items falling.

Avoid stacking palletised loads of cartons and packs that are capable of being crushed as the strength and stability of the stack cannot be maintained. Loads that are capable of being stacked directly on top of each other should generally be provided with additional packing on top of the lower palletised load, depending on the characteristics of the load and design of the pallet.

Stacking should not be so high or leaning that it poses a risk of falling over and crushing someone, a number of workers have been killed in this way. In order to prevent stacks of materials from falling it is important to re-strict the height of stacks. Generally, pallet stacks should not be stacked more than a 4:1 ratio between height of stack and the minimum depth/width of the pallets. Four loads high might also be considered a maximum due to the potential problem of crushing the material on the bottom pallet, leading to sudden collapse.

Figure 8-67: Free standing stacks and stacks in racks.
Source: RMS.

In some circumstances, dependent on the height, strength and stability of the loads, taller stacks may be built. The maximum practical height may be up to six loads high, provided that the pallet itself and the packaging of the material are designed to exceed the four-high limit. This may be the case when using inter-locking stillages that create a strong structure.

Weather conditions are an additional factor that influence the stability of stacks. The heights discussed may be suitable for stacking indoors; however stack heights may have to be reduced outdoors to take account of weather conditions, for example high winds. It is also important to consider the effects of wet weather on mate-rials that lose their strength when they become wet, which could lead to sudden collapse and fall of materials. When deciding the safe height of stacks consideration should be made to the movement of overhead equipment that could strike the stack, for example, an overhead travel-ling crane.

Heights of stacks should be limited to ensure a safe margin between the stacks and the minimum height of the overhead equipment. It can be useful to indicate maximum heights by painted markings on walls or posts.

Figure 8-68: Unstable stack.
Source: RMS.

Element 8

Adequate clearance should be maintained between rows of stacks to ensure safe stacking and withdrawal of materials. This will avoid the possibility of contacting adjacent stacks when removing materials and causing stacks to fall. Do not allow items to stick out from stacks, because if the items contact adjacent stacks when loads are removed or added they may cause the stack to fall. Ensure that there are no loose items on top of stacks, as they may be dislodged and fall when the load is moved or by strong winds. Care should be taken to avoid a single full height stack being left when materials are moved, as it will be less stable than if it were along with other stacks, particularly in bad weather.

Pallets and stillages should be visually inspected before use, particular care should be taken with disposable pallets as they tend to be of lighter construction and can easily be damaged. Check stacks periodically for stability and take corrective action where necessary.

Workers should be trained in the correct way to remove items from the stack and the dangers associated with leaving an unsafe stack. Do not remove materials from a stack by throwing down material from the top or pulling materials out from the bottom. Workers should be instructed not to climb up a stack or stand on the top of it, as the additional forces of their weight can cause the stack to become unstable, overturn and materials to fall.

Stacking in rack systems

In order to ensure stability of the rack and stacks located in them it is important to use a properly constructed rack and to secure it to the floor or wall. The rack should be matched to the load being stacked in it, ensuring it is of adequate size and weight carrying capacity. Do not exceed the safe weight carrying capacity of racks. Racks should be inspected regularly for damage due to the possible damage caused by impact or overload.

Figure 8-69: Stack being placed in racking.
Source: UK, HSE, HSG76.

It is important to ensure suitable pallets are used to carry the material placed in the racking. Disposable pallets are not normally suitable for use in racking, because they are usually of lighter construction and not strong enough. Material on pallets should be secured by straps or stretch/shrink wrap plastic, this ensures items cannot fall while being placed in storage or removed.

Care should be taken to ensure that loose material is not left on top of the load when it is stacked in the racking and that material being stored or packing around the load is not dislodged when the load is retrieved from the rack. Access to the stacked material in the racking for retrieval of part of a load must be controlled to prevent the remaining material becoming insecure and falling.

Storage

The safe storage of materials is dependent on the type of material to be stored. It is useful to plan to store heavy or large items at the bottom and smaller, lighter items, higher up. This will help with stability, reducing the risk of materials falling, and assist with manual handling. Items that are to be most frequently used should be put in an easily accessible area, this will reduce the effort related to handling and reduce the risk of materials being dislodged or falling while gaining access to the items. If there is a risk of anything falling that could injure someone, access to the storage area should be controlled and those entering should wear head protection designed for falling material.

Storage areas should be specifically designated, be clearly marked, and be controlled by a responsible person. This will help to ensure they remain organised and that material is stored in a safe manner that prevents it falling. Storage areas that are designed to be accessed by a number of workers, rather than a responsible person supervising or issuing materials, may need particular care as there can be a tendency for workers to leave materials in a poor condition, allowing them to fall and injure the next worker using the store.

Storage should be designed for the purpose and enable new stock of materials to be readily stored and ensure stock can be removed for use without disturbing other material stored and causing it to fall. The storage may involve widely used methods of storage, for example, cupboards, cabinets and shelves, or the use of specialist systems, for example, racking storage for sheet material or tiered storage bin systems for loose items.

Aisles in storage areas should be clearly marked, be of ample width for the type of storage, and be kept free from obstacles and waste materials. Generally, with heavy or awkward materials, the storage area should be designed to allow access by fork-lift trucks (FLTs) as well as workers. It should be kept tidy and suitably lit. These measures will ensure ease of movement and reduce the likelihood of contact with materials that can cause them to become unstable or fall.

CASE STUDY

A 20 year old male fork truck driver was injured when a pallet carrying containers of battery acid collapsed whilst he attempted to place it into storage raking. One of the containers fell on to the floor, burst and spayed acid onto the driver's legs causing chemical burns. On investigation a disposable pallet had been used in error. Disposable pallets are now stored in a designated location and clearly labelled not for re-use.

The material should not be stored so that it blocks fire points, extinguishers or blocks out the light. It should not be placed on soft ground or floors that may not be strong enough to support it. The material should not be stored too close to vehicle routes to avoid vehicles hitting them as they pass or turn and causing them to become unstable and fall.

The storage area should be protected against unauthorised use and should have the correct signs, for example, warning of any hazards related to the type of material stored or requiring personal protective equipment (PPE) to be worn.

Chamber	Caisson, cofferdam, or interception chamber for water.
Tank	Storage tanks for solid or liquid chemicals.
Vat	Process vessel which may be open, but by its depth confines a person.
Silo	May be an above-the-ground structure for storing cereal, crops.
Pit	Below ground, such as a chamber for a pump.
Pipe	Concrete, plastic, steel, etc. fabrication used to carry liquids or gases.
Sewer	Brick or concrete structure for the carrying of liquid waste.
Flue	Exhaust chimney for disposal of waste gases.
Well	Deep source of water.

Figure 8-71: Confined space - sewer.
Source: RMS.

8.3 Safe working in confined spaces

TYPES OF CONFINED SPACES AND WHY THEY ARE DANGEROUS

Types of confined spaces

A confined space is a substantially enclosed (though not always entirely) place which, by virtue of its enclosed nature, there is a foreseeable risk of serious injury from hazards within the space or nearby (for example, lack of oxygen). Typically confined spaces will include a chamber, tank, vat, silo, pit, pipe, sewer, flue, well or similar space.

Figure 8-72: Confined space - worker in pipe.
Source: Shutterstock.

Why they are dangerous

Confined spaces are dangerous because the hazards may not be obvious to workers until they enter the confined space and experience the effects of the hazard. Failure to appreciate the dangers associated with confined spaces has led to the deaths of many workers and often the death of some of those who have attempted to rescue them. This is because in order to rescue a

Figure 8-70: Confined space - chamber.
Source: RMS.

worker, the rescuers urgently enter the confined space, often without establishing what the hazards affecting the worker were and without taking sufficient precautions to protect themselves, and they also become affected by the hazard. In addition, because of the confined nature of the space it can make rescuing a worker very difficult and slow, which means that a worker affected by hazards suffers serious harm or dies before the rescue is completed.

Figure 8-73: Confined space - Tank entry.
Source: www.picstopin.com

Hazards of confined spaces

The main hazards associated with working within a confined space include:

- Flammable vapours/dusts or excess oxygen that could cause fire or explosion.

- Heat that could cause loss of consciousness from increased body temperature.

- Exposure to a toxic gas/fume/vapour or lack of oxygen that could cause loss of consciousness or asphyxiation.

- Increase in the level of liquid that could cause drowning.

- Free-flowing solid that if trapped by it could cause asphyxiation.

- High concentrations of dust that could cause respiratory difficulties, for example dust found in a flour silo.

- Moving parts of machinery that could cause impact, crush, entanglement, etc.

- Steam that could cause burns.

- Restrictions in the space that could cause a worker to be trapped or difficulties in escaping.

Assessing risks from a confined space

Before work in a confined space can be carried out it must be shown that it not practicable to carry out the work without entering the confined space. If it is considered necessary to enter a confined space then the employer must carry out a risk assessment to identify risks and the precautions that will be necessary to ensure a safe system of work.

An assessment of risks from a confined space should consider the risks present in the confined space, those introduced by the work done in the confined space and those created by work carried out near to the confined space that can affect those working in the confined space.

The risk assessment should, in particular, consider the hazards and how they arise, including from the following typical situations:

- Previous contents of the confined space - whether they have been removed effectively or whether some material remains as a vapour or gas.

- Residues left in the confined space or equipment attached to the confined space - for example, sediment or material trapped in seals that may release toxic or flammable material when disturbed.

- Substances and other materials allowed to enter the confined space while working in it – for example, flood water entering a sewer or process substances entering a mixing tank.

- Oxygen deficiency and oxygen enrichment – normal oxygen content of air is approximately 21%.

- Shape and dimensions of the confined space - the atmosphere may vary in parts of the confined space, for example, low lying compartments.

- Temperature changes may also cause toxic materials to be released – for example, changes may be due to introducing heat in the form of a steam lance for cleaning or welding tasks.

- Cleaning chemicals or chemicals introduced as part of the task being done in the confined space could affect the atmosphere - for example, vapours from adhesives used to glue a replacement tank lining or the products of combustion from a welding task.

- Substances or dangerous fumes entering the confined space - for example, from adjacent work or contamination being released from the ground (methane can leach into groundwater and be released or limestone may give off carbon dioxide in the presence of acid groundwater).

- The temperature in the confined space - ambient temperature may be too hot or too cold to allow ease of breathing or control a worker's body core temperature (welding activities will produce a lot of heat and with storage vessels that are located outside the temperature inside the vessels will rise significantly when exposed to sunlight).

- Mechanical equipment operating while workers are in the confined space – for example, a stirrer in a mixing vessel.

- Services connected to the confined space – for example, steam, electricity, water.

- Sources of ignition for flammable materials – in the confined space or brought in by workers, for example their equipment or portable lighting.

- Ventilation of the confined space – adequacy of natural ventilation or need to provide forced ventilation (consider monitoring the atmosphere, a source of fresh air for forced ventilation, the effects of failure of forced ventilation).

- Lighting limitations – adequacy of natural lighting or need to provide lighting (consider risks introduced by electrical lighting equipment).

It is important that the risk assessment also take account of:

- The task, as well as the materials and equipment involved.

- The people - including the competency of those involved and the suitability of those who will enter the confined space.

- What could influence safe access and egress from the confined space – for example, limitations of access routes and whether a hoist is required.

- What could influence the health and safety of the working environment.

- Reliability of control measures – need for a permit-to-work.

- Emergency arrangements – consider if a harness and lifeline are required, factors that will affect a rescue, the resources available to a rescue team and the location of external emergency services.

Precautions to include in a safe system of work for confined spaces

Similar to other risks, the priority would be to determine if entry into the confined space can be avoided, for example, by using cleaning and other equipment operated from outside the confined space (for example, water jets, long-handled tools or vibrating blockage clearing equipment) and use of viewing panels or cameras for inspection. If it is necessary to enter the confined space it is important to take into account the risks and put into place precautions to minimise the risks.

Some of the precautions to consider in a safe system of work for confined spaces are:

- Level of supervision - Supervisors should be given responsibility to make sure that the necessary precautions are taken, by checking precautions at each stage and they may need to remain present while work is carried out.

- Suitability of the workers – their competence (training and experience), size, fitness (ability to wear respiratory protective equipment) and claustrophobia.

- Isolation and lock-off of any service or material entry and exit points (such as water stop valves).

- Isolation of electrical systems and moving parts of the equipment.

- Removal of residues before entry into the space, by cleaning from outside the confined space.

- Atmosphere testing, before entry into the space and at intervals, personal monitors may also be required.

- Ventilation before and during entry into the space, forced ventilation may be needed.

- PPE requirements, including respiratory protection if required.

- Access (the way in) and egress (the way out), ladders or hoists may be required and it is important to ensure unobstructed routes for escape.

- Communication methods between those in the confined space and those outside.

- Fire precautions, deluge fire systems would usually need to be isolated so portable fire extinguishing equipment would need to be provided.

- Lighting levels, provision of additional temporary lighting. Consider flammability of the atmosphere when choosing the type of equipment.

- Equipment suitable for the confined space, extra low voltage (less than 25V) for work inside metal confined spaces, use of residual current devices, non-sparking tools.

- Control of the duration of a working period to manage fatigue and heat/cold exposure.

- Emergency rescue arrangements, for example a lifeline attached to a harness should run back to a point outside the confined space, someone to keep watch, communicate and raise the alarm.

The risk assessment will, in particular, help to identify the need for a formal administrative system such as a permit-to-work *(see 3.7 - Permit-to-work systems)*. The permit-to-work-system will typically include consideration of precautions described above and, in particular, procedures for:

- Testing the atmosphere.
- Respiratory protective equipment and other personal protective equipment for those risks that cannot be controlled by other means.
- Equipment for safe access and egress.
- Suitable and sufficient emergency rescue arrangements.

Testing the atmosphere

Testing the air will need to be carried out to check that it is free from both toxic and flammable gases/vapours and that it is fit to breathe. Testing should be carried out by a competent person using suitable detectors, which are correctly calibrated. Where the risk assessment indicates that conditions may change, or as a further precaution, continuous monitoring of the air may be needed. The appropriate choice of testing equipment will depend on the particular circumstances of the confined space and the work to be done.

Tests should be carried out by competent and experienced persons and records should be kept of the results of tests. Personal air monitoring equipment should be worn whenever appropriate to minimise the risks from hazards present as local pockets of contaminant in the confined space.

Personal protective equipment

Figure 8 - 74: Typical entrance to a confined space.
Source: RMS.

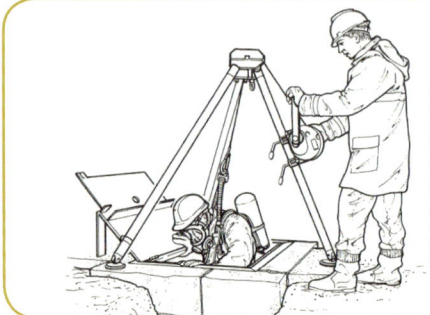

Here a worker wearing full breathing apparatus is also wearing a harness with a lanyard connected to a winch so that he can be hauled to the surface in an emergency without others having to enter the manhole to rescue him.

Figure 8-75: Access to a confined space.
Source: UK HSE, HSG150.

Where respiratory protective equipment is provided or used in connection with confined space entry (including emergency rescue), it must be suitable. Other equipment - ropes, harnesses, lifelines, resuscitating apparatus, first-aid equipment, protective clothing and other special equipment will usually need to be provided.

Safe access to and egress from confined spaces

The arrangements for safe access and egress from confined spaces include:

- Openings need to be sufficiently large and free from obstruction - to allow the passage of workers wearing the necessary protective clothing and equipment and to allow access for rescue purposes.

- Practice drills - to check that the size of openings and entry procedures are satisfactory.

- Identification of the confined space warning signs placed – to ensure the correct confined space is being worked on and to warn other that confined space work is in progress so they do not do anything that could harm those workers inside.

- An observer present at all times when work is in progress - the observer should stay in contact with the workers and be equipped to raise the alarm immediately if an emergency occurs.

Emergency arrangements

Rescuers need to be properly trained, sufficiently fit to carry out their task, on standby, and capable of using any equipment provided for rescue, for example, breathing apparatus, lifelines and firefighting equipment. Rescuers also need to be protected against the cause of the emergency.

Suitable and sufficient rescue arrangements should be provided before any worker is allowed to enter or carry out work in a confined space. The arrangements for emergency rescue will depend on the nature of the confined space, the risks identified and the likelihood of an emergency rescue being needed.

The arrangements might need to cover:

- Raising the alarm and rescue.
- Safeguarding the rescuers.
- First-aid, including resuscitation equipment.
- Special arrangements with local hospitals (for example, for foreseeable poisoning).
- Firefighting equipment.
- Public emergency services contact methods.

When a permit-to-work would not be required for a confined space

A permit-to-work would not be required for a confined space that did not present significant risks. This would typically be where a confined space was large enough for a worker to enter, but did not contain the hazards that could cause serious harm/death described earlier in this element and the work being done could also not introduce these hazards. This would be identified and confirmed through a risk assessment process.

 CASE STUDY

Prosecution of a UK company following a triple fatality which resulted from confined-space working.

A company transported waste water/sludge, using a tanker lorry, from food factories and abattoirs to various farms for spreading on agricultural land. This involved transferring the waste from the tanker into a holding tank (at the side of the field), from which it was pumped to a tractor for spreading on the agricultural site.

On occasions, a thick sludge built up in the bottom of the holding tank. This required an employee to climb onto the top of the tank and to use a high-pressure hose to dislodge and dilute the sludge. The person undertaking the task fell into the tank through an open hatch, became unconscious (the tank atmosphere was incapable of supporting life) and drowned in the liquid at the bottom of the tank.

Two men who attempted a rescue also suffered the same fate. A fourth man also tried a rescue but managed to get out.

In summary, there were no measures to prevent access to the tank, and training and risk assessments were said to be inadequate. The company and their directors were successfully prosecuted and fined.

 REVIEW

List four possible risks when working in a confined space.

Define confined space.

What should the competent person consider when developing a safe system of work for a confined space?

WHAT A LONE WORKER IS AND EXAMPLES OF LONE WORKING

Lone workers are those who work by themselves without close or direct supervision. This may mean they are with other people who are not their work colleagues, but may be customers or workers of another company. Lone workers are found in a wide range of situations. Some examples include:

People in fixed establishments where:

- Only one person works on the premises, for example, petrol stations, kiosks, shops and also home workers.

- People work separately from others, for example, in factories, warehouses, leisure centres or offices.

- People work outside normal hours, for example, cleaners, security workers, facilities management workers or contractors conducting special tasks that are better done at this time.

Mobile workers working away from their fixed base:

- Doing construction activities, plant installation, maintenance, cleaning work, electrical repairs, painting and decorating.

- Agricultural and forestry workers.

- Service workers, for example, rent collectors, postal workers, workers providing care in someone's home, drivers, doctors and district nurses.

In managing the risks from lone working it is necessary to identify the hazards of the work, assess the risks involved, and put control measures in place to avoid or minimise the risks.

PARTICULAR HAZARDS OF LONE WORKING

The hazards that lone workers may experience are essentially the same hazards that people working together may encounter. The major difference is that a lone worker may be more at risk of harm than when working with other workers. For example, a lone worker may be more likely to experience threats or actual violence whilst working. They may be more likely to do things on their own that would normally need more than one person or carrying out tasks that are outside their limit of competence.

Examples of particular hazards of lone working include:

- Over exertion - while conducting manual handling tasks.

- Overtuning of equipment – while operating equipment that needs two people to operate it safely, for example when operating a crane that needs an observer and a crane operator.

- Work-related upper limb disorders – due to having to adopt awkward postures and hold equipment in place at the same time as fitting it.

- Electric shock – due to difficulty in carrying out electrical tests.

- Struck by vehicles – due to needing someone to observe vehicles passing near the worker while they were working.

- Violence – due to the fact they are on their own allows people to attack and overcome them more easily and without a witnesses.

WHAT SHOULD BE CONSIDERED WHEN ASSESSING THE RISKS OF LONE WORKING.

Employers should take into account both normal work and foreseeable emergencies, for example fire, equipment failure, illness and injuries. When considering the risks related to lone working, particular attention should be paid to:

- The workplace, for example a railway worker inspecting a railway track would find it difficult to look out for trains coming while doing the inspection work.

- Work equipment, which could present a specific hazard to the lone worker, for example due to the use of temporary access equipment that one person would have difficulty using safely, for example the stability of a ladder.

- Access and egress could be difficult for one person, for example for a lone worker out of normal working hours where the workplace could be locked while they are working or carrying out tasks in a location that is difficult to access (collecting a water sample by accessing a slippery river bank).

- Equipment involved in the work may not be safe for one person to operate the essential controls.

- Chemicals or hazardous substances used could present a particular risk to the lone worker.

- Heavy or large objects, which could be outside the capability of one person to lift or move.

- Violence or aggression, which might be more likely because the worker is on their own, particularly if they in a work situation that takes them into someone's home, they are carrying out a role on behalf of an enforcing authority, they are carrying valuable items, the activity is carried out in a remote area or at night.

- An individual might be more vulnerable than others and be particularly at risk if they work alone, for example if they are young, pregnant, disabled, a trainee or have a pre-existing health condition.

In addition, if things go wrong there will not be other workers on hand to provide assistance, for example, provide first-aid, help someone trapped under a fallen load, come to a worker's aid when overcome by chemicals or rescue them from falling into a river/water filled tank.

CONTROL MEASURES FOR LONE WORKING

Where the risks are higher, consideration should always be made to finding alternative ways of working that avoid lone working. When the risk assessment shows that it is not possible for the work to be carried out safely by a lone worker, arrangements for providing assistance should be put in place. Work may be carried out faster, safer and more effectively by using two workers.

Establishing safe working for lone workers is no different from organising the health and safety of other workers. Employers need to know the law and standards that apply to their work activities and then assess whether the requirements can be met by people working alone. Employers need to be aware of any specific law on lone working applying in their industry; examples include supervision in diving operations, vehicles carrying explosives, fumigation work.

When the risk assessment shows that it is not possible for the work to be done safely by a lone worker, arrangements for providing assistance should be put in place. There are some high-risk activities where at least one other person may need to be present. Examples include confined space working (where a supervisor may need to be present, as well as someone dedicated to the rescue role) and electrical work at or near exposed live conductors (where at least two people are sometimes required).

Where a worker is new to a job, undergoing training, doing a job that presents specific risks, or dealing with new situations, it may be advisable for them to be accompanied when they first take up the job. It should not be left to individuals to decide whether they need assistance.

Element 8

The risk assessment should help to decide the right level and method of supervision. The extent of supervision required is a management decision and depends on the risks involved and the ability of the lone worker to identify and handle health and safety issues. Generally, the higher the risk, the greater the level of supervision required.

Where a lone worker is working at another employer's workplace, that employer should inform the lone worker's employer of any risks from being in that workplace and the control measures that should be taken. This helps the lone worker's employer to assess the risks and make appropriate arrangements with the lone worker.

Control measures for lone working may include training, supervision, monitoring, consideration of medical conditions, welfare arrangement and emergency procedures etc.

Training

When establishing control measures for lone working, attention should be paid to the importance of training, particularly where there is limited supervision. Workers should be:

- Experienced enough to understand the risks and precautions of lone working fully and to avoid panic reactions in unusual situations.

- Competent to deal with circumstances which are new, unusual or beyond the scope of training, for example, when to stop work and seek advice from a supervisor and how to handle aggression.

- Able to respond correctly to emergencies; they should also be given information on established emergency procedures and danger areas.

- Able to administer first-aid.

Supervision

When providing supervision for lone working, the following factors should be considered:

- Supervision can ensure that workers understand the risks associated with their work and that the necessary safety precautions are carried out.

- Supervision can provide guidance in situations of uncertainty.

- Supervision can be carried out when checking the progress and quality of the work. This may take the form of periodic site visits combined with discussions in which health and safety issues are raised.

- Supervision is important when a worker is new to a job, undergoing training, doing a job which presents special risks, or dealing with new situations. Such workers may need to be accompanied when they first take up the work.

- The level of supervision required should be based on the findings of a risk assessment and the competency of the worker.

Procedures for monitoring

When establishing monitoring procedures for lone working, the following methods should be considered:

- Supervisors periodically visiting and observing people working alone.

- Regular contact using either a telephone or radio.

- Automatic warning devices which operate if specific signals are not received periodically from the lone worker, for example, systems for security staff.

- Other devices designed to raise the alarm in the event of an emergency and which are operated manually or automatically by the absence of activity.

- Checking that the worker returned to their base or home on completion of a task.

Lone workers may carry mobile phones to assist with communication, the signal strength should be checked before entering a lone working situation. A mobile phone should never be relied on as the only means of communication. Lone workers should tell their manager or a colleague about any visit in advance, including its location and nature, and when they expect to arrive and leave. Afterwards, the lone working they should confirm they are safe.

Medical considerations

When determining medical considerations for lone working:

- Check that lone workers have no medical conditions that make them unsuitable for working alone.

- Consider both routine work and foreseeable emergencies that may impose additional physical and mental burdens on the individual.

- Ensure they have access to first-aid facilities, and mobile workers should carry a first-aid kit suitable for treating minor injuries.

REVIEW

What procedural controls could employers implement to help minimise the risk to lone workers?

What factors should be considered to reduce the risk to lone workers?

COMMON CAUSES OF SLIPS AND TRIPS

Uneven surfaces

Uneven surfaces can cause a worker to trip. The surface does not have to be very uneven to cause a trip, it just has to be uneven enough for the worker's footwear to catch on the surface, causing a sudden stop of movement of the foot and the worker to lose balance. Walking surfaces can become uneven due to damage, where cracks can appear and holes develop in the surface. Some surfaces may be constructed to be rough, but if care is not taken in their construction the roughness may be excessively uneven. When sites are being developed, for example for construction purposes, the ground surface may be left in an uneven state after ground-works have taken place, which will present a trip hazard for workers. In offices, the floor surface may have tiles or a floor covering, which may be unevenly laid or have edges that have become less secured by adhesive. This is often where there has been a lot of pedestrian traffic, for example at the entrance to rooms or corridors.

Figure 8-76: Trip hazard, damaged tile.
Source: RMS.

Unsuitable surfaces

Slips on an unsuitable surface can cause major injuries at work. Generally, slips occur due to a loss of grip between the shoe and the walking surface. One of the main causes of workers slipping at work is on surfaces that are not suitable for the contamination that affects them. For example, the contamination could be spilt liquids in a kitchen or rain water in the entrance of an office block. In these situations if the surface of the floor is very smooth it will allow contamination to lie on the surface and not provide footwear with sufficient grip.

A surface may also become unsuitable because it has become highly polished over a period of time due to use or cleaning processes. In addition, condensation can make a smooth floor slippery and cold weather can cause ice to form on a surface. It should be remembered that cleaning processes may leave a floor surface in a wet and slippery condition for a period of time, making it an unsuitable surface to walk on.

Figure 8-77: Slip hazard spilt liquid.
Source: RMS.

Trailing cables

Trailing cables from various forms of electrical equipment are a common cause of trips in the workplace. The nature of the cable, which may have curls and loops in it, can make it easier for someone to catch their foot in it and trip. Other factors, like poor lighting and the lack of colour differentiation between the cable and floor can make a trip more likely.

Obstructions in walkways

Obstructions in walkways can easily lead to workers tripping, particularly if they are carrying something and cannot see the obstruction. In some cases, the obstruction may not cause a worker to trip, but if they place their foot on the obstruction while walking it may roll or slide away from the foot causing the worker to slip. These obstructions could be due to such things as waste material building up from processes and spilling out onto the walkway, maintenance work taking place where parts of the equipment are on the floor and in walkways or general construction site debris. Obstructions in walkways can also be due to the bad placement of a worker's personal items in their workspace, for example, in an office a worker may badly place their bag under their desk so that it becomes an obstruction in the walkway. Other forms of obstruction could include insecure or uneven ducting covers, mats, hoses or changes in level of the walkway, like curbs and steps.

Unsuitable footwear

As previously stated, slips generally occur due to a loss of grip between the shoe and the walking surface, therefore unsuitable footwear is an important part of the cause of slips. If the surface of the sole (the bottom part) of a shoe is particularly smooth, for example because it is made of a plastic that has worn smooth or the grip of a sole made of other material has worn away, this can mean that the worker loses grip on a walking surface. Sometimes a worker's footwear can become unsuitable because it is wearing out and the stitches or glue holding the sole to the top of the shoe is failing. This can cause the sole to separate from the top of the shoe and be sufficient to increase the chances of the worker tripping. Footwear that is suitable for one workplace may be unsuitable for another, for example it may be suitable for a rough concrete floor, but not for a floor that is contaminated by oil.

MAIN CONTROL MEASURES FOR SLIPS AND TRIPS

Slip-resistant surfaces and footwear

Slip-resistant surfaces

In order to ensure the safe movement of people slip-resistant surfaces should be provided:

- At the entrance of buildings, for example, a mat that provides both slip-resistance and can absorb water brought in on footwear.

- On designated walkways.

- On changes of level, such as stairs, steps, ladders, or footholds to vehicles.

- On ramps or slopes.

- Where walkways intersect with internal transport routes and people may need to stop suddenly.

- In work areas where spills of liquids or dry contaminants are likely.

- Where liquids are poured or containers filled or stored.

- On access areas used for inspection or maintenance.

- Locations where workers need to go that are exposed to the weather and where surfaces may become covered in environmental grime or slippery growth.

- On construction sites and where outdoor work takes place, for example provision of roughened concrete walkways.

Figure 8-78: Slip resistant surface on steps.
Source: RMS.

When selecting a slip resistant surface consider:

- The consequences of slipping, a slip while holding a knife or when working at height could be severe.
- The type of contamination likely, for example, liquid or dry; water, oil or blood.
- Ability to control contamination, for example, drainage.
- Amount of use of the surface.
- The numbers and types of people using the surface, including their age, disability, etc.
- What people might be doing on the surface, for example, walking, climbing, carrying, turning, and moving fast.
- Environmental issues, weather, hot or cold.
- Level of control over footwear used.
- Ease of cleaning and the desired level of cleanliness.
- The floor surface, for example, smooth or rough profile, wear durability.
- The slip resistance rating (SRV) and roughness (Rz) needed.

Figure 8-79: Slip resistant flooring.
Source: RMS.

Floors with a rough surface are best for wet conditions. Research has shown that rough floors can be cleaned to the same level of cleanliness as smooth floors and should not conflict with food hygiene requirements, but rough surfaces may require more cleaning effort and specialist equipment.

Slip-resistant footwear

It is important to choose slip-resistant footwear that matches the circumstances it is used in. The tread pattern on the bottom of footwear (the sole) and material the sole is made from are both important for slip resistance. Generally a softer sole and close-packed tread pattern work well with fluid contaminants and indoor environments. A more open pattern works better outdoors or with solid contaminants. Some footwear may be described as 'oil-resistant' this does not usually mean 'slip-resistant', oil resistant usually means that the soles will not be damaged by oil.

Tread patterns on the sole should not become clogged with waste or debris – soles should be cleaned regularly.

Slip resistance properties of footwear can change with wear, for example, when the tread pattern becomes worn down or when the sole absorbs liquid contaminants. Therefore, a system should be put in place for checking and replacing footwear before it becomes worn and dangerous.

Maintenance

Spillage control and damage

Material and substances spilt onto floors and walkways can present a risk of slipping. Therefore, arrangements should be in place to control spills. This will include control measures to prevention spills onto floors and walkways, in the form of spill proof containers instead of open containers and the use of drip trays.

Where preventive measures are not enough to control spills, employers should anticipate where and how spills may occur and provide control measures to deal with the spills. Where there is a particular risk of large spills it may be necessary to provide drainage to draw spills away from walkways and floors.

Equipment may also have to be provided to stop spills spreading, for example, sand, special granules or absorbent rolls of material. Spill material and cleaning stations may need to be located at intervals in the workplace. It is important to ensure that spills of harmful substances are not washed into rain/storm water drains and watercourses. Covers and barriers may be necessary to prevent this and alternative drainage provided to deal with this type of substance. It is important to ensure the safe disposal of any spilled substance and any absorbent material used.

Indoor locations in the workplace may be subject to frequent wetting during normal use, for example, welfare facilities such as shower areas, or cleaning operations, for example, a car wash facility. Where the floor or pedestrian route in these areas is likely to frequently get wet, drainage should be provided.

Dust is a common hazard generated by many processes and could, if allowed, enter the atmosphere and spread throughout the workplace. Apart from the health issues that could affect workers, the dust may present a risk of slipping where it has built up on floors and walkways, particularly where the dust is large and granular. Dust that does settle on floors or walkways should be removed by means that do not cause dust to spread in the atmosphere, for example, vacuuming or wet scrubbing methods.

Cleaning and housekeeping requirements

Maintenance of floors and walkways, in the form of cleaning activities, may have to be planned when workers and other people are not moving around the workplace in order to limit risks from the cleaning activity, for example, risks of slipping on recently cleaned floors that are wet. Where this is not possible careful control should be taken of work areas affected, including the use of barriers and signs.

Employers should follow an effective cleaning regime for floors and walkways in order to maximise the surface roughness and slip resistance of floors. Consider available advice from the flooring manufacturer to ensure the regime is repeated often enough and is effective in removing layers of contamination and residues of cleaning agent. Frequent spot cleaning can supplement cleaning the whole of a floor, which can be useful where spills occur or where there are routes that are used higher than the average amount.

Maintenance of a safe workplace includes the development of good housekeeping. Good housekeeping includes following the philosophy of 'a place for everything and everything in its place.' Good housekeeping will benefit from the use of the following control measures:

- Identification of storage areas.
- Tidy storage of tools, equipment and materials.
- Marking areas to be kept clear.
- Providing areas for waste material.
- Paying particular attention to keeping emergency routes clear.

Introduce procedures for reporting defects and for dealing with obstacles on the floor that could cause workers to trip or spills that might reduce the slip resistance of surfaces. This will help to ensure high standards of housekeeping are maintained and keep floors clear of obstructions, debris and spills. Maintenance activities should include repair of such things as damaged tiles and carpets - particularly on routes, at the threshold of doorways and at changes of level. Suitable and sufficient lighting should be provided and maintained in order to preventing slips and trips, enabling people to avoid contamination that is on the floor and obstructions in the workplace.

Competent supervision of activities is essential to monitor and control any changes to the work, making sure any additional housekeeping is conducted at regular intervals to prevent slips and trips from debris created during the work.

CASE STUDY

Pictures 1 and 2, in *Figure 8-80,* respectively show the modification of floor colouring by the use of a contrasting green colour cover so that the changes in level attracted the attention and were apparent to pedestrians.

The walkways were covered in dark maroon tiles and were in good condition throughout, but close to two entrance doors there were sloping areas of walkway. The building was quite old and the changes in level were part of the building structure but, because of the dark colour of the floor, it was not obvious that there was a change in level. All of the areas were part of fire escape routes and one was at the entrance to the main reception. The control measure was to use a contrasting colour section of carpet.

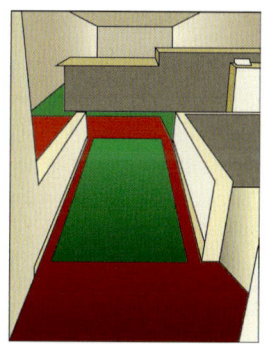

Figure 8-80: Walkway level changes.
Source: UK, HSE.

CONSIDER

Consider how well slip and trip hazards are reduced in your workplace.

8.6 Safe movement of people and vehicles in the workplace

HAZARDS TO PEDESTRIANS

Being struck by moving, flying or falling objects

Being struck by moving, flying or falling objects are hazards to pedestrians in the workplace and may also affect people in areas adjoining the workplace if precautions are not taken. Areas adjoining the workplace could include a public footpath, road, yard of a dwelling or other building beside a workplace.

The Health and Safety Executive (HSE) reported in 2017/18 that in the UK the average number of workers each year who suffered serious injuries by being struck by a moving object (including flying and falling objects) in the workplace was 7,000 and this type of hazard accounted for 23 workers being killed in the UK in 2017/18.

Examples of moving, flying or falling objects that could strike a pedestrian include:

- An object falling from a structure – for example scaffolding components, tools, roofing materials, bricks, ceiling panels that have not been securely fixed and materials that fall from over-stacked shelving.

- An object falling from lifting equipment, a vehicle or other moving equipment - including loads being lifted that are not well secured or are unstable.

- An object or material ejected while using machinery or hand tools – including parts of the machine/tool and waste material.

- The collapse of an unstable structure - including stacks of material, shelves, benches, racking and mezzanine floors not strong enough to bear the weight of objects on them.

- An object that has been deliberately thrown when carrying out a task - including throwing a scaffold coupling.

- Unstable objects falling - for example, an unsecured ladder.

Figure 8-81: Risk from falling materials.
Source: Protect-it

Some of the situations that increase the likelihood of these hazards and the risk of injury to pedestrians include:

- Stacking materials too high – sacks, pallets of materials and drums.

- Use of damaged or unsuitable pallets.

- Overloading of materials on storage systems – for example racking.

- Loose materials stacked at too steep an angle - for example, soil, grain or vegetables.

- Faulty or inappropriate means of lifting or moving materials – poor slinging methods, driving a fork-lift truck so fast that part of the load fall from the forks.

- Unstable loads on vehicles - for example, not correctly supported, tied or shrink wrapped.

- Insecure components or workpiece in moving machinery.

- Products of machining processes not contained within the machine - for example, waste material being ejected.

- Free dust from work processes or outside areas blown into the eyes of workers.

Figure 8-82: Obstructed designated pedestrian route.
Source: RMS.

The figures reported by the UK HSE are for injuries that took place in workplaces and do not include injuries involving workers on the public highway.

Situations that increase the likelihood of a pedestrian being involved in a collision with moving vehicles include:

- Restricted space that does not allow for vehicle manoeuvring and the pedestrians in an area.
- Obstructed designated pedestrians routes.
- Undefined routes that do not segregate vehicles from pedestrian - for example, people walking where fork-lift trucks operate in warehouses or heavy goods vehicles move in loading bays.
- Large vehicles reversing without anyone providing assistance.
- Disregard for rules concerning vehicles - for example, speed restrictions, competent drivers.
- Insufficient devices fitted to vehicles to warn pedestrians of their presence.
- Failure to provide recognised pedestrian crossing points.
- Lack of lighting, particularly at pedestrian crossing points and pedestrian exits.
- Lack of barriers to segregate pedestrian from vehicles.
- Absence of separate exits from buildings for pedestrian and vehicles.
- Pedestrian exits from work areas that open out onto vehicle routes.

CASE STUDY

A father of four was killed while working in the UK for an Essex based granite manufacturer. The worker was unloading a vehicle which had delivered stone slabs to the warehouse. He was moving the slabs over to A-frames to be stored when the accident happened. He was lowering a load which weighed around 3 tonnes when the pile of slabs fell and crushed him up against a wall. The worker was pronounced dead at the scene aged just 57.

The investigation had shown that the system of work which was being followed by the worker was unsafe, and therefore he was put at risk by his employers.

During the investigation it was discovered that the A-frames were not in a suitable position in the warehouse. Simple changes made to the way the storage had been designed and used could have prevented the death of the worker and saved his family from so much suffering. The employer pleaded guilty in court and was fined £20,000 with additional costs of £40,000.

Source: UK, HSE.

Collisions with moving vehicles

Pedestrian collisions with moving vehicles are one of the most common causes of fatal injuries in the workplace, typically accounting for approximately 18% of fatalities.

The Health and Safety Executive (HSE) reported in 2017/18 that in the UK the average number of workers killed each year in the workplace by a moving vehicle was 26 and 1,500 others suffered serious injuries.

REVIEW

Identify five control measures that should be followed by workers to improve the safety of pedestrians in vehicle manoeuvring areas.

What are the main causes of slips and trips in a workplace?

Pedestrians in the workplace face the risk of slips, trips and falls: outline three other hazards faced by pedestrians.

List eight control measures related to walkways.

CASE STUDY

A building contractor was working at a nuclear site in the UK when one of their employees was run over by a cherry picker (mobile elevating work platform). The injured employee was guiding the vehicle on foot when he was hit, and as a result of the accident he lost his leg.

The Health and Safety Executive prosecuted his employer as an investigation into the accident revealed a number of problems. The injured employee had been put at risk because there wasn't a safe system of work created or followed for the task he was performing. The inspectors also discovered that the employee had not received any training for the task despite having taken on the role several times a month for 14 months before the accident.

The employee was directing the cherry picker to ensure it moved safely down the one way road. Unfortunately it ran over him and he was rushed to hospital where they had no choice but to amputate his leg from above his knee joint. The company decided to plead guilty to the charges and was fined £65,000 and an additional £8,162 in court costs.

The HSE commented that the injured employee has suffered an injury that will stay with him for his entire life.

Source: UK, HSE.

Striking against fixed or stationary objects

The Health and Safety Executive (HSE) reported in 2017/18 that in the UK the average number of workers each year who suffered serious injuries by striking against a fixed or stationery object in the workplace was 2,400.

Situations that increase the likelihood of striking against fixed or stationary objects include:

- Poorly positioned machinery, equipment and furniture - avoid sharp corners protruding out.
- Insufficient space for storing tools and materials causing poor access and egress.
- Inadequate width of walkways.
- Inadequate lighting.
- Work in enclosed areas - such as restricted height in a confined spaces.
- Crane lifting accessories left in a position that workers can walk into them - such as hanging hooks/slings.
- Low overhead fixtures - such as pipework fixed at head height.

Figure 8-83: Wide enough walkways.
Source: ILO, Ergonomic checkpoints.

HAZARDS AND RISKS FROM WORKPLACE TRANSPORT OPERATIONS

Many workers die each year because of workplace transport accidents/incidents. There are also a high number of accidents/incidents that cause serious injury, for example, spinal damage, amputation and crush injuries. In addition, accidents/incidents relating to workplace transport operations cause damage to plant, equipment, infrastructure and vehicles.

Typical hazards relating to vehicle movement

Vehicle movement can present a number of hazards, some of which may be caused by driver error, for example, driving too fast. Features of the workplace can also cause vehicle movement hazards; for example, lack of signs may cause a driver to carry out an unexpected manoeuvre or to take a wrong route and have to reverse their vehicle. Other hazards may relate to the vehicle itself, including mechanical failure (for example, the brakes may fail) or the 'silent' operation of some vehicles and poor visibility around loads that are carried. Environmental conditions (for example, mud on the road, ice or snow) will also influence safe vehicle movement.

Driving too fast

The hazard of driving too fast may occur because the workload is too great for one driver to complete in the time available or there are incentives to complete the work as fast as possible, for example, the incentive created when fork-lift truck or delivery vehicle drivers are allowed to go home when the work is finished, even though their shift may have several hours left.

Often accidents/incidents occur when the driver who is driving too fast changes direction, for example, driving around a bend or steering to avoid an obstacle in the vehicle's path. A driver's inexperience may lead them to drive too fast for the characteristics of the vehicle, causing them to have to brake too late or make too sharp a turn.

Reversing

The potential for collision with pedestrians/fixed items is greatly increased when vehicles are required to reverse. This is partly because the field of vision of the driver to the rear of the vehicle may be limited by the fixed position of the vehicle mirrors, or because drivers have to adopt an awkward posture in order to be able to turn to see where they are going. In addition, large vehicles may have a significant area at their rear that the driver cannot see.

The hazard of reversing is also influenced by the driver having to operate the steering of the vehicle in the opposite way to that used when travelling forwards.

Silent operation of machinery

Vehicles that operate 'silently', typically battery-powered vehicles, present a particular hazard to pedestrians as their silent operation may mean that they are not heard in time to avoid injury. This is a significant hazard where segregation of pedestrians and vehicles is not practicable, for example, where a fork-lift truck frequently brings materials to a worker or where vehicles and pedestrians work together in warehouses. Workplace background noise or workers wearing ear muffs or ear plugs may also reduce the pedestrian's perception of such vehicles approaching.

Poor visibility

The load on a vehicle may obstruct the driver's vision and cause them to look to alternate sides of the load in order to see past it. This difficulty could mean the driver may not see pedestrians, vehicles or fixed objects sufficiently to avoid impact with them. Parts of the vehicle, for example, the lowered jib of a mobile crane, may prevent good visibility for the driver. The height of the driver above the road and the position of windows in a vehicle may create poor visibility, in particular to the rear of the vehicle. Similar difficulties can occur where the driver sits in an enclosed cab and environmental conditions like dust/sand, high humidity, fog, rain or ice make visibility through the windows of the cab poor. When vehicles move through areas with changing light levels, for example from high light levels to low light levels, it can lead to poor visibility while the driver's eyes adjust to the change. This can occur in a number of situations, for example, when moving from indoor areas to outdoor areas at night or moving from an area in bright sunlight to an area of shade during daylight hours. In addition, reliance on vehicle lights when driving outdoors at night may not be adequate to enable good visibility.

Poor visibility also relates to being able to see the vehicle. Poor light levels may make it hard to see a vehicle approaching, particularly if it does not have lights, its colour is not readily distinguishable from the background and it is moving quickly. Poor visibility can exist in many workplaces where doors, corners of buildings, and stacked materials prevent pedestrians and vehicle drivers having a clear view of vehicles approaching.

Typical risks from workplace transport operations

The main areas of risks associated with workplace transport operations are:

- Vehicle collision with other vehicles, pedestrians or fixed objects.
- Crushed or struck by a vehicle overturning.
- Injury caused due to mechanical failure of a vehicle component.
- A person being struck by a load or materials falling from a vehicle.
- Coming into contact with the inside of the vehicle while travelling in it.
- A person striking a fixed or moving object whilst travelling in a moving vehicle.

Collisions with other vehicles, pedestrians or fixed objects

People may unexpectedly appear from a part of a building/structure, or workers intent on the work they are doing may step away from where they are working to collect materials or tools. Through these actions they may step in front of vehicles and cause the vehicle to collide with them or the driver to take emergency action.

Often, the space in workplaces, such as warehouses, is restricted. To deal with this restriction, racking may be increased in height to maximise the floor space. This can then lead to poor visibility, especially at busy junctions where vehicles meet. This can cause a risk of collisions with other vehicles, pedestrians and fixed objects, such as roof and racking supports. Factors that affect the risk of collisions include:

- Inadequate lighting and direction signs.

- Inadequate signs or signals to identify the presence of vehicles.

- Drivers unfamiliar with the site.

- Needing to reverse a vehicle.

- Poor visibility - for example, sharp bends, mirror/ windscreen misted up.

- Poor identification of fixed objects - for example, overhead pipes, doorways, storage tanks, corners of buildings.

- Lack of safe crossing points on roads and vehicle routes.

- Lack of separate entrance/exit for vehicles and pedestrians.

- Lack of segregation of pedestrians and vehicles.

- Pedestrians using doors provided for vehicle only use.

- Lack of barriers to prevent pedestrians suddenly stepping from an exit/entrance into a vehicle's path.

- Poor maintenance of vehicles - for example, tyres or brakes.

- Excessive speed of vehicles.

- Lack of vehicle management - for example, use of traffic control, signaller/banksman/reversing assistant (appointed individuals who control or direct vehicles).

- Environmental conditions - for example, poor lighting, rain, mud, strong winds, dust/sand storms, snow or ice.

Figure 8-84: Poor maintenance of vehicle tyres.
Source: RMS.

Figure 8-85: Reduced risk of collision - people/vehicles.
Source: RMS.

Overturning

There are various circumstances that may cause a vehicle to overturn. They generally relate to factors that cause the centre of gravity of the vehicle to be acting so that the force created is mainly to one side of the vehicle instead of down to the ground. These uneven forces can contribute to causing vehicles to overturn, particularly when the vehicle is manoeuvring at speed, for example, going round a corner. Factors that could cause vehicles to overturn include:

- Overloading or uneven loading, for example overloading of lifting equipment attached to the vehicle.
- Insecure and unstable loads that move so their weight is not evenly distributed.
- Driving with the load elevated.
- Driving too fast, cornering at excessive speed.
- Sudden braking or acceleration.
- Hitting obstructions, including kerbs, buildings, structures or other vehicles.
- Driving across slopes.
- Driving too close to the edges of slopes, embankments or excavations.
- Driving over debris.
- Driving over soft ground or holes in the ground, such as drains.
- Poorly maintained or uneven road surfaces.
- Mechanical defects that occur because of lack of maintenance.
- Inappropriate or unequal tyre pressures, causing the weight of the load to be poorly distributed or movement of the load.

Figure 8-86: Poorly maintained road surface.
Source: Express Newspapers.

Failure of vehicle components

There is a risk of injury to people due to the mechanical failure of a part of a vehicle, in particular load-bearing parts. This may be caused by hydraulic failure of a hoist mechanism or a failure of a chain mechanism that is used to lift a load. The sudden collapse of the load may cause injury to pedestrians or others working in relation to the vehicle (loading/unloading), or ejected broken parts may strike the vehicle driver or those nearby. In addition, a vehicle may be overloaded to the point that its structure cannot hold the weight and may collapse. Heavy goods vehicles may be overloaded to the point that the trailer's load carrying capacity is exceeded and it fails. Where articulated vehicles are used there is a risk of the trailer collapsing as it is coupled or de-coupled from the tractor unit, particularly if the trailer supports are not set correctly or when the tractor does not make a good connection with the trailer and leaves the trailer unsupported as it drives away.

Figure 8-87: Trailer overloaded and therefore structurally collapsed. Source: www.worldduh.com

Struck by load

The insecurity of a load being moved by a vehicle can lead to the risk of materials falling from the vehicle or a person being struck by the load. If the load has not been secured to the vehicle properly, for example by straps, sheets or nets and this is combined with cornering or driving over an uneven road surface/object the load may fall from the vehicle. In addition, if the load extends beyond the vehicle it may hit people, equipment or structures as the vehicle moves around the site.

Coming into contact with a vehicle structure while travelling in it

As a vehicle moves there is a risk that the driver or a passenger may come into contact with the vehicle structure. A vehicle occupant is particularly vulnerable to this when the head is 'bumped' against the door frame. The risk is increased the rougher the surface that the vehicle travels over and where the person is not using driver/passenger restraint systems. If the vehicle is subject to excessive braking or comes to a sudden stop (when it strikes another object/vehicle) the occupants retain the vehicle speed and move within the cab– unless they are able to restrain themselves or are wearing seat restraints. The movement of people within the cab can prove fatal as they strike the vehicle cab interior or if they come into contact with other vehicle occupants.

Person striking a fixed object

As the vehicle is travelling along there is a risk that the driver or passenger may strike a fixed or moving item. This is not likely if an enclosed structure completely surrounds both the driver and passengers. However, many workplace vehicles have open frameworks or windows that can allow a person to put a part of their body, typically their head or hand, outside the protective structure. Many injuries have occurred to fork-lift truck drivers who, while driving forward with a load obstructing their view, lean out to look round the load and their head strikes a fixed object.

Sometimes, passengers and drivers trail an arm or a leg outside the structure of the vehicle they are travelling on

and there is a risk of it striking a fixed object as it passes, particularly while manoeuvring in areas where movement is restricted.

Typical hazards related to when vehicles are not moving

Person falling

Falling from height is a significant hazard in many transport activities conducted when the vehicle is not moving. The risk of a person falling from a vehicle can arise when loading or unloading the vehicle as this may require a person to go onto the vehicle or the top of a load and work at height. This work at height, along with slippery or uneven surfaces of the vehicle or load can create a significant risk of a person falling, particularly where there is no organised means of access to the vehicle. In addition, workers loading or unloading vehicles at height may lose awareness of where the edge of the vehicle or load is, which may lead to falls and major injury.

The risk of falling can also arise when securing the load or fitting a sheet over the load, particularly as this can involve strenuous work and the person may not have their attention on the need to avoid falling. The risk of falling when fitting a sheet is greatly increased when it is done at times when there are strong winds.

Risks of falling also occur when the driver/passenger are climbing up to enter or climbing down to exit the vehicle. There is an additional risk of falling from a vehicle when the vehicle is travelling over rough terrain, where a person in the vehicle may be dislodged by the sudden movement of the vehicle. This risk is increased where the person is not using a driver/passenger restraint system.

Materials falling

There is a hazard of objects falling on a vehicle driver during loading and unloading operations, particularly where vehicles are used to load materials at a height or to remove materials from delivery vehicles. In addition, when vehicles are being unloaded by hand, there is a hazard related to the stability of the load, as it may have moved and become unstable during transportation. A stack of material may become unstable and collapse when items are removed or added to the stack, causing serious injury or death.

Figure 8-88: Danger from pallet collapse and protected cab for driver. Source: US, HSE.

Typical workplace transport activities and their hazards

Securing loads

A number of methods may be used to secure loads to vehicles, for example, ropes, webbing and chains. It is common practice to use the hooks found on most platform vehicles as the anchor points for the load restraint systems that secure loads to the vehicle. They are usually welded or bolted to the underside of the side rail or outriggers of the vehicle. Anchor points should be firmly attached to the vehicle, either directly to the chassis or to a metal crosspiece or outrigger. Anchor points that are secured only to wooden members are unlikely to provide the restraint strength required.

The hazards relating to securing loads include work at height, slippery surfaces, weather conditions, and manual handling hazards. There is the additional hazard of a badly secured load shifting at some stage in the transport operation. Tasks to secure loads may cause the person to go onto the vehicle in order to fix straps and tighten devices to hold the load. This will present them with the risk of falling from the vehicle or from the load onto the vehicle.

The vehicle may be contaminated with dirt, dust, oil or materials spilt from previous loads, which presents a risk of slipping. Load securing equipment will usually have to be placed in position and tightened by the person securing the load. This could expose them to manual handling hazards related to moving things and over-exertion.

Sheeting

Sheeting is the covering of a load with a sheet or net to protect the load and/or restrain it in position on the vehicle.

Stages of sheeting	
Having positioned the sheets on the load, ensure all parts of the load are covered and that sheets are equal on each side. Secure the front of the rear sheet, followed by the rear of the front sheet. Do not over-tighten or the sheets will be drawn up to expose the load at the rear or at the front.	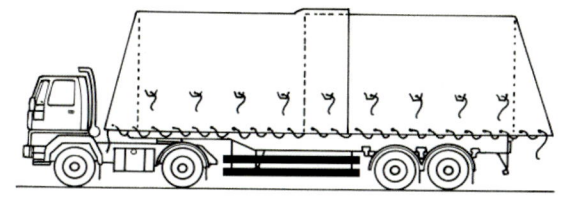 Figure 8-89 a): Ensure all parts covered and all sheets equal.
There should be no loose flaps or tears in the sheet, which might cause danger to other road users when the vehicle is moving *(see Figure 8-89 b)*. Care should be taken to avoid striking any person in the vicinity of the vehicle when throwing lashings over loads during the securing operation. Any sheeting aids or structural features of the workplace that are provided to assist in such operations should be made full use of.	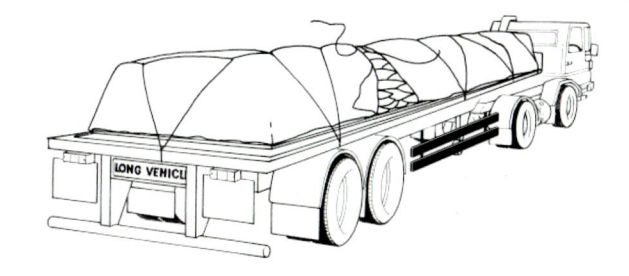 Figure 8-89 b): Use lashings to secure any loose flaps.
The next stage is to secure the front of the sheet. • Step 1, draw in surplus sheet from sides, cross over front and secure. • Step 2, draw down the remaining surplus sheet over cross-overs to form a full width flat front flap *(see Figure 8-89 c)*.	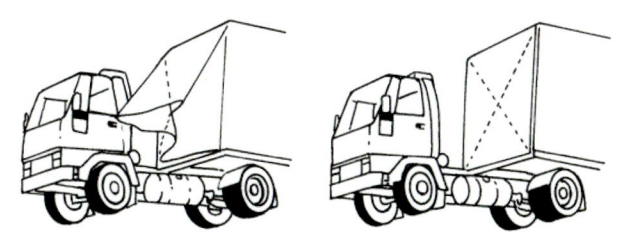 Figure 8-89 c): Secure the front of the sheet.
Having secured the front sheet, secure the sides of the rear sheet to the rearmost corners. The rear of the load should then be sheeted and folded *(see Figure 8-89 d)*. After the sheeting and roping is completed, ensure that all loose rope ends have been tied up and that the lights, reflectors, number plates and rear markings, etc, of the vehicle have not been obscured by any part of the sheet.	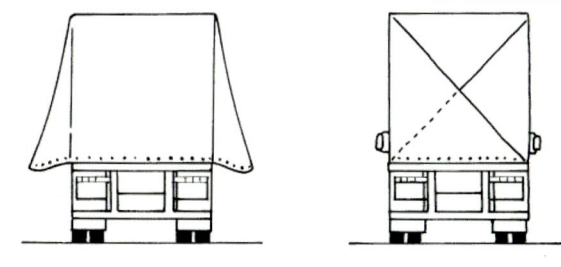 Figure 8-89 d): Secure the sides and rear of the sheet.
Source: Code of Practice: Safety of Loads on Vehicles - Department of Transport, UK	

Sheets are generally of two types:

1) Sheets or nets that only provide weather protection and protection from the effects of wind/air movement created by movement of the vehicle. They should not be used as part of a load restraint system.

2) Purpose-made load restraint sheets or nets that incorporate webbing straps to restrain the load. They are satisfactory for restraining the load up to their rated capacity, provided the straps are secured to body attachments of equivalent strength.

If the covering of a load with a sheet or net is not a mechanised process, it involves the manually unrolling of the sheet or net over the load and the sides of the vehicle, which presents hazards of falls from height, slips and manual handling. Though it may be possible to sheet some vehicles from the ground, it is not unusual for the person sheeting to have to gain access to the top of the vehicle and sometimes the top of the load, which can present significant risks of falling or slipping. As sheets are designed to be robust, they are also quite heavy and it takes a considerable amount of exertion to pull sheets over the load, leading to risk of strains and back injury.

Guidance on the correct sheeting of vehicles can be found in the UK Department of Transport Code of Practice 'Safety of Loads on Vehicles', *see Figure 8-89 (a,b,c,d) earlier* and the accompanying text taken from the code of practice. The European Commission (EC) has also produced guidelines, 'European Best Practice Guidelines on Cargo Securing for Road Transport'. The EC guidelines are primarily based on the EN 12195-1:2010 standard and also include examples of safe practices from throughout the road transport sector.

Where more than one sheet is required to cover and protect the load the rearmost sheet should be positioned first. This ensures that overlaps do not face forward allowing wind and rain etc. to get between sheets *(see Figure 8-89 a)*. The same principle should be applied to folds at the front or on the sides of the vehicle so that wind pressure will tend to close any gaps or folds in the sheet rather than open them.

Coupling

The activity of coupling a tractor unit (heavy-duty commercial vehicle within the large goods vehicle (LGV) category, also known as semi-tractor in the USA) to a trailer requires that the tractor unit is reversed towards the front of the trailer. The '5th wheel' is the connection between the tractor and the trailer. If the 5th wheel jaws are not located properly, this can lead to unexpected movement of the trailer. Though most injuries caused during coupling involve drivers or other people being run over, hit or crushed by moving vehicles, a number of non-vehicle movement based hazards exist. For example, there is a significant risk of falling during

Figure 8-90: Coupling (5th wheel).
Source: TruckNetUK.

Figure 8-91: Synchronised vehicle lifting system.
Source: Rotala PLC.

coupling, especially in the dark, as the person coupling may be less able to see slippery surfaces, obstructions or steps.

Another significant hazard is that the vehicle trailer could move or overturn. Though the ground may look flat, there may be a gradient sufficient to cause the trailer to become unstable or move under its own weight. In addition, during uncoupling the trailer goes from being supported to bearing its own weight, and this can cause the trailer to sink into the ground and overturn.

Vehicle maintenance work

There are a number of hazards and risks associated with maintenance work, including open pits (working at height/falling in), access gantries (working at height/falling off), oils and greases (slip and dermatitis) and fitting of replacement parts (manual handling/lifting, lowering, bending, etc.). One of the particular hazards associated with LGV maintenance work is the hazard of being crushed by a falling cab that is tilted to gain access to parts of the vehicle, but is not adequately propped or locked in place. Similar hazards exist at the rear of the vehicle where it has a tipping facility. Hydraulic ramps and hoists can create crush hazards as they are lowered to the ground or if there is a sudden failure of the hydraulic or mechanical system.

Figure 8-92: Tilted LGV cab and prop.
Source: UK, HSE.

There is an additional hazard that a vehicle or equipment could fall from a lift system, particularly if not located correctly or if the lift system is not raised uniformly. There are many manual handling and posture hazards associated with vehicle maintenance work, from leaning over to reach parts to the movement of large vehicle wheels. The presence of flammable liquids, in the form of fuels, oils and paints, presents a hazard of fire and explosion.

Electrical hazards are present as portable electric equipment may be used, some operating at mains voltage. Where pneumatic equipment is used it presents the hazard of noise, flying particles and possible injection of air into the body.

It may be necessary to work on part of the vehicle at height, for example, to repair the top of a tanker trailer or to refill refrigerant for a chilled foods vehicle. This work can present falling hazards that can lead to major injury or death. The risk of injury is increased when working on a vehicle that has come straight from use and is wet and slippery.

> **🔍 REVIEW**
>
> What hazards should be considered during loading and unloading of a lorry in a loading bay?

CONTROL MEASURES TO MANAGE WORKPLACE TRANSPORT

The employer, through its managers, needs to carry out an assessment of risk with regard to the safe movement of vehicles and their loads as part of the overall health and safety policy. This includes the use of vehicles such as dumper trucks, fork-lift trucks, and those vehicles used for delivery. Control measures should consider the following: safe site, safe vehicles and safe drivers.

Safe site

To ensure that the site is kept safe, the following factors should be addressed:

- Suitability of traffic routes (including site access and egress).
- Spillage control.
- Management of vehicle movements.
- Environmental considerations (visibility, gradients, changes of level, surface conditions).
- Segregating of pedestrians and vehicles and measures to be taken when segregation is not practicable.
- Protective measures for people and structures (barriers, marking signs, warnings of vehicle approach and reversing).
- Site rules (including speed limits).

Suitability of traffic routes

Suitability of traffic routes should consider the needs of the range of vehicles likely to use them and the people that might be affected by them. This will mean consideration of the needs of large vehicles to turn corners and arrangements to avoid or plan for vehicles reversing. It will also mean consideration of the needs of fork-lift trucks to ensure they are able to negotiate speed-retarding humps put in place to control large road vehicles that come on to site, and provision of cycle lanes on large sites where cycles move around with larger vehicles. The emphasis should not only be on safe access and egress into the site, from the public highway, but the provision of safe access and egress on traffic routes within the site. When establishing a safe site, consideration should be given to the following general principles:

- There should be enough traffic routes to prevent overcrowding.

- Traffic routes should be wide enough for the safe movement of the largest vehicle allowed to use them, including visiting vehicles.

- They should be made of a suitable material, and should be constructed to safely bear the weight of vehicles, and their loads, that will pass over them. Particular consideration of load carrying capacity must be made where routes take vehicles across bridges or into buildings. Upper floors and floors covering underground areas must be considered.

- Consideration should be given to vehicle weight and height restriction on routes - signs, barriers and weight checks may be necessary.

- Traffic routes should be clearly marked and signed, including directional information and markings to separate vehicles travelling in different directions and to separate pedestrians from vehicles. Markings and signs should incorporate speed limits, one-way systems, priorities and other factors normal to public roads.

- Consideration should be given to adequate lighting on routes and in loading/unloading and operating areas.

- Traffic routes should have adequate drainage. A surface gradient or road camber should be provided so that water should run off into gullies or drainage channels. Where there is a risk of oil or chemical spillage, drains should be fitted with facilities to prevent spillages escaping the site through the drainage system in an uncontrolled way.

- Separate routes and designated crossing places should be provided for pedestrians and suitable barriers installed at recognised danger spots. As far as is practicable pedestrians should be kept clear of vehicle operating areas and/or a notice displayed warning them that they are entering an operating area. Similarly, pedestrian-only zones must be clearly marked so that they are clear to drivers and preferably protected by physical barriers.

- Plan traffic routes to give the safest route between places where vehicles have to go. Avoid vehicle routes passing close to:

  - Dangerous items, unless they are well protected, for example, fuel or chemical tanks or pipes.
  - Any unprotected edge from which vehicles could fall, or where they could become unstable, such as unfenced edges of elevated weighbridges, loading bays or excavations.
  - Any unprotected and vulnerable features, for example, anything that is likely to collapse or be left in a dangerous condition if hit by a vehicle, such as scaffolding.

- Clear direction signs and marking of storage areas and buildings can help to avoid unnecessary movement, such as reversing.

- Sharp bends and overhead obstructions should be avoided where possible. Hazards that cannot be removed should be clearly marked with black and yellow diagonal stripes, for example, loading bay edges and pits. Where practicable barriers should be installed.

- Entrances and gateways must be wide enough for the traffic. Where possible, there should be enough space to allow two vehicles to pass each other without causing a blockage. Where this is not possible, traffic management systems should be in place to control the movement of vehicles from each direction, such as traffic lights, the use of a worker to control and signal to vehicles or signage to indicate who has priority. If gates or barriers are to be left open, they should be secured in position. There should be separate entrances/exits for pedestrians and vehicles, unless movement of vehicles is controlled while pedestrians are using the same entrance/exit.

- People are included in the term 'traffic'. It is important to consider pedestrians and the types of work equipment they might use, for example, pallet handlers, stackers and other handling equipment.

- Large vehicles, especially articulated vehicles and those with a drawbar trailer, often need to perform complicated manoeuvres to turn safely. This is because the trailers swing out behind the tractor unit and this often involves a larger turning circle than other vehicles.

Part 13 of the ILO Code of Practice on 'Safety and health in the iron and steel industry' deals with the subject of internal transport. Specifically, 13.2 addresses prevention and control strategies for internal transport, *see Figure 8-93.*

"13.2.2. Prevention and control

13.2.2.1. Transport routes should be planned and constructed to minimize the risk of collision and with sufficient safe clearance to allow for aisles and turns, or other types of control area. Where appropriate, maps showing the proposed route should be provided.

13.2.2.2. Transport routes should be clear of obstructions and, where possible, without irregular surfaces.

13.2.2.3. Transport routes and work areas containing transport vehicles should be visibly marked and segregated from waterways to the greatest extent possible.

13.2.2.4. The safe operating speed for vehicles should be posted and enforced."

Figure 8-93: Internal transport prevention and control strategies. Source: ILO, Code of Practice Safety and Health in the Iron and Steel Industry.

Spillage control

Material and substances spilt onto traffic routes can cause a vehicle to lose control and cause a collision. Arrangements should be in place to control spills. This will include control measures to prevention spills in the form of transporting materials in spill proof containers instead of open containers, the use of containers to hold loose items and binding containers/items together so they do not fall while being moved, for example by shrink wrapping them in plastic film.

Control measures should include materials to deal with spills. Materials to deal with spills may need to be located at intervals in the workplace, for example, sand, special granules or absorbent rolls of material. Where there is a particular risk of large spills it may be necessary to provide drainage. Arrangements should be made to remove obstacles, from traffic routes and place them in a location where they can be returned to use or disposed of, for example items that fall from vehicles.

Management of vehicle movements

Many sites are complex in nature and require the careful management of vehicles in order to ensure that they are brought onto, move around and leave the site safely. It is useful to appoint someone on site to have overall responsibility for vehicle movements.

Managers should be given formal responsibility for vehicle movements in the area or for the work they control. Where a number of organisations share a workplace, it is important that they co-operate with each other and coordinate vehicle movements.

Where materials are brought to site it may be necessary to manage deliveries to prevent too many vehicles arriving at the site at the same time, causing them to back up into the public highway. Site security arrangements play a significant part in the management of vehicles on site and will assist with controlling vehicles so that they are routed correctly and safely. It is not uncommon for vehicles that are too big or too heavy to access the roadways to be sent to site. Site security staff should be trained to identify these to prevent them accessing the site and causing harm.

Visiting drivers must be carefully managed and made aware of site rules that affect them. They must not be allowed to bring unauthorised passengers (for example, children during school holidays) onto the site. It is important to make sure that visiting drivers are aware of the workplace layout, the route they need to take, and relevant safe working practices, for example, for parking and unloading. Delivery drivers may never have visited the site before, and may only be on site for a short time. Vehicle movements should be organised so that visiting drivers do not have to enter potentially dangerous areas to move to or from their vehicles or places they need to go, such as the site office, toilet or welfare facilities.

When vehicle drivers arrive on site, it should be clear what their responsibilities are and who is in control of their activity on the site. Usually, a driver will be responsible for everything relating to the movement of their vehicle and site workers are responsible for everything that happens while the vehicle is stationary, like loading and unloading.

Vehicle movements may be organised so that most of the movements take place outside the time that pedestrians are entering/leaving the site when they start and finish work.

On a smaller scale, pedestrians may be given right of way over vehicles and if pedestrians are in an area, supervisors of vehicle movements may stop all movement until pedestrians have left the area. This could be essential if vehicles and pedestrians share an entrance/exit. The management of vehicle movements will include ensuring that those that move vehicles are trained and competent to move the specific class of vehicle. This will usually involve ensuring unauthorised people are not allowed to drive, including the use of secure control of keys or electronic access to vehicles.

The need for vehicles to reverse should be avoided where possible. A one-way system can help reduce the need to reverse and the risk of collision. In workplaces where one-way systems are not practical, it may be appropriate to use a turning point to allow vehicles to turn and drive forwards for most of the time. Turning arrangements should ideally be by means of a roundabout. If other ways of making reversing safe are not effective enough, the employer whose premises are being used may need to consider providing a competent and authorised signaller/banksman/reversing assistant in order to manage vehicle movements. They should have appropriate high-visibility equipment and use agreed hand signals.

Rules and procedures relating to vehicle movement, for example, speed limits, direction of travel and use of loading/unloading areas must be enforced to ensure compliance with requirements. It is important to have speed limits that are practicable and effective. Speed limit signs should be posted and traffic slowing measures such as speed bumps and ramps may be necessary in certain situations. Monitoring speed limit compliance is necessary, along with some kind of action against persistent offenders. Speed limits of 10-15 miles/h (15-25 km/h) are usually considered appropriate, although 5 miles/h (8 km/h) may be necessary in certain situations.

If rules on vehicle movements are difficult to enforce, physical measures such as gates, barriers, flow plates or control spikes (sprung flaps/spikes that only allow vehicles to cross in one direction) can be very effective.

Accidents/incidents can be caused where vehicles are unsafely parked as they can be an obstruction and restrict visibility. There should be clear, designated parking areas that allow the safe checking of the loads and sheeting of outgoing vehicles to ensure they are safe before leaving the site. In addition, there should be arrangements for suitable battery-charging or refuelling areas, if necessary.

Systems of inspection and maintenance of roads, routes and equipment that enables the safe movement of vehicles should take place on a planned, regular basis.

Figure 8-94: Barriers to ensure direction of travel of vehicles.
Source: RMS.

Article 16 of the ILO 'Safety and Health in Construction Convention C167 (Convention C167), 1988' establishes requirements for the safe use of vehicles and equipment on construction sites, *see Figure 8-95*.

"2) On all construction sites on which vehicles, earth-moving or materials - handling equipment are used:

a) Safe and suitable access ways shall be provided for them.

b) Traffic shall be so organised and controlled as to secure their safe operation."

Figure 8-95: Requirements for the safe use of vehicles and equipment. Source: ILO, Safety and Health in Construction Convention C167, Article 16.2

Environmental considerations

Where vehicles operate, lighting and adverse weather will make a significant impact on their safe operation. Good lighting, whether natural or artificial, is vital in promoting health and safety at work. It also has operational benefits, for example, making it easier to read labels. In all working and access areas, sufficient lighting should be provided to enable work activities to be carried out safely at all times during working hours. Where practicable, a suitable standard of lighting must be maintained so that operators of vehicles can see to operate their vehicle

and can be seen by others. It is important to avoid areas of glare or shadow that could mask the presence of a person or vehicle.

The relationship between the lighting in the work area and adjacent areas is important; large differences in luminance may cause visual discomfort or affect safety in places where there are frequent movements. For example, if vehicles travel from within buildings to the outside it is important that the light level is maintained at a roughly even level in order to give the driver's eyes time to adjust to the change in light.

Figure 8-96: Good lighting.
Source: UK HSE, HSG76.

Fixed structure hazards should be made as visible as possible with additional lighting and/or reflective strips. It is important to ensure good visibility in low levels of light by the provision of adequate lighting, particularly at crossing points and similar danger points where vehicles and pedestrians meet each other. Lighting should be located to provide good illumination of the area without causing glare to pedestrians or vehicle drivers. Care should be taken when positioning lighting structures near to roads to ensure they do not present a hazard to vehicles using the roads, particularly when the vehicles have wide loads or are turning. Where the failure of artificial lighting might expose workers to danger, emergency lighting should be provided that is automatically triggered by a failure of the normal lighting system.

Roads, gangways and aisles should have sufficient width and overhead clearance for the largest vehicle.

Attention should be paid to areas where they might meet other traffic, for example, the entrance to the site. If speed control humps are used, a bypass should be provided for trolleys and shallow-draft vehicles such as a fork-lift trucks.

Gradients and changes in ground level, such as ramps, represent a specific hazard to plant and vehicle operation. Vehicles have a limit of stability dependent on loading and their wheelbase. These conditions could put them at risk of overturning or cause damage to articulated vehicle couplings. Any gradient in a vehicle operating area should be kept as gentle as possible.

Steep gradients should be avoided as they can make driving vehicles difficult, especially if the surface is made slippery by a spill or poor weather. Slopes steeper than 1 in 10 should be avoided for most types of vehicle.

Where **changes in level** are at an edge that a vehicle might approach, and there is risk of falling, a robust barrier or similar means to demarcate the edge must be provided. Particular care must be taken at points where loading and unloading is conducted. Changes in floor level may also cause 'grounding' of part of the vehicle, which may cause the load being carried to become unstable.

In some workplaces, such as factories or chemical plants, process products may contaminate the surface condition of the road, making it difficult for vehicles to brake effectively. It is important to have a programme that anticipates this with regular cleaning of the surface as well as means of dealing with spills. Drainage systems should be fitted with facilities to prevent contaminants escaping the site through the drainage system in an uncontrolled way and reaching water-courses. In addition, the floor surface should be in good condition, free of litter and obstructions.

Figure 8-97: Changes in floor level may cause vehicle to 'ground'. Source: UK HSE, HSG76 Warehousing and storage: A guide to health and safety.

It is important to consider the adequacy of road surfaces for the adverse weather conditions that may be encountered. It may be necessary to select a rougher surface to provide a better grip for vehicle tyres. Drainage features will need to be large enough and spaced so that they can deal with the greatest expected demands on them. Provisions should be made to clear substances from the surface that might occur as a result of weather conditions, for example, mud, sand, ice or snow. These substances may make it difficult for vehicles to grip the surface or present obstructions. Provisions could include locating cleaning facilities at regular intervals along traffic routes.

Excessive ambient noise levels can mask the sound of vehicles working in the area. Therefore additional visual warning of the presence of vehicles should be used, for example, flashing lights. Ventilation should be considered when diesel-powered transport is to be used inside a building. Sufficient and suitable parking areas should

be provided away from the main work area and located where the risk of unauthorised use of the vehicles will be reduced.

Segregation

Segregating pedestrians and vehicles

Where practicable, pedestrians should be segregated from vehicles. There are degrees of segregation that are used; segregation may be total in that pedestrians are removed from the workplace where vehicles operate. This can be difficult to do unless the workplace is very controlled, for example, in some warehouse situations. Some organisations establish segregation of general pedestrians from vehicle operation in this way, but accept that drivers may have to leave their vehicle occasionally, for example, to check loads, to make adjustments to their vehicle or if it breaks down. In this sense, the drivers are authorised pedestrians and all unauthorised pedestrians are segregated from the area.

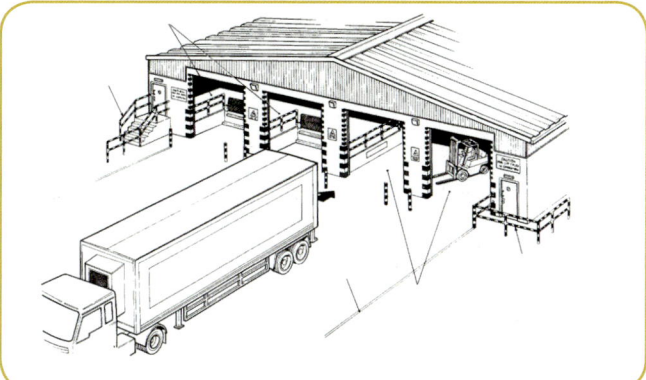

Figure 8-98: Vehicle unloading/loading showing separate pedestrian access, fencing and guarding barriers. Source: UK, HSE, HSG76.

Where possible, there should be physical segregation of pedestrians from vehicles by the means of separate doors, barriers or fencing. This is a very effective method of ensuring pedestrian safety. Barriers can be used to channel pedestrians into prescribed pedestrian routes, ensuring they are kept sufficiently clear of vehicle entrances and preventing them entering a traffic route directly from buildings.

Clearly defined and marked routes should be provided for the general movement of pedestrians at work. These should be provided for access and egress points to the workplace, car parks, and vehicle delivery routes.

Figure 8-99: Segregating pedestrians and vehicles. Source: asafe.com

Safe crossing places should be provided where pedestrians have to cross main traffic routes. In buildings where vehicles operate, separate doors and walkways should be provided for pedestrian segregation to enable them to get from building to building.

In some situations where physical segregation by barriers is not possible, segregation is based on the provision of a raised walkway for pedestrians. This allows vehicles to pass pedestrians and establishes a safe distance by provision of a sufficiently wide walkway and the raised edge of the walkway; this may be adequate for many workplaces.

Where it is not possible to have a pedestrian route with physical segregation by barriers or raised walkways segregation may be established by means of markings and signs. In this situation, segregation is established by markings on the ground that indicate safe walk routes that pedestrians can use on which they have right of way over vehicles. Though they do not provide physical segregation, they provide a degree of segregation and improved safety for pedestrians.

Where pedestrians and vehicle routes cross, appropriate crossing points should be provided and used. Where necessary, barriers could be provided to guide pedestrians to designated crossing points. Crossing points may be raised to slow vehicles and marked in a suitable way to clearly identify them. At crossing points there should be adequate visibility and open space to enable the pedestrian to see vehicles approaching. Where volumes of vehicle traffic are particularly heavy the provision of suitable bridges or subways to enable pedestrians to cross should be considered.

Figure 8-100: Pedestrian crossing point.
Source: UK HSE, HSG76.

Figure 8-101: Segregating pedestrians and vehicles.
Source: RMS.

Measures to be taken when segregation is not practicable

Where segregation is not practicable and vehicles share the same workplace as pedestrians it is important to mark the work areas and indicate main routes followed by vehicles such as fork-lift trucks. This will warn drivers to adjust their driving and be more aware of pedestrians when they are leaving a main route. Any traffic route that is used by both pedestrians and vehicles should be wide enough to enable any vehicle likely to use the route to pass pedestrians safely. Where it is not practical to make the route wide enough, passing places or traffic management systems should be provided as necessary. Audible and visual warnings of the presence of the vehicle would also assist.

Where routes used by automatic driverless vehicles that are also used by pedestrians, steps should be taken to ensure that vehicles do not trap pedestrians. The vehicles should be fitted with safeguards (sensing devices to detect humans or audible sounds to warn of their approach) to minimise the risk of injury, sufficient clearance should be provided between the vehicles and pedestrians, and care should be taken that fixtures along the route do not create trapping hazards.

Figure 8-102: Mirror fitted to improve vision.
Source: RMS.

Locations where the vision of the driver or pedestrian is restricted, at corners and similar 'blind spots', should be dealt with by the careful positioning of mirrors on walls, plant or storage or reducing the height of racking at the end of the aisle.

Where vehicles are dominant, but pedestrians need to access the same part of the workplace, similar means may be used, but for the opposite reason. In this situation the added use of high-visibility personal protective equipment that increases the ability of the pedestrian to be seen, and safety footwear including steel toe protectors, is usually needed.

Wearing high-visibility clothing is often mandatory on major construction projects and its use should be considered anywhere where there are significant risks from vehicle movement and visibility of pedestrians is necessary.

Figure 8-103: Mirror fitted to improve vision round a corner.
Source: UK HSE, HSG136.

Figure 8-104: Pedestrian's view in mirror fitted to improve vision.
Source: Aims Industrial

The reversing of large vehicles that have a restricted view should be controlled by the use of a signaller/banksman/reversing assistant to guide them. Where it is unavoidable that pedestrians will come into proximity with vehicles, pedestrians should be reminded of the hazards by briefings, site induction and signs, so they are aware at all times.

Vehicle manoeuvring areas

Factors to be considered to reduce the level of risk when pedestrians work in vehicle manoeuvring areas include:

- Defined traffic routes.

- One-way systems.

- Segregated systems for vehicular and pedestrian traffic where possible, including barriers and separate doors.

- Maintaining good visibility by the use of mirrors, transparent doors, provision of lighting and vehicle reversing cameras.

- Signs indicating where vehicles operate in the area.

- Audible warnings and flashing lights on vehicles.

- Established and enforced site rules, including speed limits.

- The provision of pedestrian safe zones in vehicle manoeuvring areas, so that workers can go to them and avoid moving vehicles. This is particularly important when vehicles are reversing and where workers might get trapped against fixed objects.

- The wearing of high-visibility clothing, foot protection.

- A good standard of housekeeping.

- Traffic control, for example, identification of "no go" or restriction on reversing areas.

- Training for, and supervision of, all concerned, competency certificates, refresher training, trained signaller/banksman/reversing assistant to direct vehicle.

- Provision of suitable battery-charging or refuelling areas if necessary.

Figure 8-105: Control of vehicle movement.
Source: RMS.

Figure 8-106: Use of banksman.
Source: RMS.

Figure 8-107: General view of site traffic.
Source: RMS.

REVIEW

What practices and procedures can be used for reducing risk when vehicles are reversing?

Protective measures for people and structures

Barriers

It is important to anticipate that drivers of vehicles might misjudge a situation and collide with structures or people whilst operating. Moving vehicles, and in particular large plant, have high impact energy when they are in contact with structures or people. It is therefore essential that vehicles be separated from people and structures.

Plant and parts of the building that are likely to be struck by vehicles should be provided with barriers that continuously surround the plant or alternatively posts can be provided at key positions. It is important to identify locations that warrant protection, for example, storage tanks or bund walls. Clearly, important plant and structures that are at the same level as the vehicle should be protected. Care should also be taken of structures at a height, such as a pipe bridge, roof truss or door lintel. All of these could be damaged by tall vehicles or those that have tipping mechanisms that raise their effective height. Because of the possible high energy involved in collisions the barrier used must be capable of resisting any likely impact. If it is only people that might contact the barrier a simple portable barrier may be adequate, but if it is heavy plant then robust barriers may need to be considered, for example, concrete structures.

Markings and signs

Though it may not always be possible to protect all structures, such as a doorway, it is important to apply *markings* to make them more visible or to install height limitation/protection barriers. In addition, *signs* warning of overhead structures or the presence of vehicles in the area will help to increase perception of hazards, raise awareness of the need to take care and avoid collisions.

Any structure that represents a height or width restriction should be readily identified for people and vehicle drivers. This will include low beams or doorways, pipe bridges and protruding scaffolds, and edges where a risk of falling exists. These markings may be by means of attaching hazard tape or painting the structure to highlight the hazard. Signs should be used to provide information, such as height restrictions, and to warn of hazards on site. Signs may be used to direct vehicles around workers at a safe distance.

Warnings of vehicle approach and reversing

There is a need to alert people to the hazard when working in or near a vehicle operating area; signs may be used to indicate the general operation of vehicles in the area. These can be supplemented by *visual and audible warning systems* that confirm the presence of a vehicle, such as a flashing light on the top of a dumper truck, *see Figure 8-109*. Warning systems may be operated by the driver, such as a horn on a dumper truck or automatically, such as an audible reversing signal on a large road vehicle. These are designed to alert people in the area in order that they can place themselves in a position of safety. They do not provide the driver with authority to reverse the vehicle or to proceed in a work area without caution.

Figure 8-108: Barriers and markings.
Source: RMS.

Figure 8-109: Visual warning on a dumper truck.
Source: RMS.

Figure 8-110: Segregation of pedestrians.
Source: RMS.

Figure 8-111: Barrier moved by pedestrians.
Source: RMS.

Site rules

It is important to establish clear and well understood site rules regarding vehicle operations. Drivers that are visiting the site should be made aware of site rules and conditions. These may have to be communicated to visiting drivers by security workers at the time they visit the site.

Site rules would often stipulate site speed limits, where the driver should be whilst vehicles are being loaded, where the keys to the vehicle should be and rules about not reversing without permission. They may also set out what access drivers have to areas of the site for their welfare, for example, availability of refreshment or toilet facilities. If there is a designated smoking area on site, this should also be clearly signed and drivers informed.

It is important that site workers and visiting drivers are able to communicate effectively; therefore arrangements should be made to communicate with visiting drivers who do not speak or have a limited vocabulary/understanding of the language used on site. Many organisations have site rules prepared to cover different languages that are illustrated with pictograms.

Figure 8-113: Correct clothing and footwear.
Source: RMS.

Where sites are made up of or join a highway, such as carriageway repairs on a motorway, it is important to identify who has priority - the vehicle or the worker. Site rules may clarify that once the site boundary is crossed, the worker has priority - this has to be clear. The site rules may be reinforced by the provision of additional signs, for example, to clarify a speed limit and the nature of the priority of workers.

Figure 8-112: Visiting vehicles.
Source: RMS.

Site rules should also be established for workplace transport that remains on site. This may mean establishing additional rules related to the types of vehicle that operate on the site.

Pedestrians should also know what the site rules for workplace transport are in order to keep themselves safe. Site rules for the safe movement of pedestrians might include such things as using pedestrian exits/entrances or crossing points, not entering hazardous areas or the need to wear personal protective equipment in hazardous areas, and not walking behind a reversing vehicle.

 CASE STUDY

A UK organisation was prosecuted and fined after a reversing vehicle at its site killed a delivery driver. The driver was delivering materials to the site and was standing by the side of his vehicle after overseeing the removal of its load when he was struck by a reversing fork-lift truck and died instantly.

The organisation had failed to carry out a suitable risk assessment for the movement of loads at the site. This would have shown the need for fork-lift trucks to avoid reversing for long distances and that delivery drivers should be removed from the danger area. They should have installed suitable barriers to prevent pedestrians gaining access to areas where vehicles were working, and established a formal system for supervising visiting drivers to the site.

Safe vehicles

When undertaking an assessment of vehicle safety, the following factors need to be addressed:

- Suitable vehicles.

- Maintenance/repair of vehicles.

- Visibility from vehicles/reversing aids.

- Driver protection and restraint systems.

Suitable vehicles

Vehicles should be safe by design, but it is important to choose the correct vehicle for the task. Consideration will need to be given to the size and weight of what needs to be moved; therefore a suitable vehicle with a sufficient safe working load should be selected. When considering the safe working load, this will include the vehicle capacity when transporting bulk liquids or slurries such as solvents, cement or silage. Vehicles should be suitable for any loads carried and it may be important that the vehicle has adequate anchor points to make sure that loads can be carried securely.

Figure 8-114: Gas operated fork-lift truck, weather protection and lights.
Source: RMS.

The vehicle will also need to be suitable for the environment in which it is to be used. This will include consideration of the space available for it to be used and whether it is to be used inside buildings, in which case battery-powered vehicles may be most suitable. The assessment of suitability must consider factors like weather conditions and provide protection from the cold, shade from heat and protection/visibility in the rain. In addition, vehicles to be used in flammable atmospheres should be suitable for work in that environment.

Vehicles have to be suitable for the terrain on which they are going to be used. For example, a standard counterbalance fork-lift truck is suitable for use in a warehouse that has smooth floors, but would be unsuitable for use on a construction site with rough-terrain. A suitable rough terrain lift truck would be required in this environment.

There should be provision of a safe way to get into and out of the driving/passenger position and any other parts of the vehicle that need to be accessed regularly.

Access features on vehicles, such as ladders, steps or walkways, should have the same basic features as site-based access systems. Vehicles should have seats and seat belts or other restraints that are safe and comfortable, where they are necessary. Guards should be fitted round dangerous parts of the vehicle, for example, power take-offs, chain drives and exposed hot exhaust pipes. Where vehicle like fork-lift trucks are used to raise loads at a height, consideration should be made to the provision of an enclosed driving position or over-head guards. These will also afford the driver some protection in the event that the vehicle should overturn.

Figure 8-115: Safe operating platform.
Source: SRTS Limited.

It is important that drivers are able to see clearly around their vehicle, to allow them to identify hazards and avoid them. Consider fitting a horn, vehicle lights, reflectors, reversing lights and possibly other warning devices, for example, rotating beacons or reversing alarms. Painting and markings in distinct colours may be necessary to make the vehicle stand out against the workplace background.

Maintenance and repair of vehicles

Maintenance

All vehicles should be well maintained and 'roadworthy', with a formal system of checks and maintenance in place. Critical items such as tyres and brakes should be kept to the same high standard on a vehicle that does not usually go outside the site (such as one used for moving trailers in a transport yard) as in one that is used on public roads. Vehicle maintenance should be planned for at regular intervals and vehicles taken out of use if critical items are not at an acceptable standard.

In addition, it is important to conduct a pre-use check of the vehicle. This is usually done by the driver as part of their taking it over for a period of use such as a work shift or day. This check would note the condition of critical items and provide a formal system to identify and consider problems that may affect the safety of the vehicle. Typical checks will include checking the pressure and condition of tyres, testing brakes, lights, horn, windscreen wipers and checking the adequacy of screen washer fluid. If the vehicle is to use attachable accessories the driver should ensure they are correctly fitted and secured. For example, for vehicles that pull trailers, the driver should check the adequacy of the trailer coupling and any hydraulic hose/electrical connections between the vehicle and trailer. If there is no designated driver and the vehicle is for general use someone should be nominated to make these checks. A record book or card would usually be used to record the checks and the findings, together with a means of reporting defects.

Repairs

Repairs should be carried out promptly to ensure safe operation of the vehicle. Where damage has been caused to the vehicle exterior body shell, this should be assessed and if it is a risk to pedestrians or other road users, the vehicle should be taken out of service until such repairs have been made.

The ILO refers to the maintenance of vehicles within Article 16 of their Safety and Health in Construction Convention C167, *see Figure 8-116*.

"Article 16

Transport, earth-moving and materials-handling equipment

1) All vehicles and earth-moving or materials-handling equipment shall:

a) Be of good design and construction taking into account as far as possible ergonomic principles.

b) Be maintained in good working order.

c) Be properly used.

d) Be operated by workers who have received appropriate training in accordance with national laws and regulations."

Figure 8-116: Internal transport vehicle safety precautions.
Source: ILO, Safety and Health in Construction Convention C167.

Visibility from vehicles/reversing aids

Driver visibility is often restricted to the side or rear of the vehicle. Side visibility can be improved greatly with the addition of extra mirrors to the sides of the vehicle. These mirrors provide a good low-level view, assisting with parking and the identification of hazards close to the vehicle sides. Vehicles that have limited driver visibility behind the vehicle may be provided with an extra supplementary mirror that improves vision. For example, a mobile crane may have limited visibility from the driving cab and a mirror in the centre/rear of the vehicle to aid the driver's vision.

A variety of other reversing aids are now in common use for vehicles that travel on ordinary roads. These include simple proximity detectors to the front and rear of the vehicle that trigger an audible alarm when the vehicle approaches an object. The audible alarm tone usually varies as the vehicle gets closer to the object. More sophisticated devices will distinguish between front and rear detection by the use of different sounds.

Figure 8-117: Audible proximity detectors.
Source: Transport support.

Figure 8-118: Reversing camera systems.
Source: Transport support.

In addition, the device may also display a schematic on a screen in the vehicle cab, indicating the precise direction of the obstruction to the driver.

Further advances in vehicle detection include the use of cameras to the rear or sides of the vehicle with a clear display available to the driver in the cab.

Control measures for reversing vehicles within the workplace include:

- Avoiding the need for vehicles to reverse (by the use of one-way and 'drive-through' systems or turning circles).
- Separation of vehicles and pedestrians as much as is practicable.
- Signs that warn of the hazard.
- Visual and audible warning systems fitted to vehicles.
- Adequate mirrors/vehicle cameras that are kept clean.

- Space to allow good visibility.
- Adequate lighting minimising sudden changes in luminance.
- Procedural measures, such as the use of trained signaller/banksman/reversing assistant.
- Refuges to protect signaller/banksman/reversing assistant.
- Appropriate site rules adequately enforced.

 **CASE STUDY**

A construction site worker was injured by a 360° excavator, which was operating in a poorly organised construction site.

The site worker was carrying out construction activities when the reversing excavator hit him and the track went over his right leg. The excavator was not fitted with devices to improve visibility from the cab, such as rear-mounted convex mirrors or closed-circuit television (CCTV), and the driver had not received formal excavator training. The excavator had been working within 3-4 metres of the injured worker on a daily basis, had knocked him once before and would often lift material over the worker's head.

After the injury, the firm reorganised the construction site and fenced off the area where workers were carrying out construction tasks. The excavator is now used in a pedestrian-free area.

Driver protection and restraint systems

Roll-over protection and restraint systems

In many work vehicle accidents/incidents drivers are injured because the vehicles do not offer protection when they roll over, or they do not restrain the drivers to prevent them falling out of the vehicles and being injured by the fall or the vehicles falling on them.

Vehicles with open cabs, such as dumper trucks, road rollers and fork-lift trucks, are examples of equipment that may present this risk.

The ILO recognises the importance of driver protection and recommends that equipment should be adapted, where practicable, to provide this protection.

New equipment purchases should include the provision of these types of protection and restraint systems. Roll-over protection can take the form of a complete, enclosed cab or of a system of bars that prevent the vehicle driver/passengers from being crushed by the vehicle in the event that it rolls over.

Figure 8-119: Seat belt status alarm beacon.
Source: Transport support.

Restraint systems typically involve some sort of adjustable seat belt that is worn to prevent the driver/passenger from being thrown from the vehicle when travelling or falling from the vehicle in the event that it rolls over. Other devices include seat belt status alarm detectors. This system is widely used on construction equipment to indicate that the operator is wearing the seat belt. It comprises a retractable lap belt, the buckle of which has a switch that is activated by the belt when fastened. This is connected to a green beacon on the roof of the vehicle to show that the operator is wearing the seat belt.

Figure 8-120: Roll bar and seat restraint.
Source: RMS.

The ILO refers to the fitting of roll-over protection systems in their Safety and Health in Construction code of practice, *see Figure 8-121.*

"6.1.9. Where appropriate, earth-moving or materials-handling equipment should be fitted with structures designed to protect the operator from being crushed, should the machine overturn, or from falling material."

Figure 8-121: Requirement for roll-over and falling material protection.
Source: ILO, Code of practice - Safety and health in construction.

Protection from falling materials

There is a risk of material falling on a vehicle driver where the vehicle is used to load and unload materials at a height or to remove materials from delivery vehicles.

Figure 8-122: Unstable stack of material, potential for falling materials. Source: RMS.

It is important that suitable protection of the driver be provided by the structural design of the vehicle. Part 6 of the ILO Code of Practice (CoP) 'Safety and health in construction' also includes reference to protection needed from falling materials, *see Figure 8-121.*

Protection should also be provided where vehicles work in close proximity to areas where there is a risk from falling materials. This may be by the provision of a frame and mesh above the driver or by the provision of a totally enclosing cab. Safety helmets alone are unlikely to provide sufficient protection but may be a useful addition where a mesh structure is provided.

 CONSIDER

Consider how transport safety might be improved in your own workplace.

Safe drivers

All drivers must be safe; the following factors will address driver safety:

- Selection and training of drivers.
- Banksman (reversing assistant).
- Management systems for assuring driver competence, including local codes of practice.

Selection and training of drivers

People are only permitted to operate plant or vehicles after they have been selected, trained and authorised to do so, or are undergoing properly organised formal training under competent supervision.

Selection

The safe usage of vehicles calls for a reasonable degree of both physical and mental fitness. The selection procedure should be devised to identify people who have been shown to be reliable and mature enough to perform their work responsibly and carefully. To avoid wasteful training for workers who lack coordination and the ability to learn, selection tests should be used.

Consideration must be given to any legal age restrictions that apply to vehicles that operate on public roads. A similar approach may be adopted for similar vehicles used on site though national laws and regulations may not be specific about age limitations in such cases.

Potential drivers should be medically examined prior to employment/training in order to assess their physical ability to cope with this type of work. Drivers should also be medically examined every five years in middle age and after sickness or accident. Points to be considered are:

General: Mobility, having full movement of trunk, neck and limbs.

Vision: Good eyesight, which may be achieved by wearing corrective lenses (spectacles), is important as operators are required to have good judgement of space and distance. In many countries, driving with vision in only one eye is legal. The question of ability to drive usually relates to field of vision. Since removing an eye reduces the visual field by only about 40%, monocular vision alone is not a barrier to driving when the other eye has a very good visual field.

Hearing: The ability to hear instructions and warning signals with each ear (to enable the direction of the source to be located) is important.

Training

It is essential that the immediate supervisors of drivers receive training in the safe operation of vehicles, and that senior management of the organisation appreciate the risks resulting from the interaction of vehicles and the workplace.

For the driver, safety must constitute an integral part of the skill-training programme and not be treated as a separate subject. The driver should be trained to a level consistent with efficient operation and care for the safety of themselves and other persons.

A comprehensive training programme has to be delivered and must consider the following topics:

- Parking and any other restrictions that may be in place, such as speed limits.

- The need to be courteous and give right of way to pedestrians and other vehicles.

- The vehicle's controls and how to drive safely both forwards and in reverse.

- Hazards that are specific to the workplace, such as height restrictions and hazardous materials.

- Safety issues associated with refuelling.

- How to carry out pre-use inspections including the procedures for reporting defects.

- Maintenance requirements.

- The issues of substance/alcohol misuse.

- How to secure the vehicle after use and the procedures for key security.

- Any legal requirements that might apply to the driver, such as limitations on driving hours.

On completion of training the worker should be issued with an authority to drive from the organisation. This will often involve a further test related to the site conditions, if the driver has been trained elsewhere. A record of all basic training, refresher training and tests should be maintained in the driver's personal documents file. Certification of training by other organisations must be checked for adequacy before the driver is given authorisation to drive.

Banksman

Banksman is a common term used for a vehicle reversing assistant. The role of the banksman is to assist the driver by extending their field of vision. This is achieved by the banksman standing in a position where they can see the area behind the vehicle and signalling to the driver to direct movement of the vehicle. The use of banksmen to control reversing operations is not a preferred option and should be subject to a full risk assessment, as it can involve putting the banksman in the potential danger area of a reversing vehicle. If a banksman has to be used, this is best achieved by correct training in the use of international hand signals to direct the driver. The driver should be instructed to stop the vehicle immediately if the banksman disappears from view.

The banksman should wear suitable high-visibility clothing and be taught to place themselves in a safe position that ensures the driver can see their signals and does not put them in danger from the reversing

vehicle. They should be aware of the limited view a driver might have from their mirrors, particularly when turning the vehicle as well as reversing, and the effect of driver position when the vehicle is left- or right-hand drive. They should take these factors into account when selecting a position to be seen and be safe.

Management systems for assuring driver competence

The trained driver

It should not be assumed that workers who join as trained drivers have received adequate training to operate safely in their new organisations. The management must ensure that they have the basic skills and receive training in the organisation's methods (local practices) and procedures for the type of work they are to undertake. They should be examined and tested before issue of an authority to drive.

A copy of the site rules and *any local codes of practice* must be given to all internal drivers and to external visitors such as delivery drivers, preferably before they arrive at the site.

Testing

On completion of training, drivers should be examined and tested to ensure that they have achieved the required standard. It is recommended that a formal test is introduced at set intervals or when there is indication of a driver not working to required standards, or following an accident/incident.

Refresher training

If high standards are to be maintained, periodic refresher training and testing should be considered.

Driver identification

Many organisations operate local codes of practice and take great care to confirm the authority they have given to drivers by the provision of a licence for that vehicle and sometimes a visible badge to confirm this. Access to vehicles must be supervised and authority checked carefully to confirm that the actual class of vehicle to be driven is within the authority given. This is important with such things as reach-lift trucks that operate differently from a standard counterbalance truck. It is essential that access to keys for vehicles is restricted to those persons who are competent to drive them.

 **REVIEW**

What factors should be included in a training programme for drivers to reduce the risk of accidents/incidents to themselves and other workers?

8.7   Work related driving

C.J.L Murray and A.D. Lopez conducted a study that analysed global harm due to a range of risk factors. In 1996 they reported in the publication 'Summary, the Global Burden of Disease' that they estimated the number of annual road traffic fatalities in the developed regions of the world to be 222,000. A study conducted in Finland by the Ministry of Labour in 1994 estimated that nearly 8% of the road traffic fatalities that occurred in Finland were work-related. Using the estimate of annual road traffic fatalities and the percentage that were work-related, it would mean that there are an estimated 17,000 work-related road traffic fatalities in the developed world annually. This is in addition to the annual workplace fatalities, estimated by the ILO to be 248,000 globally. The ratio of work-related traffic fatalities to workplace fatalities is likely to differ between countries, but it is clear that the risk of work-related driving fatalities is significant and should be managed as well as workplace risks.

The true cost of work-related driving accidents/incidents to any organisation is nearly always higher than just the costs of repairs and insurance claims. It can include personal costs, such as ill-health, time in hospital, stress on family members and possible sanctions being imposed on a driver's licence following an accident/incident. There is also the potential for drivers to have their licence revoked by the national driver licensing authority that issued them. Therefore the management of driving activities through the introduction of policies, risk assessments and developing safe systems for work-related driving is essential.

MANAGING WORK-RELATED DRIVING

Plan

Assess the risks

As with other work activities, work-related driving should be risk assessed. This will guide the employer and managers, enabling them to establish effective arrangements for controlling the driving risks. The risk assessments should be used to guide action to plan, do, check and act to manage the risks.

Risk assessments for work-related driving activities should follow the same principles as risk assessments for any other work activity, for example, by following the '5 steps' approach. They should be carried out by a competent person with practical knowledge, relating to the work-related driving activity being assessed.

The range of hazards is likely to be wide and the main areas to consider are the driver, the vehicle and the journey.

Also consider the factors that affect these hazards and increase the risk, these factors are described later in this element. When assessing the risks it is necessary to evaluate the current control measures, consider the control measures described later in this element. Talk to the drivers involved and those that manage them in order to try to determine what happens in practice. Consider different categories of driver, for example, inexperienced drivers, infrequent drivers, those that drive long distances and those that drive specific or unusual vehicles.

Step 1 - Identify the hazards
Hazards will fall into the following categories:

The driver:	Competence, fitness and health, alcohol and drug use, behaviour etc.
The vehicle:	Suitability, condition, safety equipment, ergonomic considerations, the load, security, etc.
The journey:	Route planning, scheduling, time, distance, driving hours, weather conditions, stress, volume of traffic, passengers, etc.

Step 2 - Decide who might be harmed
Obviously the driver, but this might include any passengers, other road users and/or pedestrians. Consideration should also be given to other groups who may be particularly at risk, such as young or newly qualified drivers and those driving long distances.

Step 3 - Evaluate the risk and decide on precautions
Risks may vary depending on whether driving is done at night or in the day, the type of vehicle or driving conditions. Decide the likelihood of the harm and severity (consequence) of any outcome. Consider risk factors such as the load being carried, distance travelled, driving hours, work schedules, traffic levels and weather conditions. Decide on appropriate precautions and once these have been established, decide whether the residual risk is acceptable.

Step 4 - Record the findings and implement them
Significant risk assessment findings need to be recorded and should be made available to the drivers affected.

Step 5 - Review the assessment and update if necessary
Monitor and review the assessments to ensure the risks are suitably controlled. Systems should be put in place to gather, monitor, record and analyse accidents/incidents that might affect these risk assessments. The vehicle and driver's history should also be recorded. Any changes in the route, new equipment and changes in the vehicle specifications should be reviewed and recorded. This will ensure the effectiveness of controlling the risks.

Policy

All employers who require workers to drive as part of their work should have a policy to address the risks from these activities, whether the employer provides vehicles or expects workers to drive their own vehicles for work purposes. Where it is practical, a commitment to managing work-related driving risks should be included in the employer's general health and safety policy. Where this is not practical employers may establish a separate policy to manage work-related driving risks. Without a specific policy commitment to manage these risks managers may not be given the resources to manage them effectively. It is therefore important that these risks are recognised at senior management level in the health and safety policy and a commitment made to manage them.

Taken into account by top management

In order to manage the risks of work-related driving it is important there is a top management commitment that ensures it is taken into account when business decisions are being made. This will help to ensure that those with responsibilities for managing the risk put the necessary effort and resources in place to avoid the risks where possible and minimise those that cannot be avoided. Top management should ensure that those managing the risk have the resources and authority to do this effectively.

Roles and responsibilities

In a large organisation it is likely that various departments within the organisation will have different responsibilities for the management of work-related driving risk. For example, the despatch department would often be responsible for planning journeys, the training department may be responsible for driver competence, the human resources department could be responsible for driver selection and checking the validity of driving licences, the maintenance department would usually be responsible for the condition of vehicles and ensuring their roadworthiness and the occupational health department may be responsible for carrying out routine health surveillance of drivers.

In order to establish an organised and co-ordinated approach the responsibilities of different parts of the organisation and those that manage them should be defined, including the responsibility to co-operate with each other.

Legal responsibilities of individuals on public roads

The specific legal responsibilities of individuals while driving on public roads will vary from country to country. Typical responsibilities of individuals could relate to their fitness to drive, driver competence, the condition of the vehicle and driving behaviour.

Fitness to drive

Legal responsibilities related to the driver's fitness to drive may include:

- The driver having adequate vision for driving. The requirements for adequate vision may be defined in legislation, for example, being able to read (with the aid of vision correcting lenses) a vehicle number plate, in good daylight, from a distance of approximately 20 metres.

- The driver reporting to a competent authority or doctor any health condition likely to affect their driving.

- The driver not driving while affected by fatigue. This may be linked to specific limitations on driving hours and rest periods.

- The driver not driving whilst affected by drugs or alcohol. Legal limits may be set to define when a driver is considered to be affected by drugs or alcohol.

Driver competence

Legal responsibilities related to driver competence may include:

- The driver being of a minimum age to drive the type of vehicle they are driving.

- The driver holding a valid driving licence that confirms their ability to drive the type of vehicle they are driving.

Vehicle condition

Legal responsibilities related to the vehicle condition may include:

- The driver ensuring the vehicle complies with requirements for roadworthiness. This may be a shared responsibility with the employer where the vehicle is provided by the employer. However, the driver will often be held initially responsible for the vehicle they are driving.

- The driver not overloading the vehicle.

- The driver securing any load on a vehicle correctly and ensuring it does not protrude dangerously.

Driving behaviour

Legal responsibilities related to driving behaviour may include:

- The driver driving with due care and attention for others.

- The driver not exceeding speed limits.

- The driver and passengers wearing a seat belt or suitable restraining device.

Do

Co-operation between departments

Where a number of departments are involved in the management of work-related driving it is important to ensure communication between the departments and their co-operation in providing effort to managing driving risks. If arrangements are not put in place this could seriously affect the effectiveness of the management systems. For example, failure of the maintenance department to communicate to the despatch department that a vehicle needed to be withdrawn from service for maintenance work might lead to the vehicle being driven in an unsafe condition. It is therefore essential that the structure of the organisation enables the co-operation of departments to allow for the easy and reliable interchange of work-related driving risk management information, for example related to licences to drive classes of vehicle, medical conditions, any damage to vehicles and other problems that could increase risk.

Systems to manage work-related driving

In support of their policy, employers should establish management systems to manage their work-related driving risks and managers should be trained to ensure they are competent to manage and implement the work-related driving systems.

It is important that systems are put in place to reduce work-related driving risk by ensuring that driving distances are kept to a minimum, driving hours are controlled and work schedules are organised to avoid pressure on drivers to complete their journeys in an unreasonable time. Specific systems should be established to manage fatigue and the taking of breaks from driving. This should include consideration of workers who might be expected to drive long distances as part of their work after completing a normal work day. Organisations should have systems in place to respond to inclement weather and enable drivers to take safe decisions about the effects of weather on their driving or journey. Systems should also be established to manage vehicle break-downs, emergencies, problems experienced while travelling and delays that could affect the journey.

Where relevant, systems should be established to manage circumstances where drivers provide their own vehicles for work activities, this should include:

- A requirement that if workers use their own vehicle for work activities they must maintain it in a roadworthy condition.

- A requirement that where the age of the vehicle is significant the vehicle has been confirmed to be roadworthy by an independent organisation. The requirements for initial and regular checks of roadworthiness are sometimes specified by national laws.

For example, national laws may require that vehicles undergo an initial test of roadworthiness when the vehicle is three years old and annual roadworthiness tests thereafter.

- A requirement that workers who drive their own vehicle on behalf of their employer have a current driving licence that allows them to drive the vehicle on public roads.

- A requirement that appropriate insurance for the vehicle is held by the worker and it covers the worker to drive it. The worker should present copies of certificates of insurance that cover work-related driving annually for inspection by the employer.

- A requirement that workers inform their line manager of any changes in circumstances that could affect their ability to drive on the employer's behalf. For example, changes to the vehicle the worker provides or insurance arrangements affecting it, penalty points/citations/endorsements issues for driving offences, the use of any prescription medication or changes to the worker's health that could affect ability to drive safely.

Systems of work to manage work-related driving risk should be formalised, for example, as written procedures. This would help to avoid ambiguity and ensure more consistent management of the risks.

Communication and consultation with the workforce

The arrangements put in place to manage work-related driving risk should include communication and consultation with the workforce, including those that carry out the driving. This will establish a clear understanding of the importance of managing the risk and the role that the workforce plays, this is more likely to lead to them following the systems introduced. They are also more likely to feel that they can report problems, difficulties and incidents related to work-related driving.

Adequate instruction and training

Driver competence is an essential part of managing work-related driving. It is therefore important that systems are established that ensure adequate driver instruction, training and competence for the vehicles they drive. It is important that the drivers hold a valid driving licence for the type of vehicle they drive for work activities. If drivers are expected to drive large goods vehicles (LGVs) or large passenger-carrying vehicles (PCVs) or other specialist vehicles, employers may require confirmation that the driver holds a specific licence or qualifications as proof of the driver's ability to competently drive these vehicles. Some employers carry out internal assessments of the driving skills of everyone that drives on the employer's behalf, in addition to confirming that the minimum legal requirements for driving licences are met. These driving skill assessments can be carried out in-house or by an external assessor. Some employers also offer specific training in safe driving techniques for their workers.

Systems should be established that ensure drivers are fit to carry out the work-related driving activities they do. This may involve consideration of medical conditions that could affect driving during a formal medical examination or a driver declaration process. Where drivers drive specialist vehicles, for example, a large goods vehicle, an initial medical examination is often carried out before allowing new drivers to drive the vehicles and a similar medical examination may be carried out periodically. In addition, employers may carry out tests to determine if drivers are driving or attempt to drive while influenced by alcohol or drugs.

Check

Monitoring performance

It is an essential safe working practice and may be a specific legal requirement to monitor health and safety systems related to driving, in particular, driving hours. The performance of an organisation in managing work-related driving risks should also be monitored using a range of active and reactive methods.

Information on driving schedules, including driving hours and distances can be monitored actively to determine if they conform to acceptable standards. The licences of drivers and the condition of vehicles can also be checked regularly. Arrangements for breakdowns and problems drivers experience while driving can be monitored actively by examining reports or records of how they were dealt with and any effects on safety. In addition, poor driving behaviour could be monitored by direct observation or by electronic surveillance systems.

Information on accidents/incidents linked to work-related road safety risk should also be monitored. This should include situations where the driver or other people are injured, any events that result in damage to vehicles or property and near misses. Examples of near misses would include situations where drivers fall asleep momentarily while driving, but do not lose control of their vehicle and cause harm. In addition, certain road traffic accidents/incidents may have to be reported to a competent authority under national law. The ILO's 'P155 - Protocol of 2002 to the Occupational Safety and Health Convention, 1981' (P155) encourages member nations to establish requirements for employers to record and notify the relevant competent authority about commuting accidents/incidents. P155 defines the term commuting accident in relation to work-related journeys, **see Figure 8-123.**

"The term commuting accident covers an accident resulting in death or personal injury occurring on the direct way between the place of work and:

i) The worker's principal or secondary residence.

ii) The place where the worker usually takes a meal.

iii) The place where the worker usually receives his or her remuneration."

Figure 8-123: Definition of term commuting accident.
Source: ILO, P155 Protocol of 2002 to the Occupational Safety and health Convention, 1981.

Reporting work-related driving incidents

In order to support the organisation's monitoring of the management of work-related driving risks and any necessary reports to competent authorities it is important that all workers report incidents, including near misses. Where possible this should be extended to reporting problems and difficulties that increased risk so that the organisation can take account of these and take action to reduce the risk in future activities.

Drivers must be required to record information about all incidents involving injury, whether minor or serious. A similar reporting procedure should be in place for reporting significant 'near-misses', driver and manager training regarding this should emphasis how to recognise, analyse and learn from such events. The data provided should be analysed and any changes or improvements noted. These should be communicated to those concerned and the work-related driving systems updated.

Act

Review performance and learn from experience

As with all risks, the organisation should review its performance in the management of work-related driving risks periodically and after a major work-related driving incident. This review should consider the range of factors that influence the risk, the control measures in place to minimise this risk and the effectiveness of these measures. This will inform the management of the organisation and provide an opportunity to learn from experience and determine what measures have been effective and any opportunities for improvement.

Regularly update the policy

The review of performance should lead the employer to consider any impacts on the current work-related driving policy and make any changes to the policy that would strengthen it and provide an improvement in the management of these risks.

FACTORS THAT INCREASE RISK WHEN DRIVING AT WORK

There are many factors that increase the risks of an accident/incident while driving at work and it is the responsibility of the employer to assess these based on the nature and circumstances of the work. Important considerations include distance travelled, driving hours, work schedules and stressors, such as traffic congestion and inclement weather conditions.

Distance

Road conditions have improved in most countries over the years, which has allowed greater travel distances in shorter periods of time. However the fatigue encountered by drivers often increases due to other drivers' poor performance, creating the need for a particularly high degree of concentration. LGV drivers' travel distance is often regulated by limitations on their driving hours. However, there are often no such limits on drivers of smaller vehicles and private cars. When assessing risk of any kind, an important consideration is the frequency and duration of exposure to hazards. It is logical then that the greater the distance of the journey (i.e. the duration of exposure to the hazard) the greater the risk. It is possible that scheduling of routes may extend or reduce the distance that has to be travelled and therefore affect the level of risk significantly.

Driving hours

If driving hours are excessive the driver is likely to become tired, their attention and reaction levels will fall and they are at increased risk of making errors. The driving hours may be excessive because the driver has been driving too long without a break, their cumulative hours in a day have become too much or their rest period between days has become too short. For some drivers the driving hours is in addition to other work hours, for example, for someone travelling to a meeting, these cumulative hours can have a significant effect on fatigue.

Many commercial vehicles will be fitted with tachometers (digital or analogue), which record the time spent driving, or drivers will be required to keep a record/log of driving hours. These driving hours are often limited by national legislation or international agreement. However most company-car drivers do not have this specific monitoring imposed on them. Therefore safe systems of work and appropriate training should be given to all drivers concerning the hazards and risks associated with excessive hours. It is important that cumulative hours are monitored and controlled and that breaks are taken at intervals on longer journeys.

The European Agreement concerning the work of Crews of Vehicles Engaged in International Road Transport (AETR) regulates the working and rest periods of professional drivers travelling to or through certain European countries. The agreement covers more than 40 European states and other countries, including, Azerbaijan, Croatia, Kazakhstan, Russia, Turkey, Ukraine, and Uzbekistan.

Driving hours of goods vehicles over 3.5 tonnes and passenger vehicles designed to carry nine or more passengers are regulated by AETR rules. These set limits on drivers' hours and are shown in *Figure 8-126.*

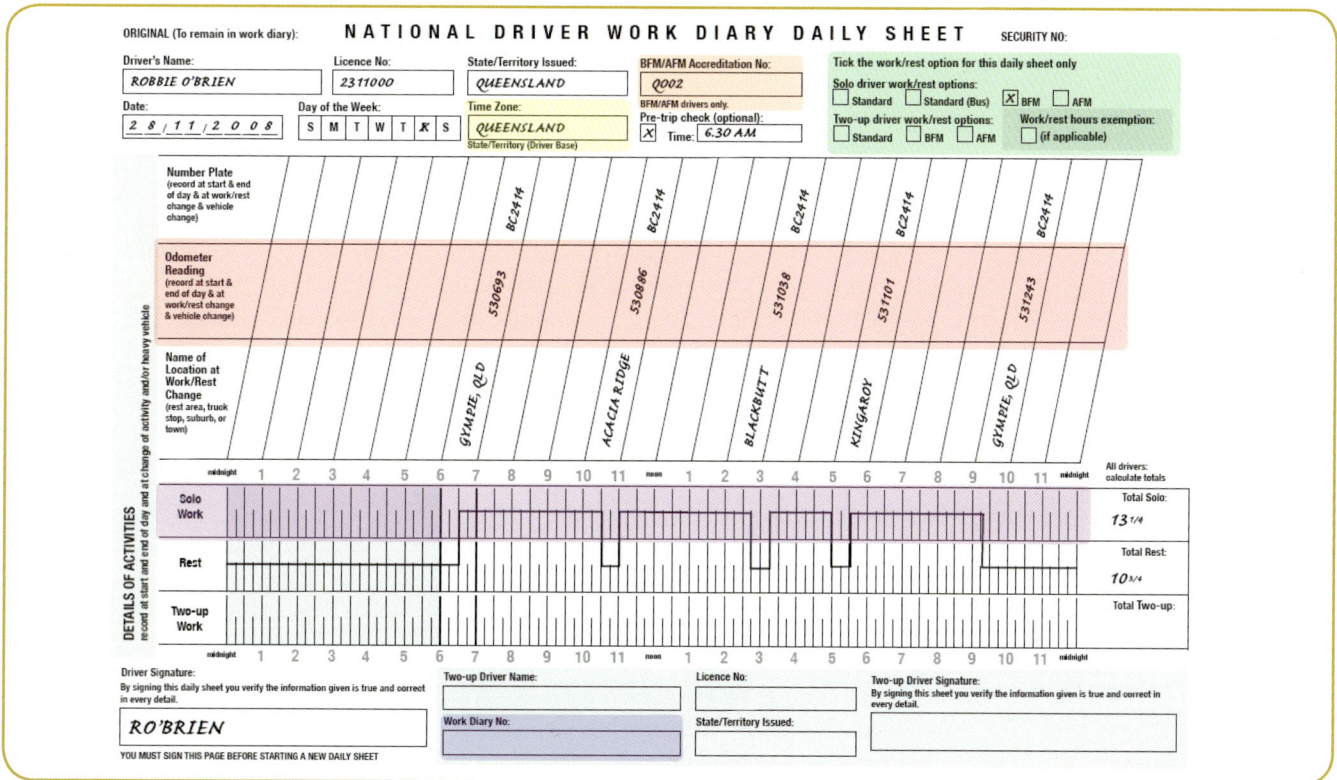

Figure 8-124: Driver diary. Source: Government of South Australia.

The responsibility for monitoring hours and taking breaks is shared between the driver and the employer.

The driving limits set by AETR follow a similar structure to the minimum requirements set out in the ILO Hours of Work and Rest Periods (Road Transport) Convention, C153 (Convention C153); limits set by the convention are added in italics, as shown in *Figure 8-126.*

Work schedules

Work schedules that are badly organised can put increased pressure on drivers to be in a place by a given time. This can lead to them being tempted to increase speed and take abrupt action to change lanes to improve their progress. This risky action can lead to higher risk of collision and reduced stopping distances. Driver fatigue issues are emphasised in the example from the government of South Australia in *Figure 8-125*.

"Fatigue is a problem on our roads. Just over 25% of heavy vehicle driver's drive fatigued at least once every four trips. It's caused by work pressures and lack of proper rest and puts a lot of people at risk."

Figure 8-125: Driver fatigue.
Source: Government of South Australia.

Work schedules should take account of periods when drivers are most likely to feel sleepy. The high-risk times are 2am to 6am and 2pm to 4pm. Employers should provide drivers with the means to stop and take a break if they feel sleepy, without the fear of recrimination.

Where possible, schedules should be organised so that breaks can naturally and easily be taken, with agreed break points if travel is going or not going as planned. Driving to and from the place of employment does not usually form part of the working day when deciding the working hours of drivers, but may be a factor in the cause of accidents/incidents if travelling home follows a long working day.

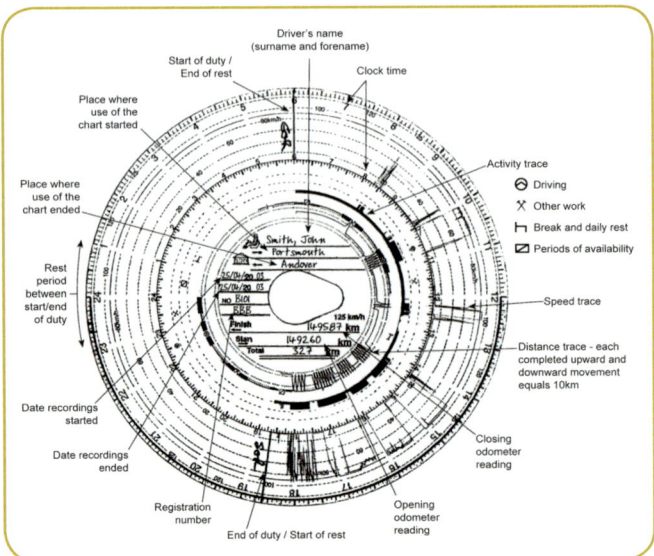

Figure 8-127: Analogue tachograph.
Source: UK Dept. of Transport.

Stress due to traffic

Driving-related stress is likely to be experienced when the demands of the road/traffic environment exceed the driver's ability to cope with or control that environment.

Rule	Time	Comment
Maximum driving limit before a break:	4 ½ hours	A break of no less than 45 minutes or divided into two periods (first 15 minutes, second 30 minutes) during the 4 ½ hours. *ILO - 5 hours.*
Daily driving limit:	9 hours	Extendable to 10 hours no more than twice a week. *ILO - 9 hours, averaged.*
Daily rest period:	11 hours	Minimum rest required, which can be reduced to 9 hours no more than three times between weekly rests. Can be taken in two periods, the first at least 3 hours and second at least 9 hours. *ILO - 10 consecutive hours in 24 hour period, reduced to 8 hours no more than twice a week.*
Weekly driving limit:	56 hours	*ILO - 48 hours, averaged.*
Two weekly driving limit:	90 hours	
Weekly rest period:	45 hours	A regular weekly rest of at least 45 hours, or a reduced weekly rest of at least 24 hours, must be started no later than the end of six consecutive 24 hour periods from the end of the last weekly rest. In any two consecutive weeks, a driver must have at least two weekly rests - one of which must be at least 45 hours long.

Figure 8-126: Vehicle driving limits. Source: AETR and ILO C153.

Sometimes driving stress can show itself as a 'road rage'. This is an irrational human mechanism that is activated to protect the sufferer from what they perceive to be actual or potential danger.

A person may often resort to irrational behaviour in order to avoid certain situations or events. Others may perceive the circumstances which aggravate the 'road rage' as being either insignificant or trivial, but to the individual concerned they are 'very real' and 'very relevant'.

Weather conditions

Weather conditions have a significant impact on the risks of driving. Sudden rainfall, sand storm or snow and fog can lead to poor visibility and sudden braking. Snow and surface water conditions can increase stopping distances. Even good weather can have a negative effect as glare from the sun can limit visibility. This is particularly the case in early morning or evening and is most significant during the winter when the sun is lower in the sky for longer.

EVALUATING THE RISKS

The driver

The level of risk is particularly affected by the driver's competence. If someone is new to driving, their skill may be adequate to provide them with a national driving licence but their experience of driving will be low.

Drivers that only drive intermittently also present a high risk as any competence they may have may decay over the time between driving.

The driver's competence has to be appropriate to the driving expected of them. An inexperienced driver driving in heavy traffic in complicated driving settings at night in the winter will present a particularly high risk. Similarly a person that is reasonably experienced in driving their small car may have difficulty when first driving a larger or faster-accelerating vehicle. Some drivers will need to adjust to driving slower vehicles and ones with a different centre of gravity.

Driver fitness may influence their ability to see well when driving at night, their ability to travel distances without breaks and may put them in a high-risk category for heart attack or other type of seizure. Pre-existing health conditions such as back injuries and late-term pregnancy could influence a driver's ability to concentrate on road conditions.

The level of training that the driver has received may affect risk in that if no training is given a driver may have developed bad driving habits and not be aware of them. Refresher driver training may help to reduce risk.

The vehicle

The size, weight, centre of gravity and power of a vehicle will all influence its functioning and therefore the risk that may arise from its use. It is important that the vehicle is suitable for the task and the driver.

A large, powerful, fast-acceleration car may be suitable for a specific task but very unsuitable for an inexperienced driver. The condition of the vehicle will have a significant effect on the level of risk. Vehicles that have poor brakes or lights, do not steer well or have poor suspension will represent a high risk in any driving situation. Vehicles with broken or missing mirrors will mean that the driver will not be able to see other road users adequately and will increase the risk when changing lanes.

Much of the **safety equipment** to prevent accidents/ incidents, such as anti-lock braking systems (ABS) should be built into most modern vehicles. Other items may be optional, such as 'run-flat' tyres, which may reduce certain road risks. Other safety equipment is designed to reduce the consequences of accidents/ incidents and can therefore reduce risk, such as airbags, head restraints, escape kits, warning triangles, fire extinguishers and high-visibility vests. The absence of safety-critical information, like the height of the vehicle, can have an immediate and significant effect on the level of risk. Consideration has to be given to **safety-critical information** related to the vehicle, including its height, width, length, weight and load carrying/towing capacity.

Vehicle Number _____ Date _____	Yes	No
Are all departmental vehicles subject to State licensing requirements equipped with the following items in good operating condition:		
Adequate rearview mirrors?	❑	❑
Safety belts?	❑	❑
Windshield wiper blades and fluid?	❑	❑
Horn?	❑	❑
Correctly adjusted headlights?	❑	❑
Brakes with adequate stopping power?	❑	❑
Emergency brake?	❑	❑
Turn/directional signals?	❑	❑
Good tires with adequate tread and correct pressure?	❑	❑
Oil and coolant levels?	❑	❑
Brake lights?	❑	❑
Taillights?	❑	❑
License plate light?	❑	❑
Tight muffler system?	❑	❑
Properly serviced fire extinguisher?	❑	❑
Intact windshield, with no cracks?	❑	❑
Is all seating in the vehicle secured to the frame?	❑	❑
Is there an Automobile Liability ID Card located in the glove compartment or elsewhere in the vehicle?	❑	❑
Are appropriate notices posted in each vehicle as a reminder that all employees and their passengers are required to wear seat belts?	❑	❑
Have all employees been instructed on safe backing practices?	❑	❑
Have employees been informed of what actions to take in the event they are involved in a vehicle accident?	❑	❑
Have employees been informed of appropriate safety guidelines when hauling loads?	❑	❑

Employee Signature _____

Supervisor's Signature _____

Figure 8-128: Typical vehicle pre-use safety checklist.
Source: www.tdi.texas.gov.

Ergonomic considerations have an effect on both comfort and ability to control the vehicle effectively. Comfort issues can increase fatigue, and ergonomic considerations like seat height adjustment can significantly affect the ability of the driver to see out of the vehicle properly.

The journey

When evaluating the risk related to journeys it is important to take into account such factors as:

- The *route* being taken - motorways are safer than smaller roads; routes using motorways will be lower in risk.

- *Scheduling* - if the journey is to be made early in the morning there might be an increased level of risk due to tiredness but this may be offset by the reduced level of traffic.

- *Sufficient time* allowed for travel - if not enough time is allowed for the journey, including an allowance for expected delays, the risk will be higher due to the increased likelihood of the driver speeding or taking unauthorised routes to make up time.

- *Weather* conditions can rapidly increase risks, conditions such as ice and snow and strong winds will have a significant effect.

WORK RELATED DRIVING CONTROL MEASURES

Avoiding work-related driving activities should be the first consideration, for example:

- The journey may not be necessary; communication may be sufficiently effective by telephone or video conference instead of driving.

- It may be possible to send goods by rail or air freight.

- Some of the driver's journey may be able to be done by rail.

Consider whether your organisation's policy on the allocation of vehicle to workers actively encourages them to drive rather than use other means of transport.

The driver

It is the employer's duty to ensure that drivers are competent, fit and in good health and capable of doing their work in a way that is safe for them. Employers must insist that all drivers produce evidence that they have a current licence to drive their class of vehicle. Without a current licence or test certificate, the vehicle insurance may be invalid. Regular assessments of driver competence and monitoring the validity of driving documentation, such as a driving licence, should be carried out. LGV drivers and

PCV drivers may have to maintain a certificate of professional competence. For example, in the UK, drivers are required to undertake 35 hours of training every 5 years.

The vehicle

It is the employer's responsibility to ensure that the vehicle is appropriate for its use. It is important that the vehicle is in a good condition, safe and suitable for the task to be carried out, including any safety equipment that is required being properly fitted and maintained. Any safety-critical information should be displayed within the cab, for example, the height or width of the vehicle.

Ergonomic factors related to the use of the vehicle should also be considered. For example, the driver's seat may require additional lumbar cushions. Whole-body vibration caused by the vehicle should be considered, for example, air-suspension seats may be required to reduce the amount of whole-body vibration received by the driver. The employer should ensure appropriate safety equipment is available and used, for example:

- Seat belts and air bags should be installed, maintained and used correctly.

- Two-wheeled vehicle users should use appropriate safety helmets and protective clothing.

- Vehicles could be fitted with speed-limiting devices and electronic trackers to monitor and control the speed of vehicles.

- Many countries require first-aid equipment to be carried in the vehicle for use in the event of an accident/incident. This helps to minimise potential consequences of driving-related injuries.

- Additional equipment may also be required, such as high-visibility clothing, a warning triangle to place in the road for if the vehicle breaks down, warm clothing or blanket for cold weather, shovel and equipment to provide grip under tyres for snow, sand or muddy conditions, portable lighting and welfare facilities.

The journey

Journey planning and scheduling are essential in ensuring the health and safety of workers who drive for work. Investing time in ensuring that journey planning is implemented as a component of the policy, will ensure that where possible, routes are planned thoroughly, schedules are realistic, and sufficient time is allocated to complete journeys safely and without fatigue. It may be necessary to plan overnight stops if the journey time extends due to bad weather or traffic conditions.

Delivery schedules should be adjusted so that unrealistic targets are not set. This will reduce the fatigue and stress of drivers and will not encourage them to drive too fast for the conditions or exceed speed limits.

Do your management systems ensure that work-related driving is managed effectively?

For example, are you confident that your vehicles are regularly inspected and serviced in accordance with manufacturers' recommendations?

ELECTRIC AND HYBRID VEHICLES

The growing environmental agenda to reduce carbon emissions has seen an increase in the development and use of electric powered vehicles; this includes battery electric, hybrid electric and plug-in hybrid electric vehicles. As this technology becomes more widespread it is important that organisations and individuals increase their awareness of the health and safety hazards which may be presented by these vehicles and that effective controls are implemented to reduce the likelihood of injuries occurring.

The main hazards presented by these vehicles can be categorised as:

- Silent operation of vehicles.

- Availability and location of charging points.

- Electric shock from high voltage components and cabling.

- Retained electrical charge in components.

- Unexpected movement of the vehicle or engine components.

- Potential for the release of explosive gases and harmful liquids from batteries.

Silent operation of vehicles

The lack of traditional engine noise from these electrically powered vehicles increases the risk of contact be-tween the vehicle and other road users. In particular pedestrians who are visually impaired and rely on engine noise to assess if it is safe to cross routes where these vehicles operate may be unaware of the presence of near silent electric vehicles. Whilst there are no current engineering controls to combat this hazard, new European Union regulations will require electric vehicle manufacturers to fit sound generators to all new vehicles to alert other road users and vulnerable pedestrians to the vehicle's presence. In the meantime, reliance remains with the vehicle driver and other road users to remain vigilant.

Availability and location of charging points

If there is limited availability of charging points the vehicles may run out of power and stop in locations that could present a danger to the occupants of the vehicles and others. This could be where there is a risk of collision with other vehicles or in a remote location that presents dangers related to the remoteness and possible vulnerability of the occupants of the vehicle to violent attack.

As the popularity of electric vehicles increases, charging points will become more prevalent. With this increase, the location of the electric vehicle charging point may present its own hazards. Close proximity to passing traffic could present the possibility of larger vehicles making contact with the charging points and causing damage and potential for exposure of live sources of electricity. Equally the electric vehicle driver may be forced to interact with passing traffic when exiting their vehicle to connect the charger.

Locations with a high amount of passing pedestrians may also be subjected to both accidental damage, whilst loading or unloading goods, and deliberate vandalism of the charging points. Drivers waiting nearby for their vehicle to charge could also find themselves facing the threat of violence, aggression or theft.

Exposure to weather extremes and proximity of the charging points to water courses, such as rivers, streams or pools, could, where drainage is inadequate, lead to flooding and ingress of water into the electrical cables sup-plying the charging points. Therefore it is really important when selecting a suitable location for charging points that consideration is given to access and egress, adequate distance from passing traffic and the size of parking bays. Effective drainage, elevation or bunding around charging points should be considered when sources of water are close by. Locations should also provide adequate lighting, suitable warning signage of the presence of electrical hazards, and visible instruction on the safe use of the charging points.

Electric shock from high voltage components and cabling

Voltages present in electric vehicles are significantly higher (up to 650 volts dc) than those found in more contemporary vehicles. Contact with live electrical parts above 110 volts dc can be fatal. It is important that manufacturers information specific to the vehicle in question is available in order to avoid hazards and actions which may expose an individual to live electrical parts.

This applies equally to drivers utilising charging points, those undertaking maintenance and repair work, incident responders (such as emergency services and vehicle recovery operatives), car wash, valeting and sales personnel.

For those performing higher risk activities such as maintenance and repair, then appropriate competence, skills and training will be required with a focus on familiarisation of the location of live electrical parts and lock off/isolation processes for electrical systems and batteries. These workers should use suitable tools and test equipment, including electrically insulated tools and wearing appropriate PPE to protect against inadvertent electric shock or electrical flashover.

Regular inspection and maintenance of vehicles and charging points is essential to ensure that any damage or faults are identified early and necessary repairs undertaken before incidents take place. Visual checks of electrical components and cabling for signs of damage should also be undertaken before any works on the vehicle are commenced. Users of charging points should only use the appropriate charging leads and adaptors and signage and information on correct use should be on display.

Retained electrical charge in components

Even when the electric vehicle is switched off, electrical energy is stored in the battery which has the potential to cause electric shock, explosion or fire. Individual components on the vehicle may also continue to retain a dangerous voltage. Manufacturer's guidelines and information for the specific vehicle must be made available to any workers working on the electric vehicle and they must comply with them. Specific training, skills and competency will be required to conduct higher risk maintenance and repair activities. These manufacturer's guidelines will include safe isolation procedures and instructions on how to discharge stored energy from individual components. Again only suitable tools should be used and appropriate PPE should be worn.

Unexpected movement of the vehicle or engine components

Electric motors or the vehicle itself may move unexpectedly due to magnetic forces within the motors. This could lead to personnel being injured due to contact with the vehicle, moving parts in the motors or individual components, or contact with live electricity. As with the other hazards already outlined, safe isolation, appropriate levels of competence and safe systems of work will be necessary. In addition to this vehicles with remote operation / proximity key systems should have the keys locked away in a secure place at a suitable distance from the vehicle to prevent the vehicle unintentionally starting up and moving; access to the keys should be controlled by the person working on the vehicle.

Potential for the release of explosive gases and harmful liquids from damaged batteries

Battery systems may contain chemicals which can be harmful if released and will also store significant amounts of energy that could cause an explosion in certain circumstances. This is more likely to become a problem if batteries are damaged, faulty or have been incorrectly modified. Therefore it is important that pre-work inspection of the batteries is conducted before work is carried out and regular maintenance and inspection schedules are in place to ensure the integrity of batteries.

Safe isolation, in accordance with manufacturer's guidance, appropriate training and competency and safe systems of work is again essential.

Battery packs are also susceptible to high temperatures, and appropriate labelling should be present to indicate maximum safe temperatures. This should be considered when carrying out operations where temperature may be adversely affected, such as painting booths or hot works. In these cases the battery should either be safely removed or insulation provided to limit temperature increases.

More general hazards that should also be considered include; slips, trips or falls of the driver or passing pedestrians whilst charging cables are used at charging points; manual handling risks of pulling charging cables into position and the potential for electrical systems within the vehicle to affect medical devices such as pacemakers.

Sources of reference

Reference information provided, in particular web links, was correct at time of publication, but may have changed.

A guide to workplace transport safety, HSG136, HSE Books, http://www.hse.gov.uk/pubns/priced/hsg136.pdf

Ambient factors in the Workplace, International Labour Organisation (ILO) Code of Practice (CoP), ISBN: 92-2-11628-X www.ilo.org/safework/info/standards-and-instruments/WCMS_107729/lang--en/index.htm

Directive 89/656/EEC – use of personal protective equipment https://osha.europa.eu/en/legislation/directives/4

Driving at work, Managing work-related road safety, INDG382, HSE Books, http://www.hse.gov.uk/pubns/indg382.pdf

Electric and hybrid vehicles, HSE Website, http://www.hse.gov.uk/mvr/topics/electric-hybrid.htm

Graphical symbols – Registered safety signs, ISO 7010:2011, International Organisation for Standardisation (ISO)

Health and Safety in Construction HSG150, 3rd Edition, HSE Books ISBN 978-0-7176-6182-2 www.hse.gov.uk/pubns/priced/hsg150.pdf

Health and Safety Toolbox, online resource, HSE www.hse.gov.uk/toolbox/index.htm

Hygiene (Commerce and Offices), ILO Convention, C120, 1964, ILO, http://www.ilo.org/dyn/normlex/en/f?p=NORMLEXPUB:12100:0::NO::P12100_ILO_CODE:C120

Hygiene (Commerce and Offices), ILO Recommendation, R120, 1964, ILO, http://www.ilo.org/dyn/normlex/en/f?p=NORMLEXPUB:12100:0::NO:12100:P12100_INSTRUMENT_ID:312458:NO

ILO Hygiene (Commerce and Offices), ILO Convention, 1964 (No. 120) - C120, http://www.ilo.org/dyn/normlex/en/f?p=NORMLEXPUB:12100:0::NO::P12100_ILO_CODE:C120

ILO Hygiene (Commerce and Offices), Recommendation, 1964 (No. 120), http://www.ilo.org/dyn/normlex/en/f?p=NORMLEXPUB:12100:0::NO:12100:P12100_INSTRUMENT_ID:312458:NO

ISO 70101:2019 – Graphical symbols – Safety colours and safety signs – Registered safety signs

Lighting at Work, HSG38, second edition 1997, HSE Books, ISBN 978-0-7176-1232-1 www.hse.gov.uk/pubns/priced/hsg38.pdf

Managing Health and Safety in Construction, Construction (Design and Management) Regulations 2015, Guidance on regulations, L153, HSE Books, ISBN: 978-0-7176-6623-3, http://www.hse.gov.uk/pubns/priced/L153.pdf

Personal Protective Equipment at Work (second edition), Personal Protective Equipment at Work Regulations 1992 (as amended), Guidance on Regulations, HSE Books, ISBN: 978-0-7176-6139-8 www.hse.gov.uk/pubns/priced/l25.pdf

Personal Protective Equipment at Work Regulations (PPER) 1992, UK, www.legislation.gov.uk (please enter the name of the Regulation)

Safe Use of Work Equipment, ACOP and guidance, L22, third edition 2008, HSE Books ISBN 978-0-7176-6295-1 www.hse.gov.uk/pubns/priced/l22.pdf

Safety and Health in Construction Convention, C167, 1988, ILO www.ilo.org/dyn/normlex/en/f?p=1000:12100:0::NO::P12100_ILO_CODE:C167

Safety and Health in Construction Recommendation, R175, 1988, ILO www.ilo.org/dyn/normlex/en/f?p=1000:12100:0::NO::P12100_ILO_CODE:R175

Safety and Health in Construction, ILO CoP, ILO Geneva 1992, ISBN: 92-2-107104-9 http://www.ilo.org/safework/info/standards-and-instruments/codes/WCMS_107826/lang--en/index.htm

Safety signs and signals, Guidance on Regulations, L64, HSE Books, ISBN: 978-0-7176-6598-3, http://www.hse.gov.uk/pubns/priced/l64.pdf

Seating at Work, HSG57, third edition 2002, HSE Books, ISBN 978-0-7176-1231-4 www.hse.gov.uk/pubns/priced/hsg57.pdf

Slips and trips, HSE, http://www.hse.gov.uk/slips/index.htm

The Construction (Design and Management) Regulations (CDM) 2015, Produced by CONIAC, published by CITB

ISBN: 978-1-85751-389-9, Industry Guidance for Clients, https://www.citb.co.uk/documents/cdm%20regs/2015/cdm-2015-clients-interactive.pdf

ISBN 978-1-85751-390-5, Industry Guidance for Principal Designers, http://www.citb.co.uk/documents/cdm%20regs/2015/cdm-2015-principal-designers-interactive.pdf

ISBN 978-1-85751-393-6, Industry Guidance for Designers, https://www.citb.co.uk/documents/cdm%20regs/2015/cdm-2015-designers-interactive.pdf

ISBN 978-1-85751-393-6, Industry Guidance for Principal Contractors, https://www.citb.co.uk/documents/cdm%20regs/2015/cdm-2015-principal-contractors-interactive.pdf

ISBN: 978-1-85751-391-2, Industry Guidance for Contractors, https://www.citb.co.uk/documents/cdm%20regs/2015/cdm-2015-contractors-interactive.pdf

ISBN 978-1-85751-394-3, Industry Guidance for Workers, https://www.citb.co.uk/documents/cdm%20regs/2015/cdm-2015-workers-interactive.pdf

The Health and Safety (Safety Signs and Signals) Regulations (SSSR) 1996, UK, www.legislation.gov.uk (please enter the name of the Regulation)

The Health and Safety (Safety Signs and Signals) Regulations 1996, Guidance on Regulations, second edition 2009, L64, HSE Books ISBN 978-0-7176-6359-0 www.hse.gov.uk/pubns/priced/l64.pdf

The Provision and Use of Work Equipment Regulations (PUWER) 1998 - Part III in particular, UK, www.legislation.gov.uk (please enter the name of the Regulation)

Welfare Facilities Recommendation, R102, 1956, http://www.ilo.org/dyn/normlex/en/f?p=1000:12100:0::NO::P12100_ILO_CODE:R102

The Work at Height Regulations (WAH) 2005 (as amended), UK, www.legislation.gov.uk (please enter the name of the Regulation)

The Workplace (Health, Safety and Welfare) Regulations (WHSWR) 1992, UK, www.legislation.gov.uk (please enter the name of the Regulation)

Understanding ergonomics at work, INDG90 (rev2), HSE Books www.hse.gov.uk/pubns/indg90.pdf

Vehicles at work, HSE, http://www.hse.gov.uk/workplacetransport/index.htm

Welfare Facilities Recommendation, R102, 1956, www.ilo.org/dyn/normlex/en/f?p=1000:12100:0::NO::P12100_ILO_CODE:R102

Work at Height Regulations 2005 (as amended) - A Brief Guide, INDG401 (rev1), HSE Books www.hse.gov.uk/pubns/indg401.pdf

Workplace health, safety and welfare, Workplace (Health, Safety and Welfare) Regulations 1992, ACOP, L24, HSE Books, ISBN 978-0-7176-0413-5 www.hse.gov.uk/pubns/priced/l24.pdf

Workplace Transport Safety - Guidance for Employers, HSG136, HSE Books, ISBN 978-0-7176-6154-1 www.hse.gov.uk/pubns/priced/hsg136.pdf

Web links to these references are provided on the RMS Publishing website for ease of use – www.rmspublishing.co.uk

STUDY QUESTIONS

1) What are the effects of extremes of temperature on the body? (8)

2) What are the control measures to reduce work-related driving risks? (8)

3) What are two design features from a vehicle intended to minimise the consequences of an overturn? (8)

4) What are the possible causes of a dumper truck overturning? (8)

5) How do non-movement related hazards result in injury to drivers? (8)

6) What measures can be used to segregate pedestrians and vehicles in the workplace? (8)

7) What control measures should you consider when planning work-related journeys? (8)

8) How can working at height be avoided? (8)

For guidance on how to answer NEBOSH questions please refer to the 'study question answer guidance' section located at the back of this guide.

Element 9

Work equipment

Contents

9.1 General requirements

TYPES OF WORK EQUIPMENT

Work equipment may be defined as any machinery, appliance, apparatus, tool or assembly of components that are arranged so that they function as a whole. Clearly, the term embraces many hand tools, power tools and machinery types.

Examples of work equipment commonly used in the workplace include:

- Air compressor.
- Automatic car wash.
- Automatic storage and retrieval equipment.
- Butcher's knife.
- Car ramp.
- Checkout machine.
- Circular saw.
- Digital image projector.
- Lifting sling.
- Liquefied Petroleum Gas (LPG) filling plant.
- Medical image scanners.
- Mobile access platform.
- Photocopier.
- Portable drill.
- Power press.
- Bench-top grinder.
- Pedestal drill.

Not work equipment:

- Livestock.
- Substances.
- Structural items (buildings).
- Private car.

Figure 9-1: Air compressor.
Source: RMS.

Figure 9-2: Vehicle lift.
Source: RMS.

Figure 9-3: Circular saw.
Source: RMS.

Figure 9-4: Photocopier.
Source: RMS.

PROVIDING SUITABLE EQUIPMENT

General requirements for suitability of work equipment

Article 16 of the ILO Convention C155 confirms that the employer has a responsibility to provide safe work equipment, *(see Figure 9-5)*.

"Article 16

1. Employers shall be required to ensure that, so far as is reasonably practicable, the workplaces, machinery, equipment and processes under their control are safe and without risk to health."

Figure 9-5: Requirements for work equipment.
Source: ILO, Occupational Safety and Health Conventions C155.

The ILO Code of Practice in the use of machinery sets out principles concerning safety and health in the use of machinery and confirms the responsibilities of employers, (***see Figure 9-6***).

"2.4.2. Buying and selecting machinery

2.4.2.1. Machinery should not be selected on purely economic or technical criteria but also carefully taking into account all factors affecting OSH and working conditions.

2.4.2.2. Employers should buy machinery for use in the workplace only if it complies with national laws and regulations and relevant international standards.

2.4.2.3. In selecting machinery, employers should ensure that they select machinery that is fit for the purpose and suitable for the specific working conditions at issue.

2.4.2.4. When second-hand machinery is introduced at the workplace, employers should ensure that it is safe and meets the technical requirements determined by national laws and regulations."

Figure 9-6: Responsibilities of employers.
Source: ILO, Code of Practice in the Use of Machinery.

When selecting work equipment the employer should ensure that it is suitable. Suitability should consider:

- Its initial integrity.
- The place where it will be used.
- The purpose for which it is used.

Integrity - The equipment should be safe through its design, construction and adaptation. This requires the equipment to be made in conformity with relevant health and safety standards. Consideration should be given to manufacturing deficiencies that leave sharp edges on the equipment where it is handled, 'home-made' tools and any equipment adapted from its original design to do a specific task.

Place - The equipment should be suitable for different environments and the risks that relate to them. For example, the environment may have high levels of moisture and humidity or be wet due to the process undertaken, or the atmosphere might be explosive. Account must also be taken of the possibility of the equipment causing a synergistic (combination) hazard in the workplace. For example, a petrol generator used in a confined space could emit toxic exhaust gases, or a hydraulic access platform may be used in a location with a low roof and present a risk of crushing the user. Similarly, the equipment may be designed for indoor use only and not have any protection from the ingress of water, which could present an electrical risk if used outdoors.

Use - The equipment should be suitable for the specific task for which it is used. Equipment used for any activity must be suitable to fulfil the exact requirements of the task. This means considering its strength, durability, power source, portability, protection against the environment, the range of tasks to be carried out, the frequency and duration of use and risks posed due to ergonomic constraints. Whenever equipment is required, the full capacity and limitation requirements should be identified and it should be confirmed that the equipment provided is suitable for the demands/limitations placed upon it. The equipment should be classed as an industrial or commercial type, designed for work activities, and not designed for use in the domestic environment.

For example, a hacksaw may be unsuitable for cutting metal straps used to secure goods to a pallet because the strap would be under tension and not restrained; a purpose-designed tool that cuts and keeps the strap safely would be more suitable than a simple saw. Similarly, a ladder may be unsuitable for work at a height where the worker needs to use both hands to carry out the task, and a scaffold or other access platform may be more suitable. Other examples of unsuitability include using a crane or fork-lift truck to lift a heavy load that was in excess of the safe working capacity of the equipment, using a swivel chair as a means of access to a shelf or using a heavy electrical tool as a hammer.

In the UK, the main requirements of the Provision and Use of Work Equipment Regulations 1998 is to ensure that the equipment used is suitable for its purpose and maintained to be safe, (***see Figure 9-7***).

"(1) Every employer shall ensure that work equipment is so constructed or adapted as to be suitable for the purpose for which it is used or provided.

(2) In selecting work equipment, every employer shall have regard to the working conditions and to the risks to the health and safety of persons which exist in the premises or undertaking in which that work equipment is to be used and any additional risk posed by the use of that work equipment.

(3) Every employer shall ensure that work equipment is used only for operations for which, and under conditions for which, it is suitable.

(4) In this regulation "suitable" means suitable in any respect which it is reasonably foreseeable will affect the health or safety of any person."

Figure 9-7: Suitability of work equipment.
Source: UK, Provision and Use of Work Equipment Regulations (PUWER) 1998.

CONFORMITY WITH BASIC HEALTH AND SAFETY STANDARDS

> "2.2. General responsibilities of designers and manufacturers
>
> 2.2.1. Machinery should be designed to be inherently safe so that hazards are eliminated. Where this is not possible, manufacturers and designers should ensure that adequate guards and safety devices are provided, so that risks are reduced to the lowest practicable level."

Figure 9-8: Responsibilities of designers and manufacturers of machinery. Source: ILO, Code of Practice in the Use of Machinery.

Those involved in the supply (including design and manufacture) of equipment have a responsibility to ensure that it is safe and healthy, so far as is reasonably practicable. This requires them to take account of all relevant standards, including basic health and safety standards.

The International Organization for Standardization (ISO) produces a number of standards that relate to work equipment. These standards have been established by international agreement and are adopted by countries globally. They set out basic health and safety standards with which designers and manufacturers should comply, for example, ISO 12100: 'Safety of machinery - General principles of design - Risk assessment and risk reduction'. ISO 12100:2010 specifies basic terminology, principles and a methodology for achieving health and safety in the design of machinery. It also specifies principles of risk assessment and risk reduction to help designers in achieving this objective. These principles are based on international knowledge and experience of the design, use, incidents, accidents and risks associated with machinery. Procedures are described for identifying hazards and estimating and evaluating risks during relevant phases of the machine life cycle, and for the elimination of hazards or sufficient risk reduction. Guidance is also given on the documentation and verification of the risk assessment and risk reduction process. ISO 12100:2010 is also intended to be used as a basis for the preparation of type-B (which cover specific safety devices and machine ergonomics) or type-C health and safety standards (which deal with specific classes of machinery and their requirements) by countries. Other ISO standards that relate to work equipment include:

- ISO 13849-1:2006 'Safety of machinery - Safety-related parts of control systems - Part 1: General principles for design'.
- ISO 13850:2006 'Safety of machinery - Emergency stop - Principles for design'.
- ISO 13856-1:2013 'Safety of machinery - Pressure-sensitive protective devices - Part 1: General principles for design and testing of pressure-sensitive mats and pressure-sensitive floors'.

In the European Union, ISO 12100:2010 has been adopted and given an EN prefix to signify this. Equipment made in conformity with harmonised European standards and meeting essential health and safety requirements may carry a CE marking to confirm this. The letters 'CE' are the abbreviation of the French phrase 'Conformité Européene', which literally means 'European Conformity'.

Figure 9-9: CE mark.
Source: European Commission.

It is a requirement of European Union legislation that employers in any member state ensure that an item of work equipment brought into use in the workplace has been designed and constructed in compliance with the directives/harmonised standards that relate to it and meets any essential health and safety requirements set for it. This includes types of work equipment that are classed as machinery.

PREVENTING ACCESS TO DANGEROUS PARTS OF MACHINERY

Dangerous parts of machinery present particularly high risks to anyone that may be exposed to them. Employers should ensure that measures are taken to:

a) Prevent access to any dangerous part of machinery or to any rotating stock-bar.

b) Stop the movement of any dangerous part of machinery or rotating stock-bar before any part of a person enters a danger zone.

See section 9.4 - Control measure for reducing risks from machinery hazards - for details of the steps to be taken to achieve these requirements.

THE NEED TO RESTRICT THE USE AND MAINTENANCE OF EQUIPMENT WITH SPECIFIC RISKS

Where the use of work equipment is likely to involve a specific risk to health or safety, employers should ensure that:

- The use of that work equipment is restricted to those persons given the task of using it.

- Repairs, modifications, maintenance or servicing of that work equipment is restricted to those workers who have been specifically chosen to perform the tasks, and also those who have received adequate training related to the tasks they have been assigned.

For example, in view of the specific risks, it would be appropriate to restrict the use and operation of a compactor, person-handling hoist, fork-lift truck, brush-cutter, tractor, abrasive wheel, nail gun, circular saw or mobile elevated work platform to those competent and authorised to use it. In the same way, maintenance should be restricted to persons who have the required competence to undertake the activity, for example, replacement of a grinding wheel, or replacement of load-bearing components of a rough-terrain fork-lift truck. Maintenance workers may not be considered to have the specific expertise or skill necessary to work on the equipment, particularly new equipment, leading to increased risk to themselves and others. It is essential that operation and maintenance of equipment with specific risks be restricted to those who have the appropriate skill and expertise.

PROVIDING INFORMATION, INSTRUCTION AND TRAINING

In most countries, there is a broad-based principle of 'duty of care' that requires workers to be involved in the prevention of health and safety risk and not to rely solely on the actions of their employer. This duty of care establishes the responsibility of workers for the health and safety of themselves and others that may be affected by their acts or omissions.

This principle of 'duty of care' is reflected in Article 19 of the ILO Occupational Safety and Health Convention C155 in general terms and specifically in Article 16 of the ILO Occupational Safety and Health Recommendation R164. The duty placed on workers is often set out at national or local level in both civil and criminal law. Users of work equipment, in particular machinery, should therefore use it in accordance with any training or instruction that they have received. They should use safety devices and protective equipment correctly and not render them inoperative.

Furthermore, users should report any faults or defects in their equipment that, if not rectified, could lead to the development of hazardous situations. The expectations placed on workers do not reduce the responsibility of the employer to ensure the use of work equipment is safe and healthy.

> "Every employee shall use any machinery, equipment, dangerous substance, transport equipment, means of production or safety device provided to him by his employer in accordance both with any training in the use of the equipment concerned which has been received by him and the instructions respecting that use which have been provided to him by the said employer in compliance with the requirements and prohibitions imposed upon that employer by or under the relevant statutory provisions."

Figure 9-10: Employee duties.
Source: UK, Regulation 14, Management of Health and Safety at Work Regulations 1999.

Whenever equipment is provided and used in the workplace, employers should ensure that workers who use it are adequately trained and are supplied with adequate health and safety information, and, where appropriate, written instructions relating to its use.

Employers should also ensure that those who supervise or manage the use of work equipment have been adequately trained and are knowledgeable regarding the necessary health and safety information and, where appropriate, have written instructions relating to its use. Training, information and instructions include those concerning the:

- Conditions and methods of use of the work equipment, including the capacities and limitations of the equipment.

- Risks that may arise from use of the equipment.

- Precautions to be taken to avoid and reduce risk, including safe operating procedures provided by the manufacturer/supplier and those drawn from experience in using the work equipment.

Training may be needed for existing workers, as well as workers using the equipment for the first time (including temporary workers), particularly if they have to use powered machinery. The greater the level of danger, the more substantial the training will need to be. For some high-risk work, such as driving fork-lift trucks, using a chainsaw or operating a crane, training should be carried out by specialist instructors. It should also be remembered that newly trained workers may obtain basic skills through training but may lack experience and judgement and require close supervision for an initial period. The requirement for refresher training at appropriate intervals should also be established. To ensure the effectiveness of training, those participating should be assessed for retention, comprehension and, where appropriate, skill.

Whilst employers should ensure that any written instructions are available to the people directly using the work equipment, they should also ensure such instructions are made available to other appropriate people. For example, maintenance instructions should be made available to the people involved in maintaining the work equipment.

Examples of training required for different workers:

Users - how to carry out pre-use checks, report defects, only to use equipment for the purpose designed. The training of someone to use a grinding machine should cover the proper methods of dressing the abrasive wheels.

Maintenance workers - safe isolation, acceptable replacement parts and adjustments in accordance with manufacturer's manuals. Maintenance workers should be given training on the specific risks related to maintenance, modification or repair work, such as stored energy in electrical systems.

Managers - be aware of the hazards and controls and maintain effective supervision.

In addition, where the use of work equipment is likely to involve a specific risk to health and safety, the employer must provide those authorised to repair, modify, maintain or service the equipment with adequate training.

WHY EQUIPMENT SHOULD BE MAINTAINED AND MAINTENANCE CONDUCTED SAFELY

Equipment to be maintained

Procedures for defective equipment

Workers should co-operate with the employer and notify it of any shortcomings in the health and safety arrangements, even when no immediate danger exists, so that the employer can take remedial action if needed. This extends to the identification of defective equipment of which any worker becomes aware; they should take reasonable steps to safeguard themselves and others.

Appropriate action includes following procedures to deal with defective equipment and, where appropriate and with the employer's consent, taking it out of use and quarantining it until it can be repaired by a competent person.

Maintenance of work equipment

Employers should ensure that work equipment is maintained throughout its working life, so that it is in efficient working order and in good repair. The manufacturer's instructions should be taken into account when maintenance work is planned and carried out.

The frequency at which maintenance activities are carried out should also take into account the intensity of use and the operating environment, for example, marine (wet, salt water) or outdoors, the variety of operations (whether the machine performs many different tasks or just one task very frequently) and risk to health and safety from malfunction or failure.

The extent and complexity of maintenance may vary from simple checks to integrated programmes for complex plant. Maintenance must, however, be effective and be targeted at the parts of the work equipment where the failure or deterioration could lead to health and safety risks.

When carrying out maintenance work on large process machinery, there are many hazards to be aware of. These include contact with dangerous moving parts of the machinery, electricity, stored energy such as heat or pressure; contact with gases, fumes and vapours and exposure to radiation and biological agents; manual handling of heavy machine parts or tools; noise and vibration, working at height or in confined spaces.

Practical measures that should be taken to protect people undertaking this type of work include: where possible, designing the machine to reduce the need to remove guards for routine maintenance and lubrication; the use of a permit-to-work system to assist in administrative controls to ensure that electrical power to the machine is isolated and locked off, and all pipelines leading to the machine are similarly isolated. This may include the need to release stored energy, for example, hydraulic pressure and, where necessary, to allow time for high-temperature equipment to cool down before maintenance starts. Where necessary, means of access such as a scaffold may have to be erected, and barriers and warning signs placed round the machinery to advise workers that maintenance work is in progress.

It is important that only skilled and competent workers undertake maintenance work and that they use appropriate tools, for example, spark-reduced tools in potentially flammable areas, a torque wrench for correct tensioning of securing bolts to pipework flanges. Consideration should be given to ensuring adequate standards of lighting and ventilation in the work area and to arrange for the work to be properly supervised and coordinated with other workers who may be present in the workplace at the time.

Where necessary, maintenance workers should be provided with appropriate personal protective equipment (PPE). Typical considerations include, for example, head protection (bump cap or helmet), eye protection (visor for arc welding, goggles for chemical splash) and fall-arrest systems (for example, harness, nets).

Inspections

For new equipment, inspections should be carried out as appropriate to the risks, for example, after installation and often with a commissioning engineer before first use. When existing equipment is modified or relocated, it should be re-inspected.

Employers should ensure that work equipment exposed to conditions causing deterioration, for example, corrosion from process materials, or significant cycles in temperature change that may result in dangerous conditions, is inspected at an appropriate frequency.

This should be:

- At suitable intervals – this may involve different degrees of inspection, including a thorough inspection by a competent person.

- Each time that exceptional unplanned circumstances occur – this could include natural phenomena (severe weather conditions such as high winds and seismic activity), accidents/incidents or a prolonged inactivity that is liable to result in damage or deterioration.

When equipment is used away from the workplace (for example, off-site) or when equipment is hired to meet a temporary requirement (for example, lifting equipment from a third party), it should be accompanied by relevant documentation indicating that a recent inspection has been carried out (many organisations use a system of labelling for portable electrical appliances, i.e. portable appliance testing (PAT). Equipment must be inspected by the user before use to identify any obvious defects and tested regularly by a competent person to ensure that it remains in a suitable condition. User checks may include guards, cables, casing integrity, cutting or machine parts and safety devices such as cut-outs. Competent persons include a maintenance worker as part of a preventive maintenance programme or it may be necessary to involve a third party, such as an insurance provider inspector, to carry out a periodic inspection of a pressure system or lifting equipment to ensure and validate its safe condition.

Inspections should address a list of identifiable health-and-safety-critical parts and will often use a check sheet as an aid to ensure all parts are considered. The purpose of an inspection check sheet is to record deterioration of specific parts, any abuse or misuse and any corrective action necessary. The results of the inspection will confirm whether or not a piece of equipment is in a safe enough condition to use. The inspection record should be kept for a suitable period and at least until the next inspection. The person carrying out the inspection must be competent. They should be capable of identifying any faults and determining their effects on health and safety. Specific requirements for the inspection, including thorough inspection (examination), of different work equipment may be specified by national law, for example, equipment used for work at height.

Maintenance can be carried out according to various systems of control. These may include a reactive approach, i.e. breakdown maintenance, or more active approaches such as planned preventive maintenance or condition-based maintenance.

Breakdown maintenance

Breakdown maintenance is concerned with repair when things go wrong, it is reactive and causes delays to the work schedule. If breakdown maintenance is the only approach used, the employer must consider all modes of failure and identify any which may put workers at risk and establish appropriate secondary health and safety controls, for example, spillage containment, respiratory protection, guards and other PPE.

It is widely accepted that a maintenance strategy based solely on repair at the time of component breakdown is neither efficient nor effective and may result in unacceptable unsafe conditions.

Planned preventive maintenance

Planned preventive maintenance (PPM) is an important strategy in maintenance and seeks to maximise the life of the component/equipment. This is done through a number of maintenance methods that are both planned and preventive. The planned method seeks to inspect and replace components on a scheduled basis. Preventive maintenance seeks to keep the condition of the component at its best by carrying out frequent care, for example lubrication, adjustment, cleaning. Although all maintenance is preventive in some respect, the primary aim of PPM is to prevent failures occurring while the equipment is in use.

Benefits of PPM

The main benefits are:

- Extended life of components.
- Assurance of reliability.
- Confirmation of condition of components.
- Reduced risk of loss-producing failure events.
- Ability to carry out work at a suitable time.
- Better utilisation of maintenance staff.
- Less standby facility required.
- Less expensive (last-minute) contracted facility required.
- Cost-effective actions.
- Demonstrates the employer has taken steps to meet the legal duties to maintain safe equipment.

Condition-based maintenance

Condition-based maintenance (CBM) is sometimes referred to as 'predictive' maintenance. Unlike PPM, which uses information provided by manufacturers to set timescales for changing components, CBM is a technique that involves monitoring the condition of the equipment and predicting equipment failure. Typical measurements include:

- Vibration analysis – rotating equipment such as compressors, pumps, motors all exhibit a certain degree of vibration. As they degrade or fall out of alignment, the amount of vibration increases. Vibration sensors can be used to detect when this becomes excessive.

- Infrared – IR cameras can be used to detect high-temperature conditions in energised equipment.

- Ultrasonic – detection of deep sub-surface defects such as boat hull corrosion.

- Acoustic – used to detect gas, liquid or vacuum leaks.

- Oil analysis – to measure the number and size of particles in a sample to determine equipment wear.

- Electrical – motor current readings using clamp-on ammeters.

- Operational performance – sensors throughout a system to measure variables such as pressure, temperature, or flow.

Maintenance action is then determined by this monitoring and prediction information. Many CBM systems are controlled by computers. CBM assumes that all equipment will deteriorate and that partial or complete loss of function will occur at some point. CBM monitors the condition or performance of plant equipment through various technologies. The data is collected, analysed, trended, and used to project equipment failures. Once the timing of equipment failure is known, action (such as replacing an oil filter when a replacement is needed not on a predetermined schedule) can be taken to prevent or delay failure. In this way, the reliability of the equipment can remain high. CBM can rely on visual inspections but often uses technology to gather, store and analyse data that may require a substantial financial investment in measuring equipment and worker up-skilling. The initial costs of implementation for such a system may be high. Also, additional competences are required to interpret and utilise the results from the system analysis correctly.

Maintenance to be conducted safely

No one should be exposed to undue risk during maintenance operations. In order to achieve this, equipment should be stopped and isolated as appropriate before work starts. If it is necessary to keep equipment running then the risks must be justified and the work adequately controlled. Controls may take the form of reducing running speed or range of movement or providing temporary guards.

Maintenance hazards

The principal sources of hazards are associated with maintenance work on:

- Conveyors.
- Lifts and hoists.
- Cranes.
- Concrete pumps.
- Live electrical equipment.
- Storage tanks.
- Hoppers.
- Chemical and degreasing plant.
- Compactors.
- X-ray machinery.

Typical hazards associated with maintenance operations include:

Mechanical:	Entanglements, machinery traps, contact, shearing traps, in-running nips, ejection, unexpected start-up.
Electrical:	Electrocution, shock, burns.
Pressure:	Unexpected pressure releases, explosion.
Physical:	Extremes of temperature, noise, vibration, dust.
Chemical:	Gases, vapours, mists, fumes.
Structural:	Obstructions, floor openings, voids.
Access:	Working at heights, confined spaces.

Figure 9-11: Conveyor.
Source: RMS.

Typical accidents/incidents associated with maintenance operations include:

- Crush, by moving machinery.
- Falls.
- Burns.
- Asphyxiation.
- Electrocution.
- Explosions.

One or more of the following factors causes maintenance accidents/incidents:

- Lack of perception of risk by managers/supervisors, often because of lack of necessary training.

- Unsafe or no system of work devised, for example, no permit-to-work system in operation, no facility to lock-off machinery and electricity supply before work starts and until the work has finished.

- No coordination between workers, and lack of communication with other supervisors or managers.

- Lack of perception of risk by workers, including failure to wear protective clothing or equipment.

Element 9

- Inadequacy of design, installation, siting of plant and equipment.

- Use of contractors with no health and safety control systems, i.e. risk assessments, method statements or who are inadequately briefed on health and safety aspects specific to the client.

- Lack of appreciation of synergistic (combined) hazards in the workplace.

Maintenance control measures

Isolation

This does not simply mean switching off the equipment using the stop button. It includes switching the equipment off at the electrical isolator for the equipment. In new workplaces, individual equipment isolators should be provided, i.e. each piece of equipment has its own isolator near to it. One isolator should not control several items of equipment as it is then impossible to isolate a single piece of equipment.

Figure 9-12: Electrical isolator with hole for padlock.
Source: RMS.

Figure 9-13: Physical isolation of valve.
Source: RMS.

Lockout and tagout

Electrical isolation by itself does not afford adequate protection, because there is nothing to prevent the isolator being switched back on or someone replacing removed fuses while the maintenance worker who isolated the equipment is still working on the equipment and is in danger.

To ensure that this does not happen, the isolator needs to be physically locked in the off position, typically using a padlock, and the key held by the worker that may be in danger if the isolation is removed. Multiple lockout devices are often used where multiple trades are carrying out work on the same equipment or plant. The multiple lockout devices are designed to accommodate a number of isolation padlocks for the different people working on the equipment, and the equipment cannot be energised until all the padlocks are removed, thereby protecting the last worker on the job. It is also a good idea to add a sign or tag – 'Do not switch on...' – on the equipment at the point of isolation. This approach is often called lockout and tagout (LOTO). Furthermore, after lockout and tagout (LOTO), maintenance workers should test the circuit before working on it, to ensure the circuit is no longer live and there is no residual energy in the circuit. Hence LOTO – lockout and tagout becomes LOTOTO – lockout, tagout and testout.

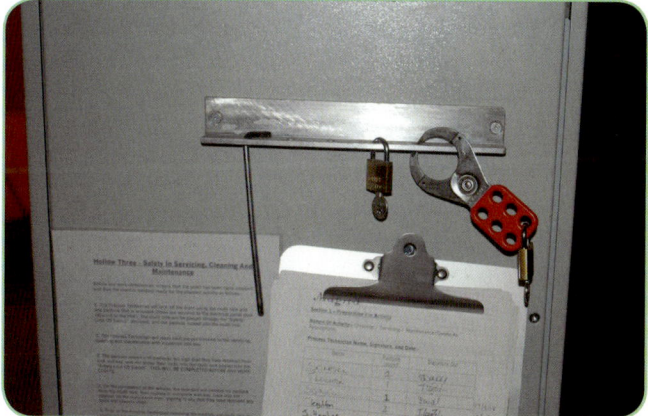
Figure 9-14: Multiple (padlock) lock-off device.
Source: RMS.

The control measures to enable maintenance work to be conducted safely include:

- Plan work in advance – provide safe access, support parts of equipment that could move or fall.

- Use written safe systems of work, method statements or permit-to-work systems as appropriate.

- Plan specific operations using method statements.

- Use physical means of isolating or locking off plant.

- Incorporate two-man working for high-risk operations.

- Integrate safety requirements in the planning of specific high-risk tasks.

- Prevent unauthorised access to the work area by using barriers and signs.

- Ensure the competence of those carrying out the work.

- Ensure the availability and use of appropriate PPE – gloves, eye protection, etc.

- Prevent fire or explosion – thoroughly clean vessels that have contained flammable solids or liquids, gases or dusts and check them thoroughly before hot work is carried out.

- Consider the need for rescue or treatment of workers in the event of an accident/incident.

EMERGENCY OPERATION CONTROLS, STABILITY, LIGHTING, MARKINGS AND WARNINGS, CLEAR WORK SPACE.

Operation and emergency controls

Controls for starting or making a significant change in operating conditions

Equipment should be fitted with a specific start-control device. It should only be possible to start equipment by positive, voluntary activation of the control device provided for that purpose. Start controls should be shrouded or otherwise protected to prevent inadvertent operation. Near each start control there should be a stop control. If it is possible to restart or change the operating conditions of the machine without using the start-control device, this should only be done where it does not lead to a hazardous situation (for example, the initiation of certain functions of machinery by the closure of an interlocking guard).

Section 4 of the ILO CoP Safety and Health in the Use of Machinery states: Control devices should be:

(a) Clearly visible and identifiable and readily distinguishable from one another by their separation, size, shape, colours or feel, and by labelling controls either with words or with unambiguous and easily recognizable symbols to identify the function or consequences of using the controls;

(b) Designed in such a way that controls for starting or stopping are clearly marked;

(c) Positioned in such a way as to be safely operated without hesitation or loss of time and without ambiguity;

(d) Designed in such a way that the movement of the control device is consistent with its effect;

(e) Located outside danger zones, except where necessary for certain control devices such as an emergency stop or a teach pendant;

(f) Positioned in such a way that their operation cannot cause additional risk;

(g) Designed or protected in such a way that the desired effect, where a hazard is involved, can be achieved only by a deliberate action;

(h) Made in such a way as to withstand any foreseeable forces; particular attention should be paid to emergency stop devices likely to be subjected to considerable forces.

Figure 9-15: Control devices for work equipment.
Source: ILO CoP Safety and Health in the Use of Machinery 2013.

For equipment that is capable of functioning in automatic mode, the starting of the equipment, restarting after a stoppage, or a change in operating conditions may be possible without intervention provided this does not lead to a hazardous situation.

Where equipment has several start-control devices and users can therefore place one another in danger, additional devices should be fitted to preclude such risks, for example, a system that requires positive start signals from all users before the equipment can start. If starting and stopping must be performed in a specific sequence in order to ensure health and safety, there should be devices that ensure that these operations are performed in the correct order. Any change in the operating conditions should only be possible by the use of a control, unless the change does not increase risk to health or safety. Examples of operating conditions include speed, pressure, temperature and power.

The controls provided should be designed and positioned so as to prevent, as far as possible, inadvertent or accidental operation. Buttons and levers should be of appropriate design, for example, including a shroud or locking facility. It should not be possible for the control to 'operate itself' due to the effects of gravity, vibration, or failure of a spring mechanism, for example.

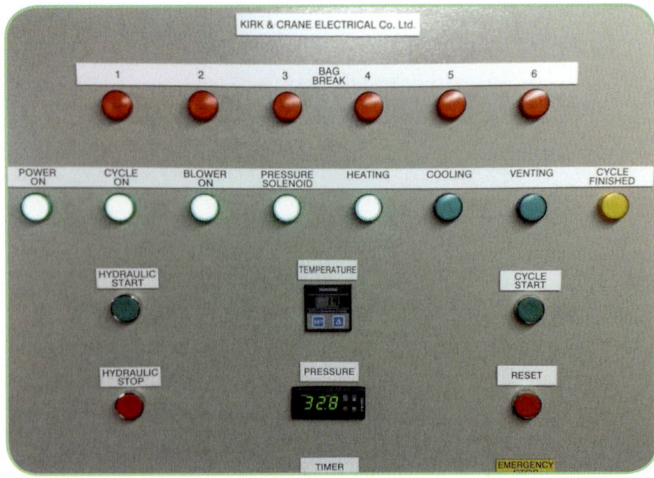

Figure 9-16: Control panel clearly marked; operation status displayed; emergency stop available.
Source: Kirk & Crane Electrical Co. Ltd.

Stop controls

The action of a stop control should bring the equipment to a safe condition in a safe manner. This acknowledges that it is not always desirable to bring all items of work equipment immediately to a complete or instantaneous stop, for example, to prevent the unsafe build-up of heat or pressure or to allow a controlled run-down of large rotating parts. Similarly, stopping the mixing mechanism of a reactor during certain chemical reactions could lead to a dangerous exothermic reaction.

The stop control should have priority over the start controls. Once the equipment or its hazardous functions have stopped, the energy supply to the actuators concerned should be cut off, unless this is undesirable for operational reasons; in these circumstances, the stop condition should be monitored and maintained. Accessible dangerous parts must be rendered stationary. However, parts of equipment that do not present a risk, such as suitably guarded cooling fans, do not need to be positively stopped and may be allowed to idle.

"Regulation 15 - Stop controls

(1) Every employer shall ensure that, where appropriate, work equipment is provided with one or more readily accessible controls the operation of which will bring the work equipment to a safe condition in a safe manner.

(2) Any control required by paragraph (1) shall bring the work equipment to a complete stop where necessary for reasons of health and safety.

(3) Any control required by paragraph (1) shall, if necessary for reasons of health and safety, switch off all sources of energy after stopping the functioning of the work equipment.

(4) Any control required by paragraph (1) shall operate in priority to any control, which starts or changes the operating conditions of the work equipment."

Figure 9-17: Stop controls for work equipment.
Source: UK, Regulation 15, Provision and Use of Work Equipment Regulations 1998.

Emergency stops are intended to effect a rapid response to potentially dangerous situations and they should not be used as functional stops during normal operation. The ILO Code of Practice for the use of machinery advises that emergency stop controls should be:

- Coloured red.

- Positioned in such a way as to be safely operated without hesitation or loss of time and without ambiguity.

- Designed in such a way that the movement of the control device is consistent with its effect.

- Located outside the danger zones, except where necessary for certain control devices such as an emergency stop (for use of maintenance workers) or a teach pendant (hand-held control).

- Positioned in such a way that their operation cannot cause additional risk.

- Designed or protected in such a way that the desired effect can be achieved only by a deliberate action.

- Made in such a way as to withstand any foreseeable forces; particular attention should be paid to emergency stop devices liable to be subjected to considerable force at the time of use.

Emergency stop controls should be easily reached and actuated. They should remain available and operational at all times, regardless of the operating mode of the equipment, for example, where special operating modes are applied to the equipment during maintenance work. Common types are mushroom-headed buttons, bars, levers, kick plates or pressure-sensitive cables.

Figure 9-18: Start and stop controls.
Source: Fine Woodworking Tools.

Once the emergency stop control has been activated, the command must remain in place until the control has been specifically overridden by a direct action to disengage the control by resetting it. Disengagement of the emergency stop control must not automatically restart the equipment, but it should require a positive action by the operator to restart it.

Figure 9-19: Control panel of CNC machining center showing emergency stop button.
Source: Shutterstock.

"Regulation 16 - Emergency stop controls

(1) Every employer shall ensure that, where appropriate, work equipment is provided with one or more readily accessible emergency stop controls unless it is not necessary by reason of the nature of the hazards and the time taken for the work equipment to come to a complete stop as a result of the action of any control provided by virtue of regulation 15(1).

(2) Any control required by paragraph (1) shall operate in priority to any control required by regulation 15(1)."

Figure 9-20: Emergency stop controls for work equipment.
Source: UK, Regulation 16, Provision and Use of Work Equipment Regulations 1998.

Controls

It should be possible to identify easily what each control does and on which equipment it takes effect. Controls should be:

- Clearly visible and identifiable.
- Readily distinguishable from one another by their separation, size, shape, colours or feel, and by labelling.
- Labelling should be either with words or unambiguous and easily recognisable symbols to identify the functions of the controls.

Controls should be so arranged that their layout, travel and resistance to operation are compatible with the action to be performed, taking account of ergonomic principles. Equipment should be fitted with such indicators (such as red and green warning lights) as may be required for safe operation. The operator should be able to read them from the control position.

Except where necessary, the employer should ensure that no control for work equipment is in a position where any person operating the control is exposed to a risk to their health. For example, an emergency stop control may be placed inside a danger zone for use of workers

during maintenance or setting robot equipment in teach mode.

From each control position, the operator should be able to ensure that nobody is in a danger zone. Alternatively, the control system should be designed and constructed in such a way that starting is prevented while someone is in a danger zone. If neither of these solutions is practicable, an acoustic and visual warning signal should be given before the equipment starts. Any persons exposed should have time to leave the danger zone or prevent the machinery from starting up.

As well as time, suitable means of avoiding the risk should be provided. This may take the form of a device used by workers at risk to prevent start-up or to warn the operator of their presence, for example the provision of a wedge or block to physically prevent the movement of equipment. Otherwise, there must be adequate provision to enable workers at risk to withdraw, for example, sufficient space or exits.

Circumstances will affect the type of warning chosen. If necessary, the employer should ensure that the equipment can only be controlled from control positions located in one or more predetermined zones or locations. Where there is more than one control position, the control system should be designed in such a way that the use of one of them prevents the use of the others, except for stop controls and emergency stops.

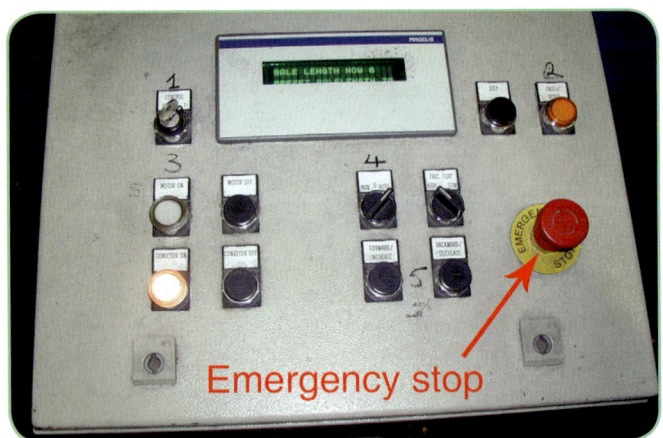

Figure 9-21: Control panel including emergency stop.
Source: RMS.

Control systems

Control systems should be designed and constructed in such a way as to ensure that as few hazardous situations as possible arise. They should be designed and constructed taking into account the following aspects:

- It should be able to withstand the intended operating stresses and external influences, taking into account foreseeable abnormal situations. External stresses include humidity, temperature, impurities, vibration and electrical fields.

- Its operation should not create any increased risk to health or safety.

- It does not impede the operation of stop and emergency stop controls.

- The loss of supply of any source of energy used by the work equipment cannot result in additional or increased risk to health or safety.

- A fault in the hardware or software of the control system does not lead to hazardous situations.

- Errors in the control system logic should not give rise to hazardous situations.

- Reasonably foreseeable human error during operation should not give rise to hazardous situations.

The ILO Code of Practice in Safety and Health in the Use of Machinery suggests that particular attention should be paid to the following when considering control systems for machinery:

- Machinery should not start unexpectedly.

- The parameters of the machinery should not change in an uncontrolled way.

- Machinery should not be prevented from stopping if the stop command has already been given.

- No moving part of the machinery or piece held by the machinery should fall or be ejected.

- Automatic or manual stopping of the moving parts, whatever they may be, should be unimpeded.

- Protective devices should remain fully effective or give a stop command.

- Safety-related parts of the control system should apply in a coherent way to the whole of an assembly of machinery and partly completed machinery.

- For cableless control, an automatic stop should be activated when correct control signals are not received, including loss of communication.

Failure of any part of the control system or its power supply should lead to a 'fail-safe' condition (more correctly and realistically called 'minimised failure to danger'), and not impede the operation of the stop or emergency stop controls. The measures that should be taken in the design and application of a control system to mitigate against the effects of failure will need to be balanced against the consequences of any failure, and the greater the risk, the more resistant the control system should be to the effects of failure.

Stability

Work equipment should be designed to be stable enough to avoid overturning, falling or uncontrolled move-ment during transportation, assembly, installation, disman-tling and use. If the shape and weight distribution of the equipment does not provide sufficient stability, appro-priate means of stabilisation must be made, including the use of anchorages and additional bracing or support structures. Where equipment has a variable reach or if the weight and centre of gravity may change during use it is important to take this into account.

Most machines used in a fixed position should be bolted or otherwise fastened down so that they do not move or rock during use. This is particularly important where the equipment is tall relative to its base or has a high centre of gravity, for example, a pedestal-mounted abrasive wheel, some diagnostic medical equipment and vertical cardboard compactors. It has long been recognised that woodworking and other machines (except those specifically designed for portable use) should be bolted to the floor or similarly secured to prevent unexpected movement.

"Regulation 20 - Stability

Every employer shall ensure that work equip-ment or any part of work equipment is stabilised by clamping or otherwise where necessary for purposes of health or safety."

Figure 9-22: Stability of work equipment.
Source: UK, Regulation 20, Provision and Use of Work Equipment Regulations 1998.

Lighting

Any workplace where a person is using work equipment must have adequate and suitable lighting for the opera-tion of the equipment, in particular machinery, so that equipment movements, controls and displays can easily be seen. Local lighting may be needed to give suffi-cient view of a dangerous process or to reduce visual fatigue. Equipment should be supplied with integral lighting suitable for the operations concerned where the absence of integral lighting would be likely to cause a risk despite ambient lighting of normal intensity.

Artificial lighting should, where possible, not produce glare, any stroboscopic effects (a visual phenomena that occurs when something that is moving appears to be represented by a series of stationary or short images – this is also known as the 'wagon wheel effect' where wagon wheels on motion pictures can appear to rotate in the opposite direction of travel) or disturbing shadows. Lighting should be suitable for the environment in which it is used. For example, it should be intrinsically safe in flammable atmospheres. Equipment should be designed and constructed so that there is no area of shadow likely to cause nuisance during its operation and no irritating dazzle. Operations such as maintenance work may not have sufficient general lighting, and equipment such as overhead cranes may temporarily obscure general lighting. In these circumstances, it may be necessary to provide equipment with local task lighting to ensure lighting levels are adequate. Internal parts of equipment requiring frequent inspection and adjustment, as well as the maintenance areas, should be provided with appropriate lighting.

"Regulation 21 - Lighting

Every employer shall ensure that suitable and sufficient lighting, which takes account of the operations to be carried out, is provided at any place where a person uses work equipment."

Figure 9-23: Lighting for work equipment.
Source: UK, Regulation 21, Provision and Use of Work Equipment Regulations 1998.

Markings and warnings

The employer should ensure that work equipment is marked in a clearly visible, legible and indelible manner with any marking appropriate for reasons of health and safety. Equipment designed and constructed for use in a potentially explosive atmosphere or similar hazard condition should be marked accordingly. Certain markings may also serve as a warning, for example, the maximum working speed, maximum working load or the contents of the equipment being of a hazardous nature (for example, the colour coding of gas bottles or service mains).

Where residual risks to health and safety remain following the provision of markings and other physical precautionary measures, warnings and warning devices should be introduced. Warnings are usually in the form of a notice, sign or similar. Examples of warnings are positive instructions (hard hats must be worn); prohibitions (no naked flames) and restrictions (do not heat above 60°C). Warning devices are active units that give out either an audible or visual signal, usually connected to the equipment in order that it operates only when a hazard exists; for example, when equipment such as a conveyor starts to operate and workers need to move to a safe position, or when a fault in the operation of unsupervised equipment may endanger people.

Warning devices are also used where the equipment is mobile; a flashing light can warn of the presence of a fork-lift truck and an audible device may warn of a vehicle reversing. Markings and warnings given by warning devices on work equipment should be unambiguous, easily perceived and easily understood. Text, signs and pictograms should only be used if they are understood in the culture in which the equipment is to be used. Consideration may have to be given to international and national standards concerning colours and safety signs/signals, for example, ISO 7010:2011 'Graphical symbols - Safety colours and safety signs used in the workplace and public areas'.

Clear unobstructed workspace

Workrooms should have enough free space to allow people to get to and from workstations and to move within the room with ease. The workspace that is necessary for the tasks to be carried out should be determined and provided to ensure safe working. Workrooms should be of sufficient height (from floor to ceiling) over most of the room to enable safe access to workstations. In older buildings with obstructions such as low beams, the obstruction should be clearly marked.

Where work equipment such as a circular saw is used in a workplace, care should be taken to ensure that adequate space exists around the equipment to make sure it is not overcrowded, and does not cause risk to users and those passing by the equipment when it is operating. It is important that the workspace allocated for users of work equipment is maintained clear and unobstructed. Obstructions could lead to slips, trips and falls, which could also involve contact with the equipment or moving parts of the equipment. Material being readied to be worked on, material being worked on and any scrap material should not be allowed to build-up in the work-space. Material such as dust, scraps of wood or metal can quickly build-up as the equipment is used. Good control of the input, output and processing of material is essential. Scrap material must be removed from the workspace at suitable intervals, so that it does not build-up and obstruct the workspace.

"Regulation 11 - Room dimensions and space

Every room where persons work shall have sufficient floor area, height and unoccupied space for purposes of health, safety and welfare."

Figure 9-24: Room dimensions and workspace. Source: UK, Regulation 11, Workplace (Health, Safety and Welfare) Regulations 1992.

REVIEW

List two categories of worker that should receive information, instruction and training on the safe use of work equipment.

What hazards are typically associated with maintenance work?

List the requirements for a stop button.

9.2 Hand-held tools

GENERAL CONSIDERATIONS FOR SELECTING HAND-HELD TOOLS (POWERED OR MANUAL)

Requirements for safe use

Workers need to be able to recognise the hazards associated with the different types of tools for their work and know the necessary safety precautions. They should be trained in the proper use of the tools they use. The tools must be suitable for the purpose and conditions (wet, dusty, flammable vapours) of use.

Requirements for safe use of hand-held tools include:

- Keep all tools in good condition with regular maintenance.

- Workers should only use tools they have been trained and authorised to use.

- Use the right tool for the work to be done – for example, use the correct size of spanner and the correct type of saw.

- Examine each tool for damage before use and do not use damaged tools.

- Operate tools according to the manufacturers' instructions.

- Use appropriate personal protective equipment.

- Do not misuse tools, for example using a screwdriver as a chisel or extending the handle of a spanner to improve leverage.

- Frequently used tools are within convenient reach.

- Adequate space is provided in the work area.

- Work benches are of a suitable height to avoid the need for awkward and uncomfortable postures.

Condition and fitness for use

Hand tools should be kept in good condition in order for them to be able to do the work required efficiently and safely, for example, sharp blades enable easy cutting with less effort, which means the worker's hands can be kept away from the blade and the tool is less likely to slip. Examples of condition and fitness for use include:

- Hammers and mallets - avoid split, broken or loose shafts and worn or chipped heads. Make sure the heads are properly secured to the shafts.

- Cutting tools – keep the cutting edge sharp and cover as much of the exposed cutting edge as is possible when in use, keep the cutting edge.

- Files - should be fitted with a proper, secure handle. Never use them as levers.

- Chisels - the cutting edge should be sharpened to the correct angle. Do not allow the head of cold chisels to spread to a mushroom shape – grind off the sides regularly. Use hand guards to prevent impact injuries when using chisels. To prevent damage to wooden handles and the potential for wood splinters, a mallet should be used for striking wood chisels.

Figure 9-25: Damaged chisel.
Source: RMS.

Suitability for purpose

Tools are designed and manufactured for specific tasks, which is why screwdrivers are made with different tip sizes and length. Many injuries happen because workers do not use a tool that is suitable for the work being done. Therefore tools should be suitable for the purpose they are to be used for. The employer should ensure that workers are provided with suitable equipment for the work being done. Consider the demands of the work activity and the suitability of the design of the equipment being used, in terms of size, shape and how appropriate it is for the task. Suitable for purpose will include:

- Quality of materials the tools are made from, manufactured to a recognised standard.

- Insulated tools for electrical work.

- Non-sparking tools for flammable atmospheres.

- The right size, for example the right size of spanner, hammer or screwdriver.

- Matching the features of the tool to the needs of the work being done, for example using pliers with serrated jaws where improved grip of items is required.

- Using low energy power tools where possible, for example battery operated or low pressure air driven tools.

- Not improvising by using a tool not designed for the work, for example, not using a screwdriver as a lever.

Location to be used in (including flammable atmosphere)

It is important to consider the location where hand-held tools are going to be used, for example whether the atmosphere is flammable, wet, dark or at a height.

When working in areas where a flammable atmosphere is present, using non-sparking tools is necessary to prevent the creation of sparks that could cause the flammable atmosphere to ignite. Non-sparking tools are made of materials that do not contain iron (non-ferrous metals) and therefore the risk of a spark being created while the tool is in use is reduced. Common materials used for non-sparking tools include brass, bronze, copper-nickel alloys and copper-aluminum alloys. Non-sparking tools can also be made of wood, leather and plastics. Some common tools that are available in a non-sparking option include hammers, chisels, pliers, screwdrivers, shovels and spanners. Specialist electrically powered hand-held tools are also available for use in flammable atmospheres.

If the location the tools are to be used in is wet it may mean that some powered tools may not be suitable, which could limit the tools that are suitable for this location to those that are manually operated, specialist electrical tools or tools operated by compressed air.

In locations that are dark supplementary lighting will be essential to ensure the worker can operate the tools safely. Where the location is cold this may require the worker to wear gloves and necessitate the use of tools designed to enable them to be gripped easily while wearing gloves. If the location of the work activity is at a height consideration should be made to the possibility of the tools being dropped a significant distance that could harm people and measures put in place to prevent this.

HAZARDS AND CONTROL MEASURES FOR A RANGE OF HAND-HELD TOOLS – MANUALLY OPERATED

Anyone who uses a hand tool may be at risk of injury, either accidentally or through misuse or equipment failure.

There is a range of injury hazards relating to the use of hand tools. For example, noise-induced hearing loss from cutting or impact tools; respiratory disease from inhalation of dust from hand sanders; eye injury from material thrown off from cutting; punctures and cuts caused by sharp equipment such as scissors, paper guillotines, knives, chisels, saws, planes and screwdrivers. Heat-producing equipment such as blowtorches and irons can cause burns and permanent scarring.

Misuse includes using the wrong tool for the job, for example, using a hammer rather than a screwdriver to drive a screw into material, or using a file as a lever, as it is brittle and may break unexpectedly.

Examples of misuse of hand tools are shown in the following schematics:

Figure 9-26: Hammer shaft split with peg holding on head and very worn hammerhead. Source: ILO.

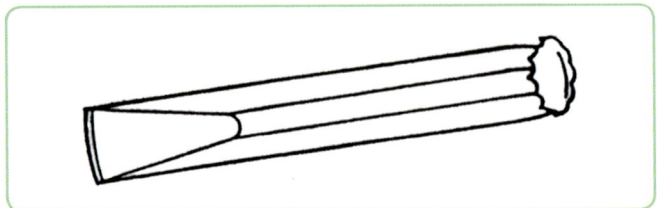

Figure 9-27: Mushroom-shaped head. Source: ILO.

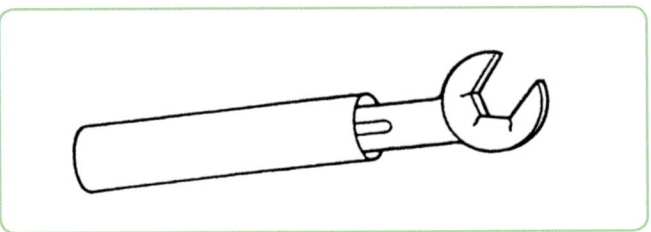

Figure 9-28: Piece of tube used as an extension. Source: ILO.

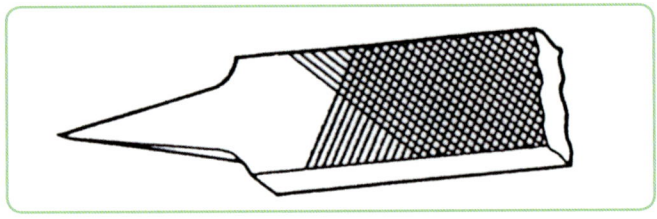

Figure 9-29: Broken file. Source: ILO.

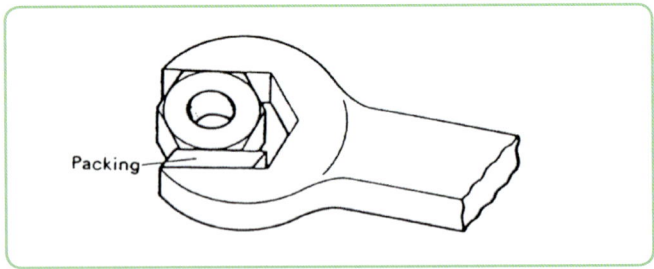

Figure 9-30: Packing material used in spanner head. Source: ILO.

Figure 9-31: File used as a lever. Source: ILO.

Hammers

Hazards

The main hazard when using a hammer is that the worker may use a hand to hold what is being hit and the hammer may hit the worker's hand or fingers instead of the intended item. Also, the hammer head could hit the intended item, but due to its shape or the angle of the impact the hammer head could slip sideways and impact the worker or twist the handle in the worker's hand, which could damage the wrist. These impacts could break a bone in the wrist, hand or fingers, as well as cause minor scrapes, cuts and bruising to other parts of the body. If the hammer head is not secure on the handle, the head could fly off when in use. The handle could become split and cause splinters or cuts to the worker's hand. If the hammer head becomes chipped, parts of it may fly off when it impacts what is being hit, this could cause eye injury or cuts.

Control measures

The following control measures will help to avoid harm from the use of hammers:

- Check the condition of the hammer – avoid split, broken or loose handles and worn or chipped heads. Heads should be properly secured to the handles.
- Place the work against a hard surface.
- Grip the handle firmly.
- Hold the hammer at the end of the handle.
- Check area is clear around you before swinging the hammer.
- Hit the surface of the work squarely with the hammer.
- Use the whole arm and elbow.
- Work in a natural position.
- Practise good hammering technique.

Files

Hazards

The main hazard when using a file is the tang, which is the pointed tip of the file that is fitted into the handle. A handle must be fitted over the tang before using the file, otherwise there is a good chance of puncturing the hand or other parts of the body with it when using the file. The handle could become split and cause splinters or cuts to the worker's hand.

If the surface of the file becomes blocked with material being filed or the file does not contact the material properly when filing, there is a risk that the file could suddenly slip forward across the surface of the material and cause damage to the worker's hand.

Files are very brittle and can snap easily if too much pressure is applied or if they are used as levers.

Control measures

These should have a proper, well-designed handle. The file should be held firmly in one hand with the fingers of the other hand used only to guide it. The material being filed should be firmly held in clamps or a vice. File strokes should be made away from the user. Take care to avoid the file slipping on the surface of the material causing sudden forward movement of the user. Files must never be used as a lever – they are very brittle and will shatter easily. They should only be cleaned by using a cleaning card and not by striking them against a solid object as this can also cause the file to shatter.

Chisels

Hazards

When a chisel is used there is a risk of injury, particularly to the eyes, from flying particles of the material being chiselled. The power required when using a chisel on metal may cause the workpiece to fly out from its holding device. Chisels have a cutting edge, which for wood chisels is particularly sharp, therefore there is a risk of cuts to the hand when handling them.

If the chisel head is allowed to become "mushroomed" pieces of the head can break off during use and cause eye injury. If a chisel is used as a screwdriver or a lever, the tip of the chisel may break suddenly and fly off, hitting the user or other workers.

Control measures

Choose a chisel large enough for the job, so the blade is used rather than only the point or corner. Never use a chisel with a dull blade – the sharper the tool, the better the performance. Chisels that are bent, cracked, or chipped should be discarded. The cutting edge should be sharpened to the correct angle. Do not allow the head of cold chisels to spread to a mushroom shape – grind off the sides regularly. Use a hand guard on the chisel and hit the chisel squarely. Chisels should not be used as a lever as they may break suddenly.

Screwdrivers

Hazards

If a worker holds a screw in their hand and the screwdriver slips out of the screw head while the worker is applying a powerful twisting motion, the screwdriver could move suddenly and stab or cut the worker's hand. A greasy handle could also cause the screwdriver to slip and lead to injury. The handle could become split and cause splinters or cuts to the worker's hand.

Repetitive use of a screwdriver, involving powerful, twisting motions, may cause health conditions, such as carpal tunnel syndrome. Most screwdrivers have metal shanks, therefore there is a risk of electric shock if the screwdriver is used near a live electrical sources.

Control measures

A screwdriver is one of the most commonly used and abused tools. The practice of using screwdrivers as punches, wedges or levers should be discouraged as this dulls blades and may cause injury if the blade fractures or slips in use. Screwdrivers should be selected so the tip fits the screw. When working on electrical equipment, screwdrivers must be equipped with insulated handles and shanks appropriate to the voltage of the equipment. Screwdrivers with split handles or damaged tips should be taken out of use and discarded safely.

Spanners

Hazards

Hazards of using a spanner may vary depending on the work being done, but could include:

- Powerful, repetitive action to turn the spanner could lead to upper limb disorders, for example damage to soft tissue in the hand, wrist and shoulder.

- The spanner slipping off the nut or other item being turned could cause the worker's hand to impact with things nearby.

- The spanner or item being turned may break with similar effects.

- The workpiece may suddenly break free from the device holding it, causing the worker to lose balance and fall.

Control measures

Avoid spanners with splayed or damaged jaws. Use ring spanners or sockets where possible, as these are less likely to slip. Discard safely any spanners that show signs of slipping. Ensure enough spanners of the correct size are available. Do not improvise by using pipes, etc., as extensions to the handle. Where necessary, use penetrating oil to loosen tight nuts.

Knives

Hazards

The most obvious hazard when using knives is the sharp blade which can easily cut the skin.

Control measures

Knives cause more disabling injuries than any other hand tool. The risks are that the hands may slip from the handle onto the blade or that the knife may strike the body or the free hand. Use knives with handle guards if possible. Make sure that the cutting stroke is always away from the body. Return the knife to a safe place and ensure it is sheathed before conducting other tasks, like moving a sack that the user has opened. Do not hold a knife with an open blade while carrying other objects.

Knives must be kept sharp and in their holders or sheaths when not in use. Dirty or oily knives should be wiped clean so that the user's grip does not slip. To clean the blade, wipe with a towel or cloth with the sharp edge turned away from the wiping hand. Foolish behaviour of any kind involving a knife (throwing, 'fencing', etc.) must not be tolerated. Where the atmosphere may be flammable, use alloy or bronze tools to prevent sparks.

HAZARDS AND CONTROL MEASURES FOR A RANGE OF HAND-HELD TOOLS - POWER OPERATED

Anyone who uses a hand-held portable power tool may be at risk of injury from the power source, equipment failure or the action of the power tool coming into contact with the worker, either by normal use, accidentally or through misuse. Injury hazards include hand-arm vibration, for example, caused by the use of hand-operated power tools.

Workers who regularly use these as part of their job may be at risk of a permanent injury known as hand-arm vibration syndrome (HAVS). In addition, many portable power tools are powered by electricity and their use in flammable atmospheres could present a risk of fire or explosion.

Electric drill

Electric drills are used for penetrating various materials and in construction are usually of a medium-to-heavy duty nature.

This equipment involves rotating shafts and tool bits, sharp tools, electricity and flying debris.

Hazards

Obvious hazards include shock and electrocution leading to possible fatalities; however, this potential is reduced by using 110 volt or battery-operated equipment. Other hazards include puncture, entanglement, noise and dust.

Control measures

Control measures include using only equipment that is suitable for the task and ensuring equipment is tested and inspected as safe to use, and the provision of suitable shut-off and isolation measures, goggles and hearing protection. Care should be taken to ensure that drill bits are kept sharp, as injuries can occur when the rotating drill bit gets stuck in material, causing the drill to kick and rotate in the operator's hands.

Sander

Hazards

Sanding equipment is available in a variety of sizes from hand-held equipment to large industrial machines. Sanders are used to provide a smooth finished surface, using a mechanical abrasive action.

Sanding operations are carried out on a wide variety of materials, including wood, minerals such as marble and man-made fibres. The main hazards associated with the sanding process are vibration and noise. Also, harm can be caused by the inhalation of respirable particles from dust. Where organic materials such as wood are being processed, fire may result from overheated surfaces or explosion from dust by-products.

Associated hazards may include electrocution if supply cables are damaged by hand-held sanders, particularly if the sander is placed on the ground whilst still rotating (orbital sanders). Other risks include trips from trailing cables, cuts from sharp surfaces and strains or sprains from manual handling of process materials.

Figure 9-32: Belt sander.
Source: Clarke International.

Control measures

When using sanding equipment, suitable personal respiratory protective equipment (RPE) is required to protect the user from dust exposure. Where possible, local exhaust ventilation equipment should be used to minimise dust in the atmosphere.

In order to protect against vibration injuries, the operator should be given regular breaks and the equipment maintained at intervals, including the renewal of sanding media to prevent the need for over-exertion by the operator. Where hand-held tools are used, suitable hand protection will also reduce injuries from vibration, cuts and manual handling.

Pre-use inspections by the operator and regular thorough examinations should be carried out to identify poten-tial electrical problems. Care should be taken to ensure that power leads do not create tripping hazards and they are positioned so that the likelihood of mechanical damage is minimised.

 **REVIEW**

What factors should be considered to reduce the risk of injury when using a hand-held hammer?

What are the risks from using hand-held power tools?

What are the control measures for the safe use of a hand-held electric drill?

9.3 Machinery hazards

MAIN MECHANICAL AND OTHER HAZARDS

This section describes the origin of machinery hazards and the ***potential consequences*** as a result of contact with mechanical and other hazards identified in ISO 12100:2010 'Safety of machinery - General principles of design - risk assessment and risk reduction', Table B1.

Mechanical hazards

Entanglement

Entanglement is the potential consequence of coming into contact with moving elements of machinery. In particular, the origin of entanglement is contact with rotating elements of machinery.

The mere fact that a machine part is revolving can consti-tute a very real hazard that can lead to entanglement. Loose clothing, jewellery, long hair, etc. increase the risk of entanglement. Examples of entanglement hazards include couplings, drill chucks/bits, flywheels, spindles, shafts (especially those with keys/bolts) and rotating tools like abrasive wheels.

Figure 9-33: Auger drill – entanglement.
Source: STIHL.

Friction and abrasion

Friction and abrasion are the potential consequences of coming into contact with moving elements of machinery. In particular, the origin of friction and abrasion is the skin coming into contact with moving elements that have a rough surface. Examples of moving elements with rough surfaces include abrasive surfaces of a sanding machine, grinding wheel or conveyor belt.

Figure 9-34: Friction and abrasion - abrasive wheel.
Source: RMS.

Cutting or severing

Cutting or severing are the consequences of coming into contact with sharp edges. The origin of cutting and severing is those sharp elements of a machine that are moving or stationery.

Saw blades, knives and even rough edges, especially when moving at high speed, can result in serious cuts and even amputation injuries. The moving elements of machinery can appear stationary due to the stroboscopic effect under certain lighting conditions, which can increase the likelihood of a worker contacting a sharp edge. Examples of cutting and severing hazards include saws blades, blades on slicing machines, abrasive cutting discs and chainsaw blades.

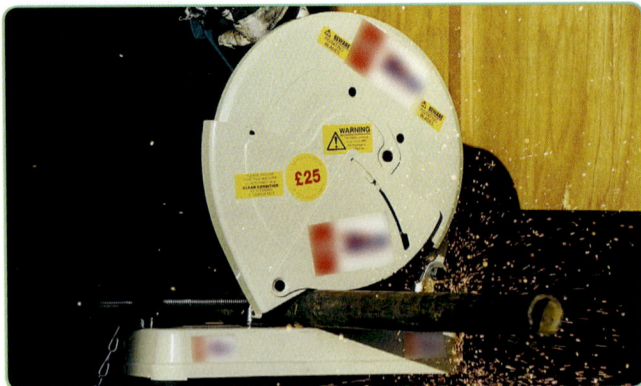

Figure 9-35: Cutting or severing - circular saw blade.
Source: Speedy Hire plc.

Shearing

Shearing is the consequence of coming in to contact with moving elements of machinery. The origin of shearing is elements of machines that move past each other or stationary objects. It differs from cutting in that the moving parts are not necessarily sharp.

Examples of shearing hazards include scissor lift platform riser and power press tools.

Figure 9-36: Shearing - scissor lift from platform riser when moving.
Source: RMS.

Stabbing and puncture

Examples of stabbing and puncture are the potential consequences of ejection of sharp items from a machine or contact with a sharp operating element of a machine.

Stabbing and puncture hazards include fixing materials deliberately ejected from a machine (such as nails fired from a nail gun), materials ejected during the machine's operating process (waste material ejected when cutting or abrading), part of a machine ejected as it wears out or breaks (part of a cutting blade) and stabbing or puncture by a drill bit or a needle of a sewing machine.

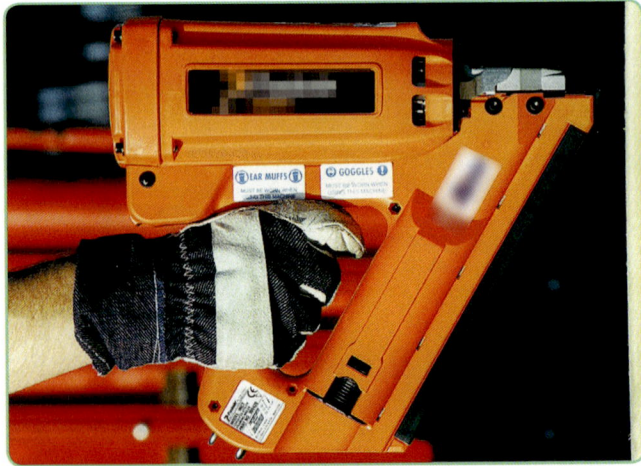

Figure 9-37: Stabbing and puncture - hand-held nail gun.
Source: Speedy Hire plc.

Figure 9-38: Stabbing and puncture - Ejected metal waste particles.
Source: RMS.

Figure 9-39: Stabbing and puncture - ejected broken disc saw blade.
Source: Water Active.

Impact

Impact is the potential consequence of a moving element of a machine or an object ejected from a machine directly striking a person. The origin of impact could be contact with those elements of a machine that strike the body, but do not penetrate or crush it, or parts of the machine, materials or tools ejected from a machine. The impact may cause the person or part of the person to be moved violently and/or local damage to the body at the site of the impact.

Examples of impact include a person being struck by the jib of a crane/excavator, a robot arm or materials being moved on a hoist or ejection of waste material from a machine or ejection of part of an abrasive wheel when it fractures.

Figure 9-40: Impact - construction work equipment.
Source: RMS.

Crushing

Crushing is the potential consequence of being caught between moving elements of a machine. The origin of crushing is part of the body being squeezed between two moving elements of a machine moving towards each other or a moving element of a machine moving towards a stationery object.

Examples of crushing hazards include the platform of a hoist closing together with the ground; moving parts of an adjustable hospital bed, tools of a power press closing together or the callipers of a spot-welding machine closing together.

Figure 9-41: Crushing - when lowering a hoist ramp to the ground.
Source: RMS.

Drawing-in or trapping

Drawing-in or trapping is the consequence of a person coming into contact with moving elements of a machine. The origin of drawing-in or trapping is contact with two rotating element that rotate towards each other or a single rotating element that rotates towards a gap in a fixed element. This movement draws in any part of the body presented to it and is therefore described as a drawing-in hazard.

Examples of drawing-in hazards are chain drives of a fork-lift truck lifting mast, V-belts on the drive from a motor to the drum of a cement mixer or the spindle of a drill, meshing gears in a lathe gearbox and conveyor belts as they pass over rollers.

Figure 9-42: Drawing-in - chain driven lifting mechanism.
Source: RMS.

Injection

Injection is the potential consequence of contact with high-pressure fluids projected from machinery.

Injection of compressed air or high-pressure fluids through the skin may lead to soft tissue injuries similar to crushing. Air entering the blood stream through the skin may be fatal. Examples of high-pressure fluid injection hazards include diesel injectors, spray painting, compressed air jets for blast cleaning the outside of a building and a high-pressure lance for cutting concrete or a hydraulic fluid leak from lifting systems.

Figure 9-43: Injection - hydraulic fluid leak from lifting system.
Source: RMS.

Summary - machinery hazards

Mechanical (example of equipment with hazards that can have these consequences)	Non-mechanical
Entanglement (Drilling machine)	Electricity
Friction and Abrasion (Grinding wheel)	Hot surfaces/fire
Cutting (Sharp edges of a circular saw)	Noise
Shear (Scissor lift mechanism)	Vibration
Stabbing and Puncture (Nail gun)	Access
Impact (Moving arm of an excavator)	Chemicals
Crushing (Platform of a hoist, ram of a forge hammer)	Radiation
Drawing-in (Conveyor belt)	Biological
Injection (High pressure hydraulic oil system)	Manual handling
	Extremes of temperature

Figure 9-44: Summary - mechanical and non-mechanical machinery hazards. Source: RMS.

Other (non mechanical) hazards

Machinery may also present other hazards. The nature of the hazard will determine the measures taken to protect people. The various sources of non-mechanical hazards include the following:

- Electricity – shock and burns.
- Hot surfaces/fire.
- Noise and vibration.
- Biological – viral and bacterial.
- High/low temperatures.
- Manual handling.
- Chemicals that are toxic, irritant, flammable, corrosive, explosive.
- Access - slips, trips and falls; obstructions and projections.
- Ionising and non-ionising radiation.

HAZARDS PRESENTED BY A RANGE OF EQUIPMENT

See also "Use of control methods related to a range of equipment" in 9.4 "Control measures for machinery".

Manufacturing/maintenance machinery

Bench-top grinder

Bench-top grinders are typically found in workshops and are suitable for indoor use. Bench-top grinders are used for sharpening tool bits (drills, chisels and blades), shaping steel and deburring cut steel components.

Mechanical hazards include friction and abrasion from contact with the moving abrasive wheel, entanglement in the rotating abrasive wheel, drawing-in between the rotating abrasive wheel and the tool rest and possible impact when parts of the wheel or workpiece break and are ejected. Other hazards are electricity from the drive motor power source, heat and noise created when the abrasive wheel grinds away the workpiece.

Figure 9-45: Abrasive wheel (grinder).
Source: RMS.

Pedestal drill

Mechanical hazards in using the equipment include failure to remove the chuck key, which will be ejected if the equipment is started and could impact the user or others. Handling the sharp edged drill bits is a cutting hazard. Failure to secure the guard covering drive pulleys after making adjustments would lead an exposed drawing-in hazard of the rotating drive belts and pulleys. When the equipment is operated the rotating chuck and drill bit present an entanglement and cutting hazard. The waste metal (swarf) cut by the drill can also present a cutting hazard. There is a risk of puncture of the worker's hand when the drill bit is lowered.

Figure 9-46: Pedestal drill.
Source: RMS.

Agricultural/horticultural machinery

Cylinder mower

Hazards are:

- Contact with revolving blades – entanglement and cutting.
- Contact with rotating drive shafts – entanglement.
- Contact with rotating drive belts or gears – drawing-in.
- Struck by stones ejected from the machine - impact.
- Collision with road traffic or pedestrians - impact.
- Overturn of machine when on inclines - crush.
- Noise and vibration.
- Manual handling.
- Combustion products from the petrol/diesel engine.
- Dust.
- Exposure to sun.
- Attacks by insects.
- Fire or explosion associated with flammable fuel.

Figure 9-47: Cylinder mower.
Source: Multihog.

Petrol-driven strimmer/brush-cutter

Hazards are:

- Contact with the moving parts of the strimmer – entanglement and cutting.
- Struck by parts of the strimmer material broken off in the operation - impact.
- Struck by fragments of the material being cut or stones - impact.
- Struck by passing traffic - impact.
- Noise and vibration.
- Fire or explosion associated with flammable fuel and hot petrol engine.
- Burns from contact with the hot casing of petrol engine.
- Exposure to combustion products from the petrol engine.
- Exposure to extreme weather conditions.
- Manual handling.
- Slips, trips and falls.

Chainsaw

The most significant mechanical hazard of using a chainsaw is the cutting or severing hazard of the moving chainsaw blade. There is a particular risk of cutting and severing injuries to the user's legs, feet, head and shoulders as the chainsaw cuts through the timber or slips to the side of what is being cut or if 'kickback' occurs.

Kickback occurs when a chain tooth at the upper quadrant of the guide bar tip cuts into wood without cutting through it. The chain cannot continue moving, and the bar is driven in an upward arc toward the operator, which present a serious risk of injury from the chainsaw blade. Kickback can result in major injuries or death.

Cutting or severing injuries can also result if the chain breaks during operation, for example due to poor maintenance or attempting to cut inappropriate materials.

In addition to the cutting and severing hazards, operation of a chainsaw presents vibration and noise hazards. The equipment is often powered by a petrol motor, which presents a fire hazard, hazard of hot petrol motor components/exhaust and exposure to combustion products from the exhaust.

Another hazard exists when heavy timber begins to fall or move when a cut is nearly complete – the chainsaw operator may be crushed by the timber or receive impact injuries.

In addition, use of the equipment may cause the operator to take up awkward postures and other non-mechanical hazards include work at height, slips, trips and falls, manual handling hazards, cuts and splinters from handling wood.

Retail machinery

Compactor

Compactors are used to make waste materials like cardboard smaller to make them easier to transport for recycling or disposal. They are commonly found in workplaces where high volumes of packaging are used, such as retail food distribution depots and supermarkets.

Mechanical hazards include crushing as the moving part of the compactor reduces the space underneath it and entanglement hazard of a moving motor shaft and drawing-in hazard of drive mechanisms, such as pulley wheels and belts. When the compactor is compacting the material hard, uncompressible items may be ejected from the material and could impact people nearby. Non-mechanical hazards include electricity, residue of hazardous substances that may be contained in material being compacted, possibly biological hazards created by the presence of vermin and manual handling issues.

Figure 9-48: Compactor.
Source: RMS.

Checkout conveyor system

Hazards include drawing-in and trapping between the rollers and the conveyor belt, also crushing between the loads being moved on the conveyor belt and stationery parts of the conveyor. The moving conveyor belt could also present a friction hazard. If the conveyor belt drive mechanism is exposed this will present drawing-in hazards and entanglement hazards of rotating motor drive shafts. Non-mechanical hazards include electrical sources, ergonomic hazards due to the operator's working position and manual handling hazards related to frequent repetitive actions.

Figure 9-49: Retail checkout conveyor.
Source: Retail Associates.

Construction site machinery

Cement mixer

A cement mixers is portable construction equipment used for mixing a variety of aggregates and cement. Hazards include:

- Drawing-in hazard of pulleys and drive belts.
- Entanglement hazard of the rotating drum and drive shaft.
- Impact hazard of materials ejected from the drum.
- Shovels and trowels entangled in the mixer blades.
- Risk of fire if fuel source is diesel or petrol.
- Exposure to fumes from petrol/diesel.
- Electricity if this the power source, particularly as water is used in the mixing process.
- Slips and trips from spilt materials.
- Manual handling when loading.
- Exposure to cement dusts and wet cement.

Figure 9-50: Cement mixer.
Source: RMS.

Bench-mounted circular saw

Cutting and severing is the main mechanical hazard, other mechanical hazards include entanglement in power drive shafts, drawing-in to drive mechanisms, such as a pulley and drive belt. Material being cut could be ejected and impact or puncture someone. Non-mechanical hazards include electricity, noise, sawdust, splinters and cuts from handling material and musculoskeletal disorders related to posture and materials handling.

Emerging technologies

Drones

A drone can refer to any unmanned mobile equipment, which would include aircraft, vessels or other vehicles. The drone is operated remotely from a ground control system (GCS), usually by a human. Perhaps the most common drones in use for work are unmanned aerial vehicles (UAVs). They are often fitted with cameras and used for site surveys, monitoring large areas of land and inspecting equipment that is difficult to gain access to, for example, wind turbines, bridges and large process structures.

One of the main hazards of drones relates to their propulsion system, UAVs used for work usually have multiple rotor blades to enable lift, movement and hovering. These blades rotate at high speed and present a cutting and severing hazard, with some risk of entanglement in the rotating blade mechanism. The other main hazard is that of impact. The drone could impact people when it is moving. The people particularly at risk are those that test the equipment, set it to work/retrieve it and those that the equipment travels near while it is operating. As the drone is operated remotely there is a risk of it moving unexpectedly towards people and striking them, perhaps because of operator error or because of extreme weather or faults in the equipment.

Loss of control of the drone could lead to a high speed impact with people or critical equipment, this could be due to failure of the GCS, failure of the drone guidance equipment or loss of the controlling signal between them.

The drone requires a power supply, which might be a battery or petrol motor. These present specific hazards – the battery is source of high energy electricity and the petrol motor has flammable liquid and an ignition source.

Driver-less vehicles

Driver-less vehicles operate autonomously, without human involvement. They are controlled by computer systems in the vehicle that make them respond to information from various sources, including sensors on the vehicle. In the workplace, at this stage in their development, they are likely to be carrying materials and equipment, rather than workers. Simple autonomous vehicles have been in use in the workplace for some time and have been used for such things as automated storage movement systems. The use of autonomous vehicles to transport goods and people on the roads is a growing area of activity.

The hazards associated with autonomous vehicles include:

- Limitations or defects in the vehicle's sensors could mean they fail to detect objects in their path, for example pedestrians, and could impact with them or crush them.

- As they are powered by electricity they are nearly silent and pose a threat of impact with pedestrians who might walk in front of the moving vehicle because they do not hear it, perhaps because they are wearing hearing protection.

- If there is a failure (error, misinterpretation, loss of signal) in the control system of the vehicle it could fail to turn a corner or stop and result in an impact with other vehicles, pedestrians, structures or other stationery objects.

- Errors in the programming may cause the vehicle to act in an unusual way, for example speed up or stop suddenly, which could cause the occupants to impact with the inside of the vehicle.

- A vehicle may also fail to recognise important information that it needs to take into account when determining safe movement, which could lead to the vehicle not take action to avoid a collision, for example, not recognising the status of traffic lights or the presence of a pedestrian crossings.

- A vehicle being maintained might unexpectedly begin to operate, causing the maintenance worker to be exposed to electrical energy sources, moving elements of the motor/drive mechanism and being struck or crushed by the vehicle.

- As with all battery operated vehicles, there are safety hazards associated with the task of battery removal/connection and battery charging.

- An autonomous vehicle may fail to appropriately respond to road surface conditions, for example changes caused by the weather (snow, ice, rain or mud) or the type of surface or contamination on the surface or objects on the surface. This could lead to lose of control of steering or unsuitable braking, with the likelihood of an uncontrolled impact.

 **REVIEW**

List three mechanical and three non-mechanical hazards associated with:

a) Cement mixer

b) Chainsaw

c) Compactor

List three different types of drawing-in incidents.

9.4 Control measures for machinery

THE PRINCIPLES OF OPERATION, ADVANTAGES AND LIMITATIONS OF CONTROL METHODS

Machinery should be designed and constructed to prevent risks. Where risks remain employers should prevent contact with hazards that could cause harm. The hazards of machinery may be mechanical or non-mechanical. In this context, the hazards are referred to as 'dangerous parts' of the machinery. There are various control methods available to prevent harm, including a range of guards and protective devices.

Guards

Guards provide a physical barrier between the hazard and people, preventing people placing themselves in the danger zone where they can contact the hazard. Guards should be used in preference to other control methods where possible.

Fixed guards

General operation of fixed guards

A fixed guard should be used whenever practicable. It should be designed so as to prevent access to the dangerous parts of the machine. A fixed guard may be designed to enable access to dangerous parts by authorised personnel for maintenance or inspection, but only when those parts have been isolated.

A fixed guard must be fitted such that it cannot be removed other than by the use of specialist tools, which are not available to users of the equipment. ISO 12100:2010 recommends that where fixed guards are held in place by fasteners, the guards should not remain in place without their fasteners.

In addition, the fasteners should remain attached to the guards when the guards are removed. A common example of a fixed guard is shown in *Figure 9-51*. Not all fixed guards are of solid construction. Some are made of mesh. The holes in a fixed guard made of mesh should be big enough to allow air circulation to cool the drive belt but positioned far enough away from the drive belt to prevent a finger from penetrating the mesh and sustaining an injury from the belt.

Distance (fixed) guards

Fixed guards do not always completely cover the danger point, the position where dangerous parts can be contacted, but place it out of normal reach. The larger the opening (to feed in material), the greater must be the distance from the opening to the danger point.

Figure 9-51: Total enclosure fixed guard.
Source: RMS.

Figure 9-52: Fixed guard – panel removed.
Source: RMS.

For example, a tree-shredding machine uses a fixed distance guard designed to prevent users reaching the dangerous parts of the machine when in use while allowing material to be fed in for shredding.

Advantages and limitations of fixed guards

The advantages of fixed guards include:

- Create a physical barrier.
- Require a tool to remove.
- No moving parts – therefore they require very little maintenance.

The limitations of fixed guards include:

- Do not disconnect power when not in place, therefore machine can still be operated without guard.
- May cause problems with visibility for inspection.
- If enclosed, may create problems with heat, which, in turn, can increase the risk of explosion.
- May not protect against non-mechanical hazards such as dust/fluids which may be ejected.

Figure 9-53: Fixed guard – mesh too big.
Source: RMS.

Interlocking guards

An interlocking guard is similar to a fixed guard, but it has a movable (usually hinged) part connected to the machine controls in such a way that if the movable part is in the open/lifted position the dangerous parts of the machine at the work point cannot operate. It can also be arranged that the action of closing the guard activates the machine and its working parts (to speed up work efficiently), for example, as is the case with the front panel of a photocopier. Interlocked guards are useful if users of the equipment need regular access to the danger area, for example, to load material in or take material out of a machine.

Everyday examples of interlocking guards are those found on domestic equipment such as dishwashers, microwave cookers and automatic washing machines. Interlocking guards should:

- As far as possible remain attached to the machinery when open.
- Be designed and constructed in such a way that they can only be operated or moved by means of an intentional action.
- Prevent the start of hazardous machinery functions until the guards are closed.
- Give a stop command whenever the guards are opened.
- Ensure that absence or failure of one of their components prevents starting, or stops the hazardous machinery functions.

Figure 9-54: Interlocking guard with viewing panel.
Source: RMS.

Where it is possible for an operator to reach the danger zone before the risk due to the hazardous machinery functions has ceased, the interlocking guard should:

- Prevent the start of hazardous machinery functions until the guard is closed and locked.
- Keep the guard closed and locked until the risk of injury from the hazardous machinery functions has ceased.

In the photograph (**see Figure 9-54**), the electrical interlock is positioned halfway down the right-hand side of the panel. The panel is made from transparent material to allow easy visual checks of the products that are manufactured by this equipment.

Advantages and limitations of interlocking guards

The advantages of interlocking guards include:

- Connected to power source, therefore machine cannot be operated with guard open.
- Allows frequent access.

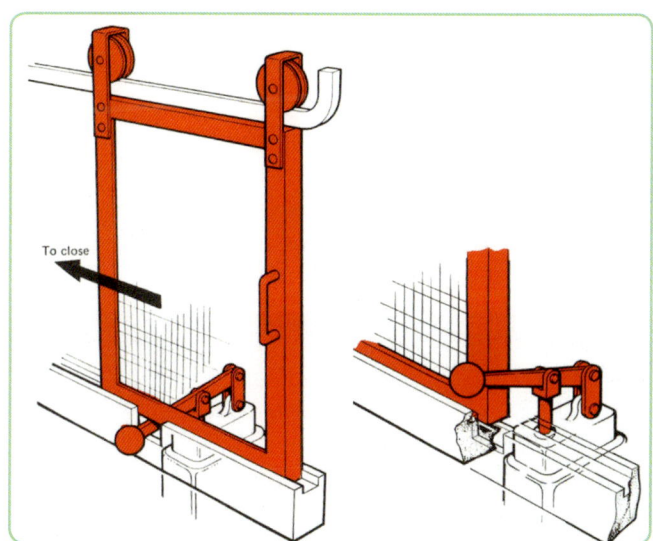

Figure 9-55: Open and closed interlock guard.
Source: BS EN ISO 12100.

The limitations of interlocking guards include:

- Have moving parts, therefore need regular maintenance.
- Can be overridden.
- If the interlock is in the form of a gate, a person can step inside and close the gate behind them (someone else could reactivate machine).
- Dangerous parts of machinery may not stop or cease to be a hazard immediately the guard is opened. It may be necessary to fit a delay timer or brake; for example, the drum on a spin drier does not stop instantly because of the momentum of the drum and contents and therefore a delay may be fitted to prevent the door being opened until the drum is stationary.

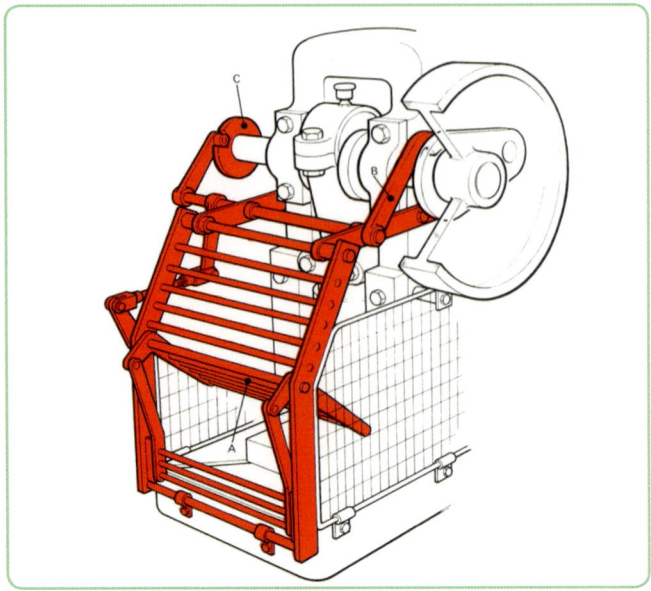

Figure 9-56: Power press – interlock guard.
Source: BS EN ISO 12100.

Self-adjusting guards

Self-adjusting guards are guards that are fixed to the moving parts of the machine, which close themselves over the dangerous parts as an integral part of the operation of the machine. They prevent accidental access by the operator but allow entry of the material to the machine in such a way that the material forms part of the guarding arrangement itself. For example, a hand-held circular saw.

Figure 9-57: Self-adjusting (fixed) guard.
Source: RMS.

Advantages and limitations of self-adjusting guards
The advantages of self-adjusting guards include:

- Close over the dangerous parts to provide protection without the operator needing to do anything.

The limitations of self-adjusting guards include:

- May obscure visibility when in use.

- Are vulnerable to damage in the operation of the equipment.

Adjustable guards

Adjustable guards are fixed guards that incorporate an adjustable element to accommodate a range of positions and size of dangerous parts of the machinery. When adjusted, it remains set in position for the duration of a particular operation.

Advantages and limitations of adjustable guards
The advantages of adjustable guards include:

- Can be adjusted by operator to provide protection.
- Protects operator where size or position of dangerous parts varies.

The limitations of adjustable guards include:

- Are reliant on the operator to adjust to the correct position.
- May obscure visibility when in use.

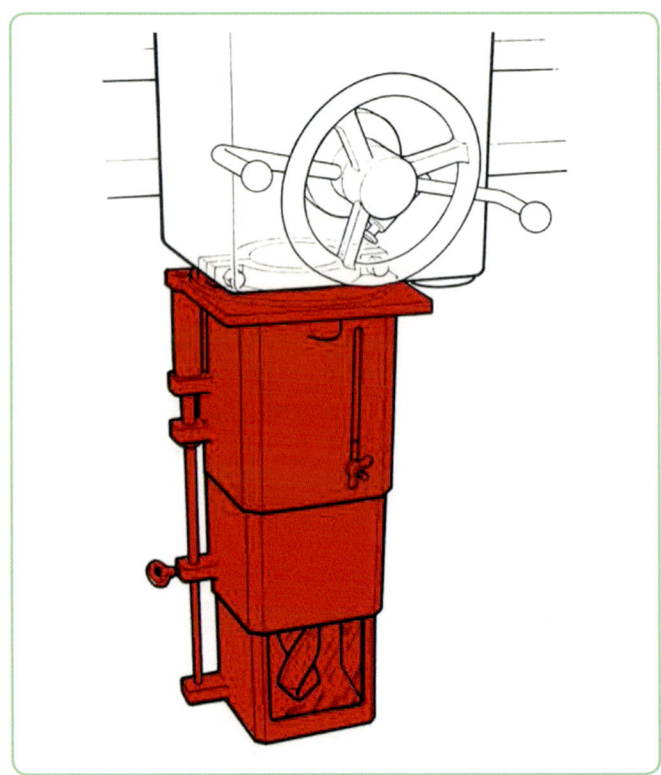

Figure 9-58: Adjustable (fixed) guard.
Source: BS EN ISO 12100.

Protective devices

Because protective devices do not provide a physical barrier to prevent the person's access to the danger zone and rely on a response mechanism they should only be used where a fixed guard is not practical. In addition, some of the protective devices are designed to keep the operator of the machine away from the hazard, do not protect other people.

Sensitive protective equipment

Sensitive protective equipment (sometimes called a trip device) has a sensing mechanism that detects that a person (or part of the person, for example, their hand) is in the danger zone (or approaching the danger zone) and at risk of harm from hazards. The sensing mechanism may be a light beam, pressure mat, rod, cable or other mechanism. The sensing mechanism causes the device to activate a further mechanism that either stops or reverses the machine, preventing further danger to the operator.

It is important to note that a sensitive protective device/ equipment is not classed as a guard. A guard is something that physically prevents access to the danger zone and the hazard, whereas a sensitive protective device/ equipment detects the person in the danger zone and responds to this.

It is essential that the sensing mechanism is positioned so that when a person approaches the danger zone there is enough time for the device/equipment to detect this and cease the action of the machine presenting a hazard before the person is harmed. This makes the location of the sensing mechanism, its sensitivity and the ability of the device/equipment to cease the hazard, critical.

Figure 9-59: Light curtain.
Source: UK HSE, HSG180.

Sensitive protective devices/equipment based on light beam/curtain sensing mechanisms have demonstrated that they can provide a high degree of protection, similar to that provided by an interlocking guard. However, an electro-mechanical telescopic mechanism of the type sometimes used with radial drills does not provide as good protection as an adjustable fixed guard.

Advantages and limitations of sensitive protective devices/equipment

The advantages of sensitive protective devices/equipment include:

- Can be used as an additional risk control measure.
- Can minimise the severity of injury

The limitations of sensitive protective devices/equipment include:

- Can be overridden.
- May not prevent harm from occurring.
- May cause production delays and increase stress in users with false 'trips'.

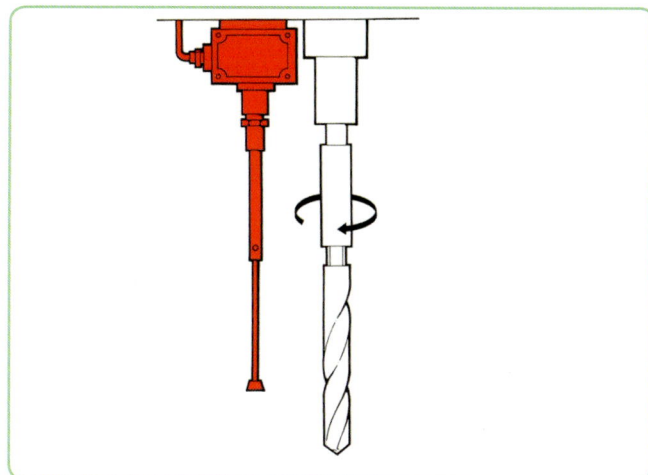

Figure 9-60: Trip device.
Source: BS EN ISO 12100.

Two-hand control device

Two-hand controls (2HC) provide a level of protection where other methods are not practicable, helping to ensure the operator's hands remain outside the danger area.

A two-hand control is a device that requires both hands to operate it. The controls must be operated simultaneously, and this helps to assure that both hands are kept away from the dangerous parts. 2HC devices protect only the operator and then only as long as a colleague does not activate one of the controls. Everyday examples of two-hand controls are a hedge trimmer and a garment press.

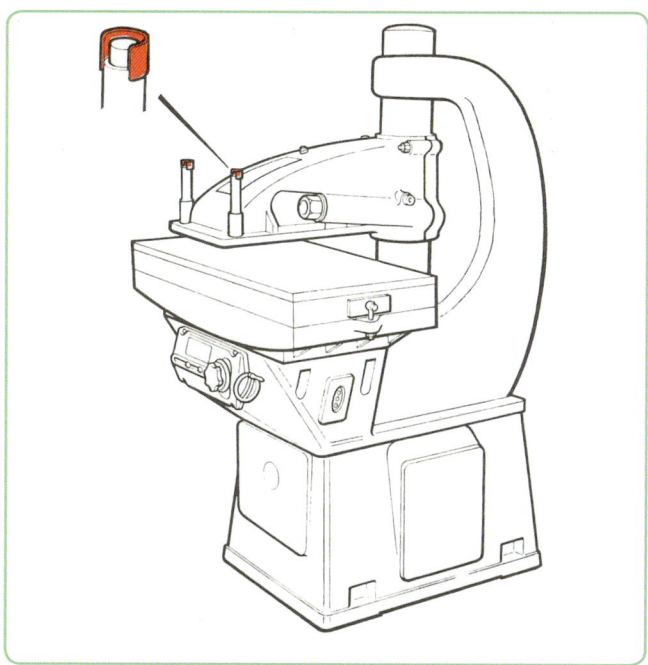

Figure 9-61: Two-hand control device.
Source: BS EN ISO 12100.

Advantages and limitations of two-hand control devices
The advantages of two-hand control devices include:

- Ensures both of the operator's hands are out of the danger area when the machine is operated.

The limitations of two-hand control devices include:

- Only protects the operator from harm.

- May limit speed of operation with delays if controls are not pressed at exactly same time.

Hold-to-run device

The principle of a hold-to-run device is that the operator has to 'hold' a button, stick or foot pedal to 'run' a piece of equipment. A common application of this device is on a domestic lawn mower or hedge trimmer (*see Figure 9-62*).

It is essential that the hold-to-run device, which when released stops the machine, is located far enough away from the danger area to prevent the operator getting access to the moving parts without releasing it.

Advantages and limitations of hold-to-run devices
The advantages of hold-to-run devices include:

- Ensures the operator is out of the danger area when the machine is operated.
- Provides distance between operator and hazard.

The limitations of hold-to-run devices include:

- Only protects the operator from harm.
- There may be residual movement of dangerous parts once the device has been released.

Figure 9-62: Hold-to-run device on hedge trimmer.
Source: RMS.

Emergency stop controls

Emergency stops are intended to effect a rapid response to potentially dangerous situations and they should not be used as functional stops during normal operation. Emergency stop controls should be easily reached and actuated. Common types are mushroom-headed buttons, bars, levers, kick plates, or pressure-sensitive cables. They can be used by the operator or other workers to stop the machine quickly, providing protection from further risk of harm. They can be useful in quickly bringing a machine to rest in situations where material has got stuck in the machine.

Advantages and limitations of emergency stop controls
The advantages of emergency stop controls include:

- Removes power immediately.
- Equipment has to be reset after use.
- Prevents accidental restarting of the equipment.

The limitations of emergency stop controls include:
- Does not prevent access to the danger area.
- May be incorrectly positioned.

Figure 9-63: Emergency stop controls.
Source: Oilybits Ltd.

Protection appliances

Protection appliances are used to hold or manipulate a work piece at a machine while keeping the operator's hands clear of the danger zone. They are commonly used in conjunction with manually-fed woodworking machines and certain other machines, such as band saws for cutting meat, where it is not possible to fully enclose the cutting tool in a guard. These appliances will normally be used in addition to guards.

Jigs, holders and push-sticks

When the methods of safeguarding explained previously are not practicable, protection appliances such as jigs, holders and push-sticks must be provided. These will help to keep the operator's hands at a safe distance from the danger zone. There is, however, no physical restraint to prevent the operator from placing their hands in danger.

Figure 9-64: Notched push-stick.
Source: Craftsmanspace.

Figure 9-65: Band saw push-stick.
Source: Toolmonger.

Advantages and limitations of protection appliances

The advantages of protection appliances include:

- Provide distance between operator and hazard.
- Inexpensive and easily replaced if damaged.
- May be shaped to suit work being carried out.

The limitations of protection appliances include:

- Harm may still occur from other non-mechanical hazards.
- Some designs can be awkward to use and may result in a lack of control.
- Adjustments have to be made, for example, when using different sizes of wood.
- Failure of the protection appliance (for example, breaking or kickback) may present an additional hazard to the operator.

Information, instruction, training and supervision

The employer should ensure that all persons who use work equipment, and any workers who supervise or manage the use of work equipment, have available to them adequate health and safety information. Where appropriate, written instructions and training pertaining to the use of the work equipment should be made available.

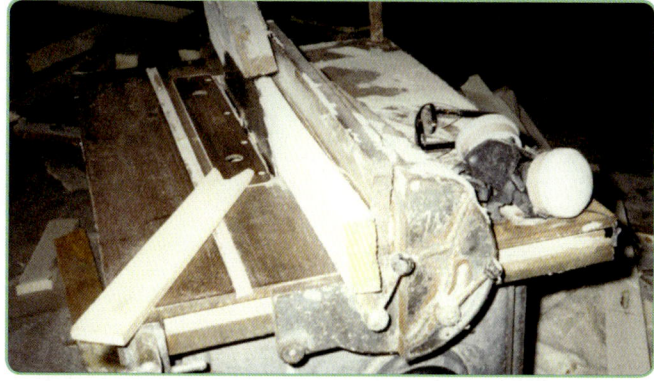

Figure 9-66: Fixed guard and push-stick used with a circular saw.
Source: Lincsafe.

This includes information and, where appropriate, written instructions that are comprehensible to those concerned with:

- The conditions in which and the methods by which the work equipment may be used.
- Foreseeable abnormal situations and the action to be taken if such a situation were to occur.
- Any conclusions to be drawn from experience in using the work equipment.

Similarly, employers should ensure that all persons who use work equipment and those who supervise or manage the use of work equipment have received training in any risks involved and precautions to be taken.

Employers should provide levels of supervision as necessary. Extra supervision will be required for new, inexperienced or less capable users of equipment.

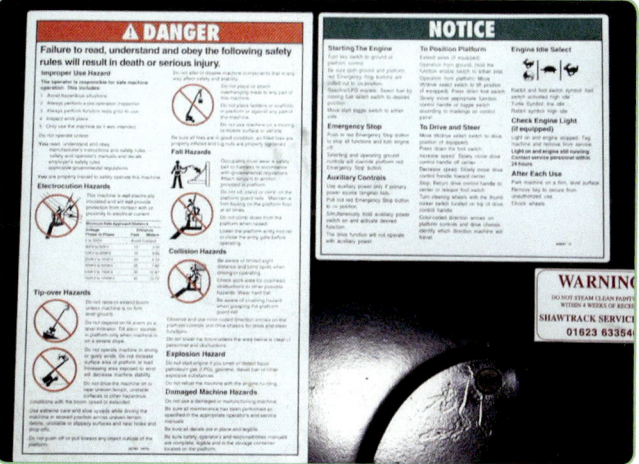

Figure 9-67: Information and instruction.
Source: RMS.

Advantages and limitations of information, training and supervision

The advantages of information, training and supervision include:

- Easy to reach a wide audience on a variety of subjects.
- Can be applied immediately and adapted to suit specific needs of the user.

The limitations of information, training and supervision include:

- Supervision may not prevent contact with the hazard.
- Relies on the person concerned to follow the instruction.
- May be misunderstood.
- Supervision is needed to a sufficient degree to ensure health and safety; a high degree of supervision may be required for some equipment.

"Regulation 11 -

Dangerous parts of machinery

(1) Every employer shall ensure that measures are taken in accordance with paragraph (2) which are effective:

(a) To prevent access to any dangerous part of machinery or to any rotating stock-bar; or

(b) To stop the movement of any dangerous part of machinery or rotating stock-bar before any part of a person enters a danger zone.

(2) The measures required by paragraph (1) shall consist of:

(a) The provision of fixed guards enclosing every dangerous part or rotating stock-bar where and to the extent that it is practicable to do so, but where or to the extent that it is not, then

(b) The provision of other guards or protection devices where and to the extent that it is practicable to do so, but where or to the extent that it is not, then

(c) The provision of jigs, holders, push-sticks or similar protection appliances used in conjunction with the machinery where and to the extent that it is practicable to do so, and

(d) The provision of information, instruction, training and supervision as is necessary.

(5) In this regulation - 'danger zone' means any zone in or around machinery in which a person is exposed to a risk to health or safety from contact with a dangerous part of machinery or a rotating stock-bar; 'stock-bar' means any part of a stock-bar which projects beyond the headstock of a lathe.

Regulation 11(2) gives the measures that an employer should take to fulfil the duty under -Regulation 11(1) a combination of measures may be necessary to satisfy Regulation 11. When deciding on the appropriate level of safeguarding, risk assessment criteria (likelihood of injury, potential severity of injury, numbers at risk) need to be considered both in relation to the normal operation of the machinery and other operations such as maintenance, repair, setting, tuning, adjustment etc."

Figure 9-68: Dangerous parts of machinery.
Source: UK, Regulation 11, Provision and Use of Work Equipment Regulations 1998.

Personal Protective Equipment

Personal protective equipment (PPE) is a last resort and should only be relied upon when other controls do not adequately control risks. The use of machinery presents a number of mechanical hazards and care has to be taken that PPE is not used in situations where it presents an increased risk of entanglement or drawing-in to machinery, such as might happen with loose overalls and gloves.

The main examples of use of PPE with regard to use of machinery are:

- Eye protection (safety spectacles/glasses, goggles and face shields) – protection for the eyes and the head from flying particles, welding glare, dust, fumes and splashes.

- Head protection (safety helmets or scalp protectors i.e. bump caps) – helmets provide protection from falling objects or the head striking fixed objects, whereas scalp protectors only provide protection from the head striking fixed objects.

- Protective clothing for the body (overalls) – protection from a wide range of hazards, including splashes of coolant used with machinery and specialist clothing for when using chainsaws. Close-fitting overalls designed to avoid entanglement may be required when working with rotating machinery.

- High-visibility clothing – enables those working in or around a dangerous area to be better seen by others working in that area. For example, where there is mobile equipment or where workers have to enter danger areas to remove or place materials.

- Gloves (chain-mail gloves and sleeves) – protection against cuts and abrasions when handling machined components, raw material or machinery cutters.

- Footwear (steel insoles and toecaps) – protection against sharp objects that might be stood on or objects dropped while being handled.

- Hearing protection for noisy machine operations.

Advantages and limitations of PPE
The advantages of PPE include:

- Easy to see if it is being worn.
- Provides protection against a variety of hazards.

The limitations of PPE include:

- Does not substitute for effective guarding of machinery.
- Only protects the user from residual hazards that guards do not deal with.
- May not give adequate protection.
- May pose additional hazards, for example, gloves becoming entangled.

USE OF CONTROL METHODS TO A RANGE OF EQUIPMENT

See also "Hazard of a range of equipment" in section 9.3 - Machinery hazards.

Manufacturing/Maintenance machinery

Bench grinder

Control measures include an adjustable fixed guard that is fitted over as much of the grinding wheel as is practicable, a work-piece rest, trained and competent users, high-impact-resistant goggles, respiratory protective equipment (depending on task) and personal hearing protection.

Pedestal drill

Appropriate precautions for pre-use are:

- Drive pulley entanglement prevented by isolation from power when adjusting speeds.
- Check that chuck key is removed before use.
- Ensure the trip device is functional.

Appropriate precautions during use are:

- Interlocking guard to cover the drive pulley and belts.
- Adjustable (fixed) guard provided round the chuck and drill bit, face shield/goggles for protection from fly-ing metal pieces from the cutting process (swarf).
- Prevent entanglement – close-fitting clothing, control of long hair and removal of jewellery.
- Puncture prevented by jigs or clamping devices.

Chainsaw

The equipment should have a hold-to-run device and sensitive protection equipment should be fitted in front of the operator's hand that will trip and stop operation if the chainsaw 'kicks-back'. The risks associated with chainsaw use mean that protective clothing, for example, forestry boots, helmet with mesh visor, chainsaw gloves, body/leg protective clothing and hearing protectors should be worn while operating them. In order to limit the effects of vibration, careful selection of equipment with low vibration output is important. Training of users is a critical element of the control strategy for use of this equipment.

Agricultural/horticultural machinery

Cylinder mower

Appropriate precautions are:

- Provision of fixed guards around any drive mechanisms, for example, chains or shafts.
- Fitting of fixed guards to enclose as much of the blades as is practicable.
- Provision of PPE (for example, ear defenders, eye protection and high-visibility clothing).
- Coning-off areas in close proximity to moving traffic.
- Training users in the operation of the machine on sloping ground and on refuelling procedures.

Petrol-driven strimmer/brush-cutter

Appropriate precautions are:

- Provision of fixed guards around any drive mechanisms and motor/engine.
- Fixed guards to enclose as much of the strimmer material or blade as is practicable.
- PPE (eye and hearing protection).
- Appropriate storage of petrol.
- Regular maintenance by authorised people, including changing of cutters.

Figure 9-69: Brush-cutter.
Source: Kawasaki-engines.

Retail machinery

Compactor

Appropriate precautions are the provision of fixed guards around the sides and back of the compactor, where access is not normally required. Interlocked guards should be fitted to the opening where the material is placed in the compactor and to access areas for removal. Good housekeeping around the compactor is essential to avoid slips, trips and falls.

Checkout conveyer system

Appropriate precautions are fixed guards around drive mechanisms, interlocked guards on panels that may need to be opened to remove material that has got stuck in the conveyor and sensitive protection equipment (trip devices) that identify items approaching the drawing-in point. The sides of the conveyor equipment should be sufficiently close to the belt to prevent abrasion contact and built up to prevent the fall of items.

Construction machinery

Cement mixer

Fixed guards must be provided around drive mechanisms. Motor covers should be closed when in use. Users of the equipment should be warned of the dangers of shovels and trowels becoming caught in the mixer blades when charging. Electrical hazards are reduced by the use of low-voltage power (for example, 110 volts compared with 230 volts mains supply) or a residual current circuit breaker, with suitable heavy-duty protected cable. If the equipment is petrol or diesel driven, hot parts, for example, the exhaust system, should be allowed to cool before refuelling. A no-smoking policy should also be in place. Other controls include manual handling training, good housekeeping of spilt materials, and avoidance of exposure to dust and fumes.

Bench-mounted circular saw

Appropriate precautions are fixed, adjustable and/or self-adjusting guards, jigs, holders, push-sticks and PPE (eye, respiratory and hearing protection).

Figure 9-70: Bench cross-cut circular saw.
Source: RMS.

Emerging technologies

Drones

The safe use of drones for the purposes of work includes the following control measures:

- Establish a clear plan for the flight – considering such things as timing, take off/landing, weather, areas of work activity and possible effects on flight, areas of potential signal interruption, task to be achieved by the drone, restrictions on the flight to protect people/property and limitations due to hazards (flammable areas and obstructions).

- Select a size of drone that suits the task to be done, smaller drones may suit congested areas.

- Obtain a permit to work/fly where necessary.

- Ensure the operator is competent and proficient in relation to the task.

- Ensure the operator has clear, direct, unaided visual line of site (VLOS), use an observer to assist if needed.

- Ensure the operator and any observer has good eyesight.

- Ensure the drone can be clearly seen and controlled safely during night flying operations, additional lighting may be required to enable this.

- Launch/recover the drone from the ground where ever possible - avoid hand launch/recovery and restrict this method to competent people.

- Fly no higher than 120 metres (400 feet) and remain below any surrounding obstacles when possible.

- Remain clear of and do not interfere with manned aircraft operations.

- Keep the drone at least 50 metres (150 feet) away from people and property, including vehicles, keep 150 metres (500ft) away from crowds and built up areas and do not fly over them - unless there is a clear work purpose and risk assessments have been made.

- Do not fly in adverse weather conditions, such as in high winds or reduced visibility.

- Do not fly under the influence of alcohol or drugs.

- Establish an emergency plan for flight propulsion failure.

Figure 9-71: Drone.
Source: Cyberhawk

Driver-less vehicles

When driver-less vehicles are used in the workplace pedestrians should be excluded from the area, wherever possible. These restricted areas should be clearly identified and suitable warnings provided. Consider interlocking access to the area so that vehicles cannot operate when pedestrians enter the area.

Where pedestrians and driver-less vehicles work in the same area they should be segregated as much as possible, for example, by separate aisles or roadways for the vehicles, separated by physical barriers. Where this is not possible clearly marked, separate routes should be provided and the vehicles must be equipped with suitable sensors that enable the detection of pedestrians in good time to enable the vehicle to stop before contact is made. Because many driver-less vehicles are almost silent in operation, consider adding visual and audible warnings devices to the vehicles to indicate that they are operating. Also consider reducing the speed of operation of the driver-less vehicles and ensure there is good lighting in the area to provide pedestrians with time to react.

Figure 9-72: Driver-less work vehicle.
Source: PSA Group

Driver-less vehicles for use in public areas and on public roads require sophisticated sensor and control systems that enable steering and braking in order to ensure they operate safely and avoid collisions with pedestrians and other vehicles.

Further general control measures for the safe operation of driver-less vehicles include:

- Only authorised persons must be permitted to control or maintain a vehicle and its operating system.

- Vehicle safety devices must not be manually overridden when the vehicle is operating in automatic or semi-automatic modes.

- The surfaces over which the vehicle operates must be maintained to ensure that the traction required for travel, steering, and braking performance can be met under the environmental conditions which may be expected on that surface.

- The environment that the vehicle operates in needs to be part of the vehicle design criteria - including temperature, humidity, ambient weather (for example, work on an exposed dock), air quality (for example, aggressive dusts or flammable atmospheres).

- Changes to the driving surface and environment must be evaluated to verify there is no adverse effect on the vehicle's safety systems.

- The vehicle manufacturer's instructions on battery removal/fitting and charging must be followed.

BASIC REQUIREMENTS FOR GUARDS AND SAFETY DEVICES

Compatibility with process

Compatibility with the material being processed – this is particularly important in the food-processing industry, where the guard material should not constitute a source of contamination of the product. It is also important that any safety device maintains its physical and mechanical properties after coming into contact with potential contaminants such as cutting fluids used in machining operations or cleaning and sterilising agents used in food-processing machinery.

In selecting an appropriate safeguard for a particular type of machinery or danger area, it should be borne in mind that a fixed guard is simple and should be used where access to the danger area is not required during operation of the machinery or for cleaning, setting or other activities. As the need for access arises and increases in frequency, the importance of safety procedures for removal of a fixed guard increases, until the frequency is such that an interlocking system should be used.

Adequate strength

Guard mounting should be compatible with the strength and duty of the guard. In selecting the material to be used for the construction of a guard, consideration should be given to the following:

- Its ability to withstand the force of ejection of parts of the machinery or material being processed, where this is a foreseeable danger.

- Its ability to provide protection against hazards identified. In many cases, the guard may fulfil a combination of functions such as prevention of access and containment of hazards. This may apply where the hazards include ejected particles, liquids, dust, fumes, radiation, noise, etc., and one or more of these considerations may govern the selection of guard materials.

Maintained

All guards must be maintained in effective order to perform their function. This will require a planned approach to checks on guards and work, such as checking the security of fixed guards.

Allow for maintenance without removal

The weight and size of the guard are factors to be considered in relation to the need to remove and replace it for routine maintenance. If a guard has to be removed in order to carry out maintenance work, it increases the risks for maintenance workers and also increases the chances that the guard will not be replaced. This is especially the case if a piece of equipment has a history of regular breakdown.

Not increase risk or restrict view

Any guard selected should not itself present a hazard such as trapping or shear points, rough or sharp edges or other hazards likely to cause injury.

Power-operated guards should be designed and constructed so that a hazard is not created by restricting the view of the process or machine by the user.

Not easily by-passed

Guards or their components must not be easily bypassed, in particular by users of the equipment. There is always a temptation to do so when under production or similar pressures. When positioning such protective devices as interlock switches it is best to locate them away from the operator and preferably within the guard.

"Regulation 11 paragraph 3 - Effective guards and devices

All guards and protection devices provided under sub-paragraphs (a) or (b) of paragraph (2) shall:

(a) Be suitable for the purpose for which they are provided.

(b) Be of good construction, sound material and adequate strength.

(c) Be maintained in an efficient state, in efficient working order and in good repair.

(d) Not give rise to any increased risk to health or safety.

(e) Not be easily bypassed or disabled.

(f) Be situated at sufficient distance from the danger zone.

(g) Not unduly restrict the view of the operating cycle of the machinery, where such a view is necessary.

(h) Be so constructed or adapted that they allow operations necessary to fit or replace parts and for maintenance work, restricting access so that it is allowed only to the area where the work is to be carried out and, if possible, without having to dismantle the guard or protection device."

Figure 9-73: Effective guards and devices.
Source: UK, Regulation 11, Provision and Use of Work Equipment Regulations 1998.

 **REVIEW**

What hazards are associated with the use of a cement mixer?

What control measures can be used to reduce the risks of injury to operators of a cement mixer?

What are the basic requirements for guarding systems?

Sources of reference

Reference information provided, in particular web links, was correct at time of publication, but may have changed.

7 ways driverless cars could fail (Forbes), https://www.forbes.com/sites/chunkamui/2016/04/08/7-driverless-doomsdays/#281914d616cf

Ambient factors in the workplace, ILO CoP, ISBN 92-2-11628-XISBN 978-92-2-111628-8 http://www.ilo.org/safework/info/standards-and-instruments/WCMS_107729/lang--en/index.htm

Buying new machinery, INDG271, HSE Books, http://www.hse.gov.uk/pubns/indg271.pdf

C155 – Occupational Safety and Health Convention, 1981. https://www.ilo.org/dyn/normlex/en/f?p=normlexpub:12100:0::no::p12100_instrument_id:312300

Directive 2006/42/EC - machinery directive https://eur-lex.europa.eu/LexUriServ/LexUriServ.do?uri=OJ:L:2006:157:0024:0086:EN:PDF

Directive 2009/104/EC - use of work equipment https://osha.europa.eu/en/legislation/directives/workplaces-equipment-signs-personal-protective-equipment/osh-directives/3

Drone safety risk: an assessment, Civil Aviation Authority, https://publicapps.caa.co.uk/docs/33/CAP1627_Jan2018.pdf

Drones offer risks, underwriting challenges, Risk & Insurance magazine, https://riskandinsurance.com/drones-offer-risks-underwriting-challenges/

Graphical symbols – Registered safety signs, ISO 7010:2011, International Organisation for Standardisation (ISO)

ISO 12100:2010 – Safety of machinery – general principles for design, risk assessment and risk reduction, International Organization for Standardization

Personal Protective Equipment at Work -Regulations 1992 (as amended). Guidance on Regulations, (L25),

HSE Books, ISBN: 978-0-7176-6139-8 http://www.hse.gov.uk/pubns/indg174.pdf

Personal Protective Equipment at Work (second edition), Personal Protective Equipment at Work Regulations 1992 (as amended), Guidance on Regulations, L25, HSE Books, ISBN: 978-0-7176-6597-6, http://www.hse.gov.uk/pubns/priced/l25.pdf

Rise of the Drones, Managing the unique risks associated with unmanned aircraft systems, Allianz, https://www.agcs.allianz.com/content/dam/onemarketing/agcs/agcs/reports/AGCS-Riseofthedrones-report.pdf

Safe Use of Work Equipment, ACoP and guidance (part III in particular), L22, third edition 2008, HSE Books, ISBN: 978-0-717662-95-1 http://www.hse.gov.uk/pubns/books/l22.htm

Safety and health in the use of machinery, ILO CoP, ISBN: 978-92-2-127726-Safety and health in the use of ma-chinery, ILO CoP http://www.ilo.org/wcmsp5/groups/public/---ed_protect/---protrav/---safework/documents/normativeinstrument/wcms_164653.pdf

Safety of machinery. General principles for design. Risk assessment and risk reduction. BS EN ISO 12100:2010,

ISBN: 978-0-580742-62-0, https://www.ilo.org/wcmsp5/groups/public/---ed_protect/---protrav/---safework/documents/normativeinstrument/wcms_164653.pdf

Safety signs and signals, Guidance on Regulations, L64 HSE Books, ISBN: 978-0-7176-6598-3, http://www.hse.gov.uk/pubns/priced/l64.pdf

Supplying new machinery, INDG270, HSE BooksSupplying new machinery, INDG270(rev1), HSE, http://www.hse.gov.uk/pubns/indg270.pdfhttp://www.hse.gov.uk/pubns/indg270.pdf

The health and safety toolbox, How to control risks at work, HSG268, HSE Books, ISBN: 978-0-7176-6587-7 , http://www.hse.gov.uk/pUbns/priced/hsg268.pdf

The Health and Safety (Safety Signs and Signals) Regulations 1996, Guidance on regulations, second edition 2009, L64, HSE Books, ISBN: 978-0-717663-59-0, http://www.hse.gov.uk/pubns/priced/l64.pdf

The Health and Safety (Safety Signs and Signals) Regulations 1996, UK, www.legislation.gov.uk (please enter the name of the Regulation)

The Personal Protective Equipment at Work - Regulations 1992, UK, www.legislation.gov.uk (please enter the name of the Regulation)

The Provision and Use of Work Equipment Regulations 1998, UK, www.legislation.gov.uk (please enter the name of the Regulation)

The Supply of Machinery (Safety) Regulations 2008, UK, www.legislation.gov.uk (please enter the name of the Regulation)

Web links to these references are provided on the RMS Publishing website for ease of use - www.rmspublishing.co.uk

STUDY QUESTIONS

1) What is the purpose of the CE mark on workplace equipment used in the European Union? (8)

2) What are the duties of workers when they discover a damaged piece of work equipment? (8)

3) What are the benefits of introducing a scheme of planned preventative maintenance for equipment in regular use? (8)

4) (a) What is the meaning of the term lock-out and tag-out (LOTO)? (2)

   (b) How does the use of LOTO systems reduce the risk to maintenance workers? (6)

5) (a) What is the purpose of emergency stops fitted to machinery? (2)

   (b) What are the design and positioning requirements for emergency stops? (6)

6) (a) Why is stability of work equipment important when in use? (2)

   (b) What factors may affect stability of work equipment? (6)

For guidance on how to answer NEBOSH questions please refer to the 'study question answer guidance' section located at the back of this guide.

Element 10

Fire

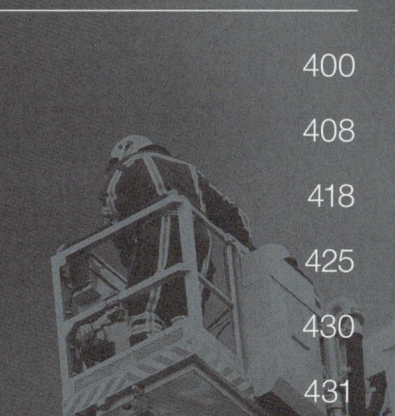

FIRE PRINCIPLES

The fire triangle

A simple approach depicts fire as having three essential components: fuel, oxygen and ignition source (heat). When the three separate parts are brought together, a fire occurs. This is often depicted by the 'fire triangle' as shown in *Figure 10-1*. This traditional approach is useful when considering the 'ingredients' needed to make a fire, and the absence of any one of these components will prevent a fire from starting. Combustion is defined as a chemical reaction during which heat energy and light energy are emitted. If the three components, fuel, oxygen and ignition come together in the right proportions, then the chemical reaction of combustion takes place.

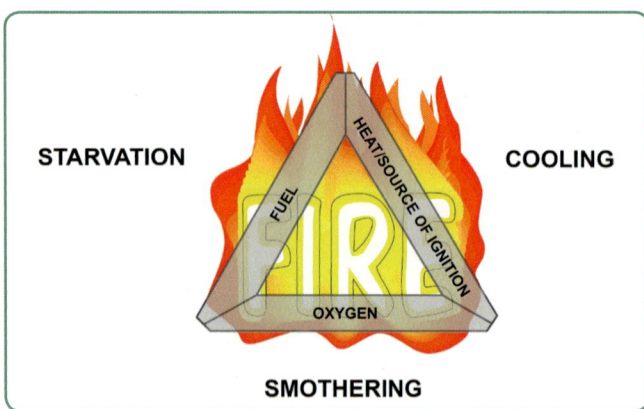

Figure 10-1: Fire triangle.
Source: RMS/Corel.

For example, when a match is struck, friction (the ignition) heats the head to a temperature at which the chemicals react and generate more heat than can escape into the air, therefore burning with a flame. The molecules in the matchstick break down and give off vapour, causing more heat to be released, propagating further chemical reactions.

Once a fire has started, a self-sustaining chain reaction begins at the surface of the fuel (solid or liquid), which turns into a vapour, and it is this that burns in the combustion process. A similar process occurs with the molecules of a gas, but because the collective surface area of the molecules is large, the combustion process is very fast. A fire may be defined as a chemical chain reaction in which fuel is reduced by reaction with an oxidiser to produce light, heat and combustion products.

For the combustion process to be maintained, all three of the component parts shown in *Figure 10-1* must remain present. If one or more of these parts of the fire is removed, the fire will be extinguished.

This can be done by:

1) **Cooling** the fire to remove the heat, for example, by use of a water or foam extinguisher.

2) **Starving** the fire of fuel, for example, isolation of a gas supply.

3) **Smothering** the fire by limiting its oxygen supply, for example, by use of a carbon dioxide or foam extinguisher.

4) **Chemical interference** of the flame reactions, for example, wet chemical extinguishing agents.

Sources of ignition

Ignition occurs when a heat source, for example, a spark, contains sufficient heat energy to cause combustion of one or more molecules of a flammable vapour or substance. Any source of heat is a possible ignition source and can be found in many forms.

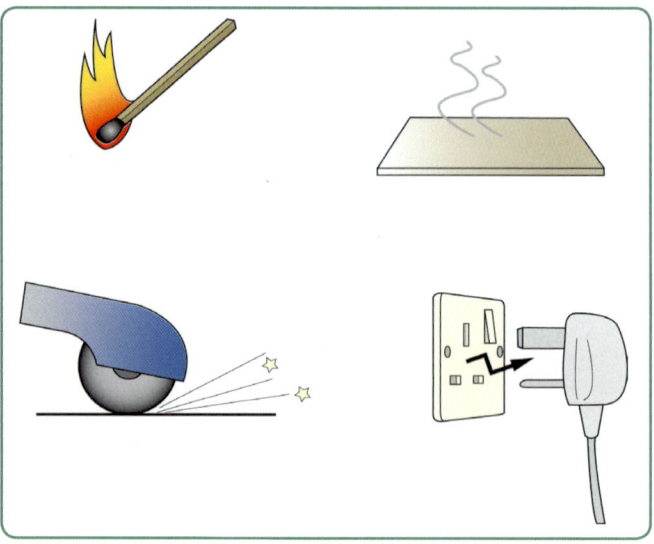

Figure 10-2: Sources of ignition.
Source: RMS/UK, Government fire guides.

Examples could be:

- Discarded smokers' materials (less of a problem in countries where smoking is prohibited in the workplace).
- Naked flames.
- Fixed or portable heaters – particularly those that use liquid fuel.
- Hot processes, for example, welding, cutting and grinding.
- Lighting.
- Deliberate ignition. Many fires in the workplace are started deliberately. The workplace may be targeted by disgruntled workers or unhappy customers or by an act of vandalism.
- Unsafe storage of flammable liquids allowing self-heating as a result of an exothermic reaction which can cause spontaneous combustion, for example, acetylene and chlorine.

Figure 10-3: Illicit smoking.
Source: FSTC Ltd.

- Cooking.
- Electrical equipment – overloading or damage to electrical circuits.
- Machinery – sparks, overheating of drive belts *(see Figure 10-5)* due to over-tightening, friction due to lack of lubrication, blunt cutting tools generating heat.
- Static electricity – most commonly from lightning strikes, although sparks from static charges are very dangerous in flammable and explosive atmospheres.

Figure 10-4: Hot lighting bulbs.
Source: UK, HSE, HSG168.

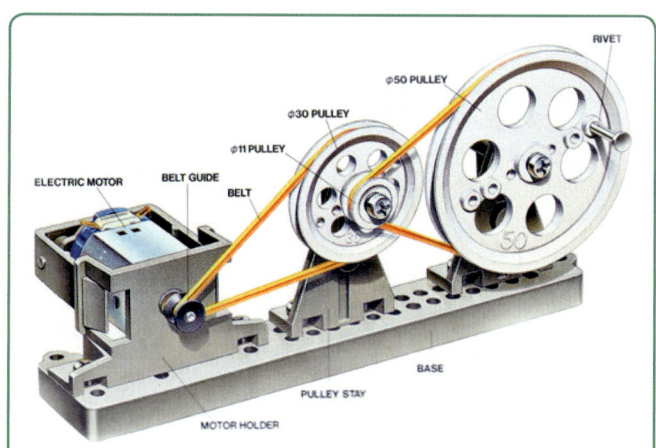

Figure 10-5: Drive belt and pulley.
Source: ServoCity.

Sources of fuel

Anything that burns is fuel for a fire. Common items that will burn in a typical workplace include:

- Flammable liquids, for example, petrol storage areas.
- Flammable gases, for example, liquefied petroleum gas (LPG) heater cylinders.
- Flammable chemicals, for example, paints and solvents.
- Wood, for example, furniture, dust from manufacturing processes.
- Plastics, rubber and foam, for example, furniture.
- Paper and card, for example, stationery cupboards, waste paper, and waste card produced in carton manufacture.
- Insulating materials, for example, walls and partition components.
- Waste materials, chemicals, waste paper, and general waste.
- Airborne metal dust, for example, aluminium metal powder.

The structure of the building should also be considered. The building itself could be made from wood or other flammable material, or it may contain flammable materials as part of the decoration, for example, wallpaper.

The words 'flammable' and 'inflammable' mean the same, i.e. that ignition takes place at normal ambient temperatures. The word 'combustible' means likely to catch fire. The terms 'highly flammable' and 'extremely flammable' indicate substances that can be ignited at low ambient temperatures. Gasoline (petrol) exposed to a spark will ignite at temperatures starting from -43°C (-45°F) and is therefore described as 'extremely flammable'.

Figure 10-6: Combustible materials in office.
Source: FSTC Ltd.

Figure 10-7: Combustible materials - waste.
Source: Shutterstock.

Sources of oxygen

The main source of oxygen for a fire is in the air around us. In an enclosed building this is provided by the ventilation system in use. This generally falls into one of two categories: natural airflow through doors, windows and other openings or mechanical air-conditioning systems and air-handling systems. In many buildings there will be a combination of systems capable of introducing/extracting air to and from the building. Leaks from cylinders or piped oxygen supplies, combined with poor ventilation, can lead to an oxygen-enriched atmosphere. Materials that ordinarily will burn only slowly will burn very vigorously in an oxygen-enriched atmosphere. The following precautions should also be considered:

- Never use oxygen instead of compressed air.
- Never use oxygen to improve air quality in a working area or confined space.
- Never use grease or oil on equipment containing oxygen.

Oxidising materials

Combustion involves the oxidation of a combustible (burnable) material. When a combustible material burns, a chemical reaction occurs in which the combustible material (fuel) combines with oxygen and gives off heat, gases, and generally light. The usual source of oxygen for combustion is air. However, oxidising materials can supply combustible materials with a source of oxygen and support a fire even when air is not present. Although most oxidising materials do not burn themselves, they can produce very flammable or explosive mixtures when combined with combustible materials.

The involvement of an oxidising material can speed up the development of a fire and make combustion more intense. Oxidising materials can cause materials that do not readily burn in air to burn rapidly in their presence. In addition, they can cause combustible materials to burn at ambient temperature without the presence of an obvious source of ignition, such as a spark or flame.

What happens when an oxidising material comes into contact with a combustible material mainly depends on the chemical stability of the oxidising material; the less stable an oxidising material is, the greater the chance that it will react in a dangerous way. Oxidising materials come in the form of gases, liquids and solids. Examples of gases include oxygen and ozone, liquids include nitric acid and chromic acid, and solids include chromate and potassium permanganate.

The United Nations (UN) has established a non-legally binding international agreement called 'Globally Harmonized System of Classification and Labelling of Chemicals (GHS)'. It has been widely accepted globally and is being established in national legislation of the countries that are adopting it. Within GHS is a classification of materials principally based on the physical hazards they present, including oxidising materials. Materials that come under the class called oxidising gases fall under a single hazard category, whilst those classed as oxidising liquids and oxidising solids have three classifications.

Oxidising gases - Annex 1 of GHS

Category 1	
Signal word: Danger	
Hazard statement May cause or intensify fire; oxidiser	

Figure 10-8: Oxidising gases.
Source: RMS (Adapted from annex 1 of GHS).

Oxidising liquids and solids - Annex 1 of GHS

Category 1	
Signal word: Danger	
Hazard statement May cause fire or explosion; strong oxidiser	
Category 2	
Signal word: Danger	
Hazard statement May intensify fire; oxidiser	
Category 3	
Signal word: Warning	
Hazard statement May cause or intensify fire; oxidiser	

Figure 10-9: Oxidising liquids and solids.
Source: RMS (Adapted from annex 1 of GHS).

CLASSIFICATION OF FIRES

The classification of fires essentially relates to the material that is being combusted and seeks to group similar materials into the same classification. There is no internationally agreed classification of fires. A common approach is set out in **Figure 10-10**, which reflects the approach taken in a number of countries, including the UK, and is based on the European Standard 'Classification of fires' – EN 2:1992.

CLASS A
Combustible solids, for example, wood, paper, textiles or plastics (usually material of an organic nature).

CLASS B
Flammable liquids or liquefiable solids – petrol, oil, paint, fat or wax.

CLASS C
Combustible gases, for example, natural gas or liquefied gases, for example, butane or propane.

CLASS D
Combustible metals such as magnesium and lithium. Such specialised fires require a specialised metal powder fire extinguisher to deal with them, and this will be required in scientific labs or where manufacturing processes involve the risk of metal fires. For example, aluminium dust or swarf can catch fire, so any process involving cutting, drilling or milling aluminium poses a potential risk.

CLASS F
Cooking oils and fats usually found in commercial kitchens such as restaurants and fast-food outlets.

ELECTRICAL FIRES
Although this is not a class of fire in the common classification, electricity is often a source of ignition and the presence of electricity will increase the risk of electric shock where water is used as the extinguishing medium.

Figure 10-10: Common classification of fires.
Source: RMS.

In Australia, the Middle East and parts of Asia, fires involving electrical equipment are given the classification E. In the United States of America (USA) a classification of fires based on five classes has been established, which is similar to the common classification. The notable difference is that Class C relates to fires involving live electrical equipment. In the USA there is no separate class for gases, which is included under Class B, and the equivalent to Class F is Class K.

CLASS	FIRES INVOLVING
A	ORDINARY COMBUSTIBLES
B	FLAMMABLE LIQUIDS
C	ENERGISED ELECTRICAL EQUIPMENT
D	COMBUSTIBLE METALS
K	COOKING OILS

Figure 10-11: United States fire classifications.
Source: USA, OSHA.

A basic understanding of the classes of fire needs to be achieved because many fire extinguishers state the classes of fire on which they may be used.

See later in this element 'International markings for portable fire extinguishers'.

PRINCIPLES OF HEAT TRANSMISSION

There are four main **methods** by which heat may be transmitted: conduction, convection, radiation and direct burning.

1) Convection

The movement of hotter gases up through the air (hot air rises and cooler or cold air falls). Convection can quickly move hot gases to another part of a building where they raise the temperature of combustible materials to a point that combustion takes place, for example, hot gases rising up a staircase through an open door.

Control measure: protection of openings by fire doors and the creation of fire-resistant compartments in buildings.

2) Conduction

Follows the principle that heat can be transmitted through solid objects. Some materials, such as metal, can absorb heat readily and transmit it to other rooms by conduction, where it can set fire to combustible items that are in contact with the heated material, for example, metal beams, ducting or pipes transmitting heat through a solid wall.

Control measure: insulation of the surface of a beam or pipe with heat-resistant materials.

3) Radiation

Transfer of heat as invisible infrared waves through the air in a similar way to light (the air or gas is not heated but solids and liquids in contact with the heat are). Any material close to a fire will absorb the heat in the form of energy waves, until the material starts to smoulder and then burn. For example, a fire in a waste container stored too near to a building may provide enough radiant heat to transfer the fire to the building.

Control measure: separation distances or fire-resistant barriers.

4) Direct burning

Combustible materials in direct contact with a naked flame, for example, process materials affected by a welding flame, or curtains, carpets and other office furnishings that may be consumed by combustion when a fire starts and enable fire to be transferred along them to other parts of a building.

Control measure: safe systems of work when doing 'hot work', separation distances between stored items and use of fire-retardant materials.

REASONS WHY FIRES SPREAD

Failure of early detection

Detection of fire spread can be delayed by:

- No detection system or fire warden patrols.
- No automatic alarm system in place.
- People not raising the alarm on discovery of a fire.
- Fire starts in an unoccupied area, remaining undetected for some time.
- Fire starts out of normal work hours, when workers are not there.
- Building material waste may be being burnt as a normal routine and smoke and other signs of fire may not be seen as unusual and ignored.
- Numerous hot working tasks conducted – therefore smells of burning ignored.
- Frequent occurrence of small, local fires caused by hot work, quickly extinguished and not seen as significant.

Absence of compartments in building structure

Fire compartmentation is the division of the building into discrete fire zones. Fire compartmentation is designed to contain the fire to within the area of the starting point of the fire. The compartmentation approach provides at least some protection for the rest of the building and its occupants even if other fire prevention systems are installed and fail.

Fire spread within a building can result from an absence of compartmentation by design or modification:

- Open-plan office (reduced compartmentation).
- Suspended ceilings create voids above.
- The structure under construction or alteration is incomplete and has reduced separation between levels and/or sections on a level.

Figure 10-12: Open-plan office.
Source: Kingsbridge College.

Figure 10-13: Suspended ceiling showing voids above.
Source: Borlaug Contracting Inc.

Figure 10-14: Fire door wedged open.
Source: RMS.

Compartments undermined

Fire spread within a building can result from compartments being undermined:

- Fire doors wedged open (*see Figure 10-14*)
- Poor maintenance of door structure, automatic closing rate or seal ineffective.
- Holes may be designed to pass through compartments and are waiting fitment of services and subsequent sealing.
- Holes cut for ducts or doorways or to provide temporary access to locate/remove equipment.
- Compartmentation may be progressively reduced in buildings under alteration, thus increasing the risk of fire spread.

Figure 10-15: Compartment undermined - holes in wall.
Source: RMS.

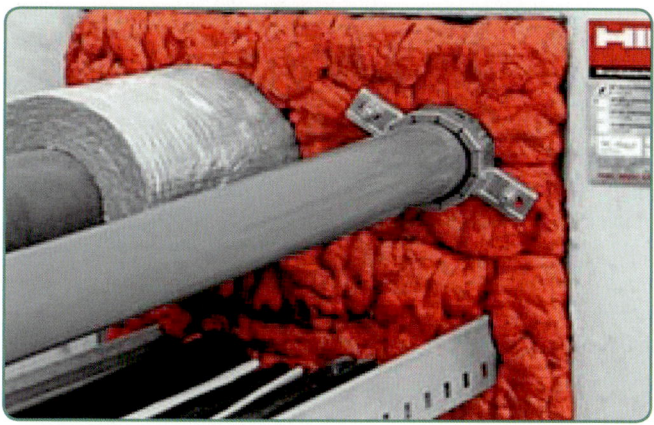

Figure 10-16: Wall gap closed by fire-resistant material.
Source: IndiaMART.

Materials inappropriately stored

Inappropriate storage of materials can cause fire spread:

- Flammable liquids not controlled – too much stored or in unsuitable containers.
- Boxes in corridors, under stairways or in access routes, obstructing safe evacuation and providing a source of fuel for the fire.

- Off-cuts of wood and sawdust left in the areas where work has taken place, increasing the risk of waste ignition.
- Flammable packing materials used in the process, such as shredded paper, polystyrene, bubble wrap allowed to accumulate, increasing the fire (combustible) load on the building.
- Pallets or waste materials inappropriately stored or discarded near to buildings, increasing risk of arson.

Figure 10-17: Materials inappropriately stored.
Source: RMS.

Figure 10-18: Materials inappropriately stored.
Source: RMS.

COMMON CAUSES AND CONSEQUENCES OF FIRE

Causes

Causes of fires in the workplace may be split into four main groups: careless actions and accidents, misuse of equipment, defective machinery or equipment and deliberate ignition (arson).

Careless actions and accidents

Careless actions and accidents relate to 'hot works' such as welding, cutting and grinding, discarded lighted cigarettes or matches, smouldering waste, unattended burning of waste materials or inappropriate electrical connections, unsafe use of gases and flammable liquids such as LPG and petrol. Careless use of chemicals that may react with other substances, for example, oxidising agents (solvents) soaked onto a cloth that give off heat as the cloth dries out.

Misuse of equipment

Misuse of equipment relates to overloading electrical circuits and/or using fuses of too high a rating. Misuse also applies when the servicing instructions are not followed or from failure to repair faulty machinery/equipment promptly. Equipment is misused when it is operated beyond its capacity, for example, a small saw blade cutting through a large section of timber, which can cause overheating. Misuse of equipment also occurs when it is stored incorrectly, for example, flammable liquids must be stored in a secure compound.

Figure 10-19: Hot work - uncontrolled welding activities can result in heat/sparks.
Source: RMS.

Figure 10-20: Misusing equipment - overloaded electrical sockets.
Source: RMS.

Defective machinery or equipment

Defective machinery or equipment may result in electrical short circuits causing arcs or sparking; an electrical earth

fault can cause local overheating. Electrical insulation failure may occur when affected by heat, damp or chemical corrosion. Mechanical, through generation of heat from friction due to wearing parts, incorrect lubrication or incorrectly tightened fan belts.

Figure 10-21: Defective electrical conductor.
Source: Shutterstock.

Deliberate ignition (arson)

Deliberate ignition is the crime of maliciously and intentionally or recklessly starting a fire or causing an explosion. These acts are often associated with; insurance fraud, aggrieved persons, concealment of another crime, political activists or vandalism. According to the Confederation of Fire Protection Associations Europe (CFPA-E) Guidelines 8:2004 'Preventing Arson – Information to Young People', arson is predominantly a problem associated with young people and young people are responsible for almost 60% of all arson fires.

Figure 10-22: Potential for deliberate ignition.
Source: RMS.

Consequences

The consequences of fires may be split into four main groups:

1) Human harm.

2) Economic effects.

3) Legal effects.

4) Environmental effects.

Human harm

The Seattle fire department reports that there are approximately 6,000 office fires in the USA each year. According to the Bureau of Labor Statistics' Census of Fatal Occupational Injuries 1992–2012, fires and explosions accounted for 3% of workplace fatalities in 2012. In the UK, fires which occur in commercial buildings result in approximately 33 deaths per year.

The Düsseldorf airport fire in 1996 caused 17 deaths and 62 injuries. The International Labour Organization (ILO) reported that in 1993, a major fire at the Kader Industrial Co. Ltd. factory in Thailand killed 188 workers. A fire in a clothing factory in Dhaka, Bangladesh, killed at least 117 workers in 2012.

Most fire deaths are not caused by burns but by smoke inhalation. Often smoke incapacitates so quickly that people are overcome and are unable to reach a safe exit.

The use of synthetic materials is now commonplace in the workplace, and when these ignite they produce toxic substances. As a fire grows inside a building, it will often consume most of the available oxygen, which in turn reduces the speed of the burning process. In addition to producing smoke, fire can incapacitate or kill by reducing oxygen levels, either by consuming the oxygen, or by displacing it with other gases. Heat is also a respiratory hazard, as superheated gases burn the respiratory tract, which often results in fatalities.

Smoke is made of components each of which can be lethal in its own way:

- **Particles**: Unburned, partially burned, and completely burned particulates can be so small they penetrate the respiratory system's protective filters and lodge in the lungs. Many products of combustion are toxic or irritating to the eyes and digestive system.
- **Vapours**: Fog-like droplets of poisonous liquid can be inhaled or absorbed through the skin.
- **Toxic gases**: The most common, carbon monoxide, can cause death, even in small quantities, as it replaces oxygen in the bloodstream. Hydrogen cyanide results from the burning of plastics, such as PVC pipe, and on inhalation interferes with cellular respiration. Phosgene is formed when products such as vinyl materials are burned. At low levels, phosgene can cause itchy eyes and a sore throat; at higher levels it can cause pulmonary oedema (fluid accumulation in the lungs) and death.

Economic effects
Commercial losses from fire are substantial, even though recent workplace injury and death figures have been low. The Düsseldorf airport fire in 1996 caused approximately €339 million worth of damage.

The fire at the Buncefield fuel depot in the UK, in December 2005, was the biggest in UK recent history. Explosions and heat from the fire caused severe damage to more than 80 buildings on the industrial estates surrounding the terminal. The cost of the damage is estimated to be

between £500 million and £1,000 million.

When fires do occur in the workplace the business is usually so badly affected it does not resume trading.

Legal effects

Article 16 of the ILO 'Convention on Occupational Safety and Health C155' sets out general responsibilities of employers, which include measures relating to fire risks, **see Figure 10-23**.

"1) Employers shall be required to ensure that, so far as is reasonably practicable, the workplaces, machinery, equipment and processes under their control are safe and without risk to health.

2) Employers shall be required to ensure that, so far as is reasonably practicable, the chemical, physical and biological substances and agents under their control are without risk to health when the appropriate measures of protection are taken."

Figure 10-23: General responsibilities of employers.
Source: ILO, Occupational Safety and Health Convention C155.

The Confederation of Fire Protection Association Europe (CFPA-Europe) is an association of national organisations in Europe concerned primarily with fire prevention and protection and also safety and security and other associated risks. It has produced a number of guidelines that are beneficial when aiding fire safety in Europe.

Where major workplace fire losses have occurred, it has generally led to the introduction of national legislation to protect workers and prevent such losses.

Environmental effects
Large uncontrolled fires create pollutants, such as smoke, that enter the atmosphere. The fire itself may cause damage to storage areas with the subsequent leakage of chemicals onto land or into watercourses, and the run-off from fire hoses may ultimately enter the water system.

Figure 10-24: Buncefield - run-off from fire hoses.
Source: BBC News.

The photographs *(see Figures 10-24 and 10-25),* show some of the damage caused by the Buncefield oil storage depot disaster in the UK in December 2005. The plume of smoke was so large it could be seen from space.

Figure 10-25: Buncefield oil storage depot disaster.
Source: Royal Chiltern Air Support Unit.

REVIEW

What are the key components of the 'fire triangle'?

Explain how a machine that has not been properly maintained may cause a fire.

What are common causes of fire in a workplace?

10.2 Preventing fire and fire spread

CONTROL MEASURES TO MINIMISE THE RISK OF FIRE STARTING IN THE WORKPLACE

Elimination and reduction of flammable and combustible materials

Where possible, employers should seek to **eliminate** the use of flammable materials in the workplace, for example, replacing adhesives that have a flammable content with those that are water based or to **substitute** the highly flammable with less flammable substances. Where this is not possible, the amount used should be **reduced** and kept to the minimum. Flammable and combustible materials in the workplace must be stored in suitable containers and minimum quantities for immediate work needs. Flammable materials not in use should be removed to a purpose-designed store in a well-ventilated area, preferably outside the building but in a secure location. Lids should be kept on containers at all times when they are not in use. Any waste containers, contaminated tools or materials in the workplace should be treated in the same way and removed to a store in fresh air until they can be dealt with.

It is important to reduce the presence of flammable and combustible materials by preventing an accumulation of waste. Waste in work areas should be removed to suitable collection points and then removed to controlled areas ready for recycling or off-site disposal. The efficient handling of waste should ensure that materials are minimised at each stage of handling – in the work area, waste points or controlled waste-handling areas. It is important to remember that containers and contaminated materials also need to be disposed of in a controlled manner so that they do not present a risk of fire. Care has to be taken to control the delivery and therefore the storage of flammable and combustible materials to site. Where possible, deliveries should be staggered to reflect the rate of use in order to minimise the amount stored on site.

Control of ignition sources

Hot work

'Hot work' is any process that can be a source of ignition, including welding, cutting, grinding, brazing and soldering processes. Hot work has been responsible for causing many fires.

One of the most tragic fires caused by hot work was Düsseldorf airport fire in 1996. The fire was started by welding on an open roadway and resulted in damage in excess of €339 million, several hundred injuries and 17 deaths. It is imperative that good safe working practices are in place. Combustible materials must be removed from the area or covered over. Consideration must be given to the effects of heat on the surrounding structure, and to where sparks, flames, hot residue or heat will travel.

Suitable fire extinguishers need to be immediately available and operatives must know how to use them. The work area must be checked thoroughly for some time after the completion of work to ensure there are no smouldering fires. A person should be appointed as a 'fire watcher' to ensure no fires result from hot work while the work is taking place and for some time after. Strong consideration should be given to the use of work permits to control hot work.

Welding and brazing

Welding and brazing activities represent a significant ignition source from the naked flame of an oxy-acetylene torch or electric arc welder and from the hot materials created by the process, for example, recently welded material that remains hot for some time, or from sparks created in the process. In addition, the equipment can represent an explosion risk if it is used incorrectly or not fitted with proper protective devices.

The following would be good practices for welding or brazing and would reduce sources of ignition from the process:

- Only use competent trained staff.
- Regulators should be of a recognised standard.
- Colour code hoses:
  - Blue - oxygen.
  - Red - acetylene.
  - Orange - propane.
- Fit non-return valves at blowpipe/torch inlet on both gas lines.
- Fit flashback arrestors incorporating cut-off valves and flame arrestors fitted to the outlet of both gas regulators.
- Use crimped hose connections not jubilee clips.
- Do not let oil or grease contaminate the oxygen supply due to explosion hazard.
- Check equipment visually before use, and check new connections with soapy water for leaks.
- Ensure the work area is well ventilated.
- Secure cylinders in upright position.
- Keep hose lengths to a minimum and check that they do not leak.
- Follow a permit-to-work system.
- Do not store standby gases that are not connected to welding apparatus in the workplace.

Figure 10-26: Hot work - welding with fire protection.
Source: American Training Resources.

Smoking

Smoking in public buildings (including the workplace) has been prohibited in some countries by national laws. Prohibition of smoking may lead to illicit smoking and extra vigilance may therefore be needed. Where general national prohibitions do not exist, it may remain necessary for the employer to prohibit smoking where hazardous materials are dealt with or where processes are carried on that involve the release of ignitable or explosive dusts or vapours or the production of readily combustible waste.

Smoking should be prohibited in stock rooms and other rooms not under continuous supervision. Any 'no

smoking' rule in these areas should be strictly enforced. Where smoking is allowed, provide easily accessible, non-combustible receptacles for cigarette ends and other smoking material and empty these daily. Smoking should cease half an hour before close-down to enable a check that smoking materials have been extinguished before people leave the workplace.

Arson

Simple, but effective, ways to deter the arsonist are by paying attention to security, both external and internal, which should encompass the following.

External security includes:

- Control of people having access to the building/site.
- Use of patrol guards.
- Lighting the premises at night – linked to closed circuit television (CCTV).
- Control of keys.
- Structural protection.
- Siting of waste containers/skips at least 8m from buildings.

Figure 10-27: Control arson by external security.
Source: RMS.

Internal security includes:

- Good housekeeping and clear access routes.
- Inspections and audits.
- Visitor supervision.
- Control of delivery and dispatch pick up areas where third parties may enter the premises.
- Control of sub-contractors.
- Control door access by keypad or electronic locks, *(see Figure 10-28).*

Further control measures to deter arson include:

- Storing flammable materials in lockable, fire-retardant cabinets.
- Storage of minimum quantities of materials on site.
- No access to ladders or other material that can be used to climb up to the roof.

Figure 10-28: Control arson by internal security.
Source: Shutterstock.

Mechanical heat

Mechanical heat, such as friction from drive belts or bearings, can be controlled by routine maintenance in which drive-belt tension is examined and the belt condition checked for signs of overheating. Bearings can be lubricated or greased as well as inspected for wear. Maintenance should also include replacement of wearing parts.

Electricity

Similar to mechanical equipment, electrical equipment must be maintained, inspected and tested to ensure circuits and their insulation has not degraded and that the system is not being overloaded. Portable and fixed electrical appliances should be checked.

Electrical equipment installed or used in flammable atmospheres must be suitable for use in that environment. There are different classes of flammable atmosphere and types of electrical equipment for use in flammable atmospheres. See also '*Use of suitable electrical equipment in flammable atmospheres' in 10.2 later in this element*.

Cooking and heating appliances

Cooking and heating appliances must not be left unattended and their use must be closely supervised. The source of energy for the appliance (electricity or gas) should be maintained, inspected and tested to ensure it functions correctly.

Systems of work

Safe systems of work to minimise fire risks in the workplace combine people, equipment, materials and the environment (the workplace) to produce the safest possible climate in which to work.

When combining these factors to make a safe system of work the following points related to fire risks can be considered.

1) Safe person

A safe person begins with raising awareness to individuals of any risk of loss resulting from outbreak of fire. Information can be provided that will identify where to raise the alarm, what the alarm sounds like, how to evacuate and where to muster, responsibility for signing in and out of the site register, fire drill procedures, trained authorised fire appointed persons, use and storage of flammable materials, good housekeeping and use of equipment producing heat or ignition (including hot processes i.e. welding). Safe systems of work must also include consideration of who is at risk, including those persons with special needs such as the young, elderly, infirm or disabled.

2) Safe materials

Safe materials begin with providing information and ensuring safe segregation and storage for materials and sources of ignition/heat. In addition, providing information on the correct way to handle materials and substances, including keeping a substance register that will detail methods of tackling a fire involving hazardous substances.

3) Safe equipment

Safe equipment begins with user information and maintenance to ensure good working and efficient order. Information should also provide the user with a safe method for use and the limitations of and risks from the equipment. Supervision may be necessary to ensure correct use and prevent misuse that may lead to short circuiting or overheating that could result in fire. Where work involves hot processes by nature (welding, grinding, casting, etc.) then permit-to-work procedures may be necessary in order to tightly control the operations.

Other equipment required in relation to fire hazards and control may include smoke or heat-detection equipment, alarm sounders/bells, alarm call-points and appropriate fire-extinguishing apparatus. It should be noted that in the event of a fire alarm, all the passenger lifts should not be used. Under normal circumstances the lift will return to the ground floor and remain in that position with the doors locked in the open position. All equipment should be regularly tested to ensure its conformity and be accompanied with a suitable certificate of validity.

4) A safe environment (workplace)

A safe workplace begins with ensuring that the fabric of the building is designed or planned in a way that will prevent ignition, suppress fire spread and allow for safe, speedy unobstructed evacuation with signs to direct people. Factors to consider will include compartmentalisation, fire-resistant materials, proper and suitable means of storage, means of detection, means of raising the alarm good housekeeping and regular monitoring and review.

In their simplest form safe systems of work to minimise fire risks might involve replacing lids on containers that contain flammable liquids in order to prevent flammable liquid vapours getting into the atmosphere of the workplace, where they could be ignited.

Other, simple safe systems of work to minimise fire risks include switching off electrical equipment that is not in use to remove it as a potential source of ignition. Systems of work for more complex situations may require written procedures to formally describe them and to help to communicate them. Because hot work presents significant fire hazards it is often controlled by safe systems of work in the form of formal written procedures involving the use of a hot work permit document.

Permit-to-work procedures

A permit-to-work is a formal documented safe system of work that is used for controlling high-risk activities. Implementation is required prior to work beginning to ensure that all precautions have been taken and are securely in place to prevent danger to the workforce.

When managed correctly, a permit-to-work prevents any mistakes or deviations through poor verbal communication by stating the specific health and safety requirements of the work activity. For fire control, a permit-to-work is typically used where there is a requirement to use flammable materials or when hot work or processes are being carried out.

Hot work permits

Hot work permits control and implement a safe system of work whenever work activities utilise heat or flame. If the risk of fire is low, it may not be necessary to implement a hot work permit; however, it should always be considered. The authorised person issues the hot work permit-to-work and will sign the document to declare that all isolations are made and remain in place throughout the duration of the work activity. Hot work permits should be issued for a specific time, for a specific place, for a specific task, and are issued to a designated competent person. In addition to this, the authorised person will make checks to ensure that all controls to be implemented by the acceptor are in place before work begins. The acceptor of the permit-to-work assumes responsibility for carrying out the work. The acceptor signs the document to declare that the terms and conditions of the permit-to-work are understood and will be complied with fully at all times by the entire work team.

Compliance with a permit-to-work system includes ensuring the required safeguards are implemented and that the work will be restricted to that stated within the permit-to-work document.

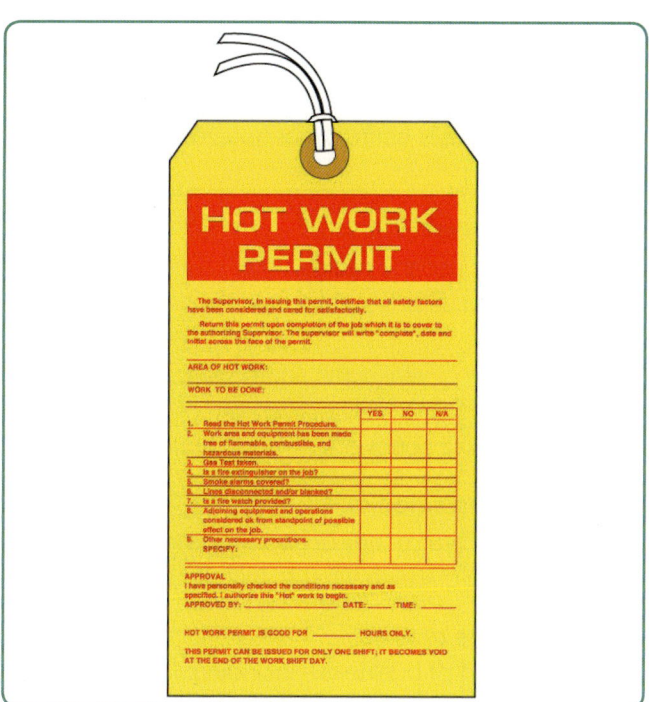

Figure 10-29: Hot work permit tag.
Source: Tufts University.

Items included in a permit-to-work document for hot work are:

- Permit issue number.
- Authorised person identification.
- Standby fire warden.
- Locations of firefighting equipment.
- Locations of flammable materials.
- Warning information sign locations.
- Emergency muster points.
- Details of the work to be carried out.
- Signature of authoriser.
- Signature of acceptor.
- Signature for work clearance/extension/handover.
- Signature for cancellation.
- Other precautions (risk assessments, method statements, personal protective equipment (PPE)).

See also - 'Permits-to-work' in 'Element 3 - Managing risk - understanding people and processes'.

Good housekeeping

Housekeeping and its effect on fire safety

'Housekeeping' means the general tidiness and order of the workplace. It should be remembered that fires need fuel, and a build-up of redundant combustible materials, rubbish and stacks of waste materials can provide that fuel. Combustible materials cannot be entirely eliminated, but they can be controlled. Any unnecessary build-up of rubbish and waste should be avoided.

If a fire starts in a neatly stacked pile of timber pallets, around which there is a clear space, the fire may be seen and extinguished before it can spread. However, if the same pile were strewn around in an untidy heap, along with adjacent rubbish, the likelihood is that fire would spread over a larger area and involve other combustible materials before it is seen. Poor housekeeping can also lead to:

- Blocked fire exits.
- Obstructed escape routes.
- Difficult access to fire alarm call-points/extinguishers/hose reels.
- Obstruction of vital signs and notices.
- A reduction in the effectiveness of automatic fire detectors and sprinklers.

Housekeeping checks

Fire prevention is a matter of good routine; the following checklists are a guide to what to look for.

List A - Routine checks

Daily at the start of business - including:

- Doors that may be used for escape purposes - unlocked and escape routes unobstructed.
- Free access to hydrants, extinguishers and fire alarm call-points.
- No dust deposits on electric motors.

List B - Routine checks

Daily at close down - including:

- Inspection of whole area of responsibility - to detect any smouldering fires, including smoking materials.
- Fire doors and shutters closed.
- All plant and equipment safely shutdown.
- Waste bins emptied (empty bins before they are full).
- No accumulation of combustible process waste, packaging materials or dust deposits.
- Safe disposal of waste, manage waste collection, keep to a minimum and store and secure from arson attack.
- Premises left secure from unauthorised access.

List C - Periodic inspection

During working hours - weekly/monthly/quarterly as decided:

- Goods neatly stored so as not to impede firefighting.
- Clear spaces around stacks of stored materials.
- Gangways kept unobstructed.
- Only minimum quantities of essential flammable and combustible material stored in work areas.
- Materials clear of light fittings.
- Smoking rules known and enforced.

STORAGE OF FLAMMABLE LIQUIDS IN WORK ROOMS AND OTHER LOCATIONS

Terms used with flammable liquids

Flash point

'Flash point' is defined as the lowest temperature at which, in a specific test apparatus, sufficient vapour is produced from a liquid sample for momentary or flash ignition to occur on the application of an ignition source.

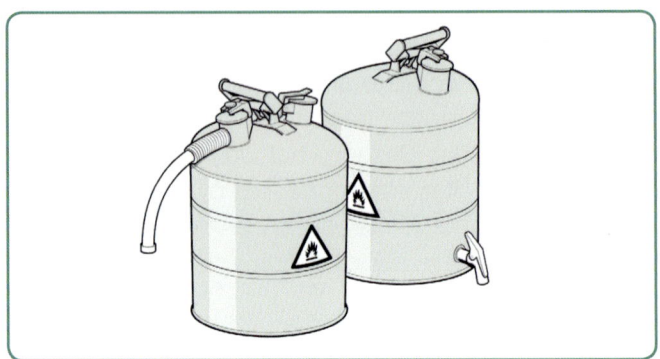

Figure 10-30: Flammable liquid containers.
Source: UK, HSE, HSG140.

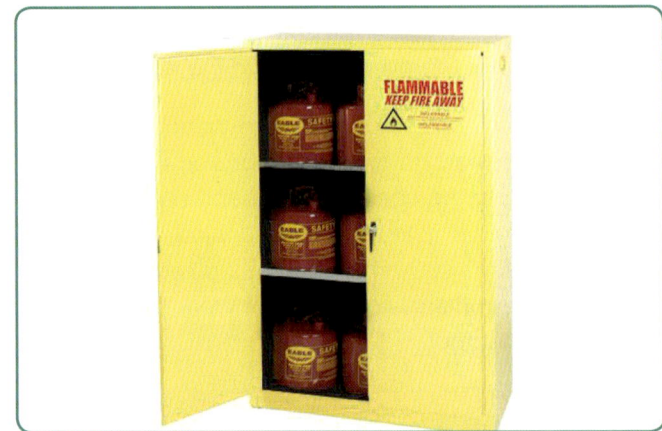

Figure 10-31: Lockable cupboard for flammable liquids.
Source: UniMac.

The flash point of a substance is the critical characteristic when deciding flammability. It must not be confused with ignition temperature, which may be considerably lower. The ignition temperature of a given substance is the measure of the minimum temperature at which the substance ignites, without the presence of an external spark or flame. Because of the fact that the material auto-ignites at this temperature range, it is also referred to as the substance's auto-ignition temperature.

Flammable

Flammability (or inflammability) is the ease with which the vapours of a substance will ignite, causing fire or combustion. Substances with vapours that will ignite at temperatures commonly encountered are considered flammable. The term flammable has various national definitions that give a temperature requirement relating to flammability.

The UN Globally Harmonized System of Classification and Labelling of Chemicals (GHS) has established an international definition of the term. This has also been reflected in the UN Recommendations on the Transport of Dangerous Goods (UNRTDG) as it relates to packaging for transport.

The GHS has established a definition that *"a flammable liquid is a liquid having a flash point of not more than 93°C"*. It further defines that a flammable liquid is classified in one of four categories.

These definitions and classifications have been adopted globally and are supported by the internationally agreed labelling under the UN GHS.

These definitions, categorisation and labelling are being introduced by many countries progressively. This has often required transitional arrangements to move to the new criteria. In time, the wider use of the GHS will make uniform the classification and other criteria for all workplace activities involving chemicals.

Label elements for flammable liquids			
Classification	Category 1	Category 2	Category 3
GHS Pictograms			
Signal word	Danger	Danger	Warning
Hazard statement	H224: Extremely flammable liquid and vapour	H225: Highly flammable liquid and vapour	H226: Flammable liquid and vapour

Figure 10-32: Flammable liquids classification and labelling.
Source: EU, Classification, Labelling and Packaging (CLP) Regulations.

The flash points of some common solvents are:

- Ethanol   +12°C.
- Toluene   +4°C.
- Methyl ethyl ketone   - 9°C.
- Acetone   -19°C.

General principles for storage and use of flammable liquids

When considering the storage or use of flammable liquids, in the UK, the HSE Guidance HSG51 'The storage of flammable liquids in containers' advises that the following safety principles should be applied. Using the acronym 'VICES' can be an aid to remembering these five principles, although there is no order of priority implied by the use of the acronym:

V Ventilation - plenty of fresh air.

I Ignition - control of ignition sources.

C Containment - suitable containers and spillage control.

E Exchange - try to use a less flammable product to do the task.

S Separation - keep storage away from process areas, by distance or a physical barrier, for example, a wall or partition.

Storage in the workplace

The objective in controlling the risk from flammable liquids is to remove all unnecessary quantities from the workplace to a recognised storage area outside the building. This may be done as part of a close-down routine at the end of the day. It is accepted that quantities of flammable liquids may need to be available in a workplace during normal working. In order to keep this to a minimum, the amount kept available in the workplace for immediate use should not exceed that which is required for one work period. In European countries the maximum recommended quantities that may be stored in suitable cabinets and bins in the workplace are:

- Liquids with a flash point below the maximum ambient temperature of the work area - no more than 50 litres.
- Other flammable liquids with a higher flash point of up to 60°C - no more than 250 litres.

Other control measures for storage in the workplace include:

- In a suitable properly labelled container, to prevent spills and sealed to prevent loss of vapour.
- In a purpose-built cabinet, bin or other storage container that is fire-resistant, with lockable doors, an inbuilt catch tray and clearly signed.
- In a designated, well-ventilated area of the workplace.
- Away from ignition sources, working or process areas.
- Capable of containing any spillage (non-spill caps and bunding) together with an appropriate spill kit.
- In a 30-minute fire-resistant structure.

Category	Flash point	Initial boiling point	Symbol	Signal word	Hazard statement
1	<23°C	≤35°C	Flame	Danger	Extremely flammable
2	<23°C	>35°C	Flame	Danger	Highly flammable
3	≥23°C ≤60°C	-	Flame	Warning	Flammable
4	≥60°C ≤93°C	-	No symbol	Warning	Combustible liquid

Figure 10-33: Flammable liquids classification.
Source: UN, Globally Harmonised System of Classification and Labelling of Chemicals.

- Provided with hazard warning signs to illustrate the flammability of the contents.
- Prohibition signs for smoking and naked flame.
- Should not contain other substances or items.
- Suitable emergency procedures.

- Fire extinguishers located nearby – consider powder type.
- Full and empty containers separated.
- Clear identification of contents.
- Area kept free of combustible materials.

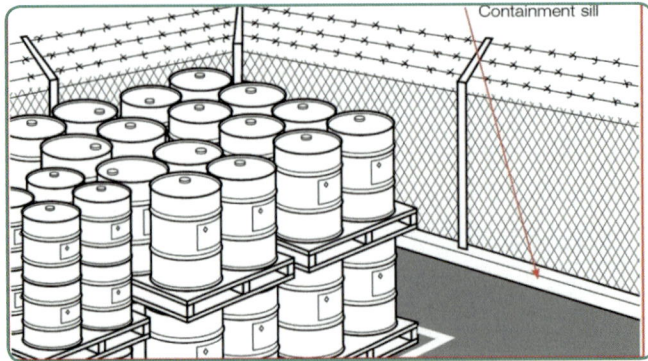

Figure 10-36: Outdoor storage of flammable materials.
Source: RMS.

Figure 10-34: Poor storage of flammable liquids.
Source: RMS.

Liquefied petroleum and other gases in cylinders

Liquefied petroleum gas (LPG) is a term that relates to gas stored in a liquefied state under pressure; common examples are propane and butane. LPG and other gas cylinders should be stored in line with the principles detailed as follows.

Figure 10-37: LPG storage.
Source: Shutterstock.

Figure 10-35: Storage of flammable materials.
Source: RMS.

Storage

Storage in the open air

Control measures for storage in the open air include:

- Storage away from potential ignition sources.
- Formal storage area on an impervious concrete pad, with a sump for spills.
- Bunded (impervious compound consisting of a solid floor and kerb or wall to enclose it) all around to take content of largest drum plus an allowance of 10%.
- Located away from other buildings.
- Secure fence and gate 2m high.
- Marked by signs warning of flammability.
- Signs prohibiting smoking or other naked flames.
- Protection from sunlight.
- If lighting is provided within store it must be flameproof.
- Provision for spill-containment materials.

Storage requirements for LPG and other gas cylinders include:

- Storage area should preferably be in clear open area outside (to ensure area is well ventilated).
- Stored in a secure compound – 2m high fence.
- Safe distance from toxic, corrosive, combustible materials, flammable liquids or general waste.
- Located away from any other building.
- If stored inside building, kept away from exit routes; consideration should be given to fire-resisting storage and forced ventilation.
- Well-ventilated area – 2.5% of total floor and wall area as vents, high and low.
- Oxygen cylinders at least 3m away from flammable gas cylinders.
- Acetylene may be stored with LPG if the quantity of LPG is less than 50 kg.

- Access to stores should be controlled to prevent LPG etc being stored in the general workplace when not in use.
- More than one exit (unlocked) may need to be available from any secure storage compound where distance to exit is greater than 12m.
- Storage compound should be locked when not in use.
- Protection from sunlight; take particular care to shield windows from direct sunlight.
- Flameproof lighting.
- Empty containers stored separately from full.
- Fire extinguishers located nearby – consider powder and water types.

Transport

Transport requirements for LPG and other gas cylinders include:

- Upright position.
- Secured to prevent falling over.
- Protection in event of accident, for example, position on vehicle.
- Transport in open vehicle preferably.
- Avoid overnight parking while loaded.
- Park in secure areas.
- Driver hazard information and warning signs.
- Driver training.
- Firefighting equipment.

General use

General requirements for use of LPG and other gas cylinders include:

- Cylinder connected for use may be stored in the general workplace; any spare cylinders must be secured in a purpose-built store until required for use.
- Fixed position to prevent falling over, or on wheeled trolley – chained.
- Well-ventilated area.
- Away from combustibles.
- Kept upright unless used on equipment specifically designed for horizontal use, for example, gas-powered lift truck.
- Handled carefully – do not drop.
- Allow to settle after transport and before use.
- Consider manual handling and injury prevention.
- Turn off cylinder before connecting or disconnecting equipment.
- Check equipment before use.
- Any smell of gas during use, turn off cylinder and investigate.
- Use correct gas regulator for equipment/task.
- Use equipment in line with manufacturers' instructions.

Use in huts

Requirements for use in huts include:

- Only allow cylinders in a hut if it is part of a heater (cabinet heater).
- Pipe into site huts from cylinder located outside where possible.
- If cylinder is outside the hut use the shortest connecting hose as possible.
- Hut to be adequately ventilated high and low.
- Heaters and cooking equipment fitted with flame failure devices.
- Turn off heater and cooking equipment and cylinder after use and overnight.
- Be aware of danger of leaks inside huts, especially overnight as a severe risk of fire or explosion may occur.
- Keep heaters and cooking equipment away from clothing and other combustibles.

Figure 10-38: Gas cylinders for huts.
Source: RMS.

STRUCTURAL MEASURES TO PREVENT SPREAD OF FIRE AND SMOKE

Properties of common building materials

Brickwork

Bricks are resistant to fire because they have already been exposed to high temperatures in the kiln where they were fired. Brickwork will usually perform well in a fire. Dependent upon the materials, workmanship, thickness, and the load carried, fire resistance of 30 minutes to 2 hours may be achieved.

Steelwork

Steel and other metals are extensively used in modern building structures. Generally they can be affected by fire at relatively low temperatures unless they are protected from the effects of the fire by some form of fire-retardant material. This may be done by encasing in concrete, fire-retardant boards or spray coatings.

Timber

Timber performs very well in fires as long as it is sufficiently large that as its outer layers become carbonised (charred), which slows the combustion process down, it retains sufficient strength to fulfil its task. Generally, timber does not fail rapidly in a fire.

Glass

Glass generally performs poorly in a fire unless it is fire-resistant glass. At high temperatures glass will melt and sag, which is why the traditional fire-resistant glass has wire within it.

Concrete

Concrete is very resistant to fire and, while heat may make the concrete spall (small sections break away), it retains its structural strength for a reasonable time period.

Structural measures to prevent spread

Measures to prevent spread of fire and smoke include:

- Fire resisting structures.
- Compartmentation to confine the fire to a predetermined size.
- Fire stopping of ducts, flues and holes in fire-resistant structures.
- Fire-resisting self-closing doors.
- Smoke seals and intumescent material (which expands when heat is applied and seals any gap between the door and the door frame) on doors.
- Early and rapid detection of a fire by use of sophisticated fire alarm systems, which may operate response systems automatically.
- Sprinklers in large compartments, in particular 'rapid response' systems to limit the size of the fire.
- Control of smoke and toxic fumes by ventilation systems, so that clear air is maintained at head height level, to enable persons to escape.

Figure 10-39: Magnetic door holder linked to alarm.
Source: RMS.

Figure 10-40: Door stop with automatic release.
Source: RMS.

Fire doors

Fire doors comprise a moving door panel and a frame that should be installed and maintained as an integral set. When the component parts fit together in an effective way they can prevent fire and smoke spread. They should have at least 30 minutes fire-resistance, the required fire resistance will depend on the location/situation where the door is installed. The door panel will usually have seals around its edges to prevent fire and smoke spread. Generally, an intumescent strip is fitted that swells when exposed to heat from a fire to seal the door. In addition, cold smoke seals may be fitted to prevent spread of smoke in situations where the temperature of the fire has not yet made the intumescent strip swell. The door is usually fitted with a self-closing device to keep the door closed when it is not being used for access. *See Figures 10-39 and 10-40*, which show mechanical means by which the door can be kept open until the sensors in the devices are activated by sound from the fire alarm system and release the door, which then closes.

Compartmentation

Compartmentation is achieved by use of compartment walls and floors which subdivide the building into smaller areas. The majority of buildings utilise traditional methods for controlling fire spread. This is mainly done by the use of fire-resisting structures and fire-resisting doors to break the building into smaller fire compartments. A fire compartment should withstand a fire for a minimum of 30 minutes, but it may require additional protection, depending upon the purpose of the structure and the use of the building.

If a fire does occur within a compartment it should be confined to that compartment by the nature of the fire-resistant materials. This should have the effect of limiting the damage done to buildings and prevent unchecked fire spread. Stairways, ducts, etc. should also form separate fire compartments to prevent vertical fire spread. In large compartments, sprinkler systems may be fitted in an attempt to limit the size of a fire, and ventilation may be provided to allow heat/smoke to escape. If additional fire safety measures such as sprinklers are installed, then the size of the fire compartments can, in general, be doubled.

Protection of openings and voids

Openings and voids in buildings include lift shafts, service ducts, voids between floors, roof voids, etc. Consideration should be given to the protection of openings and voids by the use of fire barriers such as fire shutters, cavity barriers and fire curtains. It is important that when construction or maintenance work takes place it is managed to minimise the effect on the structure being worked on to keep fire precautions intact as much as possible. This will involve planning for the reinstatement of protection of openings and voids as soon as possible after their breach to do work. The temptation to leave all breaches to the end of work and then reinstate them should be avoided – the longer that breaches are left open, the higher the risk from fires.

USE OF SUITABLE ELECTRICAL EQUIPMENT IN FLAMMABLE ATMOSPHERES

Classification of areas where explosive atmospheres may occur

The UN Common Regulatory Framework for Equipment Used in Environments with an Explosive Atmosphere supports the International Electrotechnical Commission (IEC) zone concept, established in IEC 60079–10. The zone concept established in IEC 60079–10 is widely recognised internationally and reflected in many national standards. The concept classifies hazardous locations as high, medium and low risk zones based on a standard risk-assessment methodology.

Under this zone concept, employers should classify areas where hazardous explosive atmospheres may occur into zones. The classification given to a particular zone, and its size and location, depends on the likelihood of an explosive atmosphere occurring and its persistence if it does occur. IEC 60079–10 contains descriptions of the various classifications of zones for gases and vapours and for dusts. There are three zones for gases and vapours:

- Zone 0   Flammable atmosphere highly likely to be present - may be present for long periods or even continuously.
- Zone 1   Flammable atmosphere possible but unlikely to be present for long periods.
- Zone 2   Flammable atmosphere unlikely to be present except for short periods of time - typically as a result of a process fault condition.

Similarly, there are three zones for dusts:

- Zone 20   Flammable dust cloud likely to be present continuously or for long periods.
- Zone 21   Flammable dust cloud likely to be present occasionally in normal operation.
- Zone 22   Flammable dust cloud unlikely to occur in normal operation, but if it does, will only exist for a short period.

Selection of equipment and protective systems

National legislation, for example, the European ATEX directives, governs the control of flammable atmospheres and the use of electrical equipment within them. The ATEX name is derived from the French title for the directive: 'Appareils destines á être utiliser en ATmosphères Explosibles (Devices for Use in Explosive Atmospheres).

There are two relevant directives:

- Directive 94/9/EC (also known as 'ATEX 95' or 'the ATEX Equipment Directive') on the approximation of the laws of Member States concerning equipment and protective systems intended for use in potentially explosive atmospheres.
- Directive 99/92/EC (also known as 'ATEX 137' or the 'ATEX Workplace Directive') on minimum requirements for improving the health and safety protection of workers potentially at risk from explosive atmospheres.

Areas classified into zones must be protected from sources of ignition. Electrical equipment for use in hazardous explosive atmospheres needs to be designed and constructed in such a way that it will not provide a source of ignition. Equipment intended to be used in zoned areas should be selected to meet its suitability for the zone, which is set out in IEC 60079-10, and meet the requirements of the relevant part of IEC 60079-10 that relates to the equipment type.

Zone 0 and zone 20 have the highest likelihood of an explosive atmosphere occurring and persisting; electrical equipment for this zone needs to be very well protected against providing a source of ignition. There are a number of ways in which electrical equipment can be designed to prevent ignition of explosive atmospheres, each achieving it in different ways. Different forms of electrical equipment will provide a different equipment protection level (EPL). The types of protection include:

- 'Intrinsically safe' – cannot produce a spark with sufficient energy to cause ignition – IEC symbol Ex ia (suitable for zone 0, 1, 2) or EX b (suitable for zone 1 or 2).
- 'Flameproof' – ingress of explosive atmosphere is controlled and any ignition is contained in the equipment IEC symbol Ex d (suitable for zone 1 or 2).
- 'Increased safety' equipment – does not produce sparks or hot surfaces – IEC symbol Ex e (suitable for zone 1 or 2).

 **REVIEW**

What are the storage requirements for LPG and other gas cylinders?

List the safety principles for storage and use of flammable liquids.

What are the structural features of buildings which can help prevent the spread of fire and smoke?

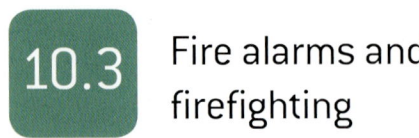

10.3 Fire alarms and firefighting

COMMON FIRE DETECTION AND ALARM SYSTEMS

Fire detection

The speed with which a fire in a building is detected is a critical factor in the determination of survival for the occupants of that building. Fires should be detected as soon as they start and building occupants alerted to the presence of the fire by the quickest possible means. It is essential that some form of detection and alarm system is used in the workplace, although the exact type of system will depend on national and local legislation and standards and the level of risk in the workplace.

Figure 10-41: Smoke detector.
Source: RMS.

Heat detection

Sensors operate by the melting of a metal (fusion detectors) or expansion of a solid, liquid or gas (thermal expansion detectors).

Radiation detection

Photoelectric cells detect the emission of infra-red/ultra-violet radiation from the fire.

Smoke detection

Smoke may be detected using ionising radiation, light scatter (smoke scatters beams of light) or obscuration (smoke entering a detector prevents light from reaching a photoelectric cell).

Flammable gas detection

Flammable gas is detected by measuring the amount of flammable gas in the atmosphere and comparing the value with a reference value.

Alarm systems

The purpose of a fire alarm is to give an early warning of a fire in a building for two reasons:

1) To increase the safety of occupants by encouraging them to escape to a place of safety.

2) To increase the possibility of early extinction of the fire, thus reducing the loss of or damage to the property.

Types of fire alarms

Alarms must make a distinctive sound, audible in all parts of the workplace. Sound levels should be 65 dB(A) or 5 dB(A) above any other noise – whichever is the greater.

Audible alarms may be supplemented by visual or tactile (vibrating) alarms where this would aid hearing-impaired workers and other people. The meaning of the alarm must be understood by all and readily differentiated from other alarms. Alarms may be manually or automatically operated.

Voice

Simplest and most effective type, but very limited because it depends on the size of the workplace and background noise levels.

Hand operated

Rotary gong, hand bell or triangle and sounder, but limited by the scale of the building.

Call-points with sounders

Standard system, operation of one call-point sounds alarm throughout workplace.

Automatic system

System as above, with added fire detection to initiate the alarm if it is not raised by a person.

Alarm call-points should be sited so that no person need travel an excessive distance to sound the alarm.

Travelling more than 30 metres (by direct measurement) or 45 metres to sound the alarm, taking into account fixtures, stock and other obstructions (actual), may be considered to be excessive. There are a number of alarm strategies that may be used depending on circumstances, including single-stage, two-stage and nominated-worker alarms.

Single-stage alarm

The alarm sounds throughout the whole of the building and signals a total evacuation.

Two-stage alarm

In certain large/high-rise buildings it may be better to evacuate the areas of high risk first, usually those closest to the fire or immediately above it. In this case, an evacuation signal is given in the affected area, together with an alert signal in other areas.

Nominated-worker alarms

In some premises, an immediate total evacuation may not be desirable, for example, nightclubs, shops, theatres, cinemas.

Figure 10-42: Easy operation alarm call-point.
Source: RMS.

Figure 10-43: Alarm point identified and well located.
Source: RMS.

A controlled evacuation by the nominated workers may be preferred so as to prevent distress and panic to the occupants. If such a system is used, the alarm must be restricted to the nominated workers and only used where there are sufficient nominated workers who have been fully trained in what to do in case of fire.

PORTABLE FIREFIGHTING EQUIPMENT

Siting

Portable firefighting equipment in the form of fire extinguishers should always be sited:

- On the line of escape routes.
- Near, but not too near, to danger points.
- Near to room exits, inside or outside according to occupancy and/or risk.
- In multi-storey buildings, at the same position on each floor, for example, at the top of stair flights or at corners in corridors.
- Where possible in groups forming fire points.
- So that no person needs to travel an excessive distance to reach an extinguisher. Travelling more than 30 metres (by direct measurement) or 45 metres (actual) taking into account fixtures, stock and other obstructions to reach an extinguisher may

be considered to be excessive. In high fire hazard areas 45 metres may be considered excessive and fire extinguishers may need to be located nearer to workers.
- National competent authorities may set travel distances dependent on the level of risk within the organisation.
- Placed on a purpose-designed floor stand or hung on a wall at a convenient height. If hung on a wall, this would usually be with the carrying handle about one metre (in the UK) from the floor to facilitate ease of handling/removal from the wall bracket.
- Away from excesses of heat, cold, dirt or dust.

Figure 10-44: Fire extinguishers - in floor stand.
Source: RMS.

Figure 10-45: Fire extinguishers - wall mounted.
Source: RMS.

Maintenance and inspection

Any firefighting equipment provided must be properly maintained and subject to examination and test at intervals so that it remains effective and available for use in an emergency.

Maintenance

Maintenance means the servicing of a fire extinguisher by a competent person. It involves thorough examination of the extinguisher (internal/external), refilling and re-pressurisation and is usually done annually.

Inspection

A monthly check (inspection) should be carried out to ensure that extinguishers are in their proper place and that they have not been discharged (the firing pin still tagged and in place), lost pressure or suffered obvious damage. It may be necessary to increase the frequency of checks for fire extinguishers located in work environments where there is less control over their use or they may be more likely to suffer damage, for example, on a construction site. Records should be kept of all visual inspections and maintenance checks that have been carried out.

Training requirements

The person responsible for ensuring fire precautions are taken must consider measures for firefighting. They should, as necessary, nominate competent persons to fight fires, and provide training and equipment accordingly. This would usually include training in how to use portable fire extinguishers, including practice in how to use them in a situation that reproduces the circumstances of a fire. Training for those who use fire extinguishers should include:

- Understanding of principles of combustion and classification of fires.
- Identification of the various types of fire extinguisher available to them.
- How to identify whether the extinguisher is appropriate to the fire and ready to use.
- Principles of use and limitations of extinguishers.
- Considerations for personal safety and the safety of others.
- How to attack fires with the appropriate extinguisher(s).
- Any specific considerations related to the environment the extinguishers are kept or used in.

Training should clarify the general and specific rules for use of portable fire extinguishers, for example:

General – aim at the seat of the fire and direct the extinguishing material across the fire to extinguish it – this is particularly appropriate for Class A fires, involving wood, paper, textiles, etc.

Specific – if using a foam portable fire extinguisher for a flammable liquid fire the foam is allowed to drop onto the fire by aiming just above it. If this is for a flammable liquid fire contained in an open tank it is possible to get good results by this process or by aiming it to the back of the tank and allowing the foam to float over the liquid. For other specific limitations or approaches to the use of individual types of extinguishing media, see the next section.

EXTINGUISHING MEDIA

In the following section, reference to a colour code means a code indicating the type of medium used in the fire extinguisher and refers to the European (BS EN 3) system of coding, which has been established for a considerable time and is widely accepted globally. *See later in this section, 'International markings for portable fire extinguishers'.*

Figure 10-46: Fire point sign. Source: BCW Office Products.

Figure 10-47: Water extinguisher colour coded red by label and sign. Source: Fire Protection Online

Water (portable fire extinguisher - colour code - red)

Water should only be used on Class A fires – those involving solids like paper and wood. Water works by cooling the burning material to below its ignition temperature, therefore removing the heat part of the fire triangle, and so putting the fire out. Water is the most common form of extinguishing medium and can be used on the majority of fires involving solid materials. It must not be used on liquid fires or in the vicinity of live electrical equipment.

Foam (portable fire extinguisher - colour code - cream)

Foam is especially useful for extinguishing Class B fires – those involving burning liquids and solids that melt and turn to liquids as they burn. Foam works in several ways to extinguish the fire, the main one being to smother the burning liquid, i.e. to stop the oxygen reaching the combustion zone. Foam can also be used to prevent flammable vapours escaping from spilled volatile liquids and also on Class A fires. It is worth noting that the modern spray foams are more efficient than water on Class A fires. *Foam must not be used in the vicinity of live electrical equipment, unless electrically rated.*

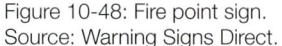

Figure 10-48: Fire point sign.
Source: Warning Signs Direct.

Figure 10-49: Foam extinguisher colour coded cream above label.
Source: Low Cost Fire.

Dry powder (portable fire extinguisher - colour code - blue)

Powder extinguishers are designed for Class A, B and C (those involving flammable gases) fires, but may only subdue Class A fires for a short while. One of the main ways in which powder works to extinguish a fire is the smothering effect, whereby it forms a thin film of powder on the burning liquid, thus excluding air. Dry powder is also excellent for the rapid knock-down (flame suppression) of flammable liquid spills and fires involving flammable gases.

It should be remembered that, if used to extinguish flammable gas fires, the source of the gas must be isolated promptly if reignition or explosion is to be prevented. Powders generally provide extinction faster than foam, but there is a greater risk of reignition. If used indoors, a powder can cause problems for the operator due to the inhalation of the powder and obscuring of vision. This type of extinguishing medium may be used on live electrical equipment.

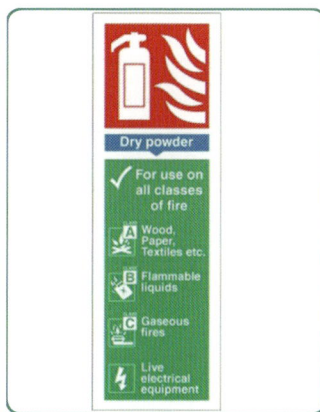

Figure 10-50: Fire point sign.
Source: Warning Signs Direct.

Figure 10-51: Powder extinguisher colour coded blue above label.
Source: The Sharpedge.

Carbon dioxide (CO_2) (portable fire extinguisher - colour code - black)

Carbon dioxide (CO_2) is safe and excellent for use on live electrical equipment. It may also be used for small Class B fires in their early stages of development, indoors or outdoors with little air movement. CO_2 replaces the oxygen in the atmosphere surrounding the fuel and thus extinguishes the fire.

Figure 10-52: Fire point sign.
Source: BCW Office Products.

Figure 10-53: CO_2 extinguisher colour coded black above label.
Source: Fire Protection Online.

CO_2 is an asphyxiant and should not be used in confined spaces. As it does not remove the heat, there is the possibility of reignition. CO_2 extinguishers are very noisy due to the rapid expansion of gas on release; this can surprise people when they operate a portable extinguisher. This expansion causes severe cooling around the discharge horn and can freeze the skin if the operator's hand is in contact with it.

As most CO_2 portable extinguishers last only a few seconds, only small fires should be tackled with this type of extinguisher. Carbon dioxide extinguishers are not suitable for Class D fires.

Figure 10-54: CO_2 extinguishing fire.
Source: Shutterstock.

Extinguishing media for specific classes of fire

Class C fires

Except in very small occurrences, a Class C fire involving gas should not normally be extinguished. If a gas leak fire is to be extinguished, the gas supply must first be isolated to prevent the gas reigniting and avoid a potential explosion.

Class D fires

Class D fires, those involving combustible metals, are a specialist type of fire and they cannot be extinguished by the use of ordinary extinguishing media. In fact, it may be dangerous to attempt to fight a metal fire with ordinary extinguishing media as the metal may explode or toxic fumes may be produced.

Metal fires can be extinguished by smothering them with dry sand. However, the sand must be absolutely dry or an explosion may occur. Other extinguishing media used may be graphite or salt. All of these extinguishing media essentially operate by the smothering principle.

Figure 10-55: Fire point sign.
Source: Midland Fire Ltd.

Figure 10-56: Wet chemical extinguisher colour coded yellow above label.
Source: Midland Fire Ltd.

Class F fires (wet chemical) (portable fire extinguisher - colour code - yellow)

Wet chemical extinguishing media have been designed to deal specifically with Class F fires, those involving cooking oil or fats where temperatures can exceed 260°C. This type of extinguishing medium congeals on top of the oil and excludes the oxygen. It may also be used on Class A fires depending upon the manufacturer's instructions.

Summary matrix - fire-extinguishing media

Figure 10-57 shows a summary of the most common fire-extinguishing media, their method of acting on fires and their effectiveness against different classes of fire. For the purposes of this summary, the classification of fires is that used in the UK/Europe and Australia/Asia.

International markings for portable fire extinguishers

The markings on portable fire extinguishers, for example, colour coding, has evolved differently in different countries. Internationally, there are several accepted categorisation methods for marking portable fire extinguishers; there is no single accepted standard.

Some are classified by their contents, i.e. the extinguishing medium, and others by the type of fire they are effective on. Some, for example, in Australia and the UK, may be categorised by the type of extinguishing media by a colour code and have additional pictogram markings to indicate the types of fires they are suited to.

Colour codes

Colour codes on extinguishers have been used for many years to identify the extinguishing media that the extinguisher contains. This has then been used to decide the class of fire on which it is effective.

Many countries use the international colour red as the dominant colour to denote fire equipment. Colour coding relating to contents has now become a supplement to this and is marked on the body of the fire extinguisher in a distinguishing way, for example, using a band or coloured label. Australia and the UK have adopted the international colour that is used to denote fire equipment. Australia and the UK recognise six classes of fire according to the fuel that is being combusted.

	Method	Class 'A'	Class 'B'	Class 'C'	Class 'D'	Electric	Class 'F'
Water	Cools	Yes	No	No	No	No	No
Spray foam	Smothers	Yes	Yes	No	No	No	No
Dry powder	Smothers & chemical	Yes	Yes	Yes & isolate	Special powders	Yes - low voltage	No
Carbon dioxide	Smothers	No	Yes - small fires	No	No	Yes	No
Wet chemical	Chemical	No	No	No	No	No	Yes
Vapourising liquids	Chemical & smothers	Special uses					

Figure 10-57: Summary matrix - fire extinguishing media.
Source: RMS.

Element 10

Australia

Type	Colour code	Class
Water	Solid red	A
Foam	Red with a blue band or label	A, B
Powder	Red with a white band or label	A, B, C, E (depending on type)
Carbon dioxide	Red with a black band or label	A (limited), B, E, F
Vapourising liquid (not halon)	Red with a yellow band or label	A, B (limited), E
Wet chemical	Red with an oatmeal band or label	A, F

Figure 10-58: Fire extinguisher colour codes and classifications - Australia. Source: Fire Protection Association Australia.

United Kingdom (UK)

Type	Colour code	Class
Water	Red	A
Foam	Red with cream as a secondary colour	A, B and sometimes E with special rating
Dry powder	Red with blue as a secondary colour	A (limited), B, C, E (depending on type)
Carbon dioxide CO_2	Red with black as a secondary colour	A (limited), B, E
Wet chemical	Red with yellow as a secondary colour	A, F
Class D powder	Red with a blue as a secondary colour	D

Figure 10-59: Fire extinguisher colour codes and classifications - United Kingdom. Source: RMS.

1) Class A fires involve organic solids such as paper and wood.

2) Class B fires involve flammable liquids.

3) Class C fires involve flammable gases.

4) Class D fires involve combustible metals.

5) Class E fires involve live electrical equipment (not strictly a class in Europe but used to denote suitability of extinguishers).

6) Class F fires involve cooking fat and oil.

Continental Europe

All extinguishers are red throughout the whole of the body of the extinguisher. No colour coding is imposed or recognised as applicable throughout Europe; the UK operates its colour coding as an allowed variation to the standard. Recently, some voluntary colour coding has appeared in Europe outside the UK.

However, it is different from the UK national addendum to EN 3, which provides for the secondary colour code, **see Figure 10-60.**

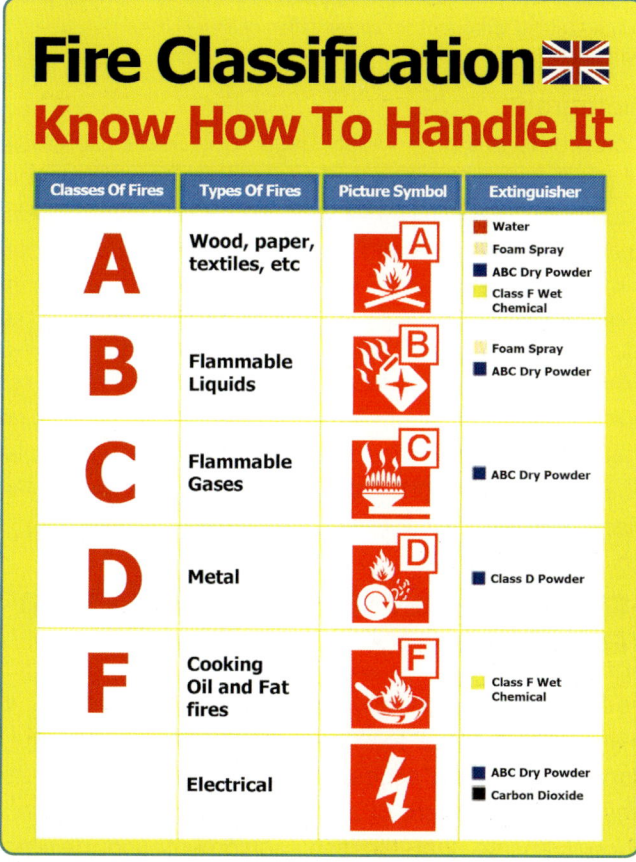

Figure 10-60: Pictograms for classification of fires - UK and Europe. Source: Safety Poster Shop.

United States

There is no official standard in the United States for the colour or colour coding of fire extinguishers, though they are typically red and may bear coloured geometric symbols, *see Figure 10-61.*

Figure 10-61: Pictograms for classification of fires - USA.
Source: Safety Poster Shop.

Pictograms

Pictograms are now widely used to indicate the class of fire that a fire extinguisher is effective on. In some countries, notably the USA, coloured geometric symbols and letters are also used with the pictograms to denote the type of fire for which the portable fire extinguisher is suitable for.

Whatever markings are applied to fire extinguishers, colour coding or symbols, the employer should ensure that workers understand their meaning and are capable of selecting the right fire extinguisher for a particular class of fire.

ACCESS FOR FIRE AND RESCUE SERVICES AND VEHICLES

As firefighting is generally carried out within buildings, it makes good sense to ensure that the fire and rescue services can gain access to a building as quickly as possible in the event of a fire. The responsible person should ensure that facilities, equipment and devices provided are maintained in an efficient state, in efficient working order and in good repair.

Vehicle access

Fire and Rescue Services need to get their appliances as close as possible to buildings to prevent time being wasted deploying firefighting equipment over extended distances. Minimum access requirements for pumping appliances and high-reach appliances will vary depending on the height, the floor area of the building and whether a fire water main is fitted. In the UK, access will be required for a minimum of 15% of the perimeter or within 45m (in the UK) of every point of the footprint of the building, up to a maximum of 100% of the perimeter, dependent upon the factors previously mentioned. Restrictions or planning consents in force at the time of the construction of the building will cause some variations between similar structures.

Access for firefighting

In low-rise buildings additional access requirements are not normally required. The means of escape is more straightforward and ladder access for firefighting is simpler. Access for firefighting appliances or other emergency vehicles must still meet the requirements given in the previous section (vehicle access).

Figure 10-62: Fire appliance – fully extended turntable ladder.
Source: www.geograph.ie.

Figure 10-63: Firefighter on turntable platform.
Source: www.geograph.ie.

In higher-rise buildings, additional facilities, including firefighting lifts, firefighting stairs and lobbies (usually called a firefighting shaft) are included as good practice. The addition of these measures allows fire and rescue services to quickly reach the floor where the fire is or to access the floor below a fire, from which an operating base can be set up. In general, buildings with floor levels over 18m, or basements more than 10m below a fire, must allow rescue service vehicle access.

REVIEW

What are two classes of fire and appropriate extinguishers for each type?

What should be included in a training course for the use of portable fire extinguishers?

10.4   Fire evacuation

MEANS OF ESCAPE

The responsible person for fire risks should make arrangements to manage risks as are appropriate to the size of the undertaking and the nature of its activities. This should include effective planning, organisation, control, monitoring and review of the preventive and protective measures. This should extend to arrangements for means of escape in the event of a fire. The means of escape must take the person to a place of total safety outside the building, and they should be able to travel the entire route by their own unaided effort. Separate arrangements will be necessary for assisting the safe escape of disabled people. Two or more escape routes may have to be provided to accommodate one being unavailable in the event of fire and to accommodate high volumes of people. The route should have no obstructions to impede the progress of people escaping so materials must not be stored on escape routes. The following general factors should be taken into consideration when planning means of escape.

Travel distances

Travel distance is a significant component of a successful means of escape plan. Travel distances are judged on the basis of how far it is to a place of safety in the open air and away from the building; the distance needs to be kept to the minimum.

The distance includes travel around obstructions in the workplace and may be greatly affected by any work in progress on a construction site. If someone is outside on a scaffold it is unlikely to be considered as a place of safety, and the distance would usually be taken as that necessary to reach the ground away from the building (for example, an assembly point). The route must be wide and short enough to allow speedy and safe evacuation.

There should normally be alternative routes leading in different directions. Everyone should be able to escape with or without assistance and before firefighters arrive. The distance between workstations and the nearest fire exit should be minimised.

Stairs

Staircases form an integral part of the means of escape from fire in most buildings. If they are to be part of the escape route, the following points must be ensured:

- Fire-resistant structure.
- Fitted with fire doors.
- Doors must not be wedged open.
- Wide enough to take the required number of people.
- Must lead direct to open air (outside the building), or to two totally separate routes of escape.
- Non-slip/trip and in good condition.
- No combustible storage within staircase.
- Adequate lighting.

Passageways

- The route should lead directly to the open air via a protected route (where necessary).
- Route to be kept unobstructed.

Escape routes	Suggested maximum travel distance	
More than one escape route provided	High fire risk	25m
	Normal fire risk	45m
	Lower fire risk	60m
Single escape route provided	High fire risk	12m
	Normal fire risk	25m
	Lower fire risk	45m

Figure 10-64: Travel distances. Source: UK, Dept of Communities.

Figure 10-65: Fire escape - hazard of falling on exit due to height from ground.
Source: RMS.

Doors

- Exit doors are to open outwards, in the direction of travel, easily (unless small numbers of people involved).
- Should be easily operated with one action by a person trying to escape.
- Should not be locked with a key or in such a way that a person inside could not open them.
- Provide fire doors along the escape route.
- Fire doors, along with fire-resistant structures, serve two purposes:

1) Prevent the spread of fire.

2) Ensure that there is means of escape for persons using the building.

- They should not be wedged open.
- Lead to open air - safety.
- May be fitted with a vision panel so that smoke and flames can be seen from the other side.

Emergency lighting

Emergency lighting should be considered if escape is likely to be required in dark conditions. This could mean any time when the sun is setting or has set and if clouds obscure natural light, not just at night time.

An emergency escape lighting system should normally cover the following:

- Each exit door from a work area.
- Escape routes.
- Intersections of corridors.
- Outside each final exit and on external escape routes.
- Emergency escape signs.
- Fire alarm call-points and firefighting equipment.
- Equipment that needs to be shutdown in an emergency.
- Lifts.
- Stairways, so that each flight receives adequate light.
- Changes in floor level.

It is not necessary to provide individual lights (luminaires) for each of the items previously listed, but there should be a sufficient overall level of light to allow them to be visible and usable. The lighting provided must be capable of providing emergency lighting for the duration of the evacuation and to assist fire rescue operatives: a typical specification for emergency lighting duration is between one and three hours.

Routine testing should take place in the form of a monthly check to ensure the emergency light works and an annual discharge test if they are maintained by a battery back-up system. Records of maintenance and inspection should be kept.

Exit and directional signs

Fire escape signs are provided to help people escape from wherever they are in a building, via a place of relative safety (the escape route) to the place of ultimate safety (the assembly area).

Fire escape signs are not usually needed on the main route into or out of a building (the one used by people for normal arrival and exit), but alternative escape routes and complicated escape routes do need to be signed.

It must not be assumed that everyone will know all safe routes through the building or that once people are out of the building they will know how to get to the assembly point. Signs directing to the assembly point will also be needed.

Figure 10-66: Fire escape sign - UK.
Source: RMS.

For information on safety signs see 'Element 3 - Managing risk - understanding people and processes'.

Exit and directional signs provided to aid escape in the event of a fire should conform to ISO 7010:2011 'Graphical Symbols – Safety Colours and Safety Signs – Registered Safety Signs'.

Element 10

This International standard has become a European Norm as PR EN 1710, which will require it to be adopted in all European Union countries.

Assembly points

The assembly point is a place of safety where people wait whilst any accident/incident is investigated, and where confirmation can be made that everyone has evacuated the premises.

Figure 10-67: Assembly point.
Source: RMS.

The main factors to consider are:

- Safe distance from building.
- Sited in a safe position.
- Not sited so that staff will be in the way of the fire and rescue service/firefighting team.
- Must be able to walk away from assembly point and back to a public road.
- Clearly signed.
- More than one provided to suit numbers and groups of people.
- Communications should be provided between assembly points.
- Measures provided to decide if evacuation successful.
- Identified person must be in charge of assembly point.
- Person to meet/brief the fire and rescue service.

EMERGENCY EVACUATION PROCEDURES

The employer should establish appropriate procedures, including fire drills, to be followed in the event of serious or imminent danger. In addition, they should nominate a sufficient number of competent persons to implement evacuation procedures.

The danger that may threaten people if an emergency occurs at work depends on many different factors; consequently, it is not possible to construct one model procedure for action in the event of fire and other emergencies for all premises.

Evacuation procedures need to reflect the type of emergency, the people affected and the premises involved. Many of the different issues to consider for different emergencies have common factors, for example, evacuation in an efficient/effective manner, an agreed assembly location (which may be different for different emergencies) and checks to ensure people are safe. These factors are considered in the following sections with regard to fire emergencies.

Fire instruction notices

Fire instruction notices should be located in conspicuous positions in all parts of the location, and adjacent to all fire alarm actuating (call) points. Printed notices should state, in concise terms, the essentials of the action to be taken upon discovering a fire and on hearing the fire alarm. It is usual to also state what someone must do when they discover a fire.

Figure 10-68: Fire instruction notice.
Source: RMS.

Fire action

The action in the event of a fire and upon discovery needs to be immediate, and a simple fire action plan should be put into effect. A good plan of action includes the following points.

On discovering a fire
- Sound the fire alarm (to warn others).
- Call the fire service.
- Go to the assembly point.

On hearing the alarm
- Leave the building by the nearest exit.
- Close doors behind you.
- Leave the building and do not re-enter.
- Go to the assembly point.

On evacuation

- Do not stop for personal belongings.
- Do not use lifts.
- Do not return to the building unless authorised to do so.
- Report to assembly point.

Fire training

Typical issues to be included in a fire-training programme relating to emergency action are:

- Fire prevention.
- Recognition of fire alarms and the actions to be taken.
- Understanding the emergency signs.
- Location of fire escape routes and assembly points.
- Consider the wording on notices that are posted and ensure that workers are instructed and trained to do what is asked of them.
- Requirements for safe evacuation (for example, non-use of lifts, do not run etc).
- Location and operation of alarm activating (call) points and other means of raising the alarm.
- How the fire service is called.
- Location, use and limitations of firefighting equipment.
- Consideration of people with special needs.
- Identity and role of fire marshals.

ROLE AND APPOINTMENT OF FIRE MARSHALS

In all premises, a person should be nominated to be responsible for coordinating the fire evacuation plan. This may be the same person who organises fire instruction and training and drills and coordinates the evacuation at the time of a fire (in the UK this person is known as the Fire Marshal). They may appoint persons such as fire marshals to assist them in fulfilling the role. This involves the appointment of certain staff to act as fire marshals to assist with evacuation.

The way in which they assist will vary between organisations, for example, some will check areas of the building in the event of a fire to ensure no person is still inside and others will lead the evacuation to show where to go.

The fire marshals' appointment should be made known to workers and they should be clearly identifiable at the time of emergency, so that those who are asked to evacuate understand the authority of the person requiring them to do so. The appointment of fire marshals helps the employer to meet the general responsibility to establish competent persons to assist with health and safety.

FIRE DRILLS AND ROLL CALL

Fire drills

A fire drill is intended to ensure, by means of training and rehearsal, that in the event of fire:

- The people who may be in danger act in a calm, orderly and efficient manner.
- Those designated with specific duties carry them out in an organised and effective manner.
- The means of escape are used in accordance with a predetermined and practised plan.

The fire drill enables all people involved in the evacuation to practice and learn under as near realistic conditions as possible and for management to demonstrate leadership. This can identify strengths and weaknesses in the evacuation procedure. Practice in the form of a drill helps people to respond quickly to the alarm and, because they have done it before, to make their way efficiently to the assembly point. It also enables a review of the plan if evacuation does not take place efficiently and effectively within the required time.

A practice fire drill should be carried out at least once a year. The drill should simulate conditions in which one or more of the escape routes from the building are obstructed. This will assist in developing an awareness of the alternative exits that can be taken and assist in ensuring people understand the unpredictability of fires.

Roll call

The traditional method of undertaking a roll call is by use of a checklist of names. Very few workplaces can now operate this system as they do not have the static workforce this system requires. Where this system can operate it will provide a speedy and efficient means of identifying who has arrived at the assembly point and who has not.

Where strict security control to a construction site is used, with signing in and out, this may make this process more viable. This requires people on site to report to their allocated assembly point and for someone (for example, a fire marshal) to confirm that they have arrived safely and to determine if anyone is missing. If it is not known exactly who is in a building, a system of fire marshals who can make a check of the building at the time of their own evacuation (without endangering their own safety) may be employed. This can assist with the process and may identify people who have not evacuated.

However, this system may not provide absolute confirmation that everyone has evacuated as there may be limited opportunity for the fire marshal to check the whole of the area allocated to them. Any doubt or confirmed missing persons should be reported to the person nominated to report to the fire service, who in turn will provide a report to the national fire and rescue service as soon as they arrive.

PROVISIONS FOR PEOPLE WITH DISABILITIES

When planning a fire evacuation system, employers need to consider who may be in the workplace, their abilities and capabilities.

Any disability, for example, hearing, vision, mental or mobility impairment, must be catered for. Some of the arrangements may involve allocating a nominated assistant(s) to support the person's speedy escape, for example, by the use of a specially designed evacuation chair to enable them to make their way out of a building down emergency exit stairs.

Figure 10-69: Evacuation chair for disabled people. Source: RMS.

Part of the provision is to make sure they are capable of knowing that an emergency exists. This may mean providing them with special alarm arrangements that cater for their disability, for example, a visual or vibrating alert for the hearing impaired.

Figure 10-70: Fire refuge sign for disabled people. Source: Safety Selector.

In some cases, disabled people may need to use a refuge area, a place of relative safety for short time periods. A refuge area is separated from the fire by a fire-resisting structure and has access via a safe route to a fire exit. It provides a temporary space for disabled people to wait for others who will help them to evacuate and often includes an intercom system, so the person using the refuge can communicate with rescuers. The refuge should be large enough to accommodate the number of wheelchair users who are likely to be using it. The refuge is also a good location for storing other emergency evacuation equipment such as 'evac-chairs' and for nominated responsible individuals, assigned with duties to help disabled persons, to meet the person they will assist to evacuate the building. Some buildings may be equipped with an evacuation lift, which has been specifically designed within a fire-resisting enclosure and has a separate power supply. Separate lift control mechanisms may be used, so that control of the lift car is by a responsible person who is aware of the evacuation plan.

BUILDING PLANS TO INCLUDE EMERGENCY ESCAPE ROUTES

Building plans should be drawn up to record emergency escape arrangements and aid the national emergency services. These will help identify the quickest and shortest route through the building, but can also be used to aid search and rescue.

Plans should clearly identify call-points, the siting of firefighting equipment/sprinklers (if fitted), fire doors and travel distances, refuge areas for the disabled, escape routes and assembly points.

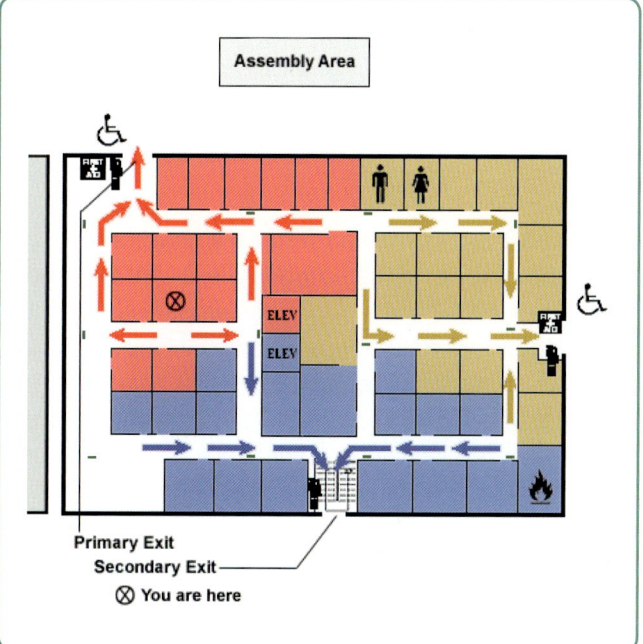

Figure 10-71: Fire evacuation plan. Source: USA, OSHA.

Figure 10-71 is taken from the US, Occupational Safety & Health Administration (OSHA's) approach to fire safety evacuation plans.

A floor plan shows the possible evacuation routes in the building. It is colour coded and uses arrows to indicate the designated exit. A room containing hazardous materials is indicated in the lower right hand corner of the building by the flame symbol. The assembly area is indicated outside the primary exit at the top of the building.

An evacuation floor plan with three exits, has the primary exit designated in the upper left by red arrows, with two main flows coming toward it indicated by bent arrows, the red rooms, and red elevator. Persons in the upper left half of the building are directed toward this exit. The secondary exit is located centrally on the adjacent outer wall on the right side of the building. Persons in the top hallway and second hallway are directed with tan arrows from the tan coloured rooms toward this exit. A male and female figure (representing restrooms) are indicated in the first tan coloured rooms in the upper hallway. The individuals should exit along the hallway toward the secondary exit at the right side of the building. Both the primary and secondary exits are marked with handicapped signs.

There is a third exit in the last hallway, centrally located in the outer wall opposite the outer wall with the primary exit and adjacent to the outer wall with the secondary exit. Persons in the third hallway are directed by blue arrows from the blue coloured rooms and blue elevator to exit out this doorway. This exit is not designated for handicapped persons as stairs are indicated.

Coloured boxes indicate a row of rooms along the outer walls, with hallways parallel to the rows of outer rooms on three sides of the building. The outer wall on the left side of the building has a hallway along the outer wall. Four sets of six coloured rooms are along the internal corridors and there are three large rooms centrally located with internal hallways connecting the top and bottom of the building.

The primary exit is marked with an arrow from the text below the map, as is the secondary exit. An X inscribed in a circle marks the position of the employee, indicated in the legend, in text 'You are here'. On the floor plan, the employee is located in the upper left hand corner in the internal set of six red coloured rooms, in the central room in the second hallway. The employee may exit the red coloured room, either to the left or right (indicated by red arrows), and then proceed toward the outer wall and the upper left primary exit.

Sources of reference

Reference information provided, in particular web links, was correct at time of publication, but may have changed.

ATEX – the current Directive 94/9/EC has been recast as Directive 2014/34/EU which was applicable from 20 April 2016 http://ec.europa.eu/growth/sectors/mechanical-engineering/atex/

94/9/EC https://eur-lex.europa.eu/legal-content/EN/TXT/?uri=CELEX:01994L0009-20130101

2014/34/EU https://eur-lex.europa.eu/legal-content/EN/TXT/?uri=CELEX:32014L0034

Fire safety, HSE Toolbox, http://www.hse.gov.uk/toolbox/fire.htm

Fire Safety Risk Assessment series, UK Home Office Publications: https://www.gov.uk/search?q=Fire+Safety+Risk+Assessment

1. Offices and shops, ISBN 978–1-8511–2815–0

2. Factories and warehouses, ISBN 978-1-8511-2816-7

3. Sleeping accommodation, ISBN 978-1-8511-2817-4

4. Residential care premises, ISBN 978-1-8511-2818-1

5. Educational premises, ISBN 978-1-8511-2819-8

6. Small and medium places of assembly, ISBN 978–1-8511-2820–4

7. Large places of assembly, ISBN 978-1-8511-2821-1

8. Theatres, cinemas and similar premises, ISBN 978-1-8511-2822-8

9. Open air events and venues, ISBN 978-1-8511-2823-5

10. Healthcare premises, ISBN 978-1-8511-2824-2

11. Transport premises and facilities, ISBN 978-1-8511-2825-9

12. Means of escape for disabled people, ISBN: 978-1-8511-2873-0

ISO 70101:2019 – Graphical symbols – Safety colours and safety signs – Registered safety signs

Safety in the use of chemicals at work, International Labour Organization (ILO) Code of Practice (CoP), ILO, 1993. ISBN: 978-9-221080-06-4 http://www.ilo.org/wcmsp5/groups/public/@ed_protect/@protrav/@safework/documents/normativeinstrument/wcms_107823.pdf

Section 6: Operational control measures and

Section 7: Design and installation

Safe use and handling of flammable liquids, HSG140,

HSE Books http://www.hse.gov.uk/pubns/priced/hsg140.pdf

Storage of flammable liquids in containers, HSG51, HSE Books, http://www.hse.gov.uk/pubns/priced/hsg51.pdf

Storage of flammable liquids in tanks, HSG176, HSE Books, http://www.hse.gov.uk/pubns/priced/hsg176.pdf

The health and safety toolbox, How to control risks at work, HSG268, HSE Books, ISBN: 978-0-7176-6587-7, http://www.hse.gov.uk/pUbns/priced/hsg268.pdf

The Regulatory Reform (Fire Safety) Order (RRFSO) 2005, UK, www.legislation.gov.uk (please enter the name of the Regulation)

Web links to these references are provided on the RMS Publishing website for ease of use – www.rmspublishing. co.uk

STUDY QUESTIONS

1) (a) How has an understanding of the principles of the fire triangle been used to develop techniques for extinguishing fires? (6)

   (b) Explain two methods, with an example for each, how a fire can be extinguished. (2)

2) What are the main methods by which fires spread through a structure? (8)

3) What are the reasons for carrying out regular reviews of fire safety measures? (8)

4) Why is good 'housekeeping' in the workplace essential to ensure safe escape in a fire? (8)

5) What should you consider before the location of a fire assembly point is decided upon? (8)

For guidance on how to answer NEBOSH questions please refer to the 'study question answer guidance' section located at the back of this guide.

This page is intentionally left blank

Element 11

Electricity

Contents

11.1   Hazards and risks

PRINCIPLES OF ELECTRICITY

Electricity is a facility that we have all come to take for granted, whether for lighting, heating, as a source of motive power or as the driving force behind the computer. Used properly, it can be of great benefit to us, but misused, it can be very dangerous and often fatal.

Electricity is used in most industries, offices and homes, and society could now not easily function without it. Despite its convenience to the user, it is of major danger (risk of electric shock, burns, fire or explosion). The normal senses of sight, hearing and smell will not detect electricity. Making contact with exposed conductors at the supply voltage of 110 volts or 230 volts can be lethal.

Unlike many other workplace accidents/incidents, the actual number of electrical accidents/incidents is small. However, with a reported 10–20 fatalities each year in the UK and 156 fatalities in the USA in 2011, the severity is high. Accidents/incidents are often caused by complacency, not just by the normally assumed ignorance. It must be recognised by everyone working with electricity that over half of all electrical fatal accidents/incidents happen to skilled/competent persons.

In order to avoid the causes of electrical injury, it is necessary to understand the basic principles of electricity, what it does to the body and what controls are necessary. Two terms appear in common usage with electricity: 'live' and 'dead'. A 'live' system (also known in some countries as a 'hot' system) is one carrying an electrical current. Once the electricity has been disconnected from its power sources, it is described as 'dead'.

Basic circuitry

The flow of electrons through a conductor is known as a current. Electric current flows due to differences in electrical 'pressure' (or potential difference, as it is often known), just as water flows through a pipe because of the pressure behind it.

Differences in electrical potential are measured in volts. In some systems, the current flows continually in the same direction. This is known as direct current (DC). However, the current may also constantly reverse its direction of flow. This is known as alternating current (AC). Most public electricity supplies are AC.

The frequency of the reversals of AC current flow varies in different countries, the most common being 50 times per second, in which case, the power supply is said to have a frequency of 50 cycles per second or 50 Hertz

(50Hz). Other countries, for example the USA, operate at 120 volts and 60Hz. DC is little used in standard distribution systems, but is sometimes used in industry for specialist applications.

Although there are slight differences in the effects under fault and shock conditions between AC and DC, the safe approach is to apply the same rules of safety for the treatment and prevention of electric shock from either type. As a current passes round a circuit under the action of an applied voltage it is impeded in its flow. This is due to the presence in the circuit of resistance, the opposition to the passage of an electrical current through that circuit (conductor). The factors that contribute to conductor resistance are measured in ohms.

Figure 11-1 Electrical hazard warning sign.
Source: Rivington Signs.

Relationship between voltage, current and resistance

There is a simple relationship between electrical pressure (volts), current (measured in amperes or milliamperes 10^{-3}) and resistance (measured in ohms) represented by Ohm's law:

Voltage (V) = current (I) multiplied by the circuit resistance (R).

Therefore,

$$V = I \times R \quad or \quad I = \frac{V}{R}$$

Hence, given any two values, the third can be calculated. Also, if one value changes, the other two values will change accordingly. This basic electrical equation can be used to calculate the current that flows in a circuit of a given resistance.

This will need to be done to determine, for example, the fuse or cable rating needed for a particular circuit. Similarly, the current that will flow through a person who touches a live conductor can be calculated.

Resistance in a circuit is dependent on many factors. Most metals, particularly copper and steel, allow current to pass very easily. These have a low resistance and are used as conductors. Other materials such as glass, plastics, rubber and textiles have a high resistance and are used as electrical protective barriers (insulators) with conductors.

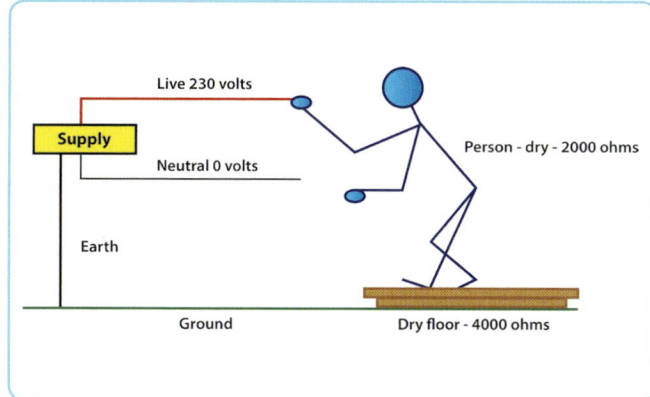

Figure 11-2: Basic electrical circuitry.
Source: RMS.

The schematic in **Figure 11-2** represents a simple electrical circuit in relation to a person. The terms earth and neutral are introduced to represent the two paths (conductors) through which the current returns to the electrical supply generator. This may be through an electrical return conductor (neutral) or literally through the ground.

If the person is on a dry concrete floor, resistance in the body to current flow will only be about 2,000 ohms and the resistance in the floor about 4,000 ohms, therefore the combined resistance to the current flowing to earth would be 6,000 ohms. Assuming the person is in contact with a live electrical supply at 230 volts, the current flowing through the person in this fault condition can be calculated.

$$I = \frac{V}{R} = \frac{230 \ Volts}{2,000 + 4,000 \ Ohms} = 0.038 \ amperes$$

The current flowing through the operator will then be about 0.04 amperes or 40mA (40 milliamperes) to earth.

At this voltage (230 volts), there is sufficient potential difference (230 volts to 0 volts) between the supply through the person and the concrete floor to earth, to enable sufficient current to pass through the person and cause a fatal shock.

ELECTRIC SHOCK AND ITS EFFECT ON THE BODY

The term electric shock is used to describe the unwanted or undesirable exposure to electricity at a detectable level (typically 1mA AC at 50Hz).

When an electric current passes through a material, the resistance to the flow of electrons dissipates energy, usually in the form of heat. If the material is human tissue and the amount of heat generated is sufficient, the tissue may be burnt. The effect is similar to damage caused by an open flame or other high-temperature source of heat, except that electricity has the ability to **burn** tissue well beneath the skin, including internal organs.

Nerve cells communicate by creating electrical signals (at very small voltages and currents) in response to the input of certain chemical compounds (neurotransmitters). If the electric shock current is of sufficient magnitude it will override the electrical impulses normally generated by the neurons, preventing both reflex and volitional (controlled by conscious choice or decision) signals from being able to operate muscles. These effects may be felt as *pain*.

Muscles may also be *triggered* by a shock current, which will cause them to *contract* involuntarily. The forearm muscles responsible for bending fingers tend to be better developed than those muscles responsible for extending fingers. If both sets of muscles attempt to contract, the 'bending' muscles will be stronger and clench the fingers into a fist.

If the conductor delivering a shock current touches the palm of the hand, the clenching action will force the hand to grasp the conductor firmly, securing contact with the conductor, and it will not be possible for the victim to release their grasp. Even when the current is stopped, the victim may not regain voluntary control over their muscles for a while, as the neurotransmitter chemistry will be in disarray. Involuntary muscle contraction is called shock-induced tetanus. Shock-induced tetanus can only be interrupted by stopping the current passing through the body.

Electric current is able to affect more than just skeletal muscles; it can also affect breathing and heart function, particularly if the path is across the chest. The diaphragm muscle controlling the lungs and the heart muscle can also be caused to be in a state of shock-induced tetanus (involuntary muscle contraction) by a shock current, leading to *respiratory failure* of the lungs and *fibrillation* of the heart or *cardiac arrest.* As discussed previously, fibrillation is a condition where all the heart muscles start moving independently in a disorganised manner, rather than in a coordinated way. It affects the ability of the heart to pump blood, resulting in brain damage and eventual cardiac arrest.

Factors influencing severity of the effects of electric shock on the body

The factors influencing the severity of the effects of electric shock on the body include the following:

- Voltage.

- Frequency and size of the current involved.

- Duration, the length of contact time (measured in milliseconds).

- The path taken through the body by the current (across the chest/heart/lungs is of significant risk).

- The electrical resistance of the skin in contact with the electrical system (perspiration, sweat) and the internal body resistance as the current flows through the body.

- The resistance of the electrical path of the current flow to earth (factors associated with the footwear/clothing being worn by the person and the conductivity of the surface they are standing on).

- The general health, gender and age of the person involved (females are more susceptible to electric shock, i.e. at lower current and time of exposure).

The amount of current that flows through the body for a given voltage will depend on the frequency of the supply voltage, on the level of the voltage that is applied and on the state of the point of contact with the body, particularly the moisture condition. The effect of electricity on the body and severity of electric shock results from a combination of the current level and the duration of the passage of that current. The voltage level is relevant mainly in that it causes the passage of the current.

Figure 11-3: Contact with high-voltage buried cable.
Source: www.safetyblog.co.uk.

Voltage

Voltage is the driving force behind the flow of electricity, in the same way that pressure in a water pipe influences the amount of water that flows. The correct name for this term is potential difference, as a voltage is the measure of the difference in electrical energy between two points. Any electrical charge that is free to move will move from the higher energy point to the lower one, taking a quantity of that energy difference with it.

Frequency

Alternating current (AC) is preferred over direct current (DC) for power generation/transmission systems and is used domestically in industry and commerce, but AC is 3 to 5 times more dangerous than DC at the same voltage and amperage. AC at frequencies of 50 to 60Hz produces a series of muscle contractions, which will, if the point of contact is the palm of the hand, cause the hand to grasp the source and prolong the exposure. DC is most likely to cause a single convulsive contraction. AC

has a greater tendency to cause fibrillation of the heart muscles, whereas DC tends to make the heart muscles stand still. Once the shock current is halted, a still heart has a better chance of regaining a normal beat pattern than one that is in fibrillation. Defibrillating equipment, used by paramedics, utilises a DC shock current to halt fibrillation to enable the heart to recover its normal beat. Though both AC and DC shocks may be fatal, more DC is required to have the same effect as AC, for example, the 'no let go' threshold for DC is reported to be 4–5 times that of AC.

Duration

For an electric shock to have an effect, a person needs to be in contact with the current for sufficient time. At low current levels, the body tolerates the current so the time is not material; however, at higher current levels, for example, 50mA, the person has to remain in contact for sufficient time for the heart to be affected, in the order of milliseconds (ms). In general, the longer a person is in contact with the current, the more harm may be caused.

Resistance

The amount of resistance in a circuit influences the amount of current that is allowed to flow, as explained by Ohm's law. It is possible for a person to be in contact with a circuit and to present sufficiently high resistance that very little current is allowed to flow through their body, (see Figure 11-2). It should be noted that the level of current flow is also dependent on the voltage; at high voltages an enormous amount of resistance is needed to ensure current flow will remain at a safe level. In an electric shock situation the human body contributes part of the resistance of the circuit, and the amount it contributes depends on the current path taken and other factors such as personal chemical make-up (a large proportion of the body is water), moisture on and the thickness of skin, and any clothing that is being worn, such as shoes and gloves.

Current path

The effect of an electric shock on the body is particularly dependent on the current path through the body. Current has to flow through from one point to another as part of a circuit. If the flow was between two points on a finger the effect on the body would be concentrated between the two points. If the current path is between one hand and another, across the chest, this means the flow will pass through major parts of the body, such as the heart, and may cause fibrillation or cardiac arrest.

In a similar way, a contact between hand and foot (feet), causing the current to travel across the chest, can have serious effects on a great many parts of the body, including the heart. These current paths tend to be the ones leading to fatal injuries.

However, a current path from hand to foot down one

side of the body may not affect the heart and therefore may not be fatal. Although many people may experience shock from 110 volt or 230 volt supplies, this may not be fatal if they are, for example, standing on or wearing some insulating material. This may be a matter of fortune and, as a rule, this sort of protection should not be relied on.

The amount of current flow through the body has a significant influence on the effect of the electric shock. At low levels of current flow, no effect may be experienced; however, at progressively higher levels of current flow the larger muscles of the body may be affected. The heart, being made of large muscles, requires a significant current to affect it. At high current levels, burns may occur at the point of current entry and exit from the body, as well as at points along the route the current takes through the body.

Common causes of electric shock

Common causes of electric shock include:

- Work on electrical circuits by unqualified persons.
- Work on live circuits.
- Replacement of fuses and light bulbs on supposedly dead circuits.
- Working on de-energised circuits that accidentally become re-energised.
- Using electrical equipment in a wet environment.
- Faults in electrical systems, which energise parts that are not normally conductors, for example, the casing of electrical equipment.

Figure 11-4: Electrical equipment near water supply.
Source: RMS.

Direct and indirect contact with electrical source

Direct

Direct contact relates to when a person makes contact with an electrical source - a charged or energised conductor that is intended to be charged or energised. In these circumstances, the electrical system is operating in its normal or proper condition. This may occur when someone is working on equipment where conductors are exposed and in the live condition.

Indirect shock

Indirect contact relates to when a person makes contact with electrical conducting material that is normally at a safe potential, but has become dangerously live through a fault condition. Conductive parts of equipment that may become live in a fault condition include the conducting casing of equipment and trunking around electrical cables. These are normally safe to touch, but under fault conditions could become dangerously live. For example, where the casing of equipment has a poor connection to earth, and when a fault occurs on the equipment, the casing may become live.

Effects of current flowing in the human body

Current (mA)	Length of time	Likely effects
0–1	Not critical	Threshold of feeling. Undetected by person.
1–15	Not critical	Threshold of cramp. Independent loosening of the hands no longer possible.
15–30	Minutes	Cramp-like pulling together of the arms, breathing difficult. Limit of tolerance.
30–50	Seconds to minutes	Strong cramp-like effects, loss of consciousness due to restricted breathing. Longer time may lead to fibrillation.
50–500	Less than one heart period (750 ms)	No fibrillation. Strong shock effects.
	Greater than one heart period	Fibrillation. Loss of consciousness. Burn marks.
Over 500	Less than one heart period	Fibrillation. Loss of consciousness. Burn marks.

Figure 11-5: Effects of current flowing in the human body.
Source: RMS.

Electrical burns

Direct burns

Direct contact with a live electrical source can allow a current to flow through the body, causing a heating effect along the route taken by the electric current as it passes through the body tissue, causing direct burns along its path.

Whilst there are likely to be burn marks on the skin at the point of contact (external burns), there may also be a deep-seated burning within the body (internal burns), which is painful and slow to heal.

External burns

When electrical current makes contact with the skin, it becomes part of the electrical circuit and can cause the point of entry to reach a high temperature, creating immediate tissue damage and charring. The point of entry tends to be limited to small areas and will often appear sunken or hollowed, whereas the exit wound is more extensively damaged and open.

Internal burns

The severity of the damage caused internally by electrical burns will depend upon the pathway along which the current flows. In the human body, the pathways of least resistance are typically, in the first instance, blood vessels, nerves and muscle, then skin, tendon and fat, and, finally, bone.

As the outer layer of skin is burnt, the resistance decreases and so the current will increase. The current flowing through the body can cause major injury to internal organs and bone marrow as it passes through them.

Indirect burns

Indirect burns can be caused when an electrical discharge (an arc or spark) occurs from a high-voltage system. This discharge contains a lot of energy, and the flash from the discharge can cause burns even without direct contact with the electrical supply.

In addition, a worker may experience indirect burns to external body parts caused by equipment coming into contact with the electrical system. For example, if while working on live equipment, the system is short circuited by an uninsulated spanner touching live and neutral, this will result in a large and sudden current flow through the spanner. This rapid discharge of energy that follows contact with high voltages not only causes the rapid melting of the spanner, but does so with such violent force that the molten particles of metal are thrown off with huge velocity. When these molten particles contact the parts of a person in the vicinity of the spanner, for example the hands or face, they can cause serious burns as the molten metal sticks to the skin.

It is not necessary to have high voltages to melt a spanner in this way – it can also occur with batteries that have sufficient stored energy, such as those on a fork-lift truck. Many workers have suffered injury in this way when servicing lift-truck batteries when a spanner has fallen out of the top pocket of their overalls.

Electrical fires and explosions

Common causes of electrical fires

Figure 11-6: Used coiled up - risk of overheating.
Source: RMS.

Much electrical equipment generates heat or produces sparks, so this equipment should not be placed where this could lead to the uncontrolled ignition of any substance. The principal causes of electrical fires are:

- Wiring with defects such as insulation failure due to age or poor maintenance or physical damage.

- Overheating of cables or other electrical equipment through overloading with currents above their design capacity (for example, a coiled extension lead will have a much lower current-carrying capacity than one that is fully uncoiled and will rapidly overheat if this is exceeded).

- Too high a fuse rating for the circuit to be protected (for example, a 13A fuse used for a circuit with a load capacity of 3A).

- Poor connections due to the effects of use/lack of maintenance or unskilled workers (for example, cables not secured by a cable grip inside a drill casing).

Figure 11-7: Max current capacity exceeded.
Source: RMS.

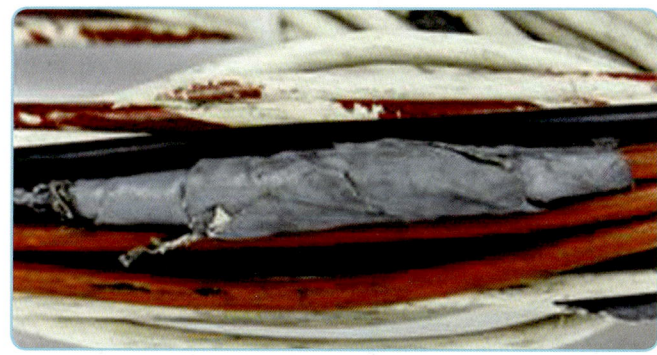

Figure 11-8: Worn cable - risk of electrical fire.
Source: RMS.

Electrical equipment may itself explode or arc violently, and it may also act as a source of ignition of flammable vapours, gases, liquids or dust through electric sparks, arcs or high surface temperatures of equipment. Other causes are heat created by poorly maintained or defective motors, heaters and lighting.

Some portable electric equipment, for example mobile phones, are powered by a battery that requires to be charged to continue to operate. During the charging process electrical energy passes from the charging device to the battery at a rate determined by the charging device. This transfer of electrical energy creates an amount of heat, but if the correct charging device is used it should remain at an acceptable level. If the incorrect charging device is used the transfer of electricity energy may take place too quickly causing the battery in the device receiving the charge to overheat. This could also occur in a situation where the mechanism that stops the charging process when the battery is at maximum charge fails and charging continues.

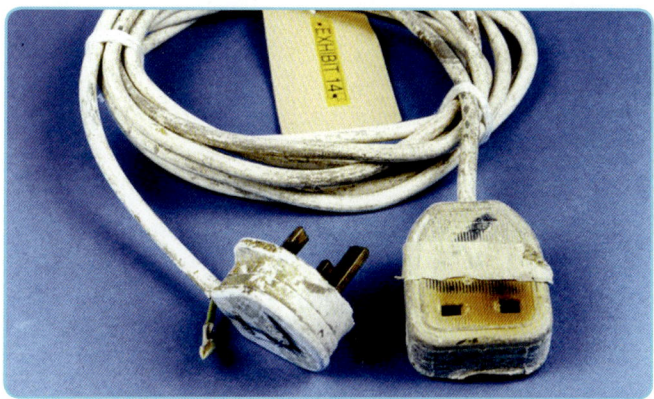

Figure 11-9: Evidence of overheating.
Source: RMS.

Static electricity

Static electricity is different from mains-power and battery-power electricity, as static electricity can be generated naturally. Static electricity is the potential difference (voltage) between surfaces resulting from friction between the surfaces. It is familiar in everyday life as the crackling sound when we remove a woollen sweater and the tiny blue sparks seen in the dark. It is also evident in the clinging together of clothing, paper or sheets of

material or the sharp shock we get when we rub and separate from dissimilar surfaces, such as when getting out of a car and then touching the bodywork. Static electricity may be generated in the following situations:

- The flow of liquids and powders through pipes, for example, when re-fuelling with petrol.
- The pouring of powders from insulating plastic bags.
- Spraying.
- The unwinding of rolled insulating foils.
- The movement of dust or liquids through air.
- The pouring of liquids, granules or powders from insulated containers.

Static electricity build-up is a potential problem where materials that are not very conductive are in contact with each other, for example, plastics and paper, and where two surfaces are rapidly separated. This is particularly dangerous when a static spark is created in a flammable or explosive atmosphere. Given the right mix of flammable material and oxygen, a static spark with sufficient energy can start a fire or explosion. The risk to persons from direct contact with static electricity is low.

Arcing

Arcing is the flow of electricity through the air from one conductor to another. The arc is a visible plasma discharge between the conductors and is caused by the electrical current ionising gases in the air. Electric arcs occur in nature in the form of lightning. The ability of electricity to arc is used industrially for welding, plasma cutting and even certain types of lighting, such as fluorescent lighting.

The dangers associated with arcing increase at the higher electrical voltages found in power distribution, where the electricity has the ability to arc distances of 10 metres or more. Being struck by the electric arc will cause an electric shock and probably significant burns. Indirect burns occur from the radiant heat given off by the arc. Damage to the eyes may result from the ultraviolet light (UV) emitted by the arc.

An arc flash is the sudden release of electrical energy through the air when a high-voltage gap exists and there is a breakdown between conductors.

Figure 11-10: Arc discharge. Source: Solar Power Planet Earth.

The best way to prevent arc flash or to protect workers in the event of an accident/incident is through effective training. In addition to being 'qualified' to local electrical standards, workers who may be exposed to arc flash hazards need to understand why arc flash occurs and how it can be prevented, and should adopt safe working practices to prevent injury.

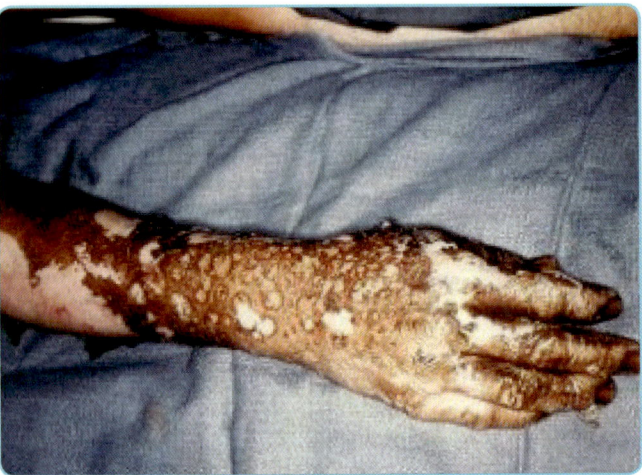

Figure 11-11: Burns from an arc flash of >1000 amps.
Source: SARI/EI.

Common causes of arc flash include:

- Insulation failure.
- Build-up of dust, impurities, and corrosion on insulating surfaces, which can provide a path for current.
- Equipment failure due to the use of sub-standard parts, improper installation, or even normal wear and tear.
- Birds, bees and rodents snapping leads at connections.
- Human error, including dropped tools, accidental contact with electrical systems, and improper work procedures.

An arc flash gives off thermal radiation (heat) and bright, intense light that can cause burns.

Figure 11-12: Equipment damaged by arc discharge.
Source: GS Engineering Consultants Inc.

Temperatures have been recorded as high as 20,000°C (36,000°F). High-voltage arcs can also produce considerable pressure waves by rapidly heating the air and creating a blast. This pressure burst can hit a worker with great force and send molten metal droplets from melted copper and aluminium electrical components great distances at extremely high velocities.

WORKPLACE ELECTRICAL EQUIPMENT

Use of unsuitable electrical equipment

Accidents/incidents can occur when electrical equipment is used that is unsuitable for the work to be done or for the environment. Electrical equipment used in work conditions where there is a high risk of damage, for example construction, mining or quarrying activities, should be of a suitable standard. If equipment is used that is only suitable for low-risk activities, for example domestic use, cable grips may become loose, casings may get damaged and cable insulation may be cut. All of these conditions could lead to users receiving an electric shock.

Portable electrical equipment that relies on an earth to limit the hazardous effects of its metal casing becoming live could be unsuitable if its arduous use is likely to lead to the earth becoming loose or getting damaged as this would present a serious risk of electric shock for the user. Similarly, if the equipment is not capable of the workload expected of it, it may become overloaded and this may cause it to overheat and cause a fire.

The use of some 230v portable electrical equipment in outdoor or wet work conditions may present a risk of electric shock that could be fatal. In such conditions, reduced-voltage equipment, such as that supplied by a transformer with an output that is centre-tapped to earth, might be more suitable, see *'Element 11.2 – Control measures'* later in this element. Similarly, the use of general electrical equipment in a flammable atmosphere could have catastrophic effects as it might ignite the atmosphere and cause an explosion. The connection of electrical equipment to an electrical supply that does not match the equipment specification, including wrong voltage and current, could overload the conductors, leading to a breakdown of insulation and electric shock or fire. A particular problem can arise through deliberate worker bad practices. Electrical equipment may be unsuitable because it has been tampered with. For example, a worker may change the fuse for a higher-rated one, and such practices can lead to a fire or electric shock.

Poorly maintained or defective electrical equipment

If electrical equipment is poorly maintained or becomes defective in use there is an increased risk of a person making contact with a live conductor that is part of the system and has become exposed. In addition, conductive parts of equipment may become live when a fault occurs. Because such faults may not be obvious, a person may use the equipment unknowingly, exposing them to a high risk of electric shock.

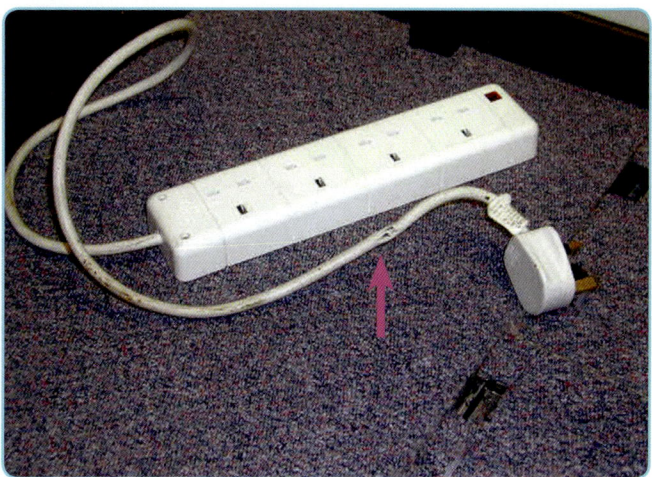

Figure 11-13: Hazard - damaged cable insulation.
Source: RMS.

The risk of fatal electric shock is increased if the equipment has a metal casing and the earth connection has become defective. Metal-cased portable electrical equipment poses a particular problem as the flexing of the cable in use can lead to the earth connection failing, often without the knowledge of the user until a fault occurs and they receive an electric shock. The practice of workers carrying portable electrical equipment by the cable or pulling it along by the cable can lead to conductors becoming loose, which can lead to a fault that could cause a fire or electric shock.

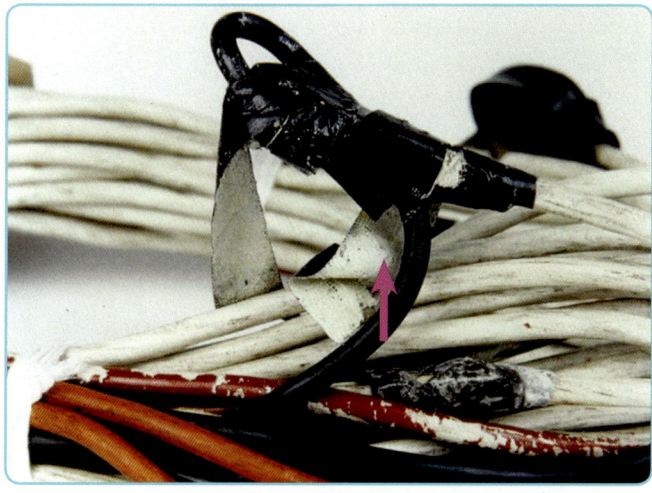

Figure 11-14: Hazard – taped joints.
Source: RMS.

Electrical equipment can become damaged in many ways, for example plugs or sockets may get broken. This could lead to live conductors being exposed or coming into contact with the outer casing of the equipment, which will significantly increase the risk of electrical shock. Cracks in insulated casings of equipment might allow moisture or liquid to enter the equipment in sufficient quantities for the moisture to conduct electricity from the live conductor to a worker in contact with the casing.

Figure 11-15: Hazard – fuse wired out.
Source: RMS.

When in use, the insulation of equipment can become less effective than expected as the plastic insulation breaks down over time or when exposed to ultraviolet light/chemicals. This means that where a live conductor comes into contact with the insulation there is a risk of electric shock to anyone touching it. Control measures are discussed in detail in '**Element 11.2 – Control measures**' later in this element.

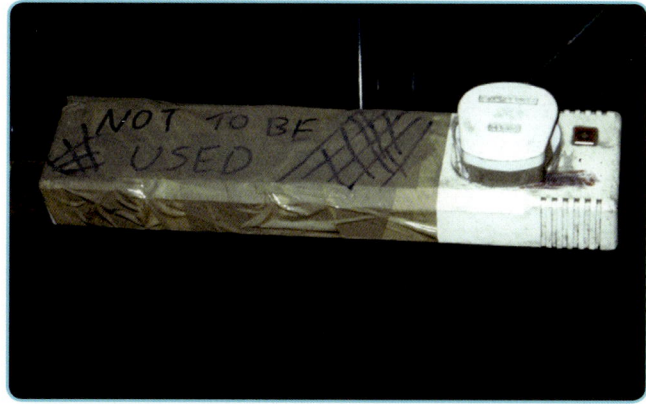

Figure 11-16: Continued use of defective equipment.
Source: RMS.

Use of electrical equipment in wet environments

Use of electrical equipment in wet conditions increases the risk of harm because the wet conditions increase the conductivity of surrounding surfaces. Where a fault exists on electrical equipment in a wet environment, it may not be necessary to make direct contact with the equipment to receive an electric shock, as the wet substance may act as a conductor, making a circuit between the faulty equipment and the person. These conditions may exist where, for example, a plasterer plasters a wall around a faulty light socket or a pressure water cleaner has a damaged cable lying in the water run-off from the cleaning operation.

SECONDARY EFFECTS

Any injury resulting indirectly from receiving an electric shock is a secondary effect. Secondary effects of electricity occur where the flow of electricity through the body causes muscle spasm or the flow/discharge causes surprise to the worker. Muscular spasm may be severe, particularly if the leg muscles are affected, causing a person to be thrown several metres.

When workers are surprised by an electric shock they receive, they may move their hand away rapidly or step back from where contact was made with the electrical source. Injuries may result from dislocation, impact with surrounding objects or fall from a height. In addition, a tool may be dropped, causing burns or impact injuries to the user or others nearby. Therefore, a broken bone, cut or bruise may all be injuries caused by 'secondary effects' from receiving an electric shock.

WORK NEAR OVERHEAD POWER LINES

Contact with live overhead power lines kills a significant number of people and causes serious injuries every year. A high proportion – about one-third – of inadvertent line contacts prove fatal. Overhead power lines may be confused with telephone lines, which can lead people not to identify the risk from contact. Lines may be hard to see at night or against a dark or very bright background.

Power lines typically carry electricity at between 110/230 and 400,000 volts; even contact with an exposed uninsulated 110/230 volt line can be fatal. Because they are generally not covered in insulation (in rural areas), direct contact can easily be made. Contact can lead to current passing through a person, and the item that contacted the line, on a path to earth. Rubber-soled shoes would not provide protection from current flow and shock. Actual contact with a power line is not necessary to result in electric shock; a close approach to the line conductors may allow an arc to take place. The risk of arcing increases as the line voltage increases.

Many occupations may require workers to perform their tasks near overhead power lines. Construction workers, truck drivers, tree service workers, mobile equipment operators, agricultural workers, and others may find themselves carrying out their work in the vicinity of live overhead power lines. They may not be trained to recog-

nise the dangers of electrocution if their bodies, equipment, tools, work materials, or vehicles come near to an over-head power line.

See also '*Protection against overhead cables*' in 11.2 later in this element.

CONTACT WITH UNDERGROUND POWER CABLES DURING EXCAVATION WORK

Underground power cables are present in most locations where excavations are conducted, in shopping areas, on construction or redevelopment sites and even in the countryside. The location of power cables may or may not be known and they may or may not be marked as power cables. Their presence is not obvious when conducting a visual site survey and so the likelihood of striking a power cable when excavating, drilling or piling is increased. The results of striking an underground power cable can include shock, electrocution, explosion and burns.

Figure 11-17: Hydraulic breaker strikes a high-voltage electric cable. Source: UK HSE, HSG47.

Figure 11-18: 11Kv power cable damaged by pneumatic drill. Source: PP Construction Safety.

As with overhead power lines, any underground service should be treated as live until confirmed dead by a power supply authority. Maps showing the location of underground power cables need to be taken only as an indication of their location, not as a guarantee of accuracy, which means that careful testing needs to take place until indicated power cables are located. A common problem when installing cables in the ground is that any attempt to lay them in straight lines is often defeated by the presence of other buried services or their natural tendency to coil out of alignment. Caution needs to be exercised as other unmarked power cables in the area may not be easily detected. See also *'Locating buried services'* in 11.2 later in this element.

WORK ON MAINS ELECTRICAL SUPPLIES

Mains electricity supplies, 110/230 volts, may be treated complacently because these are sometimes viewed as being a common and relatively low voltage. However, a number of deaths occur each year to people working on live mains electricity supplies. It is never absolutely safe to work on live electrical equipment. There are few circumstances where it is necessary to work live, and this must be done only after it has been determined that it is unreasonable for the work to be done dead. In the UK, this is a legal mandatory requirement. Even if working live can be justified, many precautions are needed to make sure that the risk is reduced to an acceptable level.

REVIEW

List four factors that may affect the severity of injury from contact with electricity.

List four common causes of electrical arc flash.

Following an electric shock, a worker may experience secondary effects which result in injury.

Explain the term secondary effects and outline two common injuries which may occur.

11.2 Control measures

PROTECTION OF CONDUCTORS

Conductors, whether live, neutral or earthed, require protection from accidental or deliberate contact, interference, misuse or even abuse. This is often in the form of insulation. Insulators are materials that do not readily conduct electricity. Some common insulator materials are glass, plastic, rubber, air, and wood.

The insulation must be in good condition. Insulation appropriate to the environment should be used to give resistance to abrasion, chemicals, heat and impact. Where conductors that are insulated are exposed to a higher risk of damage they may be further protected by a metal casing, which provides reinforcement around the insulation, or by ducting, which provides a protective location for the cable.

When conductors are covered with insulation they are often called cables or leads. Flexible cables with multi-strand conductors are required for portable tools and extension leads. Cables must be secured by the outer sheath at their point of entry into the apparatus, including plugs. The individual conductor insulation should not show through the sheath and conductors must not be exposed. Extension leads must be fused.

Temporary wiring should be used in compliance with standards and properly secured, supported and mechanically protected against damage. Taped joints in cables are not allowed and proper line connectors must be used to join the conductors in cables. Connectors should be kept to a minimum to reduce earth path impedance (i.e. the total opposition the AC current faces, so includes resistance and other factors beyond the scope of this course).

Many cables are set up on a temporary basis, but these improvised arrangements may get left for a considerable time. Care should be taken to identify true short-term temporary arrangements and those that warrant full longer-term arrangements such as placing them in trunking for better protection.

Attention to cables in offices is particularly important to avoid tripping hazards. Cable trips may loosen or damage the conductors' outer insulation sheath or pull the conductors free from terminals within. If this is not noticed, potential shock injuries may occur to workers who subsequently unplug the affected equipment. Cables routed across roads or pedestrian routes should be covered to protect them from damage. Where cables are to cross a doorway, this is usually best done by taking it around the door frame instead of trailing it across the floor. Overhead cables likely to be hit by vehicles or persons carrying ladders, pipes, etc. should be highlighted by the use of appropriate signs.

Regular examination should be made for deterioration, cuts (these are best identified by systematic bending of short sections of cable by hand, progressing along the length of the cable – this will open any cuts, revealing the conductor), kinks or bend damage (particularly near to the point of entry into the apparatus), exposed conductors, overheat or burn damage, trapping damage, insulation becoming brittle or corrosion.

STRENGTH AND CAPABILITY OF EQUIPMENT

It is critical to ensure that all electrical equipment is suitable for its purpose. Electrical equipment must be carefully selected to ensure it is suitable for the electrical system it will become part of, the task it will be expected to perform and the environment in which it will be used.

For example, if it is to be used for outdoor work on a construction site in conditions where it might get wet, equipment providing protection from the ingress of water must be selected. Many tools designed for use in a domestic situation may not be suitable for use in the potentially harsh conditions (for example, wet, high humidity, flammable atmospheres, in constant use) of a workplace.

For example, cable-entry grips may be required to be more secure and the outer protection of cables thicker on equipment designed for construction work. Part of the selection process is to determine situations where low-voltage systems can be used, for example, 110 volts in preference to 230 volts.

The strength and capability of electrical equipment should be considered when determining its suitability. This should include the fact that any electrical equipment needs to be capable of withstanding the thermal or other effects of the electrical currents that flow through it without failure to danger. This will include normal and fault currents. In order for equipment to remain safe when subjected to sustained fault conditions it may require the inclusion of protective devices that detect the fault and break the circuit containing the fault.

In addition, insulation must be sufficiently effective to enable the equipment to withstand the applied voltage and any transient over-voltages.

Electrical equipment should be used within the manufacturer's rating and the requirements of any instructions supplied. Knowledge of the electrical specification and the tests carried out by the manufacturer, based on the requirements of national or international standards, may be required. This will assist the employer in identifying the strength and capabilities of the equipment, so that it may be selected and installed appropriately.

- Hazardous environments may include:
- Weather – equipment and cables must be capable of withstanding exposure to rain, snow, ice, temperature extremes, etc.
- Natural hazards associated with animals (rats gnawing on cables), solar radiation degradation of insulation, plant growth.
- Extremes of operating temperatures and pressures.
- Corrosive environments (acids/chlorine etc.).
- Flammable substances (gases/dusts/vapours).
- Contamination due to water ingress or dusts.

Additionally, electrical equipment must be sufficiently robust to protect against foreseeable mechanical damage from the environment and from use. If cables are likely to be repeatedly coiled and uncoiled, the movement may result in damage to the flex or connectors. The cable will, therefore, need to be flexible enough to operate in these conditions.

Figure 11-19: Electric cable to the crane coiled (looping).
Source: RMS.

Figure 11-20: Electric cable to the crane uncoiled (stretched).
Source: RMS.

ADVANTAGES AND LIMITATIONS OF PROTECTIVE SYSTEMS

Fuses

A fuse is a device designed to automatically cut off the power supply to a circuit within a given time when the current flow in that circuit exceeds a given value. A fuse may be a tinned copper wire in a suitable carrier or a wire or wires in an enclosed cartridge. In effect, it is a weak link in the circuit that melts when heat is created by too high a current passing through the thin wire in the fuse case. When this happens, the circuit is broken and no more current flows. A fuse usually has a rating in the order of amperes rather than milliamperes, which means it has **limited usefulness in protecting people from electric shock.** The fuse will operate (break the circuit) relatively slowly if the current is just above the fuse rating. Using a fuse with too high a rating means that the circuit will remain intact and the equipment will draw power. This may cause it to overheat, leading to a fire, or if a fault exists, the circuit will remain live and the fault current may pass through the user of the equipment when they touch or operate it.

The following formula should be used to calculate the correct rating for a fuse:

$$\text{Current (amperes)} = \frac{\text{Power (watts)}}{\text{Voltage (volts)}}$$

For example, the correct fuse current rating for a 2 kilowatt kettle on a 230 volt supply would be:

$$\frac{2,000 \text{ W}}{230 \text{ V}} \qquad 8.69 \text{ amperes}$$

Typical fuses for domestic appliances in Europe are 3, 5, and 13 ampere ratings. The nearest fuse just above 8.69 amperes is 13 amperes.

Advantages

- Offers a good level of protection for the equipment.
- Very cheap and reliable within its limits.

Limitations

- A weak link in the circuit that melts slowly when heat is created by a fault condition. However, this usually happens too slowly to protect people.
- Easy to replace with wrong rating.
- Needs tools to replace.
- Easy to override by replacing a fuse with one of a higher rating or putting in an improvised 'fuse', such as a nail, that has a high rating.

Many countries now use circuit breakers (electro-mechanical devices) instead of fuses. The advantages of circuit breakers is that they can be readily reset, but only when the fault condition is rectified, unlike a fuse, which might be replaced by one with the wrong rating.

Earthing

A conductor called an earth wire is fitted to the system; it is connected at one end to a plate buried in the ground and at the other to the metal casing of the equipment. If for any reason a conductor touches the casing so that the equipment casing becomes 'live', the current will flow to the point of lowest potential, the earth. The path to this point (earth) is made easier as the wire is designed to have very little resistance. This may prevent electric shock provided it is used in association with a correctly rated fuse, or, better still, a residual current device (RCD), and no one is in contact with the equipment at the time the fault occurs.

Figure 11-21: Plug - foil fuse, no earth. Source: RMS.

Figure 11-22: Earthing. Source: RMS.

It should be remembered that earthing is provided where the casing can become live. If the equipment is designed so that this cannot happen, such as double-insulated equipment where the user touches non-conducting surfaces, earthing of the equipment is no advantage. In summary, earthing provides a path of least resistance for leaked current and provides protection against indirect shock.

Isolation of supply

Safety devices such as barrier guards or guarding devices are installed on systems to maintain worker safety while these systems are being operated. When non-routine activities such as maintenance, repair, or set-up, or the removal of process material blockages or misaligned feeds are carried out, these safety devices may be removed provided there are alternative methods in place to protect workers from the increased risk of injury from exposure to the unintended or inadvertent release of energy.

	Typical examples of power ratings are:	Suitable fuses at 230 volts:
Computer processor	350 Watts	3 amperes
Electric kettle	1,850–2,200 Watts	10–13 amperes
Dishwasher	1,380 Watts	10 amperes
Refrigerator	90 Watts	3 amperes

Isolation of an electrical circuit involves the removal of electrical power from the circuit or system, sometimes referred to as 'making dead' (zero voltage). Isolation of an electrical system is an excellent way of achieving safety for those that need to work on or near the system; for example, isolation of a power supply into a building that is to be refurbished or isolation of plant that is to be maintained. In its simplest form, it can mean switching off and unplugging a portable appliance at times it is not in use. Care must be taken to check that the isolation has been adequate and effective before work starts; this may include tests on the system. It is also important to ensure the isolation is secure; 'lock-out' and 'tag-out' (LOTO) systems will assist with this.

In practice, lock out is the isolation of electrical energy from the system (a machine, equipment, or process), which physically locks the system controls in a safe mode.

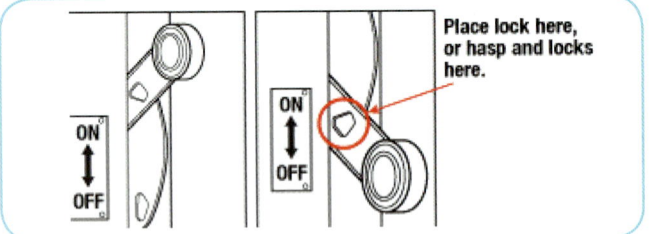

Figure 11-23: Lock out.
Source: Canadian Centre for Occupational Health & Safety.

The energy-isolating device can be a manually operated disconnect switch or a circuit breaker (not the user on–off switch). In most cases, these devices will have loops or tabs which can be locked to a stationary item in a de-energised position. The lock-out device can be any device that has the ability to secure the energy-isolating device in a safe position. See the example of the lock-and-hasp combination in **Figure 11-23**, taken from Canadian standard CSAZ460-05 (R2010) 'Control of Hazardous Energy - Lockout and Other Methods'.

Tag out is a labelling process that is used when lock out is required. The process of tagging out a system involves attaching a standardised label that includes the following information:

- The reason for lock out/tag out (for example, repair or maintenance).
- Time of application of the lock/tag.
- The name of the authorised person who attached the tag and lock to the system.
- To prevent isolated equipment start-up without the authorised individual's knowledge, the LOTO management system should specify that the only person authorised to remove it is the person(s) who placed the lock and tag onto the equipment.

Figure 11-24: Lock-out / tag-out station.
Source: Total Safes.

Figure 11-25: Lock-out / tag-out system in use.
Source: Safety Partners Ltd.

Isolation is a very effective method of protection and ensures people cannot be injured by electrical energy. A limitation of isolation is that certain types of circuit testing and fault finding often have to be carried out with the system live.

Double insulation

This is a common protection device and consists of a layer of insulation around the live electrical parts of the equipment and a second layer of insulation material around this. Commonly, the second layer of insulation is the casing of the equipment (often made from a physical damage-resistant, non-conducting plastic). When the electrical equipment casing is made from an insulator, the equipment is not normally fitted with an earth wire. To make sure that the double insulation is not impaired, the casing must not be pierced by conducting parts such as metal screws. Neither must insulating screws be used, because there is the possibility that they will be lost and will be replaced by metal screws. Any holes in the enclosure of a double-insulated appliance, such as those to allow ventilation, must be so small that fingers cannot reach live parts.

Each layer of insulation must be sufficient in its own right to give adequate protection against shock. In Europe, double-insulated equipment has to be identified by being classified at 'Class II' or have the double-insulated symbol, (**see Figure 11-36**).

In summary, double-insulated equipment has two layers of insulating material between the live parts of the equipment and the user. If a fault occurs with the live parts and a conductor touches the insulating material surrounding it, no current can pass to the user, therefore no shock occurs.

Figure 11-26: Double-Insulated 230V drill.
Source: RMS.

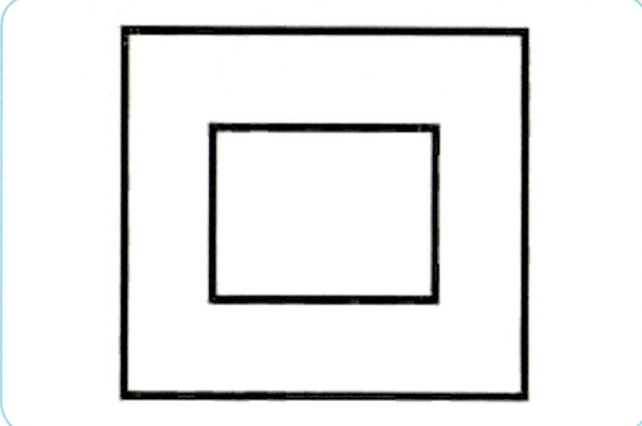

Figure 11-27: Double insulation.
Source: RMS.

Figure 11-28: Double-insulation symbol displayed on equipment.
Source: Panasonic.

Residual current device

A residual current device (RCD) is an electro-mechanical switching device that is used to automatically isolate the supply when there is a difference between the current flowing into a device and the current flowing from the device. Such a difference might result from a fault that causes current leakage, with possible fire risks or the risk of a shock when a person provides a path to earth for the current. *Figure 11-29*, schematics 1 to 5, illustrates the process of disconnection when a fault condition occurs.

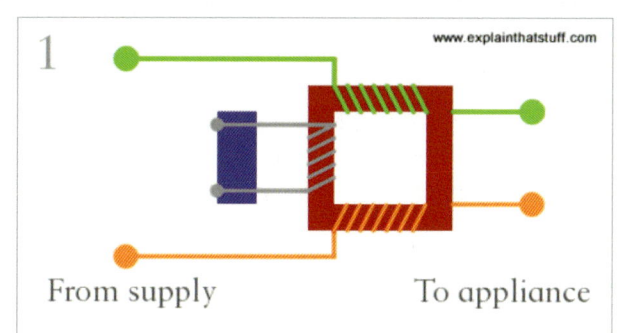

Under normal operation, when power flows from the live wire (green) through the neutral wire (orange) without loss or leakage. The magnetic field is in balance in the core, no electricity flows through the search coil (grey) or relay (blue).

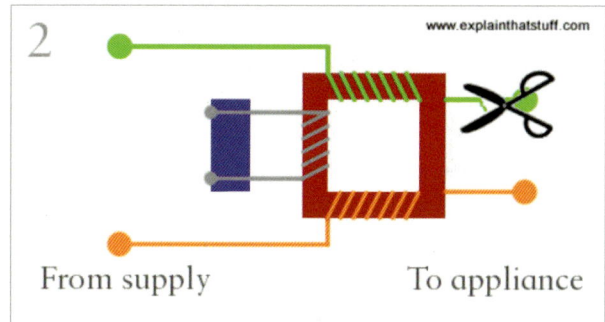

If the live wire (green) is cut (typically by fault or accident).

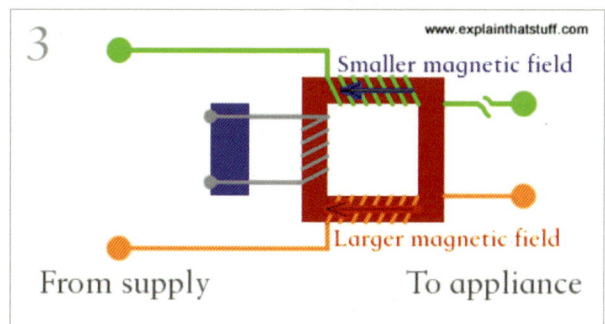

When the live is cut, there is a current imbalance in the core, more current flows through the neutral wire (orange) than through the live wire (green). The two magnetic fields no longer cancel out.

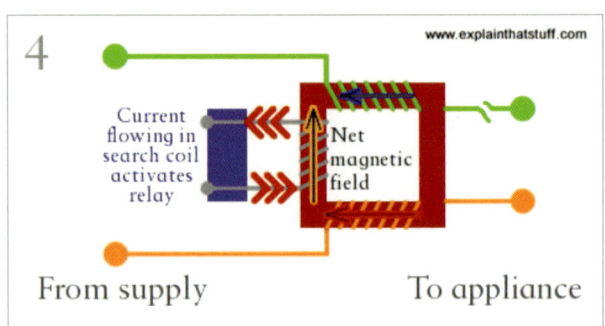

The net magnetic field in the core causes an electric current to flow in the search coil (grey), which activates the relay (blue).

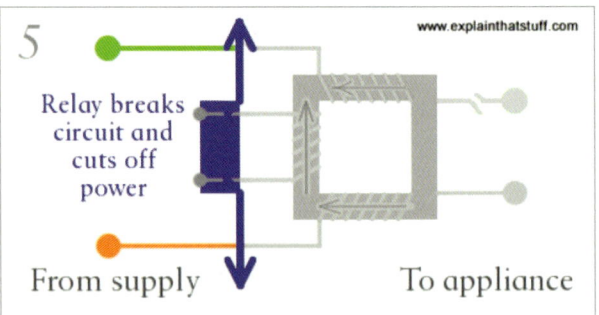

The relay snaps open, breaking the incoming circuit cables and stopping all power from flowing in as little as 30 milliseconds; much faster than the cable can be cut.

Figure 11-29: Operation of an RCD.
Source: www.explainthatstuff.com.

RCDs can be designed to operate at low currents and fast response times (usually 30 mA and 30 ms) and thus they **reduce the effect of an electric shock.** Although they do not prevent the person receiving an electric shock, they are very sensitive and operate very quickly and reduce some of the primary effects of the shock, including fibrillation. Despite this, it is still possible for a person to receive secondary injuries from a fall caused after receiving the electric shock. The equipment needs to be de-energised from time to time in order to be confident it will work properly when needed. This can easily be done by a simple test routine before use, as the equipment is plugged into the RCD.

Summary

- Rapid and sensitive.
- Difficult to defeat.
- Easy and safe to test and reset.
- Does not prevent shock, but reduces the effect of a shock.
- They can be prone to 'nuisance' tripping because they are so sensitive to fault conditions.

Figure 11-30: Portable socket with built-in RCD.
Source: Omega.

Figure 11-31: Residual current device.
Source: RS Components Ltd.

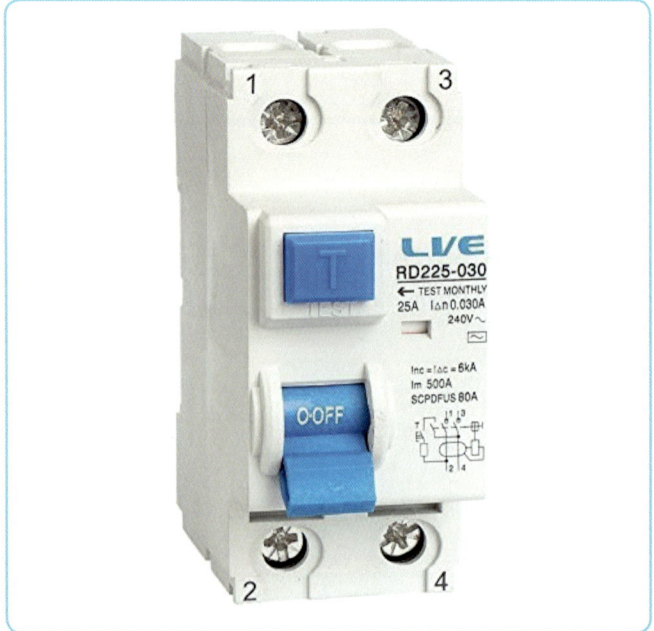

Figure 11-32: RCD/circuit breaker.
Source: RS Components Ltd.

Reduced and low voltage systems

One of the best ways to reduce the risk from electricity is to reduce the supply voltage. This is frequently achieved by the use of a transformer.

In simple terms, a transformer consists of two coils of wires around a common iron core. The first coil of wires is known as the primary winding (and is connected to the supply voltage). The second coil of wires is known as the secondary winding (and generates the output voltage). The ratio between the number of coils in the primary winding and the number of coils in the secondary winding will determine the voltage generated in the secondary winding. If the number of windings in the primary circuit is double that in the secondary, the voltage generated in the secondary circuit will be halved, for example, 230 volts AC to 115 volts AC.

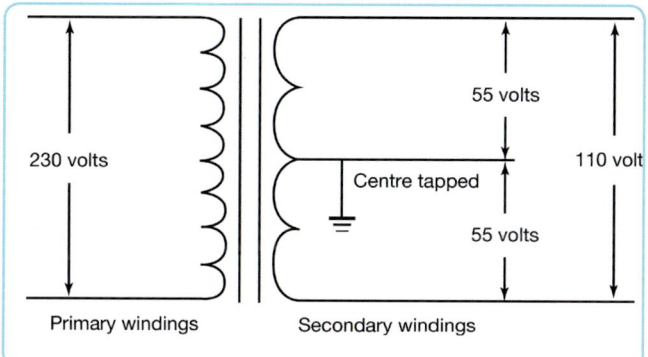

Figure 11-33: Centre-tapped transformer.
Source: RMS

The ratio of winding between the primary and secondary coils can be adjusted (**see Figure 11-33**) to generate any voltage required in the secondary circuit.

In Europe, the primary voltage is 230 volts and the secondary (from a step-down transformer) is 110 volts; when the transformer is centre-tapped to neutral (earth), the generated voltage is 55 volts AC, either side of the centre tap. In practice, this means that any voltage involved in an electrical shock to neutral (earth) will be 55 volt (much less risk to a worker if a fault condition occurs).

Using the earlier example of Ohm's law, if the voltage is 230 volts then:

$$I = \frac{V=}{R} \quad \frac{230 \text{ Volts}}{2,000 + 4,000 \text{ Ohms}} = \frac{0.038 \text{ amperes or 38 mA}}{}$$

However, if a centre tapped to earth transformer is used, then current is reduced substantially:

$$I = \frac{V}{R} \quad \frac{55 \text{ Volts}}{2,000 + 4,000 \text{ Ohms}} \quad \frac{0.009 \text{ amperes or 9 mA}}{}$$

Reference to **Figures 11-2 and 11-5** will clearly show how this **reduces the effects of electric shock on the body**.

Figure 11-34: 110V powered drill.
Source: RMS.

Figure 11-35: 110V powered drill.
Source: RMS.

An alternative to reduction in voltage by means of a transformer is to provide battery-powered equipment; this will commonly run on 12 to 24 volts, but voltages may be higher. The common method is to use a rechargeable battery to power the equipment, which eliminates the need for a cable and gives a greater flexibility of use, for example, for drills and power drivers.

The appropriate siting of the charger unit (connected to 230 volts) is the main risk associated with the use of battery-powered low-voltage equipment.

Figure 11-36: 110V centre tapped earth transformer.
Source: RMS.

Figure 11-37: Battery-powered drill - 19.2V.
Source: Charles & Hudson.

USE OF COMPETENT PERSONS

It is particularly important that anyone who undertakes electrical work is able to satisfy the requirements for competence. Those who wish to undertake electrical testing work would normally be expected to have more knowledge and to be able to demonstrate competence through the successful completion of a suitable training course. More complex electrical tasks such as motor repair or maintenance of radio frequency heating equipment should only be carried out by someone who has been trained to do them. Work on higher-voltage systems must be carried out by specialist electrical engineers trained for this purpose. Work on live electrical systems must be controlled and requires specialist competence and systems of work to ensure safety. The competent person will need to have:

- Knowledge of electricity.
- Experience of electrical work (similar to that being undertaken).
- An understanding of the hazards presented by electricity and the control measures to reduce the hazard.
- The ability to recognise dangerous situations and whether it is safe for work to continue.

A person who is learning these skills and experience will need to be supervised by a competent person.

 **CASE STUDY**

A person received severe electric shock injuries after incorrectly wiring a machine, which resulted in the machine frame becoming live. The injured person was not competent to undertake such work, yet competent persons were available. The incident occurred in the UK and the employer was fined for their failure to supervise.

USE OF SAFE SYSTEMS OF WORK

Live electrical work

Common terms used when describing live working are given in the UK Guidance note HSG85. These include:

Charged:	The item has acquired a charge either because it is live or because it has become charged by other means, such as by static or induction charging, or has retained or regained a charge due to capacitance effects, even though it may be disconnected from the rest of the system.
Live:	Equipment that is at a voltage by being connected to a source of electricity. Live parts that are uninsulated and exposed so that they can be touched either directly or indirectly by a conducting object are hazardous if the voltage exceeds 50V AC or 120V DC in dry conditions – see BSI publication PD65193 – and/or if the fault energy level is high.
Live work:	Work on or near conductors that are accessible and 'live' or 'charged'. Live work includes live testing, such as using a test instrument to measure voltage on a live power distribution or control system.

If it is necessary to work on live conductors, very strict controls must be in place and a robust administrative safe system of work must be adhered to.

These will include:

- A full justification of why there is no reasonably practicable alternative to live working.
- A 'live electrical permit-to-work', which will be valid for the stated task only.
- At least two competent persons must be present during the work and the more senior of these will sign off the permit when the work is complete and after all equipment has been restored to a fully safe condition.
- Restriction of work area to these competent people.
- Protection of the work area from other hazards, such as vehicles.
- Protection of workers from non-essential live conductors, by isolation or screening.
- Provision of information on the system and task.
- Use of suitable insulated test equipment and tools.
- Adequate lighting and clear space to work.
- First-aid service immediately available.
- Competent supervision.

Isolation

In the UK, the HSE Guidance note HSG85 defines isolated as equipment (or part of an electrical system) which is disconnected and separated by a safe distance (the isolating gap) from all sources of electrical energy in such a way that the disconnection is secure, i.e. it cannot be re-energised accidentally or inadvertently.

Any isolation method used for electrical equipment must be adequate and secure. This means that turning equipment off at a general on–off switch or other mechanism is not adequate. Isolation therefore requires dis-connection at the primary isolation mechanism for the equipment or circuit. In order to ensure a high level of security if work on the circuit or equipment takes the person away from the site of the isolator mechanism, a lock or some other means of secure isolation, such as fuse removal, should be considered.

Isolation procedures, often referred to as 'lock out / tag out', in conjunction with a permit-to-work, often form part of a safe system of work. Various competences are required to ensure full compliance with these procedures, with specialist training and supervision required.

Figure 11-38: Multi-lock system.
Source: RMS.

If multi-lock systems are used (***see Figure 11-38***) there must be a contingency plan in case someone loses a key or even leaves at the end of a shift without removing their personal padlock.

CASE STUDY

In the UK, an employee received a 650 volt AC electric shock when he picked up a cable lying on the ground that was connected to a generator and began to apply insulating tape to exposed wires. On prosecution, the employers were fined £15,000.

People who undertake work on electrical equipment or installations should be competent to do so. Those in control of work should instruct workers only to undertake work they are competent to carry out.

Locating buried services

Excavation operations should not begin until all available service location drawings have been identified and thoroughly examined. Record plans and location drawings should not be considered as totally accurate but serve only as an indication of the likelihood of the presence of services, their location and depth.

Figure 11-39: Cable Avoidance Tool. Source: Allpipe Stoppers.

It is possible for the position of an electricity supply cable to alter if previous works have been carried out in the location due to the flexibility of the cable and movement of surrounding features since original installation.

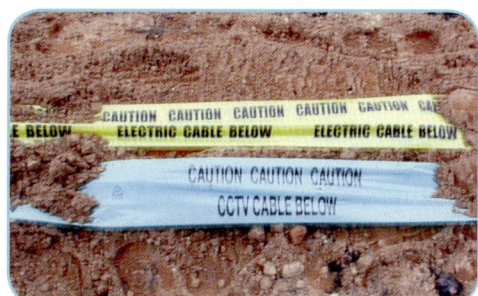

Figure 11-40: Marketing of services. Source: RMS.

Plans often show a proposed position for the services that does not translate to the ground, such that services are placed in position only approximately where the plan says.

Before groundwork is due to commence it is common and good practice to check for the presence of any of the services or hazards by using a detection device. A common detection device used is the cable avoidance tool, more commonly known as a CAT scanner. It is important that service location devices such as CAT scanners are used by competent, trained operatives to assist in the identification and marking of the actual location and position of buried services. When these have been identified, it is essential that physical markings be placed on the ground to show where the services are located.

Protection against overhead cables

Contact with live overhead lines kills people and causes serious injuries every year. It is important, where possible, to eliminate the danger by:

- Avoidance – carrying out as much work as possible remote from the lines.
- Diversion – diverting all overhead lines clear of the work area.
- Isolation – make lines dead while the work is in progress.
- Insulation.

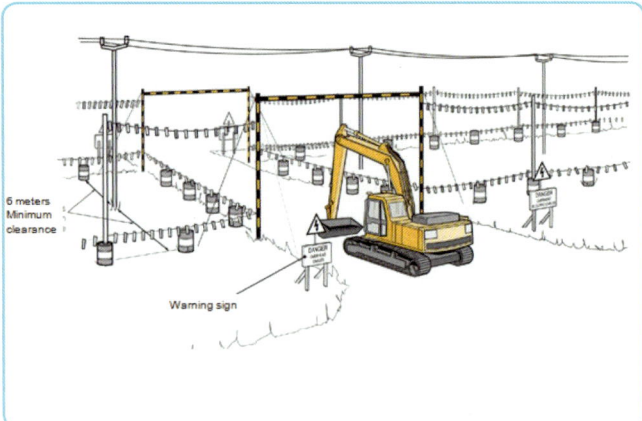

Figure 11-41: Passing beneath power lines, overhead cable protection. Source: UK HSE, HSG144.

Figure 11-42: Working near power lines, ground-level barrier. Source: UK HSE, HSG150.

If the danger cannot be eliminated, the risk must be managed by controlling access to, and work beneath, over-head power lines. For lower-voltage cables it may be possible to temporarily isolate the supply and fit insulation to prevent contact with the live conductor when power is restored. The following precautions may also be needed to manage the risk:

- Isolate the supply.
- Erect 'goal-post' barriers to define clearance distances.
- Clearly mark danger zones with signs and/or bunting.
- Ensure safe access under lines.
- Use a signaller/banksman or marshals where appropriate.
- Restrict the use of metal equipment such as ladders and scaffolds.

EMERGENCY PROCEDURES FOLLOWING AN ELECTRICAL INCIDENT

Anyone working around electrical systems should be aware of what needs to be done for a casualty of electrical shock.

If someone is lying unconscious and in contact with conductors the following actions should be taken in an order depending on the circumstances:

- Assess the situation.
- Summon help, including qualified medical support.
- If possible, shut off the power.
- Do not touch the casualty – there may be enough voltage across the body of the casualty to shock the would-be rescuer. The problem with this rule is that the source of power may not be known or easily found in time to save the casualty from shock.
- Remove the power. If possible, prove the system is discharged and dead. If this is not possible, take the following action.
- Remove the casualty from the power. It may be possible to dislodge the casualty from the circuit with a dry wooden board or piece of non-metallic material or using a jacket as a loop around the person, holding both sleeves and pulling away.
- Reassess the situation and any remaining danger to yourself and the casualty.
- Once the casualty has been disconnected from the source of electric power, the immediate medical concerns should be respiration and circulation (breathing and pulse). If the rescuer is trained in cardiopulmonary resuscitation (CPR), they should follow the appropriate steps of checking breathing (including the airway) and pulse, then applying CPR as necessary to keep the casualty's body from de-oxygenating.
- If the casualty is conscious, lay them in the recovery position and keep them warm to reduce the chances of physiological shock until qualified medical personnel arrive on the scene.
- Keep the casualty under observation for secondary effects. Cool burns with water.

Figure 11-43: First-aid sign. Source: RMS.

Further considerations:

- Do not go near the casualty until the electricity supply is proven to be off. This is especially important with overhead high-voltage lines: keep yourself and others at least 18 metres away until the electricity supply company personnel advise otherwise.

Figure 11-44: Risk of electric shock due to damage to cable resting on metal checker plate flooring. Source: RMS.

- Do not delay – after 3 minutes without blood circulation, irreversible damage can be done to the casualty.

- Do not wait for an accident to happen – train in emergency procedures and first-aid, plan procedures for an emergency (calling for help, making calls to the emergency services, meeting ambulances and leading them to the casualty) and hold emergency drills.

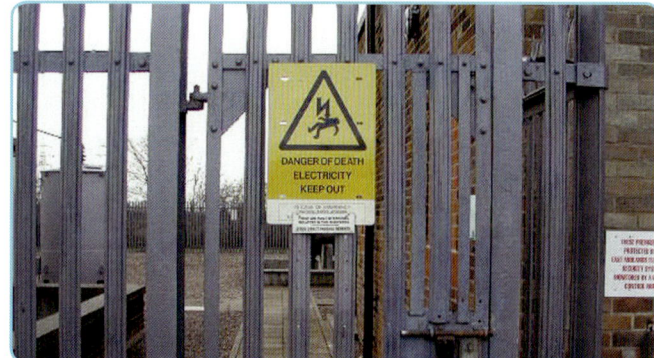

Figure 11-45: Restriction on work on live credits.
Source: RMS.

- Establish whether the accident/incident has to be reported to the competent authority under national legislation.

INSPECTION AND MAINTENANCE STRATEGIES

Maintenance

All electrical systems should be maintained by those that have control of them to an extent necessary to prevent danger. Danger can be considered to be the risk of injury from electric shock, electric burn, fire of an electrical origin, electric arcing or explosion initiated or caused by electricity. Maintenance is a general term that in practice can include visual inspection, repair, testing and replacement. Maintenance will determine whether equipment is fully serviceable or needing repair. It further suggests that cost-effective maintenance can be achieved by a combination of:

- Checks by the user.
- Visual inspections by a person appointed to do this.
- Combined inspection and tests by a competent person or by a contractor.

In order to identify what systems will need maintenance, they should be listed. This same listing can be used as a checklist recording that the appropriate checks have been done. It may also include details of the type of equipment and the checks and tests to be carried out. In the UK, Regulation 4 of the Electricity at Work Regulations 1989 states:

> "All systems shall at all times be of such construction as to prevent, so far as reasonably practicable, such danger". "As may be necessary to prevent danger, all systems shall be maintained so as to prevent, so far as reasonably practicable, such danger."

Figure 11-46: Maintenance of electrical system.
Source: UK, Regulation 4, Electricity at Work Regulations 1980.

User checks

The user of electrical equipment should be encouraged, after basic training, to look critically at apparatus and the source of power. If any defects are found, the apparatus should be marked and not be used again before examination by a competent person. Obviously, there must be a procedure by which the user brings faults to the attention of a supervisor and/or a competent person who might rectify the fault. Checks by the user are the first line of defence but should never be the only line taken. Such inspections should be aimed at identifying the following:

- Cuts, abrasions and cracks in inner and outer cable insulation.
- Damaged plugs – cracked casing or bent pins, failure of the cord grip.

- Taped or other inadequate cable joints.
- Evidence of bare wires.
- Faulty or ineffective switches.
- Overloaded sockets.
- Lack of formal testing.
- Burn marks or discolouration.
- Damaged casing.
- Loose parts or screws.
- Wet or contaminated equipment.
- Loose or damaged sockets or switches.
- Lack of circuit protection – no RCD.
- Evidence of unauthorised repairs.
- Cables trapped under furniture or in floor boxes.
- Vent holes blocked.

Formal inspection and tests

Inspection

The maintenance system should always include formal visual inspection of all portable electrical equipment and electrical tests. The factors that might affect the frequency of inspection and testing of, say, a portable drill that is used on a building site, include the nature of the work and the environmental conditions in which the drill is to be used, the frequency and duration of use, the age of the equipment, the intrinsic safety features of the equipment such as double insulation and the use of low voltage, user checks and the number of problems reported, the number and competency of the users, manufacturers' recommendations and best practice guidance, and the results of previous tests and inspections.

The inspection can be done by a member of staff who has been trained in what to look for and has basic electrical knowledge. They should know enough to avoid danger to themselves or others. Visual inspections are likely to need to look for the same types of defects as user checks, but should also include opening accessible parts of equipment/plugs (where this is possible) and inspection of fixed installations.

Opening plugs of portable equipment to check for:

- Use of correctly rated fuse.
- Effective cord grip.
- Secure and correct cable terminations.

Inspection of fixed installations for:

- Damaged or loose cable racks, conduit or cabling.
- Missing, broken or inadequately secured covers.
- Loose or faulty joints.
- Loose earth connections.
- Moisture, corrosion or contamination.
- Burn marks or discolouration.
- Open or inadequately secured panel doors.
- Ease of access to switches and isolators.
- Presence of temporary wiring.

User checks and a programme of formal visual inspections are found to pick up some 95% of faults.

Testing

Faults such as loss of earth or broken wires inside an installation or cable cannot be found by visual inspection, so some apparatus needs to have a combined inspection and test. This is particularly important for all earthed equipment and leads and plugs connected to hand-held or hand-operated equipment. The system should be tested regularly; tests may include earth continuity and impedance tests and tests of insulation material.

Frequency of inspection and testing

Deciding the frequency

Many approaches to establishing frequency suggest that inspection and testing should be done regularly. The term 'regularly' is not generally specified in terms of a fixed time interval that is applicable for all systems. A risk-based judgement must be made to decide an appropriate frequency.

In effect, the frequency will depend on the conditions in which the system is used; for example, a test of office portable equipment may be sufficient if conducted every three years, whereas equipment used on a construction site may need to be tested every three months. The electrical system as a whole, not just portable equipment, must also be tested periodically. Again, this depends on the conditions of use and may vary from 10 years to 6 months. Factors to be considered when deciding the frequency include:

- Type of equipment.
- Whether it is hand-held.
- Manufacturer's recommendations.
- Its initial integrity and soundness.
- Age.
- Working environment.
- Likelihood of mechanical damage.
- Frequency of use.
- Duration of use.
- Foreseeable use.
- Who uses it?
- Modifications or repairs.
- Past experience.

Records of inspection and testing

In order to identify what systems and equipment will need inspection and testing they should be listed.

This same listing can be used as a checklist recording that the appropriate checks, inspections and tests have been done. It would be usual to include details of the type of the equipment, its location and its age.

It is important that a cumulative record of equipment and its status is available to managers who are responsible for employees using the equipment as well as contractors or suitably qualified employees who are conducting the inspection or test.

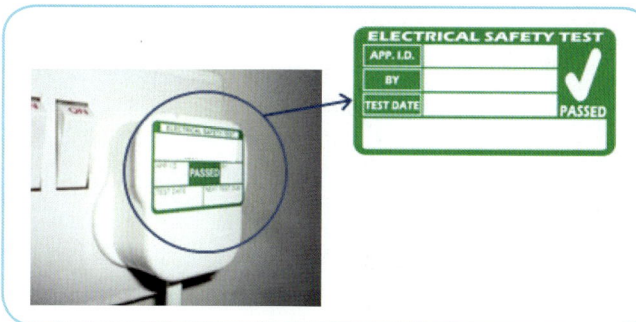

Figure 11-47: PAT labels. Source: RMS.

In addition, it is common practice to add a label to the system or part of the system (for example, portable appliances) to indicate that an inspection and/or test has taken place and its status following this. Some labels show the date that this took place; others prefer to show the date of next inspection or test.

There is a growing trend, especially in offices, for workers to bring to work their own electrically powered equipment, including calculators, radios, kettles and coffee makers.

The number of electrical accidents/incidents has grown accordingly, and fires from electrical equipment chargers left on overnight are growing in number. All such equipment should be recorded, inspected and tested by a competent person before use and at regular intervals, as if it were the property of the organisation.

 **CONSIDER**

What faults would you look for when carrying out a visual inspection of a portable electric appliance?

Advantages and limitations of portable appliance testing

The purpose of portable appliance testing (PAT) is to periodically confirm the critical aspects of the electrical integrity of portable appliances. Three levels of inspection should be included in a maintenance and inspection strategy for portable electrical appliances:

1) The first level of inspection should be carried out by the operator before the appliance is used and consists of an informal check of the condition of the appliance and its cable and plug.

2) The second level of inspection is a more formal visual inspection by an appointed person, which follows a laid-down procedure and includes other matters such as the correctness of the rating of the fuses fitted, security of cable grips, earth continuity, impedance and insulation.

3) The third level of the strategy includes the periodic combined inspection and testing of the appliance by a competent person.

It is important to keep centralised records of the results of PAT within an organisation. Such records can then be used for setting the frequency for appliance testing, to verify whether unlabelled equipment had been tested or has merely lost its label, and to provide a record of past faults on all appliances that have been reported. This approach will demonstrate that the employer is in compliance with the Regulations.

The limitation with PAT is that people tend to over-rely on the apparent assurance that the test indicates. They may be tempted to see it as a permanent assurance that the equipment is safe. This can lead to users not making their own pre-use checks of the appliance. The appliance test is of its condition at a point in time. The test does not assure that the fuse is of the correct rating or that the cable grip is sound and not loose.

Fixed installations, for example outlet sockets, should also be inspected (depending on use, typically every five years) to ensure appropriate surface insulation (no visible cracks, scorching or damage), fitted with a functioning on–off switch, correctly wired line polarity (live, neutral), and resistance of earth connection (as low as possible).

Sources of reference

Reference information provided, in particular web links, was correct at time of publication, but may have changed.

Ambient factors in the workplace, International Labour Organisation (ILO) Code of Practice (CoP), ISBN 92-2-11628-, http://www.ilo.org/safework/info/standards-and-instruments/WCMS_107729/lang--en/index.htm

ATEX - Directive 2014/34/EU, https://eur-lex.europa.eu/legal-content/EN/TXT/?uri=CELEX:32014L0034

Electricity at work, safe working practices, HSG85, HSE Books, ISBN: 978-0-7176-6581-5 http://www.hse.gov.uk/pUbns/priced/hsg85.pdf

Guidance on the management of electrical safety and safe isolation procedures for low voltage installations, Electrical Safety First in partnership with others (including the HSE), https://www.electricalsafetyfirst.org.uk/downloads/

The health and safety toolbox, How to control risks at work, HSG268, HSE Books, ISBN: 978-0-7176-6587-7, http://www.hse.gov.uk/pUbns/priced/hsg268.pdf

IET Wiring Regulations 18th Edition BS7671:2018

Maintaining Portable and Transportable Electrical Equipment, HSG107, HSE Books, ISBN: 978-0-7176-6606-5 http://www.hse.gov.uk/pubns/priced/hsg107.pdf

The Electricity at Work Regulations 1989, HSR25 HSE Books, ISBN: 978-0-7176-6636-2, UK, http://www.hse.gov.uk/pubns/priced/hsr25.pdf

Web links to these references are provided on the RMS Publishing website for ease of use - www.rmspublishing.co.uk

STUDY QUESTIONS

1) Explain the progressive effects that electrical shock may have on the body. (8)

2) Explain the purpose of electrical double insulation. (8)

3) What are typical user checks that should be carried out, prior to using a portable item of electrical equipment? (8)

4) What should you consider when determining the frequency for the inspection and testing of electrical equipment? (8)

5) What are the advantages and disadvantages of the use of a fuse as a protective device in an electrical circuit? (8)

For guidance on how to answer NEBOSH questions please refer to the 'study question answer guide' section located at the back of this guide.

Summary of ILO, OSH Conventions and Recommendations

The following abridged versions of the International Labour Organisation (ILO), occupational safety and health (OSH) conventions and recommendations are included to show how member states of the United Nations have evolved fundamental standards for the improvement in the health safety and welfare of workers over time.

International labour standards are legal instruments drawn up by the ILO and set out basic principles and rights at work. They are either **conventions**, which are legally binding international treaties that may be ratified by member states, or **recommendations**, which serve as accompanying non-binding guidelines.

ASBESTOS CONVENTION C162, 1986

Convention concerning Safety in the Use of Asbestos.

PART I. SCOPE AND DEFINITIONS

Article 1

1. This Convention applies to all activities involving exposure of workers to asbestos in the course of work.

Article 2

For the purpose of this Convention-

b) the term asbestos means the fibrous form of mineral silicates belonging to rock-forming minerals of the serpentine group, i.e. chrysotile (white asbestos), and of the amphibole group, i.e. actinolite, amosite (brown asbestos, cummingtonite-grunerite), anthophyllite, crocidolite (blue asbestos), tremolite, or any mixture containing one or more of these.

PART II. GENERAL PRINCIPLES

Article 3

1. National laws or regulations shall prescribe the measures to be taken for the prevention and control of, and protection of workers against, health hazards due to occupational exposure to asbestos.

PART III. PROTECTIVE AND PREVENTIVE MEASURES

Article 9

The national laws or regulations adopted pursuant to Article 3 of this Convention shall provide that exposure to asbestos shall be prevented or controlled by one or more of the following measures:

a) making work in which exposure to asbestos may occur subject to regulations prescribing adequate engineering controls and work practices, including workplace hygiene;

b) prescribing special rules and procedures, including authorisation, for the use of asbestos or of certain types of asbestos or products containing asbestos or for certain work processes.

Article 11

1. The use of crocidolite and products containing this fibre shall be prohibited.

Article 15

1. The competent authority shall prescribe limits for the exposure of workers to asbestos or other exposure criteria for the evaluation of the working environment.

Article 16

Each employer shall be made responsible for the establishment and implementation of practical measures for the prevention and control of the exposure of the workers he employs to asbestos and for their protection against the hazards due to asbestos.

Article 19

1. In accordance with national law and practice, employers shall dispose of waste containing asbestos in a manner that does not pose a health risk to the workers concerned, including those handling asbestos waste, or to the population in the vicinity of the enterprise.

PART IV. SURVEILLANCE OF THE WORKING ENVIRONMENT AND WORKERS' HEALTH

Article 20

1. Where it is necessary for the protection of the health of workers, the employer shall measure the concentrations of airborne asbestos dust in workplaces, and shall monitor the exposure of workers to asbestos at intervals and using methods specified by the competent authority.

For the full text of this Convention go to the ILO website:

http://www.ilo.org/dyn/normlex/en/f?p=NORML EXPUB:12100:0::NO::P12100_ILO_CODE:C162

ASBESTOS RECOMMENDATION R172, 1986

Recommendation concerning Safety in the Use of Asbestos

I. SCOPE AND DEFINITIONS

1. The provisions of the Asbestos Convention, 1986, and of this Recommendation should be applied to all activities involving a risk of exposure of workers to asbestos in the course of work.

2. Measures should be taken, in accordance with national law and practice, to afford to self-employed persons protection analogous to that provided for in the Asbestos Convention, 1986, and in this Recommendation.

3. Employment of young persons of less than 18 years of age in activities involving a risk of occupational

exposure to asbestos should receive special attention, as required by the competent authority.

II. GENERAL PRINCIPLES

4. The measures prescribed pursuant to Article 3 of the Asbestos Convention, 1986, should be so framed as to cover the diversity of risks of occupational exposure to asbestos in all branches of economic activity, and should be drawn up with due regard to Articles 1 and 2 of the Occupational Cancer Convention, 1974.

5. The competent authority should periodically review the measures prescribed, taking into account the Code of practice on safety in the use of asbestos published by the International Labour Office and other codes of practice or guides which may be established by the International Labour Office and the conclusions of meetings of experts which may be convened by it, as well as information from other competent bodies on asbestos and substitute materials.

6. The competent authority, in the application of the provisions of this Recommendation, should act after consultation with the most representative organisations of employers and workers.

III. PROTECTIVE AND PREVENTIVE MEASURES

13. (1) With a view to the effective enforcement of the national laws and regulations, the competent authority should prescribe the information to be supplied in the notifications of work with asbestos provided for in Article 13 of the Asbestos Convention, 1986.

(2) This information should include in particular the following:

a)   the type and quantity of asbestos used;

b)   the activities and processes carried out;

c)   the products manufactured;

d)   the number of workers exposed and the level and frequency of their exposure;

e)   the preventive and protective measures taken to comply with the national laws and regulations;

f)   any other information necessary to safeguard the workers' health.

IV. SURVEILLANCE OF THE WORKING ENVIRONMENT AND WORKERS' HEALTH

31. (1) For the prevention of disease and functional impairment related to exposure to asbestos, all workers assigned to work involving exposure to asbestos should be provided, as appropriate, with-

a)   a pre-assignment medical examination;

b)   periodic medical examinations at appropriate intervals;

c)   other tests and investigations, in particular chest radiographs and lung function tests, which may be necessary to supervise their state of health in relation to the occupational hazard and to identify early indicators of disease caused by asbestos.

(2) The intervals between medical examinations should be determined by the competent authority, taking into account the level of exposure and the workers' state of health in relation to the occupational hazard.

(3) The competent authority should ensure that provision is made, in accordance with national law and practice, for appropriate medical examinations to continue to be available to workers after termination of an assignment involving exposure to asbestos.

(4) The examinations, tests and investigations provided for in subparagraphs (1) and (3) above should be carried out as far as possible in working hours and should entail no cost to the worker.

(5) Where the results of medical tests or investigations reveal clinical or preclinical effects, measures should be taken to prevent or reduce exposure of the workers concerned and to prevent further deterioration of their health.

36. (1) The records of the monitoring of the working environment should be kept for a period of not less than 30 years.

(2) Records of the monitoring of exposure of workers as well as the sections of their medical files relevant to health hazards due to exposure to asbestos and chest radiographs should be kept for a period of not less than 30 years following termination of an assignment involving exposure to asbestos.

V. INFORMATION AND EDUCATION

40. The competent authority should take measures to promote the training and information of all persons concerned with respect to the prevention and control of, and protection against, health hazards due to occupational exposure to asbestos.

41. The competent authority, in consultation with the most representative organisations of employers and workers concerned, should draw up suitable educational guides for employers, workers and others.

For the full text of this Recommendation go to the ILO website:

http://www.ilo.org/dyn/normlex/en/f?p=1000:12100:0::NO::P12100_ILO_CODE:R172

CHEMICALS CONVENTION C170, 1990

Convention concerning Safety in the use of Chemicals at Work.

Asbestos and carcinogens each have their own Conventions: Asbestos Convention C162, 1986 (see page 458) and Occupational Cancer Convention C139, 1974 (see page 467).

Article 1

1. This Convention applies to all branches of economic activity in which chemicals are used.

Article 2

For the purposes of this Convention:

a) the term chemicals means chemical elements and compounds, and mixtures thereof, whether natural or synthetic;

b) the term hazardous chemical includes any chemical which has been classified as hazardous in accordance with Article 6 or for which relevant information exists to indicate that the chemical is hazardous;

PART III. CLASSIFICATION AND RELATED MEASURES

Article 6
Classification systems

1. Systems and specific criteria appropriate for the classification of all chemicals according to the type and degree of their intrinsic health and physical hazards and for assessing the relevance of the information required to determine whether a chemical is hazardous shall be established by the competent authority, or by a body approved or recognised by the competent authority, in accordance with national or international standards.

Article 7
Labelling and marking

1. All chemicals shall be marked so as to indicate their identity.

Article 8
Chemical safety data sheets

1. For hazardous chemicals, chemical safety data sheets containing detailed essential information regarding their identity, supplier, classification, hazards, safety precautions and emergency procedures shall be provided to employers.

Article 9
Responsibilities of suppliers

1. Suppliers of chemicals, whether manufacturers, importers or distributors, shall ensure that:

a) such chemicals have been classified in accordance with Article 6 on the basis of knowledge of their properties and a search of available information or assessed in accordance with paragraph 3 to determine whether they are hazardous chemicals.

PART IV. RESPONSIBILITIES OF EMPLOYERS

Article 10
Identification

1. Employers shall ensure that all chemicals used at work are labelled or marked as required by Article 7 and that chemical safety data sheets have been provided as required by Article 8 and are made available to workers and their representatives.

Article 11
Transfer of chemicals

Employers shall ensure that when chemicals are transferred into other containers or equipment, the contents are indicated in a manner which will make known to workers their identity, any hazards associated with their use and any safety precautions to be observed.

Article 12
Exposure

Employers shall:

a) ensure that workers are not exposed to chemicals to an extent which exceeds exposure limits or other exposure criteria for the evaluation and control of the working environment established by the competent authority, or by a body approved or recognised by the competent authority, in accordance with national or international standards;

Article 13
Operational control

1. Employers shall make an assessment of the risks arising from the use of chemicals at work, and shall protect workers against such risks by appropriate means,

2. Employers shall:

a) limit exposure to hazardous chemicals so as to protect the safety and health of workers;

b) provide first aid;

c) make arrangements to deal with emergencies and disposal.

Article 15
Information and training

Employers shall:

a) inform the workers of the hazards associated with exposure to chemicals used at the workplace;

b) instruct the workers how to obtain and use the information provided on labels and chemical safety data sheets;

PART V. DUTIES OF WORKERS

Article 17

1. Workers shall co-operate as closely as possible with their employers in the discharge by the employers of their responsibilities and comply with all procedures and practices relating to safety in the use of chemicals at work.

2. Workers shall take all reasonable steps to eliminate or minimise risk to themselves and to others from the use of chemicals at work.

For the full text of this Convention go to the ILO

website:http://www.ilo.org/dyn/normlex/en/f?p=1000:12100:0::NO::P12100_ILO_CODE:C170

CHEMICALS RECOMMENDATION R177, 1990

Recommendation concerning Safety in the use of Chemicals at Work.

I. GENERAL PROVISIONS

1. The provisions of this Recommendation should be applied in conjunction with those of the Chemicals Convention, 1990 (hereafter referred to as "the Convention").

2. The most representative organisations of employers and workers concerned should be consulted on the measures to be taken to give effect to the provisions of this Recommendation.

3. The competent authority should specify categories of workers who for reasons of safety and health are not allowed to use specified chemicals or are allowed to use them only under conditions prescribed in accordance with national laws or regulations.

4. The provisions of this Recommendation should also apply to such self-employed persons as may be specified by national laws or regulations.

5. The special provisions established by the competent authority to protect confidential information, under Article 1, paragraph 2(b), and Article 18, paragraph 4, of the Convention.

II. CLASSIFICATION AND RELATED MEASURES

Classification
6. The criteria for the classification of chemicals established pursuant to Article 6, paragraph 1, of the Convention should be based upon the characteristics of chemicals.

Labelling and marking
8. (1) The requirements for the labelling and marking of chemicals established pursuant to Article 7 of the Convention, should be such as to enable persons handling or using chemicals to recognise and distinguish between them both when receiving and when using them, so that they may be used safely.

(2) The labelling requirements for hazardous chemicals should, in conformity with existing national or international systems, cover:

a) the information to be given on the label including as appropriate:

i.   trade names;

ii. identity of the chemical;

iii. name, address and telephone number of the supplier;

iv. hazard symbols;

v.   nature of the special risks associated with the use of the chemical;

vi. safety precautions;

vii. identification of the batch;

viii. the statement that a chemical safety data sheet giving additional information is available from the employer;

ix. the classification assigned under the system established by the competent authority;

b) the legibility, durability and size of the label;

c) the uniformity of labels and symbols, including colours.

(3) The label should be easily understandable by workers.

(4) In the case of chemicals not covered by subparagraph (2) above, the marking may be limited to the identity of the chemical.

9. Where it is impracticable to label or mark a chemical in view of the size of the container or the nature of the package, provision should be made for other effective means of recognition such as tagging or accompanying documents. However, all containers of hazardous chemicals should indicate the hazards of the contents through appropriate wording or symbols.

Chemical safety data sheets
10. (1) The criteria for the preparation of chemical safety data sheets for hazardous chemicals should ensure that they contain essential information including, as applicable:

a) chemical product and company identification (including trade or common name of the chemical and details of the supplier or manufacturer);

b) composition/information on ingredients (in a way that clearly identifies them for the purpose of conducting a hazard evaluation).

III. RESPONSIBILITIES OF EMPLOYERS

Monitoring of exposure

11. (1) Where workers are exposed to hazardous chemicals, the employer should be required to:

a) limit exposure to such chemicals so as to protect the health of workers;

b) assess, monitor and record, as necessary, the concentration of airborne chemicals at the workplace,

Operational control within the workplace

12. (1) Measures should be taken by employers to protect workers against hazards arising from the use of chemicals at work, based upon the criteria established pursuant to Paragraph 13 below.

13. The competent authority should ensure that criteria are established for safety in the use of hazardous chemicals, including provisions covering, as applicable:

a) the risk of acute or chronic diseases due to entry into the body by inhalation, skin absorption or ingestion;

b) the risk of injury or disease from skin or eye contact;

c) the risk of injury from fire, explosion or other events resulting from physical properties or chemical reactivity;

d) the precautionary measures to be taken through:

i. the choice of chemicals that eliminate or minimise such risks;

ii. the choice of processes, technology and installations that eliminate or minimise such risks;

iii. the use and proper maintenance of engineering control measures;

iv. the adoption of working systems and practices that eliminate or minimise such risks;

v. the adoption of adequate personal hygiene measures and provision of adequate sanitary facilities;

vi. the provision, maintenance and use of suitable personal protective equipment and clothing, at no cost to the worker where the above measures have not proved sufficient to eliminate such risks;

vii. the use of signs and notices;

viii. adequate preparations for emergencies.

14. The competent authority should ensure that criteria are established for safety in the storage of hazardous chemicals, including provisions covering, as applicable:

a) the compatibility and segregation of stored chemicals;

b) the properties and quantity of chemicals to be stored;

c) the security and siting of and access to stores;

d) the construction, nature and integrity of storage containers;

e) loading and unloading of storage containers;

f) labelling and relabelling requirements;

g) precautions against accidental release, fire, explosion and chemical reactivity;

h) temperature, humidity and ventilation;

i) precautions and procedures in case of spillage;

j) emergency procedures;

k) possible physical and chemical changes in stored chemicals.

15. The competent authority should ensure that criteria consistent with national or international transport regulations are established for the safety of workers involved in the transport of hazardous chemicals.

Medical surveillance

18. (1) The employer, or the institution competent under national law and practice, should be required to arrange, through a method which accords with national law and practice, such medical surveillance of workers as is necessary.

V. Rights of workers

26. Workers should receive:

a) information on the classification and labelling of chemicals and on chemical safety data sheets in forms and languages which they easily understand;

b) information on the risks which may arise from the use of hazardous chemicals in the course of their work;

c) instruction, written or oral, based on the chemical safety data sheet and specific to the workplace if appropriate;

d) training and, where necessary, retraining in the methods which are available for the prevention and control of, and for protection against, such risks, including correct methods of storage, transport and waste disposal as well as emergency and first-aid measures.

For the full text of this Recommendation go to the ILO website:

http://www.ilo.org/dyn/normlex/
en/f?p=1000:12100:0::NO::P12100_ILO_CODE:R177

EMPLOYMENT INJURY BENEFITS CONVENTION C121, 1964

Convention concerning Benefits in the Case of Employment Injury.

Article 4

1. National legislation concerning employment injury benefits shall protect all employees, including apprentices, in the public and private sectors, including co-operatives, and, in respect of the death of the breadwinner, prescribed categories of beneficiaries.

2. Any Member may make such exceptions as it deems necessary in respect of-

   a) persons whose employment is of a casual nature and who are employed otherwise than for the purpose of the employer's trade or business;

   b) out-workers;

   c) members of the employer's family living in his house, in respect of their work for him;

   d) other categories of employees, which shall not exceed in number 10 percent of all employees other than those excluded under clauses (a) to (c).

Article 5

Where a declaration provided for in Article 2 is in force, the application of national legislation concerning employment injury benefits may be limited to prescribed categories of employees, which shall total in number not less than 75 percent of all employees in industrial undertakings, and, in respect of the death of the breadwinner, prescribed categories of beneficiaries.

Article 6

The contingencies covered shall include the following where due to an employment injury:

a) a morbid condition;

b) incapacity for work resulting from such a condition and involving suspension of earnings, as defined by national legislation;

c) total loss of earning capacity or partial loss thereof in excess of a prescribed degree, likely to be permanent, or corresponding loss of faculty; and

d) the loss of support suffered as the result of the death of the breadwinner by prescribed categories of beneficiaries.

Article 7

1. Each Member shall prescribe a definition of "industrial accident", including the conditions under which a commuting accident is considered to be an industrial accident, and shall specify the terms of such definition in its reports upon the application of this Convention submitted under Article 22 of the Constitution of the International Labour Organisation.

2. Where commuting accidents are covered by social security schemes other than employment injury schemes, and these schemes provide in respect of commuting accidents benefits which, when taken together, are at least equivalent to those required under this Convention, it shall not be necessary to make provision for commuting accidents in the definition of "industrial accident".

Article 8

Each Member shall-

a) prescribe a list of diseases, comprising at least the diseases enumerated in Schedule I to this Convention, which shall be regarded as occupational diseases under prescribed conditions; or

b) include in its legislation a general definition of occupational diseases broad enough to cover at least the diseases enumerated in Schedule I to this Convention; or

c) prescribe a list of diseases in conformity with clause (a), complemented by a general definition of occupational diseases or by other provisions for establishing the occupational origin of diseases not so listed or manifesting themselves under conditions different from those prescribed.

Article 9

1. Each Member shall secure to the persons protected, subject to prescribed conditions, the provision of the following benefits:

   a) medical care and allied benefits in respect of a morbid condition;

   b) cash benefits in respect of the contingencies specified in Article 6, clauses (b), (c) and (d).

2. Eligibility for benefits may not be made subject to the length of employment, to the duration of insurance or to the payment of contributions: Provided that a period of exposure may be prescribed for occupational diseases.

For the full text of this Recommendation go to the ILO website:

https://www.ilo.org/dyn/normlex/en/f?p=NORML EXPUB:12100:0::NO::P12100_ILO_CODE:C121

HOURS OF WORK AND REST PERIODS (ROAD TRANSPORT) CONVENTION C153, 1979

Convention concerning Hours of Work and Rest Periods in Road Transport.

Article 1

1. This Convention applies to wage-earning drivers working, whether for undertakings engaged in transport for third parties or for undertakings transporting goods or passengers for own account, on motor vehicles engaged professionally in the internal or international transport by road of goods or passengers.

2. Except as otherwise provided herein, this Convention further applies to owners of motor vehicles engaged professionally in road transport and non-wage-earning members of their families, when they are working as drivers.

Article 4

1. For the purpose of this Convention the term hours of work means the time spent by wage-earning drivers on-

   a) driving and other work during the running time of the vehicle; and

   b) subsidiary work in connection with the vehicle, its passengers or its load.

2. Periods of mere attendance or stand-by, either on the vehicle or at the workplace and during which the drivers are not free to dispose of their time as they please, may be regarded as hours of work to an extent to be prescribed in each country by the competent authority or body, by collective agreements or by any other means consistent with national practice.

Article 5

1. No driver shall be allowed to drive continuously for more than four hours without a break.

2. The competent authority or body in each country, taking into account particular national conditions, may authorise the period referred to in paragraph 1 of this Article to be exceeded by not more than one hour.

3. The length of the break referred to in this Article and, as appropriate, the way in which the break may be split shall be determined by the competent authority or body in each country.

4. The competent authority or body in each country may specify cases in which the provisions of this Article are inapplicable because drivers have sufficient breaks as a result of stops provided for in the timetable or as a result of the intermittent nature of the work.

Article 6

1. The maximum total driving time, including overtime, shall exceed neither nine hours per day nor 48 hours per week.

2. The total driving times referred to in paragraph 1 of this Article may be calculated as an average over a number of days or weeks to be determined by the competent authority or body in each country.

3. The total driving times referred to in paragraph 1 of this Article shall be reduced in the case of transport activities carried out in particularly difficult conditions. The competent authority or body in each country shall define these activities and determine the total driving times to be applied in respect of the drivers concerned.

Article 10

1. The competent authority or body in each country shall-

   a) provide for an individual control book and prescribe the conditions of its issue, its contents and the manner in which it shall be kept by the drivers; and

   a) lay down a procedure for notification of the hours worked in accordance with Article 9, paragraph 1, of this Convention and the circumstances justifying them.

2. Each employer shall-

   a) keep a record, in a form approved by the competent authority or body in each country, indicating the hours of work and of rest of every driver employed by him; and

   b) place this record at the disposal of the supervisory authorities in a manner determined by the competent authority or body in each country.

Article 15

1. This Convention shall be binding only upon those Members of the International Labour Organisation whose ratifications have been registered with the Director-General.

2. It shall come into force twelve months after the date on which the ratifications of two Members have been registered with the Director-General.

3. Thereafter, this Convention shall come into force for any Member twelve months after the date on which its ratification has been registered.

For the full text of this Convention go to the ILO website:

http://www.ilo.org/dyn/normlex/
en/f?p=1000:12100:0::NO::P12100_ILO_CODE:C153

HYGIENE (COMMERCE AND OFFICES) CONVENTION C120, 1964

Convention concerning Hygiene in Commerce and Offices.

PART I. OBLIGATIONS OF PARTIES

Article 1

This Convention applies to-

a) trading establishments;

b) establishments, institutions and administrative services in which the workers are mainly engaged in office work;

PART II. GENERAL PRINCIPLES

Article 7

All premises used by workers, and the equipment of such premises, shall be properly maintained and kept clean.

Article 8

All premises used by workers shall have sufficient and suitable ventilation, natural or artificial or both, supplying fresh or purified air.

Article 9

All premises used by workers shall have sufficient and suitable lighting; workplaces shall, as far as possible, have natural lighting.

Article 10

As comfortable and steady a temperature as circumstances permit shall be maintained in all premises used by workers.

Article 11

All workplaces shall be so laid out and work-stations so arranged that there is no harmful effect on the health of the worker.

Article 12

A sufficient supply of wholesome drinking water or of some other wholesome drink shall be made available to workers.

Article 13

Sufficient and suitable washing facilities and sanitary conveniences shall be provided and properly maintained.

Article 14

Sufficient and suitable seats shall be supplied for workers and workers shall be given reasonable opportunities of using them.

Article 15

Suitable facilities for changing, leaving and drying clothing which is not worn at work shall be provided and properly maintained.

Article 16

Underground or windowless premises in which work is normally performed shall comply with appropriate standards of hygiene.

Article 17

Workers shall be protected by appropriate and practicable measures against substances, processes and techniques which are obnoxious, unhealthy or toxic or for any reason harmful. Where the nature of the work so requires, the competent authority shall prescribe personal protective equipment.

Article 18

Noise and vibrations likely to have harmful effects on workers shall be reduced as far as possible by appropriate and practicable measures.

Article 19

Every establishment, institution or administrative service, or department thereof, to which this Convention applies shall, having regard to its size and the possible risk-

a) maintain its own dispensary or first-aid post; or

b) maintain a dispensary or first-aid post jointly with other establishments, institutions or administrative services, or departments thereof; or

c) have one or more first-aid cupboards, boxes or kits.

For the full text of this Convention go to the ILO website:

http://www.ilo.org/dyn/normlex/en/f?p=1000:12100:0::NO::P12100_ILO_CODE:C120

HYGIENE (COMMERCE AND OFFICES) RECOMMENDATION R120, 1964

III. Maintenance and Cleanliness

5. All places in which work is carried on, or through which workers may have to pass, or which contain sanitary or other facilities provided for the common use of workers, and the equipment of such places, should be properly maintained.

6. (1) All such places and equipment should be kept clean.

(2) In particular the following should be regularly cleaned:

a) floors, stairs and passages;

b) windows used for lighting, and sources of artificial lighting;

c) walls, ceilings and equipment.

7. Cleaning should be carried out--

a) by means raising the minimum amount of dust;

b) outside working hours, except in particular circumstances or where cleaning during working hours can be effected without disadvantage for the workers.

8. Cloakrooms, lavatories, washstands and, if necessary, other facilities for the common use of workers should be regularly cleaned and periodically disinfected.

9. All refuse and waste likely to give off obnoxious, toxic or harmful substances, or be a source of infection, should be made harmless, removed or isolated at the earliest possible moment; disposal should be in accordance with standards approved by the competent authority.

10. Removal and disposal arrangements for other refuse and waste should be made and sufficient receptacles for such refuse and waste should be provided in suitable places.

IV. Ventilation

11. In all places in which work is carried on, or which contain sanitary or other facilities for the common use of workers, there should be sufficient and suitable ventilation, natural or artificial or both, supplying fresh or purified air.

12. In particular-

a) apparatus ensuring natural or artificial ventilation should be so designed as to introduce a sufficient quantity of fresh or purified air per person and per hour into an area, taking into account the nature and conditions of the work;

b) arrangements should be made to remove or make harmless, as far as possible, fumes, dust and any other obnoxious or harmful impurities which may be generated in the course of work;

c) the normal speed of movement of air at fixed workstations should not be harmful to the health or comfort of the persons working there;

d) as far as possible and in so far as conditions require, appropriate measures should be taken to ensure that in enclosed premises a suitable hygrometric level in the air is maintained.

13. Where a workplace is wholly or substantially air conditioned, suitable means of emergency ventilation, natural or artificial, should be provided.

V. Lighting

14. In all places in which work is carried on, or through which workers may have to pass or which contain sanitary or other facilities provided for the common use of workers, there should be, as long as the places are likely to be used, sufficient and suitable lighting, natural or artificial, or both.

15. In particular, all practicable measures should be taken-

a) to ensure visual comfort-

i. by openings for natural lighting which are appropriately distributed and of sufficient size;

ii. by a careful choice and appropriate distribution of artificial lighting;

iii. by a careful choice of colours for the premises and their equipment;

b) to prevent discomfort or disorders caused by glare, excessive contrasts between light and shade, reflection of light and over-strong direct lighting;

c) to eliminate harmful flickering whenever artificial lighting is used.

16. Wherever sufficient natural lighting is reasonably practicable it should be adopted in preference to any other.

17. Suitable standards of natural or artificial lighting for different types of work and premises and various occupations should be fixed by the competent authority.

18. In premises where there are large numbers of workers or visitors, emergency lighting should be provided.

For the full text of this Convention go to the ILO website:

https://www.ilo.org/dyn/normlex/
en/f?p=NORMLEXPUB:55:0:::55:P55_TYPE,P55_
LANG,P55_DOCUMENT,P55_NODE:REC,en,R120,/
Document

LIST OF OCCUPATIONAL DISEASES RECOMMENDATION R194, 2002

1. In the establishment, review and application of systems for the recording and notification of occupational accidents and diseases, the competent authority should take account of the 1996 Code of practice on the recording and notification of occupational accidents and diseases, and other codes of practice or guides relating to this subject that are approved in the future by the International Labour Organization.

2. A national list of occupational diseases for the purposes of prevention, recording, notification and, if applicable, compensation should be established by the competent authority, in consultation with the most representative organizations of employers and workers, by methods appropriate to national conditions and practice, and by stages as necessary. This list should:

a) for the purposes of prevention, recording, notification and compensation comprise, at the least, the diseases enumerated in Schedule I of the Employment Injury Benefits Convention, 1964, as amended in 1980;

b) comprise, to the extent possible, other diseases contained in the list of occupational diseases as annexed to this Recommendation; and

c) comprise, to the extent possible, a section entitled "Suspected occupational diseases".

3. The list as annexed to this Recommendation should be regularly reviewed and updated through tripartite meetings of experts convened by the Governing Body of the International Labour Office. Any new list so established shall be submitted to the Governing Body for its approval, and upon approval shall replace the preceding list and shall be communicated to the Members of the International Labour Organization.

4. The national list of occupational diseases should be reviewed and updated with due regard to the most up-to-date list established in accordance with Paragraph 3 above.

5. Each Member should communicate its national list of occupational diseases to the International Labour Office as soon as it is established or revised, with a view to facilitating the regular review and updating of the list of occupational diseases annexed to this Recommendation.

6. Each Member should furnish annually to the International Labour Office comprehensive statistics on occupational accidents and diseases and, as appropriate, dangerous occurrences and commuting accidents with a view to facilitating the international exchange and comparison of these statistics.

For the full text of this Convention go to the ILO website:

https://www.ilo.org/dyn/normlex/en/f?p=NORMLEXPUB:12100:0::NO::P12100_ILO_CODE:R194

OCCUPATIONAL CANCER CONVENTION C139, 1974

Convention concerning Prevention and Control of Occupational Hazards caused by Carcinogenic Substances and Agents.

Article 1

1. Each Member which ratifies this Convention shall periodically determine the carcinogenic substances and agents to which occupational exposure shall be prohibited or made subject to authorisation or control, and those to which other provisions of this Convention shall apply.

Article 2

1. Each Member which ratifies this Convention shall make every effort to have carcinogenic substances and agents to which workers may be exposed in the course of their work replaced by non-carcinogenic substances or agents or by less harmful substances or agents.

Article 4

Each Member which ratifies this Convention shall take steps so that workers who have been, are, or are likely to be exposed to carcinogenic substances or agents are provided with all the available information on the dangers involved and on the measures to be taken.

Article 5

Each Member which ratifies this Convention shall take measures to ensure that workers are provided with such medical examinations or biological or other tests or investigations during the period of employment and thereafter as are necessary to evaluate their exposure and supervise their state of health in relation to the occupational hazards.

For the full text of this Convention go to the ILO website:

http://www.ilo.org/dyn/normlex/en/f?p=NORMLEXPUB:12100:0::NO:12100:P12100_ILO_CODE:C139

OCCUPATIONAL CANCER RECOMMENDATION R147, 1974

Recommendation concerning Prevention and Control of Occupational Hazards caused by Carcinogenic Substances and Agents.

I. GENERAL PROVISIONS

1. Every effort should be made to replace carcinogenic substances and agents to which workers may be exposed in the course of their work by non-carcinogenic substances or agents or by less harmful substances or agents; in the choice of substitute substances or agents account should be taken of their carcinogenic, toxic and other properties.

2. The number of workers exposed to carcinogenic substances or agents and the duration and degree of such exposure should be reduced to the minimum compatible with safety.

II. PREVENTIVE MEASURES

6. The competent authority should periodically determine the carcinogenic substances and agents to which occupational exposure should be prohibited or made

subject to authorisation or control.

7. In making such determinations the competent authority should give consideration to the latest information contained in the codes of practice or guides which may be established by the International Labour Office, and in the conclusions of meetings of experts which may be convened by the International Labour Office, as well as to information from other competent bodies.

III. SUPERVISION OF HEALTH OF WORKERS

11. Provision should be made, by laws or regulations or any other method consistent with national practice and conditions, for all workers assigned to work involving exposure to specified carcinogenic substances or agents to undergo as appropriate-

a) a pre-assignment medical examination;

b) periodic medical examinations at suitable intervals;

c) biological or other tests and investigations which may be necessary to evaluate their exposure and supervise their state of health in relation to the occupational hazards.

IV. INFORMATION AND EDUCATION

16. (1) The competent authority should promote epidemiological and other studies and collect and disseminate information relevant to occupational cancer risks, with the assistance as appropriate of international and national organisations, including organisations of employers and workers.

(2) It should endeavour to establish the criteria for determining the carcinogenicity of substances and agents.

V. MEASURES OF APPLICATION

22. Each Member should-

a) by laws or regulations or any other method consistent with national practice and conditions, take such steps, including the provision of appropriate penalties, as may be necessary to give effect to the provisions of this Recommendation;

b) in accordance with national practice, specify the bodies or persons on whom the obligation of compliance with the provisions of this Recommendation rests;

c) provide appropriate inspection services for the purpose of supervising the application of the provisions of this Recommendation, or satisfy itself that appropriate inspection is carried out.

For the full text of this Recommendation go to the ILO website:
http://www.ilo.org/dyn/normlex/en/f?p=1000:12100:0::NO::P12100_ILO_CODE:R147

OCCUPATIONAL HEALTH SERVICES CONVENTION C161, 1985

Convention concerning Occupational Health Services

PART I. PRINCIPLES OF NATIONAL POLICY

Article 1

For the purpose of this Convention-

a) the term occupational health services means services entrusted with essentially preventive functions and responsible for advising the employer, the workers and their representatives in the undertaking on-

i. the requirements for establishing and maintaining a safe and healthy working environment which will facilitate optimal physical and mental health in relation to work;

ii. the adaptation of work to the capabilities of workers in the light of their state of physical and mental health;

b) the term workers' representatives in the undertaking means persons who are recognised as such under national law or practice.

PART II. FUNCTIONS

Article 5

Without prejudice to the responsibility of each employer for the health and safety of the workers in his employment, and with due regard to the necessity for the workers to participate in matters of occupational health and safety, occupational health services shall have such of the following functions as are adequate and appropriate to the occupational risks of the undertaking:

a) identification and assessment of the risks from health hazards in the workplace;

b) surveillance of the factors in the working environment and working practices which may affect workers' health, including sanitary installations, canteens and housing where these facilities are provided by the employer;

c) advice on planning and organisation of work, including the design of workplaces, on the choice, maintenance and condition of machinery and other equipment and on substances used in work;

d) participation in the development of programmes for the improvement of working practices as well as testing and evaluation of health aspects of new equipment;

e) advice on occupational health, safety and hygiene and on ergonomics and individual and collective protective equipment;

f) surveillance of workers' health in relation to work;

g) promoting the adaptation of work to the worker;

h) contribution to measures of vocational rehabilitation;

i) collaboration in providing information, training and education in the fields of occupational health and hygiene and ergonomics;

j) organising of first aid and emergency treatment;

k) participation in analysis of occupational accidents and occupational diseases

PART IV. CONDITIONS OF OPERATION

Article 9

1. In accordance with national law and practice, occupational health services should be multidisciplinary. The composition of the personnel shall be determined by the nature of the duties to be performed.

2. Occupational health services shall carry out their functions in co-operation with the other services in the undertaking.

3. Measures shall be taken, in accordance with national law and practice, to ensure adequate co-operation and co-ordination between occupational health services and, as appropriate, other bodies concerned with the provision of health services.

Article 10

The personnel providing occupational health services shall enjoy full professional independence from employers, workers, and their representatives, where they exist, in relation to the functions listed in Article 5.

Article 11

The competent authority shall determine the qualifications required for the personnel providing occupational health services, according to the nature of the duties to be performed and in accordance with national law and practice.

Article 12

The surveillance of workers' health in relation to work shall involve no loss of earnings for them, shall be free of charge and shall take place as far as possible during working hours.

Article 13

All workers shall be informed of health hazards involved in their work.

Article 14

Occupational health services shall be informed by the employer and workers of any known factors and any suspected factors in the working environment which may affect the workers' health.

For the full text of this Convention go to the ILO website:

https://www.ilo.org/dyn/normlex/en/f?
p=NORMLEXPUB:12100:0::NO:121
00:P12100_INSTRUMENT_ID:312306:NO

OCCUPATIONAL HEALTH SERVICES RECOMMENDATION R171, 1985

Recommendation concerning Occupational Health Services

I. PRINCIPLES OF NATIONAL POLICY

1. Each Member should, in the light of national conditions and practice and in consultation with the most representative organisations of employers and workers, where they exist, formulate, implement and periodically review a coherent national policy on occupational health services, which should include general principles governing their functions, organisation and operation.

2. (1) Each Member should develop progressively occupational health services for all workers, including those in the public sector and the members of production co-operatives, in all branches of economic activity and all undertakings. The provision made should be adequate and appropriate to the specific health risks of the undertakings.

(2) Provision should also be made for such measures as may be necessary and reasonably practicable to make available to self-employed persons protection analogous to that provided for in the Occupational Health Services Convention, 1985, and in this Recommendation.

II. FUNCTIONS

3. The role of occupational health services should be essentially preventive.

4. Occupational health services should establish a programme of activity adapted to the undertaking or undertakings they serve, taking into account in particular the occupational hazards in the working environment as well as the problems specific to the branches of economic activity concerned.

A. Surveillance of the working environment

5. (1) The surveillance of the working environment should include-

a) identification and evaluation of the environmental factors which may affect the workers' health;

b) assessment of conditions of occupational hygiene and factors in the organisation of work which may give rise to risks for the health of workers;

c) assessment of collective and personal protective equipment;

d) assessment where appropriate of exposure of workers to hazardous agents by valid and generally accepted monitoring methods;

e) assessment of control systems designed to eliminate or reduce exposure.

(2) Such surveillance should be carried out in liaison with the other technical services of the undertaking and in co-operation with the workers concerned and their representatives in the undertaking or the safety and health committee, where they exist.

B. Surveillance of the workers' health

11. (1) Surveillance of the workers' health should include, in the cases and under the conditions specified by the competent authority, all assessments necessary to protect the health of the workers, which may include-

a) health assessment of workers before their assignment to specific tasks which may involve a danger to their health or that of others;

b) health assessment at periodic intervals during employment which involves exposure to a particular hazard to health;

c) health assessment on resumption of work after a prolonged absence for health reasons for the purpose of determining its possible occupational causes, of recommending appropriate action to protect the workers and of determining the worker's suitability for the job and needs for reassignment and rehabilitation;

d) health assessment on and after the termination of assignments involving hazards which might cause or contribute to future health impairment.

(2) Provisions should be adopted to protect the privacy of the workers and to ensure that health surveillance is not used for discriminatory purposes or in any other manner prejudicial to their interests.

14. (1) Occupational health services should record data on workers' health in personal confidential health files. These files should also contain information on jobs held by the workers, on exposure to occupational hazards involved in their work, and on the results of any assessments of workers' exposure to these hazards.

C. Information, education, training, advice

19. Occupational health services should participate in designing and implementing programmes of information, education and training on health and hygiene in relation to work for the personnel of the undertaking.

D. First aid, treatment and health programmes

23. Taking into account national law and practice, occupational health services in undertakings should provide first-aid and emergency treatment in cases of accident or indisposition of workers at the workplace and should collaborate in the organisation of first aid.

IV. CONDITIONS OF OPERATION

39 (1) The competent authority may prescribe standards for the premises and equipment necessary for occupational health services to exercise their functions.

(2) Occupational health services should have access to appropriate facilities for carrying out the analyses and tests necessary for surveillance of the workers' health and of the working environment.

For the full text of this Recommendation go to the ILO website:

http://www.ilo.org/dyn/normlex/
en/f?p=1000:12100:0::NO::P12100_ILO_CODE:R171

OCCUPATIONAL SAFETY AND HEALTH (DOCK WORK) CONVENTION C152, 1979

Article 1

For the purpose of this Convention, the term dock work covers all and any part of the work of loading or unloading any ship as well as any work incidental thereto; the definition of such work shall be established by national law or practice. The organisations of employers and workers concerned shall be consulted on or otherwise participate in the establishment and revision of this definition.

Article 4

1. National laws or regulations shall prescribe that measures complying with Part III of this Convention be taken as regards dock work with a view to--

a) providing and maintaining workplaces, equipment and methods of work that are safe and without risk of injury to health;

b) providing and maintaining safe means of access to any workplace;

c) providing the information, training and supervision necessary to ensure the protection of workers against risks of accident or injury to health arising out of or in the course of their employment;

d) providing workers with any personal protective equipment and protective clothing and any lifesaving appliances reasonably required where adequate protection against risks of accident or injury to health cannot be provided by other means;

e) providing and maintaining suitable and adequate first-aid and rescue facilities;

f) developing and establishing proper procedures to deal with any emergency situations which may arise.

2. The measures to be taken in pursuance of this Convention shall cover--

a) general requirements relating to the construction, equipping and maintenance of dock structures and other places at which dock work is carried out;

b) fire and explosion prevention and protection;

c) safe means of access to ships, holds, staging, equipment and lifting appliances;

d) transport of workers;

e) opening and closing of hatches, protection of hatchways and work in holds;

f) construction, maintenance and use of lifting and other cargo-handling appliances;

g) construction, maintenance and use of staging;

h) rigging and use of ship's derricks;

i) testing, examination, inspection and certification, as appropriate, of lifting appliances, of loose gear, including chains and ropes, and of slings and other lifting devices which form an integral part of the load;

j) handling of different types of cargo;

k) stacking and storage of goods;

l) dangerous substances and other hazards in the working environment;

m) personal protective equipment and protective clothing;

n) sanitary and washing facilities and welfare amenities;

o) medical supervision;

p) first-aid and rescue facilities;

q) safety and health organisation;

r) training of workers;

s) notification and investigation of occupational accidents and diseases.

3. The practical implementation of the requirements prescribed in pursuance of paragraph 1 of this Article shall be ensured or assisted by technical standards or codes of practice approved by the competent authority, or by other appropriate methods consistent with national practice and conditions.

For the full text of this Recommendation go to the ILO website:

https://www.ilo.org/dyn/normlex/en/f?p=NORMLEXPUB:12100:0::NO::P12100_ILO_CODE:C152

OCCUPATIONAL SAFETY AND HEALTH CONVENTION C155, 2003

Convention concerning Occupational Safety and Health and the Working Environment

PART I. SCOPE AND DEFINITIONS

Article 1

1. This Convention applies to all branches of economic activity.

PART II. PRINCIPLES OF NATIONAL POLICY

Article 4

1. Each Member shall, in the light of national conditions and practice, and in consultation with the most representative organisations of employers and workers, formulate, implement and periodically review a coherent national policy on occupational safety, occupational health and the working environment.

2. The aim of the policy shall be to prevent accidents and injury to health arising out of, linked with or occurring in the course of work, by minimising, so far as is reasonably practicable, the causes of hazards inherent in the working environment.

Article 5

The policy referred to in Article 4 of this Convention shall take account of the following main spheres of action in so far as they affect occupational safety and health and the working environment:

a) design, testing, choice, substitution, installation, arrangement, use and maintenance of the material elements of work (workplaces, working environment, tools, machinery and equipment, chemical, physical and biological substances and agents, work processes);

b) relationships between the material elements of work and the persons who carry out or supervise the work, and adaptation of machinery, equipment, working time, organisation of work and work processes to the physical and mental capacities of the workers;

c) training, including necessary further training, qualifications and motivations of persons involved, in one capacity or another, in the achievement of adequate levels of safety and health;

d) communication and co-operation at the levels of the working group and the undertaking and at all other appropriate levels up to and including the national level;

e) the protection of workers and their representatives from disciplinary measures as a result of actions properly taken by them in conformity with the policy referred to in Article 4 of this Convention.

Article 7

The situation regarding occupational safety and health and the working environment shall be reviewed at appropriate intervals, either over-all or in respect of particular areas, with a view to identifying major problems, evolving effective methods for dealing with them and priorities of action, and evaluating results.

Article 9

1. The enforcement of laws and regulations concerning occupational safety and health and the working environment shall be secured by an adequate and appropriate system of inspection.

2. The enforcement system shall provide for adequate penalties for violations of the laws and regulations.

Article 12

Measures shall be taken, in accordance with national law and practice, with a view to ensuring that those who design, manufacture, import, provide or transfer machinery, equipment or substances for occupational use-

a) satisfy themselves that, so far as is reasonably practicable, the machinery, equipment or substance does not entail dangers for the safety and health of those using it correctly;

b) make available information concerning the correct installation and use of machinery and equipment and the correct use of substances, and information on hazards of machinery and equipment and dangerous properties of chemical substances and physical and biological agents or products, as well as instructions on how known hazards are to be avoided;

c) undertake studies and research or otherwise keep abreast of the scientific and technical knowledge necessary to comply with subparagraphs (a) and (b) of this Article.

For the full text of this Convention go to the ILO website:

http://www.ilo.org/dyn/normlex/en/f?p=NORMLEX
PUB:12100:0::NO:12100:P12100_ILO_CODE:C155

OCCUPATIONAL SAFETY AND HEALTH RECOMMENDATION R164, 2006

Recommendation concerning Occupational Safety and Health and the Working Environment.

II. TECHNICAL FIELDS OF ACTION

3. As appropriate for different branches of economic activity and different types of work and taking into account the principle of giving priority to eliminating hazards at their source, measures should be taken in pursuance of the policy referred to in Article 4 of the Convention, in particular in the following fields:

a) design, siting, structural features, installation, maintenance, repair and alteration of workplaces and means of access thereto and egress therefrom;

b) lighting, ventilation, order and cleanliness of workplaces;

c) temperature, humidity and movement of air in the workplace;

d) design, construction, use, maintenance, testing and inspection of machinery and equipment liable to present hazards and, as appropriate, their approval and transfer;

e) prevention of harmful physical or mental stress due to conditions of work;

f) handling, stacking and storage of loads and materials, manually or mechanically;

g) use of electricity;

h) manufacture, packing, labelling, transport, storage and use of dangerous substances and agents, disposal of their wastes and residues, and, as appropriate, their replacement by other substances or agents which are not dangerous or which are less dangerous;

i) radiation protection;

j) prevention and control of, and protection against, occupational hazards due to noise and vibration;

k) control of the atmosphere and other ambient factors of workplaces;

l) prevention and control of hazards due to high and low barometric pressures;

m) prevention of fires and explosions and measures to be taken in case of fire or explosion;

n) design, manufacture, supply, use, maintenance and testing of personal protective equipment and protective clothing;

o) sanitary installations, washing facilities, facilities for changing and storing clothes, supply of drinking water, and any other welfare facilities connected with occupational safety and health;

p) first-aid treatment;

q) establishment of emergency plans;

r) supervision of the health of workers.

IV. ACTION AT THE LEVEL OF THE UNDERTAKING

16. The arrangements provided for in Article 19 of the Convention should aim at ensuring that workers-

a) take reasonable care for their own safety and that of other persons who may be affected by their acts or omissions at work;

b) comply with instructions given for their own safety and health and those of others and with safety and health procedures;

c) use safety devices and protective equipment correctly and do not render them inoperative;

d) report forthwith to their immediate supervisor any situation which they have reason to believe could present a hazard and which they cannot themselves correct;

e) report any accident or injury to health which arises in the course of or in connection with work.

17. No measures prejudicial to a worker should be taken by reference to the fact that, in good faith, he complained of what he considered to be a breach of statutory requirements or a serious inadequacy in the measures taken by the employer in respect of occupational safety and health and the working environment.

For the full text of this Recommendation go to the ILO website:

http://www.ilo.org/dyn/normlex/en/f? p=NORMLEXPUB:12100:0::NO:121 00:P12100_INSTRUMENT_ID:312502

RADIATION PROTECTION CONVENTION C115, 1960

Convention concerning the Protection of Workers against Ionising Radiations

PART I. GENERAL PROVISIONS

Article 2

1. This Convention applies to all activities involving exposure of workers to ionising radiations in the course of their work.

PART II. PROTECTIVE MEASURES

Article 4

The activities referred to in Article 2 shall be so arranged and conducted as to afford the protection envisaged in this Part of the Convention.

Article 5

Every effort shall be made to restrict the exposure of workers to ionising radiations to the lowest practicable level, and any unnecessary exposure shall be avoided by all parties concerned.

Article 6

1. Maximum permissible doses of ionising radiations which may be received from sources external to or internal to the body and maximum permissible amounts of radioactive substances which can be taken into the body shall be fixed in accordance with Part I of this Convention for various categories of workers.

2. Such maximum permissible doses and amounts shall be kept under constant review in the light of current knowledge.

Article 7

1. Appropriate levels shall be fixed in accordance with Article 6 for workers who are directly engaged in radiation work and are-

   a) aged 18 and over;

   b) under the age of 18.

2. No worker under the age of 16 shall be engaged in work involving ionising radiations.

Article 9

1. Appropriate warnings shall be used to indicate the presence of hazards from ionising radiations. Any information necessary in this connection shall be supplied to the workers.

2. All workers directly engaged in radiation work shall be adequately instructed, before and during such employment, in the precautions to be taken for their protection, as regards their health and safety, and the reasons therefor.

Article 11

Appropriate monitoring of workers and places of work shall be carried out in order to measure the exposure of workers to ionising radiations and radioactive substances, with a view to ascertaining that the applicable levels are respected.

Article 12

All workers directly engaged in radiation work shall undergo an appropriate medical examination prior to or shortly after taking up such work and subsequently undergo further medical examinations at appropriate intervals.

For the full text of this Convention go to the ILO website:

http://www.ilo.org/dyn/normlex/
en/f?p=1000:12100:0::NO::P12100_ILO_CODE:C115

RADIATION PROTECTION RECOMMENDATION R114, 1960

Recommendation concerning the Protection of Workers against Ionising Radiations

I. GENERAL PROVISIONS

2. (1) This Recommendation applies to all activities involving exposure of workers to ionising radiations in the course of their work.

III. COMPETENT PERSON

6. The employer should appoint a competent person to deal on behalf of the undertaking with questions of protection against ionising radiations.

IV. METHODS OF PROTECTION

7. (1) In cases where they ensure effective protection preference should be given to methods of collective protection, both physical and operational.

(2) Wherever methods of collective protection are inadequate, personal protective equipment and, as necessary, appropriate protective procedures should be used.

8. (1) All protective devices, appliances and apparatus should be so designed or modified as to fulfil their intended purpose.

9. (1) Unsealed sources should be manipulated with due regard to their toxicity.

(2) The methods of manipulation should be chosen with a view to minimising the risk of entry of radioactive substances into the body and the spread of radioactive contamination.

10. Plans should be made in advance for measures-

   a) to detect as promptly as possible any leakage from, or breakage of, a sealed source of radioactive substances which may involve a risk of radioactive contamination and;

   b) to take prompt remedial action to prevent the further spread of radioactive contamination and to apply other appropriate safety precautions,

including decontamination procedures, with, as necessary, the immediate collaboration of all authorities concerned.

11. Sources which may involve exposure of workers to ionising radiations, and the areas in which such an exposure may occur or where workers may be exposed to radioactive contamination, should be identified, in appropriate cases, by means of easily recognisable warnings.

12. All sources of radioactive substances, whether sealed or unsealed, in use or stored by an undertaking, should be appropriately recorded.

15. (2) If the loss, theft or damage is confirmed, the competent authority should be notified without delay.

V. MONITORING

17. (1) Appropriate monitoring of workers and places of work should be carried out in order to measure the exposure of workers to ionising radiations and radioactive substances, with a view to ascertaining that the applicable levels are respected.

19. Persons who carry out monitoring in pursuance of the provisions of the Radiation Protection Convention, 1960, and of this Recommendation, should be afforded adequate equipment and facilities for carrying out this work.

VI. MEDICAL EXAMINATIONS

20. All medical examinations referred to in the Radiation Protection Convention, 1960, should be carried out by a suitably qualified physician.

24. For all workers who undergo such medical examinations health records should be established and kept in accordance with the requirements of the competent authority.

VIII. CO-OPERATION OF EMPLOYERS AND WORKERS

32. Every effort should be made by both the employers and the workers to secure the closest co-operation in carrying out the measures for protection against ionising radiations.

For the full text of this Recommendation go to the ILO website:

http://www.ilo.org/dyn/normlex/
en/f?p=1000:12100:0::NO::P12100_ILO_CODE:R114

SAFETY AND HEALTH IN CONSTRUCTION CONVENTION C167, 1988

Convention concerning Safety and Health in Construction

I. SCOPE AND DEFINITIONS

Article 1

1. This Convention applies to all construction activities, namely building, civil engineering, and erection and dismantling work, including any process, operation or transport on a construction site, from the preparation of the site to the completion of the project.

Article 2

For the purpose of this Convention:

a) The term construction covers:

(i) building, including excavation and the construction, structural alteration, renovation, repair, maintenance (including cleaning and painting) and demolition of all types of buildings or structures.

(ii) civil engineering, including excavation and the construction, structural alteration, repair, maintenance and demolition of, for example, airports, docks, harbours, inland waterways, dams, river and avalanche and sea defence works, roads and highways, railways, bridges, tunnels, viaducts and works related to the provision of services such as communications, drainage, sewerage, water and energy supplies;

(iii) the erection and dismantling of prefabricated buildings and structures, as well as the manufacturing of prefabricated elements on the construction site;

b) the term construction site means any site at which any of the processes or operations described in subparagraph (a) above are carried on.

II. GENERAL PROVISIONS

Article 9

Those concerned with the design and planning of a construction project shall take into account the safety and health of the construction workers in accordance with national laws, regulations and practice.

III. PREVENTIVE AND PROTECTIVE MEASURES

Article 13
Safety of workplaces

1. All appropriate precautions shall be taken to ensure that all workplaces are safe and without risk of injury to the safety and health of workers.

2. Safe means of access to and egress from all workplaces shall be provided and maintained, and indicated where appropriate.

3. All appropriate precautions shall be taken to protect persons present at or in the vicinity of a construction site from all risks which may arise from such site.

Article 14
Scaffolds and ladders

1. Where work cannot safely be done on or from the ground or from part of a building or other permanent structure, a safe and suitable scaffold shall be provided and maintained, or other equally safe and suitable provision shall be made.

Article 15
Lifting appliances and gear

1. Every lifting appliance and item of lifting gear, including their constituent elements, attachments, anchorages and supports, shall-

a) be of good design and construction, sound material and adequate strength for the purpose for which they are used;

b) be properly installed and used;

c) be maintained in good working order;

d) be examined and tested by a competent person at such times and in such cases as shall be prescribed by national laws or regulations; the results of these examinations and tests shall be recorded;

e) be operated by workers who have received appropriate training in accordance with national laws and regulations.

Article 20
Cofferdams and caissons

1. Every cofferdam and caisson shall be-

a) of good construction and suitable and sound material and of adequate strength;

b) provided with adequate means for workers to reach safety in the event of an inrush of water or material.

2. The construction, positioning, modification or dismantling of a cofferdam or caisson shall take place only under the immediate supervision of a competent person.

3. Every cofferdam and caisson shall be inspected by a competent person at prescribed intervals.

Article 21
Work in compressed air

1. Work in compressed air shall be carried out only in accordance with measures prescribed by national laws or regulations.

2. Work in compressed air shall be carried out only

by workers whose physical aptitude for such work has been established by a medical examination and when a competent person is present to supervise the conduct of the operations.

Article 30
Personal protective equipment and protective clothing

1. Where adequate protection against risk of accident or injury to health, including exposure to adverse conditions, cannot be ensured by other means, suitable personal protective equipment and protective clothing, having regard to the type of work and risks, shall be provided and maintained by the employer, without cost to the workers, as may be prescribed by national laws or regulations.

2. The employer shall provide the workers with the appropriate means to enable them to use the individual protective equipment, and shall ensure its proper use.

3. Protective equipment and protective clothing shall comply with standards set by the competent authority taking into account as far as possible ergonomic principles.

4. Workers shall be required to make proper use of and to take good care of the personal protective equipment and protective clothing provided for their use.

Article 31
First-aid

1. The employer shall be responsible for ensuring that first- aid, including trained personnel, is available at all times. Arrangements shall be made for ensuring the removal for medical attention of workers who have suffered an accident or sudden illness.

Article 32
Welfare

1. At or within reasonable access of every construction site an adequate supply of wholesome drinking water shall be provided.

Article 33
Information and training

Workers shall be adequately and suitably-

a) informed of potential safety and health hazards to which they may be exposed at their workplace;

b) instructed and trained in the measures available for the prevention and control of, and protection against, those hazards.

For the full text of this Convention go to the ILO website:

http://www.ilo.org/dyn/normlex/ en/f?p=1000:12100:0::NO::P12100_ILO_CODE:C167

SAFETY AND HEALTH IN CONSTRUCTION RECOMMENDATION R175, 1988

Recommendation concerning Safety and Health in Construction

I. SCOPE AND DEFINITIONS

1. The provisions of the Safety and Health in Construction Convention, 1988, hereinafter referred to as the Convention and of this Recommendation should be applied in particular to:

a) building, civil engineering and the erection and dismantling of prefabricated buildings and structures, as defined in Article 2(a) of the Convention;

b) the fabrication and erection of oil rigs, and of offshore installations while under construction on shore.

III. PREVENTIVE AND PROTECTIVE MEASURES

Scaffolds

16. Every scaffold and part thereof should be of suitable and sound material and of adequate size and strength for the purpose for which it is used and be maintained in a proper condition.

17. Every scaffold should be properly designed, erected and maintained so as to prevent collapse or accidental displacement when properly used.

Lifting appliances and lifting gear

22. National laws or regulations should prescribe the lifting appliances and items of lifting gear which should be examined and tested by a competent person-

a) before being taken into use for the first time;

b) after erection on a site;

c) subsequently at intervals prescribed by such national laws or regulations;

d) after any substantial alteration or repair.

Transport, earth-moving and materials-handling equipment

30. The drivers and operators of vehicles and of earth-moving or materials-handling equipment should be persons trained and tested as required by national laws or regulations.

Excavations, shafts, earthworks, underground works and tunnels

34. Shoring or other support for any part of an excavation, shaft, earthworks, underground works or tunnel should not be erected, altered or dismantled except under the supervision of a competent person.

Health hazards

41. (1) An information system should be set up by the competent authority, using the results of international scientific research, to provide information for architects, contractors, employers and workers' representatives on the health risks associated with hazardous substances used in the construction industry.

First-aid

49. The manner in which first-aid facilities and personnel are to be provided in pursuance of Article 31 of the Convention should be prescribed by national laws or regulations drawn up after consulting the competent health authority and the most representative organisations of employers and workers concerned.

Welfare

51. In appropriate cases, depending on the number of workers, the duration of the work and its location, adequate facilities for obtaining or preparing food and drink at or near a construction site should be provided, if they are not otherwise available.

For the full text of this Recommendation go to the ILO website:

https://www.ilo.org/dyn/normlex/
en/f?p=1000:12100:0::NO::P12100_ILO_CODE:R175

WELFARE FACILITIES RECOMMENDATION R102, 1956

Recommendation concerning Welfare Facilities for Workers.

I. SCOPE

1. This Recommendation applies to manual and non-manual workers employed in public or private undertakings, excluding workers in agriculture and sea transport.

III. FEEDING FACILITIES

A. Canteens

4. Canteens providing appropriate meals should be set up and operated in or near undertakings where this is desirable, having regard to the number of workers employed by the undertaking, the demand for and prospective use of the facilities, the non-availability of other appropriate facilities for obtaining meals and any other relevant conditions and circumstances.

B. Buffets and trolleys

10. (1) In undertakings where it is not practicable to set up canteens providing appropriate meals, and in other undertakings where such canteens already exist, buffets or trolleys should be provided, where necessary and practicable, for the sale to the workers of packed meals or snacks and tea, coffee, milk and other beverages. Trolleys should not, however, be introduced into workplaces in which dangerous or harmful processes make it undesirable that workers should partake of food and drink there.

(2) Some of these facilities should be made available not only during the midday or midshift interval but also during the recognised rest pauses and breaks.

C. Messrooms and other suitable rooms

11. (1) In undertakings where it is not practicable to set up canteens providing appropriate meals, and, where necessary, in other undertakings where such canteens already exist, messroom facilities should be provided, where practicable and appropriate, for individual workers to prepare or heat and take meals provided by themselves.

D. Mobile canteens

12. In undertakings in which workers are dispersed over wide work areas, it is desirable, where practicable and necessary, and where other satisfactory facilities are not available, to provide mobile canteens for the sale of appropriate meals to the worker.

E. Other facilities

13. Special consideration should be given to providing shift workers with facilities for obtaining adequate meals and beverages at appropriate times.

F. Use of facilities

15. The workers should in no case be compelled, except as required by national laws and regulations for reasons of health, to use any of the feeding facilities provided.

IV. REST FACILITIES

A. Seats

16. (1) In undertakings where any workers, especially women and young workers, have in the course of their work reasonable opportunities for sitting without detriment to their work, seats should be provided and maintained for their use.

(2) Seats so provided should be in adequate numbers and reasonably near the work posts of the workers concerned.

B. Rest rooms

19. (1) In an undertaking where alternative facilities are not available for workers to take temporary rest during working hours, a rest room should be provided, where this is desirable, having regard to the nature of the work and any other relevant conditions and circumstances. In particular, rest rooms should be provided to meet the needs of women workers; of workers engaged on particularly arduous or special work requiring temporary rest during working hours; or of workers employed on broken shifts.

20. The facilities so provided should include at least-

a) a room in which provision suited to the climate is made for relieving discomfort from cold or heat;

b) adequate ventilation and lighting;

c) suitable seating facilities in sufficient numbers.

For the full text of this Recommendation go to the ILO website:

http://www.ilo.org/dyn/normlex/ en/f?p=1000:12100:0::NO::P12100_ILO_CODE:R102

WORKING ENVIRONMENT (AIR, POLLUTION, NOISE AND VIBRATION) CONVENTION C148, 1977

Convention concerning the Protection of Workers against Occupational Hazards in the Working Environment Due to Air Pollution, Noise and Vibration

PART I. SCOPE AND DEFINITIONS

Article 2

1. Each Member, after consultation with the representative organisations of employers and workers, where such exist, may accept the obligations of this Convention separately in respect of-

a) air pollution;

b) noise; and

c) vibration.

Article 3

For the purpose of this Convention-

a) the term air pollution covers all air contaminated by substances, whatever their physical state, which are harmful to health or otherwise dangerous;

b) the term noise covers all sound which can result in hearing impairment or be harmful to health or otherwise dangerous;

c) the term vibration covers any vibration which is transmitted to the human body through solid structures and is harmful to health or otherwise dangerous.

PART III. PREVENTIVE AND PROTECTIVE MEASURE

Article 8

(1) The competent authority shall establish criteria for determining the hazards of exposure to air pollution, noise and vibration in the working environment and, where appropriate, shall specify exposure limits on the basis of these criteria.

(3) The criteria and exposure limits shall be established, supplemented and revised regularly in the light of current national and international knowledge and data, taking into account as far as possible any increase in occupational hazards resulting from simultaneous exposure to several harmful factors at the workplace.

Article 9

As far as possible, the working environment shall be kept free from any hazard due to air pollution, noise or vibration-

a) by technical measures applied to new plant or processes in design or installation, or added to existing plant or processes; or, where this is not possible,

b) by supplementary organisational measures.

Article 10

Where the measures taken in pursuance of Article 9 do not bring air pollution, noise and vibration in the working environment within the limits specified in pursuance of Article 8, the employer shall provide and maintain suitable personal protective equipment. The employer shall not require a worker to work without the personal protective equipment provided in pursuance of this Article.

Article 13

All persons concerned shall be adequately and suitably-

a) informed of potential occupational hazards in the working environment due to air pollution, noise and vibration; and

b) instructed in the measures available for the prevention and control of, and protection against, those hazards.

For the full text of this Convention go to the ILO website:

http://www.ilo.org/dyn/normlex/ en/f?p=1000:12100:0::NO::P12100_ILO_CODE:C148

WORKING ENVIRONMENT (AIR, POLLUTION, NOISE AND VIBRATION) RECOMMENDATION R156, 1977

Recommendation concerning the Protection of Workers against Occupational Hazards in the Working Environment due to Air Pollution, Noise and Vibration

I. SCOPE

(2) Measures should be taken to give self-employed persons protection in the working environment analogous to that provided for in the Working Environment (Air Pollution, Noise and Vibration) Convention, 1977, and in this Recommendation.

II. PREVENTIVE AND PROTECTIVE MEASURES

(2) Special monitoring in relation to the exposure limits referred to in Article 8 of the Working Environment (Air Pollution, Noise and Vibration) Convention, 1977, should be undertaken in the working environment when machinery or installations are first put into use or significantly modified, or when new processes are introduced.

(3) It should be the duty of the employer to arrange for equipment used to monitor air pollution, noise and vibration in the working environment to be regularly inspected, maintained and calibrated.

(4) The workers and/or their representatives and the inspection services should be afforded access to the records of the monitoring of the working environment and to the records of inspection, maintenance and calibration of apparatus and equipment used therefor.

(5) Substances which are harmful to health or otherwise dangerous and which are liable to be airborne in the working environment should, as far as possible, be replaced by less harmful or harmless substances.

(6) Processes involving air pollution, noise or vibration in the working environment as defined in Article 3 of the Working Environment (Air Pollution, Noise and Vibration) Convention, 1977, should be replaced as far as possible by processes involving less or no air pollution, noise or vibration.

III. SUPERVISION OF THE HEALTH OF WORKERS

16. (1) The supervision of the health of workers provided for in Article 11 of the Working Environment (Air Pollution, Noise and Vibration) Convention, 1977, should include, as determined by the competent authority-

a) a pre-assignment medical examination;

b) periodic medical examinations at suitable intervals;

c) biological or other tests or investigations which may be necessary to control the degree of exposure and supervise the state of health of the worker concerned;

d) medical examinations or biological or other tests or investigations after cessation of the assignment which, when medically indicated, should be made available as of right on a regular basis and over a prolonged period.

(2) The competent authority should require that the results of any such examinations or tests be made available to the worker, and at his request to his personal physician.

17. The supervision provided for in Paragraph 16 of this Recommendation should normally be carried out in working hours and should be free of cost to the worker.

IV. TRAINING, INFORMATION AND RESEARCH

23. Employers' and workers' organisations should take positive action to carry out programmes of training and information with respect to the prevention and control of, and protection against, existing and potential occupational hazards in the working environment due to air pollution, noise and vibration.

24. Workers' representatives within undertakings should have the facilities and necessary time, without loss of pay, to play an active role in respect of the prevention and control of, and the protection against, occupational hazards in the working environment due to air pollution, noise and vibration. For this purpose, they should have the right to seek assistance from recognised experts of their choice.

25. Such measures as are necessary should be taken to secure that, in connection with the use at a workplace of a substance liable to be harmful to health or otherwise dangerous, adequate information is available on-

a) the results of any relevant tests relating to the substance; and

b) the conditions required to ensure that, when properly used, it is without danger to the health of workers.

For the full text of this Recommendation go to the ILO website:

http://www.ilo.org/dyn/normlex/en/f?p=1000:12100:0::NO::P12100_ILO_CODE:R156

This page is intentionally left blank

Assessment

Contents

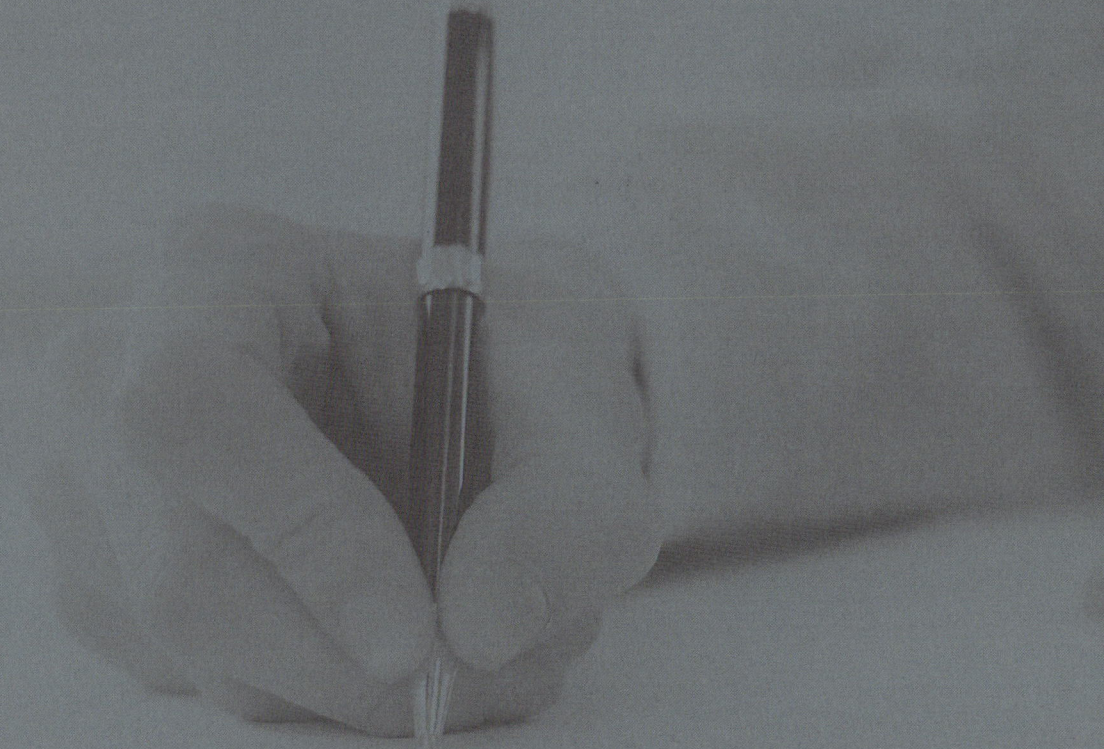

A.1 Assessments of understanding

It is understood that those using this publication may be doing so to broaden their understanding of this important topic, management of health and safety risks, wherever in the world they may be working, and others will be studying in order to obtain a specific health and safety related qualification. The approach taken by those assessing such qualifications will vary.

This element provides information on how one such qualification is assessed, the NEBOSH International General Certificate in Occupational Health and Safety.

The NEBOSH International General Certificate qualification (IG) has two unit assessments:

- Unit IG1: Management of health and safety – question paper.
- Unit IG2: Risk assessment – practical assessment.

The questions and related guidance to answers provided in this element may prove useful for those that want to assess their understanding for both units of this qualification and for more general reasons.

The example risk assessment, provided by NEBOSH within their resources section on page ***https://www. nebosh.org.uk/qualifications/international-general- certificate/*** will be of great use to those studying for the NEBOSH qualification and will prove particularly useful to those that wish to develop and use their own means of carrying out risk assessments in their workplace.

UNIT IG1: MANAGEMENT OF HEALTH AND SAFETY - EXAM PAPER

The written paper is an examination of your knowledge of elements 1 to 4 only and is 2 hours duration and contains 11 questions.

Question 1 carries 20 marks. This question should be allocated 30 minutes in total. If time (for example, five minutes) is given to reading, planning and checking, the time available for writing is 25 minutes. Learners should produce approximately 1½ sides to give themselves the best chance of obtaining good marks.

Questions 2 to 11 each carry eight marks. If time (for example, 10 minutes) is allowed for reading, planning and checking then there are eight minutes to answer each question. Learners should produce approximately ½ a page, depending on the question.

At every examination a number of learners - including some good ones - perform less well than they might because of poor examination technique. It is essential that learners practice answering questions and learn to budget their time according to the number of marks allocated to questions (and parts of questions) as shown on the paper. ***See also the 'Revision and exam guidance' section later in this study book.***

STUDY QUESTIONS

In order that learners may check their understanding of the topics covered, a number of questions have been included at the end of each element. Although elements 5 to 11 are only assessed by the risk assessment (IG2), RMS has included a selection of study questions to help reinforce learning of the required syllabus content for the NEBOSH International General Certificate qualification.

Information regarding the required content of answers to the study questions is provided on pages 412-427. Any resemblance to NEBOSH examination questions is coincidental; the questions have been created to illustrate the type of question which may be asked at the IG1 examination.

UNIT IG2 - RISK ASSESSMENT

The aim of this unit is to assess a learner's ability to complete successfully a health and safety risk assessment in their workplace and to prioritise (with justification) THREE actions. The risk assessment MUST NOT take place until you have completed your studies of the whole of the IG syllabus (elements 1 to 11).

The time allowed by NEBOSH to complete the assessment is not restricted, but it is advised that learners should aim to complete the full risk assessment documentation within three hours. This indicates an acceptable amount of time that could produce an acceptable result. The actual time taken will depend on the learner, the type of workplace and the method used to produce the risk assessment, i.e. either hand written or word processed format.

Always refer to your Learning Partner for instructions on when to carrying out the risk assessment and the deadline for receipt of your completed submission, please make sure you are clear about when and where you will carry out the risk assessment.

The risk assessment format is based on the Health and Safety Executive's (HSE) "Risk assessment. A brief guide to controlling risks in the workplace" (INDG163 rev4). It is recommended that you refer to the risk assessment guidance, examples and frequently asked questions accessible from the HSE website http://www.hse.gov. uk/risk/controlling-risks.htm

The stages of the assessment are as follows:

Stage 1. Description of the organization and the methodology used.

Stage 2. Risk assessment.

Stage 3. Prioritise three actions with justification.

Stage 4. Review, communicate and check.

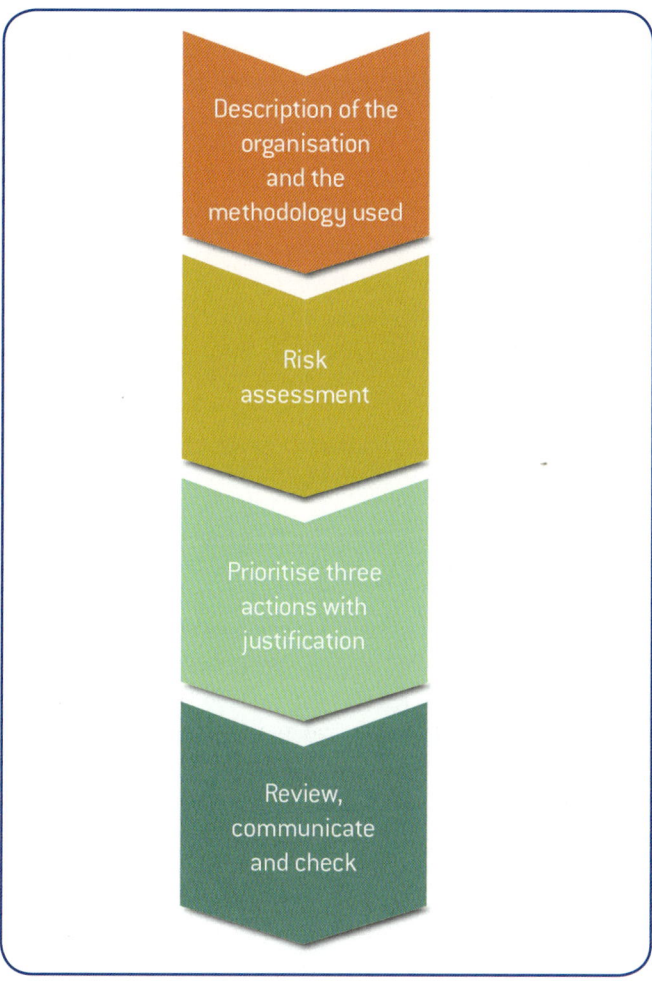

Figure A-1: Assessment stages.
Source: RMS.

The following will give you some tips on what to address at each stage.

Stage 1. Description of the organization and the methodology used

Description of the organization:

You must give a clear and concise description of the organization which will "paint a mental picture" for the examiner. You may create a pseudonym for the organization if you are concerned about confidentiality.

Methodology used:

In this section you need to describe how you carried out the assessment. As a minimum, you should include: the sources of information that you consulted; who you spoke to; and how the hazards and controls were identified.

Stage 2. Risk assessment

This is the section where you must carry out a thorough risk assessment of the organization by completing all of the columns on the risk assessment forms that will be provided for you.

Column 1 – Hazard category and hazards

You must address at least ten different hazards that cover at least five different hazard categories. Remember, hazards are usually defined as "those things that have the potential to cause harm". Hazard categories are the topic headings which can be found in elements 5 to 11 of the IG syllabus.

Column 2 - Who might be harmed and how?

At this stage you must think about those who could be affected by the hazard. Of course, this will include those workers who are doing the task but you should also consider other people who may be a greater risk i.e. contractors, pregnant workers, young people, people with disabilities, etc.

You should then describe the task that is being done and the consequences of exposure to the hazard.

Columns 3 and 4 – "What are you already doing" and "What further actions/controls are required?"

Columns 3 and 4 work together. If you identify a hazard that is well controlled, there will be a lot of information in column 3 and little in column 4. Conversely, if a hazard is not controlled adequately, there will not be much information in column 3 but column 4 will contain a lot more.

Learners should avoid generic phrases being repeatedly used, for example, 'monitor' and 'train staff'. You should give appropriate clarification by giving examples of appropriate monitoring and the type of training required.

Column 5 – Timescales for further actions to be completed

Learners should identify timescales that are both practical and realistic to ensure that appropriate actions can be taken. Where possible, you should avoid generic phrases such as "immediate" and "on-going".

Column 6 – Responsible person's job role

It is important for learners to understand that it is not their job to action all of recommendations. In this column learners should state the job title of the person responsible for ensuring that the actions are completed. The seniority of the person in the organization should reflect the importance of the action(s) allocated to them. For example, formulating policy should be the responsibility of a director whereas carrying out regular workplace inspections is the role of the office manager.

Stage 3. Prioritise three actions with justification

At this stage learners should refer to **element 1** of the syllabus and discuss the moral, legal and financial arguments for all three actions. When discussing the legal argument, you must mention specific International Conventions, ISO's and, if appropriate, any relevant local laws and legislation. Learners should also consider the likelihood and severity of loss occurring. For example, this discussion should take into account the number of workers at risk, the severity of harm that may occur, the frequency and duration of exposure to the hazard, etc.

Stage 4. Review, communicate and check

The final part of the assessment is to: set a realistic review date for the risk assessment and say why you have chosen that review date; indicate how the findings of the risk assessment are to be communicated (verbal or written up date and the methods to be used, for example, email, noticeboards) and who needs to know the information; and indicate how you will follow up on the risk assessment to check that the actions have been carried out.

An assessment pack is available from the NEBOSH website which includes everything required in order to complete the IG2 assessment including guidance, helpful hints and tips, forms to use and a completed example risk assessment.

This pack can be downloaded from *https://www.nebosh. org.uk/qualifications/international-general-certificate/*

IMPORTANT: The example risk assessment referred to is for illustrative purposes only and must not be reproduced in part, section or full. Following a full investigation by NEBOSH, where it has been proven that this example has been plagiarised by learners, risk assessments will be voided and those learners involved potentially barred from undertaking any further NEBOSH qualifications.

A.2 Study questions - answer guidance

The following are not sample NEBOSH questions/answers but are representative and have been developed by RMS Publishing. Please note these are not always complete or exhaustive answers and you are required to add appropriate detail where applicable.

ELEMENT 1: WHY WE SHOULD MANAGE WORKPLACE HEALTH AND SAFETY

1) Explain what is meant by the terms direct and indirect cost of a health and safety incident, with an example for each? (8)

Direct costs are those costs that directly relate to the accident/incident or incidence of ill-health. Direct costs may be insured or uninsured depending on the cause of the loss. In addition, many smaller costs may not be covered by insurance policies because the insurance company expects the employer to meet some of the costs of losses.

An example of an insured cost is claims for compensation made against the employer by workers and other people. This may be covered by employer's liability insurance or public liability insurance or similar insurance policies.

Indirect costs are those costs that indirectly relate to the accident/incident or incidence of ill-health. Some of these costs may be caused by a single accident/incident or incidence of ill-health or may be the result of an accumulation of occurrences over time. Direct costs may also be insured or uninsured although not many can be insured.

An example of an uninsured cost is loss of important experience that the worker had.

2) What are the social reasons for preventing accidents/incidents and ill-health in the workplace? (8)

*The **moral/social** reason to prevent harm is usually further reinforced by social expectations of behaviour, which require consideration of others that may be affected by interaction with them. In particular this includes work activities and how they may harm those involved in them or affected by them. This social expectation is often expressed in both civil law and criminal law as, without the potential for litigation or regulatory action, many employers would not act upon their implied obligation to provide protection. In many countries, it is a specific legal requirement to safeguard the health and safety of workers and others that might be affected by the organization's operations.*

3) What are responsibilities and rights of workers in the Occupational Safety and Health Convention C155? (8)

*Worker **responsibilities** include co-operating with employers with regard to obligations placed upon the employer, including reporting any situation that presents imminent risk or serious danger. In order to achieve this, workers and their representatives have rights.*

*Workers' **rights** include; they should receive adequate information and training on measures taken by the employer to secure occupational health and safety. They or their representatives should be consulted by the employer on all aspects of occupational health and safety associated with their work. Workers should not be made to work in situations of continuing imminent or serious danger, until the employer takes remedial action.*

4) What is meant by practicable duties placed on employers? (8)

Practicable duties have to be met to an extent only limited by the current state of knowledge and invention, irrespective of cost or difficulty. This duty is often used for control of the provision of important or high risk precautions that involve applying technical solutions. As technical solutions may be influenced by technical characteristics of the work the duty allows the employer to adjust the solution to meet these characteristics and achieve health and safety. In addition, technical solutions may change over time as "knowledge and invention" produces alternative and better solutions. Employers are expected to use these solutions to meet the duty, therefore the duty expects employers to keep up to date with alternative solutions and apply them promptly when they become generally available.

5) Why are ISO international standards not in any way binding on either governments or industry merely by virtue of being international standards? (8)

ISO international standards are not in any way binding on either Governments or industry merely by virtue of being international standards. This is to allow for situations where certain types of standards may conflict with social, cultural or legislative expectations and requirements. This also reflects the fact that national and international experts responsible for creating these standards do not always agree and not all proposals become standards by unanimous vote. The individual nations and their standards bodies remain the final arbiters.

ELEMENT 2: HOW HEALTH AND SAFETY MANAGEMENT SYSTEMS WORK AND WHAT THEY LOOK LIKE

1) (a) What are the five key elements in ILO 'Guidelines on Occupational Safety and Health Management Systems'? (5)

(b) Explain the requirement for two of the elements identified. (3)

For (a) The ILO 'Guidelines on Occupational Safety and Health Management Systems - ILO-OSH 2001' follow a structure that uses the following five key elements:

- Policy.
- Organising.
- Planning and implementation.
- Evaluation.
- Actions for improvement.

For (b) any two from the following is required by the question:

Policy

The health and safety policy establishes a plan which influences all the organization's activities and decisions, including those to do with the selection of resources and information, the design and operation of working systems, the design and delivery of products and services, and the control and disposal of waste.

Organization

The organization section of the policy should clearly define the roles and responsibilities of everyone in the organization. This is required to establish a positive culture that secures involvement and participation at all levels. This culture is sustained by effective communications and the promotion of competence that enables all managers and workers to make a responsible and informed contribution to the health and safety effort.

Planning and implementation

Planning and implementation should aim to minimise the risks created by work activities, products and services. They use risk assessment methods to identify hazards, decide priorities and set objectives for hazard elimination and risk reduction. Wherever possible, risks are eliminated by the careful selection and design of facilities, equipment, substances and processes or minimised by the use of physical control measures. Where this is not possible, provision of a safe system of work and personal protective equipment are used to control risks.

Evaluation

Responsibilities, accountability and authority for evaluation should be clearly allocated at different levels in the management structure. Performance evaluation is consistently used to determine the extent to which the policy and objectives have been met and risks controlled. Many organizations establish health and safety committees to evaluate and report on the on-going progress of health and safety performance.

Actions for improvement

When system non-conformities, for example, the identification of additional hazards, procedures not being followed or planned action not implemented, prompt corrective and preventive action systems should be taken. Corrective action is that action to correct the non-conformity and prevent further harm being caused by it, this tends to deal with the immediate causes of the non-conformity. Preventive action deals with the underlying and root causes of the non-conformity. This type of action reduces the likelihood of the non-conformity re-occurring in the same place or in other places.

2) What are the main benefits of introducing a recognised health and safety management system? (8)

- *Demonstrates that the organization meets the requirements of a recognised standard.*

- *Provides objective proof that the organization attaches great importance to health and safety.*

- *Ensures clear responsibilities, (communication) structures and processes throughout the entire organization.*

- *Confirms a commitment to be open to independent scrutiny.*

- *Adds credibility to the organization.*

- *Improves regard and reputation, giving a positive image of the organization.*

- *Communicates a positive message to workers, customers and other stakeholders.*

- *Confirms to regulators that recognised health and safety standards have been met.*

- *Meets customers' expectations.*

- *Communicates a readiness to ensure consistent and improving health and safety performance.*

- *Increases the reliability of the work done and raises awareness.*

- *Establishes a level of performance to be maintained over the certification period.*

- *Increases confidence in the organization and its ability to manage health and safety risk.*

- *Helps to distinguish the organization from others who do not have certification.*

3) What are the key aims an organization should commit to in their health and safety policy statement? (8)

ILO-OSH 2001 suggests the health and safety policy statement should include, as a minimum, the following key aims:

Protecting the health and safety of all members of the organization by preventing work-related injuries, ill-health, diseases and incidents. Complying with relevant National/International occupational health and safety legislation, voluntary programmes, collective agreements on health and safety and other requirements to which the organization subscribes. Ensuring that workers and their representatives are consulted and encouraged to participate actively in all elements of the health and safety management system. Continually improving the performance of the health and safety management system.

4) What is the purpose of the three key features of an effective health and safety policy? (8)

*A **general statement** of management commitment, sometimes called a statement of intent.*

The statement defines what is to be achieved. The purpose of the statement section of the health and safety policy is to set and demonstrate management's commitment to its health and safety aims and to set objectives and quantifiable targets for the organization.

*Details of the **organization** of health and safety*

The organization section of the policy clarifies who will achieve the aims. Its purpose is to define the structure, role, relationships and responsibilities of individuals. It allocates responsibilities and ensures effective delegating, reporting and accountability. The organization section will usually define the organization of worker participation through health and safety representatives, committees and direct involvement.

***Arrangements** to control risks.*

The arrangements section describes show the aims will be achieved. The purpose of the arrangement section is to detail the specific systems and procedures that establish the direction, scope and actions of the organisation-organization when managing health and safety.

5) What methods can could you used to communicate the health and safety policy d to workers and others? (8)

All affected by the policy must understand their obligations. In order to achieve this, training and briefings will be necessary, as a minimum, to ensure effective communication. For new workers this is often done as part of the induction process. The health and safety policy should be communicated in a language and medium that managers and workers readily understand. The health and safety policy should be readily accessible to all persons in their workplace and made available to external interested parties, for example visitors, neighbours, customers and suppliers. The health and safety policy statement is often posted in prominent positions in the organization as part of the communication process and can serve as a lasting reminder to managers and workers after completion of training.

6) (a) Who should sign the health and safety policy statement? (2)

(b) Why should the policy be signed and dated? (6)

For (a) The most senior accountable person in the organization should sign the policy statement.

For (b): In order to emphasise the necessary commit-

ment to the organization's aims the statement should be signed by the most senior accountable person in the organization and dated. This shows that the senior person has accepted their responsibilities. It also gives authority to the policy and demonstrates senior management commitment to health and safety. By signing the policy, the personal commitment demonstrated should mean that managers are more likely to implement the policy and workers are more likely to believe in it and follow it. Strong management commitment, shown by the application of a signature, can emphasise that health and safety is equal to other business objectives. The addition of a date to the statement will indicate the last time the statement was reviewed, which will assist with ensuring the statement remains relevant and up to date. The health and safety policy statement is often posted in prominent positions in the organization to reinforce the commitment to health and safety.

ELEMENT 3: MANAGING RISK - UNDERSTANDING PEOPLE AND PROCESSES

1) What are the additional arrangements employers may need to make to meet their responsibility to protect visitors, neighbours or members of the public from their work activities? (8)

The type of arrangements employers could make to meet their responsibility to protect visitors include; requiring prior notification of a visitor's intention to visit, so that risks related to the visit can be assessed and control measures put in place. Providing visitors with health and safety information in suitable languages, for example, information on hazards, control measures and emergency procedures. Providing visitors with an explanation of relevant site rules, for example, requirements not to enter restricted areas or to wear specified personal protective equipment (PPE). Controlling access to the site, for example, by procedures for signing in/out and issuing visitors with badges to confirm their presence on-site is approved. Providing visitors with special clothing so that they are easy to identify, enabling managers and workers to take account of their lack of awareness or knowledge of hazards. Providing visitors with a suitable person to guide them while on site, someone who knows the hazards of the site and how to keep the visitors safe and healthy.

2) What is meant by 'health and safety culture of an organization'? (8)

The health and safety culture of an organization is concerned with:

'How people feel' about how health and safety encompasses the values, beliefs, attitudes and perceptions of individuals and groups at all levels of the organiza-

tion, which are often referred to as the health and safety climate of the organization.

'What people do' within the organization includes the health and safety related activities, actions and behaviours of individuals and groups at all levels. For example, individuals making time for health and safety and giving it due priority when making decisions.

'What the organization has' is reflected in the organization's policies, operating procedures, management systems, control systems, communication and workflow systems. For example, health and safety is integrated in planning work activities and design of the workplace.

3) Explain why an understanding of individual factors is important in the workplace. (8)

It is important to know what a particular job involves so that the effects of individual factors can be minimised, especially for high hazard jobs. Not all individuals are suited for all tasks, for example, they may not have the physical strength and stamina required for activities such coal mining. Some individuals mental disposition might put them at risk, for example, their motivation, i.e. they may feel that while doing their work in a particular way presents additional risks it is worth taking the risk to make the work easier or to complete the work quicker. Some individual characteristics, such as skills and attitudes, can be modified by training, experience and involvement.

4) (a) Explain the role of worker participation in health and safety. (4)

(b) What are the benefits of participation for the employer? (4)

For (a) The role of worker participation in health and safety is to provide the employer with a wider view of how risks affect workers, their view on the effectiveness of current control measures and on proposed control measures. In addition, the role of worker participation is to show management commitment and motivate workers to work safely and healthily.

For (b) The benefits of worker participation of this type is that it has been identified as one of the most significant factors to influence health and safety behaviour and promote a positive health and safety culture in organizations. It can lead to shared health and safety values and the motivation of those involved to work together to improve health and safety.

5) What are the key stages that need to be followed in the risk assessment process? (8)

The key stages in a risk assessment process are:

- The identification of hazards relative to the work activity or task being assessed.

- *Identification of the population at risk, who might be harmed and how, particular regard should be given to young persons, those that are inexperienced, pregnant or nursing mothers and those with a disability.*

- *The evaluation of the risks from the hazards (likelihood and severity (consequence)) and deciding on precautions (adequacy of current controls and need for additional controls). Consideration of any residual risk that may remain.*

- *Recording significant findings and the implementation of them.*

- *Reviewing the risk assessment and updating it if necessary (periodically or when there is a significant change, for example, process or legislative change).*

6) (a) What are two reasons why visitors to a workplace might be at greater risk of injury than a worker? (2)

(b) What are the measures to be taken to ensure the health and safety of visitors to the workplace? (6)

For (a) Visitors may be unfamiliar with controls and the processes carried out in the workplace, their vulnerability particularly if they are disabled or young persons; they may not have been issued with or know how to correctly use personal protective equipment; a lack of knowledge of the site layout including pedestrian routes, which might not be clearly defined or adequate; be unfamiliar with the emergency procedures.

For (b) Procedures to deal with visitors to a workplace such as, visitor identification, by the issue of badges and a system requiring sign in and out; prior notification to those members of staff to be involved in the visit; the need for visitors to be escorted by a member of management or supervisory staff; the provision of information to the visitors on hazards and emergency procedures; an explanation of specific site rules, for example, the wearing of personal protective equipment; the clear marking of pedestrian routes.

7) What should you consider when developing and implementing a safe system of work for general activities? (8)

Development of a safe system of work requires a systematic approach and generally requires the involvement of a number of people in order to establish an effective system of work. The development process requires a number of stages. Identification and analysis of the task, for example, consider the risks, the complexity and layout, equipment, environment and materials. Identification of hazards and risk assessment and any issues which might affect individuals with special needs or disabilities.

Introducing controls and formulating procedures - including the definition of the safe and healthy method and the implementation of the system, procedure, method statements, permits to work. Instruction and training in the operation of the system, developing skill and knowledge and close working to support a new person or trainee. Monitoring the system, supervisory checks and feedback of improvements identified.

It is important that the development of a safe system of work involves relevant people; this could include managers, workers who will work to the system, maintenance workers, competent health and safety practitioners and specialists.

8) What is the function of a permit-to-work system? (8)

The function of a permit-to-work system is to ensure the proper authorisation of specified work, by an appointed individual competent in the task and associated hazards and risk controls systems available. This will include confirmation of the identity, nature, timing, extent and limitations of the work. Consideration of the need for minimum daily staffing levels and where there may be identical or similar work equipment in the same vicinity, how confusion is prevented. Controlling change and considering other work activities that might interact with specified work. Establish criteria to be considered when identifying hazards and what they are. Confirm through the permit in writing, that hazards have been removed, where possible and that control measures are in place to deal with residual hazards. Confirm who has control of the location and equipment relating to the work when it passes between parties. Confirm work is started, suspended, conducted, and finished safely, evidenced by time and signature for each stage of the activity. Ultimately providing a method to identify the necessary steps and sequence to be followed.

9) Why it is important to develop emergency procedures for the workplace? (8)

It is important to develop and implement emergency procedures for potentially major loss-causing events in order to bring the event under control promptly, reduce the effects of the event (on premises, equipment, materials, environment and people that might be affected) and enable a fast return to normal operations. In the absence of emergency procedures there may not be a timely or suitable response to the emergency allowing it to get out of control and cause more serious effects than if procedures were in place. Depending on the emergency, this could result in major loss of life, long term ill-health, environmental effects, significant damage to buildings and equipment and long delays in the organization returning to normal operations. Some organizations or locations where emergencies take place that are

*not effectively controlled never recover from the effects
and cease to function or trade.*

10) What are eight items that should be included in a hot
work permit-to-work? (8)

*Identify eight items included in a permit-to-work
document from the following:*

- *Permit issue number.*
- *Authorised person identification.*
- *Standby fire warden.*
- *Locations of firefighting equipment.*
- *Locations of flammable materials.*
- *Warning information sign locations.*
- *Emergency muster points.*
- *Details of the work to be carried out.*
- *Signature of authoriser.*
- *Signature of acceptor.*
- *Signature for work clearance/extension/handover.*
- *Signature for cancellation.*
- *Other precautions (risk assessments, method state-
ments, personal protective equipment (PPE)).*

11) (a) What is the purpose of first aid? (2)

   (b) Explain the role of first-aiders. (6)

*For (a) The **purpose** of first-aid is to preserve life; prevent
the condition requiring first-aid getting worse, i.e.
minimise its consequences until medical help arrives; to
promote recovery of the person requiring first-aid and
provide treatment where medical attention of a minor
injury is not required.*

*For (b) The **role** of first-aiders is to ensure good planning
is in place to manage health and safety incidents swiftly
when they occur for the foreseeable risks of the organization. It is important to have arrangements in place for
when accidents/incidents and ill-health occur. Emergencies that require first-aid treatment can happen at any
time. Therefore the provision of first-aiders is an important part of an organization's emergency arrangements
to provide a prompt first response to emergencies,
preventing injuries and illness getting worse and providing
care until the local medical emergency services respond.
Where they are trained to do so they can also treat minor
injuries that would not receive or do not require professional medical attention.*

ELEMENT 4: HEALTH AND SAFETY MONITORING AND MEASURING

1) What are three different types of inspections that
might be used in any workplace and give an example
of each one? (8)

Examples could include the following:

***General** workplace inspections - carried out by local
first-line managers and worker health and safety representatives. An example of a routine inspection would
be an inspection of an office location undertaken every
three months where standards of housekeeping were
monitored.*

***Statutory** thorough examination of equipment, for
example, boilers, lifting equipment - carried out by
specialist competent persons.*

***Preventive** maintenance inspections of specific (critical)
items - carried out by maintenance staff.*

***Pre-use** checks of equipment, for example, vehicles,
fork-lift trucks, access equipment - carried out by the
user.*

2) What are the advantages and disadvantages of the
use of a checklist when carrying out inspections? (8)

*The question requires advantages and disadvan-
tages, this question is best answered under two distinct
headings, to demonstrate a full understanding of the
requirement, and typical examples include:*

Advantages

- *Enables prior preparation and planning.*
- *Quick and easy to arrange.*
- *Brings a consistent approach.*
- *Clearly identifies standards.*
- *Thorough.*
- *Provides readymade basis for inspection report.*
- *Provides evidence for audits.*

Disadvantages

- *Does not encourage the inspector to think beyond
the scope of the checklist.*
- *Items not on checklist are not inspected.*
- *May tempt people who are not authorised/compe-
tent to carry out the inspection.*
- *Can be out of date if standards change.*
- *Inspectors might be tempted to fill in the checklist
without checking the work area/equipment.*

3) What are the functions of an accident investigation?
(8)

*Care should be taken with this style of question. Answers
should state that any investigation will vary according to
the circumstances, but it will generally include the need
to establish the causes of an accident, both immediate
and underlying; and the appropriate preventative action
to be taken. The identification of weaknesses in current
systems, including any non-compliance with statutory requirements so that standards can be improved.
Collection of statistics, by identification of trends or*

frequency of occurrence. Determination of economic losses caused by delay or failure to complete customer expectations; including making any preparations for criminal/civil action. Improve staff relations by demonstrating commitment to health and safety.

4) What is the purpose of active and reactive monitoring? (8)

There is a need for a range of both active and reactive measures to determine whether health and safety objectives have been met. A balanced approach to monitoring seeks to learn from all available sources. Hence two broad categories of monitoring are required:

Active monitoring, before the event, involves identification through regular, planned observations of workplace conditions, systems and the actions of people, to ensure that performance standards are being implemented and management controls are working, for example, workplace and plant inspections.

Reactive monitoring, after the event, involves learning from mistakes, whether they result in injuries, illness, property damage or near-misses, for example, accident investigation.

Organizations need to ensure that information from both active and reactive monitoring is used to identify situations that create risks, and to do something about them. Priority should be given where risks are greatest.

5) What are the steps to be taken when conducting an accident investigation? (8)

When conducting accident investigations the following steps should be considered:

- Step one: gathering the information.
- Step two: analysing the information .
- Step three: identifying risk control measures.
- Step four: the action plan and its implementation.

6) What actions which should be taken following an audit? (8)

The outcome from an audit should be a detailed written report of findings and recommendations to improve or maintain the health and safety management system. A structure and approach to the report should be agreed at the pre-audit stage. The final report should give a clear assessment of the overall performance of the organisationorganization, identify deficiencies and make recommendations for improvement. It should also identify the observed strengths and suggest how they can be built upon. All audit reports need to be accurately and clearly communicated. In addition to the provision of a detailed written report a verbal presentation of the report may be provided soon after the close of the audit, in order to give an early opportunity for management to learn and take action.

7) What are the differences between audits and inspections? (8)

Health and safety audits assess the health and safety system, or parts of it, to determine if the system is ensuring health and safety. One of the parts of the system that may be examined by an audit is active monitoring methods, like inspections. In this way, the audit would identify if the correct people were conducting them, using the right methods, at the right frequency and how effective they were.

Inspections usually involve the examination of the workplace, work equipment or work activities; with the purpose of identification of hazards, or conditions that can lead to hazards, and to put in controls to mitigate the hazards. It can therefore be said that inspections are concerned with hazard identification in the workplace, whereas auditing relates to the systems that manage the prevention and control of hazards.

8) (a) Who in the organization should receive reports on health and safety performance from managers? (2)

(b) Why should others in the organization also receive the reports? (6)

For (a) The results of the review of health and safety performance should be reported at senior management level. This is particularly important in situations where the review has been conducted by a work group drawn from the senior management team, as this will enable all of senior management to understand and accept the implications of the review.

For (b) The results of the review should be communicated widely in the organization and in particular to those managers that have responsibility for responding to the actions arising from the review. It is customary to include a statement of health and safety performance, along with other risks, within the annual report. Such reports should be available to all workers and other stakeholders.

9) What are the advantages and disadvantages of internally conducted audits of an organization? (8)

Advantages

- Internal audits ensure local acceptance to implement recommendations and actions improving ownership of issues found.
- The auditor often has intimate knowledge of the hazards and existing work practices.
- Auditors are aware of what might be appropriate for the industry.
- Auditors are familiar with the workforce including

their strengths and weaknesses.

- *Relatively low cost and easier to arrange.*
- *Builds internal competence.*

Disadvantages

- *Auditors may not possess auditing skills.*
- *Auditors may not be up to date with current legislation and best practice.*
- *The auditor may also be responsible for implementation of any proposed changes and this might inhibit recommendations because of the effect on their workload.*
- *Auditors may be subject to pressure from management and time constraints, causing them to carry out less verification and to make more assumptions with respect to compliance evaluation.*

10) What information should you consider when carrying out a review of health and safety performance? (8)

- *Level of compliance with relevant legal and organizational requirements.*
- *Accident and incident data, corrective and preventive actions.*
- *Inspections, tours and sampling.*
- *Absences and sickness.*
- *Quality assurance reports.*
- *Audits and other monitoring data/records/reports.*
- *External communications and complaints.*
- *Results of participation and consultation.*
- *Objectives met.*
- *Actions from previous management reviews.*
- *Legal/good practice developments.*
- *Assessing opportunities for improvement and the need for change.*

ELEMENT 5: PHYSICAL AND PSYCHOLOGICAL HEALTH

1) What circumstances may lead to occupational stress amongst workers? (8)

Stress in workers can be caused by a range of factors relating both to the organization they work in and to issues that are external to the organization but influence the worker while they are at work, for example, personal relationship problems at home. What may cause one worker to become stressed may not have the same effect on others. Pressure perceived as unacceptable to one worker may help to retain focus and motivation in another; however, when that pressure becomes excessive or unmanageable it can lead to stress. The causes of stress in the workplace may be grouped under two

categories, 'work content', for example, workload and pace, participation and control; and 'work context', for example, status, pay, the role of the worker in the organization, interpersonal relationships.

2) What are the health effects that may be caused by exposure to ionising radiation? (8)

Beyond certain thresholds, radiation can impair the functioning of tissues and/or organs and can produce acute effects such as skin redness, vomiting, hair loss, radiation burns, or acute radiation syndrome and death. These effects are more severe at higher doses and higher dose rates.

If the dose is low or delivered over a long period of time (low dose rate), the damaged cells are more likely to repair themselves successfully. However, long-term effects may still occur if the cell damage is repaired but incorporates errors in the DNA, transforming an irradiated cell that still retains its capacity for cell division. This transformation may lead to cancer after years or even decades have passed.

3) Explain four control methods to reduce noise at source. (8)

The answer should include the method and then a brief explanation of how noise reduction is achieved by the method identified. The answer should include a choice of four methods and how noise reduction can be achieved from:

Enclosure:	Total enclosure to contain noise at source by using an acoustic enclosure or haven.
Isolation:	A form of separation between the noise and the worker by distance or use of a barrier of an absorbent character (for example, an acoustic absorbent wall) in the path of noise transmission or relocation of workers into another room remote from the noise source.
Absorption:	When noise passes through porous materials (for example, foam, mineral or wool) some of its energy is absorbed, reducing the noise level.

Insulation:	Imposing a barrier (for example, a brick wall or lead sheet) between the noise source and the workers to deflect or block the noise path.
Lagging:	Insulation of pipework and fluid containers to reduce noise levels, similar to absorption.
Damping:	Mechanical vibration can be reduced through conversion into heat by damping materials, for example, use of plastic gears, rubber coating on conveyors or rollers, reinforcement of metal panels.
Silencing:	Pipes/boxes can be designed to reduce air/gas noise (for example, engine exhaust silencers or duct silencers). Silencers use change of air direction (by baffles) and expansion chambers deplete the source energy and reduce noise levels.
Work practice:	Modify material-handling processes to reduce the noise from shock and impact, for example, reducing the distance where objects fall onto hard surfaces or fixing damping material to surfaces or containers. Review frequency of maintenance programmes, for example, equipment lubrication and replacement of worn bearings.

4) Explain the technique of audiometry used in health surveillance and say why it is necessary. (8)

Audiometry is a medical testing procedure that establishes hearing sensitivity across a range of sound frequencies, which can then be monitored over time. It will assist with the identification of noise hazards and the evaluation of noise control measures.

When workers may be exposed to noise at work audiometry is used as a health surveillance technique from the start of a worker's assignment to work; to establish a base line of hearing sensitivity. Regular surveillance will then detect early signs of hearing loss and provide for early intervention to prevent further deterioration.

5) What are the typical health effects that might be shown by those who use vibrating hand-held tools regularly at work? (8)

Effects on the body usually begin with sharp tingling pains in the fingers, the affected area becomes increasingly more painful, the skin tone becomes lighter (Caucasian race people) and with continued exposure often turns white and over the same time period feeling sensitivity in the fingers decreases and progressively they become numb and result in a syndrome referred to as vibration white finger (VWF). In addition to VWF there may be a range of conditions relating to long-term damage to the circulatory system, nerves, soft tissues, bones and joints which including loss of grip and manual strength.

6) What are the control measures which should be taken when a noise survey has determined an average noise level of 83 dbA in a factory workplace? (8)

The European approach recognises the potential harm that may be done by particularly high-intensity noise experienced over a short period and sets limit and action levels for this situation in the form of peak sound pressures. The lower exposure action value on the A weighted scale is 80dB. Where a worker is likely to be exposed to noise at or above the lower exposure action level values, the employer must make hearing protection available if requested and provide the workers and their representatives with suitable and sufficient information, instruction and training. This should include, for example, the nature of risks from exposure to noise and the organizational and technical measures taken in order to comply. The area should be signed to state the noise level is in excess of 80dBA, and noise protective equipment is available on request. All workers exposed to noise levels above 80dBA should be subject to regular audiometric health surveillance.

7) What are the risks to workers from the misuse of substances at work? (8)

Apart from the general effects on the individual's health (for example, cirrhosis of the liver from alcohol abuse), the risks related to substance misuse and work is derived from the effects the substances have on the individual's biological and psychological systems. These effects create risks to the health and safety of the worker and others. Answers should be supported with examples from:

- Poor coordination and balance.
- Perception ability reduced.
- Overall state of poor health including fatigue, poor concentration and stress.
- Poor attitude, lack of adherence to rules.
- Increased risk of violence.
- Increased likelihood of transport incidents.
- Reduction in work rate which may encourage risk taking when catching up.

This could have a negative affect whilst working at height, working with equipment, tools or transport and many other activities that rely on care and accuracy.

8) How can an employer determine the size of the problem of violence at work? (8)

To answer this question, the approach should include reference to: asking workers informally, through managers and health and safety representatives. Encouraging workers to report all incidents and keep detailed records. Classify all incidents according to their actual or potential severity of outcome. Endeavor to predict circumstances where violence may arise such as those with a heavy workload or pressures to complete the job against unrealistic timescales; those dealing with complaints from members of the public. Absence records and the effectiveness of an alcohol and substance abuse policy.

ELEMENT 6: MUSCULOSKELETAL HEALTH

1) Explain, with examples, how mechanisation may reduce fatigue or injury from manual handling operations. (8)

Mechanisation involves the use of handling aids. Although this may retain some elements of manual handling, bodily forces are applied more efficiently.

- **Levers** - reduces bodily force to move a load, can avoid trapping fingers.

- **Hoists** - can support weights, allowing handler to position load.

- **Trolley** - sack truck, truck roller, hoist; reduces effort to move loads horizontally.

- **Chutes** - a way of using gravity to move loads from one place to another.

- **Handling devices** - hand-held hooks or suction pads can help when handling a load that is difficult to grasp.

2) (a) Why should periodic inspection and examination of lifting equipment be carried out?

(b) Who should be selected to do the inspection and examination?

For (a) Lifting equipment should be thoroughly examined prior to first use, after assembly and on change of location in order to ensure that it has been installed correctly and is safe to operate. The equipment should be examined as necessary to ensure that damaged or dangerously worn equipment does not remain in service.

For (b) All items of lifting equipment must be periodically examined by a competent person. A competent person is someone who has acquired, through training, qualifications or experience or a combination of these things, the knowledge and skills required to do that work competently.

3) What are the principal hazards associated with a mobile crane lifting operation? (8)

The principal hazards associated with any crane lifting operation are: overturning, which can be caused by operating outside the capabilities of the machine, uneven or weak ground (cellars or drains), outriggers not extended, insufficient counterweight and adverse weather and by striking obstructions. Overloading by exceeding the operating capacity or operating radii, or by failure of safety devices. Collision with other cranes, overhead cables or structures. Failure of load-bearing part from structural components of the crane itself or an accessory fitted to it. This may be due to overloading or degradation of the load-bearing part due to damage, use (wear) or faults (corrosion). Loss of load from failure of lifting tackle, incorrect hook fittings or slinging procedure. People in and around the crane may get entangled in or trapped by moving parts.

4) Explain the musculoskeletal disorders sprain and strain. (8)

A sprain is an injury to a ligament. The ligament is tough fibrous tissue that connects a bone to another bone. Ligament injuries typically involve overstretching or a tearing of this fibrous tissue, i.e. a 'torn ligament'.

A strain is an injury to either a muscle or a tendon that connects muscles to bones. Depending on the severity of the injury, a strain may be a simple overstretch of the muscle or tendon or it may result in a partial or complete tear. Rupture of the muscles in a section of the abdominal wall can cause a hernia.

5) What should be considered when carrying out a manual handling risk assessment? (8)

The four elements are Load, Individual, Task, and

Working Environment, each answer should be supported by examples such as:

Load, is it:

- Heavy?
- Bulky or unwieldy?
- Difficult to grasp?
- Unstable, or with contents likely to shift?

Individual

- Require unusual strength.
- Create a hazard to those who have a health problem.
- Require special knowledge or training.
- Involve clothing and footwear that presents an increased risk.

Task

- Holding a load at a distance from the trunk.
- A long carrying distance.
- Twisting the trunk.
- Excessive pushing or pulling distances.
- Stooping.

Working environment

- Space constraints preventing good posture.
- Uneven, slippery or unstable floors.
- Lighting.
- Is it a cold or hot environment?

6) (a) What are the health effects associated with carpal tunnel syndrome?

(b) Identify two examples of work tasks which may result in work-related upper limb disorders.

For (a) Carpal tunnel syndrome is the painful inflammation of the nerves and tendons that pass through the carpal tunnel formed by the carpal bones in the wrist area. The syndrome usually affects the thumb, index finger, middle finger and half of the ring finger (but may spread outside these areas).

For (b) Examples include:

- Keyboard operation.
- Assembly of small components.
- Bricklaying.
- Checkout operators.

ELEMENT 7: CHEMICAL AND BIOLOGICAL AGENTS

1) What are two forms of biological agent and give a workplace example for each. (8)

Answers should have considered any two from the following:

Fungi, are a variety of organisms that act in a parasitic manner, feeding on organic matter. Most are either harmless or positively beneficial to health; however, a number cause harm to humans and may be fatal. An example of fungi is the mould from rotten hay called aspergilla, which causes aspergillosis ('farmer's lung') a risk associated with agricultural work.

Bacteria, are single-cell organisms. Most bacteria are harmless to humans and many are beneficial. The bacteria that can cause disease are called pathogens. Examples of harmful bacteria are leptospira (causing Weil's disease), leptospirosis, a risk associated with sewage work.

Viruses, are the smallest known type of infectious agent. They invade the cells of other organisms, which they take over and where they make copies of themselves, and while not all cause disease, many of them do. Examples of viruses are hepatitis, which can cause liver damage, a risk associated with contaminated body fluids encountered in clinical work.

2) What are the practical measures that complement the body's protection mechanisms against ill-heath from agents in the workplace? (8)

Practical measures include:

- Maintain good personal hygiene.
- Do not apply cosmetics in the workplace.
- Do not allow eating or drinking in the workplace.
- Provide proper containers/storage for food and drink.
- Provide and ensure use of appropriate personal protective equipment.
- Clean contaminated protective clothing, before removing.

3) What are the limitations of information provided by manufacturers and suppliers in assessing risks to health? (8)

Individual susceptibility of workers differs by, for example, age, gender or ethnic origin, and this can limit the value of information sourced from manufacturers and guidance documents related to occupational exposure limits. Exposure history varies over the working life of an individual and current exposure may not indicate that the individual may suffer due to a cumulative effect from earlier exposures. For example, an individual may have been or be engaged in a number of processes within a variety of workplaces or personal pastimes. These limitations are reflected in the use of occupational exposure limits, where the limits are considered to be a maximum and control measures should be applied to control assessed risks to well below this limit.

4) What are the basic components of a local exhaust ventilation (LEV) system? (8)

The components of a basic system are:

- *Extractor Hood(s) to collect airborne contaminants at, or near, where they are created (the source).*
- *Ducts to carry the airborne contaminants away from the process.*
- *Air cleaner to filter and clean the extracted air.*
- *Fan must be the right size and type to deliver sufficient low pressure at the extractor hood to remove contaminates to the hood.*
- *Discharge - the safe release of cleaned, extracted air into the atmosphere.*

5) What are the factors that reduce the efficiency of a local exhaust ventilation (LEV) system? (8)

Factors might include:

- *Damaged ducting, from corrosion or impact resulting in leaks, distortion of ducts reducing flow efficiency.*
- *Alterations, or modifications, such as the addition of more extractions than the system was designed for.*
- *Incorrect extractor hood location, not close enough to source of contaminate.*
- *Poor design with too many bends in ducts, reducing flow efficiency.*
- *Blocked or defective filters.*
- *Leaving too many ports open.*
- *Process changes leading to overwhelming amounts of contamination.*
- *Fan strength or incorrect adjustment of fan.*
- *Excessive amounts of contamination.*

6) What are the circumstances where breathing apparatus may be the preferred respiratory protection rather than the use of respirators? (8)

Breathing apparatus provides a separate supply of air (including oxygen) to the user, whereas respirators purify the air by drawing it through a filter to remove contaminants, therefore cannot be used if there is insufficient oxygen in the atmosphere. In addition breathing apparatus as a high assigned protection factor (APF) and may be used in an atmosphere with high levels of airborne toxic substance.

Breathing apparatus can be worn for long periods if connected to a permanent supply of air, whereas respirators have a finite life depending on the concentration of airborne contaminant.

ELEMENT 8: GENERAL WORKPLACE ISSUES

1) What are the effects of extremes of temperature on the body? (8)

Mention of cold and hot conditions based on some of the following is required. An approach might look at the progressive effects of each as listed:

Under cold conditions:

- *Loss of concentration in mental work.*
- *Reduced manipulative powers in manual work.*
- *Chilblains.*
- *Frost burns caused by skin contact with very cold surfaces.*
- *Shivering, leading to chattering of the teeth.*
- *Hypothermia, death quickly follows.*

Under hot conditions:

- *Loss of concentration.*
- *Reduced activity rate.*
- *Discomfort and dehydration caused by sweating.*
- *Muscle cramping as a result of lost body salts during sweating.*
- *Heat exhaustion just prior to heat stroke.*
- *Heat rash or 'prickly heat'.*
- *Heat exhaustion/stroke.*
- *Fainting due to vasodilatation (widening of the blood vessels, especially the arteries, leading to increased blood flow or reduced blood pressure).*
- *Breakdown of control mechanisms, body temperatures soar, death quickly follows.*

2) What are the control measures to reduce work-related driving risks? (8)

The scope of this question requires a consideration of the need to drive, the driver, the vehicle, the journey and incident reporting, for example:

Elimination: *The journey may not be necessary; communication may be sufficiently effective by telephone or video conference instead of driving.*

Driver: *In good health and capable of doing their work in a way that is safe for them. Evidence that they have a current license to drive their class of vehicle.*

The vehicle: *Vehicle is in a good condition, safe and suitable for the task to be carried out, including any safety equipment that is required being properly fitted and maintained, for example, seat belts.*

The journey: *Journey planning and scheduling routes should be planned thoroughly, to ensure schedules are realistic, without fatigue, will not encourage them to drive too fast for the conditions or exceed speed limits.*

Incident reporting: *Drivers must report all incidents, including near-misses. Data provided should be analysed and any changes or improvements noted, communicated to those concerned and the work-related road safety procedures should be updated.*

3) What are two design features of a vehicle intended to minimise the consequences of an overturn? (8)

Roll-over protection can take the form of a complete, enclosed cab or of a system of bars that prevent the vehicle driver/passengers from being crushed by the vehicle in the event that it rolls over.

Restraint systems typically involve some sort of adjustable seat belt that is worn to prevent the driver/passenger from being thrown from the vehicle when travelling or falling from the vehicle in the event that it rolls over.

4) What are the possible causes of a dumper truck overturning? (8)

Possible causes could include: overloading or uneven loading, for example overloading of lifting equipment attached to the vehicle. Insecure and unstable loads that move so their weight is not evenly distributed. Driving with the load skip elevated. Driving too fast and cornering at excessive speed. Sudden braking or acceleration. Hitting obstructions, including kerbs, buildings, structures or other vehicles. Driving across slopes. Driving too close to the edges of slopes, embankments or excavations. Driving over debris, soft ground or holes in the ground, such as drains. Poorly maintained or uneven road surfaces. Mechanical defects that occur due to the lack of maintenance. Inappropriate or unequal tyre pressures, causing the weight of the load to be poorly distributed or movement of the load.

5) How do non-movement related hazards result in injury to drivers? (8)

The vehicle load may collapse or fall on a vehicle driver during loading and unloading operations, particularly where vehicles are used to load materials at a height. A stack of material may become unstable and collapse when items are removed or added to the stack.

A number of methods may be used to secure loads to vehicles, for example, ropes, webbing and chains, which could expose them to manual handling hazards related to moving things and over-exertion. Similarly If the covering of a load with a sheet may involve the manually unrolling of the sheet or net over the load and the sides of the vehicle, which presents hazards of falls from height, slips and manual handling.

6) What measures can be used to segregate pedestrians and vehicles in the workplace? (8)

Segregation may be total in that pedestrians are removed from the workplace where vehicles operate, for example, in some warehouse situations. Clearly defined and marked routes should be provided for the general movement of pedestrians at work. These should be provided for access and egress points to the workplace, car parks, and vehicle delivery routes. Safe crossing places should be provided where pedestrians have to cross main traffic routes. In buildings where vehicles operate, separate doors and walkways should be provided for pedestrian segregation to enable them to get from building to building. Where it is not possible to have a pedestrian route with physical segregation by barriers or raised walkways segregation may be established by means of markings and signs.

7) What control measures should you consider when planning work-related journeys? (8)

Journey planning and scheduling are essential in ensuring the health and safety of workers who drive for work. Journey planning should be implemented as a component of the policy, to ensure where possible, routes are planned thoroughly, schedules are realistic, and sufficient time is allocated to complete journeys safely and without fatigue. It may be necessary to plan overnight stops if the journey time extends due to bad weather or traffic conditions. Delivery schedules should be adjusted so that unrealistic targets are not set. This will reduce the fatigue and stress of drivers and will not encourage them to drive too fast for the conditions or exceed speed limits.

8) How can working at height be avoided? (8)

This could be achieved by using different equipment or methods of work and answers should be supported with examples such as: pre-assembly, pre-painted or pre-drilled materials such as roof trusses, either before delivery or on the ground on-site, instead of assembly at a height. Decorative wooden building panels can be pre-treated on the ground to avoid weather treatment being brushed on at a height. Long reach handling devices can be used to allow cleaning or other tasks to be conducted from the ground. Where equipment, such as light units, requires maintenance an option may be to lower it sufficiently to enable bulbs to be changed and cleaning to be conducted from the ground.

ELEMENT 9: WORK EQUIPMENT

1) What is the purpose of the CE mark on workplace equipment used in the European Union? (8)

It is a requirement of European Union legislation that employers in any member state ensure that an item of work equipment brought into use in the workplace has been designed and constructed in compliance with the directives/harmonised standards that relate to it and meets the essential health and safety requirements set for it.

2) What are the duties of workers when they discover a damaged piece of work equipment? (8)

In most countries, there is a broad-based principle of 'duty of care' that requires workers to be involved in the prevention of health and safety risk and not to rely solely on the actions of their employer. This duty of care establishes the responsibility of workers for the health and safety of themselves and others that may be affected by their acts or omissions.

Users of work equipment, in particular machinery, should therefore use it in accordance with any training or instruction that they have received. They should use safety devices and protective equipment correctly and not render them inoperative.

This includes an expectation that users report any faults or defects in their equipment that, if not rectified, could lead to the development of hazardous situations.

3) What are the benefits of introducing a scheme of planned preventive maintenance for equipment in regular use? (8)

Planned preventive maintenance (PPM) is an important strategy in maintenance and seeks to maximise the life of the component/equipment. The planned method seeks to inspect and replace components on a scheduled basis. Preventive maintenance seeks to keep the condition of the component at its best by carrying out frequent care, for example lubrication, adjustment, cleaning.

Although all maintenance is preventive in some respect, the primary aim of PPM is to prevent failures occurring while the equipment is in use. The main benefits are: extended life of components and assurance of reliability leading to confirmation of the condition of essential components. Reduced risk of loss-producing failure events. Ability to carry out work at a suitable time. Better utilisation of maintenance staff and no standby facility required. Emergency contracted out repair work reduced making plant more cost effective. Demonstrates the employer has made arrangements to meet the legal duties to maintain safe equipment.

4) (a) What is the meaning of the term lock-out and tag-out (LOTO)? (2)

(b) How does the use of LOTO systems reduce the risk to maintenance workers? (6)

For (a) Lock-out (LO) is the isolation of electrical energy

from the system (a machine, equipment, or process), and physically locks the system controls in a safe mode. Tag-out (TO) is a labelling process that is used when lock-out is required. The process of tagging out a system involves attaching a standardised label that includes certain information.

For (b) Electrical isolation by itself does not afford adequate protection, because there is nothing to prevent the isolator being switched back on or someone replacing removed fuses while the maintenance worker who isolated the equipment is still working on the equipment and in danger. To ensure that this does not happen, the isolator needs to be physically locked in the off position, typically using a padlock, and the key held by the worker that may be in danger if the isolation is removed.

5) (a) What is the purpose of emergency stops fitted to machinery? (2)

(b) What are the design and positioning requirements for emergency stops? (6)

For (a) Emergency stops are a hand-operated safety device, intended to ensure a rapid shut off response to potentially dangerous situations. Emergency stop controls should be easily reached and actuated. They should remain available and operational at all times, regardless of the operating mode of the equipment and they should not be used as functional stops during normal operation.

For (b) The design aspect of this question should be answered with reference to some examples from:

- *Coloured red.*
- *Positioned in such a way as to be safely operated without hesitation or loss of time and without ambiguity.*
- *Designed in such a way that the movement of the control device is consistent with its effect.*
- *Located outside the danger zones, except where necessary for certain control devices such as an emergency stop (for use of maintenance workers) or a teach pendant (hand-held control).*
- *Positioned in such a way that their operation cannot cause additional risk.*
- *Designed or protected in such a way that the desired effect can be achieved only by a deliberate action.*
- *Made in such a way as to withstand any foreseeable forces; particular attention should be paid to emergency stop devices liable to be subjected to considerable force at the time of use*

6) (a) Why is stability of work equipment important when in use? (2)

(b) What factors may affect stability of work equipment? (6)

For (a) Work equipment should be stable enough to avoid overturning, falling or uncontrolled movement during transportation, assembly, installation, dismantling and use. If the shape and weight distribution of the equipment does not provide sufficient stability, appropriate means of stabilisation must be made, including the use of anchorages and additional bracing or support structures. Where equipment has a variable reach or if the weight and centre of gravity may change during use it is important to take this into account.

For (b) Most machines used in a fixed position should be bolted or otherwise fastened down so that they do not move or rock during use. This is particularly important where the equipment is tall relative to its base or has a high centre of gravity, for example, a pedestal-mounted abrasive wheel, some diagnostic medical equipment and vertical cardboard compactors.

ELEMENT 10: FIRE

1) (a) How has an understanding of the principles of the fire triangle been used to develop techniques for extinguishing fires? (6)

(b) Explain two methods, with an example for each, how a fire can be extinguished. (2)

For (a) A simple approach depicts fire as having three essential components, the fire triangle: fuel, oxygen and ignition source (heat). When the three separate parts are brought together, a fire occurs. If one component is removed the fire is extinguished.

For (b) One example is to remove heat, this can be achieved by cooling and for Class A fires water can be used to achieve this. Another example is removal of the fuel, for a gas fire this would be achieved by isolation of the gas supply to the fire.

2) What are the main methods by which fires spread through a structure? (8)

The main methods by which fires spread are:

Convection, the movement of hotter gases up through the air can quickly move hot gases to another part of a building where they raise the temperature of combustible materials to a point that combustion takes place, for example, hot gases rising up a staircase through an open door.

Conduction, some materials, such as metal, can absorb heat readily and transmit it to other rooms by conduction, where it can set fire to combustible items that are in contact with the heated material, for example, metal beams, ducting or pipes transmitting heat through a solid wall.

Direct burning, this where ignited combustible materials, for example, furnishings in direct contact with other combustible materials will cause them to ignite and travel along them spreading the fire, for example, a carpet.

Radiation, is the transfer of heat as invisible infrared waves through the air in a similar way to light. Any material close to a fire will absorb the heat in the form of energy waves, and eventually ignite.

3) What are the reasons for carrying out a regular reviews of fire safety measures? (8)

Fire risk assessments and arrangements should be reviewed by the 'responsible person' if there is reason to suspect that it is no longer valid or there has been a significant change in the matters to which it relates. Examples: Changes to the workplace are proposed, for example, increased storage of flammable materials. Similarly if changes to work processes/activities are proposed, for example, introducing a new night shift. Changes to the number or type of people present are proposed, for example, public are invited on site to buy goods. Following a near-miss incident or if a fire occurs. When failure of fire precautions occurs, for example, fire-detection and alarm systems. When there is a change in legislation and periodically to include any improvements which have been identified.

4) Why is good 'housekeeping' in the workplace essential to ensure safe escape in a fire? (8)

'Housekeeping' means the general tidiness and order of the workplace. It should be remembered that fires need fuel, and a build-up of redundant combustible materials, rubbish and stacks of waste materials can provide that fuel. Combustible materials cannot be entirely eliminated, but they can be controlled. Any unnecessary build-up of rubbish and waste should be avoided. Poor housekeeping can also lead to obstruction of hose reels; of vital signage and a reduction in the effectiveness of automatic fire detectors and sprinklers.

5) What should you consider before the location of a fire assembly point is decided upon? (8)

The assembly point is a place of safety where people wait whilst any incident is investigated, and where confirmation can be made that everyone has evacuated the premises. The assembly point should be in a safe position, at a distance from buildings. The location should enable workers to be able to walk away from the assembly point and back to a public road, but sited so that staff will not be in the way of the fire service and rescue teams. More than one assembly point may be required, for example to suit numbers and location of different groups of people. Assembly points should be suitably illuminated and be clearly signed to enable them to be differentiated from each other. Communication should be provided between assembly points to enable an accurate role call to be made.

ELEMENT 11: ELECTRICITY

1) Explain the progressive effects that electrical shock may have on the body. (8)

Low voltage, will cause pain at the point of contact. Increased voltage, will cause burns to the skin on entry to and from the body and burns to tissue on the path the current flows through the body. Muscle will contract involuntary, causing the leg muscles to jump or if contact is with the palm of the hand, for the hand to clench and grip onto the conductor causing the shock. If the current path is across the chest this may lead to respiratory failure, fibrillation of the heart and cardiac arrest.

2) Explain the purpose of electrical double insulation. (8)

Double insulation is used in the design of some electrical equipment, for example, portable equipment such as a telephone charger, its purpose is to protect the user from an electrical fault condition in the equipment, receiving an electrical shock. Shock is prevented by the use of two layers of insulation to all parts of the electrical components. In the event of the failure of one layer of insulation the remaining layer will still afford protection from electric shock. In many instances the casing of the equipment is made of non-conducting material which provides the second layer of insulation protection.

3) What are typical user checks that should be carried out, prior to using a portable item of electrical equipment? (8)

- *Typical user checks might include:*
- *Evidence of formal testing*
- *Cuts, abrasions and cracks in inner and outer cable insulation.*
- *Damaged plugs – cracked casing or bent pins, failure of the cord grip.*
- *Taped or other inadequate cable joints.*
- *Faulty or ineffective switches.*
- *Overloaded sockets.*
- *Equipment that is wet or dirty.*
- *Burn marks or discolouration.*
- *Damaged casing.*
- *Loose or damaged sockets or switches.*
- *Evidence of unauthorised repairs.*

4) What should you consider when determining the frequency for the inspection and testing of electrical equipment? (8)

The frequency will depend on the conditions the system is used in, for example, a test of office portable equipment may be sufficient if conducted every three years, whereas equipment used on a construction site may need to be tested every three months. You should consider when deciding the frequency: the type of equipment, i.e whether it is hand-held, a review of manufacturer's recommendations; the suitability of Its initial integrity and soundness for the task. The working environment and its suitability in flammable areas or wet conditions. The likelihood of mechanical damage with consideration of frequency of use, duration of use, foreseeable use. Does the user need special skills (welding equipment), can it be used by young persons (butchers meat cutter). Any modifications or repairs been made.

5) What are the advantages and disadvantages of the use of a fuse as a protective device in an electrical circuit? (8)

Fuses are primarily designed to prevent circuit overload and only provide indirect protection from electric shock.

Advantages are:

- *Offers a good level of protection for the equipment.*
- *Very cheap and reliable within its limits.*

Limitations are:

- *A weak link in the circuit that melts slowly when heat is created by a fault condition.*
- *Easy to replace with wrong rating.*
- *Needs tools to replace.*
- *Easy to override by replacing a fuse with one of a higher rating or putting in an improvised 'fuse', such as a nail, that has a high rating.*

This page is intentionally left blank

Revision and examination guidance

Contents

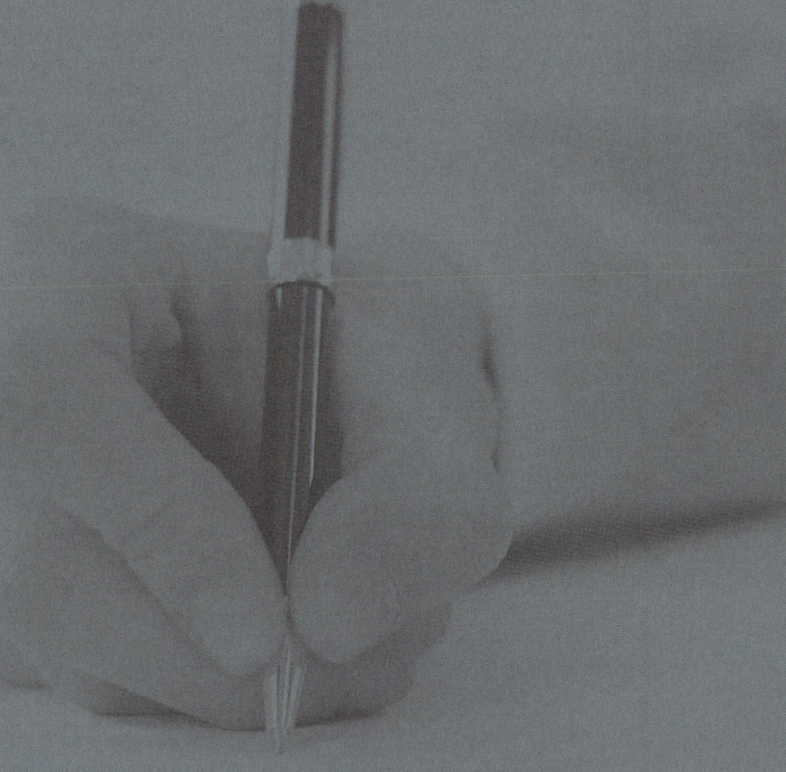

Revision and examination guidance

Being successful in exams means having more than an in-depth knowledge of the subject matter, it means also having the necessary exam technique. Studying and revising are activities in their own right and there are tips and techniques that you can adopt which will help you to be more effective and give yourself a greater chance of success.

This section is aimed at providing some useful advice on how to improve your study skills and how to revise. It will also advise you on exam technique and help you to organise your time.

Finally, it will also explain what to expect on the day of the exam.

How to study

TIME MANAGEMENT

Most people live busy lives and finding time to study and take exams can be difficult. So, do not commit to study 'all day'. Make a study plan. Be realistic and disciplined - don't plan a schedule you can't manage.

Remember, too much work can be as unproductive as too little work. A good way to start is to work for 50 minutes, then give yourself a 10 minute break. Also remember to allow breaks in your day for food, relaxation and exercise, but not all at once. Physical exercise will help to increase your concentration. Even a short, brisk walk will improve your attention span.

The best time to do difficult tasks is when you are at your most productive so schedule these topics at the start of your study period. Decide on the time of day when your concentration is at its best; for some this might be first thing in the morning, so consider rising earlier and study before going to work; for others it may be late in the evening, but always remember most people's day jobs are demanding and you may well be too tired!

Another consideration might be to plan to do some work at lunch time, whatever you decide, aim to work to a regular regime that suits you and your family. Try to give each subject equal time; do not concentrate on one subject at the expense of another.

Don't be inflexible to the plan, intrusions will occur that takes priority over your plan - always leave time at the end of each study/revision period for reviewing what you have done and what you must still do.

FINDING A SUITABLE ENVIRONMENT

It is important that you work in comfortable surroundings. Your study environment should be calm and quiet and free from distraction (i.e. e-mails, mobile phone, family, TV, Facebook).

Ensure your room has adequate lighting to prevent eye strain, is at a comfortable temperature and is well ventilated with plenty of fresh air to keep you awake.

You should also ensure that you have a good chair (an upright chair is better than an armchair) and a spacious desk to take all your books and other study aids.

REVISION AND EXAM TECHNIQUE

It may have been some time since you last prepared for and took a professional exam. Remember, success in an exam depends mainly on:

- **Revision** - your ability to remember, recall and apply the information contained in this book.
- **Exam technique** - your ability to understand the questions and write good answers in the time available.

Revision and exam technique are skills that can be learned. We will now deal with both of these skills so that you can prepare yourself for the exam. There is a saying that 'proper planning and preparation prevents a poor performance'. This was never truer than in an exam.

REVISION TIPS

You should recognise the knowledge that you already possess that is relevant to the syllabus. Don't underestimate the importance of this. But, it is never too early to start revising!

Throughout the course or learning you should have been thinking about this and noting the topics that you are finding it difficult to deal with.

- Simply reading this publication repeatedly will not normally help you to remember the information; from the start of your study revisit each days study and abstract (in handwritten format, ideally) the key learning points and essence of the subject. This process will ensure greater retention and recall because we remember better that which we do, rather than that which we have just read. This process will identify any ambiguity you might have, early in the study period and enable you to seek clarity from your tutor before training is complete.

- As the course progresses, condense the information you have abstracted on to revision cards. This is a revision technique in itself. It means that you can carry a lot of information with you and read them in a spare moment.

Other techniques, which you may consider worth trying, include: mind maps, key words and mnemonics.

MIND MAPS/KEY WORDS

Mind maps (often referred to as a spider gram) are a powerful graphic technique which provide a universal key to unlock the potential of the brain. It uses the visualisation part of the brain to retain information in such a way that it can be quickly recovered when required. Tony Buzan is recognised by many as the inventor of mind maps in the 1960's.

Mind maps are best used for a single topic, such as learning a language. The technique is used to identify all the components necessary to achieve this and possibly the order in which they should be carried out. This part is an essential aspect of revision; the fact that you have abstracted that which is required will imbed the information into the brain and enable threads of connectivity to be established simplifying recall in the future.

The mind map should start in the centre of a blank page turned sideways. The blank page, does not restrict the brain to one dimension, as with lined paper, thus enabling the brain to work in a two dimensional way and to spread out in all directions.

Connect the main branches to the centre start point, avoid straight lines and use lines which are curved as this increases the brains attention. The brain works best by association so that when the lines are connected to the second and third level branches you will understand and remember a lot more easily. Use images or sketches and colours throughout to make this more interesting for you and increase the brains stimulation and retention.

Use one key word per line; single key words give your mind map more power and flexibility. An example of an approach to learning a foreign language is illustrated in **Figure R-1** below.

MNEMONICS

The use of mnemonics is another brain association technique to aid revision and memory recall. The technique relies on the use of key words to trigger memory recall, often when there are several key headings which require recall to fully answer a question.

Commonly encountered mnemonics are often used for lists and in auditory form, such as short poems, acronyms, or memorable phrases.

A common acronym, based on People, Equipment, Material, Environment (PEME) is used to identify the considerations required to determining a safe system of work.

The acronym PEME is sufficient to trigger the answer, thus:

- A safe system of work involves the combination of People, Equipment, Materials and working Environment in the correct way to ensure a safe outcome.

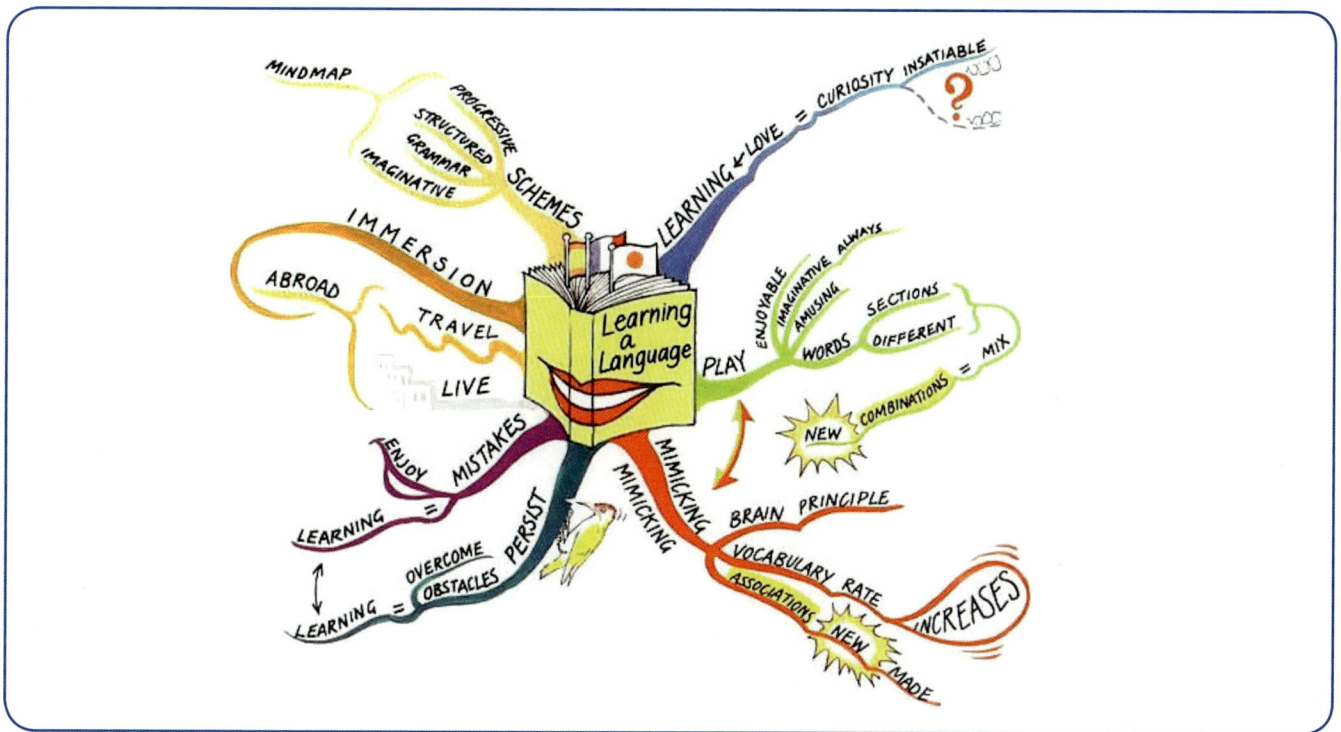

Figure R-1: Learning a language mind map. Source: Tony Buzan

R.3 Approaching the time of the examination

As the exam gets nearer, you will need to think about organising all the information from the course in preparation for revision.

We recommend that you refer to the revision checklist located at the end of this section, which contains the syllabus for your exam. If a topic is in the syllabus then it is possible that there will be an examination question on that topic.

You can assess your level of knowledge and recall against the revision checklist. Look at the content listed for each element in the revision checklist. Ask yourself the following question:

"If there is a question in the exam about that topic. Could I answer it?"

This technique will help you to identify the areas of weakness and then to prepare a plan of work which concentrates on the knowledge that you need to develop.

R.4 The lost art of written communications

From the start of your revision it is essential to practice hand writing whenever possible. Most people have become so used to using personal computers (PC's) and mobile phones for written or text communications, that they experience real difficulty in writing on the day of the exam. Do not underestimate the effort required to handwrite text continuously for two hours for each of the exams. When using a PC you can write whatever you wish in any order and then re-organise the text easily into a logical sequence; this is not possible with the written word on paper in a NEBOSH answer book. You must plan your answer or approach before you commence writing. It is important to provide sufficient written material in your answer, the examiner cannot award marks for that which is not there; again, the use of e-mails has encouraged us to use as few words as possible, leaving the reader to fill in the gaps; the examiner will not fill in the gaps, they will only mark that which is written and developed.

Attempt the questions provided within the assessment section and answer within the exam time frame (10 minutes for a short question, 30 minutes for a long question) without reference to your notes or the suggested answer. This will enable you to get an idea of working under exam conditions. You will need to practice writing answers in, for example, black ink, write your answer on lined paper, but leave each alternate line free. When you have finished; review your answer with your study book and annotate your work with, for example, a *green ink* pen, and use the lines left free in your answer. The *green ink* annotations will help with that which you may have missed or remind you of the detail not covered sufficiently. This way of doing things will help you to build understanding and memory retention.

R.5 NEBOSH General Certificate - Examination

There is a single written examination which lasts two hours and the exam paper consists of; one long answer question 1 (20 marks), and ten short answer questions 2 to 11 (8 marks each).

EXAM TECHNIQUE

Make sure that you study the exam questions provided in the NEBOSH Examiners Expectation Report and syllabus, available via the NEBOSH website. These are produced to help candidates and tutors in future exams. They are intended to be constructive and warn of the common mistakes made by people sitting these exams. A common fault is that candidates fail to pay attention to the command word in each question.

UNDERSTANDING THE ACTION VERBS

In this context, the phrase 'action verb' is used to refer to the words specifically associated with the learning outcomes and assessment objectives of the NEBOSH Certificate qualifications. The syllabus learning outcomes are concerned with what students can do at the end of a learning activity, to assess this, action verbs are used.

The most common 'action verbs' used in certificate examination questions are in the table on the following page.

Certificate questions will predominantly assess **knowledge, application and comprehension**.

Knowledge requires an ability to recall or remember facts without necessarily understanding them. Action verbs used in knowledge based questions include **LIST**. **Application** is the skill of being able to take knowledge and apply it in different contexts and circumstances in order to understand why and where problems and issues arise. The important thing to remember is that, whatever

Action Verb	Definition
List	To give reference to an item, this could be its name or title. NB: Normally a word or phrase will be sufficient, provided the reference is clear.
Why/how	To show for what cause, reason, or purpose. E.g. why did you do it? Why is it a good idea to use a permit-to-work system for some work activities?
What	What is used in questions when you are being asked for specific information about something. E.g. What are an employer's health and safety responsibilities?
Explain	To provide an understanding. To make an idea or relationship clear. E.g. Explain what an organisation should do to provide effective first-aid.
Describe	To give a detailed written account of the distinctive features of a subject. The account should be factual, without any attempt to explain. When describing a subject (or object) a test of sufficient detail would be that another person would be able to visualise what you are describing.

Source: RMS

the context - a transport company, a communications centre or an oil refinery - the principles being assessed are the same, but will have different implications given the different industry or issue being considered. Action verbs used to assess application include **EXPLAIN**. **Comprehension** requires an ability to understand and interpret learned information. Action verbs used in comprehension based questions include **EXPLAIN**.

In every question the skills required by the specific action verbs are also shown in the marks allocated for the question. In general there are going to be more marks available for application and comprehension skill questions than for knowledge based questions.

It is important to read the whole question and to understand what the question requires as the action verb on its own will need to be reinforced by the remainder of the question.

The need to understand the meaning of the 'action word' and to read the question carefully is emphasised in the comments below that are typical of those used in NEBOSH examiner's reports.

"It was disappointing to note that some candidates again misread the question and provided outlines of the duties of employers rather than employees."

"Failure to read the question correctly or misinterpreting what the question was asking, for example, a question could ask for a range of hazards, yet the candidate would give control measures as the answer."

"It is evident that in some cases candidates feel obliged to give the appearance of a broad answer by repeating the same point in a number of different ways. Repeating points within the answer will not gain further marks and will also waste time......"

Source: NEBOSH examiners report.

TIME MANAGEMENT

Good time management during the examination is an essential component for success.

There are 11 questions in total on each paper:

* Question 1 = 20 marks and questions 2-11 = 8 marks each.

You should allow yourself about twenty-five to thirty minutes for the twenty mark question and eight to ten minutes for the short answer questions. An easy rule is to allow one minute for each mark available. There will be a clock in the room but most candidates often find it useful to take their own watch into the exam and have it visible on their desk to help keep track of time. NEBOSH questions are often split into parts to help signpost the answer required and to give you an idea of how much time should be spent on each part of the question.

ATTEMPT ALL QUESTIONS

Good exam technique includes making sure you attempt all the questions to maximise your chances of a pass standard.

If you **DO NOT** attempt all the questions and you average 50% of the available marks for those attempted, as you can see from the following table your margin for a pass is significantly reduced.

No. of questions attempted:	For example:	If you only average 50% of the total available marks this will equal:
11	One 20 mark + Ten 8 mark questions	50 marks - Pass
10	One 20 mark + Nine 8 mark questions	46 marks - Pass
8	One 20 mark + Eight 8 mark questions	42 marks - Referred

The fewer the questions attempted, the higher the average mark for each that must be obtained for a pass standard to be achieved overall. It is important to remember that most marks are achieved in the earlier part of the available time spent on any question; do not spend too much time on any one question, certainly do not exceed the time available for each.

If you finish with time to spare go back over the questions and add extra material if you feel it is relevant.

ATTEMPTING QUESTIONS

You must attempt all 11 questions but you may answer them in any order. Start each question on a new page in the answer book. Put the question number at the start of each page. Tick off questions on the answer book front cover to indicate your progress. Make sure that you read each question very carefully. Read through the whole question before you begin your answer. With questions that comprise of more than one part, it is normal for the parts to be related and you must make sure that you do not answer the sections in the wrong order. It is not uncommon for candidates to miss marks simply by failing to answer the question that was set.

A typical, NEBOSH Examiners Report comment:

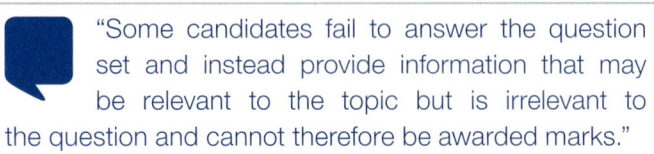
"Some candidates fail to answer the question set and instead provide information that may be relevant to the topic but is irrelevant to the question and cannot therefore be awarded marks."

So, be sure that you answer the question that has been asked and not the one that you wish had been asked!

MAKE IT EASIER FOR THE EXAMINER

Make it as easy as possible for the examiner to find places to award you marks. Keep writing legible - remember the examiner cannot award marks for that which cannot be read. If a question consists of different sub-sections then ensure each sub section is correctly numbered and answered against the appropriate number, within your answer. Do not expect the examiners to give marks if the appropriate number and answer are not clearly connected. A general all-embracing answer, may cover the question requirements, but it is not the examiners job to make it fit any specific question numbering or example requirements. Beware, this is a common trap that candidates fall into when rushing through the exam questions. Take care with abbreviations; PPE, SSW, explain the words at least once using brackets to introduce the abbreviation first before using again, for example, personal protective equipment (PPE), safe system of work (SSW). This rule applies for each question answered.

Marks will not be awarded if the examiner cannot understand the relevance of the sentence that you have written, poor grammar or spelling will not be penalised, as long as your answers are clear and can be understood. It often helps to keep sentences relatively short, but take care not to assume that the examiner will fill in any gaps in the information provided. As discussed earlier avoid using generic references such as, train, monitor, risk assess, etc.

An example is often required: **'train all workers'**, would communicate better if it were more specific to the question, such as **'workers should be trained in the correct disposal technique for waste oil, including hazards that are present and control measures that are in place'**. Pay close attention to the number of marks available for each question or part of a question - this usually indicates how many key pieces of information the examiner expects to see in your answer.

An example can be used to illustrate an idea and demonstrate that you understand the requirements of the question. If you start to run out of time, write your answers in bullet point or checklist style, rather than failing to answer a question at all.

On the following page is an example of how an answer to a question might need to be developed to show depth of understanding for a question requiring an explanation of the use of personal protective equipment (PPE).

Use a similar approach to develop the following statements:

- Provide training.
- Provide supervision.
- Provide safe access and egress.
- Provide suitable fire extinguishers.

Your answer	Too general	Better, but generic	Clear explanation of type	Why given	Improved	Done
Provide PPE	X					
Provide suitable PPE	X					
Provide eye protection		X				
Provide chemical resistant goggles			X			
Provide chemical resistant goggles to protect the worker				X		
Provide chemical resistant goggles to protect the workers eyes from chemical splash					✓	
Provide chemical resistant goggles to protect the workers eyes from chemical splash when dispensing acid from a container						✓

Finally:

Before the exam you should check the following:

- Know where the exam is to take place.
- Arrive in good time.
- Bring your examination entry confirmation, which includes your candidate number. The invigilator will have a seating plan and your place in the examination room is allocated to you.
- Bring photographic proof of identity.
- Bring pens, pencils, ruler, etc. (Remember these must be in a clear plastic bag or wallet).
- Bags, books, reference materials must be left out of reach to the side, rear or front of the room.
- Mobile phones or any other electronic devices may **NOT** be used. Please ensure devices are switched off and placed in your bag (out of reach) or handed to the invigilator to keep for the duration of the examination.
- Know that bilingual dictionaries (in paper format, with no personal annotations) approved by NEBOSH may be used.
- Check with your approved course provider if refreshments are provided and if you may bring sweets into the examination room.

Good luck, but always remember:

"The greatest barrier to success is the fear of failure."

Source: Sven Goran Eriksson.

R.6 Revision checklist

The revision checklist is designed to help those wishing to take the NEBOSH International General Certificate. By now you should have been trained in the requirements necessary to fulfil the syllabus requirements and have a good understanding of the action verbs, which are used at examination.

The following revision checklist is taken from elements 1 to 4 of this publication. The purpose of the checklist is for you to review the topics against the action verbs. Column 1 is the element and sub element number. Column 2 is the element and sub element topics. Column 3 and 4 are for you to identify if you are generally comfortable at that level (Ok) or not (require more revision – Rev).

	Element topics	Ok	Rev
1.0	**Why we should manage workplace health and safety**		
1.1	Morals and money		
	Moral expectations of good standards of health and safety.		
	The financial costs of incidents (insured and uninsured costs)		
1.2	Regulating health and safety		
	What enforcement agencies do and what happens if you don't comply		
	The part played by international standards (like ISO 45001)		
	The International Labour Organisation's (ILO) Convention C155 and Recommendation R164		
	Employers' responsibilities (C155 Article 16 and R164 recommendation 10)		
	Workers responsibilities and rights (C155 Article 19 and R164 recommendation 16)		
	Where you can find information on national standards.		
1.3	Who does what in organisations		
	Roles of directors/managers/supervisors		
	How top management can demonstrate commitment by:		
	Making resources available to design, implement and maintain the occupational health and safety management system		
	Defining roles and responsibilities		
	Appointing senior managers with specific responsibility for health and safety		
	Appointing competent people (internal and external, including specialists) to help the organisation meet its health and safety obligations		
	Reviewing health and safety performance		
	Responsibilities of organisations who share a workplace to work together on health and safety issues (C155 Article 17, R164 Recommendation 11)		
	How clients and contractors should work together:		
	The duties they owe each other ('Safety and health in construction', ILO Code of Practice - Chapter 2)		
	Effective planning and co-ordination of contracted work		
	Pre-selection and management of contractors		
	Overall aims of the organisation in terms of health and safety performance		
2.0	**How health and safety management systems work and what they look like**		
2.1	What they are and the benefits they bring		
	The basics of a health and safety management system: the 'Plan, Do, Check, Act' model (see ISO 45001:2018 and ILO-OSH2001)		
	The benefits of having a formal/certified health and safety management system.		
2.2	What good health and safety management systems look like		
	The occupational health and safety policy (see clause 5.2 ISO 45001:2018):		
	Role		
	Typical content		
	Proportionate to the needs of the organisation		
	Responsibilities – all workers at all levels of an organisation have responsibility for health and safety		
	Practical arrangements for making it work:		
	The importance of stating the organisation's arrangements for planning and organising, controlling hazards, consultation, communication, monitoring compliance, assessing effectiveness		

	Element topics	Ok	Rev
	Keeping it current: when you might need to review the health and safety management system, including passage of time, technological, organisational or legal changes, and results of monitoring.		
3.0	**Managing risk - understanding people and processes**		
3.1	Health and safety culture		
	Meaning of the term 'health and safety culture'		
	Relationship between health and safety culture and health and safety performance		
	Indicators of an organisation's health and safety culture:		
	Incidents, absenteeism, sickness rates, staff turnover, level of compliance with health and safety rules and procedures, complaints about working conditions		
	Influence of peers on health and safety culture.		
3.2	Improving health and safety culture		
	Gaining commitment of management		
	Promoting health and safety standards by leadership and example and appropriate use of disciplinary procedures		
	Competent workers		
	Good communication within the organisation:		
	Benefits and limitations of different methods of communication (verbal, written and graphic)		
	Use and effectiveness of noticeboards and health and safety media		
	Co-operation and consultation with the workforce and contractors, including:		
	Benefits of worker participation (including worker feedback)		
	The role of health and safety committees		
	When training is needed:		
	Induction (key health and safety topics to be covered)		
	Job change		
	Process change		
	Introduction of new legislation		
	Introduction of new technology		
3.3	How human factors influence behaviour positively or negatively		
	Organisational factors, including: culture, leadership, resources, work patterns, communications		
	Job factors, including: task, workload, environment, display and controls, procedures		
	Individual factors, including: competence, skills, personality, attitude and risk perception		
	Link between individual, job and organisational factors		
3.4	Assessing risk		
	Meaning of hazard, risk, risk profiling and risk assessment		
	Risk profiling: What is involved? Who should be involved? The risk profiling process		
	Purpose of risk assessment and the 'suitable and sufficient' standard it needs to reach (see HSG65: 'Managing for health and safety')		
	A general approach to risk assessment (5 steps):		
	Identify hazards:		
	Cources and form of harm; sources of information to consult; use of task analysis, legislation, manufacturers' information, incident data, guidance		

	Element topics	Ok	Rev
	Identify people at risk:		
	Including workers, operators, maintenance staff, cleaners, contractors, visitors, public		
	Evaluate risk (taking account of what you already do) and decide if you need to do more:		
	Likelihood of harm and probable severity		
	Possible acute and chronic health effects		
	Risk rating		
	Principles to consider when controlling risk (section 3.10.1 ILO-OSH 2001 – 'Guidelines on occupational safety and health management systems')		
	Practical application of the principles – applying the general hierarchy of control (clause 8.1.2 of ISO 45001:2018)		
	Application based on prioritisation of risk		
	Use of guidance; sources and examples of legislation		
	Applying controls to specified hazards		
	Residual risk; acceptable/tolerable risk levels		
	Distinction between priorities and timescales		
	Record significant findings		
	Reasons for review		
	Application of risk assessment for specific types of risk and special cases:		
	Examples of when they are required, including fire, DSE, manual handling, hazardous substances, noise		
	Why specific risk assessment methods are used for certain risks – to enable proper, systematic consideration of all relevant issues that contribute to the risk		
	Special case applications to young people, expectant and nursing mothers; also consideration of disabled workers and lone workers.		
3.5	Management of change		
	Typical types of change faced in the workplace and the possible impact of such change, including: construction works, change of process, change of equipment, change in working practices		
	Managing the impact of change:		
	Communication and co-operation		
	Risk assessment		
	Appointment of competent people		
	Segregation of work areas		
	Amendment of emergency procedures		
	Welfare provision		
	Review of change (during and after).		
3.6	Safe systems of work for general work activities		
	Why workers should be involved when developing safe systems of work		
	Why procedures should be recorded/written down		
	The differences between technical, procedural and behavioural controls		
	Developing a safe system of work:		
	Analysing tasks, identifying hazards and assessing risks		

	Element topics	Ok	Rev
	Introducing controls and formulating procedures		
	Instruction and training in how to use the system		
	Monitoring the system		
3.7	Permit-to-work systems		
	Meaning of a permit-to-work system		
	Why permit-to-work systems are used		
	How permit-to-work systems work and are used		
	When to use a permit-to-work system, including: hot work, work on non-live (isolated) electrical systems, machinery maintenance, confined spaces, work at height.		
3.8	Emergency procedures		
	Why emergency procedures need to be developed		
	What to include in an emergency procedure (see HSG268: 'The health and safety toolbox')		
	Why people need training in emergency procedures		
	Why emergency procedures need to be tested		
	What to consider when deciding on first aid needs in a workplace:		
	Shift patterns		
	Location of site		
	Activities carried out		
	Number of workers		
	Location relative to hospitals/emergency services		
4.0	**Health and safety monitoring and measuring**		
4.1	Active and reactive monitoring		
	The differences between active and reactive monitoring		
	Active monitoring methods (health and safety inspections, sampling and tours) and their usefulness:		
	Differences between the methods; frequency; competence and objectivity of people doing them; use of checklists; allocation of responsibilities and priorities for action		
	Reactive monitoring measures and their usefulness:		
	Data on accidents, dangerous occurrences, near misses, ill-health, complaints by workforce, and enforcement action and incident investigations		
	Why lessons need to be learnt from beneficial and adverse events		
	The difference between leading and lagging indicators		
4.2	Investigating incidents		
	The different levels of investigations: minimal, low, medium and high (see HSG245)		
	Basic incident investigation steps		
	Step one: gathering the information		
	Step two: analysing the information		
	Step three: identifying risk control measures		
	Step four: the action plan and its implementation		
	How occupational accidents and diseases are recorded and notified by the organisation (Recording and notification of occupational accidents and diseases, ILO Code of Practice – chapters 4–7)		
4.3	Health and safety auditing		

	Element topics	Ok	Rev
	Definition of the term 'audit' (clause 3.32, ISO 45001:2018)		
	Why health and safety management systems should be audited, including:		
	Negative: identifying failing of a management system		
	Positive: organisational learning and assurance		
	Difference between audits and inspections		
	Types of audit: product/services, process, system		
	Advantages and disadvantages of external and internal audits		
	The audit stages:		
	Notification of the audit and timetable for auditing		
	Pre-audit preparations, including competent audit team, time and resources required		
	Information gathering		
	Information analysis		
	Completion of audit report		
4.4	Review of health and safety performance		
	Why health and safety performance should be reviewed		
	What the review should consider:		
	Level of compliance with relevant legal and organisational requirements		
	Accident and incident data, corrective and preventive actions		
	Inspections, tours and sampling		
	Absences and sickness		
	Quality assurance reports		
	Audits		
	Monitoring data/records/reports		
	External communications and complaints		
	Results of participation and consultation		
	Objectives met		
	Actions from previous management reviews		
	Legal/good practice developments		
	Assessing opportunities for improvement and the need for change		
	Reporting on health and safety performance		
	Feeding review outputs into action and development plans as part of continuous improvement		

INDEX

This page is intentionally left blank